About the Cover

The cover image illustrates a sphidron, which is a circular disc formed by spiral arms. Sphidrons are a variety of spidrons, which were invented by Daniel Erdely in 1979 as a homework assignment for Erno Rubik, inventor of the Rubik's Cube.

▶ Meet the Artist

The cover art was generated by Paul Nylander, a mechanical engineer with strong programming and mathematical skills that he uses to design complex engineering systems. Nylander says that he always enjoyed science and art as a hobby. When he was in high school, he had some aptitude for math, but programming was difficult for him. However, he became much more interested in programming when he began studying computer graphics. Most of Nylander's artwork is created in Mathematica, POV-Ray, and C++.

You can find a short bio and description of his work at: http://virtualmathmuseum.org/mathart/ArtGalleryNylander/ Nylanderindex.html.

Cover: Sphidron. By Daniel Erdely and Paul Nylander (bugman123.com)

GLENCOE

ALGEBRA 1

McGraw Hill

mheducation.com/prek-12

Send all inquiries to:
McGraw-Hill Education
8787 Orion Place
Columbus, OH 43240

ISBN: 978-0-07-903989-7
MHID: 0-07-903989-8

Printed in the United States of America.

5 6 7 8 9 LWI 23 22 21 20 19

Understanding by Design® is a registered trademark of the
Association for Supervision and Curriculum Development
("ASCD").

McGraw-Hill is committed to providing instructional materials in
Science, Technology, Engineering, and Mathematics (STEM) that
give all students a solid foundation, one that prepares them for
college and careers in the 21st century.

Contents in Brief

Authors

Our lead authors ensure that the Macmillan/McGraw-Hill and Glencoe/McGraw-Hill mathematics programs are truly vertically aligned by beginning with the end in mind — success in Algebra 1 and beyond. By "backmapping" the content from the high school programs, all of our mathematics programs are well articulated in their scope and sequence.

LEAD AUTHORS

John A. Carter, Ph.D.

Mathematics Teacher
WINNETKA, ILLINOIS

Areas of Expertise:
Using technology and manipulatives to visualize concepts; mathematics achievement of English-language learners

Gilbert J. Cuevas, Ph.D.

Professor of Mathematics Education, Texas State University— San Marcos
SAN MARCOS, TEXAS

Areas of Expertise:
Applying concepts and skills in mathematically rich contexts; mathematical representations; use of technology in the development of geometric thinking

Roger Day, Ph.D., NBCT

Mathematics Department Chairperson, Pontiac Township High School
PONTIAC, ILLINOIS

Areas of Expertise:
Understanding and applying probability and statistics; mathematics teacher education

In Memoriam
Carol Malloy, Ph.D.

Dr. Carol Malloy was a fervent supporter of mathematics education. She was a Professor at the University of North Carolina, Chapel Hill, NCTM Board of Directors member, President of the Benjamin Banneker Association (BBA), and 2013 BBA Lifetime Achievement Award for Mathematics winner. She joined McGraw-Hill in 1996. Her influence significantly improved our programs' focus on real-world problem solving and equity. We will miss her inspiration and passion for education.

PROGRAM AUTHORS

Berchie Holliday, Ed.D.

National Mathematics Consultant
SILVER SPRING, MARYLAND

Areas of Expertise:
Using mathematics to model and understand real-world data; the effect of graphics on mathematical understanding

Beatrice Moore Luchin

Mathematics Consultant
HOUSTON, TEXAS

Areas of Expertise:
Mathematics literacy; working with English language learners

CONTRIBUTING AUTHORS

Dinah Zike FOLDABLES

Educational Consultant Dinah-Might Activities, Inc.
SAN ANTONIO, TEXAS

Jay McTighe

Educational Author and Consultant
COLUMBIA, MARYLAND

Consultants and Reviewers

These professionals were instrumental in providing valuable input and suggestions for improving the effectiveness of the mathematics instruction.

LEAD CONSULTANT

Viken Hovsepian

Professor of Mathematics
Rio Hondo College
WHITTIER, CALIFORNIA

CONSULTANTS

MATHEMATICAL CONTENT

Grant A. Fraser, Ph.D.
Professor of Mathematics
California State University, Los Angeles
LOS ANGELES, CALIFORNIA

Arthur K. Wayman, Ph.D.
Professor of Mathematics Emeritus
California State University, Long Beach
LONG BEACH, CALIFORNIA

GIFTED AND TALENTED

Shelbi K. Cole
Research Assistant
University of Connecticut
STORRS, CONNECTICUT

COLLEGE READINESS

Robert Lee Kimball, Jr.
Department Head, Math and Physics
Wake Technical Community College
RALEIGH, NORTH CAROLINA

DIFFERENTIATION FOR ENGLISH-LANGUAGE LEARNERS

Susana Davidenko
State University of New York
CORTLAND, NEW YORK

Alfredo Gómez
Mathematics/ESL Teacher
George W. Fowler High School
SYRACUSE, NEW YORK

GRAPHING CALCULATOR

Ruth M. Casey
T³ National Instructor
FRANKFORT, KENTUCKY

Jerry Cummins
Former President
National Council of Supervisors of Mathematics
WESTERN SPRINGS, ILLINOIS

MATHEMATICAL FLUENCY

Robert M. Capraro
Associate Professor
Texas A&M University
COLLEGE STATION, TEXAS

PRE-AP

Dixie Ross
Lead Teacher for Advanced Placement Mathematics
Pflugerville High School
PFLUGERVILLE, TEXAS

READING AND WRITING

ReLeah Cossett Lent
Author and Educational Consultant
MORGANTON, GEORGIA

Lynn T. Havens
Director of Project CRISS
KALISPELL, MONTANA

REVIEWERS

Sherri Abel
Mathematics Teacher
Eastside High School
TAYLORS, SOUTH CAROLINA

Kelli Ball, NBCT
Mathematics Teacher
Owasso 7th Grade Center
OWASSO, OKLAHOMA

Cynthia A. Burke
Mathematics Teacher
Sherrard Junior High School
WHEELING, WEST VIRGINIA

Patrick M. Cain, Sr.
Assistant Principal
Stanhope Elmore High School
MILLBROOK, ALABAMA

Robert D. Cherry
Mathematics Instructor
Wheaton Warrenville South
   High School
WHEATON, ILLINOIS

Tammy Cisco
8th Grade Mathematics/
   Algebra Teacher
Celina Middle School
CELINA, OHIO

Amber L. Contrano
High School Teacher
Naperville Central High School
NAPERVILLE, ILLINOIS

Catherine Creteau
Mathematics Department
Delaware Valley Regional
   High School
FRENCHTOWN, NEW JERSEY

Glenna L. Crockett
Mathematics Department Chair
Fairland High School
FAIRLAND, OKLAHOMA

Jami L. Cullen
Mathematics Teacher/Leader
Hilltonia Middle School
COLUMBUS, OHIO

Franco DiPasqua
Director of K–12 Mathematics
West Seneca Central Schools
WEST SENECA, NEW YORK

Kendrick Fearson
Mathematics Department Chair
Amos P. Godby High School
TALLAHASSEE, FLORIDA

Lisa K. Gleason
Mathematics Teacher
Gaylord High School
GAYLORD, MICHIGAN

Debra Harley
Director of Math & Science
East Meadow School District
WESTBURY, NEW YORK

Tracie A. Harwood
Mathematics Teacher
Braden River High School
BRADENTON, FLORIDA

Bonnie C. Hill
Mathematics Department Chair
Triad High School
TROY, ILLINOIS

Clayton Hutsler
Teacher
Goodwyn Junior High School
MONTGOMERY, ALABAMA

Gureet Kaur
7th Grade Mathematics Teacher
Quail Hollow Middle School
CHARLOTTE, NORTH CAROLINA

Rima Seals Kelley, NBCT
Mathematics Teacher/
   Department Chair
Deerlake Middle School
TALLAHASSEE, FLORIDA

Holly W. Loftis
8th Grade Mathematics Teacher
Greer Middle School
LYMAN, SOUTH CAROLINA

Katherine Lohrman
Teacher, Math Specialist,
   New Teacher Mentor
John Marshall High School
ROCHESTER, NEW YORK

Carol Y. Lumpkin
Mathematics Educator
Crayton Middle School
COLUMBIA, SOUTH CAROLINA

Ron Mezzadri
Supervisor of Mathematics K–12
Fair Lawn Public Schools
FAIR LAWN, NEW JERSEY

Bonnye C. Newton
SOL Resource Specialist
Amherst County Public Schools
AMHERST, VIRGINIA

Kevin Olsen
Mathematics Teacher
River Ridge High School
NEW PORT RICHEY, FLORIDA

Kara Painter
Mathematics Teacher
Downers Grove South
   High School
DOWNERS GROVE, ILLINOIS

Sheila L. Ruddle, NBCT
Mathematics Teacher,
   Grades 7 and 8
Pendleton County Middle/
   High School
FRANKLIN, WEST VIRGINIA

Angela H. Slate
Mathematics Teacher/Grade 7,
   Pre-Algebra, Algebra
LeRoy Martin Middle School
RALEIGH, NORTH CAROLINA

Cathy Stellern
Mathematics Teacher
West High School
KNOXVILLE, TENNESSEE

Dr. Maria J. Vlahos
Mathematics Division Head
   for Grades 6–12
Barrington High School
BARRINGTON, ILLINOIS

Susan S. Wesson
Mathematics Consultant/
   Teacher (Retired)
Pilot Butte Middle School
BEND, OREGON

Mary Beth Zinn
High School Mathematics Teacher
Chippewa Valley High Schools
CLINTON TOWNSHIP, MICHIGAN

Digital Tools to Enhance Your Learning

Students today have an unprecedented access to and appetite for technology and new media. You see technology as your friend and rely on it to study, work, play, relax, and communicate.

You are accustomed to the role that computers play in today's world. You are the first generation whose primary educational tool is a computer or a cell phone. The eStudentEdition gives you access to your math curriculum anytime, anywhere.

Investigate

Sketchpad Discover concepts using The Geometer's Sketchpad®.

Vocabulary tools include fun Vocabulary Review Games.

Tools enhance understanding through exploration.

The Geometer's Sketchpad

The Geometer's Sketchpad gives you a tangible, visual way to see the math in action through dynamic model manipulation of lines, shapes, and functions.

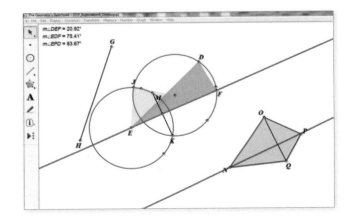

eToolkit

eToolkit helps you to use virtual manipulative's to extend your learning beyond the classroom by modifying concrete models in a real-time, interactive format focused on problem-based learning.

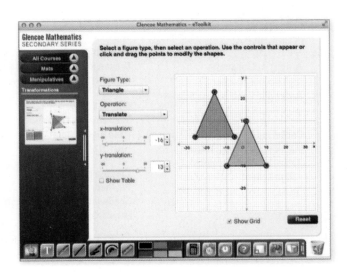

Learn

LearnSmart
Topic based online assessment

Animations
illustrate key concepts through step-by-step tutorials and videos.

Tutors
See and hear a teacher explain how to solve problems.

Calculator Resources
provides other calculator keystrokes for each Graphing Technology Lab.

Practice

Self-Check Practice
allows students to check their understanding and send results to their teacher.

eBook
Interactive learning experience with links directly to assets

eStudentEdition

Use your eStudentEdition to access your print text 24/7. This interactive eBook gives you access to all the resources that help you learn the concepts. You can take notes, highlight, digitally write on the pages, and bookmark where you are.

Interactive Student Guide

Interactive Student Guide is a dynamic resource to help you meet the challenges of content standards.

This guide works together with the student edition to ensure that you can reflect on comprehension and application, apply math concepts to the real world, and internalize concepts to develop "second nature" recall.

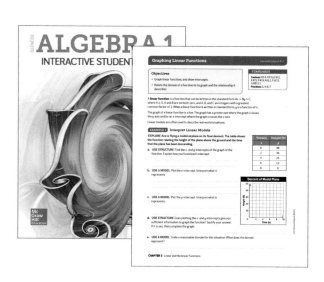

Adaptive Learning

LEARNSMART®

LearnSmart is your online test-prep solution with adaptive capability and resources for end-of-course assessment. Through a series of adaptive questions, LearnSmart identifies areas where you need to focus your learning. It provides you with learning resources such as slides, videos, kaleidoscopes, and label games to encourage you to review material again. This adaptive learning will help increase your likelihood that you will retain your new knowledge.

LearnSmart can be used as a self-study tool or it can be assigned to you by your teacher through your ConnectED platform.

ALEKS®

Through a cycle of assessment and learning, *ALEKS determines the topics that you are ready to learn next and develops a personalized learning path for you. ALEKS provides you with real-time, actionable data that informs you what you need to learn. It will help you review and master the skills needed to be successful in your math class. Use your ALEKS Pie to see a snapshot of your current progress of the course. Click on the Pie slice to see the number of topics mastered and ready to learn next.

Use the timeline to watch your progress toward your learning goals, and if needed you can toggle between English and Spanish translations of the content and interface.

*Ask your teacher if you have access to ALEKS.

McGraw-Hill Education

CHAPTER 0
Preparing for Algebra

TABLE OF CONTENTS

Wim van den Heever/Tetra Images/Alamy

Name _____ Date _____ Period _____

Scavenger Hunt

Let's Get Started!

To help you find the information you need quickly, use the Scavenger Hunt below to learn where things are located in each chapter.

1. What is the title of Chapter 1?

2. How can you tell what you'll learn in Lesson 1-1?

3. There is a Real-World Career featured in Lesson 1-1. What is the career?

4. How many examples are presented in Lesson 1-2?

Worksheets help to explain key concepts, let you practice your skills, and offer opportunities for extending the lessons. Find them in the Resources in ConnectED.

Go Online!
connectED.mcgraw-hill.com

CHAPTER 1
Expressions and Functions

Go Online!
connectED.mcgraw-hill.com

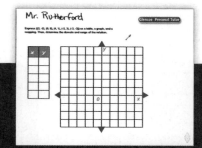

Personal Tutors let you hear a real teacher discuss each step to solving a problem. There is a Personal Tutor for each example, just a click away in ConnectED.

CHAPTER 2
Linear Equations

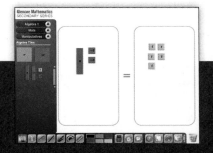

 Using the algebra tiles in the **Virtual Manipulatives** can be helpful as you study this chapter. Find them in the eToolkit in ConnectED.

 *Go Online!*
connectED.mcgraw-hill.com

CHAPTER 3
Linear and Nonlinear Functions

Radius Images/Alamy

Go Online!
connectED.mcgraw-hill.com

With the **Graphing Tools** in ConnectED, you can explore how changing the values affects the graph of a linear function.

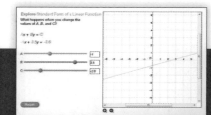

CHAPTER 4
Equations of Linear Functions

altrendo images/Media Bakery

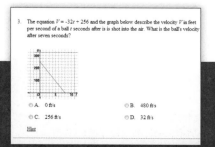

3. The equation $V = -32t + 256$ and the graph below describe the velocity V in feet per second of a ball t seconds after is is shot into the air. What is the ball's velocity after seven seconds?

A. 0 ft/s
B. 480 ft/s
C. 256 ft/s
D. 32 ft/s

Hint

Review concepts with quick **Self-Check Quizzes** in ConnectED. Use them to check your own progress as you complete each lesson.

Go Online!
connectED.mcgraw-hill.com

CHAPTER 5
Linear Inequalities

ASSESSMENT

Juniors Bildarchiv GmbH/Alamy

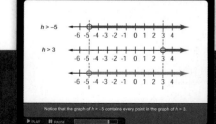

Go Online!
connectED.mcgraw-hill.com

Animations demonstrate Key Concepts and topics from the chapter. Click to watch animations in ConnectED.

CHAPTER 6
Systems of Linear Equations and Inequalities

ASSESSMENT

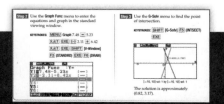

Graphing Calculator Keystrokes help you make the most of your graphing calculator. Find the keystrokes for your calculator in the Resources in ConnectED.

Go Online!
connectED.mcgraw-hill.com

CHAPTER 7
Exponents and Exponential Functions

Purestock/SuperStock

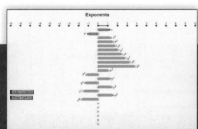

Go Online!
connectED.mcgraw-hill.com

Geometer's Sketchpad® gives you a tangible, visual way to learn mathematics. Investigate exponents and exponential functions with sketches in ConnectED.

CHAPTER 8
Polynomials

Spaces Images/Blend Images

Worksheets help to explain key concepts, let you practice your skills, and offer opportunities for extending the lessons. Find them in the Resources in ConnectED.

Go Online!
connectED.mcgraw-hill.com

CHAPTER 9
Quadratic Functions and Equations

Getty Images

Go Online!
connectED.mcgraw-hill.com

With **Graphing Calculator** Easy Files™, your graphing calculator can do more than graph! Practice vocabulary in English or Spanish with Lesson Vocabulary Review Files. Or review with a 5-Minute Check. Ask your teacher to assign them to you in ConnectED.

CHAPTER 10
Statistics

ASSESSMENT

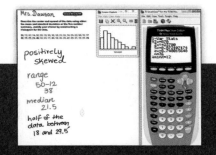

Personal Tutors that use graphing calculator technology show you every step to solving problems with this powerful tool. Find them in the Resources in ConnectED.

Go Online!
connectED.mcgraw-hill.com

Student Handbook

Built-In Workbook

Reference

(MP) Standards for Mathematical Practice

Glencoe Algebra 1 exhibits these practices throughout the entire program. All of the Standards for Mathematical Practice will be covered in each chapter. The MP icon notes specific areas of coverage.

Mathematical Practices	What does it mean?
1. Make sense of problems and persevere in solving them.	Solving a mathematical problem takes time. Use a logical process to make sense of problems, understand that there may be more than one way to solve a problem, and alter the process if needed.
2. Reason abstractly and quantitatively.	You can start with a concrete or real-world context and then represent it with abstract numbers or symbols (decontextualize), find a solution, then refer back to the context to check that the solution makes sense (contextualize).
3. Construct viable arguments and critique the reasoning of others.	Sound mathematical arguments require a logical progression of statements and reasons. Mathematically proficient students can clearly communicate their thoughts and defend them.
4. Model with mathematics.	Modeling links classroom mathematics and statistics to everyday life, work, and decision-making. High school students at this level are expected to apply key takeaways from earlier grades to high-school level problems.
5. Use appropriate tools strategically.	Certain tools, including estimation and virtual tools are more appropriate than others. You should understand the benefits and limitations of each tool.
6. Attend to precision.	Precision in mathematics is more than accurate calculations. It is also the ability to communicate with the language of mathematics. In high school mathematics, precise language makes for effective communication and serves as a tool for understanding and solving problems.
7. Look for and make use of structure.	Mathematics is based on a well-defined structure. Mathematically proficient students look for that structure to find easier ways to solve problems.
8. Look for and express regularity in repeated reasoning.	Mathematics has been described as the study of patterns. Recognizing a pattern can lead to results more quickly and efficiently.

FOLDABLES® by Dinah Zike

Folding Instructions

The following pages offer step-by-step instructions to make the Foldables® study guides.

Layered-Look Book

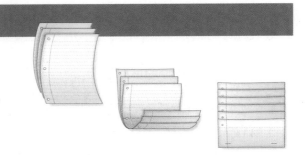

1. Collect three sheets of paper and layer them about 1 cm apart vertically. Keep the edges level.

2. Fold up the bottom edges of the paper to form 6 equal tabs.

3. Fold the papers and crease well to hold the tabs in place. Staple along the fold. Label each tab.

Shutter-Fold and Four-Door Books

1. Find the middle of a horizontal sheet of paper. Fold both edges to the middle and crease the folds. Stop here if making a shutter-fold book. For a four-door book, complete the steps below.

2. Fold the folded paper in half, from top to bottom.

3. Unfold and cut along the fold lines to make four tabs. Label each tab.

Concept-Map Book

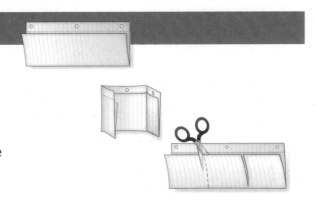

1. Fold a horizontal sheet of paper from top to bottom. Make the top edge about 2 cm shorter than the bottom edge.

2. Fold width-wise into thirds.

3. Unfold and cut only the top layer along both folds to make three tabs. Label the top and each tab.

Vocabulary Book

1. Fold a vertical sheet of notebook paper in half.

2. Cut along every third line of only the top layer to form tabs. Label each tab.

Pocket Book

1. Fold the bottom of a horizontal sheet of paper up about 3 cm.

2. If making a two-pocket book, fold in half. If making a three-pocket book, fold in thirds.

3. Unfold once and dot with glue or staple to make pockets. Label each pocket.

Bound Book

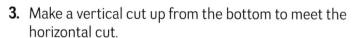

1. Fold several sheets of paper in half to find the middle. Hold all but one sheet together and make a 3-cm cut at the fold line on each side of the paper.

2. On the final page, cut along the fold line to within 3-cm of each edge.

3. Slip the first few sheets through the cut in the final sheet to make a multi-page book.

Top-Tab Book

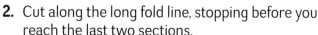

1. Layer multiple sheets of paper so that about 2–3 cm of each can be seen.

2. Make a 2–3-cm horizontal cut through all pages a short distance (3 cm) from the top edge of the top sheet.

3. Make a vertical cut up from the bottom to meet the horizontal cut.

4. Place the sheets on top of an uncut sheet and align the tops and sides of all sheets. Label each tab.

Accordion Book

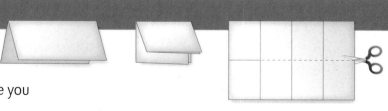

1. Fold a sheet of paper in half. Fold in half and in half again to form eight sections.

2. Cut along the long fold line, stopping before you reach the last two sections.

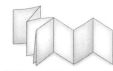

3. Refold the paper into an accordion book. You may want to glue the double pages together.

NOW

Chapter 0 contains lessons on topics from previous courses. You can use this chapter in various ways.

- Begin the school year by taking the Pretest. If you need additional review, complete the lessons in this chapter. To verify that you have successfully reviewed the topics, take the Posttest.

- As you work through the text, you may find that there are topics you need to review. When this happens, complete the individual lessons that you need.

- Use this chapter for reference. When you have questions about any of these topics, flip back to this chapter to review definitions or key concepts.

(MP) WHY

Use the Mathematical Practices to complete the activity.

Use Tools Use the Internet or another source to find how fast cheetahs run. How long will it take a cheetah to run 100 miles?

Make Sense Use the KWL chart from ConnectED to organize the information.

Topic		
What I Know	What I Want to Know	What I Learned
A cheetah can run 72 miles per hour. This means a cheetah runs 72 miles in 1 hour.	How long will it take a cheetah to run 100 mi?	An equation can be used to solve for an unknown value.

Model With Mathematics Construct an equation to solve for the missing number.

Discuss Compare your equation with those created by your classmates. What is the best way to solve this problem?

Wim van den Heever/Tetra Images/Alamy

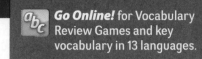
Go Online to Guide Your Learning

Organize

Throughout this text, you will be invited to use Foldables to organize your notes.

Why should you use them?

- They help you organize, display, and arrange information.
- They make great study guides, specifically designed for you.
- They give you a chance to improve your math vocabulary.

How should you use them?

- Write general information — titles, vocabulary terms, concepts, questions, and main ideas — on the front tabs of your Foldable.
- Write specific information — ideas, your thoughts, answers to questions, steps, notes, and definitions — under the tabs.

When should you use them?

- Set up your Foldable as you begin a chapter, or when you start learning a new concept.
- Write in your Foldable every day.
- Use your Foldable to review for homework, quizzes, and tests.

New Vocabulary

English		Español
integer	p. P7	entero
absolute value	p. P11	valor absolute
opposites	p. P11	opuestos
reciprocal	p. P18	recíproco
perimeter	p. P23	perímetro
circle	p. P24	círculo
diameter	p. P24	diámetro
center	p. P24	centro
circumference	p. P24	circunferencia
radius	p. P24	radio
area	p. P26	area
volume	p. P29	volumen
surface area	p. P31	area de superficie
probability	p. P33	probabilidad
sample space	p. P33	espacio muestral
complements	p. P33	complementos
tree diagram	p. P34	diagrama de árbol
odds	p. P35	probabilidades

Explore & Explain

 Tools

Investigate circumference and area of circles using the **2-D Figures** tool. How does changing the radius or diameter affect the circumference?

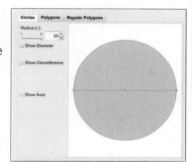

Determine whether you need an estimate or an exact answer. Then solve.

1. **SHOPPING** Addison paid $1.59 for gum and $1.29 for a package of notebook paper. She gave the cashier a $5 bill. If the tax was $0.14, how much change should Addison receive?

2. **DISTANCE** Luis rode his bike 1.2 miles to his friend's house, then 0.7 mile to the convenience store, then 1.9 miles to the library. If he rode the same route back home, about how far did he travel in all?

Find each sum or difference.

3. $20 + (-7)$

4. $-15 + 6$

5. $-9 - 22$

6. $18.4 - (-3.2)$

7. $23.1 + (-9.81)$

8. $-5.6 + (-30.7)$

Find each product or quotient.

9. $11(-8)$

10. $-15(-2)$

11. $63 \div (-9)$

12. $-22 \div 11$

Replace each ● with <, >, or = to make a true sentence.

13. $\frac{7}{20}$ ● $\frac{2}{5}$

14. 0.15 ● $\frac{1}{8}$

15. Order 0.5, $-\frac{1}{7}$, -0.2, and $\frac{1}{3}$ from least to greatest.

Find each sum or difference. Write in simplest form.

16. $\frac{5}{6} + \frac{2}{3}$

17. $\frac{11}{12} - \frac{3}{4}$

18. $\frac{1}{2} + \frac{4}{9}$

19. $-\frac{3}{5} + \left(-\frac{1}{5}\right)$

20. If $\frac{3}{5}$ of the students in Mrs. Hudson's class are girls, what fraction of the class are boys?

Find each product or quotient.

21. $2.4(-0.7)$

22. $-40.5 \div (-8.1)$

23. $15.9(1.2)$

24. $-6.5 \div 0.5$

Name the reciprocal of each number.

25. $\frac{4}{11}$

26. $-\frac{3}{7}$

Find each product or quotient. Write in simplest form.

27. $\frac{2}{21} \div \frac{1}{3}$

28. $\frac{1}{5} \cdot \frac{3}{20}$

29. $\frac{6}{25} \div \left(-\frac{3}{5}\right)$

30. $\frac{1}{9} \cdot \frac{3}{4}$

31. $-\frac{2}{21} \div \left(-\frac{2}{15}\right)$

32. $2\frac{1}{2} \cdot \frac{2}{15}$

Express each percent as a fraction in simplest form.

33. 20%

34. 7.5%

35. 1.4%

36. 37%

Use the percent proportion to find each number.

37. 18 is what percent of 72?

38. 35 is what percent of 200?

39. 24 is 60% of what number?

40. **TEST SCORES** James answered 14 items correctly on a 16-item quiz. What percent did he answer correctly?

41. **BASKETBALL** Emily made 75% of the baskets that she attempted. If she made 9 baskets, how many attempts did she make?

Find the perimeter and area of each figure.

42.
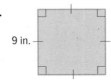
9 in.

43.
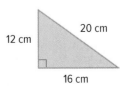
20 cm
12 cm
16 cm

44. A parallelogram has side lengths of 7 inches and 11 inches. Find the perimeter.

45. GARDENS Find the perimeter of the garden.

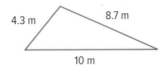

4.3 m   8.7 m

10 m

46. The perimeter of a square is 16 centimeters. What is the length of one side?

Find the circumference and area of each circle. Round to the nearest tenth.

47.

2 m

48.

16 cm

49. BIRDS The floor of a birdcage is a circle with a circumference of about 47.1 inches. What is the diameter of the birdcage floor? Round to the nearest inch.

Find the volume and surface area of each rectangular prism given the measurements below.

50. $\ell = 3$ cm, $w = 1$ cm, $h = 3$ cm

51. $\ell = 6$ ft, $w = 2$ ft, $h = 5$ ft

52. $\ell = 5$ in., $w = 4$ in., $h = 2$ in.

53. Find the volume and surface area of the rectangular prism.

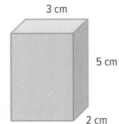

3 cm

5 cm

2 cm

54. A can of soup has a volume of 20π cubic inches. The diameter of the can is 4 inches. What is the height of the can?

One pencil is randomly selected from a case containing 3 red, 4 green, 2 black, and 6 blue pencils. Find each probability.

55. P(green)

56. P(red or blue)

57. Use a tree diagram to find the sample space for the event *a die is rolled, and a coin is tossed*. State the number of possible outcomes.

A die is rolled. Find each probability.

58. rolling a multiple of 5

59. rolling a number divisible by 2

60. rolling a number divisible by 3

One coin is randomly selected from a jar containing 20 pennies, 15 nickels, 3 dimes, and 12 quarters. Find the odds of each outcome. Write in simplest form.

61. a penny

62. a penny or nickel

63. a dime

64. a nickel or quarter

65. A coin is tossed 50 times. The results are shown in the table. Find the experimental probability of heads. Write as a fraction in simplest form.

Lands Face-Up	Number of Times
heads	22
tails	28

66. A die is rolled 100 times. The results are shown in the table. Find the experimental probability of rolling a 6. Write as a fraction in simplest form.

Number Rolled	Number of Times
1	17
2	15
3	14
4	18
5	16
6	20

LESSON 1
Plan for Problem Solving

∴ Objective

- Use the four-step problem-solving plan.

New Vocabulary

four-step problem-solving plan

defining a variable

Mathematical Practices

1 Make sense of problems and persevere in solving them.

Using the **four-step problem-solving plan** can help you solve any word problem. Each step of the plan is important.

Step 1 **Understand the Problem**

To solve a verbal problem, first read the problem carefully and explore what the problem is asking.
- Identify what information is given.
- Identify what you need to find.

Step 2 **Plan the Solution**

One strategy you can use is to write an equation. Choose a variable to represent one of the unspecified numbers in the problem. This is called **defining a variable**. Then use the variable to write expressions for the other unspecified numbers in the problem.

Step 3 **Solve a Problem**

Use the plan or strategy you chose in Step 2 to find a solution to the problem.

Step 4 **Check the Solution**

Check your answer in the context of the original problem.
- Are the steps that you took to solve the problem effective?
- Is your answer reasonable?

Example 1   Use the Four-Step Plan

FLOORS Ling's hallway is 10 feet long and 4 feet wide. He paid $200 to tile his hallway floor. How much did Ling pay per square foot for the tile?

Understand We are given the measurements of the hallway and the total cost of the tile. We are asked to find the cost of each square foot of tile.

Plan Write an equation. Let f represent the cost of each square foot of tile. The area of the hallway is 10×4 or 40 ft^2.

40	times	the cost per square foot	equals	200.
40	·	f	=	200

Solve $40 \cdot f = 200$. Find f mentally by asking, "What number times 40 is 200?"
$f = 5$
The tile cost $5 per square foot.

Check If the tile costs $5 per square foot, then 40 square feet of tile costs $5 \cdot 40$ or $200. The answer makes sense.

When an exact value is needed, you can use estimation to check your answer.

Example 2   Use the Four-Step Plan

TRAVEL Emily's family drove 254.6 miles. Their car used 19 gallons of gasoline. Describe the car's gas mileage.

Understand We are given the total miles driven and how much gasoline was used. We are asked to find the gas mileage of the car.

Plan Write an equation. Let G represent the car's gas mileage.

gas mileage = number of miles ÷ number of gallons used

$G = 254.6 ÷ 19$

Solve $G = 254.6 ÷ 19$

$= 13.4$ mi/gal

The car's gas mileage is 13.4 miles per gallon.

Check Use estimation to check your solution.

260 mi ÷ 20 gal = 13 mi/gal

Since the solution 13.4 is close to the estimate, the answer is reasonable. Dividing total miles by the gallons of gas used provides the gas mileage.

Exercises

Determine whether you need an estimate or an exact answer. Then use the four-step problem-solving plan to solve.

1. **DRIVING** While on vacation, the Jacobson family drove 312.8 miles the first day, 177.2 miles the second day, and 209 miles the third day. About how many miles did they travel in all?

2. **PETS** Ms. Hernandez boarded her dog at a kennel for 4 days. It cost $28.90 per day, and she had a coupon for $5 off. What was the final cost for boarding her dog?

3. **MEASUREMENT** William is using a 1.75-liter container to fill a 14-liter container of water. About how many times will he need to fill the smaller container?

4. **SEWING** Fabric costs $12.05 per yard. The drama department needs 18 yards of the fabric for their new play. About how much should they expect to pay?

5. **FINANCIAL LITERACY** The table shows donations to help purchase a new tree for the school. How much money did the students donate in all?

Number of Students	Amount of Each Donation
20	$2.50
15	$3.25

6. **SHOPPING** Is $12 enough to buy a half gallon of milk for $2.49, a bag of apples for $4.35, and four cups of yogurt that cost $0.89 each? Explain.

Real Numbers

:: Objective

- Classify and use real numbers.

New Vocabulary

positive number
negative number
natural number
whole number
integer
rational number
square root
principal square root
perfect square
irrational number
real number
graph
coordinate

A number line can be used to show the sets of natural numbers, whole numbers, integers, and rational numbers. Values greater than 0, or **positive numbers**, are listed to the right of 0, and values less than 0, or **negative numbers**, are listed to the left of 0.

natural numbers: 1, 2, 3, ...

whole numbers: 0, 1, 2, 3, ...

integers: ... , −3, −2, −1, 0, 1, 2, 3, ...

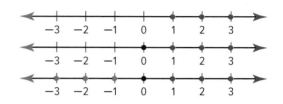

rational numbers: numbers that can be expressed in the form $\frac{a}{b}$, where a and b are integers and $b \neq 0$

A **square root** is one of two equal factors of a number. For example, one square root of 64, written as $\sqrt{64}$, is 8 since $8 \cdot 8$ or 8^2 is 64. The nonnegative square root of a number is the **principal square root**. Another square root of 64 is −8 since $(-8) \cdot (-8)$ or $(-8)^2$ is also 64. A number like 64, with a square root that is a rational number, is called a **perfect square**. The square roots of a perfect square are rational numbers.

A number such as $\sqrt{3}$ is the square root of a number that is not a perfect square. It cannot be expressed as a terminating or repeating decimal; $\sqrt{3} \approx 1.73205....$ Numbers that cannot be expressed as terminating or repeating decimals, or in the form $\frac{a}{b}$, where a and b are integers and $b \neq 0$, are called **irrational numbers**. Irrational numbers and rational numbers together form the set of **real numbers**.

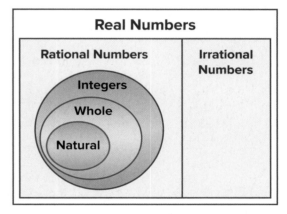

Example 1   Classify Real Numbers

Name the set or sets of numbers to which each real number belongs.

a. $\frac{5}{22}$

Because 5 and 22 are integers and $5 \div 22 = 0.2272727...$ or $0.2\overline{27}$, which is a repeating decimal, this number is a rational number.

b. $\sqrt{81}$

Because $\sqrt{81} = 9$, this number is a natural number, a whole number, an integer, and a rational number.

c. $\sqrt{56}$

Because $\sqrt{56} = 7.48331477...$, which is not a repeating or terminating decimal, this number is irrational.

To **graph** a set of numbers means to draw, or plot, the points named by those numbers on a number line. The number that corresponds to a point on a number line is called the **coordinate** of that point. The rational numbers and the irrational numbers complete the number line.

Example 2   Graph and Order Real Numbers

Graph each set of numbers on a number line. Then order the numbers from least to greatest.

a. $\left\{ \frac{5}{3}, -\frac{4}{3}, \frac{2}{3}, -\frac{1}{3} \right\}$

$$-\frac{5}{3} \quad -\frac{4}{3} \quad -1 \quad -\frac{2}{3} \quad -\frac{1}{3} \quad 0 \quad \frac{1}{3} \quad \frac{2}{3} \quad 1 \quad \frac{4}{3} \quad \frac{5}{3} \quad 2 \quad \frac{7}{3}$$

From least to greatest, the order is $-\frac{4}{3}$, $-\frac{1}{3}$, $\frac{2}{3}$, and $\frac{5}{3}$.

b. $\left\{ 6\frac{4}{5}, \sqrt{49}, 6.\overline{3}, \sqrt{57} \right\}$

Express each number as a decimal. Then order the decimals.

$6\frac{4}{5} = 6.8$ $\sqrt{49} = 7$ $6.\overline{3} = 6.33333333...$ $\sqrt{57} = 7.5468344...$

$$6.\overline{3} \qquad 6\frac{4}{5} \quad \sqrt{49} \qquad \sqrt{57}$$

$$6.0 \quad 6.2 \quad 6.4 \quad 6.6 \quad 6.8 \quad 7.0 \quad 7.2 \quad 7.4 \quad 7.6 \quad 7.8 \quad 8.0$$

From least to greatest, the order is $6.\overline{3}$, $6\frac{4}{5}$, $\sqrt{49}$, and $\sqrt{57}$.

c. $\left\{ \sqrt{20}, 4.7, \frac{12}{3}, 4\frac{1}{3} \right\}$

$\sqrt{20} = 4.47213595...$ $4.7 = 4.7$ $\frac{12}{3} = 4.0$ $4\frac{1}{3} = 4.33333333...$

$$\frac{12}{3} \qquad 4\frac{1}{3} \quad \sqrt{20} \qquad 4.7$$

$$3.8 \quad 3.9 \quad 4.0 \quad 4.1 \quad 4.2 \quad 4.3 \quad 4.4 \quad 4.5 \quad 4.6 \quad 4.7 \quad 4.8$$

From least to greatest, the order is $\frac{12}{3}$, $4\frac{1}{3}$, $\sqrt{20}$, and 4.7.

Any repeating decimal can be written as a fraction.

Example 3   Write Repeating Decimals as Fractions

Write $0.\overline{7}$ as a fraction in simplest form.

Step 1 $N = 0.777...$ Let N represent the repeating decimal.
$10N = 10(0.777...)$ Since only one digit repeats, multiply each side by 10.
$10N = 7.777...$ Simplify.

Step 2 Subtract N from $10N$ to eliminate the part of the number that repeats.

$$10N = 7.777...$$
$$-(N = 0.777...)$$

$9N = 7$ Subtract.

$\frac{9N}{9} = \frac{7}{9}$ Divide each side by 9.

$N = \frac{7}{9}$ Simplify.

Perfect squares can be used to simplify square roots of rational numbers.

Study Tip

Perfect Squares Keep a list of perfect squares in your notebook. Refer to it when you need to simplify a square root.

Key Concept Perfect Square

Words   Rational numbers with square roots that are rational numbers.

Examples   25 is a perfect square since $\sqrt{25} = 5$.

144 is a perfect square since $\sqrt{144} = 12$.

Example 4   Simplify Roots

Simplify each square root.

a. $\sqrt{\dfrac{4}{121}}$

$\sqrt{\dfrac{4}{121}} = \sqrt{\left(\dfrac{2}{11}\right)^2}$ $2^2 = 4$ and $11^2 = 121$

$= \dfrac{2}{11}$ Simplify.

b. $-\sqrt{\dfrac{49}{256}}$

$-\sqrt{\dfrac{49}{256}} = -\sqrt{\left(\dfrac{7}{16}\right)^2}$ $7^2 = 49$ and $16^2 = 256$

$= -\dfrac{7}{16}$

You can estimate roots that are not perfect squares.

Example 5   Estimate Roots

Estimate each square root to the nearest whole number.

a. $\sqrt{15}$

Find the two perfect squares closest to 15. List some perfect squares.

1, 4, 9, 16, 25, 36, ...

15 is between 9 and 16.

$9 < 15 < 16$ Write an inequality.

$\sqrt{9} < \sqrt{15} < \sqrt{16}$ Take the square root of each number.

$3 < \sqrt{15} < 4$ Simplify.

```
        3              4
    ←───┼──────────────┼───→
       √9            √15 √16
```

Since 15 is closer to 16 than 9, the best whole-number estimate for $\sqrt{15}$ is 4.

b. $\sqrt{130}$

Find the two perfect squares closest to 130. List some perfect squares.

81, 100, 121, 144

130 is between 121 and 144.

| $121 <$ | 130 | < 144 | Write an inequality. |

$$121 < 130 < 144 \qquad \text{Write an inequality.}$$
$$\sqrt{121} < \sqrt{130} < \sqrt{144} \qquad \text{Take the square root of each number.}$$
$$11 < \sqrt{130} < 12 \qquad \text{Simplify.}$$

Study Tip

Draw a Diagram
Graphing points on a number line can help you analyze your estimate for accuracy.

Since 130 is closer to 121 than to 144, the best whole number estimate for $\sqrt{130}$ is 11.

CHECK $\sqrt{130} \approx 11.4018$ Use a calculator.

Rounded to the nearest whole number, $\sqrt{130}$ is 11. So the estimate is valid.

Exercises

Name the set or sets of numbers to which each real number belongs.

1. $-\sqrt{64}$

2. $\dfrac{8}{3}$

3. $\sqrt{28}$

4. $\dfrac{56}{7}$

5. $-\sqrt{22}$

6. $\dfrac{36}{6}$

7. $-\dfrac{5}{12}$

8. $\dfrac{18}{3}$

9. $\sqrt{10.24}$

10. $\dfrac{-54}{19}$

11. $\sqrt{\dfrac{82}{20}}$

12. $-\dfrac{72}{8}$

Graph each set of numbers on a number line. Then order the numbers from least to greatest.

13. $\left\{ \dfrac{7}{5}, -\dfrac{3}{5}, \dfrac{3}{4}, -\dfrac{6}{5} \right\}$

14. $\left\{ \dfrac{1}{2}, -\dfrac{7}{9}, \dfrac{1}{9}, -\dfrac{4}{9} \right\}$

15. $\left\{ 2\dfrac{1}{4}, \sqrt{7}, 2.\overline{3}, \sqrt{8} \right\}$

16. $\left\{ \dfrac{4}{5}, \sqrt{2}, 0.\overline{1}, \sqrt{3} \right\}$

17. $\left\{ -3.5, -\dfrac{15}{5}, -\sqrt{10}, -3\dfrac{3}{4} \right\}$

18. $\left\{ \sqrt{64}, 8.8, \dfrac{26}{3}, 8\dfrac{2}{7} \right\}$

Write each repeating decimal as a fraction in simplest form.

19. $0.\overline{5}$

20. $0.\overline{4}$

21. $0.\overline{13}$

22. $0.\overline{21}$

Simplify each square root.

23. $-\sqrt{25}$

24. $\sqrt{361}$

25. $\pm\sqrt{36}$

26. $\sqrt{0.64}$

27. $\pm\sqrt{1.44}$

28. $-\sqrt{6.25}$

29. $\sqrt{\dfrac{16}{49}}$

30. $\sqrt{\dfrac{169}{196}}$

31. $\sqrt{\dfrac{25}{324}}$

Estimate each root to the nearest whole number.

32. $\sqrt{112}$

33. $\sqrt{252}$

34. $\sqrt{415}$

35. $\sqrt{670}$

Operations with Integers

∴ Objective

- Add, subtract, multiply, and divide integers.

New Vocabulary

absolute value

opposites

additive inverses

An integer is any number from the set {..., −3, −2, −1, 0, 1, 2, 3, ...}. You can use a number line to add integers.

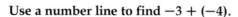

Example 1   Add Integers with the Same Sign

Use a number line to find −3 + (−4).

Step 1 Draw an arrow from 0 to −3.

Step 2 Draw a second arrow 4 units to the left to represent adding −4.

$$-3 + (-4) = -7$$

The second arrow ends at −7. So, −3 + (−4) = −7.

You can also use absolute value to add integers. The **absolute value** of a number is its distance from 0 on the number line.

Same Signs (+ + or − −)		Different Signs (+ − or − +)	
$3 + 5 = 8$	3 and 5 are positive. Their sum is positive.	$3 + (-5) = -2$	−5 has the greater absolute value. Their sum is negative.
$-3 + (-5) = -8$	−3 and −5 are negative. Their sum is negative.	$-3 + 5 = 2$	5 has the greater absolute value. Their sum is positive.

Example 2   Add Integers Using Absolute Value

Find −11 + (−7).

$$-11 + (-7) = -(|-11| + |-7|)$$ Add the absolute values. Both numbers are negative, so the sum is negative.

$$= -(11 + 7)$$ Absolute values of nonzero numbers are always positive.

$$= -18$$ Simplify.

Every positive integer can be paired with a negative integer. These pairs are called **opposites**. A number and its opposite are **additive inverses**. Additive inverses can be used when you subtract integers.

Example 3   Subtract Positive Integers

Find 18 − 23.

$$18 - 23 = 18 + (-23)$$ To subtract 23, add its inverse.

$$= -(|-23| - |18|)$$ Subtract the absolute values. Because $|-23|$ is greater than $|18|$, the result is negative.

$$= -(23 - 18)$$ Absolute values of nonzero numbers are always positive.

$$= -5$$ Simplify.

Study Tip

Products and Quotients
The product or quotient of two numbers having the *same sign* is positive. The product or quotient of two numbers having *different signs* is negative.

	Same Signs (+ + or − −)		Different Signs (+ − or − +)	
3(5) = 15	3 and 5 are positive. Their product is positive.	3(−5) = −15	3 and −5 have different signs. Their product is negative.	
−3(−5) = 15	−3 and −5 are negative. Their product is positive.	−3(5) = −15	−3 and 5 have different signs. Their product is negative.	

Example 4   Multiply and Divide Integers

Find each product or quotient.

a. $4(-5)$

$4(-5) = -20$ different signs ⟶ negative product

b. $-51 \div (-3)$

$-51 \div (-3) = 17$ same sign ⟶ positive quotient

c. $-12(-14)$

$-12(-14) = 168$ same sign ⟶ positive product

d. $-63 \div 7$

$-63 \div 7 = -9$ different signs ⟶ negative quotient

Exercises

Find each sum or difference.

1. $-8 + 13$

2. $11 + (-19)$

3. $-19 - 8$

4. $-77 + (-46)$

5. $12 - 34$

6. $41 + (-56)$

7. $50 - 82$

8. $-47 - 13$

9. $-80 + 102$

Find each product or quotient.

10. $5(18)$

11. $60 \div 12$

12. $-12(15)$

13. $-64 \div (-8)$

14. $8(-22)$

15. $54 \div (-6)$

16. $30(14)$

17. $-23(5)$

18. $-200 \div 2$

19. WEATHER The outside temperature was −4°F in the morning and 13°F in the afternoon. By how much did the temperature increase?

20. DOLPHINS A dolphin swimming 24 feet below the ocean's surface dives 18 feet straight down. How many feet below the ocean's surface is the dolphin now?

21. MOVIES A movie theater gave out 50 coupons for $3 off each movie. What is the total amount of discounts provided by the theater?

22. WAGES Emilio earns $11 per hour. He works 14 hours a week. His employer withholds $32 from each paycheck for taxes. If he is paid weekly, what is the amount of his paycheck?

23. FINANCIAL LITERACY Talia is working on a monthly budget. Her monthly income is $500. She has allocated $200 for savings, $100 for vehicle expenses, and $75 for clothing. How much is available to spend on entertainment?

Adding and Subtracting Rational Numbers

Objective

- Compare and order, add and subtract rational numbers.

You can use different methods to compare rational numbers. One way is to compare two fractions with common denominators. Another way is to compare decimals.

Example 1   Compare Rational Numbers

Replace ● with <, >, or = to make $\frac{2}{3}$ ● $\frac{5}{6}$ a true sentence.

Method 1 Write the fractions with the same denominator.

The least common denominator of $\frac{2}{3}$ and $\frac{5}{6}$ is 6.

$$\frac{2}{3} = \frac{4}{6}$$

$$\frac{5}{6} = \frac{5}{6}$$

Since $\frac{4}{6} < \frac{5}{6}$, $\frac{2}{3} < \frac{5}{6}$.

Method 2 Write as decimals.

Write $\frac{2}{3}$ and $\frac{5}{6}$ as decimals. You may want to use a calculator.

2 ÷ 3 ENTER .6666666667

So, $\frac{2}{3} = 0.\overline{6}$.

5 ÷ 6 ENTER .8333333333

So, $\frac{5}{6} = 0.8\overline{3}$.

Since $0.\overline{6} < 0.8\overline{3}$, $\frac{2}{3} < \frac{5}{6}$.

You can order rational numbers by writing all of the fractions as decimals.

Example 2   Order Rational Numbers

Order $5\frac{2}{9}$, $5\frac{3}{8}$, 4.9, and $-5\frac{3}{5}$ from least to greatest.

$$5\frac{2}{9} = 5.\overline{2} \qquad\qquad 5\frac{3}{8} = 5.375$$

$$4.9 = 4.9 \qquad\qquad -5\frac{3}{5} = -5.6$$

$-5.6 < 4.9 < 5.\overline{2} < 5.375$. So, from least to greatest, the numbers are $-5\frac{3}{5}$, 4.9, $5\frac{2}{9}$, and $5\frac{3}{8}$.

To add or subtract fractions with the same denominator, add or subtract the numerators and write the sum or difference over the denominator.

Example 3   Add and Subtract Like Fractions

Find each sum or difference. Write in simplest form.

a. $\dfrac{3}{5} + \dfrac{1}{5}$

$$\dfrac{3}{5} + \dfrac{1}{5} = \dfrac{3+1}{5}$$ The denominators are the same. Add the numerators.

$$= \dfrac{4}{5}$$ Simplify.

b. $\dfrac{7}{16} - \dfrac{1}{16}$

$$\dfrac{7}{16} - \dfrac{1}{16} = \dfrac{7-1}{16}$$ The denominators are the same. Subtract the numerators.

$$= \dfrac{6}{16}$$ Simplify.

$$= \dfrac{3}{8}$$ Rename the fraction.

c. $\dfrac{4}{9} - \dfrac{7}{9}$

$$\dfrac{4}{9} - \dfrac{7}{9} = \dfrac{4-7}{9}$$ The denominators are the same. Subtract the numerators.

$$= -\dfrac{3}{9}$$ Simplify.

$$= -\dfrac{1}{3}$$ Rename the fraction.

> **Study Tip**
>
> **Mental Math** If the denominators of the fractions are the same, you can use mental math to determine the sum or difference.

To add or subtract fractions with unlike denominators, first find the least common denominator (LCD). Rename each fraction with the LCD, and then add or subtract. Simplify if possible.

Example 4   Add and Subtract Unlike Fractions

Find each sum or difference. Write in simplest form.

a. $\dfrac{1}{2} + \dfrac{2}{3}$

$$\dfrac{1}{2} + \dfrac{2}{3} = \dfrac{3}{6} + \dfrac{4}{6}$$ The LCD for 2 and 3 is 6. Rename $\dfrac{1}{2}$ as $\dfrac{3}{6}$ and $\dfrac{2}{3}$ as $\dfrac{4}{6}$.

$$= \dfrac{3+4}{6}$$ Add the numerators.

$$= \dfrac{7}{6} \text{ or } 1\dfrac{1}{6}$$ Simplify.

b. $\dfrac{3}{8} - \dfrac{1}{3}$

$$\dfrac{3}{8} - \dfrac{1}{3} = \dfrac{9}{24} - \dfrac{8}{24}$$ The LCD for 8 and 3 is 24. Rename $\dfrac{3}{8}$ as $\dfrac{9}{24}$ and $\dfrac{1}{3}$ as $\dfrac{8}{24}$.

$$= \dfrac{9-8}{24}$$ Subtract the numerators.

$$= \dfrac{1}{24}$$ Simplify.

c. $\dfrac{2}{5} - \dfrac{3}{4}$

$$\dfrac{2}{5} - \dfrac{3}{4} = \dfrac{8}{20} - \dfrac{15}{20}$$ The LCD for 5 and 4 is 20. Rename $\dfrac{2}{5}$ as $\dfrac{8}{20}$ and $\dfrac{3}{4}$ as $\dfrac{15}{20}$.

$$= \dfrac{8-15}{20}$$ Subtract the numerators.

$$= -\dfrac{7}{20}$$ Simplify.

You can use a number line to add rational numbers.

Example 5   Add Decimals

Use a number line to find $2.5 + (-3.5)$.

Step 1 Draw an arrow from 0 to 2.5.

Step 2 Draw a second arrow 3.5 units to the left.

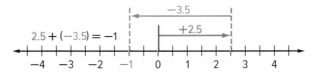

The second arrow ends at -1.

So, $2.5 + (-3.5) = -1$.

You can also use absolute value to add rational numbers.

Same Signs $(+ + \text{ or } - -)$		Different Signs $(+ - \text{ or } - +)$	
$3.1 + 2.5 = 5.6$	3.1 and 2.5 are positive, so the sum is positive.	$3.1 + (-2.5) = 0.6$	3.1 has the greater absolute value, so the sum is positive.
$-3.1 + (-2.5) = -5.6$	-3.1 and -2.5 are negative, so the sum is negative.	$-3.1 + 2.5 = -0.6$	-3.1 has the greater absolute value, so the sum is negative.

Example 6   Use Absolute Value to Add Rational Numbers

Find each sum.

a. $-13.12 + (-8.6)$

$$-13.12 + (-8.6) = -(|-13.12| + |-8.6|)$$ Both numbers are negative, so the sum is negative.

$$= -(13.12 + 8.6)$$ Absolute values of nonzero numbers are always positive.

$$= -21.72$$ Simplify.

b. $\dfrac{7}{16} + \left(-\dfrac{3}{8}\right)$

$$\dfrac{7}{16} + \left(-\dfrac{3}{8}\right) = \dfrac{7}{16} + \left(-\dfrac{6}{16}\right)$$ The LCD is 16. Replace $-\dfrac{3}{8}$ with $-\dfrac{6}{16}$.

$$= \left(\left|\dfrac{7}{16}\right| - \left|-\dfrac{6}{16}\right|\right)$$ Subtract the absolute values. Because $\left|\dfrac{7}{16}\right|$ is greater than $\left|-\dfrac{6}{16}\right|$, the result is positive.

$$= \dfrac{7}{16} - \dfrac{6}{16}$$ Absolute values of nonzero numbers are always positive.

$$= \dfrac{1}{16}$$ Simplify.

To subtract a negative rational number, add its inverse.

Example 7	Subtract Decimals

Find $-32.25 - (-42.5)$.

$-32.25 - (-42.5) = -32.25 + 42.5$ To subtract -42.5, add its inverse.

$ = |42.5| - |-32.25|$ Subtract the absolute values. Because $|42.5|$ is greater than $|-32.25|$, the result is positive.

$ = 42.5 - 32.25$ Absolute values of nonzero numbers are always positive.

$ = 10.25$ Simplify.

Exercises

Replace each ● with $<$, $>$, or $=$ to make a true sentence.

1. $-\dfrac{5}{8}$ ● $\dfrac{3}{8}$ **2.** $\dfrac{4}{5}$ ● 0.71 **3.** $\dfrac{5}{6}$ ● 0.875

4. 1.2 ● $1\dfrac{2}{9}$ **5.** $\dfrac{8}{15}$ ● $0.5\overline{3}$ **6.** $-\dfrac{7}{11}$ ● $-\dfrac{2}{3}$

Order each set of rational numbers from least to greatest.

7. $3.8, 3.06, 3\dfrac{1}{6}, 3\dfrac{3}{4}$ **8.** $2\dfrac{1}{4}, 1\dfrac{7}{8}, 1.75, 2.4$

9. $0.11, -\dfrac{1}{9}, -0.5, \dfrac{1}{10}$ **10.** $-4\dfrac{3}{5}, -3\dfrac{2}{5}, -4.65, -4.09$

Find each sum or difference. Write in simplest form.

11. $\dfrac{2}{5} + \dfrac{1}{5}$ **12.** $\dfrac{3}{9} + \dfrac{4}{9}$ **13.** $\dfrac{5}{16} - \dfrac{4}{16}$

14. $\dfrac{6}{7} - \dfrac{3}{7}$ **15.** $\dfrac{2}{3} + \dfrac{1}{3}$ **16.** $\dfrac{5}{8} + \dfrac{7}{8}$

17. $\dfrac{4}{3} + \dfrac{4}{3}$ **18.** $\dfrac{7}{15} - \dfrac{2}{15}$ **19.** $\dfrac{1}{3} - \dfrac{2}{9}$

20. $\dfrac{1}{2} + \dfrac{1}{4}$ **21.** $\dfrac{1}{2} - \dfrac{1}{3}$ **22.** $\dfrac{3}{7} + \dfrac{5}{14}$

23. $\dfrac{7}{10} - \dfrac{2}{15}$ **24.** $\dfrac{3}{8} + \dfrac{1}{6}$ **25.** $\dfrac{13}{20} - \dfrac{2}{5}$

Find each sum or difference. Write in simplest form if necessary.

26. $-1.6 + (-3.8)$ **27.** $-32.4 + (-4.5)$ **28.** $-38.9 + 24.2$

29. $-9.16 - 10.17$ **30.** $26.37 + (-61.1)$ **31.** $72.5 - (-81.3)$

32. $43.2 + (-27.9)$ **33.** $79.3 - (-14)$ **34.** $1.34 - (-0.458)$

35. $-\dfrac{1}{6} - \dfrac{2}{3}$ **36.** $\dfrac{1}{2} - \dfrac{4}{5}$ **37.** $-\dfrac{2}{5} + \dfrac{17}{20}$

38. $-\dfrac{4}{5} + \left(-\dfrac{1}{3}\right)$ **39.** $-\dfrac{1}{12} - \left(-\dfrac{3}{4}\right)$ **40.** $-\dfrac{7}{8} - \left(-\dfrac{3}{16}\right)$

41. GEOGRAPHY About $\dfrac{7}{10}$ of the surface of Earth is covered by water. The rest of the surface is covered by land. How much of Earth's surface is covered by land?

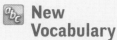

New Vocabulary

multiplicative inverses

reciprocals

The product or quotient of two rational numbers having the *same sign* is positive. The product or quotient of two rational numbers having *different signs* is negative.

Example 1   Multiply and Divide Decimals

Find each product or quotient.

a. $7.2(-0.2)$

different signs ⟶ negative product

$7.2(-0.2) = -1.44$

b. $-23.94 \div (-10.5)$

same sign ⟶ positive quotient

$-23.94 \div (-10.5) = 2.28$

To multiply fractions, multiply the numerators and multiply the denominators. If the numerators and denominators have common factors, you can simplify before you multiply by canceling.

Example 2   Multiply Fractions

Find each product.

a. $\dfrac{2}{5} \cdot \dfrac{1}{3}$

$\dfrac{2}{5} \cdot \dfrac{1}{3} = \dfrac{2 \cdot 1}{5 \cdot 3}$ Multiply the numerators.
Multiply the denominators.

$= \dfrac{2}{15}$ Simplify.

b. $\dfrac{3}{5} \cdot 1\dfrac{1}{2}$

$\dfrac{3}{5} \cdot 1\dfrac{1}{2} = \dfrac{3}{5} \cdot \dfrac{3}{2}$ Write $1\dfrac{1}{2}$ as an improper fraction.

$= \dfrac{3 \cdot 3}{5 \cdot 2}$ Multiply the numerators.
Multiply the denominators.

$= \dfrac{9}{10}$ Simplify.

c. $\dfrac{1}{4} \cdot \dfrac{2}{9}$

$\dfrac{1}{4} \cdot \dfrac{2}{9} = \dfrac{1}{\underset{2}{\cancel{4}}} \cdot \dfrac{\overset{1}{\cancel{2}}}{9}$ Divide by the GCF, 2.

$= \dfrac{1 \cdot 1}{2 \cdot 9}$ or $\dfrac{1}{18}$ Multiply the numerators.
Multiply the denominators and simplify.

Example 3   Multiply Fractions with Different Signs

Find $\left(-\dfrac{3}{4}\right)\left(\dfrac{3}{8}\right)$.

$\left(-\dfrac{3}{4}\right)\left(\dfrac{3}{8}\right) = -\left(\dfrac{3}{4} \cdot \dfrac{3}{8}\right)$ different signs ⟶ negative product

$= -\left(\dfrac{3 \cdot 3}{4 \cdot 8}\right)$ or $-\dfrac{9}{32}$ Multiply the numerators.
Multiply the denominators and simplify.

Two numbers whose product is 1 are called **multiplicative inverses** or **reciprocals**.

Example 4   Find the Reciprocal

Name the reciprocal of each number.

a. $\frac{3}{8}$

$\frac{3}{8} \cdot \frac{8}{3} = 1$ The product is 1.

The reciprocal of $\frac{3}{8}$ is $\frac{8}{3}$.

b. $2\frac{4}{5}$

$2\frac{4}{5} = \frac{14}{5}$ Write $2\frac{4}{5}$ as $\frac{14}{5}$.

$\frac{14}{5} \cdot \frac{5}{14} = 1$ The product is 1.

The reciprocal of $2\frac{4}{5}$ is $\frac{5}{14}$.

To divide one fraction by another fraction, multiply the dividend by the reciprocal of the divisor.

Example 5   Divide Fractions

Find each quotient.

a. $\frac{1}{3} \div \frac{1}{2}$

$\frac{1}{3} \div \frac{1}{2} = \frac{1}{3} \cdot \frac{2}{1}$ Multiply $\frac{1}{3}$ by $\frac{2}{1}$, the reciprocal of $\frac{1}{2}$.

$= \frac{2}{3}$ Simplify.

b. $\frac{3}{8} \div \frac{2}{3}$

$\frac{3}{8} \div \frac{2}{3} = \frac{3}{8} \cdot \frac{3}{2}$ Multiply $\frac{3}{8}$ by $\frac{3}{2}$, the reciprocal of $\frac{2}{3}$.

$= \frac{9}{16}$ Simplify.

c. $\frac{3}{4} \div 2\frac{1}{2}$

$\frac{3}{4} \div 2\frac{1}{2} = \frac{3}{4} \div \frac{5}{2}$ Write $2\frac{1}{2}$ as an improper fraction

$= \frac{3}{4} \cdot \frac{2}{5}$ Multiply $\frac{3}{4}$ by $\frac{2}{5}$, the reciprocal of $2\frac{1}{2}$.

$= \frac{6}{20}$ or $\frac{3}{10}$ Simplify.

d. $-\frac{1}{5} \div \left(-\frac{3}{10}\right)$

$-\frac{1}{5} \div \left(-\frac{3}{10}\right) = -\frac{1}{5} \cdot \left(-\frac{10}{3}\right)$ Multiply $-\frac{1}{5}$ by $-\frac{10}{3}$, the reciprocal of $-\frac{3}{10}$.

$= \frac{10}{15}$ or $\frac{2}{3}$ Same sign ⟶ positive quotient; simplify.

Study Tip

Use Estimation You can justify your answer by using estimation. $\frac{3}{8}$ is close to $\frac{1}{2}$ and $\frac{2}{3}$ is close to 1. So, the quotient is close to $\frac{1}{2}$ divided by 1 or $\frac{1}{2}$.

Find each product or quotient. Round to the nearest hundredth if necessary.

1. $6.5(0.13)$

2. $-5.8(2.3)$

3. $42.3 \div (-6)$

4. $-14.1(-2.9)$

5. $-78 \div (-1.3)$

6. $108 \div (-0.9)$

7. $0.75(-6.4)$

8. $-23.94 \div 10.5$

9. $-32.4 \div 21.3$

Find each product. Simplify before multiplying if possible.

10. $\frac{3}{4} \cdot \frac{1}{5}$

11. $\frac{2}{5} \cdot \frac{3}{7}$

12. $-\frac{1}{3} \cdot \frac{2}{5}$

13. $-\frac{2}{3} \cdot \left(-\frac{1}{11}\right)$

14. $2\frac{1}{2} \cdot \left(-\frac{1}{4}\right)$

15. $3\frac{1}{2} \cdot 1\frac{1}{2}$

16. $\frac{2}{9} \cdot \frac{1}{2}$

17. $\frac{3}{2} \cdot \left(-\frac{1}{3}\right)$

18. $\frac{1}{3} \cdot \frac{6}{5}$

19. $-\frac{9}{4} \cdot \frac{1}{18}$

20. $\frac{11}{3} \cdot \frac{9}{44}$

21. $\left(-\frac{30}{11}\right) \cdot \left(-\frac{1}{3}\right)$

22. $-\frac{3}{5} \cdot \frac{5}{6}$

23. $\left(-\frac{1}{3}\right)\left(-7\frac{1}{2}\right)$

24. $\frac{2}{7} \cdot 4\frac{2}{3}$

Name the reciprocal of each number.

25. $\frac{6}{7}$

26. $\frac{1}{22}$

27. $-\frac{14}{23}$

28. $2\frac{3}{4}$

29. $-5\frac{1}{3}$

30. $3\frac{3}{4}$

Find each quotient.

31. $\frac{2}{3} \div \frac{1}{3}$

32. $\frac{16}{9} \div \frac{4}{9}$

33. $\frac{3}{2} \div \frac{1}{2}$

34. $\frac{3}{7} \div \left(-\frac{1}{5}\right)$

35. $-\frac{9}{10} \div 3$

36. $\frac{1}{2} \div \frac{3}{5}$

37. $2\frac{1}{4} \div \frac{1}{2}$

38. $-1\frac{1}{3} \div \frac{2}{3}$

39. $\frac{11}{12} \div 1\frac{2}{3}$

40. $4 \div \left(-\frac{2}{7}\right)$

41. $-\frac{1}{3} \div \left(-1\frac{1}{5}\right)$

42. $\frac{3}{25} \div \frac{2}{15}$

43. PIZZA A large pizza at Pizza Shack has 12 slices. If Bobby ate $\frac{1}{4}$ of the pizza, how many slices of pizza did he eat?

44. MUSIC Samantha practices the flute for $4\frac{1}{2}$ hours each week. How many hours does she practice in a month?

45. BAND How many band uniforms can be made with $131\frac{3}{4}$ yards of fabric if each uniform requires $3\frac{7}{8}$ yards?

46. CARPENTRY How many boards, each 2 feet 8 inches long, can be cut from a board 16 feet long if there is no waste?

47. SEWING How many 9-inch ribbons can be cut from $1\frac{1}{2}$ yards of ribbon?

- Use and apply the percent proportion.

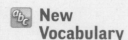

New Vocabulary

percent

percent proportion

A **percent** is a ratio that compares a number to 100. To write a percent as a fraction, express the ratio as a fraction with a denominator of 100. Fractions should be expressed in simplest form.

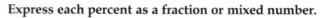

Example 1   Percents as Fractions

Express each percent as a fraction or mixed number.

a. 79%

$$79\% = \frac{79}{100} \qquad \text{Definition of percent}$$

b. 107%

$$107\% = \frac{107}{100} \qquad \text{Definition of percent}$$

$$= 1\frac{7}{100} \qquad \text{Simplify.}$$

c. 0.5%

$$0.5\% = \frac{0.5}{100} \qquad \text{Definition of percent}$$

$$= \frac{5}{1000} \qquad \text{Multiply the numerator and denominator by 10 to eliminate the decimal.}$$

$$= \frac{1}{200} \qquad \text{Simplify.}$$

In the **percent proportion**, the ratio of a part of something to the whole (base) is equal to the percent written as a fraction.

part $\longrightarrow$
whole $\longrightarrow$ $\dfrac{a}{b} = \dfrac{p}{100}$ $\longleftarrow$ percent

percent   whole   part
$\downarrow$ $\downarrow$ $\downarrow$
Example: 25% of 40 is 10.

You can use the percent proportion to find the part.

Example 2   Find the Part

40% of 30 is what number?

$$\frac{a}{b} = \frac{p}{100} \qquad \text{The percent is 40, and the base is 30. Let } a \text{ represent the part.}$$

$$\frac{a}{30} = \frac{40}{100} \qquad \text{Replace } b \text{ with 30 and } p \text{ with 40.}$$

$$100a = 30(40) \qquad \text{Find the cross products.}$$

$$100a = 1200 \qquad \text{Simplify.}$$

$$\frac{100a}{100} = \frac{1200}{100} \qquad \text{Divide each side by 100.}$$

$$a = 12 \qquad \text{Simplify.}$$

The part is 12. So, 40% of 30 is 12.

You can also use the percent proportion to find the percent of the base.

Example 3   Find the Percent

SURVEYS Kelsey took a survey of students in her lunch period. 42 out of the 70 students Kelsey surveyed said their family had a pet. What percent of the students had pets?

$\dfrac{a}{b} = \dfrac{p}{100}$ The part is 42, and the base is 70. Let p represent the percent.

$\dfrac{42}{70} = \dfrac{p}{100}$ Replace a with 42 and b with 70.

$4200 = 70p$ Find the cross products.

$\dfrac{4200}{70} = \dfrac{70p}{70}$ Divide each side by 70.

$60 = p$ Simplify.

The percent is 60, so $\dfrac{60}{100}$ or 60% of the students had pets.

Study Tip

Percent Proportion In percent problems, the whole, or base usually follows the word *of*.

Example 4   Find the Whole

67.5 is 75% of what number?

$\dfrac{a}{b} = \dfrac{p}{100}$ The percent is 75, and the part is 67.5. Let b represent the base.

$\dfrac{67.5}{b} = \dfrac{75}{100}$ Replace a with 67.5 and p with 75.

$6750 = 75b$ Find the cross products.

$\dfrac{6750}{75} = \dfrac{75b}{75}$ Divide each side by 75.

$90 = b$ Simplify.

The base is 90, so 67.5 is 75% of 90.

Exercises

Express each percent as a fraction or mixed number in simplest form.

1. 5%   **2.** 60%   **3.** 11%

4. 120%   **5.** 78%   **6.** 2.5%

7. 0.6%   **8.** 0.4%   **9.** 1400%

Use the percent proportion to find each number.

10. 25 is what percent of 125?   **11.** 16 is what percent of 40?

12. 14 is 20% of what number?   **13.** 50% of what number is 80?

14. What number is 25% of 18?   **15.** Find 10% of 95.

16. What percent of 48 is 30?   **17.** What number is 150% of 32?

18. 5% of what number is 3.5?   **19.** 1 is what percent of 400?

20. Find 0.5% of 250.   **21.** 49 is 200% of what number?

22. 15 is what percent of 12?   **23.** 36 is what percent of 24?

24. **BASKETBALL** Madeline usually makes 85% of her shots in basketball. If she attempts 20, how many will she likely make?

25. **TEST SCORES** Brian answered 36 items correctly on a 40-item test. What percent did he answer correctly?

26. **CARD GAMES** Juanita told her dad that she won 80% of the card games she played yesterday. If she won 4 games, how many games did she play?

27. **SOLUTIONS** A glucose solution is prepared by dissolving 6 milliliters of glucose in 120 milliliters of pure solution. What is the percent of glucose in the resulting solution?

28. **DRIVER'S ED** Kara needs to get a 75% on her driving education test in order to get her license. If there are 35 questions on the test, how many does she need to answer correctly?

29. **HEALTH** The U.S. Food and Drug Administration requires food manufacturers to label their products with a nutritional label. The label shows the information from a package of macaroni and cheese.

    a. The label states that a serving contains 3 grams of saturated fat, which is 15% of the daily value recommended for a 2000-Calorie diet. How many grams of saturated fat are recommended for a 2000-Calorie diet?

    b. The 470 milligrams of sodium (salt) in the macaroni and cheese is 20% of the recommended daily value. What is the recommended daily value of sodium?

    c. For a healthy diet, the National Research Council recommends that no more than 30 percent of the total Calories come from fat. What percent of the Calories in a serving of this macaroni and cheese come from fat?

Nutrition Facts		
Serving Size 1 cup (228g)		
Servings per container 2		
Amount per serving		
Calories 250 Calories from Fat 110		
		%Daily value*
Total Fat 12g		18%
Saturated Fat 3g		15%
Cholesterol 30mg		10%
Sodium 470mg		20%
Total Carbohydrate 31g		10%
Dietary Fiber 0g		0%
Sugars 5g		
Protein 5g		
Vitamin A 4%	•	Vitamin C 2%
Calcium 20%	•	Iron 4%

30. **TEST SCORES** The table shows the number of points each student in Will's study group earned on a recent math test. There were 88 points possible on the test. Express all answers to the nearest tenth of a percent.

Name	Will	Penny	Cheng	Minowa	Rob
Score	72	68	81	87	75

    a. Find Will's percent correct on the test.

    b. Find Cheng's percent correct on the test.

    c. Find Rob's percent correct on the test.

    d. What was the highest percentage? The lowest?

31. **PET STORE** In a pet store, 15% of the animals are hamsters. If the store has 40 animals, how many of them are hamsters?

Perimeter

● Find the perimeter of two-dimensional figures.

 New Vocabulary

perimeter
circle
diameter
circumference
center
radius

Perimeter is the distance around a figure. Perimeter is measured in linear units.

Rectangle

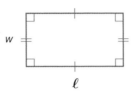

$P = 2(\ell + w)$ or
$P = 2\ell + 2w$

Parallelogram

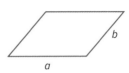

$P = 2(a + b)$ or
$P = 2a + 2b$

Square

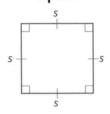

$P = 4s$

Triangle

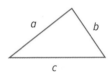

$P = a + b + c$

Example 1	Perimeters of Rectangles and Squares

Find the perimeter of each figure.

a. a rectangle with a length of 5 inches and a width of 1 inch

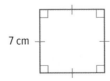

$P = 2(\ell + w)$ Perimeter formula

$= 2(5 + 1)$ $\ell = 5, w = 1$

$= 2(6)$ Add.

$= 12$ The perimeter is 12 inches.

b. a square with a side length of 7 centimeters

$P = 4s$ Perimeter formula

$= 4(7)$ Replace *s* with 7.

$= 28$ The perimeter is 28 centimeters.

Example 2 — Perimeters of Parallelograms and Triangles

Find the perimeter of each figure.

a.

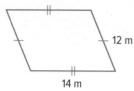

12 m

14 m

$P = 2(a + b)$ Perimeter formula

$= 2(14 + 12)$ $a = 14, b = 12$

$= 2(26)$ Add.

$= 52$ Multiply.

The perimeter of the parallelogram is 52 meters.

b.

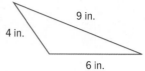

9 in.

4 in.

6 in.

$P = a + b + c$ Perimeter formula

$= 4 + 6 + 9$ $a = 4, b = 6, c = 9$

$= 19$ Add.

The perimeter of the triangle is 19 inches.

A **circle** is the set of all points in a plane that are the same distance from a given point.

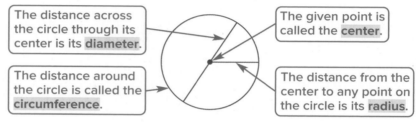

The distance across the circle through its center is its **diameter**.

The given point is called the **center**.

The distance around the circle is called the **circumference**.

The distance from the center to any point on the circle is its **radius**.

The formula for the circumference of a circle is $C = \pi d$ or $C = 2\pi r$.

Example 3 — Circumference

Find each circumference to the nearest tenth.

a. The radius is 4 feet.

$C = 2\pi r$ Circumference formula

$= 2\pi(4)$ Replace r with 4.

$= 8\pi$ Simplify.

The exact circumference is 8π feet.

8 [π] [ENTER] 25.13274123

The circumference is about 25.1 feet.

b. The diameter is 15 centimeters.

$C = \pi d$ Circumference formula

$= \pi(15)$ Replace d with 15.

$= 15\pi$ Simplify.

≈ 47.1 Use a calculator to evaluate 15π.

The circumference is about 47.1 centimeters.

c.

3 m

$C = 2\pi r$ Circumference formula

$= 2\pi(3)$ Replace r with 3.

$= 6\pi$ Simplify.

≈ 18.8 Use a calculator to evaluate 6π.

The circumference is about 18.8 meters.

Find the perimeter of each figure.

1.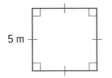
   5 m

2.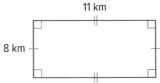
   11 km
   8 km

3.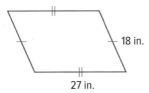
   18 in.
   27 in.

4.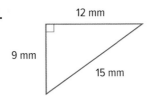
   12 mm
   9 mm
   15 mm

5. a square with side length 8 inches

6. a rectangle with length 9 centimeters and width 3 centimeters

7. a triangle with sides 4 feet, 13 feet, and 12 feet

8. a parallelogram with side lengths $6\frac{1}{4}$ inches and 5 inches

9. a quarter-circle with a radius of 7 inches

Find the circumference of each circle. Round to the nearest tenth.

10.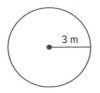
    3 m

11.
    10 in.

12.
    12 cm

13. **GARDENS** A square garden has a side length of 5.8 meters. What is the perimeter of the garden?

14. **ROOMS** A rectangular room is $12\frac{1}{2}$ feet wide and 14 feet long. What is the perimeter of the room?

15. **CYCLING** The tire for a 10-speed bicycle has a diameter of 27 inches. Find the distance traveled in 10 rotations of the tire. Round to the nearest tenth.

16. **GEOGRAPHY** Earth's circumference is approximately 25,000 miles. If you could dig a tunnel to the center of the Earth, how long would the tunnel be? Round to the nearest tenth mile.

Find the perimeter of each figure. Round to the nearest tenth.

17.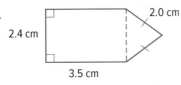
    2.0 cm
    2.4 cm
    3.5 cm

18.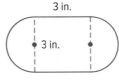
    3 in.
    3 in.

19.
    4 ft

20.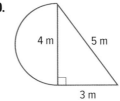
    4 m   5 m
    3 m

Area is the number of square units needed to cover a surface. Area is measured in square units.

Rectangle

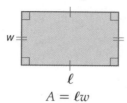

$A = \ell w$

Parallelogram

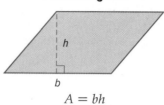

$A = bh$

Square

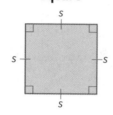

$A = s^2$

Triangle

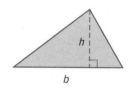

$A = \frac{1}{2}bh$

Example 1   Areas of Rectangles and Squares

Find the area of each figure.

a. a rectangle with a length of 7 yards and a width of 1 yard

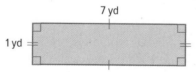

$A = \ell w$ Area formula

$= 7(1)$ $\ell = 7, w = 1$

$= 7$ The area of the rectangle is 7 square yards.

b. a square with a side length of 2 meters

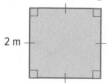

$A = s^2$ Area formula

$= 2^2$ $s = 2$

$= 4$ The area is 4 square meters.

Example 2   Areas of Parallelograms and Triangles

Find the area of each figure.

a. a parallelogram with a base of 11 feet and a height of 9 feet

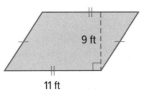

$A = bh$ Area formula

$\quad = 11(9)$ $b = 11, h = 9$

$\quad = 99$ Multiply.

The area is 99 square feet.

b. a triangle with a base of 12 millimeters and a height of 5 millimeters

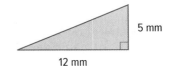

$A = \frac{1}{2}bh$ Area formula

$\quad = \frac{1}{2}(12)(5)$ $b = 12, h = 5$

$\quad = 30$ Multiply.

The area is 30 square millimeters.

The formula for the area of a circle is $A = \pi r^2$.

Example 3   Areas of Circles

Find the area of each circle to the nearest tenth.

a. a radius of 3 centimeters

$A = \pi r^2$ Area formula

$\quad = \pi(3)^2$ Replace r with 3.

$\quad = 9\pi$ Simplify.

$\quad \approx 28.3$ Use a calculator to evaluate 9π.

The area is about 28.3 square centimeters.

> **Study Tip**
>
> **Mental Math** You can use mental math to check your solutions. Square the radius and then multiply by 3.

b. a diameter of 21 meters

$A = \pi r^2$ Area formula

$\quad = \pi(10.5)^2$ Replace r with 10.5.

$\quad = 110.25\pi$ Simplify.

$\quad \approx 346.4$ Use a calculator to evaluate 110.25π.

The area is about 346.4 square meters.

Example 4   Estimate Area

Estimate the area of the polygon if each square represents 1 square mile.

One way to estimate the area is to count each square as one unit and each partial square as a half unit, no matter how large or small.

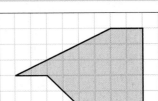

$A \approx$ squares $+$ partial squares

$\quad \approx 21(1) + 8(0.5)$ 21 whole squares and 8 partial squares

$\quad \approx 21 + 4$ or 25

The area of the polygon is about 25 square miles.

Find the area of each figure.

1.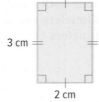

3 cm

2 cm

2.

6 in.

3.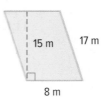

15 m   17 m

8 m

Find the area of each figure. Round to the nearest tenth if necessary.

4. a triangle with a base 12 millimeters and height 11 millimeters

5. a square with side length 9 feet

6. a rectangle with length 8 centimeters and width 2 centimeters

7. a triangle with a base 6 feet and height 3 feet

8. a quarter-circle with a diameter of 4 meters

9. a semi-circle with a radius of 3 inches

Find the area of each circle. Round to the nearest tenth.

10.

5 in.

11.

2 ft

12.

2 km

13. The radius is 4 centimeters.

14. The radius is 7.2 millimeters.

15. The diameter is 16 inches.

16. The diameter is 25 feet.

17. **CAMPING** The square floor of a tent has an area of 49 square feet. What is the side length of the tent?

Estimate the area of each polygon in square units.

18.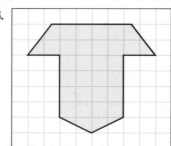

19.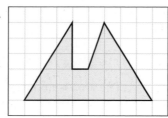

20. **ART** The Blueprints at Addison Circle is a sculpture that took 18,000 worker-hours to construct using over 410,000 pounds of steel. The piece of art is placed within a traffic circle with a diameter of 133 feet. Find the area of the circle. Round to the nearest tenth of a square foot.

Find the area of each figure. Round to the nearest tenth.

21.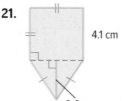

4.1 cm

2.6 cm

22.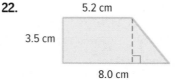

5.2 cm

3.5 cm

8.0 cm

23.

2.9 cm

1.2 cm

- Find the volumes of rectangular prisms and cylinders.

 **New Vocabulary**

volume

Volume is the measure of space occupied by a solid. Volume is measured in cubic units.

To find the volume of a rectangular prism, multiply the length times the width times the height. The formula for the volume of a rectangular prism is shown below.

$$V = \ell \cdot w \cdot h$$

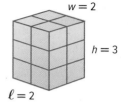

$w = 2$

$h = 3$

$\ell = 2$

The prism at the right has a volume of $2 \cdot 2 \cdot 3$ or 12 cubic units.

Example 1 Volumes of Rectangular Prisms

Find the volume of each rectangular prism.

a. The length is 8 centimeters, the width is 1 centimeter, and the height is 5 centimeters.

$V = \ell \cdot w \cdot h$ Volume formula
$= 8 \cdot 1 \cdot 5$ Replace ℓ with 8, w with 1, and h with 5.
$= 40$ Simplify.

The volume is 40 cubic centimeters.

b.

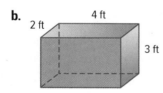

4 ft

2 ft

3 ft

The prism has a length of 4 feet, width of 2 feet, and height of 3 feet.

$V = \ell \cdot w \cdot h$ Volume formula
$= 4 \cdot 2 \cdot 3$ Replace ℓ with 4, w with 2, and h with 3.
$= 24$ Simplify.

The volume is 24 cubic feet.

The volume of a solid is the product of the area of the base and the height of the solid. For a cylinder, the area of the base is πr^2. So the volume is $V = \pi r^2 h$.

Example 2 Volume of a Cylinder

Find the volume of the cylinder.

$V = \pi r^2 h$ Volume of a cylinder

$= \pi(3^2)6$ $r = 3, h = 6$

$= 54\pi$ Simplify.

≈ 169.6 Use a calculator.

3 in.

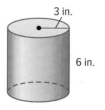

6 in.

The volume is about 169.6 cubic inches.

Find the volume of each rectangular prism given the length, width, and height.

1. $\ell = 5$ cm, $w = 3$ cm, $h = 2$ cm

2. $\ell = 10$ m, $w = 10$ m, $h = 1$ m

3. $\ell = 6$ yd, $w = 2$ yd, $h = 4$ yd

4. $\ell = 2$ in., $w = 5$ in., $h = 12$ in.

5. $\ell = 13$ ft, $w = 9$ ft, $h = 12$ ft

6. $\ell = 7.8$ mm, $w = 0.6$ mm, $h = 8$ mm

Find the volume of each rectangular prism.

7.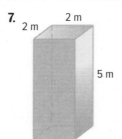

8.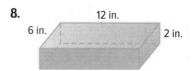

9. GEOMETRY A cube measures 3 meters on a side. What is its volume?

10. AQUARIUMS An aquarium is 8 feet long, 5 feet wide, and 5.5 feet deep. What is the volume of the aquarium?

11. COOKING What is the volume of a microwave oven that is 18 inches wide by 10 inches long with a depth of $11\frac{1}{2}$ inches?

12. BOXES A cardboard box is 32 inches long, 22 inches wide, and 16 inches tall. What is the volume of the box?

13. SWIMMING POOLS A children's rectangular pool holds 480 cubic feet of water. What is the depth of the pool if its length is 30 feet and its width is 16 feet?

14. BAKING A rectangular cake pan has a volume of 234 cubic inches. If the length of the pan is 9 inches and the width is 13 inches, what is the height of the pan?

15. GEOMETRY The volume of the rectangular prism at the right is 440 cubic centimeters. What is the width?

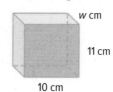

Find the volume of each cylinder. Round to the nearest tenth.

16.

17.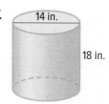

18.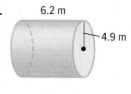

19. FIREWOOD Firewood is usually sold by a measure known as a *cord*. A full cord may be a stack 8 × 4 × 4 feet or a stack 8 × 8 × 2 feet.

a. What is the volume of a full cord of firewood?

b. A "short cord" of wood is 8 × 4 × the length of the logs. What is the volume of a short cord of $2\frac{1}{2}$-foot logs?

c. If you have an area that is 12 feet long and 2 feet wide in which to store your firewood, how high will the stack be if it is a full cord of wood?

∷Objective

● Find the surface areas of rectangular prisms and cylinders.

New Vocabulary

surface area

Surface area is the sum of the areas of all the surfaces, or faces, of a solid. Surface area is measured in square units.

Key Concept Surface Area

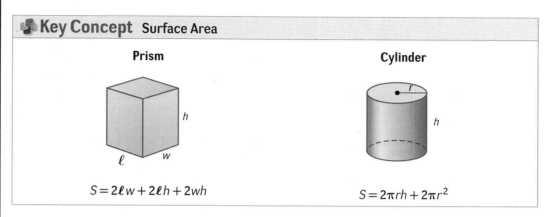

Prism

$$S = 2\ell w + 2\ell h + 2wh$$

Cylinder

$$S = 2\pi rh + 2\pi r^2$$

Example 1   Find Surface Areas

Find the surface area of each solid. Round to the nearest tenth if necessary.

a.

5 m

1 m

3 m

The prism has a length of 3 meters, width of 1 meter, and height of 5 meters.

$S = 2\ell w + 2\ell h + 2wh$	Surface area formula
$= 2(3)(1) + 2(3)(5) + 2(1)(5)$	$\ell = 3,\ w = 1,\ h = 5$
$= 6 + 30 + 10$	Multiply.
$= 46$	Add.

The surface area is 46 square meters.

b.

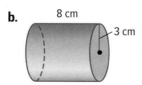

8 cm

3 cm

The height is 8 centimeters and the radius of the base is 3 centimeters. The surface area is the sum of the area of each base, $2\pi r^2$, and the area of the side, given by the circumference of the base times the height or $2\pi rh$.

$S = 2\pi rh + 2\pi r^2$	Formula for surface area of a cylinder
$= 2\pi(3)(8) + 2\pi(3^2)$	$r = 3,\ h = 8$
$= 48\pi + 18\pi$	Simplify.
$\approx 207.3 \text{ cm}^2$	Use a calculator.

Find the surface area of each rectangular prism given the measurements below.

1. $\ell = 6$ in., $w = 1$ in., $h = 4$ in

2. $\ell = 8$ m, $w = 2$ m, $h = 2$ m

3. $\ell = 10$ mm, $w = 4$ mm, $h = 5$ mm

4. $\ell = 6.2$ cm, $w = 1$ cm, $h = 3$ cm

5. $\ell = 7$ ft, $w = 2$ ft, $h = \frac{1}{2}$ ft

6. $\ell = 7.8$ m, $w = 3.4$ m, $h = 9$ m

Find the surface area of each solid.

7.

2 m
2 m
5 m

8.

4 ft
2 ft
3 ft

9.

12 in.
6 in.
2 in.

10.

5 mm
8 mm
1.2 mm

11.

4.5 in.
12.5 in.

12.

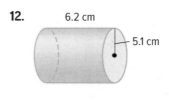
6.2 cm
5.1 cm

13. GEOMETRY What is the surface area of a cube with a side length of 2 meters?

14. GIFTS A gift box is a rectangular prism 14 inches long, 5 inches wide, and 4 inches high. If the box is to be covered in fabric, how much fabric is needed if there is no overlap?

15. BOXES A new refrigerator is shipped in a box 34 inches deep, 66 inches high, and $33\frac{1}{4}$ inches wide. What is the surface area of the box in square feet? Round to the nearest square foot. (*Hint:* 1 ft^2 = 144 in^2)

16. PAINTING A cabinet is 6 feet high, 3 feet wide, and 2 feet long. The entire outside surface of the cabinet is being painted except for the bottom. What is the surface area of the cabinet that is being painted?

17. SOUP A soup can is 4 inches tall and has a diameter of $3\frac{1}{4}$ inches. How much paper is needed for the label on the can? Round your answer to the nearest tenth.

18. CRAFTS For a craft project, Sarah is covering all the sides of a box with stickers. The length of the box is 8 inches, the width is 6 inches, and the height is 4 inches. If each sticker has a length of 2 inches and a width of 4 inches, how many stickers does she need to cover the box?

Simple Probability and Odds

Eyewire/Getty Images

::Objective

- Find the probability and odds of simple events.

New Vocabulary

probability
sample space
equally likely
complements
tree diagram
Fundamental Counting Principle
odds

The **probability** of an event is the ratio of the number of favorable outcomes for the event to the total number of possible outcomes. When you roll a die, there are six possible outcomes: 1, 2, 3, 4, 5, or 6. This list of all possible outcomes is called the **sample space**.

When there are n outcomes and the probability of each one is $\frac{1}{n}$, we say that the outcomes are **equally likely**.

For example, when you roll a die, the 6 possible outcomes are equally likely because each outcome has a probability of $\frac{1}{6}$. The probability of an event is always between 0 and 1, inclusive. The closer a probability is to 1, the more likely it is to occur.

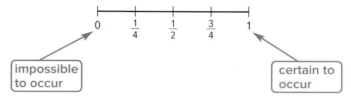

impossible to occur

certain to occur

Example 1   Find Probabilities

A die is rolled. Find each probability.

a. rolling a 1 or 5

There are six possible outcomes. There are two favorable outcomes, 1 and 5.

$$\text{probability} = \frac{\text{number of favorable outcomes}}{\text{total number of possible outcomes}} = \frac{2}{6}$$

So, $P(1 \text{ or } 5) = \frac{2}{6}$ or $\frac{1}{3}$.

b. rolling an even number

Three of the six outcomes are even numbers. So, there are three favorable outcomes.

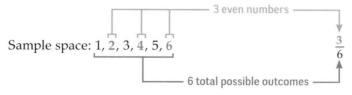

3 even numbers

Sample space: 1, 2, 3, 4, 5, 6 $\frac{3}{6}$

6 total possible outcomes

So, $P(\text{even number}) = \frac{3}{6}$ or $\frac{1}{2}$.

The events for rolling a 1 and for *not* rolling a 1 are called **complements**.

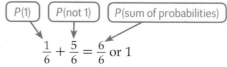

$P(1)$ $P(\text{not 1})$ $P(\text{sum of probabilities})$

$$\frac{1}{6} + \frac{5}{6} = \frac{6}{6} \text{ or } 1$$

The sum of the probabilities for any two complementary events is always 1.

Example 2   Find Probabilities

A bowl contains 5 red chips, 7 blue chips, 6 yellow chips, and 10 green chips. One chip is randomly drawn. Find each probability.

a. blue

There are 7 blue chips and 28 total chips.

$$P(\text{blue chip}) = \frac{7}{28} \qquad \longleftarrow \text{number of favorable outcomes}$$
$$\longleftarrow \text{number of possible outcomes}$$
$$= \frac{1}{4}$$

The probability can be stated as $\frac{1}{4}$, 0.25, or 25%.

b. red or yellow

There are 5 + 6 or 11 chips that are red or yellow.

$$P(\text{red or yellow}) = \frac{11}{28} \qquad \longleftarrow \text{number of favorable outcomes}$$
$$\longleftarrow \text{number of possible outcomes}$$
$$\approx 0.39$$

The probability can be stated as $\frac{11}{28}$, about 0.39, or about 39%.

c. not green

There are 5 + 7 + 6 or 18 chips that are not green.

$$P(\text{not green}) = \frac{18}{28} \qquad \longleftarrow \text{number of favorable outcomes}$$
$$\longleftarrow \text{number of possible outcomes}$$
$$= \frac{9}{14} \text{ or about } 0.64$$

The probability can be stated as $\frac{9}{14}$, about 0.64, or about 64%.

> **Study Tip**
>
> **Alternate Method** A chip drawn will either be green or not green. So, another method for finding P(not green) is to find P(green) and subtract that probability from 1.

One method used for counting the number of possible outcomes is to draw a **tree diagram**. The last column of a tree diagram shows all of the possible outcomes.

Example 3   Use a Tree Diagram to Count Outcomes

School baseball caps come in blue, yellow, or white. The caps have either the school mascot or the school's initials. Use a tree diagram to determine the number of different caps possible.

> **Study Tip**
>
> **Counting Outcomes** When counting possible outcomes, make a column in your tree diagram for each part of the event.

Color	Design	Outcomes
blue	mascot	blue, mascot
	initials	blue, initials
yellow	mascot	yellow, mascot
	initials	yellow, initials
white	mascot	white, mascot
	initials	white, initials

The tree diagram shows that there are 6 different caps possible.

This example is an illustration of the **Fundamental Counting Principle**, which relates the number of outcomes to the number of choices.

Key Concept — Fundamental Counting Principle

Words	If event *M* can occur in *m* ways and is followed by event *N* that can occur in *n* ways, then the event *M* followed by *N* can occur in *m* • *n* ways.
Example	If there are 4 possible sizes for fish tanks and 3 possible shapes, then there are 4 • 3 or 12 possible fish tanks.

Example 4 — Use the Fundamental Counting Principle

a. An ice cream shop offers one, two, or three scoops of ice cream from among 12 different flavors. The ice cream can be served in a wafer cone, a sugar cone, or in a cup. Use the Fundamental Counting Principle to determine the number of choices possible.

There are 3 ways the ice cream is served, 3 different servings, and there are 12 different flavors of ice cream.

Use the Fundamental Counting Principle to find the number of possible choices.

number of scoops	number of flavors	number of serving options	number of choices of ordering ice cream
3 •	12 •	3 =	108

So, there are 108 different ways to order ice cream.

b. Jimmy needs to make a 3-digit password for his log-on name on a Web site. The password can include any digit from 0-9, but the digits may not repeat. How many possible 3-digit passwords are there?

If the first digit is a 4, then the next digit cannot be a 4.

We can use the Fundamental Counting Principle to find the number of possible passwords.

1st digit	2nd digit	3rd digit	number of passwords
10 •	9 •	8 =	720

So, there are 720 possible 3-digit passwords.

Study Tip

Odds The sum of the number of successes and the number of failures equals the size of the sample space, or the number of possible outcomes.

The **odds** of an event occurring is the ratio that compares the number of ways an event can occur (successes) to the number of ways it cannot occur (failures).

Example 5 — Find the Odds

Find the odds of rolling a number less than 3.

There are six possible outcomes; 2 are successes and 4 are failures.

So, the odds of rolling a number less than 3 are $\frac{1}{2}$ or 1:2.

One coin is randomly selected from a jar containing 70 nickels, 100 dimes, 80 quarters, and 50 one-dollar coins. Find each probability.

1. *P*(quarter)

2. *P*(dime)

3. *P*(quarter or nickel)

4. *P*(value greater than $0.10)

5. *P*(value less than $1)

6. *P*(value at most $1)

One of the polygons below is chosen at random. Find each probability.

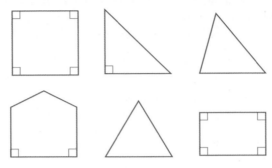

7. *P*(triangle)

8. *P*(pentagon)

9. *P*(not a quadrilateral)

10. *P*(more than 2 right angles)

Use a tree diagram to find the sample space for each event. State the number of possible outcomes.

11. The spinner at the right is spun and two coins are tossed.

12. At a restaurant, you choose two sides to have with breakfast. You can choose white or whole wheat toast. You can choose sausage links, sausage patties, or bacon.

13. How many different 3-character codes are there using A, B, or C for the first character, 8 or 9 for the second character, and 0 or 1 for the third character?

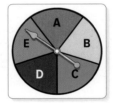

A bag is full of different colored marbles. The probability of randomly selecting a red marble from the bag is $\frac{1}{8}$. The probability of selecting a blue marble is $\frac{13}{24}$. Find each probability.

14. *P*(not red)

15. *P*(not blue)

Find the odds of each outcome if a computer randomly picks a letter in the name THE UNITED STATES OF AMERICA.

16. the letter *A*

17. the letter *T*

18. a vowel

19. a consonant

Margaret wants to order a sub at the local deli.

20. Find the number of possible orders of a sub with one topping and one dressing option.

21. Find the number of possible ham subs with mayonnaise, any combination of toppings or no toppings at all.

22. Find the number of possible orders of a sub with any combination of dressing and/or toppings.

Subs	
ham, salami, roast beef, turkey, bologna, pepperoni	
Dressing	Toppings
mayonnaise, mustard, vinegar, oil	lettuce, onions, peppers, olives

Determine whether you need an estimate or an exact answer. Then use the four-step problem-solving plan to solve.

1. **DISTANCE** Fabio rode his scooter 2.3 miles to his friend's house, then 0.7 mile to the grocery store, then 2.1 miles to the library. If he rode the same route back home, about how far did he travel in all?

2. **SHOPPING** The regular price of a T-shirt is $19.99. It is on sale for 15% off. Sales tax is 6%. If you give the cashier a $20 bill, how much change will you receive?

Find each sum or difference.

3. $-31 + (-4)$

4. $48 - 55$

5. $-71 - (-10)$

6. $31 - 42.9$

7. $-11.5 + 8.1$

8. $-0.38 - (-1.06)$

Find each product or quotient.

9. $-21(-5)$

10. $-81 \div (-3)$

11. $-120 \div 8$

12. $-39 \div -3$

Replace each ● with <, >, or = to make a true sentence.

13. $-0.62 ● -\frac{6}{7}$

14. $\frac{12}{44} ● \frac{8}{11}$

15. Order $4\frac{4}{5}$, 4.85, $2\frac{5}{8}$, and 2.6 from least to greatest.

Find each sum or difference. Write in simplest form.

16. $\frac{1}{7} + \frac{5}{7}$

17. $\frac{7}{8} - \frac{1}{8}$

18. $\frac{1}{6} + \left(-\frac{1}{2}\right)$

19. $-\frac{1}{12} - \left(-\frac{3}{4}\right)$

20. Sara and Gabe are sharing a sheet of stickers. Sara has $\frac{2}{7}$ of the sheet. Gabe has $\frac{1}{4}$ of the sheet. What fraction of the sheet do Sara and Gabe have together?

Find each product or quotient.

21. $-1.2(9.3)$

22. $-20.93 \div (-2.3)$

23. $10.5 \div (-1.2)$

24. $(-3.4)(-2.8)$

Name the reciprocal of each number.

25. 6

26. $1\frac{2}{5}$

27. $-2\frac{3}{7}$

28. $-\frac{1}{2}$

29. $\frac{4}{3}$

30. $5\frac{1}{3}$

Find each product or quotient. Write in simplest form.

31. $\frac{2}{5} \cdot \frac{5}{9}$

32. $\frac{4}{5} \div \frac{1}{5}$

33. $-\frac{7}{8} \cdot 2$

34. $\frac{1}{3} \div 2\frac{1}{4}$

35. $-6 \cdot \left(-\frac{3}{4}\right)$

36. $\frac{7}{18} \div \left(-\frac{14}{15}\right)$

37. **PICNIC** Joseph is mixing $5\frac{1}{2}$ gallons of orange drink for his class picnic. Every $\frac{1}{2}$ gallon requires 1 packet of orange drink mix. How many packets of orange drink mix does Joseph need?

Express each percent as a fraction in simplest form.

38. 6%

39. 140%

Use the percent proportion to find each number.

40. 50% of what number is 31?

41. What number is 110% of 51?

42. Find 8% of 95.

43. **SOLUTIONS** A solution is prepared by dissolving 24 milliliters of saline in 150 milliliters of pure solution. What is the percent of saline in the pure solution?

44. **SHOPPING** Marta got 60% off a pair of shoes. If the shoes cost $15.90 (before sales tax), what was the original price of the shoes?

Find the perimeter and area of each figure.

45.

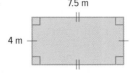

46.

47. A parallelogram has a base of 20 millimeters and a height of 6 millimeters. Find the area.

48. GARDENS Find the perimeter of the garden.

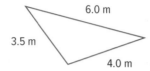

49. The area of a square is 25 square centimeters. What is the length of one side?

50. The area of a rectangle is 40 square millimeters. The length of one of the sides is 8 millimeters. What is the length of the other side?

Find the circumference and area of each circle. Round to the nearest tenth.

51.

25 in.

52.

3.5 cm

53. PARKS A park has a circular area for a fountain that has a circumference of about 16 feet. What is the radius of the circular area? Round to the nearest tenth.

Find the volume and surface area of each rectangular prism given the measurements below.

54. $\ell = 1.5$ m, $w = 3$ m, $h = 2$ m

55. $\ell = 4$ in., $w = 1$ in., $h = \frac{1}{2}$ in.

56. Find the volume and surface area of the rectangular prism.

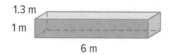

1.3 m
1 m
6 m

57. A cylinder has a height of 6 centimeters and a base radius of 5 centimeters. What is the surface area of the cylinder? Round to the nearest tenth.

58. A cylinder has a height of 8 inches and a base radius of 3 inches. What is the volume of the cylinder? Round to the nearest tenth.

One marble is randomly selected from a jar containing 3 red, 4 green, 2 black, and 6 blue marbles. Find each probability.

59. P(red or blue)

60. P(green or red)

61. P(not black)

62. P(not blue)

63. A movie theater is offering snack specials. You can choose a small, medium, large, or jumbo popcorn with or without butter, and soda or bottled water. Use a tree diagram to find the sample space for the event. State the number of possible outcomes.

A die is rolled. Find each probability.

64. rolling a 1 or a 6

65. rolling any number but 3

66. rolling a mutliple of 2

67. rolling an odd number

One coin is randomly selected from a jar containing 20 pennies, 15 nickels, 3 dimes, and 12 quarters. Find the odds of each outcome. Write in simplest form.

68. a dime

69. a value less than $0.25

70. a value greater than $0.10

71. a value less than $0.05

72. SCHOOL In a science class, each student must choose a lab project from a list of 15, write a paper on one of 6 topics, and give a presentation about one of 8 subjects. How many ways can students choose to do their assignments?

73. GAMES Marcos has been dealt seven different cards. How many different ways can he play his cards if he is required to play one card at a time?

74. CAMP Jeannette brought 4 pairs of pants, 2 skirts, and 8 shirts to summer camp. How many different ways can she make an outfit out of one shirt and either one pair of pants or one skirt?

75. DOGS Miles is taking care of 5 dogs. He has 5 different dog treats. How many different ways can he give each dog one treat?

CHAPTER 1
Expressions and Functions

THEN

You have learned how to perform operations on whole numbers.

NOW

In this chapter, you will:

- Write algebraic expressions and use the order of operations.
- Define appropriate quantities for descriptive modeling.
- Represent and interpret relations and functions.
- Interpret the graphs of functions.

(MP) WHY

SCUBA DIVING A scuba diving store rents air tanks and wet suits. An algebraic expression can be written to represent the total cost for a group to rent this equipment.

Use the Mathematical Practices to complete the activity.

1. **Use Tools** Use the Internet or another source to find the rental fees for scuba diving equipment.

2. **Sense-Making** Make a table that displays the cost of renting the equipment for 1 through 7 days.

3. **Model With Mathematics** Use the Line Graph tool in ConnectED to plot each data point. Let x be the number of days and y be the rental cost.

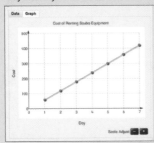

4. **Discuss** Suppose a scuba rental charges $200 per week. Compare this fee to the data you found.

Go Online to Guide Your Learning

Explore & Explain		Organize

 The Geometer's Sketchpad

Visualize the order of operations using the **Order of Operations** sketch in Lesson 1-2.

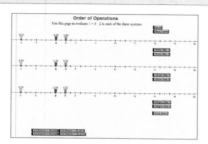

 Foldables

Get organized! Create an Expressions and Functions Foldable before you start the chapter to arrange your notes on expressions and functions.

Tools

Explore relations and functions with the **Mapping** tool. How can you tell if a relation is a function?

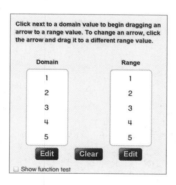

Collaborate

 Chapter Project

In the **Want to Be Your Own Boss?** project, you will use what you have learned about expressions to represent quantities that relate to business.

eBook
Interactive Student Guide

Before starting the chapter, answer the **Chapter Focus** preview questions. Check your answers as you complete each lesson. At the end of the chapter, try the **Performance Task**.

Focus

 LEARNSMART

Need help studying? Complete the **Analyze** and **Interpret Functions** domains in LearnSmart to review for the chapter test.

ALEKS

You can use the **Arithmetic Readiness**, **Real Numbers**, and **Linear Equations** topics in ALEKS to find out what you know about expressions and functions and what you are ready to learn.*

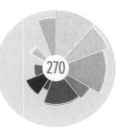

* Ask your teacher if this is part of your program.

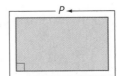
Get Ready for the Chapter

Connecting Concepts

Concept Check

Review the concepts used in this chapter by answering each question below.

1. A fraction is in simplest form if the numerator and denominator have no common factors. If you know that 54 out of 180 customers choose cookie-dough ice cream as their favorite flavor, how can you express this information as a fraction in simplest form?

2. Explain what it means to find the perimeter of a geometric figure.

3. How do you find the perimeter of a triangle?

4. How do you evaluate $P = 2\ell + 2w$ for the rectangle shown?

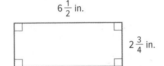

$6\frac{1}{2}$ in.

$2\frac{3}{4}$ in.

5. How can you use $P = 2\ell + 2w$ to find the amount of fencing needed to enclose a rectangular garden that measures 6 meters by 4 meters?

6. Describe what method you would use to evaluate a numeric expression.

7. Describe the first step to cut a board measuring 72 feet into three equal pieces.

New Vocabulary

English		Español
algebraic expression	p. 5	expression algebraica
variable	p. 5	variable
term	p. 5	término
power	p. 5	potencia
coefficient	p. 26	coeficiente
metric	p. 33	métrico
accuracy	p. 34	exactitud
relation	p. 42	relacíon
domain	p. 42	domino
range	p. 42	rango
independent variable	p. 44	variable independiente
dependent variable	p. 44	variable dependiente
function	p. 49	función
intercept	p. 58	intersección
line symmetry	p. 59	simetría
end behavior	p. 59	comportamiento final

Performance Task Preview

You can use the concepts and skills in this chapter to solve problems about determining how far horizontally a rock will fall from a cliff and how long it takes to fall. Understanding how to make and graph a table of values for $d = 2t$ will help you finish the Performance Task at the end of the chapter.

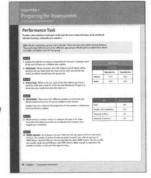

MP **In this Performance Task you will:**

• make sense of problems

• reason abstractly

• model with mathematics

Review Vocabulary

additive inverse inverso aditivo a number and its opposite

multiplicative inverse inverso multiplicativo two numbers with a product of 1

perimeter perímetro the distance around a geometric figure

Variables and Expressions

::Then	::Now	::Why?
• You performed operations on integers.	**1** Write verbal expressions for algebraic expressions. **2** Write algebraic expressions for verbal expressions.	• Cassie and her friends are at a baseball game. The stadium is running a promotion where hot dogs are $0.10 each. Suppose d represents the number of hot dogs Cassie and her friends eat. Then $0.10d$ represents the cost, in dollars, of the hot dogs they eat.

 New Vocabulary
algebraic expression
variable
term
factor
product
power
exponent
base

 Mathematical Practices
4 Model with mathematics.

1 **Write Verbal Expressions** An **algebraic expression** consists of sums and/or products of numbers and variables. In the algebraic expression $0.10d$, the letter d is called a variable. In algebra, **variables** are symbols used to represent unspecified numbers or values. Any letter may be used as a variable.

$$0.10d \qquad 2x + 4 \qquad 3 + \frac{z}{6} \qquad p \cdot q \qquad 4cd \div 3mn$$

A **term** of an expression may be a number, a variable, or a product or quotient of numbers and variables. For example, $0.10d$, $2x$, and 4 are each terms.

> A term that contains x or other letters is sometimes referred to as a *variable term*. → $2x + 4$ ← A term that does not have a variable is a *constant term*.

In a multiplication expression, the quantities being multiplied are **factors**, and the result is the **product**. A raised dot or set of parentheses is often used to indicate a product. Here are several ways to represent the product of x and y.

$$xy \qquad x \cdot y \qquad x(y) \qquad (x)y \qquad (x)(y)$$

An expression like x^n is called a **power**. The word *power* can also refer to the exponent. The **exponent** indicates the number of times the base is used as a factor. In an expression of the form x^n, the **base** is x. The expression x^n is read "x to the nth power." When no exponent is shown, it is understood to be 1. For example, $a = a^1$.

base · exponent · x^n

Study Tip
Using Your Text Notice that new terms are listed at the beginning of the lesson and also highlighted in context.

Example 1 Write Verbal Expressions

Write a verbal expression for each algebraic expression.

a. $3x^4$
three times x to the fourth power

b. $5z^4 + 16$
5 times z to the fourth power plus sixteen

 Guided Practice

1A. $16u^2 - 3$

1B. $\frac{1}{2}a + \frac{6b}{7}$

Jupiterimages/Stockbyte/Getty Images

2 Write Algebraic Expressions
Another important skill is translating verbal expressions into algebraic expressions.

Key Concept Translating Verbal to Algebraic Expressions

Operation	Verbal Phrases
Addition	more than, sum, plus, increased by, added to
Subtraction	less than, subtracted from, difference, decreased by, minus
Multiplication	product of, multiplied by, times, of
Division	quotient of, divided by

Go Online!

Personal Tutors for each example let you follow along as a teacher solves a problem. Pause and rewind as you need.

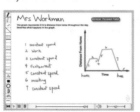

Example 2 Write Algebraic Expressions

Write an algebraic expression for each verbal expression.

a. a number t more than 6

The words *more than* suggest addition.
Thus, the algebraic expression is $6 + t$ or $t + 6$.

b. 10 less than the product of 7 and f

Less than implies subtraction, and *product* suggests multiplication.
So the expression is written as $7f - 10$.

c. two thirds of the volume v

The word *of* with a fraction implies that you should multiply.
The expression could be written as $\frac{2}{3}v$ or $\frac{2v}{3}$.

Guided Practice

2A. the product of p and 6   **2B.** one third of the area a

Variables can represent quantities that are known and quantities that are unknown. They are also used in formulas, expressions, and equations.

Real-World Example 3 Write an Expression

SPORTS MARKETING Mr. Martinez orders 250 key chains printed with his athletic team's logo and 500 pencils printed with its Web address. Write an algebraic expression that represents the cost of the order.

Let k be the cost of each key chain and p be the cost of each pencil. Then the cost of the key chains is $250k$, and the cost of the pencils is $500p$. The cost of the order is represented by $250k + 500p$.

Real-World Career

SPORTS MARKETING
Sports marketers promote and manage athletes, teams, facilities, and sports-related businesses and organizations. A minimum of a bachelor's degree in sports management or business administration is preferred.

Guided Practice

3. COFFEE SHOP Katie bakes 40 pastries and makes coffee for 200 people. Write an algebraic expression to represent the cost of this situation.

Image Source/Getty Images

 **Go Online!** for a Self-Check Quiz

Example 1 Write a verbal expression for each algebraic expression.

1. $2m$

2. $\frac{2}{3}r^4$

3. $a^2 - 18b$

Example 2 Write an algebraic expression for each verbal expression.

4. the sum of a number and 14

5. 6 less a number t

6. 7 more than 11 times a number

7. 1 minus the quotient of r and 7

8. two fifths of the square of a number j

9. n cubed increased by 5

Example 3 **10. GROCERIES** Mr. Bailey purchased some groceries that cost d dollars. He paid with a $50 bill. Write an expression for the amount of change he will receive.

Practice and Problem Solving Extra Practice is on page R1.

Example 1 Write a verbal expression for each algebraic expression.

11. $4q$

12. $\frac{1}{8}y$

13. $15 + r$

14. $w - 24$

15. $3x^2$

16. $\frac{r^4}{9}$

(17) $2a + 6$

18. $r^4 \cdot t^3$

Example 2 Write an algebraic expression for each verbal expression.

19. x more than 7

20. a number less 35

21. 5 times a number

22. one third of a number

23. f divided by 10

24. the quotient of 45 and r

25. three times a number plus 16

26. 18 decreased by 3 times d

27. k squared minus 11

28. 20 divided by t to the fifth power

Example 3 **29. GEOMETRY** The volume of a cylinder is π times the radius r squared multiplied by the height h. Write an expression for the volume.

30. FINANCIAL LITERACY Jocelyn makes x dollars per hour working at the grocery store and n dollars per hour babysitting. Write an expression that describes her earnings if she babysat for 25 hours and worked at the grocery store for 15 hours.

Write a verbal expression for each algebraic expression.

31. $25 + 6x^2$

32. $6f^2 + 5f$

33. $\frac{3a^5}{2}$

34. **MP SENSE-MAKING** A local gym membership costs $20 per month plus additional activity charges. If x is the number of exercise classes taken above the planned amount and y is the number of massages taken above the planned amount, interpret each expression.

a. $15x$

b. $55y$

c. $15x + 55y + 20$

35 **DREAMS** It is believed that about $\frac{3}{4}$ of our dreams involve people that we know.

    a. Write an expression to describe the number of dreams that feature people you know if you have d dreams.

    b. Use the expression you wrote to predict the number of dreams that include people you know out of 28 dreams.

36. **SPORTS** In football, a touchdown is awarded 6 points and the team then may attempt a kick for a point after a touchdown.

    a. Write an expression that describes the number of points scored on touchdowns T and points after touchdowns p by one team in a game.

    b. If a team wins a football game 27-0, write an equation to represent the possible number of touchdowns and points after touchdowns by the winning team.

    c. If a team wins a football game 21-7, how many possible number of touchdowns and points after touchdowns were scored during the game by both teams?

37. **MULTIPLE REPRESENTATIONS** In this problem, you will explore the multiplication of powers with like bases.

    a. **Tabular** Copy and complete the table.

10^2	$\times$	10^1	$=$	$10 \times 10 \times 10$	$=$	10^3
10^2	$\times$	10^2	$=$	$10 \times 10 \times 10 \times 10$	$=$	10^4
10^2	$\times$	10^3	$=$	$10 \times 10 \times 10 \times 10 \times 10$	$=$	?
10^2	$\times$	10^4	$=$	?	$=$	?

    b. **Algebraic** Write an equation for the pattern in the table.

    c. **Verbal** Make a conjecture about the exponent of the product of two powers with like bases.

H.O.T. Problems   Use **H**igher-**O**rder **T**hinking Skills

38. **MP REASONING** Explain the differences between an algebraic expression and a verbal expression.

39. **OPEN ENDED** Define a variable to represent a real-life quantity, such as time in minutes or distance in feet. Then use the variable to write an algebraic expression to represent one of your daily activities. Describe in words what your expression represents, and explain your reasoning.

40. **ERROR ANALYSIS** Consuelo and James are writing an algebraic expression for *three times the sum of n squared and 3*. Is either of them correct? Explain your reasoning.

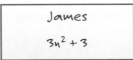

Consuelo	James
$3(n^2 + 3)$	$3n^2 + 3$

41. **CHALLENGE** For the cube, x represents a positive whole number. Find the value of x such that the volume of the cube and 6 times the area of one of its faces have the same value.

42. **WRITING IN MATH** Describe how to write an algebraic expression from a real-world situation. Include a definition of algebraic expression in your own words.

43. Which of the following best represents "five more than the product of 7 and a number t?" (MP) 2

⚪ **A** $5 > 7t$

⚪ **B** $7t + 5$

⚪ **C** $5t + 7$

⚪ **D** $5 \cdot 7t$

44. The volume of this cube can be expressed as 5^3.

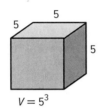

$V = 5^3$

Which expression can be used to find the volume of a cube with edges that are x units shorter? (MP) 4

⚪ **A** $(x - 5)^3$

⚪ **B** $5^3 - x^3$

⚪ **C** $(5 - x)^3$

⚪ **D** $5 - x$

45. Elsie buys a pizza for $16 and several bottles of water for $2 each. Let C represent the total amount of money that Elsie spends and let w represent how many bottles of water she buys. (MP) 2, 4

a. Which equation best represents this situation?

⚪ **A** $C = 2 + 16w$

⚪ **B** $C = 16 + 2 + w$

⚪ **C** $C = 16 + 2w$

⚪ **D** $C = 2(w + 16)$

b. Write an expression for the total amount of money Elsie spends if she buys 2 pizzas and w bottles of water.

46. Which equation best describes the data in the table? (MP) 2

x	8	4	2
y	2	−2	−4

⚪ **A** $y = x \div 4$

⚪ **B** $y = -0.5x$

⚪ **C** $y = x + 6$

⚪ **D** $y = x - 6$

47. Which equation best represents the verbal description "the quotient of a and 2"? (MP) 2

⚪ **A** $\dfrac{a}{2}$

⚪ **B** $\dfrac{2}{a}$

⚪ **C** $a + 2$

⚪ **D** $2a$

48. **MULTI-STEP** Karima earns $400 per week plus a holiday bonus of $250. Write an expression to represent the total amount Karima earns after working w weeks. (MP) 2, 4

a. Write an equation to represent this situation.

b. Find the amount Karima earns after she works 4, 12, and 36 weeks.

49. Which situation(s) can be represented by the expression $24 - n$? Choose all that apply. (MP) 4

☐ **A** Grace pays 24 dollars plus n dollars in tax for a shirt.

☐ **B** Sarah takes n bottles of water out of a case of 24 bottles.

☐ **C** Eliana has n fewer games than Joshua, who has 24 games.

☐ **D** Mazzy bought n concert tickets for 24 dollars each.

☐ **E** Kayleigh is n years younger than her cousin Elizabeth, who is 24 years old.

Order of Operations

	Ticket	Price ($)
	Adult	78.95
	Child	68.95

⋮ Then

- You expressed algebraic expressions verbally.

⋮ Now

1 Evaluate numerical expressions by using the order of operations.

2 Evaluate algebraic expressions by using the order of operations.

⋮ Why?

- The admission prices for an aquarium theme park are shown in the table. If four adults and three children go to the park, the expression below represents the cost of admission for the group.

  $4(78.95) + 3(68.95)$

 New Vocabulary
evaluate
order of operations

MP **Mathematical Practices**
7 Look for and make use of structure.

1 Evaluate Numerical Expressions To find the cost of admission, the expression $4(78.95) + 3(68.95)$ must be evaluated. To **evaluate** an expression means to find its value.

Example 1   Evaluate Expressions

Evaluate 3^5.

$3^5 = 3 \cdot 3 \cdot 3 \cdot 3 \cdot 3$ Use 3 as a factor 5 times.

$= 243$ Multiply

▶ **Guided Practice**

1A. 2^4 **1B.** 4^5 **1C.** 7^3

The numerical expression that represents the cost of admission contains more than one operation. The rule that lets you know which operation to perform first is called the **order of operations**.

⚙ Key Concept Order of Operations

Step 1	Evaluate expressions inside grouping symbols.
Step 2	Evaluate all powers.
Step 3	Multiply and/or divide from left to right.
Step 4	Add and/or subtract from left to right.

Example 2   Order of Operations

Evaluate $16 - 8 \div 2^2 + 14$.

$$16 - 8 \div 2^2 + 14 = 16 - 8 \div 4 + 14$$ Evaluate powers.
$$= 16 - 2 + 14$$ Divide 8 by 4.
$$= 14 + 14$$ Subtract 2 from 16.
$$= 28$$ Add 14 and 14.

▶ **Guided Practice**

2A. $3 + 42 \cdot 2 - 5$ **2B.** $20 - 7 + 8^2 - 7 \cdot 11$

When one or more grouping symbols are used, evaluate within the innermost grouping symbols first.

Example 3   Expressions with Grouping Symbols

Evaluate each expression.

a. $4 \div 2 + 5(10 - 6)$

$$4 \div 2 + 5(10 - 6) = 4 \div 2 + 5(4)$$ Evaluate inside parentheses.
$$= 2 + 5(4)$$ Divide 4 by 2.
$$= 2 + 20$$ Multiply 5 by 4.
$$= 22$$ Add 2 to 20.

b. $6\left[32 - (2 + 3)^2\right]$

$$6\left[32 - (2 + 3)^2\right] = 6\left[32 - (5)^2\right]$$ Evaluate innermost expression first.
$$= 6[32 - 25]$$ Evaluate power.
$$= 6[7]$$ Subtract 25 from 32.
$$= 42$$ Multiply.

c. $\dfrac{2^3 - 5}{15 + 9}$

$$\frac{2^3 - 5}{15 + 9} = \frac{8 - 5}{15 + 9}$$ Evaluate the power in the numerator.

$$= \frac{3}{15 + 9}$$ Subtract 5 from 8 in the numerator.

$$= \frac{3}{24} \text{ or } \frac{1}{8}$$ Add 15 and 9 in denominator, and simplify.

Guided Practice

3A. $5 \cdot 4(10 - 8) + 20$ **3B.** $15 - \left[10 + (3 - 2)^2\right] + 6$ **3C.** $\dfrac{(4 + 5)^2}{3(7 - 4)}$

2 **Evaluate Algebraic Expressions** To evaluate an algebraic expression, replace the variables with their values. Then find the value of the numerical expression using the order of operations.

Example 4   Evaluate an Algebraic Expression

Evaluate $3x^2 + \left(2y + z^3\right)$ if $x = 4$, $y = 5$, $z = 3$.

$3x^2 + \left(2y + z^3\right)$
$$= 3(4)^2 + (2 \cdot 5 + 3^3)$$ Replace x with 4, y with 5, and z with 3.
$$= 3(4)^2 + (2 \cdot 5 + 27)$$ Evaluate 3^3.
$$= 3(4)^2 + (10 + 27)$$ Multiply 2 by 5.
$$= 3(4)^2 + (37)$$ Add 10 to 27.
$$= 3(16) + 37$$ Evaluate 4^2.
$$= 48 + 37$$ Multiply 3 by 16.
$$= 85$$ Add 48 to 37.

Guided Practice

Evaluate each expression.

4A. $a^2(3b + 5) \div c$ if $a = 2$, $b = 6$, $c = 4$ **4B.** $5d + (6f - g)$ if $d = 4$, $f = 3$, $g = 12$

Real-World Example 5   Write and Evaluate an Expression

ENVIRONMENTAL STUDIES Science on a Sphere (SOS)® demonstrates the effects of atmospheric storms, climate changes, and ocean temperature on the environment. The volume of a sphere is four thirds of π multiplied by the radius r to the third power.

a. Write an expression that represents the volume of a sphere.

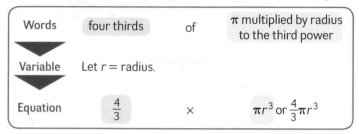

Words	four thirds	of	π multiplied by radius to the third power
Variable	Let r = radius.		
Equation	$\frac{4}{3}$	×	πr^3 or $\frac{4}{3}\pi r^3$

b. Find the volume of the 3-foot radius sphere used for SOS.

$V = \frac{4}{3}\pi r^3$ Volume of a sphere

$\quad = \frac{4}{3}\pi(3)^3$ Replace r with 3.

$\quad = \left(\frac{4}{3}\right)\pi(27)$ Evaluate $3^3 = 27$.

$\quad = 36\pi$ Multiply $\frac{4}{3}$ by 27.

The volume of the sphere is 36π cubic feet.

▶ **Guided Practice**

5. FOREST FIRES According to the California Department of Forestry, an average of 539.2 fires each year are started by burning debris, while campfires are responsible for an average of 129.1 fires each year.

A. Write an algebraic expression for the total number of fires in f years due to debris fires and campfires.

B. How many total fires would there be over a 5-year period?

Check Your Understanding   ◯ = Step-by-Step Solutions begin on page R11.

Go Online! for a Self-Check Quiz

Examples 1–3 **Evaluate each expression.**

1. 9^2

2. 4^4

3. 3^5

4. $30 - 14 \div 2$

5 $5 \cdot 5 - 1 \cdot 3$

6. $(2 + 5)4$

7. $[8(2) - 4^2] + 7(4)$

8. $\dfrac{11 - 8}{1 + 7 \cdot 2}$

9. $\dfrac{(4 \cdot 3)^2}{9 + 3}$

Example 4   **Evaluate each expression if $a = 4$, $b = 6$, and $c = 8$.**

10. $8b - a$

11. $2a + (b^2 \div 3)$

12. $\dfrac{b(9 - c)}{a^2}$

Example 5   **13. BOOKS** Akira bought one new book for $20 and three used books for $4.95 each. Write and evaluate an expression to find how much money the books cost.

14. (MP) **REASONING** Koto purchased food for herself and her friends. She bought 4 cheeseburgers for $3.99 each, 3 French fries for $1.79 each, and 4 drinks for $5.16. Write and evaluate an expression to find how much the food cost.

Examples 1–3   **Evaluate each expression.**

15. 7^2

16. 14^3

17. 2^6

18. $35 - 3 \cdot 8$

19. $18 \div 9 + 2 \cdot 6$

20. $10 + 8^3 \div 16$

21. $24 \div 6 + 2^3 \cdot 4$

22. $(11 \cdot 7) - 9 \cdot 8$

23. $29 - 3(9 - 5)$

24. $(12 - 6) \cdot 5^2$

25. $3^5 - (1 + 10^2)$

26. $108 \div [3(9 + 3^2)]$

27. $[(6^3 - 9) \div 23]4$

28. $\dfrac{8 + 3^3}{12 - 7}$

29. $\dfrac{(1 + 6)9}{5^2 - 4}$

Example 4   **Evaluate each expression if $g = 2$, $r = 3$, and $t = 11$.**

30. $g + 6t$

31. $7 - gr$

32. $r^2 + (g^3 - 8)^5$

(33) $(2t + 3g) \div 4$

34. $t^2 + 8rt + r^2$

35. $3g(g + r)^2 - 1$

Example 5   **36. GEOMETRY** Write an algebraic expression to represent the area of the triangle. Then evaluate it to find the area when $h = 12$ inches.

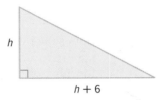

37. AMUSEMENT PARKS In 2013, there were 3344 amusement parks and arcades. Suppose by 2018 this number decreased by 148. Write and evaluate an expression to find the number of amusement parks and arcades in 2018.

38. MP STRUCTURE Marcos sells tickets at a university's athletic ticket office. If p represents a preferred season ticket, b represents a blue zone ticket, and g represents a general admission ticket, interpret and then evaluate the following expressions.

a. $45b$ **b.** $15p + 35g$ **c.** $6p + 11b + 22g$

University Football Ticket prices	
Preferred Season Ticket	$100
Blue Zone	$80
General Admission	$70

Evaluate each expression.

39. 4^2

40. 12^3

41. 3^6

42. 11^5

43. $(3 - 4^2)^2 + 8$

44. $23 - 2(17 + 3^3)$

45. $3[4 - 8 + 4^2(2 + 5)]$

46. $\dfrac{2 \cdot 8^2 - 2^2 \cdot 8}{2 \cdot 8}$

47. $25 + \left[(16 - 3 \cdot 5) + \dfrac{12 + 3}{5}\right]$

48. $7^3 - \dfrac{2}{3}(13 \cdot 6 + 9)4$

Evaluate each expression if $a = 8$, $b = 4$, and $c = 16$.

49. $a^2bc - b^2$

50. $\dfrac{c^2}{b^2} + \dfrac{b^2}{a^2}$

51. $\dfrac{2b + 3c^2}{4a^2 - 2b}$

52. $\dfrac{3ab + c^2}{a}$

53. $\left(\dfrac{a}{b}\right)^2 - \dfrac{c}{a - b}$

54. $\dfrac{2a - b^2}{ab} + \dfrac{c - a}{b^2}$

55. SALES One day, 28 small and 12 large merchant spaces were rented. Another day, 30 small and 15 large spaces were rented. Write and evaluate an expression to show the total rent collected.

56. SHOPPING Isabel is shopping for back-to-school clothes. She bought 3 skirts, 2 pairs of jeans, and 4 sweaters. Write and evaluate an expression to find how much she spent, not including sales tax.

Clothing	
skirt	$35.99
jeans	$49.99
sweater	$32.99

57. PYRAMIDS The pyramid at the Louvre has a square base with a side of 35.42 meters and a height of 21.64 meters. The Great Pyramid in Egypt has a square base with a side of 230 meters and a height of 146.5 meters. The expression for the volume of a pyramid is $\frac{1}{3}Bh$, where B is the area of the base and h is the height.

  a. Draw both pyramids and label the dimensions.

  b. Write a verbal expression for the difference in volume of the two pyramids.

  c. Write an algebraic expression for the difference in volume of the two pyramids. Find the difference in volume.

58. FINANCIAL LITERACY Ginger is determining her monthly expenses. She has monthly rent r, monthly utilities u, weekly food expense f, and weekly auto expense a. Assume there are 4 weeks in a month.

  a. Write an algebraic expression to represent her spending in one month.

  b. Suppose her monthly rent is $550, her monthly utilities are $115, her weekly food expenses are $75, and her weekly auto expenses are $125. Determine her total monthly expenses.

H.O.T. Problems   Use **H**igher-**O**rder **T**hinking Skills

59. ERROR ANALYSIS Tara and Curtis are simplifying $[4(10) - 3^2] + 6(4)$. Is either of them correct? Explain your reasoning.

Tara
$[4(10) - 3^2] + 6(4)$
$= [4(10) - 9] + 6(4)$
$= 4(1) + 6(4)$
$= 4 + 6(4)$
$= 4 + 24$
$= 28$

Curtis
$[4(10) - 3^2] + 6(4)$
$= [4(10) - 9] + 6(4)$
$= (40 - 9) + 6(4)$
$= 31 + 6(4)$
$= 31 + 24$
$= 55$

60. MP REASONING Explain how to evaluate $a[(b - c) \div d] - f$ if you were given values for $a, b, c, d,$ and f. How would you evaluate the expression differently if the expression was $a \cdot b - c \div d - f$?

61. MP PERSEVERANCE Write an expression using the whole numbers 1 to 5 using all five digits and addition and/or subtraction to create a numeric expression with a value of 3.

62. OPEN ENDED Write an expression that uses exponents, at least three different operations, and two sets of parentheses. Explain the steps you would take to evaluate the expression.

63. WRITING IN MATH Choose a geometric formula and explain how the order of operations applies when using the formula.

64. WRITING IN MATH Equivalent expressions have the same value. Are the expressions $(30 + 17) \times 10$ and $10 \times 30 + 10 \times 17$ equivalent? Explain why or why not.

65. The smallest circle in the figure has a radius of 2 inches. Which equation gives the area A of the shaded part of the target? Recall that $A = \pi r^2$. 2

6 in.

  A $A = 4\pi$

  B $A = 36\pi$

  C $A = 60\pi$

  D $A = 68\pi$

66. Edgar buys a apples, b bananas, and c cantaloupes at the farmer's market. The prices at the market are shown in the table.

Fruit	Price Each
Apples	$0.50
Bananas	$0.20
Cantaloupes	$1.50

Edgar has a coupon for a free apple. If he gets 5 apples, 4 bananas, and 2 cantaloupes, how much money does he spend? **MP** 2

  A $2.20

  B $4.50

  C $5.80

  D $6.30

67. Evaluate $27 \div 3 + (12 - 4)$. **MP** 7

  A 3

  B 16

  C 17

  D 38

68. Evaluate the expression $\dfrac{6a - b^2}{2} + \left(\dfrac{b}{a}\right)^2$ if $a = 2$ and $b = 4$. **MP** 7

  A 2

  B 6

  C 10.25

  D 36

Find the total number of grams of protein she consumed. **MP** 4

  A 24

  B 37.5

  C 39.5

  D 42

69. The table shows the prices of various fruits and vegetables. Find the total cost of four oranges, three cucumbers, and one head of lettuce. **MP** 4

Oranges	Cucumbers	Lettuce
$1.00, buy one get one free	3 for $1.00	$2.50 each

  A $5.50

  B $7.50

  C $4.00

  D $5.00

70. Maya has one bran muffin, 16 ounces of orange juice, 3 ounces of sunflower seeds, 2 slices of turkey, and half of a cup of spinach.

Food	Protein (g)
bran muffin (1)	3
orange juice (8 oz)	2
sunflower seeds (1 oz)	2
turkey (1 slice)	12
spinach (1 cup)	5

71. **MULTI-STEP** Consider the expression $6 + 12 \div 3 - 2 \times 4$. **MP** 1

  a. What is the value of the expression?

  b. Use grouping symbols so the expression has a value greater than the value from part **a.**

  c. Use grouping symbols so the expression has a value less than the value from part **a.**

Properties of Numbers

- You used the order of operations to simplify expressions.

1 Recognize the properties of equality and identity properties.

2 Recognize the Commutative and Associative Properties.

- Natalie lives 32 miles away from the mall. The distance from her house to the mall is the same as the distance from the mall to her house. This is an example of the Reflexive Property.

New Vocabulary

equivalent expressions
additive identity
multiplicative identity
multiplicative inverse
reciprocal

Mathematical Practices

2 Reason abstractly and quantitatively.

3 Construct viable arguments and critique the reasoning of others.

1 **Properties of Equality and Identity** The expressions $4k + 8k$ and $12k$ are called **equivalent expressions** because they represent the same number. The properties below allow you to write an equivalent expression for a given expression.

Key Concept   Properties of Equality

Property	Words	Symbols	Examples
Reflexive Property	Any quantity is equal to itself.	For any number a, $a = a$.	$5 = 5$ $4 + 7 = 4 + 7$
Symmetric Property	If one quantity equals a second quantity, then the second quantity equals the first.	For any numbers a and b, if $a = b$, then $b = a$.	If $8 = 2 + 6$, then $2 + 6 = 8$.
Transitive Property	If one quantity equals a second quantity and the second quantity equals a third quantity, then the first qantity equals the third quantity.	For any numbers a, b, and c, if $a = b$ and $b = c$, then $a = c$.	If $6 + 9 = 3 + 12$ and $3 + 12 = 15$, then $6 + 9 = 15$.
Substitution Property	A quantity may be substituted for its equal in any expression.	If $a = b$, then a may be replaced by b in any expression.	If $n = 11$, then $4n = 4 \cdot 11$.

The sum of any number and 0 is equal to the number. Thus, 0 is called the **additive identity**.

Key Concept   Addition Properties

Property	Words	Symbols	Examples
Additive Identity	For any number a, the sum of a and 0 is a.	$a + 0 = 0 + a = a$	$2 + 0 = 2$ $0 + 2 = 2$
Additive Inverse	A number and its opposite are additive inverses of each other.	$a + (-a) = 0$	$3 + (-3) = 0$ $-4 + 4 = 0$

Ned Frisk/Blend Images

There are also special properties associated with multiplication. Consider the following equations.

$$4 \cdot n = 4$$

The solution of the equation is 1. Since the product of any number and 1 is equal to the number, 1 is called the **multiplicative identity**.

$$6 \cdot m = 0$$

The solution of the equation is 0. The product of any number and 0 is equal to 0. This is called the **Multiplicative Property of Zero**.

Two numbers whose product is 1 are called **multiplicative inverses** or **reciprocals**. Zero has no reciprocal because any number times 0 is 0.

Key Concept Multiplication Properties

Property	Words	Symbols	Examples
Multiplicative Identity	For any number a, the product of a and 1 is a.	$a \cdot 1 = a$ $1 \cdot a = a$	$14 \cdot 1 = 14$ $1 \cdot 14 = 14$
Multiplicative Property of Zero	For any number a, the product of a and 0 is 0.	$a \cdot 0 = 0$ $0 \cdot a \ 0$	$9 \cdot 0 = 0$ $0 \cdot 9 = 0$
Multiplicative Inverse	For every number $\frac{a}{b}$, where $a, b \neq 0$, there is exactly one number $\frac{b}{a}$ such that the product of $\frac{a}{b}$ and $\frac{b}{a}$ is 1.	$\frac{a}{b} \cdot \frac{b}{a} = 1$ $\frac{b}{a} \cdot \frac{a}{b} = 1$	$\frac{4}{5} \cdot \frac{5}{4} = \frac{20}{20}$ or 1 $\frac{5}{4} \cdot \frac{4}{5} = \frac{20}{20}$ or 1

Example 1   Evaluate Using Properties

Evaluate $7(4 - 3) - 1 + 5 \cdot \frac{1}{5}$. Name the property used in each step.

$$7(4 - 3) - 1 + 5 \cdot \frac{1}{5} = 7(1) - 1 + 5 \cdot \frac{1}{5} \qquad \text{Substitution: } 4 - 3 = 1$$

$$= 7 - 1 + 5 \cdot \frac{1}{5} \qquad \text{Multiplicative Identity: } 7 \cdot 1 = 7$$

$$= 7 - 1 + 1 \qquad \text{Multiplicative Inverse: } 5 \cdot \frac{1}{5} = 1$$

$$= 6 + 1 \qquad \text{Substitution: } 7 - 1 = 6$$

$$= 7 \qquad \text{Substitution: } 6 + 1 = 7$$

▶ **Guided Practice**

Name the property used in each step.

1A. $2 \cdot 3 + (4 \cdot 2 - 8)$

$= 2 \cdot 3 + (8 - 8)$?

$= 2 \cdot 3 + (0)$?

$= 6 + 0$?

$= 6$?

1B. $7 \cdot \frac{1}{7} + 6(15 \div 3 - 5)$

$= 7 \cdot \frac{1}{7} + 6(5 - 5)$?

$= 7 \cdot \frac{1}{7} + 6(0)$?

$= 1 + 6(0)$?

$= 1 + 0$?

$= 1$?

2 Use Commutative and Associative Properties

Nikki walks 2 blocks to her friend Sierra's house. They walk another 4 blocks to school. At the end of the day, Nikki and Sierra walk back to Sierra's house, and then Nikki walks home.

The distance from Nikki's house to school	equals	the distance from the school to Nikki's house.
2 + 4	=	4 + 2

This is an example of the **Commutative Property** for addition.

Key Concept Commutative Property

Words	The order in which you add or multiply numbers does not change their sum or product.
Symbols	For any numbers a and b, $a + b = b + a$ and $a \cdot b = b \cdot a$.
Examples	$4 + 8 = 8 + 4$ $7 \cdot 11 = 11 \cdot 7$

An easy way to find the sum or product of numbers is to group, or associate, the numbers using the **Associative Property**.

Key Concept Associative Property

Words	The way you group three or more numbers when adding or multiplying does not change their sum or product.
Symbols	For any numbers a, b, and c, $(a + b) + c = a + (b + c)$ and $(ab)c = a(bc)$.
Examples	$(3 + 5) + 7 = 3 + (5 + 7)$ $(2 \cdot 6) \cdot 9 = 2 \cdot (6 \cdot 9)$

Real-World Example 2   Apply Properties of Numbers

PARTY PLANNING Eric makes a list of items that he needs to buy for a party and their costs. Find the total cost of these items.

Party Supplies	
Item	**Cost ($)**
balloons	9.75
decorations	18.50
food	53.25
beverages	22.50

Balloons		Decorations		Food		Beverages
9.75	+	18.50	+	53.25	+	22.50

$9.75 + 18.50 + 53.25 + 22.50$
$= 9.75 + 53.25 + 18.50 + 22.50$ Commutative (+)
$= (9.75 + 53.25) + (18.50 + 22.50)$ Associative (+)
$= 63.00 + 41.00$ Substitution
$= 104.00$ Substitution

The total cost is $104.00

Guided Practice

2. **FURNITURE** Rafael is buying furnishings for his first apartment. He buys a couch for $450, lamps for $55.50, a rug for $43.50, and a table for $75. Find the total cost of these items.

Example 3   Use Multiplication Properties

Evaluate 5 • 7 • 4 • 2 using the properties of numbers. Name the property used in each step.

$$5 \cdot 7 \cdot 4 \cdot 2 = 5 \cdot 2 \cdot 7 \cdot 4 \qquad \text{Commutative } (\times)$$
$$= (5 \cdot 2) \cdot (7 \cdot 4) \qquad \text{Associative } (\times)$$
$$= 10 \cdot 28 \qquad \text{Substitution}$$
$$= 280 \qquad \text{Substitution}$$

▶ **Guided Practice**

Evaluate each expression using the properties of numbers. Name the property used in each step.

3A. $2.9 \cdot 4 \cdot 10$ **3B.** $\frac{5}{3} \cdot 25 \cdot 3 \cdot 2$

Go Online! for a Self-Check Quiz

Check Your Understanding

 = Step-by-Step Solutions begin on page R11.

Example 1   Evaluate each expression. Name the property used in each step.

1. $(1 \div 5)5 \cdot 14$ **2.** $6 + 4(19 - 15)$ **3.** $5(14 - 5) + 6(3 + 7)$

Example 2   **4. FINANCIAL LITERACY** Carolyn has 9 quarters, 4 dimes, 7 nickels, and 2 pennies, which can be represented as $9(25) + 4(10) + 7(5) + 2$. Evaluate the expression to find how much money she has. Name the property used in each step.

Examples 3   Evaluate each expression using the properties of numbers. Name the property used in each step.

5. $23 + 42 + 37$ **6.** $2.75 + 3.5 + 4.25 + 1.5$

7. $3 \cdot 7 \cdot 10 \cdot 2$ **8.** $\frac{1}{4} \cdot 24 \cdot \frac{2}{3}$

Practice and Problem Solving

Extra Practice is on page R1.

Example 1   Evaluate each expression. Name the property used in each step.

9 $3(22 - 3 \cdot 7)$ **10.** $7 + (9 - 3^2)$

11. $\frac{3}{4}[4 \div (7 - 4)]$ **12.** $[3 \div (2 \cdot 1)]\frac{2}{3}$

13. $2(3 \cdot 2 - 5) + 3 \cdot \frac{1}{3}$ **14.** $6 \cdot \frac{1}{6} + 5(12 \div 4 - 3)$

Example 2   **15. GEOMETRY** The expression $2 \cdot \frac{22}{7} \cdot 14^2 + 2 \cdot \frac{22}{7} \cdot 14 \cdot 7$ represents the approximate surface area of the cylinder at the right. Evaluate this expression to find the approximate surface area. Name the property used in each step.

7 in.

14 in.

16. **REASONING** A traveler checks into a hotel on Friday and checks out the following Tuesday morning. Use the table to find the total cost of the room including tax.

Hotel Rates Per Day		
Day	Room Charge	Sales Tax
Monday–Friday	$99	$12.87
Saturday–Sunday	$87	$11.31

Examples 2–3 Evaluate each expression using properties of numbers. Name the property used in each step.

17. $25 + 14 + 15 + 36$

18. $11 + 7 + 5 + 13$

19. $3\frac{2}{3} + 4 + 5\frac{1}{3}$

20. $4\frac{4}{9} + 7\frac{2}{9}$

21. $4.3 + 2.4 + 3.6 + 9.7$

22. $3.25 + 2.2 + 5.4 + 10.75$

23. $12 \cdot 2 \cdot 6 \cdot 5$

24. $2 \cdot 8 \cdot 10 \cdot 2$

25. $0.2 \cdot 4.6 \cdot 5$

26. $3.5 \cdot 3 \cdot 6$

27. $1\frac{5}{6} \cdot 24 \cdot 3\frac{1}{11}$

28. $2\frac{3}{4} \cdot 1\frac{1}{8} \cdot 32$

29. SCUBA DIVING The sign shows the equipment rented or sold by a scuba diving store.

    a. Write two expressions to represent the total sales to rent 2 wet suits, 3 air tanks, 2 dive flags, and selling 5 underwater cameras.

    b. What are the total sales?

30. COOKIES Bobby baked 2 dozen chocolate chip cookies, 3 dozen sugar cookies, and a dozen oatmeal raisin cookies. How many total cookies did he bake?

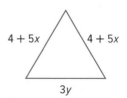

Evaluate each expression if $a = -1$, $b = 4$, and $c = 6$.

31 $4a + 9b - 2c$

32. $-10c + 3a + a$

33. $a - b + 5a - 2b$

34. $8a + 5b - 11a - 7b$

35. $3c^2 + 2c + 2c^2$

36. $3a - 4a^2 + 2a$

37. FOOTBALL A football team is on the 35-yard line. The quarterback is sacked at the line of scrimmage. The team gains 0 yards, so they are still at the 35-yard line. Which identity or property does this represent? Explain.

Find the value of x. Then name the property used.

38. $8 = 8 + x$

39. $3.2 + x = 3.2$

40. $10x = 10$

41. $\frac{1}{2} \cdot x = \frac{1}{2} \cdot 7$

42. $x + 0 = 5$

43. $1 \cdot x = 3$

44. $5 \cdot \frac{1}{5} = x$

45. $2 + 8 = 8 + x$

46. $x + \frac{3}{4} = 3 + \frac{3}{4}$

47. $\frac{1}{3} \cdot x = 1$

48. GEOMETRY Write an expression to represent the perimeter of the triangle. Then find the perimeter if $x = 2$ and $y = 7$.

49. SPORTS Tickets to a baseball game cost $35 each plus a $5.50 handling charge per ticket. If Sharon has a coupon for $10 off and orders 4 tickets, how much will she be charged?

50. **MP** **PRECISION** The table shows prices on children's clothing.

    a. Interpret the expression $5(17.99) + 2(11.99) + 7(14.99)$.

    b. Write and evaluate three different expressions that represent 8 pairs of shorts and 8 tops.

    c. If you buy 8 shorts and 8 tops, you receive a discount of 15%. Find the greatest and least amount of money you can spend on the 16 items at the sale.

Shorts	Polos	T-shirts
$16.99	$17.99	$15.99
$14.99	$13.99	$11.99

51. MULTI-STEP George is designing ledges for the octagonal (8-sided) gazebo that his brother is building. All of the sides are equal in length, and each ledge needs to be 18 inches shorter than the sides.

   a. What is the minimum length of wood George should purchase if his brother decides that the perimeter of the gazebo will be 64 feet?

   b. What was your solution process?

   c. What assumptions did you make?

52. MULTIPLE REPRESENTATIONS You can use *algebra tiles* to model and explore algebraic expressions. The rectangular tile has an area of x, with dimensions 1 by x. The small square tile has an area of 1, with dimensions 1 by 1.

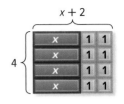

   a. Concrete Make a rectangle with algebra tiles to model the expression $4(x + 2)$ as shown. What are the dimensions of this rectangle? What is its area?

   b. Analytical What are the areas of the green region and of the yellow region?

   c. Verbal Complete this statement: $4(x + 2) = \ ?$. Write a convincing argument to justify your statement.

53 GEOMETRY A **proof** is an argument in which each statement you make is supported by a true statement. It is given that $\overline{AB} \cong \overline{CD}$, $\overline{AB} \cong \overline{BD}$, and $\overline{AB} \cong \overline{AC}$. Pedro wants to prove $\triangle ADB \cong \triangle ADC$. To do this, he must show that $\overline{AD} \cong \overline{AD}$, $\overline{AB} \cong \overline{DC}$ and $\overline{BD} \cong \overline{AC}$.

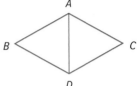

   a. Copy the figure and label $\overline{AB} \cong \overline{CD}$, $\overline{AB} \cong \overline{BD}$, and $\overline{AB} \cong \overline{AC}$.

   b. Use the Reflexive and Transitive Properties to prove $\triangle ADB \cong \triangle ADC$.

   c. If AC is x centimeters, write an equation for the perimeter of $ACDB$.

H.O.T. Problems   Use Higher-Order Thinking Skills

54. OPEN ENDED Write two equations showing the Transitive Property of Equality. Justify your reasoning.

55. MP CONSTRUCT ARGUMENTS Explain why 0 has no multiplicative inverse.

56. MP CONSTRUCT ARGUMENTS The sum of any two whole numbers is always a whole number. So, the set of whole numbers {0, 1, 2, 3, 4, … } is said to be closed under addition. This is an example of the **Closure Property**. State whether each statement is *true* or *false*. If false, justify your reasoning.

   a. The set of whole numbers is closed under subtraction.

   b. The set of whole numbers is closed under multiplication.

   c. The set of whole numbers is closed under division.

57. MP CONSTRUCT ARGUMENTS Does the Commutative Property *sometimes*, *always* or *never* hold for subtraction? Explain your reasoning.

58. MP REASONING Explain whether 1 can be an additive identity. Give an example to justify your answer.

59. WHICH ONE DOESN'T BELONG? Identify the equation that does not belong with the other three. Explain your reasoning.

$x + 12 = 12 + x$	$7h = h \cdot 7$	$1 + a = a + 1$	$(2j)k = 2(jk)$

60. WRITING IN MATH Determine whether the Commutative Property applies to division. Justify your answer.

61. Renee wants to solve the equation $\frac{3}{4}b = 6$. To do this, she will use the Multiplicative Inverse Property. Which of the following equations best illustrates this property? **MP** 3

- **A** $8 + 7 = 7 + 8$
- **B** $a = a \cdot 1$
- **C** $a \cdot 0 = 0$
- **D** $\frac{c}{d} \cdot \frac{d}{c} = 1 \ (c \neq 0, d \neq 0)$

62. Abassi will use the Additive Identity Property to solve an equation. Which of the following best illustrates the Additive Identity Property? **MP** 3

- **A** $a \cdot 1 = a$
- **B** $b + 0 = b$
- **C** $c + (-c) = 0$
- **D** $d + 1 = d + 1$

63. When a number is tripled, its value increases by 10. What is the original number? **MP** 2

- **A** 5
- **B** 10
- **C** 15
- **D** 30

64. Which property justifies rewriting the equation $\frac{1}{6} \cdot 6 + z = 8$ as $1 + z = 8$? **MP** 3

- **A** Additive Identity Property
- **B** Multiplicative Identity Property
- **C** Multiplicative Inverse Property
- **D** Substitution

65. A company creates mobile apps for a smartphone. When the app was free, they had 880 downloads. After the price was set to $0.99, they had d downloads. The company receives $0.70 in revenue for each app that is sold for $0.99. Which equation gives the average revenue R for all downloads of this app? **MP** 2, 4

- **A** $R = \dfrac{0.7d}{880 + d}$
- **B** $R = 0.7(880 - d)$
- **C** $R = 0.7d$
- **D** $R = \dfrac{0.7}{880 + d}$

66. Which of the following properties justifies the following? **MP** 2

$$a - \frac{1}{2} = 5 - \frac{1}{2}$$
$$a = 5$$

- **A** Additive Identity Property
- **B** Associative Property
- **C** Commutative Property
- **D** Reflexive Property

67. MULTI-STEP Consider the rectangle shown below. **MP** 2

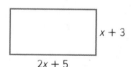

a. Which expression represents the perimeter of the rectangle?

- **A** $(x + 3) + (2x + 5)$
- **B** $(x + 3)(2x + 5)$
- **C** $(x + 3) + (2x + 5) + (x + 3) + (2x + 5)$
- **D** $(x + 3)(2x + 5)(x + 3)(2x + 5)$

b. What is the value of the perimeter if $x = 2$?

Distributive Property

::Then	::Now	::Why?

Then
- You explored Associative and Commutative Properties.

Now
1. Use the Distributive Property to evaluate expressions.
2. Use the Distributive Property to simplify expressions.

Why?
John burns approximately 420 Calories per hour by surfing. The chart below shows the time he spent surfing in one week.

Day	Mon	Tue	Wed	Thu	Fri	Sat	Sun
Time (h)	1	$\frac{1}{2}$	0	1	0	2	$2\frac{1}{2}$

To determine the total number of Calories that he burned surfing that week, you can use the Distributive Property.

New Vocabulary

like terms
simplest form
coefficient

Mathematical Practices

1 Make sense of problems and presevere in solving them.

8 Look for and express regularity in repeated reasoning.

1 Evaluate Expressions There are two methods you could use to calculate the number of Calories John burned surfing. You could find the total time spent surfing and then multiply by the Calories burned per hour. Or you could find the number of Calories burned each day and then add to find the total.

Method 1 Rate Times Total Time

$$420\left(1 + \frac{1}{2} + 1 + 2 + 2\frac{1}{2}\right)$$
$$= 420(7)$$
$$= 2940$$

Method 2 Sum of Daily Calories Burned

$$420(1) + 420\left(\frac{1}{2}\right) + 420(1) + 420(2) + 420\left(2\frac{1}{2}\right)$$
$$= 420 + 210 + 420 + 840 + 1050$$
$$= 2940$$

Either method gives the same total of 2940 Calories burned. This is an example of the **Distributive Property**.

Key Concept Distributive Property

Symbol
For any numbers a, b, and c,
$a(b + c) = ab + ac$ and $(b + c)a = ba + ca$ and
$a(b - c) = ab - ac$ and $(b - c)a = ba - ca$.

Examples

$3(2 + 5) = 3 \cdot 2 + 3 \cdot 5$ $4(9 - 7) = 4 \cdot 9 - 4 \cdot 7$
$3(7) = 6 + 15$ $4(2) = 36 - 28$
$21 = 21$ $8 = 8$

The Symmetric Property of Equality allows the Distributive Property to be written as follows.

$$\text{If } a(b + c) = ab + ac, \text{ then } ab + ac = a(b + c).$$

Real-World Example 1   Distribute Over Addition

SPORTS A group of 7 adults and 6 children are going to a University of South Florida Bulls baseball game. Use the Distributive Property to write and evaluate an expression for the total ticket cost.

Understand You need to find the cost of each ticket and then find the total cost.

Plan 7 + 6 or 13 people are going to the game, so the tickets are $2 each.

Solve Write an expression that shows the product of the cost of each ticket and the sum of adult tickets and children's tickets.

$2(7 + 6) = 2(7) + 2(6)$ Distributive Property

$\qquad = 14 + 12$ Multiply.

$\qquad = 26$ Add.

The total cost is $26.

Check The total number of tickets needed is 13, and they cost $2 each. Multiply 13 by 2 to get 26. Therefore, the total cost of tickets is $26.

USF Bulls Baseball Tickets	
Ticket	Cost ($)
Adult Single Game	5
Children Single Game (12 and under)	3
Groups of 10 or more Single Game	2
Senior Single Game (65 and over)	3

Source: USF

Guided Practice

1. **SPORTS** A group of 3 adults, an 11-year-old, and 2 children under 10 years old are going to a baseball game. Write and evaluate an expression to determine the cost of tickets for the group.

You can use the Distributive Property to make mental math easier.

Example 2   Mental Math

Use the Distributive Property to rewrite $7 \cdot 49$. Then evaluate.

$7 \cdot 49 = 7(50 - 1)$ Think: $49 = 50 - 1$

$\qquad = 7(50) - 7(1)$ Distributive Property

$\qquad = 350 - 7$ Multiply.

$\qquad = 343$ Subtract.

Guided Practice

Use the Distributive Property to rewrite each expression. Then evaluate.

2A. $304(15)$ **2B.** $44 \cdot 2\frac{1}{2}$

2C. $210(5)$ **2D.** $52(17)$

2 **Simplify Expressions** You can use algebra tiles to investigate how the Distributive Property relates to algebraic expressions.

The rectangle at the right has 3 x-tiles and 6 1-tiles. The area of the rectangle is $x + 1 + 1 + x + 1 + 1 + x + 1 + 1$ or $3x + 6$. Therefore, $3(x + 2) = 3x + 6$.

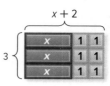

Example 3   Algebraic Expressions

Rewrite each expression using the Distributive Property. Then simplify.

a. $7(3w - 5)$

$7(3w - 5) = 7 \cdot 3w - 7 \cdot 5$ Distributive Property
$\qquad\qquad = 21w - 35$ Multiply.

b. $(6v^2 + v - 3)4$

$(6v^2 + v - 3)4 = 6v^2(4) + v(4) - 3(4)$ Distributive Property
$\qquad\qquad\qquad = 24v^2 + 4v - 12$ Multiply.

▸ **Guided Practice**

3A. $(8 + 4n)2$ **3B.** $-6(r + 3g - t)$

3C. $(3x^2 - 5x + 2)(-5)$ **3D.** $2(7 - 4m^2)$

Like terms are terms that contain the same variables, with corresponding variables having the same power.

 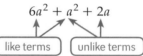

The Distributive Property and the properties of equality can be used to show that $4k + 8k = 12k$. In this expression, $4k$ and $8k$ are like terms.

$4k + 8k = (4 + 8)k$ Distributive Property
$\qquad\quad = 12k$ Substitution

An expression is in **simplest form** when it contains no like terms or parentheses.

Example 4   Combine Like Terms

a. Simplify $17u + 25u$.

$17u + 25u = (17 + 25)u$ Distributive Property
$\qquad\qquad = 42u$ Substitution

b. Simplify $6t^2 + 3t - t$.

$6t^2 + 3t - t = 6t^2 + (3 - 1)t$ Distributive Property
$\qquad\qquad\quad = 6t^2 + 2t$ Substitution

▸ **Guided Practice**

Simplify each expression. If not possible, write *simplified*.

4A. $6n - 4n$ **4B.** $b^2 + 13b + 13$

4C. $4y^3 + 2y - 8y + 5$ **4D.** $7a + 4 - 6a^2 - 2a$

Go Online!

Look for the Tools icons for places where the tools in the eToolkit may be useful. Log into ConnectED to use the tools.

Example 5   Write and Simplify Expressions

Use the expression *twice the difference of 3x and y increased by five times the sum of x and 2y.*

a. Write an algebraic expression for the verbal expression.

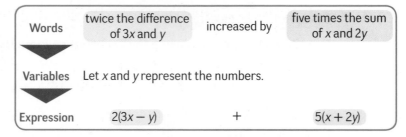

| Words | twice the difference of $3x$ and y | increased by | five times the sum of x and $2y$ |

Variables Let x and y represent the numbers.

| Expression | $2(3x - y)$ | $+$ | $5(x + 2y)$ |

b. Simplify the expression, and indicate the properties used.

$$2(3x - y) + 5(x + 2y) = 2(3x) - 2(y) + 5(x) + 5(2y) \quad \text{Distributive Property}$$
$$= 6x - 2y + 5x + 10y \quad \text{Multiply.}$$
$$= 6x + 5x - 2y + 10y \quad \text{Commutative } (+)$$
$$= (6x + 5x) + (-2y + 10y) \quad \text{Associative Property}$$
$$= (6 + 5)x + (-2 + 10)y \quad \text{Distributive Property}$$
$$= 11x + 8y \quad \text{Substitution}$$

▶ **Guided Practice**

5. Use the expression *5 times the difference of q squared and r plus 8 times the sum of 3q and 2r.*

  A. Write an algebraic expression for the verbal expression.

  B. Simplify the expression, and indicate the properties used.

The **coefficient** of a term is the numerical factor. For example, in $6ab$, the coefficient is 6, and in $\frac{x^2}{3}$, the coefficient is $\frac{1}{3}$. In the term y, the coefficient is 1 since $1 \cdot y = y$ by the Multiplicative Identity Property.

Concept Summary Properties of Numbers

The following properties are true for any numbers a, b, and c.

Properties	Addition	Multiplication
Commutative	$a + b = b + a$	$ab = ba$
Associative	$(a + b) + c = a + (b + c)$	$(ab)c = a(bc)$
Identity	0 is the identity. $a + 0 = 0 + a = a$	1 is the identity. $a \cdot 1 = 1 \cdot a = a$
Zero	—	$a \cdot 0 = 0 \cdot a = 0$
Distributive	$a(b + c) = ab + ac$ and $(b + c)a = ba + ca$	
Substitution	If $a = b$, then a may be substituted for b.	

Study Tip

Compare and Contrast
Noticing the similarities and differences in the addition and multiplication properties can help you learn these terms.

Check Your Understanding   ◯ = Step-by-Step Solutions begin on page R11.

Go Online! for a
Self-Check Quiz

Example 1
1. **PILOT** A pilot at an air show charges $55 per passenger for rides. If 12 adults and 15 children ride in one day, write and evaluate an expression to describe the situation.

Example 2
Use the Distributive Property to rewrite each expression. Then evaluate.

2. $14(51)$

3. $6\frac{1}{9}(9)$

Example 3
Use the Distributive Property to rewrite each expression. Then simplify.

4. $2(4 + t)$

5. $(2g^2 + 9g - 3)6$

Example 4
Simplify each expression. If not possible, write *simplified*.

6. $15m + m$

7. $3x^3 + 5y^3 + 14$

8. $(5m + 2m)10$

Example 5
Write an algebraic expression for each verbal expression. Then simplify, indicating the properties used.

9. 4 times the sum of 2 times x and six

10. one half of 4 times y plus the quantity of y and 3

Practice and Problem Solving   Extra Practice is on page R1.

Example 1
11. **TIME MANAGEMENT** Margo uses colors to track her activities on a calendar. Red represents homework, yellow represents work, and green represents track practice. In a typical week, she has 5 red items, 3 yellow items, and 4 green items. How many activities does Margo do in 4 weeks?

12. **MP PERSEVERANCE** The Red Cross is holding blood drives in two locations. In one day, Center 1 collected 715 pints and Center 2 collected 1035 pints. Write and evaluate an expression to estimate the total number of pints of blood donated over a 3-day period.

Example 2
Use the Distributive Property to rewrite each expression. Then evaluate.

13. $(4 + 5)6$

14. $7(13 + 12)$

15. $6(6 - 1)$

16. $(3 + 8)15$

17. $14(8 - 5)$

18. $(9 - 4)19$

19. $4(7 - 2)$

20. $7(2 + 1)$

21. $7 \cdot 497$

22. $6(525)$

23. $36 \cdot 3\frac{1}{4}$

24. $\left(4\frac{2}{7}\right)21$

Example 3
Use the Distributive Property to rewrite each expression. Then simplify.

25. $2(x + 4)$

26. $(5 + n)3$

27. $(2 - 3m^2)(-5)$

28. $8(x^2 - 9x + 5)$

Example 4
Simplify each expression. If not possible, write *simplified*.

29. $13r + 5r$

30. $3x^3 - 2x^2$

31. $7m + 7 - 5m$

32. $5z^2 + 3z + 8z^2$

33. $(2 - 4n)17$

34. $11(4d + 6)$

35. $7m + 2m + 5p + 4m$

36. $3x + 7(3x + 4)$

37. $4(fg + 3g) + 5g$

Example 5
Write an algebraic expression for each verbal expression. Then simplify, indicating the properties used.

38. the product of 5 and m squared, increased by the sum of the square of m and 5

39. 7 times the sum of a squared and b minus 4 times the sum of a squared and b

40. GEOMETRY Find the perimeter of an isosceles triangle with side lengths of $5 + x$, $5 + x$, and xy. Write in simplest form.

41 GEOMETRY A regular hexagon measures $3x + 5$ units on each side. What is the perimeter in simplest form?

Simplify each expression.

42. $6x + 4y + 5x$

43. $3m + 5g + 6g + 11m$

44. $4a + 5a^2 + 2a^2 + a^2$

45. $5k + 3k^3 + 7k + 9k^3$

46. $6d + 4(3d + 5)$

47. $2(6x + 4) + 7x$

48. FOOD Kenji is picking up take-out food for his study group.

   a. Interpret the expression
   $4(4.49) + 3(2.29) + 3(1.99) + 5(1.49)$.

   b. How much would it cost if Kenji bought four of each item on the menu?

Menu	
Item	Cost ($)
sandwich	4.49
cup of soup	2.29
side salad	1.99
drink	1.49

Use the Distributive Property to rewrite each expression. Then simplify.

49. $\left(\dfrac{1}{3} - 2b\right)27$

50. $4(8p + 4q - 7r)$

51. $6(2c - cd^2 + d)$

Simplify each expression. If not possible, write *simplified*.

52. $6x^2 + 14x - 9x$

53. $4y^3 + 3y^3 + y^4$

54. $a + \dfrac{a}{5} + \dfrac{2}{5}a$

55. MULTIPLE REPRESENTATIONS The area of the model is $2(x - 4)$ or $2x - 8$. The expression $2(x - 4)$ is in *factored form*.

   a. Geometric Use algebra tiles to form a rectangle with area $2x + 6$. Use the result to write $2x + 6$ in factored form.

   b. Tabular Use algebra tiles to form rectangles to represent each area in the table. Record the factored form of each expression.

   c. Verbal Explain how you could find the factored form of an expression.

Area	Factored Form
$2x + 6$	
$3x + 3$	
$3x - 12$	
$5x + 10$	

H.O.T. Problems   Use Higher-Order Thinking Skills

56. MP PERSEVERANCE Use the Distributive Property to simplify $6x^2[(3x - 4) + (4x + 2)]$.

57. MP REASONING Should the Distributive Property be a property of multiplication, addition, or both? Explain your answer.

58. WRITING IN MATH Why is it helpful to represent verbal expressions algebraically?

59. WRITING IN MATH Use the data about surfing on page 23 to explain how the Distributive Property can be used to calculate quickly. Also, compare the two methods of finding the total Calories burned.

60. An expression is shown below.

$6ab^2 + 9a^2b$

Which of the following shows an equivalent expression? **MP** 7

○ **A** $9a^2b^2$

○ **B** $3ab(3b + 2a)$

○ **C** $3ab(2b + 3a)$

○ **D** $3a^2b^2(2b + 3a)$

61. Which of the following is equivalent to the expression $(5m - 3)12$? **MP** 7

○ **A** $5m - 36$

○ **B** $17m + 9$

○ **C** $60m - 36$

○ **D** $60m + 36$

62. Which of the following is equivalent to the expression $3m - 7m^2 + 5m + m^2$? **MP** 7

      I. $-6m^2 + 8m$

     II. $2m(4m - 3)$

    III. $2m(4 - 3m)$

○ **A**   I only

○ **B**   I and II

○ **C**   I and III

○ **D**   III only

63. If $2x(3 + 4a) = -16x + 6x$, what is the value of a? **MP** 2

○ **A** -4

○ **B** -2

○ **C** 2

○ **D** 4

64. The diagram shows the side lengths of a quadrilateral.

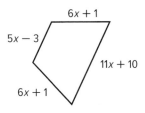

Which expression represents the perimeter of the figure? **MP** 7

○ **A** $22x + 8$

○ **B** $27x + 9$

○ **C** $28x + 9$

○ **D** $9(3x + 1)$

65. What is the simplified form of the expression $3y^2 + 7 - y^2 - 5 - 2y^2 - 1$? **MP** 1

○ **A** 0

○ **B** 1

○ **C** 2

○ **D** $6y^2 + 1$

66. Which of the following is equivalent to $2y + 5(3y - 1)$? **MP** 7

○ **A** $5y + 4$

○ **B** $5y - 5$

○ **C** $17y - 1$

○ **D** $17y - 5$

67. MULTI-STEP Admission to the movies is x dollars, and a bag of popcorn costs $4. **MP** 4

a. Write an expression to represent total cost for two people to go to the movies three times and sharing a bag of popcorn each time.

b. What is the total cost in the scenario of part **a** if the admission is $11 per person?

Algebra Lab
Operations with Rational Numbers

Is the product or sum of two rational numbers also a rational number?

Rational numbers are the set of numbers expressed in the form of a fraction $\frac{a}{b}$, where a and b are integers and $b \neq 0$.

A set is **closed** under an operation if for any members in the set, the result of the operation is also in the set.

A set may be closed under one operation but another set may not be closed under the same operation. For example, in the set of real numbers, subtracting two real numbers always results in a real number, so the set of real numbers is closed under subtraction.

However, in the set of whole numbers, the subtraction of two whole numbers is not always a whole number. For example, $12 - 15 = -3$, which is not a whole number. So the set of whole numbers is not closed under subtraction.

Mathematical Practices

MP 2 Reason abstractly and quantitatively

Activity 1 Review Hierarchy and Relationships of Sets of Numbers

Use the Venn diagram to review the hierarchy and relationships of the different sets of numbers in the real number system.

1A Complete the table. For the two rational numbers a and b, determine ab and $a + b$.

a	b	ab	$a + b$
1	2		
$\frac{1}{2}$	$\frac{1}{3}$		
0.4	0.325		
$\frac{1}{4}$	12		
0.5	0.5		
-4	0.5		
$\frac{1}{6}$	-24		
-3	-1		

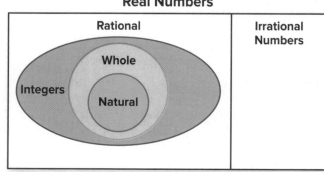

Real Numbers

Rational   Irrational Numbers

Integers   Whole   Natural

1B Are any of the sums or products not rational? Explain.

1C Make a conjecture about whether the set of rational numbers is closed under multiplication and addition.

Listing examples is not sufficient to show that the set of rational numbers is closed under multiplication and addition. To prove this is true, it must be shown that these properties are true for all rational numbers.

To prove that the set of rational numbers is closed under multiplication, it must be shown that if a and b are any rational numbers, then the product ab will always be a rational number.

Prove that the set of rational numbers is closed under addition.

Assume that a and b are two rational numbers, and that $a = \frac{m}{n}$ and $b = \frac{p}{q}$ in which m, n, p, and q are integers.

$$a + b = \frac{m}{n} + \frac{p}{q}$$ Substitution

$$= \frac{q}{q}\left(\frac{m}{n}\right) + \frac{n}{n}\left(\frac{p}{q}\right)$$ Multiply by $\frac{q}{q} = \frac{n}{n} = 1$ to get like denominators.

$$= \frac{mq}{nq} + \frac{np}{nq}$$ Multiply.

$$= \frac{mq + np}{nq}$$ Combine like terms.

The set of integers is closed under multiplication. Therefore, nq, mq, and np are all integers. Because the integers are closed under addition and mq and np are integers, $mq + np$ is an integer. Both the numerator and denominator are integers.

Because $a + b$ can be written as a fraction in which the numerator and denominator are both integers, the set of rational numbers is closed under addition.

Justify each step to prove that the set of rational numbers is closed under multiplication.

Given: a and b are rational numbers.

Prove: The set of rational numbers is closed under multiplication.

3A $a = \frac{m}{n}$ and $b = \frac{p}{q}$, in which m, n, p, and q are integers, and n, $q \neq 0$.

3B $ab = \left(\frac{m}{n}\right)\left(\frac{p}{q}\right)$

3C $= \left(\frac{mp}{nq}\right)$

3D How does $ab = \frac{mp}{nq}$ prove that the set of rational numbers are closed under multiplication?

Work cooperatively. Given that a and b are rational numbers, prove each statement.

1. The set of rational number is closed under subtraction.

2. The set of rational numbers is closed under division.

Write a verbal expression for each algebraic expression.
(Lesson 1-1)

1. $21 - x^3$

2. $3m^5 + 9$

Write an algebraic expression for each verbal expression. (Lesson 1-1)

3. five more than s squared

4. four times y to the fourth power

5. CAR RENTAL The XYZ Car Rental Agency charges a flat rate of $39 per day plus $0.47 per mile driven. Write an algebraic expression for the rental cost of a car for x days that is driven y miles. (Lesson 1-1)

Evaluate each expression. (Lesson 1-2)

6. $24 \div 3 - 2 \cdot 3$

7. $5 + 2^2$

8. $4(3 + 9)$

9. $36 - 2(1 + 3)^2$

10. $\dfrac{40 - 2^3}{4 + 3(2^2)}$

11. PARKS The costs of tickets to a local amusement park are shown. Write and evaluate an expression to find the total cost for 5 adults and 8 children. (Lesson 1-2)

12. MULTIPLE CHOICE Write an algebraic expression to represent the perimeter of the rectangle shown below. Then evaluate it to find the perimeter when $w = 8$ cm. (Lesson 1-2)

A 37 cm

B 232 cm

C 74 cm

D 45 cm

Evaluate each expression. Name the property used in each step. (Lesson 1-3)

13. $(8 - 2^3) + 21$

14. $3(1 \div 3) \cdot 9$

15. $[5 \div (3 \cdot 1)]\dfrac{3}{5}$

16. $18 + 35 + 32 + 15$

17. $0.25 \cdot 7 \cdot 4$

Use the Distributive Property to rewrite each expression. Then evaluate. (Lesson 1-4)

18. $3(5 + 2)$

19. $(9 - 6)12$

20. $8(7 - 4)$

Use the Distributive Property to rewrite each expression. Then simplify. (Lesson 1-4)

21. $4(x + 3)$

22. $(6 - 2y)7$

23. $-5(3m - 2)$

24. CAR WASH A car wash chain has three locations.

a. Use the information in the table below to write and evaluate an expression to estimate the total number of car washes sold over a 4-day period. (Lesson 1-4)

Location	Daily Car Washes
Location 1	145
Location 2	211
Location 3	184

b. **MP** What mathematical practice did you use to solve this problem?

25. Use the Distributive Property to rewrite $(8 - 3p)(-2)$. (Lesson 1-4)

Descriptive Modeling and Accuracy

∴Then

- You used numbers to model a real-world situation.

∴Now

1 Define appropriate quantities for descriptive modeling.

2 Choose appropriate levels of accuracy.

∴Why?

- Numbers and equations are used to describe and model real-world situations and to make accurate measurements. How accurate a measurement should be depends on the level of accuracy needed or desired. For example, the parts for a jet engine may need to be made to the nearest ten-thousandth of a meter or less.

New Vocabulary
metric
debt-to-income ratio
accuracy

Mathematical Practices
4 Model with mathematics.
6 Attend to precision.

1 **Descriptive Modeling** When using numbers to model a real-world situation, it is often helpful to have a metric. A **metric** is a rule for assigning a number to some characteristic or attribute. For example, teachers use metrics to determine grades. Each teacher determines the appropriate metric for assessing a student's performance and assigning a grade.

A mortgage companies use a **debt-to-income ratio** as a metric to determine if a person qualifies for a loan. The debt-to-income ratio is calculated as *how much the person owes per month* divided by *how much the person earns per month*. Many mortgage companies use a debt-to-income ratio of 0.36 or less to determine candidates for loans. Debt-to-income ratio is only one metric used by mortgage companies. The type of metric used or the value required to qualify for a loan varies by company.

Real-World Example 1   Modeling in Consumer Loans

LOANS Ryan is applying for a home loan. His monthly expenses are $1165, his monthly income is $3650, and his projected monthly mortgage for a new home is $1068. Banks determine whether a person qualifies for a home loan based on a metric. At National Road Bank, Ryan's debt-to-income ratio must be 0.36 or less to qualify for a loan, and at New Savings Bank his mortgage-to-income ratio must be 0.28 or less.

a. Determine whether Ryan qualifies for a home loan at National Road Bank.

$$\text{Debt-to-income ratio} = \frac{\text{monthly expenses}}{\text{monthly income}}$$
$$= \frac{1165}{3650}$$
$$\approx 0.319178$$

Because the ratio is less than 0.36, he does qualify for a loan at National Road Bank.

b. Determine whether Ryan qualifies for a home loan at New Savings Bank.

$$\text{Mortgage-to-income ratio} = \frac{\text{monthly mortgage}}{\text{monthly income}}$$
$$= \frac{1068}{3650}$$
$$\approx 0.292603$$

Because the ratio is greater than 0.28, he does not qualify for a loan at New Savings Bank.

c. **Compare the results of the two metrics. How effective are each of the metrics as measures of whether Ryan can afford to buy a house?**

Ryan qualifies for a home loan at National Road Bank, but at New Savings Bank, he does not. The two metrics use different factors to determine whether someone would qualify for a loan. A more effective metric might use both monthly expenses and mortgage.

> ### Guided Practice

LOANS **Luciana has monthly expenses of $975, a monthly income of $2465, and a projected monthly mortgage payment of $684.**

1A. Determine whether Luciana qualifies for a home loan at National Road Bank.

1B. Determine whether Luciana qualifies for a home loan at New Savings Bank.

1C. Compare the results of the two metrics. How could Luciana adjust her finances so that she qualifies at both banks?

2 Appropriate Levels of Accuracy All measurements taken in the real world are approximations. The greater the care with which a measurement is taken, the more accurate it will be. **Accuracy** refers to how close a measured value comes to the actual or desired value. For example, an exact fraction is more accurate than a rounded decimal.

Example 2	Accuracy in Measurement

CONSUMER PRODUCTS **Suppose you measure the width of a laptop screen and record the measurement in meters, centimeters, and millimeters.**

a. **Determine which measurement is rounded to the nearest whole number.**

The measurement in millimeters is rounded to the nearest whole number.

b. **Determine which measurement is rounded to the nearest half, tenth, or smaller.**

The measurement in centimeters is rounded to the nearest half. The measurement in meters is rounded to the nearest tenth.

c. **Determine which measurement is the most appropriate for measuring the screen.**

Usually, the most appropriate measure is the most reasonable for the size of the screen. So the measurement in centimeters is the most appropriate.

d. **Determine which unit of measure is the most accurate.**

Usually, the smaller the unit of measure, the more accurate the measure. So the measurement in millimeters is the most accurate.

Real-World Link

When manufacturers and retailers describe the size of a laptop screen, they report the length of the diagonal of the screen rather than the height or width. Common laptop screen sizes range from 11 inches to 17 inches.

> **Guided Practice**

2A. An area rug has dimensions given in inches and centimeters. Which unit of measure is more accurate? Explain.

2B. One website indicates that the distance to the Sun is 150 billion meters. Another website lists the distance as 93,000,000 miles. Determine which unit is more appropriate for measuring the distance to the sun: meters or miles. Explain.

Example 3   Decide Where to Round

Determine where to round in each situation.

a. Elan has \$13 that he wants to divide among his 6 siblings. When he types 13 ÷ 6 into his calculator, the number that appears is 2.166666667. Where should Elan round? Explain.

Since Elan is rounding money, the smallest increment is a penny, so round to the hundredths place. This will give him 2.17, and \$2.17 × 6 = \$13.02. Elan will be two pennies short, so round to \$2.16. Since \$2.16 × 6 = \$12.96, Elan can give each of his siblings \$2.16.

b. Dante's mother brings him a dozen cookies, but before she leaves, she eats one and tells Dante he has to share with his two sisters. Dante types 11 ÷ 3 into his calculator and gets 3.666666667. Where should Dante round? Explain.

After each sibling receives 3 cookies, there are two cookies left. In this case, it is more accurate to convert the decimal portion to a fraction and give each sibling $\frac{2}{3}$ of a cookie.

c. Eva measures the dimensions of a box as 8.7, 9.52, and 3.16 inches. She multiplies these three numbers to find the measure of the volume. The result shown on her calculator is 261.72384. Where should Eva round? Explain.

Eva should round to the tenths place, 261.7, because she was only accurate to the tenths place with one of her measures.

> **Guided Practice**

3A. Jessica wants to divide \$23 six ways. Her calculator shows 3.833333333. Where should she round? Explain.

3B. Ms. Harris wants to share 2 pizzas among 6 people. Her calculator shows 0.333333333. Where should she round?

3C. The measurements of an aquarium are 12.9, 7.67, and 4.11 inches. The measure of the volume is given by the product 406.65573. Where should the number be rounded?

For most real-world measurements, a decision must be made about the level of accuracy needed or desired.

Example 4   Find an Appropriate Level of Accuracy

For each situation, determine the appropriate level of accuracy.

a. **Jon needs to buy a shade for the window opening shown, but the shades are only available in whole increments. What size shade should he buy to cover the window entirely?**

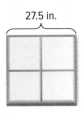

27.5 in.

He should buy the 28-inch shade because it is enough to cover the glass.

b. **Tom is buying flea medicine for his dog. The amount of medicine depends on the dog's weight. The medicine is available in packages that are sold in 10-pound increments. How accurate does Tom need to be to buy the correct medicine?**

He needs to be accurate to within 10 pounds.

c. **Tyrone is building a jet engine. How accurate do you think he needs to be with his measurements?**

He needs to be very accurate, perhaps to a thousandth of an inch.

▶ **Guided Practice**

4A. Matt's table is missing a leg. He wants to cut a piece of wood to replace the leg. How accurate do you think he needs to be with his measurement?

4B. You are estimating the height of a small child. Which unit of measure should you use: 1 foot, 1 inch, or $\frac{1}{16}$ inch?

4C. Curt is measuring the driving distance from one city to another. How accurate do you think he needs to be with his measurement?

Watch Out!

MP Modeling Some real-world situations may be modeled well by more than one unit of measurement. For instance, a box may be measured in centimeters or millimeters, but the context of the situation will help you to determine the better unit of measure.

Go Online! for a Self-Check Quiz

Check Your Understanding   ◯ = Step-by-Step Solutions begin on page R11.

Example 1

1. **GRADING** A teacher uses a metric to determine her students' quarter grades. The total number of points possible is 450. Students with at least 90% of the total points will receive an A for the quarter. Students with 89-80% will receive a B, students with 79-70% will receive a C, and students with 69-60% will receive a D. The table shows the number of points for each student. Determine the number of students who have earned each grade.

432	388	345	330
419	361	342	328
404	360	340	300
398	359	333	289
398	359	332	288

Example 2

2. Charges for water consumption in a home are recorded with a water meter. Last month, the meter read "879," and this month the meter read "932." Which unit of measure is most likely the unit for these measurements: gallons, hundreds of gallons, thousands of gallons?

Example 3

3. Kiri wants to share 3 pies among 11 people. Her calculator shows 0.2727272727. Determine where the rounding should occur and give the rounded answer.

Example 4

4. A nurse is administering medicine to a patient based on his weight. How accurate do you think she needs to be with her measurement of the medicine?

Practice and Problem Solving

Extra Practice is on page R1.

Example 1

5. Amaranda's piano teacher allows his students to pick out a prize based on how much time they've recorded in a weekly practice log. They earn a pencil for practicing 0–120 minutes, a music download for practicing 121–180 minutes, and a movie ticket for practicing 181 minutes or more. Amaranda's practice logs for 9 weeks showed 125, 130, 115, 95, 80, 175, 185, 160, and 110 minutes. How many music downloads did Amaranda earn?

Example 2

6. Jayson is the new supply manager for the football team. The order form used by last year's manager says they use 30 units of sport drink per game. Which unit of measure is most likely: ounces, cups, or gallons?

Example 3

(7) Evan's calculator shows 137.2582774 as the volume of his soccer ball. Evan measured the radius of the ball to be 3.2 inches. Determine to what place the rounding of the volume should occur and give the rounded answer.

8. Sam wants to divide $111 seven ways. Where should the rounding occur? What is the rounded answer?

Example 4

9. You are estimating the length of your school's basketball court. Which unit of measure should you use: 1 foot, 1 inch, or $\frac{1}{16}$ inch?

10. MARBLES The graph shows the mass of a set of marbles related to the number of marbles in the set. A line is drawn to represent all amounts of marbles.

a. Describe the line in terms of *accuracy*.

b. What is the approximate mass of 7 marbles?

c. Explain why some points are above the line and some points are below the line.

d. Make a conjecture about the approximate mass of 20 marbles. Explain your reasoning.

Mass of Marbles

11. PROMOTIONS Company ABC's human resources department uses the Data Quality Score as a metric to determine promotions for its sales team.

$$\text{Data Quality Score} = \frac{\text{Number of Positive Calls}}{\text{Total Number of Calls}} + \frac{\text{Actual Deals Closed}}{\text{Projected Number of Deals}}$$

	Number of Positive Calls	Total Number of Calls	Actual Deals Closed	Projected Number of Deals
Mr. Menendez	160	240	18	40
Ms. O'Toole	188	232	24	52
Ms. Randall	159	212	31	79
Mr. Fraser	162	238	19	48

a. Find the data score for each applicant.

b. Who should get a promotion? Justify your reasoning.

12. Which is the best estimate for the length of a guitar: 1 foot, 1 meter, or 1 kilometer?

13. Which is the best estimate for the volume of a bathtub: 35 ounces, 35 cups, 35 gallons?

14. Which is the best estimate for the weight of an elephant: 7.5 pounds, 7.5 tons, or 7.5 ounces?

15. Which is the best estimate for the distance from Earth to Earth's moon: 240 miles, 240 thousand miles, 240 million miles?

16. FINANCIAL LITERACY Cameron wants to buy a car. His monthly expenses are shown in the spreadsheet.

◇	A	B	C
1	**Type of Debt**	**Expenses**	**Salary**
2	Rent	900	4140
3	Credit Cards	250	
4	Mortgage	1410	
5	Home Equity Loan	225	
6	**Total**	**2785**	

Sheet 1 / Sheet 2 / Sheet 3

a. Write an expression to find the debt-to-income ratio that takes into account all monthly expenses.

b. The car loan officer will loan Cameron the money as long as the debt-to-income ratio is below 0.75. Will Cameron get approved for a loan? Explain.

c. Before taking the loan, Cameron checks another car loan company. This company's loan metric is calculated by finding the ratio of his car payment to his monthly salary. The company uses 0.15 as its ideal ratio to find the maximum monthly payment Cameron can afford. Theoretically, what is the maximum car payment Cameron can afford?

17. CONSUMER SCIENCE A consumer watch agency wants to test the accuracy of the advertised weight of bags of fertilizer having a label weight of 50 pounds.

a. Describe a way to test the weights.

b. What is an appropriate level of accuracy for their tests? Justify your reasoning.

c. How can their results guide future production of bags of fertilizer?

18. SPACE Which unit of measure is the most appropriate for measuring the distance from Earth to the star Polaris: feet, kilometers, light-years? Explain.

19 FINANCIAL LITERACY The Internal Revenue Service (IRS) of the U.S. government uses formulas to assess income tax for individuals. They provide tax-rate guidelines that include metrics. For example, for the 2015 tax year, a single person earning over $9,225 but less than $37,450 is expected to owe income tax using the following metric: $Tax = 922.50 + 0.15(x - 9225)$, where x is the amount of taxable income.

a. Use this metric to find the income tax for a single person earning $20,000 of taxable income.

b. Use the Internet to research another IRS metric. Describe your findings.

20. BASEBALL In baseball, earned run averages (ERAs) are used to compare pitchers. The spreadsheet shows the midyear ERAs for some starting pitchers of Major League Baseball teams during the 2016 season.

a. The ERA is calculated by dividing the total number of earned runs by the quotient of the number of innings pitched and 9 or the number of full games pitched. The metric below represents this calculation.

$$ERA = \frac{ER}{IP \div 9} = \frac{\text{Number of Earned Runs}}{\text{Number of Innings Pitched} \div 9}$$

At the midyear of the 2016 season, Jacob DeGrom of the New York Mets had 93 innings pitched for 27 earned runs. Find his ERA.

◇	A	B	C	D	E	
1	**Pitcher**	**Team**	**Games Won**	**Games Lost**	**ERA**	
2	Estrada, M.	Toronto Blue Jays	5	3	2.93	
3	Arrieta, J.	Chicago Cubs	12	4	2.68	
4	Wright, S.	Boston Red Sox	10	5	2.68	
5	Cueto, J.	San Francisco Giants	13	1	2.47	
6	Fernandez, J.	Miami Dolphins	11	4	2.52	
7	Teheran, J.	Atlanta Braves	3	8	2.96	

Sheet 1 / Sheet 2 / Sheet 3 /

b. Summarize the data and evaluate its effectiveness for modeling the data.

21. COLLEGE ATHLETICS Some universities use a metric called the Academic Index, which considers a student's G.P.A. and SAT scores, to qualify high school athletic recruits. Use the expression to determine whether the student athlete qualifies at a university that requires an Academic Index of 185 or greater.

$$2\left[\frac{\left(\dfrac{\text{Reading Score} + \text{Writing Score}}{2}\right) + \text{Math Score}}{20}\right] + \text{G. P. A. Value}$$

G.P.A.	G.P.A. Value
4.0	80
3.8	78
3.6	75
3.4	71
3.2	69
3.0	67

a. Determine whether a student with a high school G.P.A. of 3.2 and SAT scores of 620 in reading, 640 in writing, and 670 in math will qualify for the university based on the Academic Index.

b. What other attributes of high school recruits do you think universities might consider when creating metrics to determine qualification?

22. **COMPUTERS** Each year, the estimated number of computer tablet users in the U.S. is increasing, as shown in the table. The numbers for 2017 and 2018 are projections.

a. Is it accurate to say that in 2019, the number of tablets will be greater than 177 million? Explain.

b. The percent of increase in tablet users is calculated using the model
$$\% \text{ Increase} = \frac{\text{New Amount} - \text{Old Amount}}{\text{Old Amount}}.$$ Complete the table with the percent of increase from 2015 to 2018.

Year	Number of Tablet Users (millions)	Percent of Increase
2014	149	—
2015	159	
2016	166	
2017	172	
2018	177	

c. Make a conjecture about the number of tablet users in 2019. Justify your reasoning.

Determine the level of accuracy for each situation. Explain your reasoning.

23. A sports reporter states that there are 660,430 gallons of water in an Olympic-sized swimming pool.

24. A science magazine reported that there are, on average, 37 trillion cells that make up the human body.

25. An information station at a zoo states that male gorillas weigh 152 kilograms.

26. One study estimated that about 205 billion emails are sent worldwide each day.

H.O.T. Problems   Use **H**igher-**O**rder **T**hinking Skills

27. **OPEN ENDED** The Environmental Protection Agency (EPA) has a complex way to determine the fuel economy data for a car.

a. Describe an experiment that may be done to test whether the fuel economy rating of a new car is accurate.

b. The combined miles per gallon (MPG) is sometimes found by finding a weighted average of 55% of the city MPG rating and 45% of the highway MPG. The equation below is a metric for the combined MPG.
$$\text{Combined MPG} = \frac{1}{\frac{0.55}{\text{CITY}} + \frac{0.45}{\text{HWY}}}$$
Find the combined miles per gallon for a car with ratings CITY = 30 mpg and HWY = 35 mpg.

28. **WRITING IN MATH** The thickness of a page in a textbook was measured with a micrometer to be 0.018 centimeter. The micrometer's measurement has an *uncertainty* of plus or minus 0.001 centimeter.

a. Describe the meaning of *uncertainty* in the context of this problem.

b. How accurate does the thickness of the page need to be when publishing the textbook? Explain.

c. State a rule that you could use to express the thickness of the page in the most accurate terms.

29. MULTI-STEP The spreadsheet shows Viola's monthly expenses and salary. (MP) 1, 4

   a. Which metric(s) gives ratios below 0.3 that could be used by the lending institution to assess its risk for lending money to an individual? Select all that apply.

◇	A	B	C
1	**Type of Debt**	**Expenses**	**Salary**
2	Rent	950	4950
3	Credit Cards	220	
4	Home Equity Loan	400	
5	Student Loan	150	
6	**Total**	**1720**	
7			

Sheet 1 / Sheet 2 / Sheet 3

☐ **A** $\text{Ratio} = \dfrac{\text{total monthly debt}}{\text{monthly salary}}$

☐ **B** $\text{Ratio} = \dfrac{\text{total loan payments}}{\text{monthly salary}}$

☐ **C** $\text{Ratio} = \dfrac{\text{rent} + \text{home equity loan}}{\text{monthly salary}}$

☐ **D** $\text{Ratio} = \dfrac{\text{monthly salary} - \text{total monthly debt}}{\text{monthly salary}}$

   b. The lending institution has decided to no longer use the ratio shown in D. Use the value you calculated in D to make a conjecture about why they may have made this decision.

   c. Suppose Viola has no monthly debt when she applies for a car loan. How does this affect the ratios and the lending institution's decision about making the loan?

30. The on-base percentage (OBP) in baseball is a measure of how often a batter reaches a base.

The formula to find the OBP is

$\dfrac{H + BB + HBP}{AB + BB + HBP + SF}$, where H = hits, BB = base on balls, HBP = hit by pitch, AB = at bats, and SF = the number of times a runner reaches a base after a sacrifice fly. In 2015, Joey Votto of the Cincinnati Reds led the Major Leagues for OBP, with H = 171, BB = 143, HBP = 5, AB = 545, and SF = 2. What was his on-base percentage? (MP) 1, 4, 6

31. Suppose you measure a poster to hang on a wall and record the measurement in centimeters, meters, and millimeters. (MP) 4, 6

   a. Which measurement is the most appropriate for measuring the poster?

   b. Which unit of measure is the most accurate?

32. Raj made a profit of $890 selling T-shirts. He wants to share the profit equally among three charities. How much should he give to each charity? (MP) 4, 6

   ○ **A** $296   ○ **C** $296.67

   ○ **B** $296.66   ○ **D** $297

33. Martina measures the dimensions of a crate as 10.45, 13.275, and 12.2 inches. She multiplies these three numbers to find the volume. The result shown on her calculator is 1692.42975. Which measurement is rounded appropriately? (MP) 4, 6

   ○ **A** 1692   ○ **C** 1692.43

   ○ **B** 1692.4   ○ **D** 1692.430

34. Darrell is tiling the floor of his bathroom with 18-inch tiles. To the nearest fraction of an inch, how accurate do you think he needs to be with his measurements? Explain. (MP) 4, 6

35. The table below shows the number of grains of rice and their mass. (MP) 4, 6

Number of Grains of Rice	Mass (grams)
100	1.75
200	3.652
400	7.05
800	14.8731
1600	28.2

   a. Which measurement is the most accurate?

   b. Which measurement is the least accurate?

   c. Predict the range of mass for 3200 grains of rice. Explain your prediction.

Relations

:·Then

- You defined appropriate quantities for descriptive modeling.

:·Now

1. Represent relations.
2. Interpret graphs of relations.

:·Why?

- The deeper in the ocean you are, the greater pressure is on your body. This is because there is more water over you. The force of gravity pulls the water weight down, creating a greater pressure.

  The equation that relates the total pressure of the water to the depth is $P = rgh$, where

  P = the pressure,
  r = the density of water,
  g = the acceleration due to gravity, and
  h = the height of water above you.

New Vocabulary

coordinate system
coordinate plane
x- and y-axes
origin
ordered pair
x- and y-coordinates
relation
mapping
domain
range
independent variable
dependent variable

MP Mathematical Practices

1 Make sense of problems and persevere in solving them.

1 Represent a Relation This relationship between the depth and the pressure exerted can be represented by a line on a coordinate grid.

A **coordinate system** is formed by the intersection of two number lines, the *horizontal axis* and the *vertical axis*.

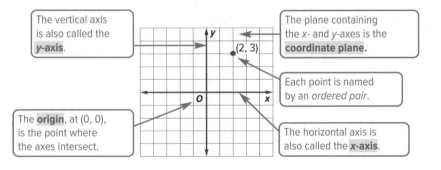

The vertical axis is also called the **y-axis**.

The plane containing the x- and y-axes is the **coordinate plane**.

Each point is named by an *ordered pair*.

The **origin**, at (0, 0), is the point where the axes intersect.

The horizontal axis is also called the **x-axis**.

A point is represented on a graph using ordered pairs.

- An **ordered pair** is a set of numbers, or *coordinates*, written in the form (x, y).

- The x-value, called the **x-coordinate**, represents the horizontal placement of the point.

- The y-value, or **y-coordinate**, represents the vertical placement of the point.

A set of ordered pairs is called a **relation**. A relation can be represented in several different ways: as an equation, in a graph, with a table, or with a mapping.

A **mapping** illustrates how each element of the *domain* is paired with an element in the *range*. The set of the first numbers of the ordered pairs is the **domain**. The set of second numbers of the ordered pairs is the **range** of the relation. This mapping represents the ordered pairs $(-2, 4), (-1, 4), (0, 6) (1, 8)$, and $(2, 8)$.

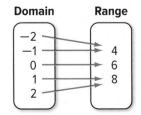

Domain	Range
−2	
−1	4
0	6
1	8
2	

Study the different representations of the same relation below.

Ordered Pairs	Table	Graph	Mapping

Ordered Pairs

(1, 2)
(−2, 4)
(0, −3)

Table

x	y
1	2
−2	4
0	−3

Graph

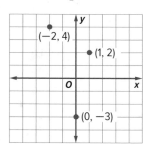

Mapping

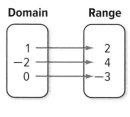

The x-values of a relation are members of the domain and the y-values of a relation are members of the range. In the relation above, the domain is {−2, 1, 0} and the range is {−3, 2, 4}.

Example 1 **Representations of a Relation**

a. Express {(2, 5), (−2, 3), (5, −2), (−1, −2)} as a table, a graph, and a mapping.

Table
Place the x-coordinates into the first column of the table. Place the corresponding y-coordinates in the second column of the table.

x	y
2	5
−2	3
5	−2
−1	−2

Graph
Graph each ordered pair on a coordinate plane.

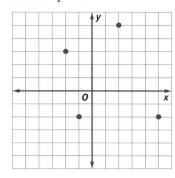

Mapping
List the x-values in the domain and the y-values in the range. Draw arrows from the x-values in the domain to the corresponding y-values in the range.

Domain   Range

Domain	Range
2	5
−2	3
5	−2
−1	

b. Determine the domain and the range of the relation.

The domain of the relation is {2, −2, 5, −1}. The range of the relation is {5, 3, −2}.

▶ **Guided Practice**

1A. Express {(4, −3), (3, 2), (−4, 1), (0, −3)} as a table, graph, and mapping.

1B. Determine the domain and range.

In a relation, the value of the variable that determines the output is called the **independent variable**. The variable with a value that is dependent on the value of the independent variable is called the **dependent variable**. The domain contains values of the independent variable. The range contains the values of the dependent variable.

Real-World Example 2   Independent and Dependent Variables

Identify the independent and dependent variables for each relation.

a. **DANCE** **The dance committee is selling tickets to the Fall Ball. The more tickets that they sell, the greater the amount of money they can spend for decorations.**

The number of tickets sold is the independent variable because it is unaffected by the money spent on decorations. The money spent on decorations is the dependent variable because it depends on the number of tickets sold.

b. **MOVIES** **Generally, the average price of going to the movies has steadily increased over time.**

Time is the independent variable because it is unaffected by the cost of attending the movies. The price of going to the movies is the dependent variable because it is affected by time.

Guided Practice

Identify the independent and dependent variables for each relation.

2A. The air pressure inside a tire increases with the temperature.

2B. As the amount of rain decreases, so does the water level of the river.

2 Graphs of a Relation A relation can be graphed without a scale on either axis. These graphs can be interpreted by analyzing their shape.

Example 3   Analyze Graphs

The graph represents the distance Francesca has ridden on her bike. Describe what happens in the graph.

As time increases, the distance increases until the graph becomes a horizontal line.

So, time is increasing but the distance remains constant. At this section Francesca stopped. Then she continued to ride her bike.

Bike Ride

Guided Practice

Describe what is happening in each graph.

3A. **Driving to School**

3B. **Change in Income**

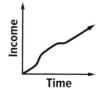

Check Your Understanding   ◯ = Step-by-Step Solutions begin on page R11.

✓ **Go Online!** for a Self-Check Quiz

Example 1 Express each relation as a table, a graph, and a mapping. Then determine the domain and range.

1. {(4, 3), (−2, 2), (5, −6)}

2. {(5, −7), (−1, 4), (0, −5), (−2, 3)}

Example 2 Identify the independent and dependent variables for each relation.

3. Increasing the temperature of a compound inside a sealed container increases the pressure inside a sealed container.

4. Mike's cell phone is part of a family plan. If he uses more data than his share, then there is less data available for the rest of his family.

5. Julian is buying concert tickets for himself and his friends. If he buys more concert tickets, the cost is greater.

6. A store is having a sale over Labor Day weekend. When there are more purchases, the profits are greater.

Example 3 **MP** **MODELING** Describe what is happening in each graph.

7. The graph represents the distance the track team runs during a practice.

8. The graph represents revenues generated through an online store.

Practice and Problem Solving   Extra Practice is on page R1.

Example 1 Express each relation as a table, a graph, and a mapping. Then determine the domain and range.

9. {(0, 0), (−3, 2), (6, 4), (−1, 1)}

10. {(5, 2), (5, 6), (3, −2), (0, −2)}

11. {(6, 1), (4, −3), (3, 2), (−1, −3)}

12. {(−1, 3), (3, −6), (−1, −8), (−3, −7)}

13. {(6, 7), (3, −2), (8, 8), (−6, 2), (2, −6)}

14. {(4, −3), (1, 3), (7, −2), (2, −2), (1, 5)}

Example 2 Identify the independent and dependent variables for each relation.

15 The Spanish classes are having a fiesta lunch. Each student who attends is to bring a Spanish side dish or dessert. The more students who attend, the more food there will be available.

16. The faster you drive your car, the longer it will take to come to a complete stop.

Example 3 **MP** **MODELING** Describe what is happening in each graph.

17. The graph represents the height of a bungee jumper.

18. The graph represents the sales of lawn mowers.

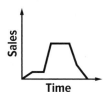

MP MODELING Describe what is happening in each graph.

19 The graph represents the value of a rare baseball card.

20. The graph represents the distance covered on an extended car ride.

For Exercises 21–23, use the graph at the right.

21. Name the ordered pair at point *A* and explain what it represents.

22. Name the ordered pair at point *B* and explain what it represents.

23. Identify the independent and dependent variables for the relation.

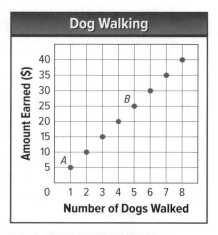

For Exercises 24–26, use the graph at the right.

24. Name the ordered pair at point *C* and explain what it represents.

25. Name the ordered pair at point *D* and explain what it represents.

26. Identify the independent and dependent variables.

Express each relation as a set of ordered pairs. Describe the domain and range.

27.

Buying Aquarium Fish	
Number of Fish	Total Cost
1	$2.50
2	$4.50
5	$10.50
8	$16.50

28.

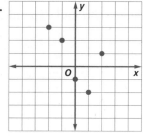

Express the relation in each table, mapping, or graph as a set of ordered pairs.

29.

x	y
4	−1
8	9
−2	−6
7	−3

30.

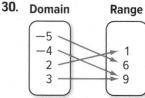

31.

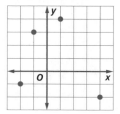

32. SPORTS In a triathlon, athletes swim 2.4 miles, bicycle 112 miles, and run 26.2 miles. Their total time includes transition time from one activity to the next. Which graph best represents a participant in a triathlon? Explain.

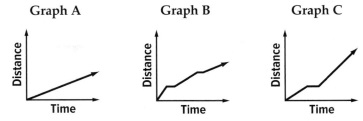

Graph A   Graph B   Graph C

Draw a graph to represent each situation.

33. ANTIQUES A grandfather clock that is over 100 years old has increased in value from when it was first purchased.

34. CAR A car depreciates in value. The value decreases quickly in the first few years.

35. REAL ESTATE A house typically increases in value over time.

36. EXERCISE An athlete alternates between running and walking during a workout.

37 PHYSIOLOGY A typical adult has about 2 pounds of water for every 3 pounds of body weight. This can be represented by the equation $w = 2\left(\dfrac{b}{3}\right)$, where w is the weight of water in pounds and b is the body weight in pounds.

  a. Make a table to show the relation between body and water weight for people weighing 100, 105, 110, 115, 120, 125, and 130 pounds. Round to the nearest tenth if necessary.

  b. What are the independent and dependent variables?

  c. State the domain and range, and then graph the relation.

  d. Reverse the independent and dependent variables. Graph this relation. Explain what the graph indicates in this circumstance.

H.O.T. Problems   Use **H**igher-**O**rder **T**hinking Skills

38. OPEN ENDED Describe a real-life situation that can be represented using a relation and discuss how one of the quantities in the relation depends on the other. Then represent the relation in three different ways.

39. CHALLENGE Describe a real-world situation where it is reasonable to have a negative number included in the domain or range.

40. 🅜🅟 **PRECISION** Compare and contrast dependent and independent variables.

41. CHALLENGE The table presents a relation. Graph the ordered pairs. Then reverse the y-coordinate and the x-coordinate in each ordered pair. Graph these ordered pairs on the same coordinate plane. Graph the line $y = x$. Describe the relationship between the two sets of ordered pairs.

x	y
0	1
1	3
2	5
3	7

42. WRITING IN MATH Use the data about the pressure of water on page 42 to explain the difference between dependent and independent variables.

43. The points on this graph represent a relation. What are the domain and range of the relation? **MP** 1

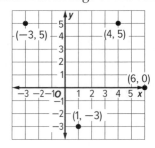

- ○ **A** domain: {−3, 0, 5}; range: {−3, 1, 4, 6}
- ○ **B** domain: {−3, 1, 1, 4, 6}; range: {−3, 0, 5, 5}
- ○ **C** domain: {−3, 1, 4, 6}; range: {−3, 0, 5, 5}
- ○ **D** domain: {−3, 1, 4, 6}; range: {−3, 0, 5}

44. A mapping is shown.

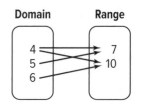

Which set of ordered pairs represents the same relation as the mapping? **MP** 1

- ○ **A** {(7, 4), (7, 5), (10, 4), (10, 6)}
- ○ **B** {(4, 7), (5, 7), (6, 10)}
- ○ **C** {(4, 7), (4, 10), (5, 7), (6, 10)}
- ○ **D** {(4, 7), (4, 10), (5, 7), (6, 7), (6, 10)}

45. Which situation is modeled by the graph? **MP** 1, 4

- ○ **A** A jogger stops to take a drink of water and then continues to jog.

- ○ **B** Ron walks to his friend's house. They drive to the mall and then return to Ron's house.

- ○ **C** Kim walks to the bus stop, waits for several minutes, and then takes the bus directly to work.

- ○ **D** Mark rides his bike to the park and takes a short break before playing basketball for an hour.

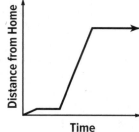

46. Three of the representations below show the same relation. Which answer choice shows a relation that is different from the other three? **MP** 1

- ○ **A**

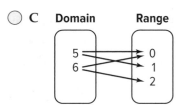

x	5	5	6	6
y	0	1	0	2

- ○ **B** {(5, 0), (5, 1), (6, 0), (6, 2)}

- ○ **C** Domain   Range

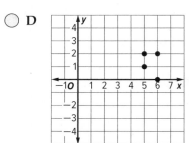

- ○ **D**

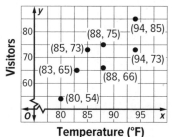

47. MULTI-STEP Data of the number of visitors to a beach and the daily temperature were recorded over one week. The data are shown in the graph. **MP** 1, 4

a. Identify the independent and dependent variables.

Independent: [_____]

Dependent: [_____]

b. What are the domain and range of the relation?

- ○ **A** domain: {80, 94}; range: {54, 85}

- ○ **B** domain: {80, 83, 85, 88, 94}; range: {54, 65, 66, 73, 75, 85}

- ○ **C** domain: {54, 65, 66, 73, 75, 85}; range: {80, 83, 85, 88, 94}

- ○ **D** domain: {80, 83, 83, 85, 88, 94, 94}; range: {54, 65, 66, 73, 73, 75, 85}

Functions

- You solved equations with elements from a replacement set.

1 Determine whether a relation is a function.

2 Find linear equations.

- The distance a car travels from when the brakes are applied to the car's complete stop is the stopping distance. This includes time for the driver to react. The faster a car is traveling, the longer the stopping distance. The stopping distance is a function of the speed of the car.

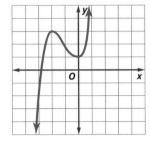

Stopping Distance of a Passenger Car

Stopping Distance (ft) vs Speed (mph)

 New Vocabulary

function
discrete function
continuous function
vertical line test
function notation
nonlinear function

MP **Mathematical Practices**

3 Construct viable arguments and critique the reasoning of others.

1 **Identify Functions** A **function** is a relationship between input and output. In a function, there is exactly one output for each input.

Key Concept   Function

Words   A function is a relation in which each element of the domain is paired with *exactly* one element of the range.

Examples

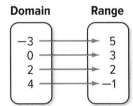

Domain   Range

−3 ——→ 5
0 ——→ 3
2 ——→ 2
4 ——→ −1

Example 1   Identify Functions

Determine whether each relation is a function. Explain.

a. Domain   Range

−2 ——→ −3
0 ——→ 6
3 ——→ 9
4

For each member of the domain, there is only one member of the range. So this mapping represents a function. It does not matter if more than one element of the domain is paired with one element of the range.

b.

Domain	1	3	5	1
Range	4	2	4	−4

The element 1 in the domain is paired with both 4 and −4 in the range. So, when x equals 1 there is more than one possible value for y. This relation is not a function.

▶ **Guided Practice**

1. {(2, 1), (3, −2), (3, 1), (2, −2)}

A function that consists of points that are not connected is a **discrete function**. A function graphed with an unbroken line or curve is a **continuous function**.

Real-World Example 2   Draw Graphs

ICE SCULPTING At an ice sculpting competition, each sculpture's height was measured to make sure that it was within the regulated height range of 0 to 6 feet. The measurements were as follows: Team 1, 4 feet; Team 2, 4.5 feet; Team 3, 3.2 feet; Team 4, 5.1 feet; Team 5, 4.8 feet.

a. Make a table of values showing the relation between the ice sculpting team and the height of their sculpture.

Team Number	1	2	3	4	5
Height (ft)	4	4.5	3.2	5.1	4.8

b. State the domain and range of the relation. Then determine whether it is a function.

The domain of the relation is {1, 2, 3, 4, 5} because this set represents values of the independent variable. It is unaffected by the heights.

The range of the relation is {4, 4.5, 3.2, 5.1, 4.8} because this set represents values of the dependent variable. This value depends on the team number.

For each member in the domain, there is only one member in the range, so the relation is a function.

c. Write the data as a set of ordered pairs. Then graph the data.

Use the table. The team number is the independent variable and the height of the sculpture is the dependent variable. Therefore, the ordered pairs are (1, 4), (2, 4.5), (3, 3.2), (4, 5.1), and (5, 4.8).

Because the team numbers and their corresponding heights cannot be between the points given, the points should not be connected.

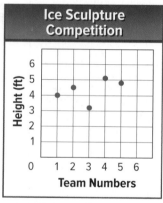

d. State whether the relation is *discrete* or *continuous*. Explain your reasoning.

Because the points are not connected, the relation is discrete.

Guided Practice

2. A bird feeder will hold up to 3 quarts of seed. The feeder weighs 2.3 pounds when empty and 13.4 pounds when full.

A. Make a table that shows the bird feeder with 0, 1, 2, and 3 quarts of seed in it weighing 2.3, 6, 9.7, 13.4 pounds respectively.

B. State the domain and range of the relation. Then determine whether it is a function.

C. Write the data as a set of ordered pairs. Then graph the data.

D. State whether the relation is *discrete* or *continuous*. Explain your reasoning.

Real-World Link

The Icehotel, located in the Arctic Circle in Sweden, is a hotel made out of ice. The ice insulates the igloo-like hotel so the temperature is at least −8°C.

Source: Icehotel

You can use the **vertical line test** to see if a graph represents a function. If a vertical line intersects the graph more than once, then the graph is not a function. Otherwise, the relation is a function.

Function	Not a Function	Function

Recall that an equation is a representation of a relation. Equations can also represent functions. Every solution of the equation is represented by a point on a graph. The graph of an equation is the set of all its solutions, which often forms a curve or a line.

Example 3   Equations as Functions

Determine whether $-3x + y = 8$ is a function.

First make a table of values. Then graph the equation.

x	−1	0	1	2
y	5	8	11	14

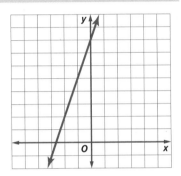

Connect the points with a smooth graph to represent all of the solutions of the equation.
The graph is a line. To use the vertical line test, place a pencil at the left of the graph to represent a vertical line. Slowly move the pencil across the graph.

For any value of x, the vertical line passes through no more than one point on the graph. So, the graph and the equation represent a function.

Guided Practice Determine whether each relation is a function.

3A. $4x = 8$

3B. $4x = y + 8$

A function can be represented in different ways.

Concept Summary Representations of a Function

Table	Mapping	Equation	Graph
	Domain Range		

x	y
−2	1
0	−1
2	1

Domain: −2, 0, 2 → Range: 1, −1

$f(x) = \frac{1}{2}x^2 - 1$

2 Find Function Values

Equations that are functions can be written in a form called **function notation**. For example, consider $y = 3x - 8$.

Equation	Function Notation
$y = 3x - 8$	$f(x) = 3x - 8$

In a function, x represents the elements of the domain, and $f(x)$ represents the elements of the range. The graph of $f(x)$ is the graph of the equation $y = f(x)$. Suppose you want to find the value in the range that corresponds to the element 5 in the domain. This is written $f(5)$ and is read f of 5. The value $f(5)$ is found by substituting 5 for x in the equation.

Example 4   Function Values

For $f(x) = -4x + 7$, find each value.

a. $f(2)$

$$f(2) = -4(2) + 7 \qquad x = 2$$
$$= -8 + 7 \qquad \text{Multiply.}$$
$$= -1 \qquad \text{Add.}$$

b. $f(-3) + 1$

$$f(-3) + 1 = [-4(-3) + 7] + 1 \qquad x = -3$$
$$= 19 + 1 \qquad \text{Simplify.}$$
$$= 20 \qquad \text{Add.}$$

▶ **Guided Practice**

For $f(x) = 2x - 3$, find each value.

4A. $f(1)$ **4B.** $6 - f(5)$

4C. $f(-2)$ **4D.** $f(-1) + f(2)$

A function with a graph that is not a straight line is a **nonlinear function**.

Example 5   Nonlinear Function Values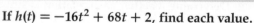

If $h(t) = -16t^2 + 68t + 2$, find each value.

a. $h(4)$

$$h(4) = -16(4)^2 + 68(4) + 2 \qquad \text{Replace } t \text{ with 4.}$$
$$= -256 + 272 + 2 \qquad \text{Multiply.}$$
$$= 18 \qquad \text{Add.}$$

b. $2[h(g)]$

$$2[h(g)] = 2[-16(g)^2 + 68(g) + 2] \qquad \text{Replace } t \text{ with } g.$$
$$= 2(-16g^2 + 68g + 2) \qquad \text{Simplify.}$$
$$= -32g^2 + 136g + 4 \qquad \text{Distributive Property}$$

▶ **Guided Practice**

If $f(t) = 2t^3$, find each value.

5A. $f(4)$ **5B.** $3[f(t)] + 2$

5C. $f(-5)$ **5D.** $f(-3) - f(1)$

Check Your Understanding ⬤ = Step-by-Step Solutions begin on page R11.

✓ **Go Online!** for a Self-Check Quiz

Examples 1–3 Determine whether each relation is a function. Explain.

1. Domain   Range

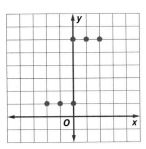

2.

Domain	Range
2	6
5	7
6	9
6	10

3. $\{(2, 2), (-1, 5), (5, 2), (2, -4)\}$

4. $y = \frac{1}{2}x - 6$

5.

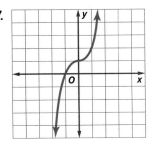

6.

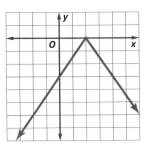

7.

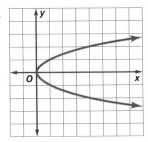

8.

Example 2

9. SCHOOL ENROLLMENT The table shows the total enrollment in U.S. public schools.

School Year	2012-13	2013-14	2014-15	2015-16
Enrollment (in thousands)	49,771	50,044	50,132	50,268

Source: *National Center for Education Statistics*

a. Write a set of ordered pairs representing the data in the table if x is the number of school years since 2012-13.

b. Draw a graph showing the relationship between the year and enrollment.

c. Describe the domain and range of the data. Is this a function?

10. ⓂⓅ REASONING The cost of printing digital pictures is given by $y = 0.13x$, where x is the number of pictures that you print and y is the cost in dollars.

a. Determine the domain and range of this relation. Why is the relation a function?

b. Write the equation in function notation. Interpret the function in terms of the context.

c. Find $f(5)$ and $f(12)$. What do these values represent?

Examples 4–5 If $f(x) = 6x + 7$ and $g(x) = x^2 - 4$, find each value.

11 $f(-3)$

12. $f(m)$

13. $f(r - 2)$

14. $g(5)$

15. $g(a) + 9$

16. $g(-4t)$

17. $f(q + 1)$

18. $f(2) + g(2)$

19. $g(-b)$

Example 1   Determine whether each relation is a function. Explain.

20. Domain   Range   **21.** Domain   Range   **22.**

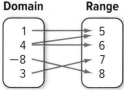

Domain	Range
4	6
−5	3
6	−3
−5	5

23.

Domain	Range
−4	2
3	−5
4	2
9	−7
−3	−5

24.

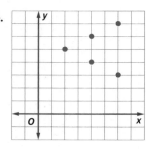

25.

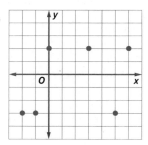

Example 2   **26.** (MP) **SENSE-MAKING** The table shows the median home prices in the United States, from 2011 to 2013.

Year	Median Home Price (S)
2011	175,600
2012	180,300
2013	208,000

a. Write a set of ordered pairs representing the data in the table.

b. Draw a graph showing the relationship between the year and price.

c. What is the domain and range for this data? Is the relation a function?

Example 3   Determine whether each relation is a function.

27. $\{(5, -7), (6, -7), (-8, -1), (0, -1)\}$ **28.** $\{(4, 5), (3, -2), (-2, 5), (4, 7)\}$

29. $y = -8$ **30.** $x = 15$

31. $y = 3x - 2$ **32.** $y = 3x + 2y$

Examples 4–5 If $f(x) = -2x - 3$ and $g(x) = x^2 + 5x$, find each value.

33. $f(-1)$ **34.** $f(6)$ **35.** $g(2)$

36. $g(-3)$ **37.** $g(-2) + 2$ **38.** $f(0) - 7$

39. $f(4y)$ **40.** $g(-6m)$ **41.** $f(c - 5)$

42. $f(r + 2)$ **43.** $5[f(d)]$ **44.** $3[g(n)]$

45 **EDUCATION** Mr. Blankenship determined that his class averages $f(z)$ can be represented as a function of the class size z by $f(z) = -\frac{5}{8}z + 87$.

a. Graph this function. Interpret the function in terms of the context.

b. What is the class size that corresponds to a class average of 72?

c. What is the domain and range of this function?

Determine whether each relation is a function.

46.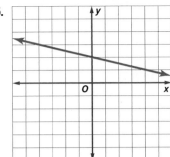

47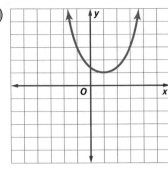

48. BABYSITTING Christina earns $7.50 an hour babysitting.

   a. Write an algebraic expression to represent the money Christina will earn if she works h hours.

   b. Choose five values for the number of hours Christina can babysit. Create a table with h and the amount of money she will make during that time.

   c. Use the values in your table to create a graph.

   d. Does it make sense to connect the points in your graph with a line? Why or why not?

H.O.T. Problems   Use Higher-Order Thinking Skills

49. OPEN ENDED Write a set of three ordered pairs that represent a function. Choose another display that represents this function.

50. MP REASONING The set of ordered pairs {(0, 1), (3, 2), (3, −5), (5, 4)} represents a relation between x and y. Graph the set of ordered pairs. Determine whether the relation is a function. Explain.

51. CHALLENGE Consider $f(x) = -4.3x - 2$. Write $f(g + 3.5)$ and simplify by combining like terms.

52. WRITE A QUESTION A classmate graphed a set of ordered pairs and used the vertical line test to determine whether it was a function. Write a question to help her decide if the same strategy can be applied to a mapping.

53. MP PERSEVERANCE If $f(3b - 1) = 9b - 1$, find one possible expression for $f(x)$.

54. ERROR ANALYSIS Corazon thinks $f(x)$ and $g(x)$ are representations of the same function. Maggie disagrees. Who is correct? Explain your reasoning.

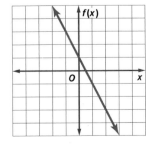

x	g(x)
−1	1
0	−1
1	−3
2	−5
3	−7

55. ⓔ WRITING IN MATH How can you determine whether a relation represents a function?

56. A student collected data on the cost of T-shirts at six stores. The table shown represents the price paid p for the number of shirts purchased n.

n	5	3	5	3	4	6
p	35	24	36	21	28	40

For the six stores, is the price paid a function of the number of shirts purchased? MP 6

○ **A** Yes, because different people shopped at different stores.

○ **B** Yes, because the total cost was more than one dollar per shirt.

○ **C** No, because there are different total costs for the same number of shirts.

○ **D** No, because the total price is not related to the number of items purchased.

57. If $g(x) = \dfrac{x^2 - 3}{10}$, what is the value of $g(1)$? MP 1

○ **A** $-\dfrac{2}{5}$

○ **B** $-\dfrac{1}{5}$

○ **C** $-\dfrac{1}{10}$

○ **D** $\dfrac{2}{5}$

58. For the function $y = 15x - 4$, assume the domain is only values of x from 0 to 5. What is the range of the function? MP 6

○ **A** All values from 15 to 20

○ **B** All values from $\dfrac{4}{15}$ to $\dfrac{3}{5}$

○ **C** All values from -4 to 71

○ **D** The two values -4 and 71

59. Which statement best describes how to determine when a graph represents a function? MP 6

○ **A** At least one vertical line intersects the function.

○ **B** Every horizontal line intersects the function.

○ **C** Every vertical line intersects the function exactly one time.

○ **D** Every vertical line intersects the function no more than one time.

60. Which of the following best describes the relation shown in the graph? MP 1

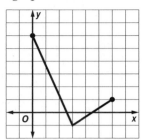

○ **A** Domain: $0 \le x \le 6$; Range: $-1 \le y \le 6$; the relation is a function

○ **B** Domain: $0 \le x \le 6$; Range: $-1 \le y \le 6$; the relation is not a function

○ **C** Domain: $-1 \le x \le 6$; Range: $0 \le y \le 6$; the relation is a function

○ **D** Domain: $-1 \le x \le 6$; Range: $0 \le y \le 6$; the relation is not a function

61. MULTI-STEP Consider the function $f(x) = x^2 + 2x - 5$. MP 2, 6

a. Find each value.

$f(-3) = $ [_____]

$f(0) = $ [_____]

$f(1) = $ [_____]

$f(-a) = $ [_____]

b. Is the relation a function? [_____]

c. Is the relation *discrete* or *continuous*?

[_____]

Representing Functions

You can use TI-Nspire Technology to explore the different ways to represent a function.

Mathematical Practices

Use Approriate Tools Strategically

Activity

Work cooperatively. Graph $f(x) = 2x + 3$ **on the TI-Nspire graphing calculator.**

Step 1 Add a new **Graphs** page.

Step 2 Enter $2x + 3$ in the entry line.

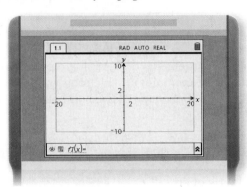

Represent the function as a table.

Step 3 Select the **Show Table** option from the **View** menu to add a table of values on the same display.

Step 4 Press **ctrl** and **tab** to toggle from the table to the graph. On the graph side, select the line and move it. Notice how the values in the table change.

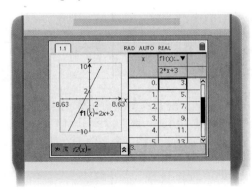

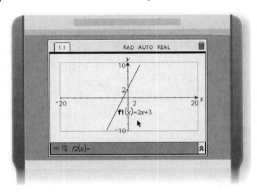

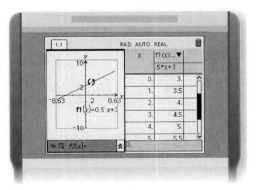

Cooperatively Analyze the Results

MP TOOLS Graph each function. Make a table of five ordered pairs that also represents the function.

1. $g(x) = -x - 3$

2. $h(x) = \frac{1}{3}x + 3$

3. $f(x) = -\frac{1}{2}x - 5$

4. $f(x) = 3x - \frac{1}{2}$

5. $g(x) = -2x + 5$

6. $h(x) = \frac{1}{5}x + 4$

Interpreting Graphs of Functions

::Then	::Now	::Why?
• You identified functions and found function values.	**1** Interpret intercepts and symmetry of graphs of functions. **2** Interpret positive, negative, increasing, and decreasing behavior, extrema, and end behavior of graphs of functions.	• Sales of video games, including hardware, software, and accessories, have increased at times and decreased at other times over the years. Annual retail video game sales in the U.S. from 2000 to 2009 can be modeled by the graph of a nonlinear function.

 New Vocabulary

intercept
x- and *y*-intercepts
line symmetry
positive
negative
increasing
decreasing
extrema
relative maximum
relative minimum
end behavior

 Mathematical Practices

1 Make sense of problems and persevere in solving them.

1 **Interpret Intercepts and Symmetry** To interpret the graph of a function, estimate and interpret key features. The **intercepts** of a graph are points where the graph intersects an axis. The *y*-coordinate of the point at which the graph intersects the *y*-axis is called a ***y*-intercept**. Similarly, the *x*-coordinate of the point at which a graph intersects the *x*-axis is called an ***x*-intercept**.

Real-World Example 1 Interpret Intercepts

PHYSICS The graph shows the height *y* of an object as a function of time *x*. Identify the function as *linear* or *nonlinear*. Then estimate and interpret the intercepts.

Linear or Nonlinear: Since the graph is a curve and not a line, the graph is nonlinear.

y-Intercept: The graph intersects the *y*-axis at about (0, 15), so the *y*-intercept of the graph is about 15. This means that the object started at an initial height of about 15 meters above the ground.

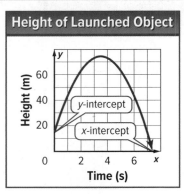

Height of Launched Object

x-**Intercept(s):** The graph intersects the *x*-axis at about (7.4, 0), so the *x*-intercept is about 7.4. This means that the object struck the ground after about 7.4 seconds.

▶ **Guided Practice**

1. The graph shows the temperature *y* of a medical sample thawed at a controlled rate. Identify the function as *linear* or *nonlinear*. Then estimate and interpret the intercepts.

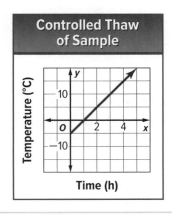

Controlled Thaw of Sample

A graph possesses **line symmetry** in the *y*-axis or some other vertical line if each half of the graph on either side of the line matches exactly. The graphs of most real-world functions do not exhibit symmetry over the entire domain. However, many have symmetry over smaller portions of the domain that are worth analyzing.

Real-World Example 2 Interpret Symmetry

PHYSICS An object is launched. The graph shows the height *y* of the object as a function of time *x*. Describe and interpret any symmetry.

The right half of the graph is the mirror image of the left half in approximately the line $x = 3.5$ between approximately $x = 0$ and $x = 7$.

In the context of the situation, the symmetry of the graph tells you that the time it took the object to go up is equal to the time it took to come down.

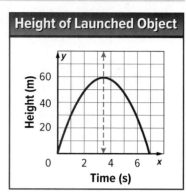

▶ **Guided Practice**

2. Describe and interpret any symmetry exhibited by the graph in Guided Practice 1.

---

2 **Interpret Extrema and End Behavior** Interpreting a graph also involves estimating and interpreting where the function is increasing, decreasing, positive, or negative, and where the function has any extreme values, either high or low.

Go Online!

Taking notes in your eStudent Edition allows you to keep your notes with each lesson to reference anytime.

Key Concepts Positive, Negative, Increasing, Decreasing, Extrema, and End Behavior

A function is **positive** where its graph lies *above* the *x*-axis, and **negative** where its graph lies *below* the *x*-axis.

A function is **increasing** where the graph goes *up* and **decreasing** where the graph goes *down* when viewed from left to right.

The points shown are the locations of relatively high or low function values called **extrema**. Point *A* is a **relative minimum**, since no other nearby points have a lesser *y*-coordinate. Point *B* is a **relative maximum**, since no other nearby points have a greater *y*-coordinate.

End behavior describes the values of a function at the positive and negative extremes in its domain.

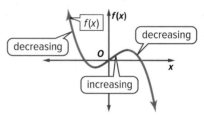

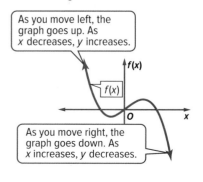

As you move left, the graph goes up. As *x* decreases, *y* increases.

As you move right, the graph goes down. As *x* increases, *y* decreases.

Study Tip

End Behavior The end behavior of some graphs can be described as approaching a specific *y*-value. In this case, a portion of the graph looks like a horizontal line.

VIDEO GAMES U.S. retail sales of video games from 2000 to 2009 can be modeled by the function graphed at the right. Estimate and interpret where the function is positive, negative, increasing, and decreasing, the *x*-coordinates of any relative extrema, and the end behavior of the graph.

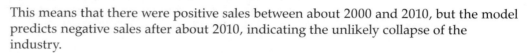

Positive: between about $x = -0.6$ and $x = 10.4$

Negative: for about $x < -0.6$ and $x > 10.4$

This means that there were positive sales between about 2000 and 2010, but the model predicts negative sales after about 2010, indicating the unlikely collapse of the industry.

Real-World Link

The first successful commercially sold portable video game system was released in 1989 and sold for $120.

Source: *PCWorld*

Increasing: for about $x < 1.5$ and between about $x = 3$ and $x = 8$

Decreasing: between about $x = 2$ and $x = 3$ and for about $x > 8$

This means that sales increased from about 2000 to 2002, decreased from 2002 to 2003, increased from 2003 to 2008, and have been decreasing since 2008.

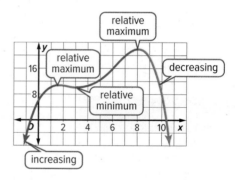

Study Tip

Constant A function is *constant* where the graph does not go up or down as the graph is viewed from left to right.

Relative Maximums: at about $x = 1.5$ and $x = 8$

Relative Minimum: at about $x = 3$

The extrema of the graph indicate that the industry experienced two relative peaks in sales during this period: one around 2002 of approximately $10.5 billion and another around 2008 of approximately $22 billion. A relative low of $10 billion in sales came in about 2003.

End Behavior:
As *x* increases or decreases, the value of *y* decreases.

The end behavior of the graph indicates negative sales several years prior to 2000 and several years after 2009, which is unlikely. This graph appears to only model sales well between 2000 and 2009 and can only be used to predict sales in 2010.

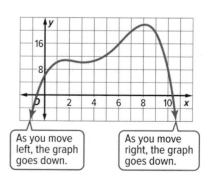

> **Guided Practice**

3. Estimate and interpret where the function graphed in Guided Practice 1 is positive, negative, increasing, or decreasing, the *x*-coordinate of any relative extrema, and the end behavior of the graph.

Real-World Example 4   Compare Properties of Functions

TENNIS A tennis player hits a forehand shot at a height of 2.8 feet and the ball travels 29 feet, when it reaches a height of about 10 feet. Then, the height of the ball decreases until it hits the ground 58 feet from where it was hit.

She then hits a backhand shot. The path of the tennis ball is shown in the graph at the right.

Compare the key features of her forehand and backhand shots.

Backhand

x-intercept

The forehand shot has an *x*-intercept of 58.
The backhand shot has an *x*-intercept of 70.
The tennis ball travels 12 feet farther during the backhand shot.

y-intercept

The forehand has a *y*-intercept of 2.8. The backhand has a *y*-intercept of about 2.5. The tennis ball is about 0.3 foot higher at the beginning of the forehand shot.

Extrema

The forehand has a maximum height of 10 feet when $x = 29$. The backhand has a maximum height of 7 feet when $x = 35$. The maximum height of the tennis ball is 3 feet higher during the forehand shot.

Guided Practice

4. Compare when the forehand and backhand shots are increasing and decreasing and when the shots are positive and negative.

Check Your Understanding

◯ = Step-by-Step Solutions begin on page R11.

Go Online! for a Self-Check Quiz

Examples 1–3   **MP** **SENSE-MAKING** Identify the function graphed as *linear* or *nonlinear*. Then estimate and interpret the intercepts of the graph, any symmetry, where the function is positive, negative, increasing, and decreasing, the *x*-coordinate of any relative extrema, and the end behavior of the graph.

1.

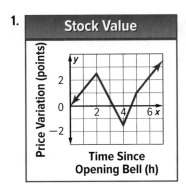

Stock Value

Time Since Opening Bell (h)

2.

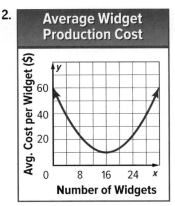

Average Widget Production Cost

Number of Widgets

3.

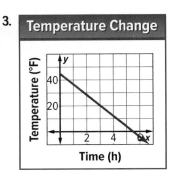

Temperature Change

Time (h)

Example 4

4. Use the description and graph to compare the fuel economy of two cars, where y is the miles per gallon and x is speed in miles per hour.

Car A

The fuel efficiency increases for speeds up to 25 mph when it reaches a relative maximum efficiency of 53 mpg. The fuel efficiency then decreases for speeds between 25 mph and 41 mph, when it gets down to 37 mpg. Above 41 mph, efficiency increases again until it reaches 48 mpg at 60 mph. Finally, the fuel efficiency rapidly decreases for speeds greater than 60 mph until leveling off at 17 mpg.

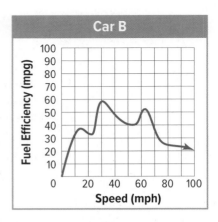

Practice and Problem Solving

Extra Practice is on page R1.

Example 1–3   Identify the function graphed as *linear* or *nonlinear*. Then estimate and interpret the intercepts of the graph, any symmetry, where the function is positive, negative, increasing, and decreasing, the x-coordinate of any relative extrema, and the end behavior of the graph.

5. Vehicle Depreciation

6. Lawn Mowing Service

7. Company Advertising

8. Web Site Traffic

9. Medicine Concentration

10. Pendulum Swing Time

Example 4   **11.** In 2011, China's urban population exceeded the rural population for the first time in the country's history. Use the description and graph to compare the key features of the urban and rural populations of China.

Rural Population

About 745,440,000 people lived in the rural regions of China in 2005. Since then, the rural population of China has been steadily decreasing to 618,660,000 in 2014.

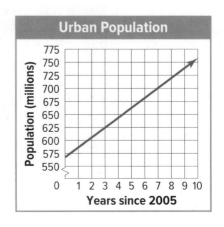

12. FERRIS WHEEL At the beginning of a Ferris wheel ride, a passenger cart is located at the same height as the center of the wheel. The position y in feet of this cart relative to the center t seconds after the ride starts is given by the function graphed at the right. Identify and interpret the key features of the graph. (*Hint:* Look for a pattern in the graph to help you describe its end behavior.)

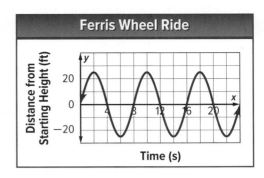

Ferris Wheel Ride

Distance from Starting Height (ft)

Time (s)

Sketch a graph of a function that could represent each situation. Identify and interpret the intercepts of the graph, where the graph is increasing and decreasing, and any relative extrema.

13. the height of a corn plant from the time the seed is planted until it reaches maturity 120 days later

14. the height of a football from the time it is punted until it reaches the ground 2.8 seconds later

15. the balance due on a car loan from the date the car was purchased until it was sold 4 years later

Sketch graphs of functions with the following characteristics.

16. The graph is linear with an x-intercept at -2. The graph is positive for $x < -2$, and negative for $x > -2$.

17. A nonlinear graph has x-intercepts at -2 and 2 and a y-intercept at -4. The graph has a relative minimum of -4 at $x = 0$. The graph is decreasing for $x < 0$ and increasing for $x > 0$.

18. A nonlinear graph has a y-intercept at 2, but no x-intercepts. The graph is positive and increasing for all values of x.

19. A nonlinear graph has x-intercepts at -8 and -2 and a y-intercept at 3. The graph has relative minimums at $x = -6$ and $x = 6$ and a relative maximum at $x = 2$. The graph is positive for $x < -8$ and $x > -2$ and negative between $x = -8$ and $x = -2$. As x decreases, y increases and as x increases, y increases.

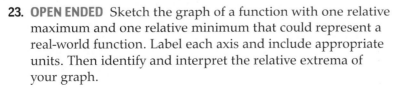

H.O.T. Problems   Use **H**igher-**O**rder **T**hinking Skills

20. **MP** **CRITIQUE ARGUMENTS** Katara thinks that all linear functions have exactly one x-intercept. Desmond thinks that a linear function can have at most one x-intercept. Is either of them correct? Explain your reasoning.

21. **CHALLENGE** Describe the end behavior of the graph shown.

22. **MP** **CRITIQUE ARGUMENTS** Determine whether the following statement is *true* or *false*. Explain.

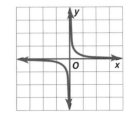

Functions have at most one y-intercept.

23. **OPEN ENDED** Sketch the graph of a function with one relative maximum and one relative minimum that could represent a real-world function. Label each axis and include appropriate units. Then identify and interpret the relative extrema of your graph.

24. **WRITING IN MATH** Describe how you would identify the key features of a graph described in this lesson using a table of values for a function.

25. Which of the following best describes the graph? MP 1

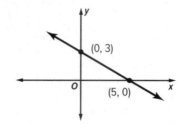

- **A** The x-intercept is 3; the y-intercept is 5; the graph is positive for $x < 5$; the graph is negative for $x > 5$.

- **B** The x-intercept is 5; the y-intercept is 3; the graph is positive for $x < 5$; the graph is negative for $x > 5$.

- **C** The x-intercept is 5; the y-intercept is 3; the graph is positive for $x > 5$; the graph is negative for $x < 5$.

- **D** The x-intercept is 5; the y-intercept is 3; the graph is positive for $x > 0$; the graph is negative for $x < 0$.

26. Determine the maximum number of y-intercepts a function can have. Justify your response. MP 1

27. Which of the following best describes the graph? MP 1

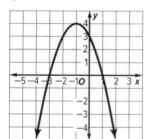

- **A** The x-intercepts are -3 and 1, the y-intercept is 3, and the axis of symmetry is $y = 4$.

- **B** The x-intercepts are -3 and 1, the y-intercept is 3, and the axis of symmetry is $x = -1$.

- **C** The x-intercepts are -1 and 3, the y-intercept is 3, and the axis of symmetry is $y = 4$.

- **D** The x-intercepts are -3 and 1, the y-intercept is 4, and the axis of symmetry is $x = -1$.

28. Which statement best describes the graph shown? MP 1

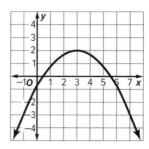

- **A** The graph is linear.

- **B** The graph is nonlinear.

- **C** There are two y-intercepts.

- **D** The graph is increasing.

29. A linear function has an x-intercept at $(-3, 0)$ and a y-intercept at $(0, -2)$. Which of the following must be true of the graph of the function? MP 1, 2

- **A** The graph is positive for $x < -3$.

- **B** The graph is increasing for $x > 0$.

- **C** The graph has another x-intercept at $(3, 0)$.

- **D** As x increases, y increases.

30. MULTI-STEP Sarah throws a ball in the air. The graph shows the height of the ball as a function of time.

Nathan also throws a ball in the air. His throw begins at a height of 2 feet. The ball travels up to 30 feet in 2 seconds and then comes back down to the ground after 4 seconds. MP 1, 2

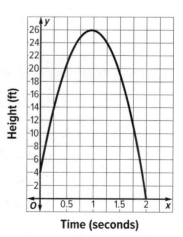

a. What are the x-intercepts of each ball's path?

b. Do both balls start from the same height?

c. Which ball goes higher in the air?

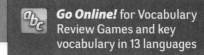

Go Online! for Vocabulary Review Games and key vocabulary in 13 languages

Study Guide

Key Concepts

Order of Operations (Lesson 1-2)

- Evalute expressions inside grouping symbols.
- Evaluate all powers.
- Multiply and/or divide in order from left to right.
- Add or subtract in order from left to right.

Properties of Equality (Lessons 1-3 and 1-4)

- For any numbers a, b, and c:

  Reflexive: $a = a$

  Symmetric: If $a = b$, then $b = a$.

  Transitive: If $a = b$ and $b = c$, then $a = c$.

  Substitution: If $a = b$, then a may be replaced by b in any expression.

  Distributive: $a(b + c) = ab + ac$ and
                $a(b - c) = ab - ac$

  Commutative: $a + b = b + a$ and $ab = ba$

  Associative: $(a + b) + c = a + (b + c)$ and
              $(ab)c = a(bc)$

Descriptive Modeling and Accuracy (Lesson 1-5)

- A metric is a rule for assigning a number to some characteristic or attribute.

Relations, Functions, and Interpreting Graphs of Functions (Lessons 1-6 through 1-8)

- Relations and functions can be represented by ordered pairs, a table, a mapping, or a graph.
- Use the vertical line test to determine if a relation is a function.
- End behavior describes the long-term behavior of a function on either end of its graph.
- Points where the graph of a function crosses an axis are called intercepts.
- A function is positive on a portion of its domain where its graph lies above the x-axis, and negative on a portion where its graph lies below the x-axis.

FOLDABLES® Study Organizer

Use your Foldable to review the chapter. Working with a partner can be helpful. Ask for clarification of concepts as needed.

Key Vocabulary

accuracy (p. 34)	metric (p. 33)
algebraic expression (p. 5)	ordered pair (p. 42)
base (p. 5)	order of operations (p. 10)
coefficient (p. 26)	origin (p. 42)
coordinate system (p. 42)	power (p. 5)
dependent variable (p. 44)	range (p. 42)
domain (p. 42)	reciprocal (p. 17)
end behavior (p. 59)	relation (p. 42)
exponent (p. 5)	relative maximum (p. 59)
function (p. 49)	relative minimum (p. 59)
independent variable (p. 44)	simplest form (p. 25)
intercept (p. 58)	term (p. 5)
like terms (p. 25)	variables (p. 5)
line symmetry (p. 59)	vertical line test (p. 51)

Vocabulary Check

State whether each sentence is *true* or *false*. If *false*, replace the underlined term to make a true sentence.

1. A <u>coordinate system</u> is formed by two intersecting number lines.

2. An <u>exponent</u> indicates the number of times the base is to be used as a factor.

3. An expression is <u>in simplest form</u> when it contains like terms and parentheses.

4. In an expression involving multiplication, the quantities being multiplied are called <u>factors</u>.

Concept Check

5. How does the Order of Operations apply to evaluating variable expressions?

6. How does the vertical line test determine whether a relation is a function?

7. Describe the difference between the additive inverse and the additive identity.

Lesson-by-Lesson Review

1-1 Variables and Expressions

Write a verbal expression for each algebraic expression.

8. $h - 7$ **9.** $3x^2$ **10.** $5 + 6m^3$

Write an algebraic expression for each verbal expression.

11. a number increased by 9

12. two thirds of a number d to the third power

13. 5 less than four times a number

Evaluate each expression.

14. 2^5 **15.** 6^3 **16.** 4^4

17. **BOWLING** Fantastic Pins Bowling Alley charges $4.75 for shoe rental plus $4.25 for each game. Write an expression representing the cost to rent shoes and bowl g games.

Example 1

Write a verbal expression for $4x + 9$.

nine more than four times a number x

Example 2

Write an algebraic expression for *the difference of twelve and two times a number cubed.*

Variable   Let x represent the number.

Expression $12 - 2x^3$

Example 3

Evaluate 3^4.

The base is 3 and the exponent is 4.

$3^4 = 3 \cdot 3 \cdot 3 \cdot 3$ Use 3 as a factor 4 times.

$\quad = 81$ Multiply.

1-2 Order of Operations

Evaluate each expression.

18. $24 - 4 \cdot 5$ **19.** $15 + 3^2 - 6$

20. $7 + 2(9 - 3)$ **21.** $8 \cdot 4 - 6 \cdot 5$

22. $\left[(2^5 - 5) \div 9\right]11$ **23.** $\dfrac{11 + 4^2}{5^2 - 4^2}$

Evaluate each expression if $a = 4$, $b = 3$, and $c = 9$.

24. $c + 3a$

25. $5b^2 \div c$

26. $(a^2 + 2bc) \div 7$

27. **ICE CREAM** The cost of a one-scoop sundae is $4.75, and the cost of a two-scoop sundae is $5.25. Write and evaluate an expression to find the total cost of 3 one-scoop sundaes and 2 two-scoop sundaes.

Example 4

Evaluate the expression $3(9 - 5)^2 \div 8$.

$3(9 - 5)^2 \div 8 = 3(4)^2 \div 8$ Work inside parentheses.

$\qquad\qquad = 3(16) \div 8$ Evaluate 4^2.

$\qquad\qquad = 48 \div 8$ Multiply.

$\qquad\qquad = 6$ Divide.

Example 5

Evaluate the expression $(5m - 2n) \div p^2$ if $m = 8$, $n = 4$, $p = 2$.

$(5m - 2n) \div p^2$

$\quad = (5 \cdot 8 - 2 \cdot 4) \div 2^2$ Replace m with 8, n with 4, and p with 2.

$\quad = (40 - 8) \div 2^2$ Multiply.

$\quad = 32 \div 2^2$ Subtract.

$\quad = 32 \div 4$ Evaluate 2^2.

$\quad = 8$ Divide.

1-3 Properties of Numbers

Evaluate each expression using properties of numbers. Name the property used in each step.

28. $18 \cdot 3(1 \div 3)$

29. $[5 \div (8 - 6)]\frac{2}{5}$

30. $(16 - 4^2) + 9$

31. $2 \cdot \frac{1}{2} + 4(4 \cdot 2 - 7)$

32. $18 + 41 + 32 + 9$

33. $7\frac{2}{5} + 5 + 2\frac{3}{5}$

34. $8 \cdot 0.5 \cdot 5$

35. $5.3 + 2.8 + 3.7 + 6.2$

36. SCHOOL SUPPLIES Monica needs to purchase a binder, a textbook, a calculator, and a workbook for her algebra class. The binder costs \$9.25, the textbook \$72.50, the calculator \$49.99, and the workbook \$15.00. Find the total cost for Monica's algebra supplies.

Example 6

Evaluate $6(4 \cdot 2 - 7) + 5 \cdot \frac{1}{5}$. Name the property used in each step.

$6(4 \cdot 2 - 7) + 5 \cdot \frac{1}{5}$

$= 6(8 - 7) + 5 \cdot \frac{1}{5}$ Substitution

$= 6(1) + 5 \cdot \frac{1}{5}$ Substitution

$= 6 + 5 \cdot \frac{1}{5}$ Multiplicative Identity

$= 6 + 1$ Multiplicative Inverse

$= 7$ Substitution

1-4 Distributive Property

Use the Distributive Property to rewrite each expression. Then evaluate.

37. $(2 + 3)6$

38. $5(18 + 12)$

39. $8(6 - 2)$

40. $(11 - 4)3$

41. $-2(5 - 3)$

42. $(8 - 3)4$

Rewrite each expression using the Distributive Property. Then simplify.

43. $3(x + 2)$

44. $(m + 8)4$

45. $6(d - 3)$

46. $-4(5 - 2t)$

47. $(9y - 6)(-3)$

48. $-6(4z + 3)$

49. TUTORING Write and evaluate an expression for the number of tutoring lessons Mrs. Green gives in 4 weeks.

Tutoring Schedule	
Day	Students
Monday	3
Tuesday	5
Wednesday	4

Example 7

Use the Distributive Property to rewrite the expression $5(3 + 8)$. Then evaluate.

$5(3 + 8) = 5(3) + 5(8)$ Distributive Property

$= 15 + 40$ Multiply.

$= 55$ Simplify.

Example 8

Rewrite the expression $6(x + 4)$ using the Distributive Property. Then simplify.

$6(x + 4) = 6 \cdot x + 6 \cdot 4$ Distributive Property

$= 6x + 24$ Simplify.

Example 9

Rewrite the expression $(3x - 2)(-5)$ using the Distributive Property. Then simplify.

$(3x - 2)(-5)$

$= (3x)(-5) - (2)(-5)$ Distributive Property

$= -15x + 10$ Simplify.

1-5 Descriptive Modeling and Accuracy

Determine which unit of measure you should use for each situation.

50. Estimating the length of a hockey rink: inches, feet, miles

51. Measuring the amount of water in a community center pool: gallons, hundreds of gallons, thousands of gallons

52. Estimating the difference in weight between two cows: ounces, pounds, tons

Determine where to round each problem.

53. Teresa splits $14 between six friends.

54. Anton divides 22 loaves of bread into 8 boxes.

55. HEIGHT Jordan's height is written as 142 centimeters +/- 3 centimeters. How could you express Jordan's height as a range?

56. LOANS Mateo wants to take out a small business loan. His monthly expenses are shown in the table. He will qualify for the loan if his debt-to-income ratio is less than 0.35. Determine his debt-to-income ratio. Does Mateo qualify for the loan?

Type of Debt	Expenses	Salary
Rent	1850	5673
Car Payment	200	
Credit Card	50	
Student Loan	100	

57. CATS The body mass index (BMI) of a cat is an indicator of the animal's fitness. A cat's BMI is calculated by using the circumference around its rib cage and the length of its back leg from knee to ankle in the formula,

$$BMI = \frac{\frac{circumference}{0.7062} - leg\ length}{0.9156} - leg\ length.$$

Find the BMI of the cats.

Cat	Circumference of Rib Cage (in.)	Leg Length (in.)
Missy	17.5	4.0
Oscar	15.0	3.5
Whiskers	20.3	3.8

Example 10

LOANS Brandon wants to buy a car. His monthly expenses are shown in the spreadsheet. Write an expression to determine his debt-to-income ratio. He will qualify for a car loan if his debt-to-income ratio is below 0.55.

Type of Debt	Expenses	Salary
Rent	1150	4375
Credit cards	585	
Student loan	475	
Telephone/Internet	105	

Add all his expenses, and then divide this amount by his salary.

Expenses = 1150 + 585 + 475 + 105 = $2315

Debt-to-income ratio $= \frac{2315}{4375} \cong 0.5291429$

Brandon's debt-to-income ratio is about 0.53, which is less than 0.55, so he will qualify for the car loan.

Example 11

Charlotte has $13 she wants to share evenly with her three friends. She divides 13 ÷ 3 on her calculator and gets 4.33333333. Where should Charlotte round?

Since Charlotte is rounding money, the smallest increment is a penny, so she should round to the hundredths place. This gives her 4.33. $4.33 × 3 = $12.99, so Charlotte will have one extra penny. She can give each friend $4.33.

1-6 Relations

Express each relation as a table, a graph, and a mapping. Then determine the domain and range.

58. {(1, 3), (2, 4), (3, 5), (4, 6)}

59. {(−1, 1), (0, −2), (3, 1), (4, −1)}

60. {(−2, 4), (−1, 3), (0, 2), (−1, 2)}

Express the relation shown in each table, mapping, or graph as a set of ordered pairs.

61.

x	y
5	3
3	−1
1	2
−1	0

62.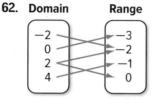

63. GARDENING On average, 7 plants grow for every 10 seeds of a certain type planted. Make a table to show the relation between seeds planted and plants growing for 50, 100, 150, and 200 seeds. Then state the domain and range and graph the relation.

Example 12

Express the relation {(−3, 4), (1, −2), (0, 1), (3, −1)} as a table, a graph, and a mapping.

Table

Place the x-coordinates into the first column. Place the corresponding y-coordinates in the second column.

x	y
−3	4
1	−2
0	1
3	−1

Graph

Graph each ordered pair on a coordinate plane.

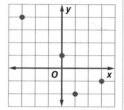

Mapping

List the x-values in the domain and the y-values in the range. Draw arrows from the x-values in set X to the corresponding y-values in set Y.

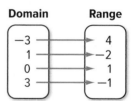

1-7 Functions

Determine whether each relation is a function.

64.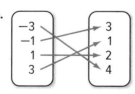

65.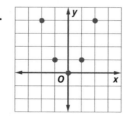

66. {(8, 4), (6, 3), (4, 2), (2, 1), (6, 0)}

If $f(x) = 2x + 4$ and $g(x) = x^2 - 3$, find each value.

67. $f(-3)$ **68.** $g(2)$ **69.** $f(0)$

70. $g(-4)$ **71.** $f(m + 2)$ **72.** $g(3p)$

73. FOOTBALL After three quarters, Omar had rushed 45 yards. If Omar averages 9 yards per carry in the fourth quarter and there is enough time to run 6 more plays before the end of the game, the equation $g(x) = 45 + 9x$ can be used to represent the situation, where x is the number of carries. Graph this function and find its domain and range.

Example 13

Determine whether $2x - y = 1$ represents a function.

First make a table of values. Then graph the equation.

x	y
−1	−3
0	−1
1	1
2	3
3	5

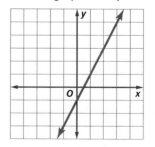

Using the vertical line test, it can be shown that $2x - y = 1$ does represent a function.

1-8 Interpreting Graphs of Functions

74. Identify the function graphed as *linear* or *nonlinear*. Then estimate and interpret the intercepts of the graph, any symmetry, where the function is positive, negative, increasing and decreasing, the *x*-coordinate of any relative extrema, and the end behavior of the graph.

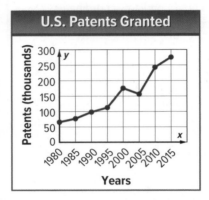

U.S. Patents Granted

75. Compare the practice runs of John and Dante during cross country practice.

John's Run

Dante: He leaves the high school, runs at a steady pace for 18 minutes until he reaches the middle school, which is about 3.2 miles away. Dante runs 10 laps around the track at the middle school, which takes him 14 minutes. Then, he runs back to the high school to complete his 55-minute run.

Example 14

POPULATION The population of American Samoa from 2004 to 2012 can be modeled by the graph shown. Estimate and interpret where the function is increasing and decreasing, the *x*-coordinates of any relative extrema, and the end behavior of the graph.

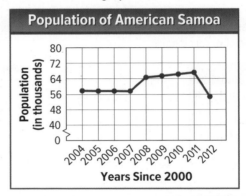

Population of American Samoa

The population remained relatively constant from 2004 to 2007, increased from 2007 to 2011, and decreased from 2011 to 2012. The relative maximum of the graph indicates that the population peaked in 2011. The end behavior indicates a decline in population from 2011 to 2012.

Write an algebraic expression for each verbal expression.

1. six more than a number

2. twelve less than the product of three and a number

3. four divided by the difference between a number and seven

Evaluate each expression.

4. $32 \div 4 + 2^3 - 3$

5. $\dfrac{(2 \cdot 4)^2}{7 + 3^2}$

6. **MULTIPLE CHOICE** Find the value of the expression $a^2 + 2ab + b^2$ if $a = 6$ and $b = 4$.

   A 68

   B 92

   C 100

   D 121

Evaluate each expression. Name the property used in each step.

7. $13 + (16 - 4^2)$

8. $\dfrac{2}{9}[9 \div (7 - 5)]$

9. $37 + 29 + 13 + 21$

Rewrite each expression using the Distributive Property. Then simplify.

10. $4(x + 3)$

11. $(5p - 2)(-3)$

12. **MOVIE TICKETS** A company operates three movie theaters. The chart shows the typical number of tickets sold each week at the three locations. Write and evaluate an expression for the total typical number of tickets sold by all three locations in four weeks.

Location	Tickets Sold
A	438
B	374
C	512

13. To find the volume of a box, a student multiplies the dimensions 7.5 centimeters, 8.9 centimeters, and 13.7 centimeters. Explain why 914.475 cubic centimeters is not an appropriate answer.

14. A teacher wants to compare the academic work of two classes. Which average for each class would you suggest that he use? Explain the reasons for your choice.

   A the average of students' quiz grades

   B the average of students' final grades

   C the average number of student absences

Express the relation shown in each table, mapping, or graph as a set of ordered pairs.

15.

x	y
−2	4
1	2
3	0
4	−2

16.
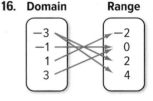
Domain   Range

17. Determine the domain and range for the relation $\{(2, 5), (-1, 3), (0, -1), (3, 3), (-4, -2)\}$.

18. Determine whether the relation $\{(2, 3), (-1, 3), (0, 4), (3, 2), (-2, 3)\}$ is a function.

If $f(x) = 5 - 2x$ and $g(x) = x^2 + 7x$, find each value.

19. $g(3)$

20. $f(-6y)$

21. Identify the function graphed as *linear* or *nonlinear*. Then estimate and interpret the intercepts of the graph, any symmetry, where the function is positive, negative, increasing, and decreasing, the x-coordinate of any relative extrema, and the end behavior of the graph.

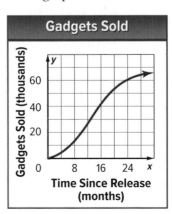

Performance Task

Provide a clear solution to each part of the task. Be sure to show all of your work, include all relevant drawings, and justify your answers.

ZOO Micah is planning a group trip to the zoo. There are two zoos within driving distance. They each have different prices for different age groups. Micah's job is to determine which zoo offers the better price for his group.

Part A

1. Explain how Micah can write an expression for the price of going to each of the zoos if there are x children and y adults.

2. **MODELING** Micah estimates that 250 children and 20 adults will be going to the zoo. Determine the total cost for each zoo and decide which zoo Micah should have the group visit.

Zoo Pricing		
Group Rates		
	Westside Zoo	Southside Zoo
Children	$3.75	$3.25
Adults	$5.75	$5.75

Part B

3. **STRUCTURE** While at the zoo, each of the 250 children purchase a drink for $1.50 and a meal for $4.25. Use the Distributive Property to show how you could determine the total cost.

Part C

4. **REASONING** There were four different sections to visit at the zoo. Micah measured how far the group walked in each section.

   Explain how you could use the properties of real numbers to determine the total distance walked.

Path	Distance (miles)
Birds	0.48
Reptiles, Fish, Amphibians	0.53
Mammals of Africa	1.27
Mammals of Americas	1.32

Part D

5. Micah wants to create a metric to compare the zoos in his state. Describe what data you would use to evaluate and compare zoos. Explain your reasoning.

Part E

6. **SENSE-MAKING** According to the zoo's Web site, the zoo opens at 10 a.m. and closes at 8 p.m. The number of visitors to the zoo peaks around 2 p.m. with an average of 4800 visitors. Around 4:30 p.m., the zoo typically has about 2000 visitors. The zoo usually has another peak around 6:00 p.m. with 3100 visitors. Make a graph to represent the number of visitors in the zoo from open to close.

Test-Taking Strategy

Test-Taking Tip

Finding the Domain
The domain of a relation is the set of inputs of the relation. The domain is found by simply listing the first elements of each ordered pair in set notation.

Example

Read the problem. Eliminate any unreasonable answers. Then use the information in the problem to solve.

What is the domain of the following relation? {(1, 3), (−6, 4), (8, 5)}

- **A** {3, 4, 5}
- **B** {−6, 1, 8}
- **C** {−6, 1, 3, 4, 5, 8}
- **D** {1, 3, 4, 5, 8}

Step 1 What do you need to find?
the domain of the relation

Step 2 Is there enough information given to solve the problem? yes

Step 3 What information, if any, is not needed to solve the problem?
The y-coordinates, because they have nothing to do with the domain

Step 4 Are there any obvious wrong answers? If so, which one(s)? Explain.
Yes; A, C, and D are wrong because they contain the y-coordinates.

Step 5 What is the correct answer? B

Apply the Strategy

Read the problem. Eliminate any unreasonable answers. Then use the information in the problem to solve.

Refer to the relation in the table below.

x	−6	−2	0	?	3	5
y	−1	8	3	−3	4	0

Consider the missing x-value in the table. Which of the following values would result in the relation not being a function?

- **A** −1
- **B** 3
- **C** 7
- **D** 8

Answer the questions below.

a. What do you need to find?

b. Is there enough information given to solve the problem?

c. What information, if any, is not needed to solve the problem?

d. Are there any obvious wrong answers? If so, which one(s)? Explain.

e. What is the correct answer?

Read each question. Then fill in the correct answer on the answer document provided by your teacher or on a sheet of paper.

1. A box of fruit has 12 plums and 15 bananas. The cost of each plum is p and the cost of each banana is 25 cents. What expression gives the total cost for all of the fruit in the box?

   ○ **A** $12 \cdot 0.25 + 15p$

   ○ **B** $12 + p + 0.25 + b$

   ○ **C** $12p + 15 \cdot 0.25$

   ○ **D** $12 \cdot 15 + p \cdot 0.25$

2. A park has the dimensions indicated in the diagram.

   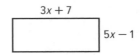

   $3x + 7$

   $5x - 1$

   Which expression represents the perimeter of the park?

   ○ **A** $= 8x + 6$

   ○ **B** $= 16x + 12$

   ○ **C** $= 6(10x + 2)$

   ○ **D** $= 15x^2 - 7$

3. Which of the following does not represent a function?

   ○ **A** $y = x + 2$

   ○ **B** $\{(3, 5), (5, 8), (8, 10), (3, 12)\}$

   ○ **C**

x	1	2	3	4
y	−4	0	−4	0

   ○ **D** $f(x) = 4 - 3x^2$

4. What is $2x^2 - 14$ equal to when $x = -3$?

   ○ **A** −32

   ○ **B** −20

   ○ **C** 4

   ○ **D** 22

5. The diagram shows the dimensions of a rectangle. Each side is measured in centimeters.

   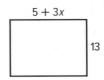

   $5 + 3x$

   13

   Which expression represents the area of the rectangle in square centimeters?

   ○ **A** $10 + 32x$

   ○ **B** $18 + 3x$

   ○ **C** $36 + 6x$

   ○ **D** $65 + 39x$

6. The table shows a relation.

x	−3	5	5	6	7
y	4	−2	3	6	−2

   Which set of ordered pairs represents the same relation?

   ○ **A** $\{-3, -2, 3, 4, 5, 6, 7\}$

   ○ **B** $\{(-3, 4), (6, 6), (7, -2)\}$

   ○ **C** $\{(4, -3), (-2, 5), (3, 5), (6, 6), (-2, 7))\}$

   ○ **D** $\{(-3, 4), (5, -2), (5, 3), (6, 6), (7, -2)\}$

7. A given function is $f(x) = 3x + 5$ and the domain of $f(x)$ is $\{-5, -3, 1, 3, 5\}$. Find the range of $f(x)$.

Test-Taking Tip

Question 6 The *x*-values represent the domain values, or the first coordinates, of a relation. In a table, the top or left row generally has the first coordinates of the ordered pairs.

8. For the function $y = 15 - 6x$, suppose the domain is only values of x from 10 to 20. What is the range of the function?

   A All values from 45 to 105

   B All values from $-\frac{5}{6}$ to $\frac{5}{6}$

   C All values from -105 to -45

   D The two values -105 and -45

9. Which statement best describes the symmetry of the graph?

   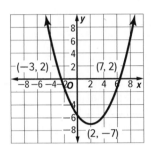

   A The graph is a mirror image of itself over the x-axis.

   B The graph is a mirror image of itself over the line $y = -7$.

   C The graph is a mirror image of itself over the line $x = 2$.

   D The graph does not exhibit any symmetry.

10. At a coffee shop, a small cup contains 8 fluid ounces and a large cup contains 14 fluid ounces. How many fluid ounces are there in 15 small cups and 20 large cups?

    A 370 fluid ounces

    B 400 fluid ounces

    C 412 fluid ounces

    D 782 fluid ounces

11. What is the value of x for the equation $3 + 7 = 7 + x$, and what property is used to identify it?

    A 3, Associative Property (+)

    B 10, Associative Property (+)

    C 3, Commutative Property (+)

    D 10, Commutative Property (+)

12. What is a simplified form of the expression $(15k + 2) - (k^2 - 2)$?

    A $14k + 4$ **C** $-k^2 + 15k + 4$

    B $-k^2 + 15k$ **D** $k^2 + 15k + 4$

13. If $h(t) = -16t^2 + 70t + 5$, what is $h(3)$?

    A 71

    B 167

    C 359

    D 2519

14. Jim is measuring a friend's height in centimeters. Do you think that he should measure to the nearest millimeter, the nearest centimeter, or the nearest 10 centimeters?

15. The ordered pairs $\{(-2, 2), (1, 2), (3, -1), (5, 2), (5, 4)\}$ represent a relation that is not a function. Which change would make it a function?

    A Replace $(5, 2)$ with $(3, 2)$.

    B Replace $(5, 4)$ with $(1, 5)$.

    C Replace $(5, 2)$ with $(5, -3)$.

    D Replace $(5, 4)$ with $(4, 5)$.

Need Extra Help?

If you missed Question...	1	2	3	4	5	6	7	8	9	10	11	12	13	14	15
Go to Lesson...	1-1	1-4	1-7	1-2	1-4	1-6	1-5	1-7	1-8	1-2	1-3	1-4	1-7	1-5	1-7

CHAPTER 2
Linear Equations

THEN

You learned to simplify algebraic expressions.

NOW

In this chapter, you will:

- Create equations that describe relationships.
- Solve linear equations in one variable.
- Solve proportions.
- Use formulas to solve real-world problems.

 WHY

SHOPPING Knowing how much something will cost before you check out is important. There may be discounts on an item, but you also need to add the sales tax.

Use the Mathematical Practices to complete the activity.

1. **Use Tools** Use the Internet or another source to find the sales tax in your state.

2. **Apply Math** If an item you want to purchase is $25 and is on sale for 15% off, what is the sale price? Calculate the sales tax. How much will it cost?

Discount		
Original Cost = $25.00		

Discount = 15% of $25.00 = $3.75

Result:

Selling Price = $25.00 − $3.75 = $21.25

3. **Discuss** Compare your solution method to that of your classmates. Did you use the same method? Can you think of a different way to solve this problem?

Go Online to Guide Your Learning

Explore & Explain

 eToolkit

Use the **Algebra Tiles** tool to model equations in Lessons 2-2 and 2-3..

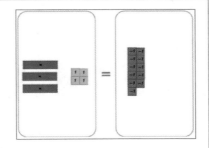

 The Geometer's Sketchpad

Visualize multi-step equations using the **Explore Linear Equations by Balancing** sketch. What does it mean when the balance is uneven?

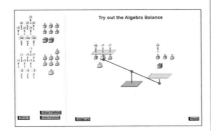

eBook

Interactive Student Guide

Before starting the chapter, answer the **Chapter Focus** preview questions. Check your answers as you complete each lesson. At the end of the chapter, try the **Performance Task**.

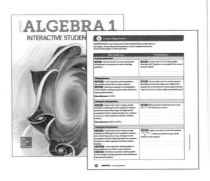

Organize

 **Foldables**

Make this Foldable to help you organize your notes about linear equations. Begin with 4 sheets of grid paper folded in half along the width. Unfold each sheet and tape to form one long piece. Label each page with the lesson number as shown. Refold to form a booklet.

Collaborate

 Chapter Project

In the **UPC ABCs** project, you will use what you have learned about expressions to complete a project.

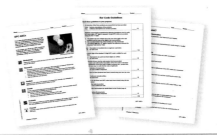

Focus

 LEARNSMART

Need help studying? Complete the **Relationships Between Quantities and Reasoning with Equations** domain in LearnSmart to review for the chapter test.

 ALEKS

You can use the **Linear Equations** topic in ALEKS to find out what you know about linear equations and what you are ready to learn.*

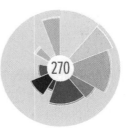

* Ask your teacher if this is part of your program.

Get Ready for the Chapter

Go Online! for Vocabulary Review Games and key vocabulary in 13 languages.

Connecting Concepts

Concept Check

Review the concepts used in this chapter by answering the questions below.

1. What is the translation of the following phrase when written as an algebraic expression: the difference between two times b and eleven?

2. What phrase describes the following algebraic expression: $8w + 9$?

3. In the expression $(9 - 4)^2 + 3$, which operation would be completed first?

4. In the expression $72 \div 9 + 3 \cdot 2^3$, which operation would be completed last?

5. The results of a survey are shown in the table. What percent of the people surveyed prefer strawberry ice cream?

Favorite Flavor	Number of Responses
vanilla	82
chocolate	76
strawberry	42

6. Based on the survey results shown in the table, what percent of the people surveyed prefer vanilla ice cream?

7. Given the following equation, what is the first step to take for solving?
$$\frac{32}{40} = \frac{p}{100}$$

Performance Task Preview

You can use the concepts and skills in the chapter to solve temperature equation problems.

Understanding linear equations will help you finish the Performance Task at the end of the chapter.

MP **In this Performance Task you will:**

- reason abstractly and quantitatively
- attend to precision
- make use of structure

New Vocabulary

English		Español
formula	p. 80	fórmula
solve an equation	p. 87	resolver una ecuación
equivalent equations	p. 87	ecuaciones equivalentes
linear equation	p. 87	ecuación lineal
multi-step equation	p. 95	ecuación de varios pasos
consecutive integers	p. 96	enteros consecutivos
number theory	p. 96	teoría del número
identity	p. 102	identidad
ratio	p. 115	razón
proportion	p. 115	proporción
means	p. 116	medios
extremes	p. 116	extremos
rate	p. 117	tasa
unit rate	p. 117	tasa unitaria
scale	p. 118	escala
scale model	p. 118	modelo de escala
literal equation	p. 123	ecuación literal
dimensional analysis	p. 124	análisis dimensional

Review Vocabulary

algebraic expression expresion algebraica **an expression consisting of one or more numbers and variables along with one or more arithmetic operations**

coordinate system sistema de coordenadas the grid formed by the intersection of two number lines, the horizontal axis and the vertical axis

function función a relation in which each element of the domain is paired with exactly one element of the range

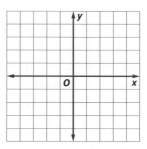

LESSON 1

Writing Equations

Then
- You evaluated and simplified algebraic expressions.

Now
1. Translate sentences into equations.
2. Translate equations into sentences.

Why?
- The Daytona 500 is widely considered to be the most important event of the NASCAR circuit. The distance around the track is 2.5 miles, and the race is a total of 500 miles. We can write an equation to determine how many laps it takes to finish the race.

 New Vocabulary
formula

 Mathematical Practices
2 Reason abstractly and quantitatively.

1 Write Equations To write an equation, identify the unknown for which you are looking and assign a variable to it. Then, write the sentence as an equation. Look for key words such as *is*, *is as much as*, *is the same as*, or *is identical to* that indicate where you should place the equals sign.

Consider the Daytona 500 example above.

Words	The length of each lap times the number of laps is the length of the race.
Variable	Let ℓ represent the number of laps in the race.
Equation	$2.5 \times \ell = 500$

Example 1   Translate Sentences into Equations

Translate each sentence into an equation.

a. Seven times a number squared is five times the difference of *k* and *m*.

Seven   times   *n* squared   is   five   times   the difference of *k* and *m*.

$$7 \cdot n^2 = 5 \cdot (k - m)$$

The equation is $7n^2 = 5(k - m)$.

b. Fifteen times a number subtracted from 80 is 25.

You can rewrite the verbal sentence so it is easier to translate. *Fifteen times a number subtracted from 80 is 25* is the same as *80 minus 15 times a number is 25*.
Let *n* represent the number.

80   minus   15   times   a number   is   25.

$$80 - 15 \cdot n = 25$$

The equation is $80 - 15n = 25$.

> ### Guided Practice
>
> **1A.** Two plus the quotient of a number and 8 is the same as 16.
>
> **1B.** Twenty-seven times *k* is *h* squared decreased by 9.

Translating sentences to algebraic expressions and equations is a valuable skill in solving real-world problems.

Real-World Example 2 Use the Four-Step Problem-Solving Plan

AIR TRAVEL Refer to the information at the left. In how many days will 261,000 flights have occurred in the United States?

Understand There are about 87,000 flights per day in the United States. We need to find how many days it will take for 261,000 flights to occur.

Plan Write an equation. Let d represent the number of days needed.

87,000	times	the number of days	equals	261,000.
87,000	$\cdot$	d	$=$	261,000

Solve $87,000\,d = 261,000$ Find d by asking, "What number times 87,000 is 261,000?"

$d = 3$

Check Check your answer by substituting 3 for d in the equation.

$87,000(3) \overset{?}{=} 261,000$ Substitute 3 for d.

$261,000 = 261,000$ ✓   Multiply.

The answer makes sense and works for the original problem.

▸ **Guided Practice**

2. GOVERNMENT There are 40 members in the Florida Senate. This is 80 fewer than the number in the Florida House of Representatives. How many members are in the Florida House of Representatives?

A rule for the relationship between certain quantities is called a **formula**. These equations use variables to represent numbers and form general rules.

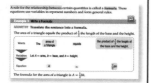

Example 3 Write a Formula

GEOMETRY Translate the sentence into a formula.

The area of a triangle equals the product of $\frac{1}{2}$ the length of the base and the height.

Words	The	area of a triangle	equals	the product of $\frac{1}{2}$ the length of the base and the height.

Variables	Let A = area, b = base, and h = height.

Equation	A	$=$	$\frac{1}{2}bh$

The formula for the area of a triangle is $A = \frac{1}{2}bh$.

▸ **Guided Practice**

3. GEOMETRY Translate the sentence into a formula.
In a right triangle, the square of the measure of the hypotenuse c is equal to the sum of the squares of the measures of the legs, a and b.

Ilene MacDonald/Alamy

Math HistoryLink

Ahmes (about 1680–1620 B.C.)
Ahmes was the Egyptian mathematician and scribe who copied the Rhind Mathematical Papyrus. The papyrus contains 87 algebra problems of the same type. The first set of problems asks how to divide n loaves of bread among 10 people.

2 **Write Sentences from Equations** If you are given an equation, you can write a sentence or create your own word problem.

| **Example 4** | **Translate Equations into Sentences** |

Translate each equation into a sentence.

a. $6z - 15 = 45$

| $6z$ | $-$ | 15 | $=$ | 45 |
| Six times z | minus | fifteen | equals | forty-five. |

b. $y^2 + 3x = w$

| | y^2 | $+$ | $3x$ | $=$ | w |
| The sum of y squared | | and | three times x | is | w. |

Guided Practice

4A. $15 = 25u^2 + 2$ **4B.** $\frac{3}{2}r - t^3 = 132$

When given a set of information, you can create a problem that relates a story.

| **Example 5** | **Write a Problem** |

Write a problem based on the given information.

$t =$ the time that Maxine drove during each of her turns; $t + 4 =$ the time that Tia drove during each of her turns; $2t + (t + 4) = 28$

Sample problem:
Maxine and Tia went on a trip, and they took turns driving. During her turn, Tia drove 4 hours more than Maxine. Maxine took 2 turns, and Tia took 1 turn. Together they drove for 28 hours. How many hours did Maxine drive?

Guided Practice

5. $p =$ Beth's salary; $0.1p =$ bonus; $p + 0.1p = 525$

Go Online! for a Self-Check Quiz

Check Your Understanding   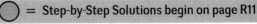 ◯ = Step-by-Step Solutions begin on page R11

Example 1   **Translate each sentence into an equation.**

1. Three times r less than 15 equals 6.

2. The sum of q and four times t is equal to 29.

③ A number n squared plus 12 is the same as the quotient of p and 4.

4. Half of j minus 5 is the sum of k and 13.

5. The sum of 8 and three times k equals the difference of 5 times k and 3.

6. Three fourths of w plus 5 is one half of w increased by nine.

7. The quotient of 25 and t plus 6 is the same as twice t plus 1.

8. Thirty-two divided by y is equal to the product of three and y minus four.

Example 2

9. FINANCIAL LITERACY There is $2000 in Trent's bank account. He wishes to increase his account to a total of $3000 by depositing $40 per week from his paycheck. Write and solve an equation to find how many weeks he needs to reach his goal.

10. **MP** **MODELING** Miguel is earning extra money by painting houses. He charges a $250 fee plus $32 per can of paint needed to complete the job. Write and use an equation to find how many cans of paint he needs for a $410 job.

Translate each sentence into a formula.

Example 3

11. The perimeter of a regular pentagon is 5 times the length of each side.

12. The area of a circle is the product of π and the radius r squared.

13. Four times π times the radius squared is the surface area of a sphere.

14. One third the product of the length of the side squared and the height is the volume of a pyramid with a square base.

Example 4

Translate each equation into a sentence.

15. $7m - q = 23$

16. $6 + 9k + 5j = 54$

17. $3(g + 8) = 4h - 10$

18. $6d^2 - 7f = 8d + f^2$

Example 5

Write a problem based on the given information.

19. g = gymnasts on a team; $3g = 45$

20. c = cost of a notebook; $0.25c$ = markup; $c + 0.25c = 3.75$

Practice and Problem Solving

Extra Practice is on page R2.

Example 1

Translate each sentence into an equation.

21. The difference of f and five times g is the same as 25 minus f.

22. Three times b less than 100 is equal to the product of 6 and b.

23. Four times the sum of 14 and c is a squared.

Example 2

24. ENDANGERED SPECIES There are 91 threatened and endangered plant and animal species in Texas. Write and solve an equation to find the number of threatened or endangered animals if there are 28 threatened and endangered plants.

25. GARDENING A flat of plants contains 12 plants. Yoshi wants a garden that has three rows with 10 plants per row. Write and solve an equation for the number of flats Yoshi should buy.

Example 3

Translate each sentence into a formula.

26. The perimeter of a rectangle is equal to 2 times the length plus twice the width.

27 Celsius temperature C is five ninths times the difference of the Fahrenheit temperature F and 32.

28. The density of an object is the quotient of its mass and its volume.

29. Simple interest is computed by finding the product of the principal amount p, the interest rate r, and the time t.

Example 4

Translate each equation into a sentence.

30. $j + 16 = 35$

31. $4m = 52$

32. $7(p + 23) = 102$

33. $r^2 - 15 = t + 19$

34. $\frac{2}{5}v + \frac{3}{4} = \frac{2}{3}x^2$

35. $\frac{1}{3} - \frac{4}{5}z = \frac{4}{3}y^3$

Example 5

Write a problem based on the given information.

36. q = quarts of strawberries; $2.50q = 10$

37. p = the principal amount; $0.12p$ = the interest charged; $p + 0.12p = 224$

38. c = number of concert tickets; $10 + 35.50c = 258.50$

39. p = the number of players in the game; $5p + 7$ = number of cards in a deck

For Exercises 40–43, match each sentence with an equation.

A. $g^2 = 2(g - 10)$ **B.** $\frac{1}{2}g + 32 = 15 + 6g$ **C.** $g^3 = 24g + 4$ **D.** $3g^2 = 30 + 9g$

40. One half of g plus thirty-two is as much as the sum of fifteen and six times g.

41. A number g to the third power is the same as the product of 24 and g plus 4.

42. The square of g is the same as two times the difference of g and 10.

43. The product of 3 and the square of g equals the sum of thirty and the product of nine and g.

44. MULTI-STEP Tim has a two-and-a-half hour meeting on the 15th floor of a downtown office building and has to decide where to park. He can park at a meter in front of the building for $0.25 per 15 minutes with a maximum of 2 hours. He can park in a lot 5 minutes away that charges a flat daily fee of $15. Or, he can park in a garage 15 minutes away for $4 per hour.

  a. Where should Tim park?

  b. Explain and evaluate your solution process.

  c. What assumptions did you make?

45 SHOPPING Pilar bought 17 items for her camping trip, including tent stakes, packets of drink mix, and bottles of water. She bought 3 times as many packets of drink mix as tent stakes. She also bought 2 more bottles of water than tent stakes. Write and solve an equation to discover how many tent stakes she bought.

46. MULTIPLE REPRESENTATIONS In this problem, you will explore how to translate relations with powers.

x	2	3	4	5	6
y	5	10	17	26	37

  a. Verbal Write a sentence to describe the relationship between x and y in the table.

  b. Algebraic Write an equation that represents the data in the table.

  c. Graphical Graph each ordered pair and draw the function. Describe the graph as discrete or continuous.

H.O.T. Problems   Use Higher-Order Thinking Skills

47. OPEN ENDED Write a problem about your favorite television show that uses the equation $x + 8 = 30$.

48. MP REASONING The surface area of a three-dimensional object is the sum of the areas of the faces. If ℓ represents the length of the side of a cube, write a formula for the surface area of the cube.

49. CHALLENGE Given the perimeter P and width w of a rectangle, write a formula to find the length ℓ.

50. WRITING IN MATH How can you translate a verbal sentence into an algebraic equation?

51. The lengths of two adjacent sides of a rectangle are $2x + 3$ and $x - 7$, and its perimeter is 40. Which answer choice shows an equation that can be used to find x? (MP) 2

    ○ **A** $2x + 3 = x - 7$

    ○ **B** $(2x + 3) + (x - 7) = 40$

    ○ **C** $2(x + 3 - 7) = 40$

    ○ **D** $2(2x + 3) + 2(x - 7) = 40$

52. This equation describes a number n:

$$n + (n + 1) = 7$$

  a. Which answer choice shows a verbal sentence of the equation? (MP) 2

    ○ **A** The sum of a number and one more than the number is 7.

    ○ **B** The sum of two numbers is 7.

    ○ **C** The sum of a number and 7 is equal to one more than the number.

    ○ **D** The sum of a number and 7 is one more than the number.

  b. What is the value of n in the equation?

    ○ **A** 2

    ○ **B** 3

    ○ **C** 4

    ○ **D** 5

53. Which equation best represents the relationship between the x- and y-values? (MP) 2

x	−2	2	4	6	10
y	7	−5	−11	−17	−29

    ○ **A** $y = x + 9$

    ○ **B** $y = x^2 + 3$

    ○ **C** $y = -3x + 1$

    ○ **D** $y = x - 7$

54. A package contains 8 hot dog buns. A shopper wants to purchase enough packages to have a total of 120 hot dog buns. Which answer choice shows an equation for the situation and the solution to that equation? (MP) 2, 4

    ○ **A** $8 \times n = 120$; $n = 15$

    ○ **B** $8 \times 120 = n$; $n = 960$

    ○ **C** $8 + n = 120$; $n = 112$

    ○ **D** $\frac{8}{n} = 120$; $n = \frac{1}{15}$

55. The triangle shown has a perimeter of 46 inches. Which equation can be used to find the value of x? (MP) 2

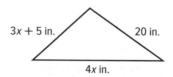

    ○ **A** $4x = 46$

    ○ **B** $4x = 20$

    ○ **C** $3x + 5 = 4x$

    ○ **D** $7x + 25 = 46$

56. MULTI-STEP (MP) 2

  a. Write an equation that says the length of a rectangle is 4 more than six times the value of x.

  b. Write an equation that says the width of a rectangle is 1 less than three times the value of x.

  c. If the perimeter of the rectangle is 96, write an equation for the perimeter of the rectangle.

  d. Simplify the right side of the equation by using the distributive property.

  e. Combine the like terms of the right side of the equation.

  f. What is the solution x of the equation?

You can use **algebra tiles** to model solving equations. To **solve an equation** means to find the value of the variable that makes the equation true. An tile represents the variable x. The $\boxed{1}$ tile represents a positive 1. The $\boxed{-1}$ tile represents a negative 1. And, the $\boxed{-x}$ tile represents the variable negative x. The goal is to get the x-tile by itself on one side of the mat by using the rules stated below.

Mathematical Practices

MP 5 Use appropriate tools strategically.
8 Look for and express regularity in repeated reasoning.

Rules for Equation Models When Adding or Subtracting:

- You can remove or add the same number of identical algebra tiles to each side of the mat without changing the equation.

- One positive tile and one negative tile of the same unit are called a zero pair. Since $1 + (-1) = 0$, you can remove or add zero pairs to either side of the equation mat without changing the equation.

Activity 1 Addition Equation

Work cooperatively. Use an equation model to solve $x + 3 = -4$.

> **Step 1** Model the equation. Place 1 x-tile and 3 positive 1-tiles on one side of the mat. Place 4 negative 1-tiles on the other side of the mat.

> **Step 2** Isolate the x-term. Add 3 negative 1-tiles to each side. The resulting equation is $x = -7$.

$$x + 3 = -4$$
$$x + 3 + (-3) = -4 + (-3)$$
$$x = -7$$

Activity 2 Subtraction Equation

Work cooperatively. Use an equation model to solve $x - 2 = 1$.

> **Step 1** Model the equation. Place 1 x-tile and two -1 tiles on one side of the mat. Place 1 positive 1-tile on the other side of the mat.

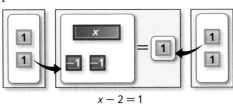

$$x - 2 = 1$$
$$x - 2 + 2 = 1 + 2$$

> **Step 2** Isolate the x-term. Add two positive 1-tiles to each side of the mat. The resulting equation is $x = 3$.

$$x - 2 = 1$$
$$x - 2 + 2 = 1 + 2$$
$$x = 3$$

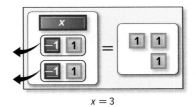

$$x = 3$$

(continued on the next page)

Model and Analyze

Work cooperatively. Solve each equation.

1. $x + 4 = 9$

2. $x + (-3) = -4$

3. $x + 7 = -2$

4. $x + (-2) = 11$

5. WRITING IN MATH If $a = b$, what can you say about $a + c$ and $b + c$? about $a - c$ and $b - c$?

When solving multiplication equations, the goal is still to get the *x*-tile by itself on one side of the mat by using the rules for dividing.

Rules for the Equation Models When Dividing:

- You can group the tiles on each side of the equation mat into an equal number of groups without changing the equation.

- You can place an equal grouping on each side of the equation mat without changing the equation.

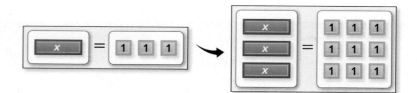

Activity 3   Multiplication Equations

Work cooperatively. Use an equation model to solve $3x = 12$.

Step 1 Model the equation. Place 3 *x*-tiles on one side of the mat. Place 12 positive 1-tiles on the other side of the mat.

Step 2 Isolate the *x*-term. Separate the tiles into 3 equal groups to match the 3 *x*-tiles. Each *x*-tile is paired with 4 positive 1-tiles. The resulting equation is $x = 4$.

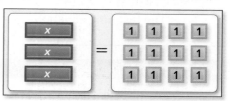

$$3x = 12$$
$$\frac{3x}{3} = \frac{12}{3}$$
$$x = 4$$

Model and Analyze

Work cooperatively. Solve each equation.

6. $5x = -15$

7. $-3x = -9$

8. $4x = 8$

9. $-6x = 18$

10. MAKE A CONJECTURE How would you use algebra tiles to solve $\frac{x}{4} = 5$? Discuss the steps you would take to solve this equation algebraically.

Solving One-Step Equations

∷Then	∷Now	∷Why?

- You translated sentences into equations.

1 Solve equations by using addition and subtraction.

2 Solve equations by using multiplication and division.

- A record for the most snow angels made at one time was set in Michigan when 3784 people participated. North Dakota had 8910 people register to break the record. To determine how many more people North Dakota had than Michigan, solve the equation $3784 + x = 8910$.

 New Vocabulary

solve an equation
equivalent equations
linear equation

 Mathematical Practices

1 Make sense of problems and persevere in solving them.

2 Reason abstractly and quantitatively.

1 **Solve Equations Using Addition or Subtraction** In an equation, the variable represents the number that satisfies the equation. To **solve an equation** means to find the value of the variable that makes the equation true.

The process of solving an equation requires assuming that the original equation has a solution and isolating the variable (with a coefficient of 1) on one side of the equation. Each step in this process results in equivalent equations. **Equivalent equations** have the same solution. An equation in which no variables has an exponent other than 1 is a **linear equation**. Some linear equations can be solved using the Addition Property of Equality.

🔄 Key Concept   Addition Property of Equality

Words	If an equation is true and the same number is added to each side of the equation, the resulting equivalent equation is also true.
Symbols	For any real numbers a, b, and c, if $a = b$, then $a + c = b + c$.
Examples	$14 = 14$ $-3 = -3$
	$14 + 3 = 14 + 3$ $+9 = +9$
	$17 = 17$ $6 = 6$

Example 1   Solve by Adding

Solve $c - 22 = 54$.

Horizontal Method		**Vertical Method**

$c - 22 = 54$	Original equation	$c - 22 = 54$
$c - 22 + 22 = 54 + 22$	Add 22 to each side.	$+ 22 = + 22$
$c = 76$	Simplify.	$c = 76$

To check that 76 is the solution, substitute 76 for c in the original equation.

CHECK $c - 22 = 54$ Original equation

          $76 - 22 \stackrel{?}{=} 54$ Substitute 76 for c.

               $54 = 54 ✓$ Subtract.

▶ **Guided Practice**

1A. $113 = g - 25$ **1B.** $j - 87 = -3$

Go Online!

Log into ConnectED to watch an **Animation** demonstrating how to solve a one-step equation with algebra tiles.

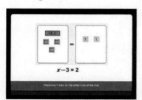

Similar to the Addition Property of Equality, the **Subtraction Property of Equality** can also be used to solve equations. Subtracting a value is equivalent to adding the opposite of the value.

Key Concept Subtraction Property of Equality

Words	If an equation is true and the same number is subtracted from each side of the equation, the resulting equivalent equation is also true.
Symbols	For any real numbers a, b, and c, if $a = b$, then $a - c = b - c$.

Examples

$$87 = 87$$
$$87 - 17 = 87 - 17$$
$$70 = 70$$

$$13 = 13$$
$$-28 = -28$$
$$\overline{-15 = -15}$$

Example 2   Solve by Subtracting

Solve $63 + m = 79$.

Horizontal Method

$63 + m = 79$	Original equation
$63 - 63 + m = 79 - 63$	Subtract 63 from each side.
$m = 16$	Simplify.

Vertical Method

$$63 + m = 79$$
$$\underline{-63 = -63}$$
$$m = 16$$

Study Tip

MP **Reasoning** When solving equations you can use either the horizontal method or the vertical method. Both methods will produce the same result.

To check that 16 is the solution, replace m with 16 in the original equation.

CHECK	$63 + m = 79$	Original equation
	$63 + 16 \stackrel{?}{=} 79$	Substitution, $m = 16$
	$79 = 79 \checkmark$	Simplify.

▶ **Guided Practice**

2A. $27 + k = 30$　　　　　　　　　**2B.** $-12 = p + 16$

2 **Solve Equations Using Multiplication or Division** In the equation $\frac{x}{3} = 9$, the variable x is divided by 3. To solve for x, undo the division by multiplying each side by 3. This is an example of the **Multiplication Property of Equality**.

Key Concept Multiplication Property of Equality

Words	If an equation is true and each side is multiplied by the same nonzero number, the resulting equation is equivalent.
Symbols	For any real numbers a, b, and c, $c \neq 0$, if $a = b$, then $ac = bc$.
Example	If $x = 5$, then $3x = 3 \cdot 5$ or 15.

Key Concept Division Property of Equality

Words	If an equation is true and each side is divided by the same nonzero number, the resulting equation is equivalent.
Symbols	For any real numbers a, b, and c, $c \neq 0$, if $a = b$, then $\frac{a}{c} = \frac{b}{c}$.
Example	If $x = -20$, then $\frac{x}{5} = \frac{-20}{5}$ or -4.

The reciprocal of a number can be used to solve equations.

Example 3 Solve by Multiplying or Dividing

Solve each equation.

a. $\frac{2}{3}q = \frac{1}{2}$

$\frac{2}{3}q = \frac{1}{2}$ Original equation

$\frac{3}{2}\left(\frac{2}{3}\right)q = \frac{3}{2}\left(\frac{1}{2}\right)$ Multiply each side by $\frac{3}{2}$, the reciprocal of $\frac{2}{3}$.

$q = \frac{3}{4}$ Simplify.

b. $39 = -3r$

$39 = -3r$ Original equation

$\frac{39}{-3} = \frac{-3r}{-3}$ Divide each side by -3.

$-13 = r$ Simplify.

Guided Practice

3A. $\frac{3}{5}k = 6$

3B. $-\frac{1}{4} = \frac{2}{3}b$

We can also use reciprocals and properties of equality to solve real-world problems.

Real-World Example 4 Solve by Multiplying

SURVEYS Of a group of 13- to 15-year-old girls surveyed, 225, or about $\frac{9}{20}$ said they text while they watch television. How many girls were surveyed?

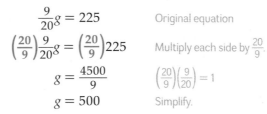

Words	Nine twentieths times those surveyed	is	225.
Variable	Let g = the number of girls surveyed.		
Equation	$\frac{9}{20}g$	=	225

$\frac{9}{20}g = 225$ Original equation

$\left(\frac{20}{9}\right)\frac{9}{20}g = \left(\frac{20}{9}\right)225$ Multiply each side by $\frac{20}{9}$.

$g = \frac{4500}{9}$ $\left(\frac{20}{9}\right)\left(\frac{9}{20}\right) = 1$

$g = 500$ Simplify.

Five hundred girls were surveyed.

Guided Practice

4. **STAINED GLASS** Allison is making a stained glass window. Her pattern requires that one fifth of the glass should be blue. She has 288 square inches of blue glass. If she intends to use all of her blue glass, how much glass will she need for the entire project?

Examples 1–3   **Solve each equation. Check your solution.**

1. $g + 5 = 33$

2. $104 = y - 67$

3. $\frac{2}{3} + w = 1\frac{1}{2}$

4. $-4 + t = -7$

5. $a + 26 = 35$

6. $-6 + c = 32$

7. $1.5 = y - (-5.6)$

8. $3 + g = \frac{1}{4}$

9. $x + 4 = \frac{3}{4}$

10. $\frac{t}{7} = -5$

11. $\frac{a}{36} = \frac{4}{9}$

12. $\frac{2}{3}n = 10$

13. $\frac{8}{9} = \frac{4}{5}k$

14. $12 = \frac{x}{-3}$

15. $-\frac{r}{4} = \frac{1}{7}$

Example 4

16. **FUNDRAISING** The television show "Idol Gives Back" raised money for relief organizations. During this show, viewers could call in and vote for their favorite performer. The parent company contributed $5 million for the 50 million votes cast. What did they pay for each vote?

17. **MP** **REASONING** Hana decides to buy her cat a bed from an online fund that gives $\frac{7}{8}$ of her purchase to shelters that care for animals. How much of Hana's money went to the animal shelter?

Practice and Problem Solving   Extra Practice is on page R2.

Examples 1–3   **Solve each equation. Check your solution.**

18. $v - 9 = 14$

19. $44 = t - 72$

20. $-61 = d + (-18)$

21. $18 + z = 40$

22. $-4a = 48$

23. $12t = -132$

24. $18 - (-f) = 91$

25 $-16 - (-t) = -45$

26. $\frac{1}{3}v = -5$

27. $\frac{u}{8} = -4$

28. $\frac{a}{6} = -9$

29. $-\frac{k}{5} = \frac{7}{5}$

30. $\frac{3}{4} = w + \frac{2}{5}$

31. $-\frac{1}{2} + a = \frac{5}{8}$

32. $-\frac{t}{7} = \frac{1}{15}$

33. $-\frac{5}{7} = y - 2$

34. $v + 914 = -23$

35. $447 + x = -261$

36. $-\frac{1}{7}c = 21$

37. $-\frac{2}{3}h = -22$

38. $\frac{3}{5}q = -15$

39. $\frac{n}{8} = -\frac{1}{4}$

40. $\frac{c}{4} = -\frac{9}{8}$

41. $\frac{2}{3} + r = -\frac{4}{9}$

Example 4

42. **CATS** A domestic cat can run at speeds of 27.5 miles per hour when chasing prey. A cheetah can run 42.5 miles per hour faster when chasing prey. How fast can the cheetah go?

43. **CLEANING** Ben and Patrick work the closing shift at the same restaurant but on different evenings. The average time it takes Ben to close the restaurant is 125 minutes. This is 20 minutes more than the time it takes Patrick to close. Write and solve an equation to find the amount of time t it takes Patrick to close the restaurant.

Solve each equation. Check your solution.

44. $\frac{x}{9} = 10$

45. $\frac{b}{7} = -11$

46. $\frac{3}{4} = \frac{c}{24}$

47. $\frac{2}{3} = \frac{1}{8}y$

48. $\frac{2}{3}n = 14$

49. $\frac{3}{5}g = -6$

50. $4\frac{1}{5} = 3p$

51. $-5 = 3\frac{1}{2}x$

52. $6 = -\frac{1}{2}n$

53. $-\frac{2}{5} = -\frac{z}{45}$

54. $-\frac{8}{24} = \frac{5}{12}$

55. $-\frac{v}{5} = -45$

Write an equation for each sentence. Then solve the equation.

56. Six times a number is 132.

57. Two thirds equals negative eight times a number.

58. Five elevenths times a number is 55.

59. Four fifths is equal to ten sixteenths of a number.

60. Three and two thirds times a number equals two ninths.

61 Four and four fifths times a number is one and one fifth.

62. **MP** **PRECISION** Adelina is comparing prices for two brands of health and energy bars at the local grocery store. She wants to get the best price for each bar.

a. Write an equation to find the price for each bar of the Feel Great brand.

b. Write an equation to find the price of each bar for the Super Power brand.

c. Which bar should Adelina buy? Explain.

63. **TRANSPORTATION** The following description appeared on a news Web site after the world's largest passenger plane was introduced. "That airline will see the A380 transporting some 555 passengers, 139 more than a similarly set-up 747." How many passengers will a similarly set-up 747 transport?

Write an equation for each scenario.

64. **FUEL** In 2004, approximately 5 million cars and trucks were classified as flex-fuel, which means they could run on gasoline or ethanol. In 2016, that number increased to about 20 million. How many more cars and trucks were flex-fuel in 2016?

65. **CHEERLEADING** At a certain cheerleading competition, the maximum time per team, including the set up, is 3 minutes. The Ridgeview High School squad's performance time is 2 minutes and 34 seconds. How much time does the squad have left for their set up?

66. **COMIC BOOKS** An X-Men #1 comic book in mint condition recently sold for $45,000. An Action Comics #63 (Mile High), also in mint condition, sold for $15,000. How much more did the X-Men comic book sell for than the Action Comics book?

67. **MOVIES** A certain movie made $1.6 million in ticket sales. Its sequel made $0.8 million in ticket sales. How much more did the first movie make than the sequel?

68. **CAMERAS** An electronics store sells a certain digital camera for $299. This is $\frac{2}{3}$ of the price that a photography store charges. What is the cost of the camera at the photography store?

69 **BLOGS** In a recent year, 57 million American adults read online blogs. However, 45 million fewer American adults say that they maintain their own blog. How many American adults maintain a blog?

70. CAREERS According to the Bureau of Labor and Statistics, approximately 144,400,000 people were employed in the United States in 2013.

   a. The number of people in production occupations times 17.5 is the number of working people. Write an equation to represent the number of people employed in production occupations in 2013. Then solve the equation.

   b. The number of people in repair occupations is 3,100,000 less than the number of people in production occupations. How many people are in repair occupations?

71. DANCES Student Council has a budget of $2500 for the homecoming dance. So far, they have spent $450 dollars for music.

   a. Write an equation to represent the amount of money left to spend. Then solve the equation.

   b. They then spent $325 on decorations. Write an equation to represent the amount of money left. Then solve the equation.

   c. If the Student Council spent their entire budget, write an equation to represent how many $15 tickets they must sell to make a profit. Then solve the equation.

H.O.T. Problems   Use Higher-Order Thinking Skills

72. WHICH ONE DOESN'T BELONG? Identify the equation that does not belong with the other three. Explain your reasoning.

$$n + 14 = 27$$ $$12 + n = 25$$ $$n - 16 = 29$$ $$n - 4 = 9$$

73. OPEN ENDED Write an equation involving addition and demonstrate two ways to solve it.

74. (MP) REASONING For which triangle is the height not $4\frac{1}{2}b$, where b is the length of the base?

75. (MP) STRUCTURE Determine whether each sentence is *sometimes*, *always*, or *never true*. Explain your reasoning.

   a. $x + x = x$ **b.** $x + 0 = x$

Triangle	Base (cm)	Height (cm)
$\triangle ABC$	3.8	17.1
$\triangle MQP$	5.4	24.3
$\triangle RST$	6.3	28.5
$\triangle TRW$	1.6	7.2

76. (MP) REASONING Determine the value for each statement below.

   a. If $x - 7 = 14$, what is the value of $x - 2$?

   b. If $t + 8 = -12$, what is the value of $t + 1$?

77. CHALLENGE Solve each equation for x. Assume that $a \neq 0$.

   a. $ax = 12$ **b.** $x + a = 15$ **c.** $-5 = x - a$ **d.** $\frac{1}{a}x = 10$

78. WRITING IN MATH Consider the Multiplication Property of Equality and the Division Property of Equality. Explain why they can be considered the same property. Which one do you think is easier to use?

79. A rectangle is divided into 6 equal-size strips. The area of the shaded part of the rectangle is 60 square units. **MP** 1,2

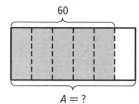

60

$A = ?$

a. Which equation can be used to find the total area, A, of the rectangle ?

○ **A** $60 = \frac{5}{6} \cdot A$

○ **B** $\frac{5}{6} \cdot 60 = A$

○ **C** $60 = \frac{6}{5} \cdot A$

○ **D** $\frac{5}{6} + A = 60$

b. What is the area of the rectangle?

○ **A** $A = 50$ square units

○ **B** $A = 59\frac{1}{6}$ square units

○ **C** $A = 72$ square units

○ **D** $A = 132$ square units

80. What is the solution to the equation $17 - x = 5$? **MP** 1,2

○ **A** $x = -12$

○ **B** $x = 3.2$

○ **C** $x = 12$

○ **D** $x = 22$

81. Consider the equation $\frac{5}{3} \cdot p = 30$. **MP** 1,2

a. What is the reciprocal of the coefficient?

b. What is the solution of the equation?

○ **A** $p = 18$

○ **B** $p = 28\frac{1}{3}$

○ **C** $p = 31\frac{2}{3}$

○ **D** $p = 50$

82. The sum of a number q and 23 is 105. Which answer choice shows both the equation that represents the sentence and the correct solution to that equation? **MP** 1,2

○ **A** $q = 23 + 105; q = 128$

○ **B** $q + 23 = 105; q = 128$

○ **C** $q + 23 = 105; q = 82$

○ **D** $q + 23 = q + 105;$ no solution

83. MULTI-STEP At a campsite in July, 120 families stayed for three or more days. The 120 families represent three-eighths of all the families who stayed at the campsite in July. **MP** 1,2

a. Write an equation to represent the situation. Let $t =$ the total number of families.

b. How many families stayed at the campsite in July?

○ **A** 15

○ **B** 45

○ **C** 300

○ **D** 320

c. In August, 120 families stayed for three more days and represent 40% of all the families who stayed at the campsite in August. Write an equation to represent this situation. Let $t =$ the total number of families.

d. How many families stayed at the campsite in August?

○ **A** 12

○ **B** 48

○ **C** 300

○ **D** 420

e. Describe how you could solve the equation from Part c by using multiplication.

f. How many more families stayed at the campsite in July than in August?

Algebra Lab
Solving Multi-Step Equations

You can use algebra tiles to model solving multi-step equations.

Mathematical Practices

MP **7** Look for and make use of structure.

Activity

Work cooperatively. Use an equation model to solve $4x + 3 = -5$.

Step 1 Model the equation.

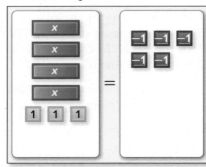

$$4x + 3 = -5$$

Place 4 x-tiles and 3 positive 1-tiles on one side of the mat. Place 5 negative 1-tiles on the other side.

Step 2 Isolate the x-term.

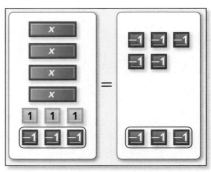

$$4x + 3 - 3 = -5 - 3$$

Since there are 3 positive 1-tiles with the x-tiles, add 3 negative 1-tiles to each side to form zero pairs.

Step 3 Remove zero pairs.

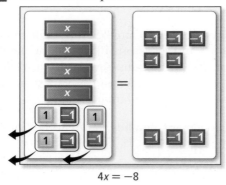

$$4x = -8$$

Group the tiles to form zero pairs and remove the zero pairs.

Step 4 Group the tiles.

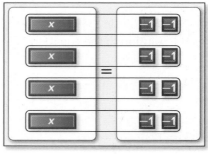

$$\frac{4x}{4} = \frac{-8}{4}$$
$$x = -2$$

Separate the remaining tiles into 4 equal groups to match the 4 x-tiles. Each x-tile is paired with 2 negative 1-tiles. The resulting equation is $x = -2$.

Model

Work cooperatively. Use algebra tiles to solve each equation.

1. $3x - 7 = -10$ **2.** $2x + 5 = 9$ **3.** $5x - 7 = 8$ **4.** $-7 = 3x + 8$

5. $5 + 4x = -11$ **6.** $3x + 1 = 7$ **7.** $11 = 2x - 5$ **8.** $7 + 6x = -11$

9. What would be your first step in solving $8x - 29 = 67$?

10. What steps would you use to solve $9x + 14 = -49$?

Solving Multi-Step Equations

:: Then

- You solved one-step equations.

:: Now

1. Solve equations involving more than one operation.

2. Solve equations involving consecutive integers.

:: Why?

- The Tour de France is the premier cycling event in the world. The map shows the 2013 Tour de France course. If the length of the shortest portion of the race can be represented by k, the expression $6k + 50.5$ is the length of the longest stage or 242.5 kilometers. This can be described by the equation $6k + 50.5 = 242.5$.

 New Vocabulary

multi-step equation
consecutive integers
number theory

 Mathematical Practices

8 Look for and express regularity in repeated reasoning.

1 Solve Multi-Step Equations Since the above equation requires more than one step to solve, it is called a **multi-step equation**. To solve this equation, we must undo each operation by working backward.

| **Example 1** | Solve Multi-Step Equations |

Solve each equation. Check your solution.

a. $11x - 4 = 29$

$$11x - 4 = 29 \qquad \text{Original equation}$$
$$11x - 4 + 4 = 29 + 4 \qquad \text{Add 4 to each side.}$$
$$11x = 33 \qquad \text{Simplify.}$$
$$\frac{11x}{11} = \frac{33}{11} \qquad \text{Divide each side by 11.}$$
$$x = 3 \qquad \text{Simplify.}$$

b. $\dfrac{a + 7}{8} = 5$

$$\frac{a + 7}{8} = 5 \qquad \text{Original equation}$$
$$8\left(\frac{a + 7}{8}\right) = 8(5) \qquad \text{Multiply each side by 8.}$$
$$a + 7 = 40 \qquad \text{Simplify.}$$
$$\underline{-7 = -7} \qquad \text{Subtract 7 from each side.}$$
$$a = 33 \qquad \text{Simplify.}$$

You can check your solutions by substituting the results back into the original equations.

> **Guided Practice**

Solve each equation. Check your solution.

1A. $2a - 6 = 4$ **1B.** $\dfrac{n + 1}{-2} = 15$

Real-World Example 2 Write and Solve a Multi-Step Equation

SHOPPING Hiroshi is buying a pair of water skis that are on sale for $\frac{2}{3}$ of the original price. After he uses a $25 gift certificate, the total cost before taxes is $185. What was the original price of the skis? Write an equation for the problem. Then solve the equation.

Words	Two thirds	of	the price	minus	25	is	185.

Variable   Let p = original price of the skis.

Equation	$\frac{2}{3}$	•	p	$-$	25	$=$	185

$$\frac{2}{3}p - 25 = 185 \qquad \text{Original equation}$$

$$\frac{2}{3}p - 25 + 25 = 185 + 25 \qquad \text{Add 25 to each side.}$$

$$\frac{2}{3}p = 210 \qquad \text{Simplify.}$$

$$\frac{3}{2}\left(\frac{2}{3}p\right) = \frac{3}{2}(210) \qquad \text{Multiply each side by } \frac{3}{2}.$$

$$p = 315 \qquad \text{Simplify.}$$

The original price of the skis was $315.

Guided Practice

2A. RETAIL A department store has sold $\frac{3}{5}$ of their formal dresses, but 10 were returned. Now the store has 62 formal dresses. How many were there originally?

2B. READING Len read $\frac{3}{4}$ of a graphic novel over the weekend. Monday, he read 22 more pages. If he has read 220 pages, how many pages does the book have?

Go Online!

Learn about two-step equations with a **BrainPOP®** **Animation** in the ConnectED Resources for this lesson.

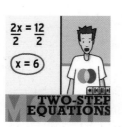

2 Solve Consecutive Integer Problems **Consecutive integers** are integers in counting order, such as 4, 5, and 6 or n, $n + 1$, and $n + 2$. Counting by two will result in *consecutive even integers* if the starting integer n is even and *consecutive odd integers* if the starting integer n is odd.

Concept Summary Consecutive Integers

Type	Words	Symbols	Example
Consecutive Integers	Integers that come in counting order.	$n, n + 1, n + 2, \ldots$	$\ldots, -2, -1, 0, 1, 2, \ldots$
Consecutive Even Integers	Even integer followed by the next even integer.	$n, n + 2, n + 4, \ldots$	$\ldots, -2, 0, 2, 4, \ldots$
Consecutive Odd Integers	Odd integer followed by the next odd integer.	$n, n + 2, n + 4, \ldots$	$\ldots, -1, 1, 3, 5, \ldots$

Number theory is the study of numbers and the relationships between them.

Example 3   Solve a Consecutive Integer Problem

NUMBER THEORY Write an equation for the following problem. Then solve the equation and answer the problem.

Find three consecutive odd integers with a sum of −51.

Let n = the least odd integer.

Then $n + 2$ = the next greater odd integer, and $n + 4$ = the greatest of the three integers.

Words	The sum of three consecutive odd integers	is	−51.
Equation	$n + (n + 2) + (n + 4)$	=	−51.

$$n + (n + 2) + (n + 4) = -51 \qquad \text{Original equation}$$
$$3n + 6 = -51 \qquad \text{Combine like terms.}$$
$$\underline{-6 = -6} \qquad \text{Subtract 6 from each side.}$$
$$3n = -57 \qquad \text{Simplify.}$$
$$\frac{3n}{3} = \frac{-57}{3} \qquad \text{Divide each side by 3.}$$
$$n = -19 \qquad \text{Simplify.}$$

$n + 2 = -19 + 2$ or -17 $n + 4 = -19 + 4$ or -15
The consecutive odd integers are −19, −17, and −15.

CHECK −19, −17, and −15 are consecutive odd integers.
      $-19 + (-17) + (-15) = -51$ ✓

Guided Practice

3. Write an equation for the following problem. Then solve the equation and answer the problem.

   Find three consecutive integers with a sum of 21.

Check Your Understanding   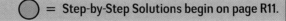 = Step-by-Step Solutions begin on page R11.

Example 1   Solve each equation. Check your solution.

1) $3m + 4 = -11$ **2.** $12 = -7f - 9$ **3.** $-3 = 2 + \frac{a}{11}$

4. $\frac{3}{2}a - 8 = 11$ **5.** $8 = \frac{x - 5}{7}$ **6.** $\frac{c + 1}{-3} = -21$

Example 2   **7. NUMBER THEORY** Fourteen decreased by twice a number equals −32. Write an equation for this situation and then find the number.

8. BASEBALL Among the career home run leaders for Major League Baseball, Hank Aaron has 181 fewer than twice the number of Chipper Jones. Hank Aaron hit 755 home runs. Write an equation for this situation. How many home runs did Chipper Jones hit in his career?

Example 3   Write an equation and solve each problem.

9. Find three consecutive odd integers with a sum of 75.

10. Find three consecutive integers with a sum of −36.

Example 1   **Solve each equation. Check your solution.**

11. $3t + 7 = -8$ **12.** $8 = 16 + 8n$ **13.** $-34 = 6m - 4$

14. $9x + 27 = -72$ **15.** $\dfrac{y}{5} - 6 = 8$ **16.** $\dfrac{f}{-7} - 8 = 2$

17. $1 + \dfrac{r}{9} = 4$ **18.** $\dfrac{k}{3} + 4 = -16$ **19.** $\dfrac{n-2}{7} = 2$

20. $14 = \dfrac{6+z}{-2}$ **21.** $-11 = \dfrac{a-5}{6}$ **22.** $\dfrac{22-w}{3} = -7$

Example 2   **(23)** **FINANCIAL LITERACY** A tablet data provider offers different plans for data usage. The plan Raul chose has a monthly fee of \$29.99 per month for 5 GB of data with an additional cost of \$10 for each gigabyte over 5.

   a. Write an expression for Raul's monthly data costs.

   b. Raul learns that his data provider is offering an unlimited data plan for \$59.99 per month. Find the minimum number of gigabytes Raul could use per month to make the unlimited plan cheaper than his current plan. Explain your solution process.

Example 3   **Write an equation and solve each problem.**

24. Fourteen less than three fourths of a number is negative eight. Find the number.

25. Seventeen is thirteen subtracted from six times a number. What is the number?

26. Find three consecutive even integers with the sum of -84.

27. Find three consecutive odd integers with the sum of 141.

28. Find four consecutive integers with the sum of 54.

29. Find four consecutive integers with the sum of -142.

Solve each equation. Check your solution.

30. $-6m - 8 = 24$ **31.** $45 = 7 - 5n$

32. $\dfrac{2b}{3} + 6 = 24$ **33.** $\dfrac{5x}{9} - 11 = -51$

34. $65 = \dfrac{3}{4}c - 7$ **35.** $9 + \dfrac{2}{3}x = 81$

36. $-\dfrac{5}{2} = \dfrac{3}{4}z + \dfrac{1}{2}$ **37.** $\dfrac{5}{6}k + \dfrac{2}{3} = \dfrac{4}{3}$

38. $-\dfrac{1}{5} - \dfrac{4}{9}a = \dfrac{2}{15}$ **39.** $-\dfrac{3}{7} = \dfrac{3}{4} - \dfrac{b}{2}$

Write an equation and solve each problem.

40. **MP** **REASONING** The ages of three brothers are consecutive integers with the sum of 96. How old are the brothers?

 41. **VOLCANOES** Moving lava can build up and form beaches at the coast of an island. The growth of an island in a seaward direction may be modeled as $8y + 2$ centimeters, where y represents the number of years that the lava flows. An island has expanded 60 centimeters seaward. How long has the lava flowed?

Solve each equation. Check your solution.

42. $-5x - 4.8 = 6.7$

43 $3.7q + 26.2 = 111.67$

44. $0.6a + 9 = 14.4$

45. $\frac{c}{2} - 4.3 = 11.5$

46. $9 = \dfrac{-6p - (-3)}{-8}$

47. $3.6 - 2.4m = 12$

48. If $7m - 3 = 53$, what is the value of $11m + 2$?

49. If $13y + 25 = 64$, what is the value of $4y - 7$?

50. If $-5c + 6 = -69$, what is the value of $6c - 15$?

51. MULTI-STEP Tickets to the NRH$_2$O Family Water Park in North Richland Hills are $25.99 per day. The park offers a season pass for $89.99, which includes a 15% discount at Riverfront Pizza & Treats. A *Greatest Value* season pass is also offered for $119.99. Some of the extra benefits are a 20% discount at Riverfront, a souvenir cup with free refills, and five meal vouchers. The McCauley family plans to spend 3 days at the park this year.

  a. Which is the better deal for the McCauley's?

  b. Explain your solution process.

  c. What assumptions would you make?

52. SHOPPING At The Family Farm, you can pick your own fruits and vegetables.

  a. The cost of a bag of potatoes is $1.50 less than $\frac{1}{2}$ of the price of apples. Write and solve an equation to find the cost of potatoes.

  b. The price of each zucchini is 3 times the price of winter squash minus $7. Write and solve an equation to find the cost of zucchini.

  c. Write an equation to represent the cost of a pumpkin using the cost of the blueberries.

The Family Farm	
Fruit	Price ($)
Apples	6.99/bag
Pumpkins	6.50 each
Blueberries	3.49/qt
Winter squash	3.29 each

H.O.T. Problems   Use Higher-Order Thinking Skills

53. OPEN ENDED Write a problem that can be modeled by the equation $2x + 40 = 60$. Then solve the equation and explain the solution in the context of the problem.

54. CHALLENGE Solve each equation for x. Assume that $a \neq 0$.

  a. $ax + 7 = 5$

  b. $\frac{1}{a}x - 4 = 9$

  c. $2 - ax = -8$

55. Ⓜ️ **REASONING** Determine whether each equation has a solution. Justify your answer.

  a. $\dfrac{a + 4}{5 + a} = 1$

  b. $\dfrac{1 + b}{1 - b} = 1$

  c. $\dfrac{c - 5}{5 - c} = 1$

56. Ⓜ️ **REGULARITY** Determine whether the following statement is *sometimes, always,* or *never* true. Explain your reasoning.

  The sum of three consecutive odd integers equals an even integer.

57. WRITING IN MATH Write a paragraph explaining the order of the steps that you would take to solve a multi-step equation.

Preparing for Assessment

58. A student wrote a list of six consecutive integers. If the sum of the first integer and sixth integer is 1, which integer is the least of the six integers? **MP** 7

- A −2
- B 1
- C 3
- D 5

59. MULTI-STEP The sum of five consecutive integers is −10. **MP** 7

a. If n is the least integer, use mathematical notation to write the five integers.

b. Write and solve an equation to find the value of n.

c. Which integer is the greatest of the five integers?

60. The sum of three consecutive even integers is 54. What is the middle integer? **MP** 7

- A 16
- B 17
- C 18
- D 20

61. Use the equation $-\dfrac{5x}{3} + 8 = 13$ to answer the following questions: **MP** 7

a. What inverse operation should be used first to solve the equation?

b. What is the solution to the equation?

- A −12.6
- B $-8\frac{1}{3}$
- C −3
- D 3

62. Which of the following equations is equivalent to $\dfrac{3x - 4}{4} = 5x + 2$? **MP** 7

- A −2x = 3
- B −2x = 6
- C −17x = 6
- D −17x = 12

63. The scale factor for two similar triangles is 2 : 3. The perimeter of the smaller triangle is 56 cm. What is the perimeter of the larger triangle? **MP** 2

- A 37 cm
- B 84 cm
- C 112 cm
- D 168 cm

64. Ken's fitness membership cost $100 to join and a $25 monthly fee. Ken also used a coupon to get 10% off the monthly fee. **MP** 4

a. Which equation models the total cost of Ken's fitness membership (C) after a number of months (m)?

b. If Ken spent $730, did he stay at the club for at least 2 years? Choose any TRUE statements about the solution process and answer to this question.

- A Substitute 730 for C.
- B To isolate m, add 100 to each side of the equation.
- C To isolate m, subtract 100 from each side of the equation.
- D Divide each side of the equation by 22.50.
- E Ken did not stay at the club at least 2 years since m = 23.
- F Ken did stay at the club at least 2 years since m = 28.

Solving Equations with the Variable on Each Side

:: **Then**	:: **Now**	:: **Why?**

You solved multi-step equations.

1 Solve equations with the variable on each side.

2 Solve equations involving grouping symbols.

The equation $y = 1.3x + 19$ represents the number of times Americans eat in their cars each year, where x is the number of years since 1985, and y is the number of times that they eat in their car. The equation $y = -1.3x + 93$ represents the number of times Americans eat in restaurants each year, where x is the number of years since 1985, and y is the number of times that they eat in a restaurant.

The equation $1.3x + 19 = -1.3x + 93$ represents the year when the number of times Americans eat in their cars will equal the number of times Americans eat in restaurants.

New Vocabulary

identity

Mathematical Practices

1 Make sense of problems and persevere in solving them.

3 Construct viable arguments and critique the reasoning of others.

4 Model with mathematics.

7 Look for and make use of structure.

1 Variables on Each Side To solve an equation that has variables on each side, use the Addition or Subtraction Property of Equality to write an equivalent equation with the variable terms on one side.

Example 1   Solve an Equation with Variables on Each Side

Solve $2 + 5k = 3k - 6$. Check your solution.

$2 + 5k = 3k - 6$	Original equation
$\underline{ -3k = -3k}$	Subtract $3k$ from each side.
$2 + 2k = -6$	Simplify.
$\underline{-2 = -2}$	Subtract 2 from each side.
$2k = -8$	Simplify.
$\dfrac{2k}{2} = \dfrac{-8}{2}$	Divide each side by 2.
$k = -4$	Simplify.

CHECK $2 + 5k = 3k - 6$	Original equation
$2 + 5(-4) \overset{?}{=} 3(-4) - 6$	Substitution, $k = -4$
$2 + -20 \overset{?}{=} -12 - 6$	Multiply.
$-18 = -18 \checkmark$	Simplify.

▶ **Guided Practice**

Solve each equation. Check your solution.

1A. $3w + 2 = 7w$ **1B.** $5a + 2 = 6 - 7a$

1C. $\dfrac{x}{2} + 1 = \dfrac{1}{4}x - 6$ **1D.** $1.3c = 3.3c + 2.8$

2 Grouping Symbols If equations contain grouping symbols such as parentheses or brackets, use the Distributive Property first to remove the grouping symbols.

Example 2   Solve an Equation with Grouping Symbols

Solve $6(5m - 3) = \frac{1}{3}(24m + 12)$.

$6(5m - 3) = \frac{1}{3}(24m + 12)$	Original equation
$30m - 18 = 8m + 4$	Distributive Property
$30m - 18 - 8m = 8m + 4 - 8m$	Subtract $8m$ from each side.
$22m - 18 = 4$	Simplify.
$22m - 18 + 18 = 4 + 18$	Add 18 to each side.
$22m = 22$	Simplify.
$\frac{22m}{22} = \frac{22}{22}$	Divide each side by 22.
$m = 1$	Simplify.

> **Study Tip**
>
> **Solving an Equation**
> You may want to eliminate the terms with a variable from one side before eliminating a constant.

▶ **Guided Practice**

Solve each equation. Check your solution.

2A. $8s - 10 = 3(6 - 2s)$ **2B.** $7(n - 1) = -2(3 + n)$

Some equations may have no solution. That is, there is no value of the variable that will result in a true equation. Some equations are true for all values of the variables. These are called **identities**.

Example 3   Find Special Solutions

Solve each equation.

a. $5x + 5 = 3(5x - 4) - 10x$

$5x + 5 = 3(5x - 4) - 10x$	Original equation
$5x + 5 = 15x - 12 - 10x$	Distributive Property
$5x + 5 = 5x - 12$	Simplify.
$\underline{-5x \qquad = -5x}$	Subtract $5x$ from each side.
$5 \neq -12$	

Since $5 \neq -12$, this equation has no solution.

b. $3(2b - 1) - 7 = 6b - 10$

$3(2b - 1) - 7 = 6b - 10$	Original equation
$6b - 3 - 7 = 6b - 10$	Distributive Property
$6b - 10 = 6b - 10$	Simplify.
$0 = 0$	Subtract $6b - 10$ from each side.

Since the expressions on each side of the equation are the same, this equation is an identity. It is true for all values of b.

Go Online!

Use the algebra tiles **Virtual Manipulatives** in the eToolkit to visualize solving equations. If the mat has the same tiles on both sides at some point in the solution, the equation is an identity.

▶ **Guided Practice**

3A. $7x + 5(x - 1) = -5 + 12x$ **3B.** $6(y - 5) = 2(10 + 3y)$

The steps for solving an equation can be summarized as follows.

🔧 Key Concept Steps for Solving Equations

Step 1 Simplify the expressions on each side. Use the Distributive Property as needed.

Step 2 Use the Addition and/or Subtraction Properties of Equality to get the variables on one side and the numbers without variables on the other side. Simplify.

Step 3 Use the Multiplication or Division Property of Equality to solve.

There are many situations in which you must simplify expressions with grouping symbols in order to solve an equation.

Example 4 Write an Equation

Find the value of x so that the figures have the same area.

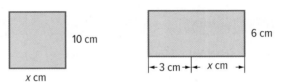

A 3	**C** 6.5
B 4.5	**D** 7

StudyTip

MP Tools There is often more than one way to solve a problem. In this example, you can write an algebraic equation and solve for x. Or you can substitute each answer choice into the formulas to find the correct answer.

Read the Item

The area of the first rectangle is $10x$, and the area of the second is $6(3 + x)$. The equation $10x = 6(3 + x)$ represents this situation.

Solve the Item

A $10x = 6(3 + x)$

$10(3) \stackrel{?}{=} 6(3 + 3)$

$30 \stackrel{?}{=} 6(6)$

$30 \neq 36$ ✗

B $10x = 6(3 + x)$

$10(4.5) \stackrel{?}{=} 6(3 + 4.5)$

$45 \stackrel{?}{=} 6(7.5)$

$45 = 45$ ✓

Since the value 4.5 results in a true statement, you do not need to check 6.5 and 7. The answer is B.

Guided Practice

4. Find the value of x so that the figures have the same perimeter.

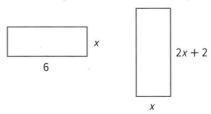

A 1.5	**B** 2	**C** 3.2	**D** 4

Check Your Understanding

◯ = Step-by-Step Solutions begin on page R11

Examples 1–3 Solve each equation. Check your solution.

1. $13x + 2 = 4x + 38$

2. $\frac{2}{3} + \frac{1}{6}q = \frac{5}{6}q + \frac{1}{3}$

3. $6(n + 4) = -18$

4. $7 = -11 + 3(b + 5)$

5. $5 + 2(n + 1) = 2n$

6. $7 - 3r = r - 4(2 + r)$

7. $14v + 6 = 2(5 + 7v) - 4$

8. $5h - 7 = 5(h - 2) + 3$

Example 4

9. **MULTIPLE CHOICE** Find the value of x so that the figures have the same perimeter.

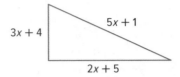

A 4   **B** 5   **C** 6   **D** 7

Practice and Problem Solving

Extra Practice is on page R2.

Examples 1–3 Solve each equation. Check your solution.

10. $7c + 12 = -4c + 78$

11. $2m - 13 = -8m + 27$

12. $9x - 4 = 2x + 3$

⑬ $6 + 3t = 8t - 14$

14. $\frac{b - 4}{6} = \frac{b}{2}$

15. $\frac{5v - 4}{10} = \frac{4}{5}$

16. $8 = 4(r + 4)$

17. $6(n + 5) = 66$

18. $5(g + 8) - 7 = 103$

19. $12 - \frac{4}{5}(x + 15) = 4$

20. $3(3m - 2) = 2(3m + 3)$

21. $6(3a + 1) - 30 = 3(2a - 4)$

Example 4

22. **GEOMETRY** Find the value of x so the rectangles have the same area.

23. **NUMBER THEORY** Four times the lesser of two consecutive even integers is 12 less than twice the greater number. Find the integers.

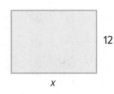
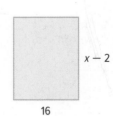

24. ⓂⓅ **SENSE-MAKING** Two times the least of three consecutive odd integers exceeds three times the greatest by 15. What are the integers?

Solve each equation. Check your solution.

25. $2x = 2(x - 3)$

26. $\frac{2}{5}h - 7 = \frac{12}{5}h - 2h + 3$

27. $-5(3 - q) + 4 = 5q - 11$

28. $2(4r + 6) = \frac{2}{3}(12r + 18)$

29. $\frac{3}{5}f + 24 = 4 - \frac{1}{5}f$

30. $\frac{1}{12} + \frac{3}{8}y = \frac{5}{12} + \frac{5}{8}y$

31. $\frac{2m}{5} = \frac{1}{3}(2m - 12)$

32. $\frac{1}{8}(3d - 2) = \frac{1}{4}(d + 5)$

33. $6.78j - 5.2 = 4.33j + 2.15$

34. $14.2t - 25.2 = 3.8t + 26.8$

35. $3.2k - 4.3 = 12.6k + 14.5$

36. $5[2p - 4(p + 5)] = 25$

37. **NUMBER THEORY** Three times the lesser of two consecutive even integers is 6 less than six times the greater number. Find the integers.

38. **MONEY** Chris has saved twice the number of quarters that Nora saved plus 6. The number of quarters Chris saved is also five times the difference of the number of quarters and 3 that Nora has saved. Write and solve an equation to find the number of quarters they each have saved.

39. **BUSINESS** A company that produces custom bike license plates spends $5525 per month in building overhead plus $2.50 per license plate. If the plates sell for $5.99, how many plates must the company sell each month before making a profit?

40. **BLOGS** The table shows the number of subscribers to a specific travel blog for two states for a recent year. How long will it take for the numbers of subscribers to be the same?

State	Blog Subscribers (thousands)	New Subscribers Each Year (thousands)
Nevada	3765	325
Texas	3842	292

41. **MULTIPLE REPRESENTATIONS** In this problem, you will explore $2x + 4 = -x - 2$.

   a. **Graphical** Make a table of values with five ordered pairs for $y = 2x + 4$ and $y = -x - 2$. Graph the ordered pairs from the tables.

   b. **Algebraic** Solve $2x + 4 = -x - 2$.

   c. **Verbal** Explain how the solution you found in part **b** is related to the intersection point of the graphs in part **a**.

H.O.T. Problems   Use **H**igher-**O**rder **T**hinking Skills

42. **MP REASONING** Solve $5x + 2 = ax - 1$ for x. Assume that $a \neq 0$. Describe each step.

43. **CHALLENGE** Write an equation with the variable on each side of the equals sign, at least one fractional coefficient, and a solution of -6. Discuss the steps you used.

44. **OPEN ENDED** Create an equation with at least two grouping symbols for which there is no solution.

45. **MP CRITIQUE ARGUMENTS** Determine whether each solution is correct. If the solution is not correct, describe the error and give the correct solution.

   a.
   $$2(g + 5) = 22$$
   $$2g + 5 = 22$$
   $$2g + 5 - 5 = 22$$
   $$2g = 17$$
   $$g = 8.5$$

   b.
   $$5d = 2d - 18$$
   $$5d - 2d = 2d - 18 - 2d$$
   $$3d = -18$$
   $$d = -6$$

   c.
   $$-6z + 13 = 7z$$
   $$-6z + 13 - 6z = 7z - 6z$$
   $$13 = z$$

46. **CHALLENGE** Find the value of k for which each equation is an identity.

   a. $k(3x - 2) = 4 - 6x$ b. $15y - 10 + k = 2(ky - 1) - y$

47. **WRITING IN MATH** Compare and contrast solving equations with variables on both sides of the equation to solving one-step or multi-step equations with a variable on one side of the equation.

48. The two figures, a regular hexagon and a regular pentagon, have the same perimeter. What is the length of a side of the pentagon? **MP** 1

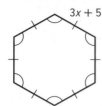 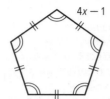

$3x + 5$

$4x - 1$

○ **A** 17.5 units

○ **B** 23 units

○ **C** 57.5 units

○ **D** 69 units

49. What is the solution to the equation $6(7b + 4) = -60$? **MP** 3

○ **A** -10

○ **B** -2

○ **C** $-1\frac{11}{21}$

○ **D** $-\frac{6}{7}$

50. Three times the lesser of two consecutive odd integers is eleven more than two times the greater of the two integers. What are the two integers? **MP** 1

○ **A** 13 and 14

○ **B** 15 and 16

○ **C** 15 and 17

○ **D** 17 and 19

51. A cell phone company charges $4 per GB beyond 20 GB in a month. The monthly plan is $59.99, and the bill for January is $75.99. How many GB were used in that month? **MP** 1, 4

○ **A** 4

○ **B** 16

○ **C** 22

○ **D** 24

52. MULTI-STEP A square has side length $4x - 3$, and an equilateral triangle has a side length $5x - 3$. **MP** 1, 3, 4, 7

 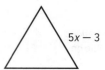

$4x - 3$

$5x - 3$

a. If the figures have the same perimeter, then write an equation to show the equal perimeters.

b. Explain the steps to solve the equation.

c. What is the length of the side of the square?

d. What is the length of the side of the triangle?

e. What is the area of the square?

53. On Monday, Jayla raised $30 and Carlos raised $50. After that, Jayla raised $12 per day and Carlos raised $7 per day. After how many days will they have raised the same amount? **MP** 1

○ **A** 0.25 day

○ **B** 4 days

○ **C** 15 days

○ **D** 16 days

54. Find three consecutive multiples of 6 with a sum that is four times the least of the numbers. **MP** 1

55. Triple a number n minus 5 is 9 more than half the number. **MP** 1

a. Write an equation for the sentence.

b. What is the number n?

56. For a rectangle with a perimeter of 20 cm, the width of the rectangle is one-third its length. Use an equation to find the length and width of the rectangle. **MP** 1

Solving Equations Involving Absolute Value

:·Then

- You solved equations with the variable on each side.

:·Now

1. Evaluate absolute value expressions.
2. Solve absolute value equations.

:·Why?

A telephone survey was conducted to determine the library activities of people 16 to 29 years old in the U.S. According to the survey, 74% browse the shelves. The survey had a margin of error of ±2.3. This means that the results could be 2.3 points higher or lower. So the percent of people who browse could be as high as 76.3% and as low as 71.7%.

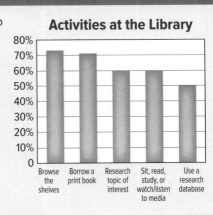

Activities at the Library

 Mathematical Practices

1 Make sense of problems and persevere in solving them.

3 Construct viable arguments and critique the reasoning of others.

4 Model with mathematics.

7 Look for and make use of structure.

1 Absolute Value Expressions Expressions with absolute values define an upper and lower range in which a value must lie. Expressions involving absolute value can be evaluated using the given value for the variable.

Example 1   Expressions with Absolute Value

Evaluate $|m + 6| - 14$ if $m = 4$.

$$|m + 6| - 14 = |4 + 6| - 14 \qquad \text{Replace } m \text{ with 4.}$$
$$= |10| - 14 \qquad 4 + 6 = 10$$
$$= 10 - 14 \qquad |10| = 10$$
$$= -4 \qquad \text{Simplify.}$$

▸ Guided Practice

1. Evaluate $23 - |3 - 4x|$ if $x = 2$.

---

2 Absolute Value Equations The margin of error in the example at the top of the page is an example of absolute value. The distance between 74 and 76.3 on a number line is the same as the distance between 71.7 and 74.

There are three types of open sentences involving absolute value, $|x| = n$, $|x| < n$, and $|x| > n$. In this lesson, we will consider only the first type. Look at the equation $|x| = 4$. This means that the distance between 0 and x is 4.

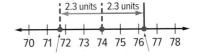

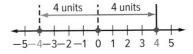

If $|x| = 4$, then $x = -4$ or $x = 4$. Thus, the solution set is $\{-4, 4\}$.

For each absolute value equation, we must consider both cases. To solve an absolute value equation, first isolate the absolute value on one side of the equals sign if it is not already by itself.

🔑 Key Concept Absolute Value Equations

Words   When solving equations that involve absolute values, there are two cases to consider.

   Case 1 The expression inside the absolute value symbol is positive or zero.

   Case 2 The expression inside the absolute value symbol is negative.

Symbols For any real numbers a and b, if $|a| = b$ and $b \geq 0$, then $a = b$ or $a = -b$.

Example $|d| = 10$, so $d = 10$ or $d = -10$.

Example 2   Solve Absolute Value Equations

Solve each equation. Then graph the solution set.

a. $|f + 5| = 17$

$|f + 5| = 17$ Original equation

Case 1

$$f + 5 = 17$$
$$f + 5 - 5 = 17 - 5 \quad \text{Subtract 5 from each side.}$$
$$f = 12 \quad \text{Simplify.}$$

Case 2

$$f + 5 = -17$$
$$f + 5 - 5 = -17 - 5$$
$$f = -22$$

b. $|b - 1| = -3$

$|b - 1| = -3$ means the distance between b and 1 is -3. Since distance cannot be negative, the solution is the empty set $\varnothing$.

▶ **Guided Practice**

2A. $|y + 2| = 4$

2B. $|3n - 4| = -1$

Absolute value equations occur in real-world situations that describe a range within which a value must lie.

Real-World Example 3   Solve an Absolute Value Equation

SNAKES The temperature of an enclosure for a pet snake should be about 80°F, give or take 5°. Find the maximum and minimum temperatures.

You can use a number line to solve.

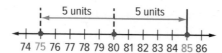

The distance from 80 to 75 is 5 units.
The distance from 80 to 85 is 5 units.

The solution set is {75, 85}. The maximum and minimum temperatures are 85° and 75°.

Ariel Skelley/Blend Images/Getty Images

Guided Practice

3. **ICE CREAM** Ice cream should be stored at 5°F with an allowance for 5°. Write and solve an equation to find the maximum and minimum temperatures at which ice cream should be stored.

When given two points on a graph, you can write an absolute value equation for the graph.

Example 4   Write an Absolute Value Equation

Write an equation involving absolute value for the graph.

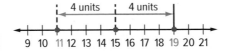

Find the point that is the same distance from 11 and from 19. This is the midpoint between 11 and 19, which is 15.

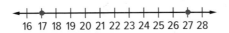

The distance from 15 to 11 is 4 units.
The distance from 15 to 19 is 4 units.

So an equation is $|x - 15| = 4$.

Guided Practice

4. Write an equation involving absolute value for the graph.

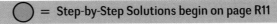

Check Your Understanding   ◯ = Step-by-Step Solutions begin on page R11

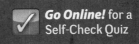
Go Online! for a Self-Check Quiz

Example 1   Evaluate each expression if $f = 3$, $g = -4$, and $h = 5$.

    1. $|3 - h| + 13$ **2.** $16 - |g + 9|$ **3.** $|f + g| - h$

Example 2   Solve each equation. Then graph the solution set.

    4. $|n + 7| = 5$ **5.** $|3z - 3| = 9$ **6.** $|4n - 1| = -6$

    7. $|b + 4| = 2$ **8.** $|2t - 4| = 8$ **9.** $|5h + 2| = -8$

Example 3   **10. FINANCIAL LITERACY** For a company to invest in a product, they need to receive a 12% return on investment (ROI) plus or minus 3%. Write an equation to find the least and the greatest ROI they need to receive.

Example 4   Write an equation involving absolute value for each graph.

12.

Example 1   Evaluate each expression if $a = -2$, $b = -3$, $c = 2$, $x = 2.1$, $y = 3$, and $z = -4.2$.

13 $|2x + z| + 2y$ **14.** $4a - |3b + 2c|$ **15.** $-|5a + c| + |3y + 2z|$

16. $-a + |2x - a|$ **17.** $|y - 2z| - 3$ **18.** $3|3b - 8c| - 3$

19. $|2x - z| + 6b$ **20.** $-3|z| + 2(a + y)$ **21.** $-4|c - 3| + 2|z - a|$

Example 2   Solve each equation. Then graph the solution set.

22. $|n - 3| = 5$ **23.** $|f + 10| = 1$ **24.** $|v - 2| = -5$

25. $|4t - 8| = 20$ **26.** $|8w + 5| = 21$ **27.** $|6y - 7| = -1$

28. $\left|\frac{1}{2}x + 5\right| = -3$ **29.** $|-2y + 6| = 6$ **30.** $\left|\frac{3}{4}a - 3\right| = 9$

Example 3   **31. SURVEY** The circle graph at the right shows the results of a survey that asked, "How likely is it that you will be rich some day?" If the margin of error is ±4%, what is the range of the percent of teens who say it is very likely that they will be rich?

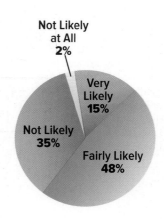

32. CHEERLEADING For competition, the cheerleading team is preparing a dance routine that must last 4 minutes, with a variation of ±5 seconds.

   a. Find the least and greatest possible times for the routine in minutes and seconds.

   b. Find the least and greatest possible times in seconds.

Example 4   Write an equation involving absolute value for each graph.

33.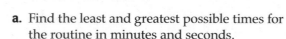

34.
```
-10-8-6-4-2  0  2  4  6  8 10
```

35.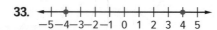

36.
```
-7-6-5-4-3-2-1  0  1  2  3
```

Solve each equation. Then graph the solution set.

37. $\left|-\frac{1}{2}b - 2\right| = 10$ **38.** $|-4d + 6| = 12$ **39.** $|5f - 3| = 12$

40. $2|h| - 3 = 8$ **41.** $4 - 3|q| = 10$ **42.** $\frac{4}{|p|} + 12 = 14$

43. MP SENSE-MAKING The 4 × 400 relay is a race where 4 runners take turns running 400 meters, or one lap around the track.

   a. If a runner runs the first leg in 52 seconds plus or minus 2 seconds, write an equation to find the fastest and slowest times.

   b. If the runners of the second and third legs run their laps in 53 seconds plus or minus 1 second, write an equation to find the fastest and slowest times.

   c. Suppose the runner of the fourth leg is the fastest on the team. If he runs an average of 50.5 seconds plus or minus 1.5 seconds, what are the team's fastest and slowest times?

44. FASHION To allow for a model's height, a designer is willing to use models that require him to change hems either up or down 2 inches. The length of the skirts is 20 inches.

  a. Write an absolute value equation that represents the length of the skirts.

  b. What is the range of the lengths of the skirts?

  c. If a 20-inch skirt was fitted for a model that is 5 feet 9 inches tall, will the designer use a 6-foot-tall model?

45. (MP) **PRECISION** Speedometer accuracy can be affected by many details such as tire diameter and axle ratio. For example, there is variation of ±3 miles per hour when calibrated at 50 miles per hour.

  a. What is the range of actual speeds of the car if calibrated at 50 miles per hour?

  b. A speedometer calibrated at 45 miles per hour has an accepted variation of ±1 mile per hour. What can we conclude from this?

Write an equation involving absolute value for each graph.

46.

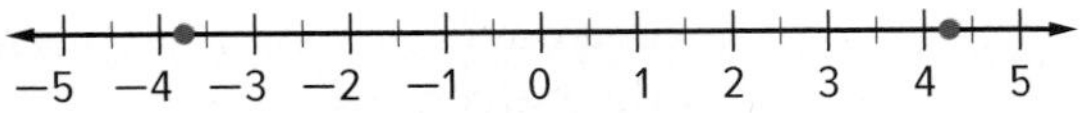

47.

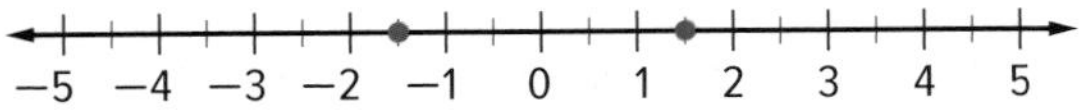

48.

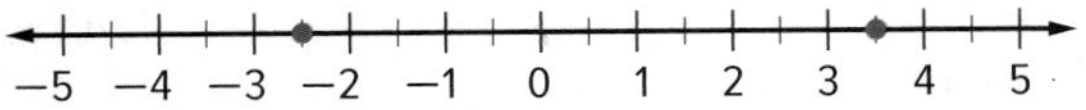

49.

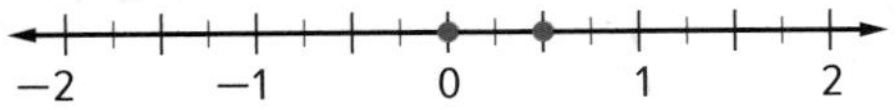

50.

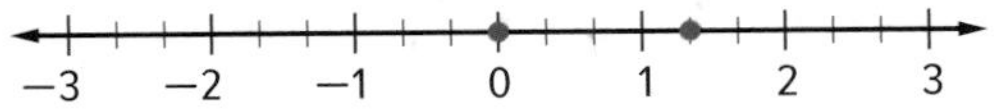

51.

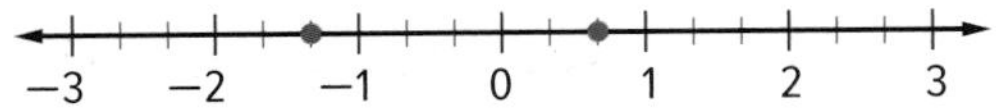

52. PHOTOS A smartphone is advertised to hold up to 1250 photos plus or minus 150, depending on quality.

  a. Write an absolute value equation that represents the number of photos that can be stored on the smartphone.

  b. What is the range in photos that can be stored on the smartphone?

  c. Graph the possible solutions on a number line.

53 ACOUSTICS The Red Rocks Amphitheater located in the Red Rock Park near Denver, Colorado, is the only naturally occurring amphitheater. The acoustic qualities here are such that a maximum of 20,000 people, plus or minus 1000, can hear natural voices clearly.

  a. Write an equation involving an absolute value that represents the number of people that can hear natural voices at Red Rocks Amphitheater.

  b. Find the maximum and minimum number of people that can hear natural voices clearly in the amphitheater.

  c. What is the range of people in part **b**?

54. BOOK CLUB The members of a book club agree to read within ten pages of the last page of a chapter. The chapter ends on page 203.

    a. Write an absolute value equation that represents the pages where club members could stop reading.

    b. Write the range of the pages where the club members could stop reading.

55. SCHOOL Teams from Washington and McKinley High Schools are competing in an academic challenge. A correct response on a question earns 10 points and an incorrect response loses 10 points. A team earns 0 points on an unattempted question. There are 5 questions in the math section.

    a. What are the maximum and minimum scores a team can earn on the math section?

    b. Suppose the Jasper team has 160 points at the start of the math section. Write and solve an equation that represents the maximum and minimum scores the team could have at the end of the math section.

    c. What are all of the possible scores that a school can earn on the math section?

H.O.T. Problems   Use Higher-Order Thinking Skills

56. OPEN ENDED Describe a real-world situation that could be represented by the absolute value equation $|x - 4| = 10$.

MP STRUCTURE Determine whether the following statements are *sometimes*, *always*, or *never* true, if c is an integer. Explain your reasoning.

57. The value of $|x + 1|$ is greater than zero.

58. The solution of $|x + c| = 0$ is greater than 0.

59. The inequality $|x| + c < 0$ has no solution.

60. The value of $|x + c| + c$ is greater than zero.

61. MP REASONING Explain why an absolute value can never be negative.

62. CHALLENGE Use the sentence $x = 7 \pm 4.6$.

    a. Describe the values of x that make the sentence true.

    b. Translate the sentence into an equation involving absolute value.

63. ERROR ANALYSIS Alex and Wesley are solving $|x + 5| = -3$. Is either of them correct? Explain your reasoning.

Alex	Wesley
$|x+5| = 3$ or $|x+5| = -3$ $x + 5 = 3$ $x + 5 = -3$ $\underline{-5 \quad -5}$ $\underline{-5 \quad -5}$ $x = -2$ $x = -8$	$|x + 5| = -3$ The solution is $\emptyset$.

64. WRITING IN MATH Explain why there are either two, one, or no solutions for absolute value equations. Demonstrate an example of each possibility.

65. The distance between an unknown length and 7 inches is 2 inches.

   a. Write an absolute value equation for the unknown length. **MP** 1

     ○ **A** $|x - 2| = 7$

     ○ **B** $|x + 2| = 7$

     ○ **C** $|x - 7| = 2$

     ○ **D** $|x + 7| = 2$

   b. What are the two possible lengths for the unknown? **MP** 1

     [   ]

66. What are the solutions to the equation $|3x - 1| = -5$? **MP** 1

     ○ **A** $x = \frac{5}{3}, x = -\frac{5}{3}$

     ○ **B** $x = 2, x = -\frac{4}{3}$

     ○ **C** $x = 5, x = -5$

     ○ **D** No solution

67. What are the solutions to the equation $|3m - 7| = 2$? **MP** 1

     ○ **A** $\frac{2}{3}, -\frac{2}{3}$

     ○ **B** $3, -3$

     ○ **C** $\frac{7}{3}, 2$

     ○ **D** $3, \frac{5}{3}$

68. The two solutions to an absolute value equation are -2 and 15. Which absolute value equation has those solutions? **MP** 1

     ○ **A** $|x - 6.5| = -8.5$

     ○ **B** $|x - 6.5| = 8.5$

     ○ **C** $|x - 8.5| = 6.5$

     ○ **D** $|x + 2| = 15$

69. MULTI-STEP A sports arena can hold 18,500 people, plus or minus 1200 people. **MP** 1, 3, 4, 6

   a. Write an absolute value equation to represent the number of people that can attend an event.

     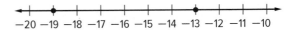

   b. Explain the steps to solve the equation.

   c. What is the maximum number of people?

   d. What is the minimum number of people?

70. The graph below shows two points on a number line.

<!-- number line from -20 to -10 with points at -19 and -13 -->
$$\overset{\text{−20 −19 −18 −17 −16 −15 −14 −13 −12 −11 −10}}{\longleftrightarrow}$$

Which equation best represents the graph? **MP** 1

     ○ **A** $|x + 3| = 16$

     ○ **B** $|x - 16| = 3$

     ○ **C** $|x + 15| = 3$

     ○ **D** $|x + 16| = 3$

71. The distance of x from -2 is 8. **MP** 1

   a. Write an equation to model x. [   ]

   b. What is the minimum value for x?

     [   ]

   c. What is the maximum value for x?

     [   ]

72. What are the two solutions of the equation, $|3x - 1| = 5$? **MP** 1

73. Given that $|x - 1| + |y - 3| = 0$, find the values of x and y. **MP** 1

     [   ]

74. Two numbers are each a distance of 3 units away from another number. The other number is 5. Write an absolute value equation to represent the two numbers. **MP** 2

Translate each sentence into an equation. (Lesson 2-1)

1. The sum of three times a and four is the same as five times a.

2. One fourth of m minus six is equal to two times the sum of m and 9.

3. The product of five and w is the same as w to the third power.

4. **MARBLES** Drew has 50 red, green, and blue marbles. He has six more red marbles than blue marbles and four fewer green marbles than blue marbles. Write and solve an equation to determine how many blue marbles Drew has. (Lesson 2-2)

Solve each equation. Check your solution. (Lesson 2-2)

5. $p + 8 = 13$

6. $-26 = b - 3$

7. $\dfrac{t}{6} = 3$

8. **MULTIPLE CHOICE** Solve the equation $\dfrac{3}{5}a = \dfrac{1}{4}$. (Lesson 2-2)

   A -3 **C** $\dfrac{5}{12}$

   B $\dfrac{3}{20}$ **D** 2

Solve each equation. Check your solution. (Lesson 2-3)

9. $2x + 5 = 13$

10. $-21 = 7 - 4y$

11. $\dfrac{m}{6} - 3 = 8$

12. $-4 = \dfrac{d + 3}{5}$

13. **FISH** The average length of a yellow-banded angelfish is 12 inches. This is 4.8 times as long as an average common goldfish. (Lesson 2-3)

   a. Write an equation you could use to find the length of the average common goldfish.

   b. What is the length of an average common goldfish?

   c. ⓂⓅ Which mathematical practice did you use to solve this problem?

Write an equation and solve each problem. (Lesson 2-3)

14. Three less than three fourths of a number is negative 9. Find the number.

15. Thirty is twelve added to six times a number. What is the number?

16. Find four consecutive integers with a sum of 106.

Solve each equation. Check your solution. (Lesson 2-4)

17. $8p + 3 = 5p + 9$

18. $\dfrac{3}{4}w + 6 = 9 - \dfrac{1}{4}w$

19. $\dfrac{z + 6}{3} = \dfrac{2z}{4}$

20. **PERIMETER** Find the value of x so that the triangles have the same perimeter. (Lesson 2-4)

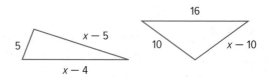

21. **PRODUCTION** ABC Sporting Goods Company produces baseball gloves. Their fixed monthly production cost is $16,000 with a per glove cost of $15. XYZ Sporting Goods Company also produces baseball gloves. Their fixed monthly production cost is $20,000 with a per glove cost of $13. Find the number of gloves produced monthly, so that the total monthly production cost is the same for both companies. (Lesson 2-4)

Evaluate each expression if $x = -4$, $y = 7$, and $z = -9$. (Lesson 2-5)

22. $|3x - 2| + 2y$

23. $|-4y + 2z| - 7z$

24. **MULTIPLE CHOICE** Solve $|6m - 3| = 9$. (Lesson 2-5)

   A $\{2\}$ **C** $\{-3, 6\}$

   B $\{-1, 2\}$ **D** $\{-3, 3\}$

25. **COFFEE** Some say to brew an excellent cup of coffee, you must have a brewing temperature of 200° F, plus or minus 5 degrees. Write and solve an equation describing the maximum and minimum brewing temperatures for an excellent cup of coffee. (Lesson 2-5)

Ratios and Proportions

::Then

- You evaluated percents by using a proportion.

::Now

1. Compare ratios.
2. Solve proportions.

::Why?

- Ratios allow us to compare many items by using a common reference. The table below shows the number of restaurants a certain popular fast food chain has per 10,000 people in the United States as well as other countries. This allows us to compare the number of these restaurants using an equal reference.

Countries	United States	Canada	Australia	Japan	United Kingdom	Taiwan
Number of Restaurants per 10,000 People	0.413	0.337	0.313	0.282	0.180	0.145

New Vocabulary

ratio
proportion
means
extremes
rate
unit rate
scale
scale model

MP Mathematical Practices

1 Make sense of problems and persevere in solving them.

3 Construct viable arguments and critique reasoning of others.

4 Model with mathematics.

7 Look for and make use of structure.

1 Ratios and Proportions The comparison between the number of restaurants and the number of people is a ratio. A **ratio** is a comparison of two numbers by division. The ratio of x to y can be expressed in the following ways.

$$x \text{ to } y \qquad x : y \qquad \frac{x}{y}$$

Suppose you wanted to determine the number of restaurants per 100,000 people in Australia. Notice that this ratio is equal to the original ratio.

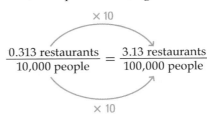

$$\frac{0.313 \text{ restaurants}}{10,000 \text{ people}} = \frac{3.13 \text{ restaurants}}{100,000 \text{ people}}$$

An equation stating that two ratios are equal is called a **proportion**. So, we can state that $\frac{0.313}{10,000} = \frac{3.13}{100,000}$ is a proportion.

Example 1   Determine Whether Ratios Are Equivalent

Determine whether $\frac{2}{3}$ and $\frac{16}{24}$ are equivalent ratios. Write *yes* or *no*. Justify your answer.

When expressed in simplest form, the ratios are equivalent.

Guided Practice

Determine whether each pair of ratios are equivalent ratios. Write *yes* or *no*. Justify your answer.

1A. $\frac{6}{10}, \frac{2}{5}$

1B. $\frac{1}{6}, \frac{5}{30}$

Ken Karp/McGraw-Hill Education

There are special names for the terms in a proportion.

1.5 and 1.2 are called the **means**. They are the middle terms of the proportion.

$$0.2 : 1.5 \;=\; 1.2 : 9.0$$

0.2 and 9.0 are called the **extremes**. They are the first and last terms of the proportion.

🔑 **Key Concept** Means-Extremes Property of Proportion

Words   In a proportion, the product of the extremes is equal to the product of the means.

Symbols   If $\frac{a}{b} = \frac{c}{d}$ and $b, d \neq 0$, then $ad = bc$.

Examples   Since $\frac{2}{4} = \frac{1}{2}$, $2(2) = 4(1)$ or $4 = 4$.

Another way to determine whether two ratios form a proportion is to use cross products. If the cross products are equal, then the ratios form a proportion.

This is the same as multiplying the means and multiplying the extremes.

Example 2   **Cross Products**

Use cross products to determine whether each pair of ratios forms a proportion.

a. $\frac{2}{3.5}, \frac{8}{14}$

$\frac{2}{3.5} \stackrel{?}{=} \frac{8}{14}$ Original proportion

$2(14) \stackrel{?}{=} 3.5(8)$ Cross products

$28 = 28$ ✓   Simplify.

The cross products are equal, so the ratios form a proportion.

b. $\frac{0.3}{1.5}, \frac{0.5}{2.0}$

$\frac{0.3}{1.5} \stackrel{?}{=} \frac{0.5}{2.0}$ Original proportion

$0.3(2.0) \stackrel{?}{=} 1.5(0.5)$ Cross products

$0.6 \neq 0.75$ ✗   Simplify.

The cross products are not equal, so the ratios do not form a proportion.

▶ **Guided Practice**

2A. $\frac{0.2}{1.8}, \frac{1}{0.9}$ **2B.** $\frac{15}{36}, \frac{35}{42}$

2 Solve Proportions To solve proportions, use cross products.

Example 3   Solve a Proportion

Solve each proportion. If necessary, round to the nearest hundredth.

a. $\dfrac{x}{10} = \dfrac{3}{5}$

$\dfrac{x}{10} = \dfrac{3}{5}$ Original proportion

$x(5) = 10(3)$ Find the cross products.

$5x = 30$ Simplify.

$\dfrac{5x}{5} = \dfrac{30}{5}$ Divide each side by 5.

$x = 6$ Simplify.

b. $\dfrac{x-2}{14} = \dfrac{2}{7}$

$\dfrac{x-2}{14} = \dfrac{2}{7}$ Original proportion

$(x-2)7 = 14(2)$ Find the cross products.

$7x - 14 = 28$ Simplify.

$7x = 42$ Add 14 to each side.

$x = 6$ Divide each side by 7.

▶ **Guided Practice**

3A. $\dfrac{r}{8} = \dfrac{25}{40}$ **3B.** $\dfrac{x+4}{5} = \dfrac{3}{8}$

The ratio of two measurements having different units of measure is called a **rate**. For example, a price of $9.99 per 10 songs is a rate. A rate that tells how many of one item is being compared to 1 of another item is called a **unit rate**.

Real-World Example 4   Rate of Growth

RETAIL In the past two years, a retailer has opened 232 stores. If that rate remains constant, how many stores will open in the next 3 years?

Understand Let r represent the number of retail stores.

Plan   Write a proportion for the problem.

$$\dfrac{232 \text{ retail stores}}{2 \text{ years}} = \dfrac{r \text{ retail stores}}{3 \text{ years}}$$

Solve $\dfrac{232}{2} = \dfrac{r}{3}$ Original proportion

$232(3) = 2r$ Find the cross products.

$696 = 2r$ Simplify.

$\dfrac{696}{2} = \dfrac{2r}{2}$ Divide each side by 2.

$348 = r$ Simplify.

The retailer will open 348 stores in 3 years.

Check   The retailer opens $232 \div 2$ or 116 stores per year and $3 \times 116 = 348$.

4. **EXERCISE** It takes 7 minutes for Isabella to walk around the gym track twice. At this rate, how many times can she walk around the track in a half hour?

A rate called a **scale** is used to make a **scale model** of something too large or too small to be convenient at actual size.

Real-World Example 5   Scale and Scale Models

MOUNTAIN TRAIL The Ramsey Cascades Trail is about $1\frac{1}{8}$ inches long on a map with scale 3 inches = 10 miles. What is the actual length of the trail?

Let ℓ represent the actual length.

$$\text{scale} \longrightarrow \frac{3}{10} = \frac{1\frac{1}{8}}{\ell} \longleftarrow \text{scale}$$
$$\text{actual} \longrightarrow \qquad\quad \longleftarrow \text{actual}$$

$$3(\ell) = 1\frac{1}{8}(10) \qquad \text{Find the cross products.}$$

$$3\ell = \frac{45}{4} \qquad \text{Simplify.}$$

$$3\ell \div 3 = \frac{45}{4} \div 3 \qquad \text{Divide each side by 3.}$$

$$\ell = \frac{15}{4} \text{ or } 3\frac{3}{4} \qquad \text{Simplify.}$$

The actual length is about $3\frac{3}{4}$ miles.

Real-World Link

The Great Smoky Mountains National Park in Tennessee is home to several waterfalls. The Ramsey Cascades is 100 feet tall. It is the tallest in the park.

Source: National Park Service

▶ **Guided Practice**

5. **AIRPLANES** On a model airplane, the scale is 5 centimeters = 2 meters. If the model's wingspan is 28.5 centimeters, what is the actual wingspan?

Check Your Understanding   ◯ = Step-by-Step Solutions begin on page R11

✓ *Go Online!* for a Self-Check Quiz

Examples 1–2   **Determine whether each pair of ratios are equivalent ratios. Write *yes* or *no*.**

1. $\frac{3}{7}, \frac{9}{14}$

2. $\frac{7}{8}, \frac{42}{48}$

3 $\frac{2.8}{4.4}, \frac{1.4}{2.1}$

Example 3   **Solve each proportion. If necessary, round to the nearest hundredth.**

4. $\frac{n}{9} = \frac{6}{27}$

5. $\frac{4}{u} = \frac{28}{35}$

6. $\frac{3}{8} = \frac{b}{10}$

Example 4   7. **RACE** Jennie ran the first 6 miles of a marathon in 58 minutes. If she is able to maintain the same pace, how long will it take her to finish the 26.2 miles?

Example 5   8. **(MP) PRECISION** On a map, the Grand Canyon and Las Vegas are about 12 inches apart. If the scale is 1 inch = 22.5 miles, how far apart are the places?

Examples 1–2 Determine whether each pair of ratios are equivalent ratios. Write *yes* or *no*.

9. $\dfrac{9}{11}, \dfrac{81}{99}$

10. $\dfrac{3}{7}, \dfrac{18}{42}$

11. $\dfrac{8.4}{9.2}, \dfrac{8.8}{9.6}$

12. $\dfrac{4}{3}, \dfrac{6}{8}$

13. $\dfrac{29.2}{10.4}, \dfrac{7.3}{2.6}$

14. $\dfrac{39.68}{60.14}, \dfrac{6.4}{9.7}$

Example 3 Solve each proportion. If necessary, round to the nearest hundredth.

15. $\dfrac{3}{8} = \dfrac{15}{a}$

16. $\dfrac{t}{2} = \dfrac{6}{12}$

17. $\dfrac{4}{9} = \dfrac{13}{q}$

18. $\dfrac{15}{35} = \dfrac{g}{7}$

19. $\dfrac{7}{10} = \dfrac{m}{14}$

20. $\dfrac{8}{13} = \dfrac{v}{21}$

21. $\dfrac{w}{2} = \dfrac{4.5}{6.8}$

22. $\dfrac{1}{0.19} = \dfrac{12}{n}$

23. $\dfrac{2}{0.21} = \dfrac{8}{n}$

24. $\dfrac{2.4}{3.6} = \dfrac{k}{1.8}$

25. $\dfrac{t}{0.3} = \dfrac{1.7}{0.9}$

26. $\dfrac{7}{1.066} = \dfrac{z}{9.65}$

27. $\dfrac{x-3}{5} = \dfrac{6}{10}$

28. $\dfrac{7}{x+9} = \dfrac{21}{36}$

29. $\dfrac{10}{15} = \dfrac{4}{x-5}$

Example 4 30. **CAR WASH** The B-Clean Car Wash washed 128 cars in 3 hours. At that rate, how many cars can they wash in 8 hours?

Example 5 31. **GEOGRAPHY** On a map of Florida, the distance between Jacksonville and Tallahassee is 2.6 centimeters. If 2 centimeters = 120 miles, what is the distance between the two cities?

32. **MP PRECISION** An artist used interlocking building blocks to build a scale model of Kennedy Space Center, Florida. In the model, 1 inch equals 1.67 feet of an actual space shuttle. The model is 110.3 inches tall. How tall is the actual space shuttle? Round to the nearest tenth.

33. **MENU** On Monday, a restaurant made $545 from selling 110 hamburgers. If they sold 53 hamburgers on Tuesday, how much did they make?

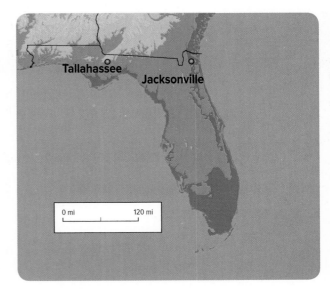

Solve each proportion. If necessary, round to the nearest hundredth.

34. $\dfrac{6}{14} = \dfrac{7}{x-3}$

35. $\dfrac{7}{4} = \dfrac{f-4}{8}$

36. $\dfrac{3-y}{4} = \dfrac{1}{9}$

37. $\dfrac{4v+7}{15} = \dfrac{6v+2}{10}$

38. $\dfrac{9b-3}{9} = \dfrac{5b+5}{3}$

39. $\dfrac{2n-4}{5} = \dfrac{3n+3}{10}$

40. **ATHLETES** At Piedmont High School, 3 out of every 8 students are athletes. If there are 1280 students at the school, how many are not athletes?

41. **BRACES** Two out of five students in the ninth grade have braces. If there are 325 students in the ninth grade, how many have braces?

42. **PAINT** Joel used a half gallon of paint to cover 84 square feet. He has 932 more square feet to paint. How many additional gallons of paint should he purchase?

43 **MOVIE THEATERS** Use the table at the right.

a. Write a ratio of the number of indoor theaters to the total number of theaters for each year.

b. Do any two of the ratios you wrote for part **a** form a proportion? If so, explain the real-world meaning of the proportion.

44. **DIARIES** In a survey, 36% of the students said that they kept an electronic diary. There were 900 students who kept an electronic diary. How many students were in the survey?

Year	Indoor	Drive-In	Total
2005	37,040	648	37,688
2006	37,765	650	38,415
2007	38,159	635	38,794
2008	38,201	633	38,834
2009	38,605	628	39,233
2010	38,902	618	39,520
2011	38,974	606	39,580

Source: North American Theater Owners

45. **MULTIPLE REPRESENTATIONS** In this problem, you will explore how changing the lengths of the sides of a shape by a factor changes the perimeter of that shape.

a. **Geometric** Draw a square *ABCD*. Draw a square *MNPQ* with sides twice as long as *ABCD*. Draw a square *FGHJ* with sides half as long as *ABCD*.

b. **Tabular** Complete the table below using the appropriate measures.

ABCD		MNPQ		FGHJ	
Side length		Side length		Side length	
Perimeter		Perimeter		Perimeter	

c. **Verbal** Make a conjecture about the change in the perimeter of a square if the side length is increased or decreased by a factor.

H.O.T. Problems   Use **H**igher-**O**rder **T**hinking Skills

46. **MP STRUCTURE** An agricultural survey determined that 353,917 milk cows produced 768 million pounds of milk. Divide one of these numbers by the other and explain the meaning of the result.

47. **MP REASONING** Compare and contrast ratios and rates.

48. **CHALLENGE** If $\frac{a+1}{b-1} = \frac{5}{1}$ and $\frac{a-1}{b+1} = \frac{1}{1}$, find the value of $\frac{b}{a}$. $\left(\textit{Hint: } \text{Choose values of } a \text{ and } b \text{ for which the proportions are true and evaluate } \frac{b}{a}.\right)$

49. **WRITING IN MATH** On a road trip, Marcus reads a highway sign and then looks at his gas gauge.

Marcus's gas tank holds 10 gallons and his car gets 32 miles per gallon at his current speed of 65 miles per hour. If he maintains this speed, will he make it to Atlanta without having to stop and get gas? Explain your reasoning.

50. **WRITING IN MATH** Describe how businesses can use ratios. Write about a real-world situation in which a business would use a ratio.

51. A scale drawing of a car is $3\frac{1}{4}$ inches long. If the scale is 1 inch = $3\frac{1}{2}$ feet, what is the length of the actual car? **MP** 1, 4

- ○ **A** $9\frac{1}{8}$ feet
- ○ **B** $10\frac{1}{8}$ feet
- ○ **C** $11\frac{3}{8}$ feet
- ○ **D** $11\frac{3}{8}$ inches

52. For what value of p is the ratio $\frac{5p-1}{11}$ equal to the ratio $\frac{3-5p}{6}$? **MP** 1

- ○ **A** $-\frac{39}{25}$
- ○ **B** $\frac{2}{5}$
- ○ **C** $\frac{39}{85}$
- ○ **D** $\frac{34}{35}$

53. The state of Colorado is almost rectangular, and the ratio of its north-south length to its east-west length is $\frac{14}{19}$. If its east-west length is about 380 miles, what is its north-south length?

a. Write a proportion for this relationship using x as the unknown length. **MP** 4

b. Find the actual north-south length. **MP** 1

- ○ **A** 206 miles
- ○ **B** 280 miles
- ○ **C** 394 miles
- ○ **D** 399 miles

54. What is the solution of the proportion $\frac{5x+3}{5} = \frac{7-2x}{3}$? **MP** 1

55. **MULTI-STEP** At a school, the school population is $\frac{2}{5}$ boys. There are 450 students at the school who are boys. **MP** 1, 3, 4, 7

a. Write a proportion to represent the number of girls in the school.

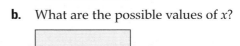

b. Explain the steps to solve the equation.

c. How many girls are in the school?

d. How many total students are in the school?

56. The ratio of an integer to 4 is the same as the ratio of 9 to that integer. **MP** 1

a. Write an equation to model this situation.

b. What are the possible values of x?

57. The ratio of girls to boys in a class is 4:3. There are 3 more girls than boys in the class. **MP** 1

a. Write a proportion to model this situation.

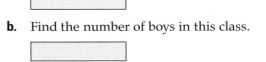

b. Find the number of boys in this class.

58. A bag contains quarters and dimes in a ratio of 3:5. If there is $6 in quarters in the bag, how many dimes are there? **MP** 1

59. If a map has a scale of 3 inches = 150 miles, how many inches represents 200 miles? **MP** 3

60. Two consecutive multiples of 6 are a ratio of 8:9. Find the sum of the two numbers. **MP** 1

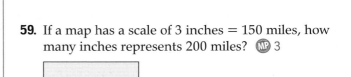

Literal Equations and Dimensional Analysis

∷ Then

- You solved equations with variables on each side.

∷ Now

1. Solve equations for given variables.
2. Use formulas to solve real-world problems.

∷ Why?

- A research station run by the National Oceanic and Atmospheric Administration (NOAA) observes waves to determine the strength of a storm. To do this, they measure the vertical distance from high point to low point and the horizontal distance between adjacent wave crests.

  The formula $v = f \cdot \lambda$ describes the relation between speed, frequency, and wavelength, where
  - v is the speed of a wave,
  - f is the frequency of the wave, and
  - λ (lambda) is wavelength of the wave.

 New Vocabulary
literal equation
dimensional analysis
unit analysis

 Mathematical Practices
4 Model with mathematics.
6 Attend to precision.
7 Look for and make use of structure.

1 Solve for a Specific Variable Some equations such as the one above contain more than one variable. At times, you will need to solve these equations for one of the variables.

Example 1   Solve for a Specific Variable

Solve $4m - 3n = 8$ for m.

$4m - 3n = 8$	Original equation
$4m - 3n + 3n = 8 + 3n$	Add $3n$ to each side.
$4m = 8 + 3n$	Simplify.
$\dfrac{4m}{4} = \dfrac{8 + 3n}{4}$	Divide each side by 4.
$m = \dfrac{8}{4} + \dfrac{3}{4}n$	Simplify.
$m = 2 + \dfrac{3}{4}n$	Simplify.

▶ **Guided Practice**

Solve each equation for the variable indicated.

1A. $15 = 3n + 6p$, for n **1B.** $\dfrac{k-2}{5} = 11j$, for k

1C. $28 = t(r + 4)$, for t **1D.** $a(q - 8) = 23$, for q

Sometimes we need to solve equations for a variable that is on both sides of the equation. When this happens, you must get all terms with that variable onto one side of the equation. It is then helpful to use the Distributive Property to isolate the variable for which you are solving.

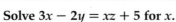

Example 2 Solve for a Specific Variable

Solve $3x - 2y = xz + 5$ for x.

$3x - 2y = xz + 5$	Original equation
$3x - 2y + 2y = xz + 5 + 2y$	Add $2y$ to each side.
$3x - xz = xz - xz + 5 + 2y$	Subtract xz from each side.
$3x - xz = 5 + 2y$	Simplify.
$x(3 - z) = 5 + 2y$	Distributive Property
$\dfrac{x(3 - z)}{3 - z} = \dfrac{5 + 2y}{3 - z}$	Divide each side by $3 - z$.
$x = \dfrac{5 + 2y}{3 - z}$	Simplify.

Since division by 0 is undefined, $3 - z \neq 0$ so $z \neq 3$.

Guided Practice

Solve each equation for the variable indicated.

2A. $d + 5c = 3d - 1$, for d **2B.** $6q - 18 = qr + t$, for q

---

2 **Use Formulas** An equation that involves several variables is called a formula or **literal equation**. To solve a literal equation, apply the process of solving for a specific variable.

Real-World Example 3 Use Literal Equations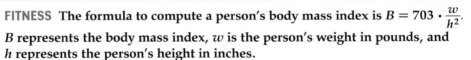

FITNESS The formula to compute a person's body mass index is $B = 703 \cdot \dfrac{w}{h^2}$. B represents the body mass index, w is the person's weight in pounds, and h represents the person's height in inches.

a. Solve the formula for w.

$B = 703 \cdot \dfrac{w}{h^2}$	Formula for body mass index
$\dfrac{B}{703} = \dfrac{703}{703} \cdot \dfrac{w}{h^2}$	Divide each side by 703.
$\dfrac{Bh^2}{703} = \dfrac{w}{h^2}\,(h^2)$	Multiply each side by h^2.
$\dfrac{Bh^2}{703} = w$	Simplify.

b. What is the weight of a person who is 64 inches tall and has a body mass index of 21.45?

$\dfrac{Bh^2}{703} = w$	Formula for weight
$\dfrac{87{,}859.2}{703} = w$	$B = 21.45, h = 64$
$125 \approx w$	Use a calculator.

The person weighs about 125 pounds.

Guided Practice

3. ELECTRICITY Ohm's Law states that $I = \dfrac{V}{R}$, where I is current in amperes, V is volts, and R is resistance in ohms.

    A. Solve the formula for R.

    B. Find the resistance of a circuit with 22 volts and 0.5 amp.

When using formulas, you may want to use dimensional analysis. **Dimensional analysis** or **unit analysis** is the process of carrying units throughout a computation.

Example 4 Use Dimensional Analysis

RUNNING A 10K run is 10 kilometers long. If 1 meter = 1.094 yards, use dimensional analysis to find the length of the race in miles. (*Hint*: 1 mi = 1760 yd)

Since the given conversion relates meters to yards, first convert 10 kilometers to meters. Then multiply by the conversion factor so the unit meters is divided out. To convert from yards to miles, multiply by $\frac{1 \text{ mi}}{1760 \text{ yd}}$.

length of run	$\times$	kilometers to meters	$\times$	meters to yards	$\times$	yards to miles
10 km	$\times$	$\dfrac{1000 \text{ m}}{1 \text{ km}}$	$\times$	$\dfrac{1.094 \text{ yd}}{1 \text{ m}}$	$\times$	$\dfrac{1 \text{ mi}}{1760 \text{ yd}}$

Notice how the units cancel, leaving the unit to which you are converting.

$$10 \text{ km} \times \frac{1000 \text{ m}}{1 \text{ km}} \times \frac{1.094 \text{ yd}}{1 \text{ m}} \times \frac{1 \text{ mi}}{1760 \text{ yd}} = \frac{10,940 \text{ mi}}{1760}$$

$$\approx 6.2 \text{ mi}$$

A 10K race is approximately 6.2 miles.

Study Tip

MP **PRECISION** As you plan your method of solution, think about what the question is asking and what units of measure will apply to the solution.

Guided Practice

4. A car travels a distance of 100 feet in about 2.8 seconds. What is the velocity of the car in miles per hour? Round to the nearest whole number.

Check Your Understanding   ◯ = **Step-by-Step Solutions begin on page R11**

Go Online! for a Self-Check Quiz

Examples 1–2 **Solve each equation or formula for the variable indicated.**

 1. $5a + c = -8a$, for a

2. $7h + f = 2h + g$, for g

3. $\dfrac{k + m}{-7} = n$, for k

4. $q = p(r + s)$, for p

Example 3

5. PACKAGING A soap company wants to use a cylindrical container to hold their new liquid soap.

 a. Solve the formula for h.

 b. What is the height of a container if the volume is 56.52 cubic inches and the radius is 1.5 inches? Round to the nearest tenth.

$V = \pi r^2 h$

Example 4

6. SHOPPING Scott found a rare comic book on an online auction site priced at 35 Australian dollars. If the exchange rate is \$1 U.S. = \$1.09 Australian, find the cost of the comic book in United States dollars. Round to the nearest cent.

7. MP **PRECISION** A fisheye lens has a minimum focus range of 13.5 centimeters. If 1 centimeter is equal in length to about 0.39 inch, what is the minimum focus range of the lens in feet?

Examples 1–2 **Solve each equation or formula for the variable indicated.**

8. $u = vw + z$, for v

9 $x = b - cd$, for c

10. $fg - 9h = 10j$, for g

11. $10m - p = -n$, for m

12. $r = \frac{2}{3}t + v$, for t

13. $\frac{5}{9}v + w = z$, for v

14. $\frac{10ac - x}{11} = -3$, for a

15. $\frac{df + 10}{6} = g$, for f

Example 3 **16. YO-YOS** The largest yo-yo in the world is 37.8 feet in circumference. The formula for the circumference of a circle is $C = 2\pi r$, where C represents circumference and r represents radius.

   a. Solve the formula for r.

   b. Find the radius of the yo-yo?

17. PHYSICS Acceleration is the measure of how fast a velocity is changing. The formula for acceleration is $a = \frac{v_f - v_i}{t}$. a represents the acceleration rate, v_f is the final velocity, v_i is the initial velocity, and t represents the time in seconds.

   a. Solve the formula for v_f.

   b. What is the final velocity of a runner who is accelerating at 2 feet per second squared for 3 seconds with an initial velocity of 4 feet per second?

Example 4 **18. SWIMMING** If each lap in a pool is 100 meters long, how many laps equal one mile? Round to the nearest tenth. (*Hint*: 1 foot ≈ 0.3048 meter)

19. MP **PRECISION** How many liters of gasoline are needed to fill a 13.2-gallon tank? There are about 1.06 quarts per 1 liter. Round to the nearest tenth.

Solve each equation or formula for the variable indicated.

20. $-14n + q = rt - 4n$, for n

21. $18t + 11v = w - 13t$, for t

22. $ax + z = aw - y$, for a

23. $10c - f = -13 + cd$, for c

Select an appropriate unit from the choices below and convert the rate to that unit.

| ft/s | mph | mm/s | km/s |

24. a car traveling at 36 ft/s

25. a snail moving at 3.6 m/h

26. a person walking at 3.4 mph

27. a satellite moving at 234,000 m/min

28. BASEBALL The formula for a pitcher's earned run average, or ERA, is $a = \frac{9r}{p}$. In this formula, a is the earned run average, r is earned runs, and p is innings pitched.

   a. Solve the formula for r.

   b. Find the earned runs for a pitcher with an ERA of 2.63 who has pitched 89 innings.

Write an equation and solve for the variable indicated.

29. Seven less than a number t equals another number r plus 6. Solve for t.

30. Ten plus eight times a number a equals eleven times number d minus six. Solve for a.

31. Nine tenths of a number g is the same as seven plus two thirds of another number k. Solve for k.

32. Three fourths of a number p less two is five sixths of another number r plus five. Solve for r.

33. **GIFTS** Ashley has 214 square inches of paper to wrap a gift box. The surface area S of the box can be found by using the formula $S = 2w(\ell + h) + 2\ell h$, where w is the width of the box, ℓ is the length of the box, and h is the height. If the length of the box is 7 inches and the width is 6 inches, how tall can Ashley's box be?

34. **DRIVING** A car is driven x miles a year and averages m miles per gallon.

   a. Write a formula for g, the number of gallons used in a year.

   b. If the average price of gas is p dollars per gallon, write a formula for the total gas cost c in dollars for driving this car each year.

   c. Car A averages 15 miles per gallon on the highway, while Car B averages 35 miles per gallon on the highway. If you average 15,000 miles each year, how much money would you save on gas per week by using Car B instead of Car A if the cost of gas averages $3 per gallon? Explain.

H.O.T. Problems   Use Higher-Order Thinking Skills

35. **CHALLENGE** The circumference of an NCAA women's basketball is 29 inches, and the rubber coating is $\frac{3}{16}$ inch thick. Use the formula $v = \frac{4}{3}\pi r^3$, where v represents the volume and r is the radius of the inside of the ball, to determine the volume of the air inside the ball. Round to the nearest whole number.

36. **MP REASONING** Select an appropriate unit to describe the highway speed of a car and the speed of a crawling caterpillar. Should the same unit be used for both? Explain.

37. **ERROR ANALYSIS** Sandrea and Fernando are solving $4a - 5b = 7$ for b. Is either of them correct? Explain.

Sandrea	Fernando
$4a - 5b = 7$	$4a - 5b = 7$
$-5b = 7 - 4a$	$5b = 7 - 4a$
$\dfrac{-5b}{-5} = \dfrac{7 - 4a}{-5}$	$\dfrac{5b}{5} = \dfrac{7 - 4a}{5}$
$b = \dfrac{7 - 4a}{-5}$	$b = \dfrac{7 - 4a}{5}$

38. **OPEN ENDED** Write a formula for A, the area of a geometric figure such as a triangle or rectangle. Then solve the formula for a variable other than A.

39. **MP PERSEVERANCE** Solve each equation or formula for the variable indicated.

   a. $n = \dfrac{x + y - 1}{xy}$ for x

   b. $\dfrac{x + y}{x - y} = \dfrac{1}{2}$ for y

40. **ⓔ WRITING IN MATH** Why is it helpful to be able to represent a literal equation in different ways?

41. The equation for the point-slope form of a line is $y - y_1 = m(x - x_1)$. What is the equation when solved for x? **MP** 7

- **A** $\dfrac{y - y_1 - mx_1}{m} = x$
- **B** $\dfrac{y - y_1 + mx_1}{m} = x$
- **C** $\dfrac{y - y_1 - x_1}{m} = x$
- **D** $\dfrac{y - y_1 + x_1}{m} = x$

42. A formula for the total surface area S of a box is $S = 2ab + 2bc + 2ac$. Which of the following represents this formula solved for c? **MP** 7

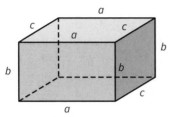

- **A** $c = \dfrac{S - 2ab}{2a + 2b}$
- **B** $c = \dfrac{S - 2ab - 2bc}{2a}$
- **C** $c = \dfrac{S - ab}{a + b}$
- **D** $c = \dfrac{S - 2ab}{a + b}$

43. A formula for the area of a kite is $A = \frac{1}{2}d_1 d_2$, where d_1 and d_2 represent the length of the diagonals. Which of the following represents this formula solved for d_2? **MP** 7

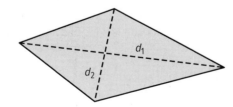

- **A** $d_2 = 2Ad_1$
- **B** $d_2 = \dfrac{A}{2d_1}$
- **C** $d_2 = 2A - d_1$
- **D** $d_2 = \dfrac{2A}{d_1}$

44. A car is traveling at 60 miles per hour. What is the speed in feet per second? **MP** 4, 7

- **A** $\frac{1}{60}$ foot per second
- **B** 1 foot per second
- **C** 88 feet per second
- **D** 3600 feet per second

45. MULTI-STEP The formula for the amount of money in an account after one year is $A = P(1 + r)$, where A is the account balance, P is the principal, and r is the interest rate. **MP** 3, 4, 7

a. Solve the equation for P.

b. Explain how to solve the equation for P.

c. Solve the equation for r.

d. Explain how to solve the equation for r.

46. The formula for the volume of a cone is $V = \dfrac{\pi r^2 h}{3}$ **MP** 7

a. Solve this formula for h.

b. Solve this formula for r.

47. The conversion formula between degrees Fahrenheit and degrees Celsius for temperature is given as $F = \frac{9}{5}C + 32$. Solve this formula for C. **MP** 4, 7

48. Solve the following equation for y. **MP** 7

$$\frac{1}{x^2} + \frac{1}{y} = 4$$

49. Solve the following equation for x. **MP** 3

$$y = \frac{2x + 2}{3}$$

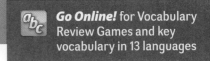
Go Online! for Vocabulary Review Games and key vocabulary in 13 languages

Study Guide

Key Concepts

Writing Equations (Lesson 2-1)

- Identify the unknown you are looking for and assign a variable to it. Then, write the sentence as an equation.

Solving Equations (Lessons 2-2 to 2-4)

- Addition and Subtraction Properties of Equality:
  If an equation is true and the same number is added to or subtracted from each side, the resulting equation is true.

- Multiplication and Division Properties of Equality:
  If an equation is true and each side is multiplied or divided by the same nonzero number, the resulting equation is true.

- Steps for Solving Equations:

  Step 1 Simplify the expression on each side. Use the Distributive Property as needed.

  Step 2 Use the Addition and/or Subtraction Properties of Equality to get the variables on one side and the numbers without variables on the other side.

  Step 3 Use the Multiplication or Division Property of Equality to solve.

Absolute Value Equations (Lesson 2-5)

- For any real numbers a and b, if $|a| = b$ and $b \geq 0$, then $a = b$ or $a = -b$.

Ratios and Proportions (Lesson 2-6)

- The Means-Extremes Property of Proportion states that in a proportion, the product of the extremes is equal to the product of the means.

Literal Equations and Dimensional Analysis (Lesson 2-7)

- An equation that involves several variables is called a formula or literal equation.

- Dimensional analysis, or unit analysis, is the process of carrying units throughout a computation.

Key Vocabulary

consecutive integers (p. 96)	number theory (p. 96)
dimensional analysis (p. 124)	proportion (p. 115)
equivalent equations (p. 87)	rate (p. 117)
extremes (p. 116)	ratio (p. 115)
formula (p. 80)	scale (p. 118)
identity (p. 102)	scale model (p. 118)
linear equation (p. 87)	solve an equation (p. 87)
literal equation (p. 123)	unit analysis (p. 124)
means (p. 116)	unit rate (p. 117)
multi-step equations (p. 95)	

Vocabulary Check

State whether each sentence is *true* or *false*. If *false*, replace the underlined term to make a true sentence.

1. To <u>solve an equation</u> means to find the value of the variable that makes the equation true.

2. The numbers 10, 12, and 14 are an example of <u>consecutive even integers</u>.

3. A(n) <u>equation</u> is a comparison of two numbers by division.

4. An equation stating that two ratios are equal is called a(n) <u>proportion</u>.

5. To write an equation to solve a problem, identify the unknown for which you are looking and assign a(n) <u>number</u> to it.

6. The <u>absolute value</u> of any number is simply the distance the number is away from zero on a number line.

FOLDABLES Study Organizer

Use your foldable to review the chapter. Working with a partner can be helpful. Ask for clarification of concepts as needed.

Concept Check

7. What is the name for an equation that involves several variables?

8. What is dimensional analysis?

Lesson-by-Lesson Review

2-1 Writing Equations

Translate each sentence into an equation.

9. The sum of five times a number x and three is the same as fifteen.

10. Four times the difference of b and six is equal to b squared.

11. One half of m cubed is the same as four times m minus nine.

Translate each equation into a sentence.

12. $3p + 8 = 20$

13. $h^2 - 5h + 6 = 0$

14. $\frac{3}{4}w^2 + \frac{2}{3}w - \frac{1}{5} = 2$

15. **FENCING** Adrianne wants to create an outdoor rectangular kennel. The length will be three feet more than twice the width. Write and use an equation to find the length and the width of the kennel if Adrianne has 54 feet of fencing.

Example 1

Translate the following sentence into an equation.

Six times the sum of a number n and four is the same as the difference between two times n to the second power and ten.

$6(n + 4) = 2n^2 - 10$

Example 2

Translate $3d^2 - 9d + 8 = 4(d + 2)$ into a sentence.

Three times a number d squared minus nine times d increased by eight is equal to four times the sum of d and two.

2-2 Solving One-Step Equations

Solve each equation. Check your solution.

16. $x - 9 = 4$

17. $-6 + g = -11$

18. $\frac{5}{9} + w = \frac{7}{9}$

19. $3.8 = m + 1.7$

20. $\frac{a}{12} = 5$

21. $8y = 48$

22. $\frac{2}{5}b = -4$

23. $-\frac{t}{16} = -\frac{7}{8}$

24. **AGE** Max is four years younger than his sister Brenda. Max is 16 years old. Write and solve an equation to find Brenda's age.

Example 3

Solve $x - 13 = 9$. Check your solution.

$$x - 13 = 9 \qquad \text{Original equation}$$
$$x - 13 + 13 = 9 + 13 \qquad \text{Add 13 to each side.}$$
$$x = 22 \qquad -13 + 13 = 0 \text{ and } 9 + 13 = 22$$

To check that 22 is the solution, substitute 22 for x in the original equation.

CHECK $\quad x - 13 = 9 \qquad$ Original equation
$$22 - 13 \stackrel{?}{=} 9 \qquad \text{Substitute 22 for } x.$$
$$9 = 9 \checkmark \qquad \text{Subtract.}$$

2-3 Solving Multi-Step Equations

Solve each equation. Check your solution.

25. $2d - 4 = 8$

26. $-9 = 3t + 6$

27. $14 = -8 - 2k$

28. $\frac{n}{4} - 7 = -2$

29. $\frac{r+4}{3} = 7$

30. $-18 = \frac{9-a}{2}$

31. $6g - 3.5 = 8.5$

32. $0.2c + 4 = 6$

33. $\frac{f}{3} - 9.2 = 3.5$

34. $4 = \frac{-3u - (-7)}{-8}$

35. CONSECUTIVE INTEGERS Find three consecutive odd integers with a sum of 63.

36. CONSECUTIVE INTEGERS Find three consecutive integers with a sum of −39.

Example 4

Solve $7y - 9 = 33$. Check your solution.

$7y - 9 = 33$	Original equation
$7y - 9 + \mathbf{9} = 33 + \mathbf{9}$	Add 9 to each side.
$7y = 42$	Simplify.
$\frac{7y}{7} = \frac{42}{7}$	Divide each side by 7.
$y = 6$	Simplify.

CHECK	$7y - 9 = 33$	Original equation
	$7(\mathbf{6}) - 9 \stackrel{?}{=} 33$	Substitute 6 for y.
	$42 - 9 \stackrel{?}{=} 33$	Multiply.
	$33 = 33$ ✓	Subtract.

2-4 Solving Equations with the Variable on Each Side

Solve each equation. Check your solution.

37. $8m + 7 = 5m + 16$

38. $2h - 14 = -5h$

39. $21 + 3j = 9 - 3j$

40. $\frac{x-3}{4} = \frac{x}{2}$

41. $\frac{6r-7}{10} = \frac{r}{4}$

42. $3(p + 4) = 33$

43. $-2(b - 3) - 4 = 18$

44. $4(3w - 2) = 8(2w + 3)$

Write an equation and solve each problem.

45. Find the sum of three consecutive odd integers if the sum of the first two integers is equal to twenty-four less than four times the third integer.

46. TRAVEL Mr. Jones drove 480 miles to a business meeting. His travel time to the meeting was 8 hours and from the meeting was 7.5 hours. Find his rate of travel for each leg of the trip.

Example 5

Solve $9w - 24 = 6w + 18$.

$9w - 24 = 6w + 18$	Original equation
$9w - 24 - \mathbf{6w} = 6w + 18 - \mathbf{6w}$	Subtract $6w$ from each side.
$3w - 24 = 18$	Simplify.
$3w - 24 + \mathbf{24} = 18 + \mathbf{24}$	Add 24 to each side.
$3w = 42$	Simplify.
$\frac{3w}{3} = \frac{42}{3}$	Divide each side by 3.
$w = 14$	Simplify.

Example 6

Write an equation to find three consecutive integers such that three times the sum of the first two integers is the same as thirteen more than four times the third integer.

Let x, $x + 1$, and $x + 2$ represent the three consecutive integers.

$3(x + x + 1) = 4(x + 2) + 13$

2-5 Solving Equations Involving Absolute Value

Evaluate each expression if $m = -8$, $n = 4$, and $p = -12$.

47. $|3m - n|$

48. $|-2p + m| - 3n$

49. $-3|6n - 2p|$

50. $4|7m + 3p| + 4n$

Solve each equation. Then graph the solution set.

51. $|x - 6| = 11$

52. $|-4w + 2| = 14$

53. $\left|\frac{1}{3}d - 6\right| = 15$

54. $\left|\frac{2b}{3} + 8\right| = 20$

Example 7

Solve $|y - 9| = 16$. Then graph the solution set.

Case 1

$y - 9 = 16$	Original equation
$y - 9 + 9 = 16 + 9$	Add 9 to each side.
$y = 25$	Simplify.

Case 2

$y - 9 = -16$	Original equation
$y - 9 + 9 = -16 + 9$	Add 9 to each side.
$y = -7$	Simplify.

The solution set is $\{-7, 25\}$.

Graph the points on a number line.

$$\overset{\quad\quad -10 \; -5 \quad\; 0 \quad\; 5 \quad\; 10 \;\; 15 \;\; 20 \;\; 25 \;\; 30}{\longleftarrow\!+\!\!-\!\bullet\!-\!+\!\!-\!+\!\!-\!+\!\!-\!+\!\!-\!+\!\!-\!+\!\!-\!\bullet\!-\!+\!\longrightarrow}$$

2-6 Ratios and Proportions

Determine whether each pair of ratios are equivalent ratios. Write *yes* or *no*.

55. $\frac{27}{45}, \frac{3}{5}$ **56.** $\frac{18}{32}, \frac{3}{4}$

Solve each proportion. If necessary, round to the nearest hundredth.

57. $\frac{4}{9} = \frac{a}{45}$

58. $\frac{3}{8} = \frac{21}{t}$

59. $\frac{9}{12} = \frac{g}{16}$

60. CONSTRUCTION A new gym is being built at Greenfield Middle School. The length of the gym as shown on the builder's blueprints is 12 inches. Find the actual length of the new gym.

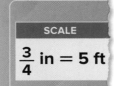

SCALE

$\frac{3}{4}$ in = 5 ft

Example 8

Determine whether $\frac{7}{9}$ and $\frac{42}{54}$ are equivalent ratios. Write *yes* or *no*. Justify your answer.

First, simplify each ratio. $\frac{7}{9}$ is already in simplest form.

$$\frac{42}{54} = \frac{42 \div 6}{54 \div 6} = \frac{7}{9}$$

When expressed in simplest form, the ratios are equivalent. The answer is yes.

Example 9

Solve $\frac{r}{8} = \frac{3}{4}$. If necessary, round to the nearest hundredth.

$\frac{r}{8} = \frac{3}{4}$	Original equation
$r(4) = 3(8)$	Find the cross products.
$4r = 24$	Simplify.
$\frac{4r}{4} = \frac{24}{4}$	Divide each side by 4.
$r = 6$	Simplify.

2-7 Literal Equations and Dimensional Analysis

Solve each equation or formula for the variable indicated.

61. $3x + 2y = 9$, for y

62. $P = 2\ell + 2w$, for ℓ

63. $-5m + 9n = 15$, for m

64. $14w + 15x = y - 21w$, for w

65. $m = \frac{2}{5}y + n$, for y

66. $7d - 3c = f + 2d$, for d

67. GEOMETRY The formula for the area of a trapezoid is $A = \frac{1}{2}h(a + b)$, where h represents the height and a and b represent the lengths of the bases. Solve for h.

68. $A = \frac{1}{2}h(b_1 + b_2)$, for b_1

69. $y = 5x - 6$, for x

70. $\frac{xy + 4}{z} = 9$, for y

71. $ax + by = c$, for x

72. $I = PRT$, for R

73. $V = \frac{1}{3}\pi r^2 h$, for h

74. $A = \frac{a + b + c}{3}$, for b

75. $y = mx + b$, for m

76. $6f - 2(f + g) = 10g - 4h$, for f.

77. $D = \frac{C - S}{n}$, for C

78. $2w - y = 7w - 2z$, for w

79. The formula $C = 6p + 200$ gives C, the total cost in dollars of hosting a birthday party for p guests at a local ice cream parlor.

 a. Solve the formula for p.

 b. Isabella can spend \$350 for her birthday party. How many people can she invite?

Example 10

Solve $6p - 8n = 12$ for p.

$6p - 8n = 12$	Original equation
$6p - 8n + 8n = 12 + 8n$	Add $8n$ to each side.
$6p = 12 + 8n$	Simplify.
$\frac{6p}{6} = \frac{12 + 8n}{6}$	Divide each side by 6.
$\frac{6p}{6} = \frac{12}{6} + \frac{8}{6}n$	Simplify.
$p = 2 + \frac{4}{3}n$	Simplify.

Example 11

Solve $\frac{6p + n}{3} = p + 4$ for p.

$\frac{6p + n}{3} = p + 4$	Original equation
$6p + n = 3(p + 4)$	Multiply each side by 3.
$6p + n = 3p + 12$	Distributive Property.
$6p - 3p + n = 3p - 3p + 12$	Subtract $3p$ from each side.
$3p + n = 12$	Simplify.
$3p + n - n = 12 - n$	Subtract n from each side.
$3p = 12 - n$	Simplify.
$\frac{3p}{3} = \frac{12 - n}{3}$	Divide each side by 3.
$p = \frac{12 - n}{3}$	Simplify.

Go Online! for another Chapter Test

Translate each sentence into an equation.

1. The sum of six and four times d is the same as d minus nine.

2. Three times the difference of two times m and five is equal to eight times m to the second power increased by four.

Solve each equation. Check your solutions.

3. $x - 5 = -11$

4. $\frac{2}{3} = w + \frac{1}{4}$

5. $\frac{t}{6} = -3$

Solve each equation. Check your solution.

6. $2a - 5 = 13$

7. $\frac{p}{4} - 3 = 9$

8. **MULTIPLE CHOICE** At Luigi's Pizza, the price of a large pizza is determined by $P = 9 + 1.5x$, where x represents the number of toppings added to a cheese pizza. Daniel spent $13.50 on a large pizza. How many toppings did he get?

   A 0

   B 1

   C 3

   D 5

Solve each equation. Check your solution.

9. $5y - 4 = 9y + 8$

10. $3(2k - 2) = -2(4k - 11)$

11. **GEOMETRY** Find the value of x so that the figures have the same perimeter.

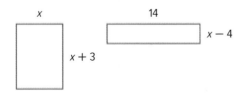

12. The sum of four consecutive integers is −62. What is the least of the four integers?

   A −17

   C 14

   B −14

   D 17

13. Evaluate the expression $|3t - 2u| + 5v$ if $t = 2$, $u = -5$, and $v = -3$.

14. Evaluate the expression $a + |4b - 2c|$ if $a = -4$, $b = -1$, and $c = 2$.

Solve each equation. Then graph the solution set.

15. $|p - 4| = 6$

16. $|2b + 5| = 9$

Solve each proportion. If necessary, round to the nearest hundredth.

17. $\frac{a}{3} = \frac{16}{24}$

18. $\frac{9}{k + 3} = \frac{3}{5}$

19. **MULTIPLE CHOICE** Akiko uses 2 feet of thread for every three squares that she sews for her quilt. How many squares can she sew if she has 38 feet of thread?

   A 19

   B 57

   C 76

   D 228

20. **TELEVISION** Ms. Garcia watched 6 commercials during a 24-minute program. At that rate, how many commercials would she watch in an hour?

21. Solve $5x - 3y = 9$ for y.

22. Solve $A = \frac{1}{2}bh$ for h.

23. The formula $d = rt$ represents the distance d that an object moving at speed r travels in time t.

   a. Solve the equation for r.

   b. What is the average speed of a car that travels 240 miles in 5 hours?

24. **MAPS** On a map, the distance between New York City and Philadelphia is 10.125 inches. If 2 inches equals 16 miles, what is the approximate distance between the two cities?

25. **WATER** Tonya drinks the recommended 8 cups of water per day. How many gallons of water does she drink per year?

Performance Task

Provide a clear solution to each part of the task. Be sure to show all of your work, include all relevant drawings, and justify your answers.

TEMPERATURE The relationship between temperatures in Kelvin (K), Celsius (°C), and Fahrenheit (°F) can be used to convert between different temperature units. A temperature in Kelvin, T_K, is 273 more than the temperature in Celsius, T_C. A temperature in Celsius is $\frac{5}{9}$ times the difference of the temperature in Fahrenheit, T_F and 32.

Part A

1. Write a formula for the relationship between a temperature in Kelvin, T_K, and the temperature in Celsius, T_C.

2. Write a formula for the relationship between a temperature in Celsius, T_C, and the temperature in Fahrenheit, T_F.

Part B

3. **Sense-Making** Use the formulas from Part A to convert 298K to temperatures in Celsius and Fahrenheit.

Part C

The formulas from Part A can be rearranged to more easily convert from Kelvin to Celsius or Celsius to Fahrenheit.

4. **Structure** Solve the formula that relates T_K and T_C for T_C. What inverse operation is needed?

5. **Arguments** Solve the formula that relates T_C and T_F for T_F. What are the steps needed to solve?

Part D

During an experiment, the temperature of a substance increases by 3°F every 7 minutes.

6. Write and solve a proportion to find how long it will take for the temperature of the substance to increase 15°F.

7. **Precision** Find the rate at which the temperature is increasing °F per second. How much will the temperature increase in 30 seconds? Round to the nearest thousandth.

Part E

The thermometer used in the experiment shows 59°F and is accurate within 0.5°F.

8. Write and solve an equation to find the minimum and maximum temperature of the substance in °F.

9. If the thermometer is accurate within 0.3°C, find the maximum and minimum temperature of the substance in °C.

Test-Taking Strategy

Example

Read the problem. Identify what you need to know. Then use the information in the problem to solve.

Ashley is 3 years older than her sister, Tina. Combined, the sum of their ages is 27 years. How old is Ashley?

A 1　　　　　　　　　　**C** 15

B 4　　　　　　　　　　**D** 12

Step 1　What do you need to find? Ashley's age

Step 2　Is there enough information given to solve the problem? yes

Step 3　What information, if any, is not needed to solve the problem? none

Step 4　Are there any obvious wrong answers? If so, which one(s)? Explain. yes, A because it makes Tina's age negative and B because the sum is too small

Step 5　What is the correct answer?

Let a represent Ashley's age. Then Tina's age is $a - 3$, since she is 3 years younger than Ashley.

Solve the equation for a.

$$a + (a - 3) = 27 \qquad \text{Original equation.}$$
$$2a - 3 = 27 \qquad \text{Add like terms.}$$
$$2a = 30 \qquad \text{Add 3 to each side.}$$
$$a = 15 \qquad \text{Divide each side by 2.}$$

Since we let a represent Ashley's age, we know that she is 15 years old.

Test-Taking Tip

Assigning a Variable

A crucial step in solving an algebra problem is assigning a variable to an unknown. This is a prerequisite to writing and solving an equation that models the problem situation.

Apply the Strategy

Read the problem. Identify what you need to know. Then use the information in the problem to solve.

Orlando has $1350 in the bank. He wants to increase his balance to a total of $2550 by depositing $40 each week from his paycheck. How many weeks will he need to save in order to reach his goal?

A 5　　　　　　　　　　**C** 40

B 0　　　　　　　　　　**D** 30

Answer the questions below.

a. What do you need to find?

b. Is there enough information given to solve the problem?

c. What information, if any, is not needed to solve the problem?

d. Are there any obvious wrong answers? If so, which one(s)? Explain.

e. What is the correct answer?

Read each question. Then fill in the correct answer on the answer document provided by your teacher or on a sheet of paper.

1. These ordered pairs represent a relation that is not a function:

    {(–1, 3), (2, 3), (4, 0), (6, 3), (6, 5)}

    Which of the following would make the relation a function?

    ○ **A** Replace (6, 3) with (4, 3).

    ○ **B** Replace (6, 5) with (2, 6).

    ○ **C** Replace (6, 3) with (6, −2).

    ○ **D** Replace (6, 5) with (5, 6).

2. Which statement describes the end behavior of this graph?

    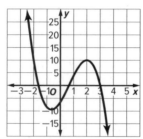

    ○ **A** As x decreases, the value of y increases; as x increases, the value of y decreases.

    ○ **B** As x decreases, the value of y decreases; as x increases, the value of y increases.

    ○ **C** As x decreases, the value of y increases; as x increases, the value of y increases.

    ○ **D** As x decreases, the value of y decreases; as x increases, the value of y decreases.

3. Which of the four answer choices illustrates the Multiplicative Identity Property?

    ○ **A** $a \cdot 1 = a$

    ○ **B** $b + 0 = b$

    ○ **C** $c + (-c) = 0$

    ○ **D** $d + 1 = d + 1$

4. If $3r + 1 = -5$, what is the value of the expression $5r + 3$?

5. The perimeter P of a regular octagon is equal to 8 times the length s of each side. Select all of the equations that represent that relationship.

    ☐ **A** $s = 8P$

    ☐ **B** $8P = s$

    ☐ **C** $P = s + 8$

    ☐ **D** $P = 8 + s$

    ☐ **E** $P = 8s$

    ☐ **F** $8s = P$

6. What is the solution to the equation $\frac{3}{8}a = 24$?

    ○ **A** $a = 8$

    ○ **B** $a = 9$

    ○ **C** $a = 64$

    ○ **D** $a = 23\frac{5}{8}$

7. The sum of four consecutive numbers is −34. What is the greatest of the four numbers?

    ○ **A** −10

    ○ **B** −8.5

    ○ **C** −7

    ○ **D** 10

8. What is the solution of the equation $4.5 - 6s = 9$?

    ○ **A** $s = -6.00$

    ○ **B** $s = -0.75$

    ○ **C** $s = 0.75$

    ○ **D** $s = 2.25$

Test Taking Tip

Question 7 Remember the difference between *consecutive numbers* and *consecutive odd* (or *even*) *numbers.*

9. The figures below represent two fields. One is a square and the other is a regular hexagon. If the two fields have the same perimeter, what is the length of a side of the hexagonal field?

4n − 3

2n + 1

- **A** 4.5
- **B** 10
- **C** 15
- **D** 60

10. The graph below shows two points on a number line.

Which equation involving absolute value represents the graph?

- **A** $|x - 7| = 5$
- **B** $|x - 12| = 2$
- **C** $|x - 7| = 2$
- **D** $|x - 5| = 7$

11. The ratio $\frac{3r - 2}{5}$ is equal to the ratio $\frac{5 + 2r}{8}$. What is the value of r?

- **A** −26
- **B** $\frac{27}{14}$
- **C** $\frac{41}{14}$
- **D** 7

12. Select all of the equations that illustrate the Additive Inverse Property.

- ☐ **A** $\frac{3}{5} \cdot \frac{5}{3} = 1$
- ☐ **B** $7 + (-7) = 0$
- ☐ **C** $4 + 0 = 4$
- ☐ **D** $6 + (-5) = (-5) + 6$
- ☐ **E** $(-x) + x = 0$
- ☐ **F** $x + 0 = x$

13. A rocket is traveling at 352 feet per second. What is that rate in miles per hour?

- **A** 240 mph
- **B** 412 mph
- **C** 516.27 mph
- **D** 14,400 mph

14. A bicyclist is traveling at 36 mph. What is that speed in feet per second?

- **A** 24.5 ft/sec
- **B** 52.8 ft/sec
- **C** 72 ft/sec
- **D** 3,168 ft/sec

15. If $y = 3x - 7$, which equation is also true?

- **A** $x = 3y - 7$
- **B** $x = 3y + 7$
- **C** $x = \frac{y - 7}{3}$
- **D** $x = \frac{y + 7}{3}$

Need Extra Help?

If you missed Question...	1	2	3	4	5	6	7	8	9	10	11	12	13	14	15
Go to Lesson...	1-7	1-8	1-3	2-3	2-1	2-2	2-3	2-3	2-4	2-5	2-6	1-3	2-6	2-6	2-7

CHAPTER 3
Linear and Nonlinear Functions

THEN
You solved linear equations algebraically.

NOW
In this chapter you will:

- Identify linear equations, intercepts, and zeros.
- Graph and write linear equations.
- Use rate of change to solve problems.

(MP) WHY

PARK ATTENDANCE In 2014, Joshua Tree National Park set a new record for attendance, with 1.6 million visitors. Annual attendance figures that increase steadily each year can be modeled with a linear function.

Use the Mathematical Practices to complete the activity.

1. **Use Tools** Use the Internet or another source to find the attendance data for your favorite park.

2. **Model With Mathematics** Use the Line Graph tool in ConnectED to plot your data.

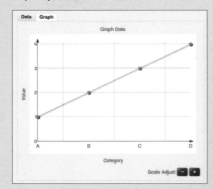

3. **Apply Math** How can this data be used by park management?

4. **Discuss** Compare your graph to the graphs your classmates created. Which park will have the greatest attendance 2 years from now?

Go Online to Guide Your Learning

Explore & Explain

 Graphing Tools

Investigate using the **Standard Form of a Linear Function** tool. How does changing the values in the equation affect the graph?

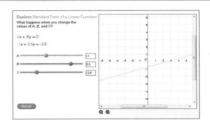

The Geometer's Sketchpad

Visualize and model slope using the **Slope of a Line** sketch in Lesson 3-3.

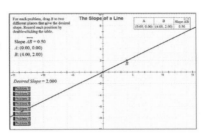

eBook

Interactive Student Guide

Before starting the chapter, answer the **Chapter Focus** preview questions. Check your answers as you complete each lesson. At the end of the chapter, try the **Performance Task**.

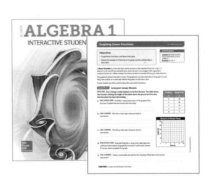

Organize

 Foldables

Get organized! Create a **Linear Functions Foldable** before you start the chapter to arrange your notes on graphing relations and functions.

Collaborate

 Chapter Project

In the **Service with a Smile** project, you will act as an entrepreneur to develop a business plan.

Focus

 LEARNSMART

Need help studying? Complete the **Analyze** and **Interpret Functions** domains in LearnSmart to review for the chapter test.

 ALEKS

You can use the **Functions and Lines** topic in ALEKS to find out what you know about linear functions and what you are ready to learn.*

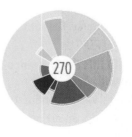

* Ask your teacher if this is part of your program.

Get Ready for the Chapter

Connecting Concepts

Concept Check

Review the concepts used in this chapter by answering each question below.

1. The origin of something is where it begins. How does this definition relate to the origin of the coordinate plane?

2. In which quadrant are the x-values negative and y-values positive?

3. Point P has coordinates (2, 5). Name another point that has the same y-coordinate.

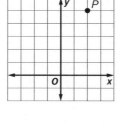

4. What is the inverse operation for subtraction?

5. Explain what step you would do first to solve $2x - 5 = 14$.

6. What does it mean to solve an equation for a given variable?

7. Explain what it means to evaluate an expression.

8. What does it mean to find the absolute value of a number?

New Vocabulary

English		Español
linear equation	p. 143	ecuación lineal
standard form	p. 143	forma estándar
constant	p. 143	constante
x-intercept	p. 144	intersección x
y-intercept	p. 144	intersección y
linear function	p. 151	función lineal
parent function	p. 151	críe la función
family of graphs	p. 151	la familia de gráficas
root	p. 151	raíz
rate of change	p. 160	tasa de cambio
slope	p. 162	pendiente
transformation	p. 181	transformación
arithmetic sequence	p. 190	sucesión arithmética
step function	p. 197	funcion escalonada
absolute value function	p. 205	función del valor absoluto

Performance Task Preview

You can use the concepts and skills in this chapter to solve problems about running your own lawn-care business. Understanding linear functions will help you finish the Performance Task at the end of the chapter.

MP **In this Performance Task you will:**

- model with mathematics
- construct an argument
- make use of structure

Review Vocabulary

origin origen
the point where the two axes in a coordinate plane intersect with coordinates (0, 0)

x-axis eje x the horizontal number line on a coordinate plane

y-axis eje y the vertical number line on a coordinate plane

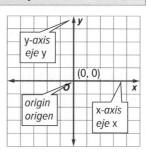

Analyzing a graph can help you learn about the relationship between two quantities. A **linear function** is a function for which the graph is a line. There are four types of linear graphs Let's analyze each type.

Mathematical Practices
MP 7 Look for and make use of structure.

Activity 1 Line that Slants Up

Work cooperatively. Analyze the function graphed at the right.

a. Describe the domain, range, and end behavior.

b. Describe the intercepts and any maximum or minimum points.

c. Identify where the function is positive, negative, increasing, and decreasing.

d. Describe any symmetry.

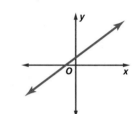

a. The domain and range are all real numbers or $\{-\infty < x < \infty\}$ and $\{-\infty < y < \infty\}$. As you move left, the graph goes down. So as x decreases, y decreases. As you move right, the graph goes up. So as x increases, y increases.

b. There is one x-intercept and one y-intercept. There are no maximum or minimum points.

c. The function value is 0 at the x-intercept. The function values are negative to the left of the x-intercept and positive to the right. The function goes up from left to right, so it is increasing on the entire domain.

d. The graph is not symmetric with respect to any vertical line.

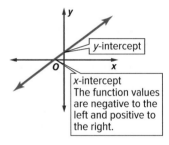

Lines that slant down from left to right have some different key features.

Activity 2 Line that Slants Down

Work cooperatively. Analyze the function graphed at the right.

a. Describe the domain, range, and end behavior.

b. Describe the intercepts and any maximum or minimum points.

c. Identify where the function is positive, negative, increasing, and decreasing.

d. Describe any symmetry.

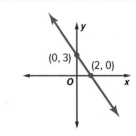

a. The domain and range are all real numbers or $\{-\infty < x < \infty\}$ and $\{-\infty < y < \infty\}$. As you move left, the graph goes up. So as x decreases, y increases. As you move right, the graph goes down. So as x increases, y decreases.

b. There is one x-intercept and one y-intercept. There are no maximum or minimum points.

c. The function values are positive to the left of the x-intercept and negative to the right.
The function goes down from left to right, so it is decreasing on the entire domain.

d. The graph is not symmetric with respect to any vertical line.

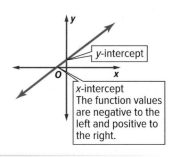

Analyzing Linear Graphs *Continued*

Horizontal lines represent special functions called **constant functions**.

Activity 3 Horizontal Line

Work cooperatively. Analyze the function graphed at the right.

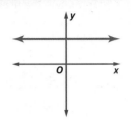

a. The domain is all real numbers, and the range is one value. As you move left or right, the graph stays constant. So as x decreases or increases, y is constant.

b. The graph does not intersect the x-axis, so there is no x-intercept. The graph has one y-intercept. There are no maximum or minimum points.

c. The function values are all positive. The function is constant on the entire domain.

d. The graph is symmetric about any vertical line.

Vertical lines represent linear relations that are *not* functions.

Activity 4 Vertical Line

Work cooperatively. Analyze the relation graphed at the right.

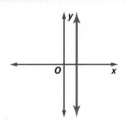

a. The domain is one value, and the range is all real numbers. This relation is not a function. Because you cannot move left or right on the graph, there is no end behavior.

b. There is one x-intercept and no y-intercept. There are no maximum or minimum points.

c. The y-values are positive above the x-axis and negative below. Because you cannot move left or right on the graph, the relation is neither increasing nor decreasing.

d. The graph is symmetric about itself.

Analyze the Results

1. Compare and contrast the key features of lines that slant up and lines that slant down.

2. How would the key features of a horizontal line below the x-axis differ from the features of a line above the x-axis?

3. Consider lines that pass through the origin.

    a. How do the key features of a line that slants up and passes through the origin compare to the key features of the line in Activity 1?

    b. Compare the key features of a line that slants down and passes through the origin to the key features of the line in Activity 2.

    c. Describe a horizontal line that passes through the origin and a vertical line that passes through the origin. Compare their key features to those of the lines in Activities 3 and 4.

4. **MP** **TOOLS** Place a pencil on a coordinate plane to represent a line. Move the pencil to represent different lines and evaluate each conjecture.

    a. *True* or *false*: A line can have more than one x-intercept.

    b. *True* or *false*: If the end behavior of a line is that as x increases, y increases, then the function values are increasing over the entire domain.

    c. *True* or *false*: Two different lines can have the same x- and y-intercepts.

Sketch a linear graph that fits each description.

5. as x increases, y decreases

6. one x-intercept and one y-intercept

7. has symmetry

8. is not a function

Graphing Linear Functions

:: Then

- You represented relationships among quantities using equations.

:: Now

1 Identify linear equations, intercepts, and zeros.

2 Graph linear equations.

:: Why?

- Recycling one ton of waste paper saves an average of 17 trees, 7000 gallons of water, 3 barrels of oil, and about 3.3 cubic yards of landfill space.

  The relationship between the amount of paper recycled and the number of trees saved can be expressed with the equation $y = 17x$, where y represents the number of trees and x represents the tons of paper recycled.

 New Vocabulary

linear equation
standard form
constant
x-intercept
y-intercept

 Mathematical Practices

8 Look for and express regularity in repeated reasoning.

---

1 **Linear Equations and Intercepts** A **linear equation** is an equation that forms a line when it is graphed. Linear equations are often written in the form $Ax + By = C$. This is called the **standard form** of a linear equation. In this equation, C is called a **constant**, or a number. Ax and By are variable terms.

🔑 Key Concept Standard Form of a Linear Equation

Words	The standard form of a linear equation is $Ax + By = C$, where $A \geq 0$, A and B are not both zero, and A, B, and C are integers with a greatest common factor of 1.
Examples	In $3x + 2y = 5$, $A = 3$, $B = 2$, and $C = 5$. In $x = -7$, $A = 1$, $B = 0$, and $C = -7$.

Example 1 Identify Linear Equations

Determine whether each equation is a linear equation. Write the equation in standard form.

a. $y = 4 - 3x$

Rewrite the equation so that it appears in standard form.

$y = 4 - 3x$	Original equation
$y + 3x = 4 - 3x + 3x$	Add $3x$ to each side.
$3x + y = 4$	Simplify.

The equation is now in standard form where $A = 3$, $B = 1$, and $C = 4$. This is a linear equation.

b. $6x - xy = 4$

Since the term xy has two variables, the equation cannot be written in the form $Ax + By = C$. Therefore, this is not a linear equation.

> **Guided Practice**

1A. $\frac{1}{3}y = -1$

1B. $y = x^2 - 4$

A linear equation can be represented on a coordinate graph. The x-coordinate of the point at which the graph of an equation crosses the x-axis is an **x-intercept**. The y-coordinate of the point at which the graph crosses the y-axis is called a **y-intercept**.

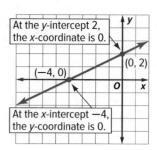

At the y-intercept 2, the x-coordinate is 0.

(0, 2)

(−4, 0)

At the x-intercept −4, the y-coordinate is 0.

The graph of a linear equation has at most one x-intercept and one y-intercept, unless it is the equation $x = 0$ or $y = 0$, in which case every number is a y-intercept or an x-intercept, respectively.

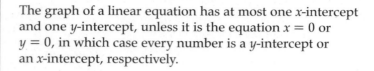

Example 2   Find Intercepts from a Graph

Find the x- and y-intercepts of the line graphed at the right.

A x-intercept is 0; y-intercept is 30.

B x-intercept is 20; y-intercept is 30.

C x-intercept is 20; y-intercept is 0.

D x-intercept is 30; y-intercept is 20.

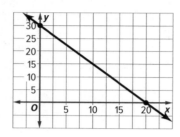

Read the Item

We need to determine the x- and y-intercepts of the line in the graph.

Solve the Item

Step 1 Find the x-intercept. Look for the point where the line crosses the x-axis.

The line crosses at (20, 0). The x-intercept is 20 because it is the x-coordinate of the point where the line crosses the x-axis.

Step 2 Find the y-intercept. Look for the point where the line crosses the y-axis.

The line crosses at (0, 30). The y-intercept is 30 because it is the y-coordinate of the point where the line crosses the y-axis.

Thus, the correct answer is B.

Guided Practice

2. **HEALTH** Find the x- and y-intercepts of the graph.

   F x-intercept is 0; y-intercept is 150.

   G x-intercept is 150; y-intercept is 0.

   H x-intercept is 150; no y-intercept.

   J No x-intercept; y-intercept is 150.

Gym Membership

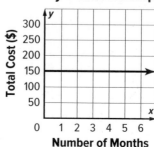

Total Cost ($)

Number of Months

Example 3   Find Intercepts from a Table

SWIMMING POOL A swimming pool is being drained at a rate of 720 gallons per hour. The table shows the function relating the volume of water in a pool and the time in hours that the pool has been draining.

Draining a Pool	
Time (h)	Volume (gal)
x	**y**
0	10,080
2	8640
6	5760
10	2880
12	1440
14	0

a. **Find the *x*- and *y*-intercepts of the graph of the function.**

x-intercept $= 14$ 14 is the value of x when $y = 0$.
y-intercept $= 10{,}080$ 10,080 is the value of y when $x = 0$.

b. **Graph the function. Describe what the intercepts mean in this situation.**

Plot the point (14, 0) on the graph. The *x*-intercept 14 means that after 14 hours, the water has a volume of 0 gallons, or the pool is completely drained.

Plot the point (0, 10,080). Connect the points. The *y*-intercept 10,080 means that the pool contained 10,080 gallons of water at time 0, or before it started to drain. This is shown in the graph.

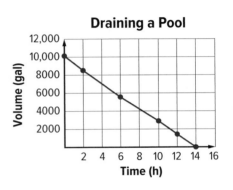

Draining a Pool

Guided Practice

3. **DRIVING** The table shows the function relating the distance to an amusement park in miles and the time in hours the Torres family has driven. Find the *x*- and *y*-intercepts. Describe what the intercepts mean in this situation.

Time (h)	Distance (mi)
0	248
1	186
2	124
3	62
4	0

2 Graph Linear Equations By first finding the *x*- and *y*-intercepts, you have the ordered pairs of two points through which the graph of the linear equation passes. This information can be used to graph the line because only two points are needed to graph a line.

Example 4   Graph by Using Intercepts

Graph $2x + 4y = 16$ by using the *x*- and *y*-intercepts.

To find the *x*-intercept, let $y = 0$.

$2x + 4y = 16$ Original equation

$2x + 4(0) = 16$ Replace y with 0.

$2x = 16$ Simplify.

$x = 8$ Divide each side by 2.

The *x*-intercept is 8. This means that the graph intersects the *x*-axis at (8, 0).

To find the y-intercept, let $x = 0$.

$$2x + 4y = 16 \quad \text{Original equation}$$
$$2(0) + 4y = 16 \quad \text{Replace } x \text{ with } 0.$$
$$4y = 16 \quad \text{Simplify.}$$
$$y = 4 \quad \text{Divide each side by 4.}$$

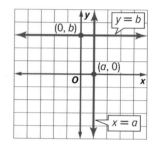

The y-intercept is 4. This means the graph intersects the y-axis at $(0, 4)$.

Plot these two points and then draw a line through them.

> **Study Tip**
>
> **Equivalent Equations**
> Rewriting equations by solving for y may make it easier to find values for y.
> $$-x + 2y = 3 \rightarrow y = \frac{x+3}{2}$$

▶ **Guided Practice**

Graph each equation by using the x- and y-intercepts.

4A. $-x + 2y = 3$ **4B.** $y = -x - 5$

Note that the graph in Example 4 has both an x- and a y-intercept. Some lines have an x-intercept and no y-intercept or vice versa. The graph of $y = b$ is a horizontal line that only has a y-intercept (unless $b = 0$). The intercept occurs at $(0, b)$. The graph of $x = a$ is a vertical line that only has an x-intercept (unless $a = 0$). The intercept occurs at $(a, 0)$.

Every ordered pair that makes an equation true represents a point on the graph. So, the graph of an equation represents all of its solutions. Any ordered pair that does not make the equation true represents a point that is not on the line.

Example 5   **Graph by Making a Table**

Graph $y = \frac{1}{3}x + 2$. Then state the domain and the range.

The domain is all real numbers, which can be written as $D = \{-\infty < x < \infty\}$. Select values from the domain and make a table. Create ordered pairs and graph them. The range is also all real numbers, or $R = \{-\infty < y < \infty\}$.

> **Study Tip**
>
> **MP Sense-Making** When the x-coefficient is a fraction, select numbers from the domain that are multiples of the denominator. It will simplify your calculations.

x	$\frac{1}{3}x + 2$	y	(x, y)
-3	$\frac{1}{3}(-3) + 2$	1	$(-3, 1)$
0	$\frac{1}{3}(0) + 2$	2	$(0, 2)$
3	$\frac{1}{3}(3) + 2$	3	$(3, 3)$
6	$\frac{1}{3}(6) + 2$	4	$(6, 4)$

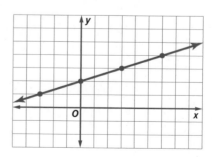

▶ **Guided Practice**

Graph each equation by making a table. Then state the domain and the range.

5A. $2x - y = 2$ **5B.** $x = 3$ **5C.** $y = -2$

Example 1 Determine whether each equation is a linear equation. Write *yes* or *no*. If yes, write the equation in standard form.

1. $x = y - 5$ **2.** $-2x - 3 = y$ **3.** $-4y + 6 = 2$ **4.** $\frac{2}{3}x - \frac{1}{3}y = 2$

Examples 2–3 Find the *x*- and *y*-intercepts of the graph of each linear function. Describe what the intercepts mean.

5.

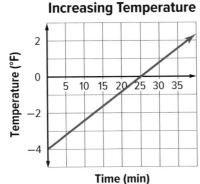

Increasing Temperature

6.

Position of Scuba Diver	
Time (s)	Depth (m)
x	*y*
0	−24
3	−18
6	−12
9	−6
12	0

Example 4 Graph each equation by using the *x*- and *y*-intercepts.

7. $y = 4 + x$ **8.** $2x - 5y = 1$

Example 5 Graph each equation by making a table. Then state the domain and range.

9. $x + 2y = 4$ **10.** $-3 + 2y = -5$ **11.** $y = 3$

12. **MP** **REASONING** The equation $15x + 30y = 180$ represents the number of children *x* and adults *y* who can attend the rodeo for $180.

   a. Use the *x*- and *y*-intercepts to graph the equation.

   b. Describe what these values mean.

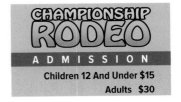

CHAMPIONSHIP RODEO
ADMISSION
Children 12 And Under $15
Adults $30

Practice and Problem Solving

Extra Practice is on page R3.

Example 1 Determine whether each equation is a linear equation. Write *yes* or *no*. If yes, write the equation in standard form.

13 $5x + y^2 = 25$ **14.** $8 + y = 4x$ **15.** $9xy - 6x = 7$

16. $4y^2 + 9 = -4$ **17.** $12x = 7y - 10y$ **18.** $y = 4x + x$

Example 2 Find the *x*- and *y*-intercepts of the graph of each linear function.

19.

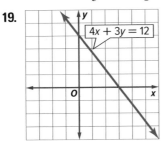

$4x + 3y = 12$

20.

x	*y*
−3	−1
−2	0
−1	1
0	2
1	3

Example 3 Find the *x*- and *y*-intercepts of each linear function. Describe what the intercepts mean.

21.   **Descent of Eagle**

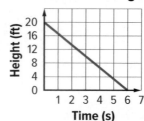

22.

Eva's Distance from Home	
Time (min)	Distance (mi)
x	*y*
0	4
2	3
4	2
6	1
8	0

Example 4 Graph each equation by using the *x*- and *y*-intercepts.

23. $y = 4 + 2x$ **24.** $5 - y = -3x$ **25.** $x = 5y + 5$

26. $x + y = 4$ **27.** $x - y = -3$ **28.** $y = 8 - 6x$

Example 5 Graph each equation by making a table. Then state the domain and range.

29. $x = -2$ **30.** $y = -4$ **31.** $y = -8x$

32. $3x = y$ **33.** $y - 8 = -x$ **34.** $x = 10 - y$

35 **TV RATINGS** The number of people who watch a singing competition can be given by $p = 0.15v$, where *p* represents the number of people in millions who saw the show and *v* is the number of potential viewers in millions.

   a. Make a table of values for the points (v, p).

   b. Graph the equation.

   c. Use the graph to estimate the number of people who saw the show if there are 14 million potential viewers.

   d. Explain why it would not make sense for *v* to be a negative number.

Determine whether each equation is a linear equation. Write *yes* or *no*. If yes, write the equation in standard form.

36. $x + \dfrac{1}{y} = 7$ **37.** $\dfrac{x}{2} = 10 + \dfrac{2y}{3}$

38. $7n - 8m = 4 - 2m$ **39.** $3a + b - 2 = b$

40. $2r - 3rt + 5t = 1$ **41.** $\dfrac{3m}{4} = \dfrac{2n}{3} - 5$

42. FINANCIAL LITERACY James earns a monthly salary of $1200 and a commission of $125 for each car he sells.

   a. Graph an equation that represents how much James earns in a month in which he sells *x* cars.

   b. Use the graph to estimate the number of cars James needs to sell in order to earn $5000.

Graph each equation. Then state the domain and range.

43. $2.5x - 4 = y$ **44.** $1.25x + 7.5 = y$ **45.** $y + \dfrac{1}{5}x = 3$

46. $\dfrac{2}{3}x + y = -7$ **47.** $5x - 3 = 2x + 6$ **48.** $3y - 7 = 4x + 1$

49. MP REASONING Mrs. Johnson is renting a car for vacation and plans to drive a total of 800 miles. A rental car company charges $195 for the week including 700 miles and $0.32 for each additional mile. If Mrs. Johnson has only $210 to spend on the rental car, can she afford to rent a car? Explain your reasoning.

50. AMUSEMENT PARKS An amusement park charges $70 for admission before 6 P.M. and $25 for admission after 6 P.M. On Saturday, the park took in a total of $28,000.

    a. Write an equation that represents the number of admissions that may have been sold. Let x represent the admissions sold before 6 P.M., and let y represent the admissions sold after 6 P.M.

    b. Graph the equation.

    c. Find the x- and y-intercepts of the graph. What does each intercept represent?

Find the x-intercept and y-intercept of the graph of each equation.

51 $5x + 3y = 15$ **52.** $2x - 7y = 14$ **53.** $2x - 3y = 5$

54. $6x + 2y = 8$ **55.** $y = \frac{1}{4}x - 3$ **56.** $y = \frac{2}{3}x + 1$

57. AP EXAMS The percent of high school graduates who scored 3+ on AP Exams can be modeled by $s = 26{,}811t + 246{,}402$, where s is the number of high school students and t represents time in years since 2000.

    a. Graph the equation and identify the t- and s-intercepts of the graph.

    b. Use the graph to estimate the number of students receiving 3+ on the AP Exams in 2022.

58. MULTIPLE REPRESENTATIONS In this problem, you will explore x- and y-intercepts of graphs of linear equations.

    a. Graphical If possible, use a straightedge to draw a line on a coordinate plane with each of the following characteristics.

x- and y-intercept	x-intercept, no y-intercept	exactly 2 x-intercepts	no x-intercept, y-intercept	exactly 2 y-intercepts

    b. Analytical For which characteristics were you able to create a line and for which characteristics were you unable to create a line? Explain.

    c. Verbal What must be true of the x- and y-intercepts of a line?

H.O.T. Problems   Use Higher-Order Thinking Skills

59. **(MP) REGULARITY** Copy and complete each table. State whether any of the tables show a linear relationship. Explain.

Perimeter of a Square	
Side Length	Perimeter
1	
2	
3	
4	

Area of a Square	
Side Length	Area
1	
2	
3	
4	

Volume of a Cube	
Side Length	Volume
1	
2	
3	
4	

60. **(MP) REASONING** Compare and contrast the graphs of $y = 2x + 1$ with the domain $\{1, 2, 3, 4\}$ and $y = 2x + 1$ with the domain of all real numbers.

OPEN-ENDED Give an example of a linear equation of the form $Ax + By = C$ for each condition. Then describe the graph of the equation.

61. $A = 0$ **62.** $B = 0$ **63.** $C = 0$

64. WRITING IN MATH Explain how to find the x-intercept and y-intercept of a graph and summarize how to graph a linear equation.

65. What are the *x*- and *y*-intercepts of the line shown in the graph? **MP** 2

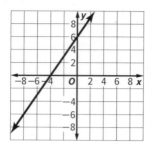

- ○ **A** *x*-intercept is −4; *y*-intercept is −6.
- ○ **B** *x*-intercept is −4; *y*-intercept is 6.
- ○ **C** *x*-intercept is 6; *y*-intercept is −4.
- ○ **D** *x*-intercept is 6; *y*-intercept is 6.

66. Which of the following shows $y = -\dfrac{5}{8}x + \dfrac{3}{2}$ written in standard form? **MP** 2

- ○ **A** $-5x - 8y = -12$
- ○ **B** $5x + 8y = 24$
- ○ **C** $5x + 8y = 12$
- ○ **D** $8y = -5x + 12$

67. Which of the following equations has the same *y*-intercept as the line shown in the graph? **MP** 1, 2

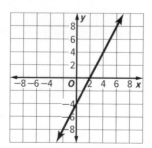

- ○ **A** $x - y = 4$
- ○ **B** $3x - y = 6$
- ○ **C** $2x + y = 4$
- ○ **D** $x + y = -2$

68. **MULTI-STEP** A candle burns as shown in the graph. **MP** 1, 2, 8

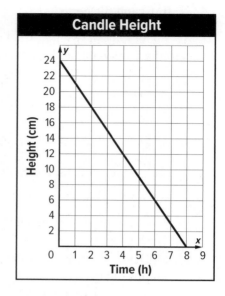

a. Which of the following statements are true?

- ☐ **A** The graph is linear.
- ☐ **B** The graph is nonlinear.
- ☐ **C** The function is increasing.
- ☐ **D** The function is decreasing.
- ☐ **E** The function is neither increasing nor decreasing.
- ☐ **F** The function is positive.
- ☐ **G** The function is negative.

b. What is the *x*-intercept?

c. What is the *y*-intercept?

d. What do the intercepts represent?

e. If the height of the candle is 8 centimeters, approximately how long has the candle been burning?

- ○ **A** 0 hours
- ○ **B** $5\frac{1}{2}$ hours
- ○ **C** 8 hours
- ○ **D** 24 hours

Zeros of Linear Functions

:: **Then**

- You graphed linear equations by using tables and finding intercepts.

:: **Now**

1 Find zeros of linear functions.

2 Model linear functions.

:: **Why?**

- The cost of braces can vary widely. The graph shows the balance of the cost of treatments as payments are made. This is modeled by the function $b = -85p + 5100$, where p represents the number of $85 payments made and b is the remaining balance.

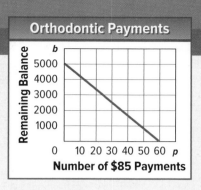

Orthodontic Payments

New Vocabulary

linear function
parent function
family of graphs
root
zeros

Mathematical Practices

4 Model with mathematics.

1 Find Zeros A **linear function** is a function for which the graph is a line. The simplest linear function is $f(x) = x$ and is called the **parent function** of the family of linear functions. A **family of graphs** is a group of graphs with one or more similar characteristics.

Key Concept Linear Equation

Parent function:	$f(x) = x$
Type of graph:	line
Domain:	all real numbers
Range:	all real numbers

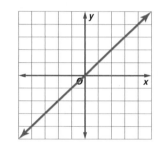

The solution or **root** of an equation is any value that makes the equation true. A linear equation has at most one root. You can find the root of an equation by graphing its related function. To write the related function for an equation, replace 0 with $f(x)$.

Linear Equation	Related Function
$2x - 8 = 0$	$f(x) = 2x - 8$ or $y = 2x - 8$

Values of x for which $f(x) = 0$ are called **zeros** of the function f. The zero of a function is located at the x-intercept of the function. The root of an equation is the value of the x-intercept. So:

- 4 is the x-intercept of $2x - 8 = 0$.

- 4 is the solution of $2x - 8 = 0$.

- 4 is the root of $2x - 8 = 0$.

- 4 is the zero of $f(x) = 2x - 8$.

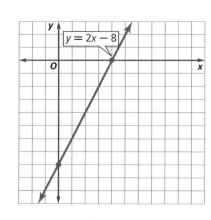

$y = 2x - 8$

Example 1   Find Zeros of a Linear Function Graphically

Find the zero of each linear function by graphing.

a. $f(x) = 3x + 3$

Make a table to graph the function.

x	$f(x) = 3x + 3$	$f(x)$	$(x, f(x))$
−2	$f(-2) = 3(-2) + 3$	−3	$(-2, -3)$
0	$f(0) = 3(0) + 3$	3	$(0, 3)$
1	$f(1) = 3(1) + 3$	6	$(1, 6)$

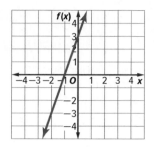

The graph intersects the x-axis at −1. So, the zero is −1.

Verify algebraically that −1 is the zero of $f(x) = 3x + 3$.

$$f(x) = 3x + 3 \qquad \text{Original equation}$$
$$f(-1) = 3(-1) + 3 \qquad \text{Substitution}$$
$$f(-1) = -3 + 3 \qquad \text{Multiply.}$$
$$f(-1) = 0 \qquad \text{Simplify.}$$

b. $f(x) = 6$

The graph is a horizontal line at $y = 6$. The line does not intersect the x-axis. This function has no zero.

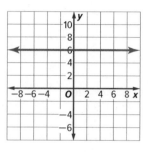

▶ **Guided Practice**

1A. $f(x) = \frac{2}{5}x + 6$

1B. $f(x) = -1.25x + 3$

Reading Math

set-builder notation
$\{x \,|\, x \geq 20\}$ is read the set of all numbers x such that x is greater than or equal to 20.

2 **Linear Function Models** Graphing may provide only an estimate. In these cases, solve algebraically to find the exact solution.

A more concise way of writing a solution set is to use **set-builder notation**. For example, the set-builder notation of a solution set that is all numbers greater than or equal to 5 is $\{x \,|\, x \geq 5\}$.

Real-World Example 2   Find Zeros of a Linear Function Algebraically

MOVIES Willow is downloading a 2-hour movie. The function $y = 1.5 - 0.0625x$ represents the number of GB left to download after x minutes. How long will it take her to download the entire movie? Identify the domain and range.

Make a table of values. Then use the table to graph the function.

x	$y = 1.5 - 0.0625x$	y	(x, y)
0	$y = 1.5 - 0.0625(0)$	1.5	$(0, 1.5)$
10	$y = 1.5 - 0.0625(10)$	0.875	$(10, 0.875)$
20	$y = 1.5 - 0.0625(20)$	0.25	$(20, 0.25)$

Movie Download

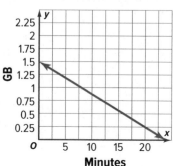

The graph appears to intersect the *x*-axis between 22 and 25.

Solve algebraically to find the exact value.

$$y = 1.5 - 0.0625x$$ Original equation

$$0 = 1.5 - 0.0625x$$ Related function

$$0 + 0.0625x = 1.5 - 0.0625x + 0.0625x$$ Add 0.0625*x* to each side.

$$0.0625x = 1.5$$ Simplify.

$$\frac{0.0625x}{0.0625} = \frac{1.5}{0.0625}$$ Divide each side by 0.0625.

$$x = 24$$ Simplify.

The zero of this function is 24. Thus, it will take Willow 24 minutes to download the movie.

The domain is $\{x \mid 0 \le x \le 24\}$, where *x* is a real number, since the amount of time can be any increment from 0 to 24 minutes. The range is $\{y \mid 0 \le y \le 1.5\}$, where *y* is a real number, since the amount of movie to download can be any number between 0 GB and 1.5 GB.

▶ **Guided Practice**

2. SCUBA Che has a 120-CF tank of gas for a 6000 feet dive. The equation $y = 120 - 2x$ describes the amount of air in his tank after *x* minutes. Find the zero and describe what it means in the context of this situation. Identify the domain and range.

In the previous example, the function was continuous, meaning that the domain can be represented by any real number in the range. That is not always the case; sometimes, the domain can only be represented by discrete values.

Real-World Example 3   Estimate a Zero by Graphing

CARNIVAL RIDES Emily is going to a local carnival. The function $m = 20 - 0.75r$ represents the amount of money *m* she has left after *r* rides. Find the zero and describe what it means in the context of this situation. Identify the domain and range.

Graph the function.

The graph appears to intersect the *r*-axis at 27.

Use algebra to find the *r*-intercept, which is about 26.67. The zero of this function is about 26.67. Since Emily cannot ride part of a ride, she can ride 26 rides before she will run out of money.

The domain is $\{x \mid 0 \le r \le 26\}$, where *r* is an integer. She can ride from 0 to 26 rides. The range is $\{m \mid 0 \le m \le 20\}$, since she can spend from $0 to $20.

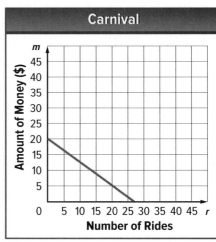

Carnival

▶ **Guided Practice**

3. FINANCIAL LITERACY Antoine's class is selling candy to raise money for a class trip. They paid $45 for the candy, and they are selling each candy bar for $1.50. The function $y = 1.50x - 45$ represents their profit *y* when they sell *x* candy bars. Find the zero and describe what it means in the context of this situation. Identify the domain and range.

Real-World Career

Entertainment Manager
An entertainment manager supervises tech tests, calls show cues, schedules performances and performers, coaches employees and guest talent, and manages expenses. Entertainment managers need a college degree in a field such as communication or theater.

Check Your Understanding ◯ = Step-by-Step Solutions begin on page R11.

Go Online! for a Self-Check Quiz

Example 1   Find the zero of each function by graphing.

1. $f(x) = -2x + 6$ **2.** $f(x) = -x - 3$

3. $f(x) = 4x - 2$ **4.** $f(x) = 9x + 3$

5. $f(x) = 13$ **6.** $f(x) = -13$

7. $f(x) = -5$ **8.** $f(x) = 2$

Examples 2–3   **9. NEWSPAPERS** The function $w = 30 - \frac{3}{4}n$ represents the weight w in pounds of the papers in Tyrone's newspaper delivery bag after he delivers n newspapers. Find the zero and explain what it means in the context of this situation. Find the domain and range.

Practice and Problem Solving
Extra Practice is on page R3.

Example 1   Find the zero of each function by graphing.

10. $f(x) = -x - 5$ **11.** $f(x) = x + 3$ **12.** $f(x) = 16$

13. $f(x) = 11$ **14.** $f(x) = 4x - 36$ **15.** $f(x) = 7x + 10$

16. $f(x) = 2x + 22$ **17** $f(x) = 7$ **18.** $f(x) = -15x$

19. $f(x) = 31$ **20.** $f(x) = 6x - 8$ **21.** $f(x) = -232$

Examples 2–3   **22. BUDGETING** Sean budgets $160 a month to cover school lunches and any additional dining expenses. The function $y = 160 - 4x$ represents the amount of money m remaining after spending $4 on x meals. Find the zero and explain what it means in the context of the situation. Find the domain and range.

23. GIFT CARDS For her birthday Kwan receives a $50 gift card to download songs. The function $m = -0.99d + 50$ represents the amount of money m that remains on the card after a number of songs d are downloaded. Find the zero and explain what it means in the context of this situation.

Find the zero of each function.

24. $f(x) = 4x + 8$ **25.** $f(x) = -16 - 2x$ **26.** $f(x) = 25 - 5x$

27. $f(x) = 10 - 3x$ **28.** $f(x) = 15 + 6x$ **29.** $f(x) = 13x + 34$

30. $f(x) = 22x - 10$ **31.** $f(x) = 25x - 17$ **32.** $f(x) = \frac{1}{2} + \frac{2}{3}x$

33. $f(x) = \frac{3}{4} - \frac{2}{5}x$ **34.** $f(x) = 13x + 117$ **35.** $f(x) = 24x - 72$

36. SEA LEVEL Parts of New Orleans lie 0.5 meter below sea level. After d days of rain the equation $w = 0.3d - 0.5$ represents the water level w in meters. Find the zero, and explain what it means in the context of this situation.

 37. **MP** **MODELING** An artist completed an ice sculpture when the temperature was $-10°C$. The equation $t = 1.25h - 10$ shows the temperature h hours after the sculpture's completion. If the artist completed the sculpture at 8:00 A.M., at what time will it begin to melt?

Find the zero of each function by graphing. Verify your answer algebraically.

38. $f(x) = 15 - x$ **39.** $f(x) = -6 - 2x$ **40.** $f(x) = -2x + 4$

41. $f(x) = -10x - 20$ **42.** $f(x) = \frac{5}{2}x - 5$ **43.** $f(x) = -\frac{8}{3}x + 3$

44. HAIR PRODUCTS Chemical hair straightening makes curly hair straight and smooth. The percent of the process left to complete is modeled by $p = -12.5t + 100$, where t is the time in minutes that the solution is left on the hair and p represents the percent of the process left to complete.

   a. Find the zero of this function.

   b. Make a graph of this situation.

   c. Explain what the zero represents in this context.

   d. State the possible domain and range of this function.

45 **MUSIC DOWNLOADS** In this problem, you will investigate the change between two quantities.

   a. Copy and complete the table.

Number of Songs Downloaded	Total Cost ($)	Total Cost Number of Songs Downloaded
2	4	
4	8	
6	12	

   b. As the number of songs downloaded increases, how does the total cost change?

   c. Interpret the value of the total cost divided by the number of songs downloaded.

H.O.T. Problems   Use **H**igher-**O**rder **T**hinking Skills

46. ERROR ANALYSIS Clarissa and Koko solve $3x + 5 = 2x + 4$ by graphing the related function. Is either of them correct? Explain your reasoning.

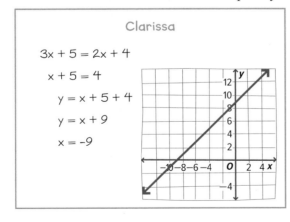

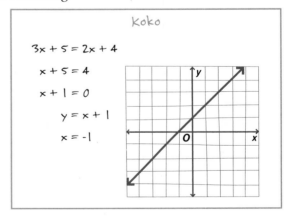

47. CHALLENGE Find the solution of $\frac{2}{3}(x + 3) = \frac{1}{2}(x + 5)$ by graphing. Verify your solution algebraically.

48. ⓂⓅ TOOLS Explain when it is better to solve an equation using algebraic methods and when it is better to solve by graphing.

49. OPEN-ENDED Write a linear equation that has a root of $-\frac{3}{4}$. Write its related function.

50. WRITING IN MATH Summarize how to solve a linear equation algebraically and graphically.

51. What is the zero of the function shown in the graph? 2

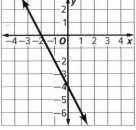

- ○ **A** −2
- ○ **B** 2
- ○ **C** 4
- ○ **D** −4

52. Which of the following functions has a zero at 4? MP 2

- ○ **A** $y = 3x + 4$
- ○ **B** $y = 4x - 2$
- ○ **C** $2x - y = -4$
- ○ **D** $x - 2y = 4$

53. Which is the best estimate for the zero of the linear function represented in the table? MP 2

- ○ **A** between 0 and 1
- ○ **B** between 2 and 3
- ○ **C** between 1 and 2
- ○ **D** between 3 and 4

x	y
0	5
1	3
2	1
3	−1
4	−3

54. The function $y = -0.00177x + 212$ estimates the boiling point of water, where y is the temperature in degrees Fahrenheit and x is the altitude in feet above sea level. What is the zero of this function? What does it represent in the context of the situation? MP 1, 4

- ○ **A** 212; the boiling point at 0 feet above sea level
- ○ **B** 119,774; the estimated number of feet above sea level when the boiling point is 0°F
- ○ **C** 0; the altitude when the boiling point is 212°F
- ○ **D** −0.00177; the change in temperature for each increase of one foot in altitude

55. Mikayla borrowed $100 from her mom. The function $y = -8x + 100$ represents her outstanding balance y after x weekly payments. What is the zero of the function? MP 2

56. Staci is walking to her friend Angie's house. The function $y = -0.5x + 10$ represents Staci's distance y in blocks from Angie's house at any time x in minutes. How long does it take Staci to walk to Angie's house? MP 2

- ○ **A** −10 minutes
- ○ **B** 5 minutes
- ○ **C** 10 minutes
- ○ **D** 20 minutes

57. The x-intercept and the y-intercept of a linear function are each equal to the same nonzero number. What is the slope of the linear function? MP 2

58. MULTI-STEP The freshman class plans to sell sweatshirts and T-shirts with the school mascot printed on the front. The function $y = -1.5x + 22.5$ represents the possible numbers of sweatshirts x and T-shirts y the class could order for $180. MP 1, 4

a. What is the zero of the function? What does it represent in the context of the situation?

- ○ **A** 22; the greatest number of T-shirts that the class can buy
- ○ **B** −1.5; the coefficient of x in the function
- ○ **C** 15; the greatest number of sweatshirts that the class can buy
- ○ **D** 180; the amount of money the class plans to spend

b. What is the domain of the function in the context of the situation?

c. What is the range of the function in the context of the situation?

The power of a graphing calculator is the ability to graph different types of equations accurately and quickly. By entering one or more equations in the calculator you can view features of a graph, such as the *x*-intercept, *y*-intercept, the origin, intersections, and the coordinates of specific points.

Mathematical Practices

MP **5** Use appropriate tools strategically.

Often linear equations are graphed in the **standard viewing window**, which is [−10, 10] by [−10, 10] with a scale of 1 on each axis. To quickly choose the standard viewing window on a TI-83/84 Plus, press ⎡ ZOOM ⎤ 6.

Activity 1   Graph a Linear Equation

Work cooperatively. Graph $3x - y = 4$.

Step 1 Enter the equation in the Y= list.

- The Y= list shows the equation or equations that you will graph.

- Equations must be entered with the *y* isolated on one side of the equation. Solve the equation for *y*, then enter it into the calculator.

$3x - y = 4$	Original equation
$3x - y - 3x = 4 - 3x$	Subtract 3x from each side.
$-y = -3x + 4$	Simplify.
$y = 3x - 4$	Multiply each side by -1.

KEYSTROKES: ⎡ Y= ⎤ 3 ⎡ X,T,θ,n ⎤ ⎡ − ⎤ 4

Step 2 Graph the equation in the standard viewing window.

- Graph the selected equation.

KEYSTROKES: ⎡ ZOOM ⎤ 6

> The equals sign appears shaded for graphs that are selected to be displayed.

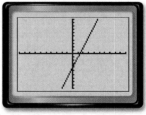

[−10, 10] scl: 1 by [−10, 10] scl: 1

Sometimes a complete graph is not displayed using the standard viewing window. A **complete graph** includes all of the important characteristics of the graph on the screen including the origin and the *x*- and *y*-intercepts. Note that the graph above is a complete graph because all of these points are visible.

When a complete graph is not displayed using the standard viewing window, you will need to change the viewing window to accommodate these important features. Use what you have learned about intercepts to help you choose an appropriate viewing window.

(continued on the next page)

Activity 2   Graph a Complete Graph

Work cooperatively. Graph $y = 5x - 14$.

Step 1 Enter the equation in the Y= list and graph in the standard viewing window.

- Clear the previous equation from the Y= list. Then enter the new equation and graph.

     KEYSTROKES: [Y=] [CLEAR] 5 [X,T,θ,n] [−] 14 [ZOOM] 6

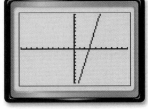

[−10, 10] scl: 1 by [−10, 10] scl: 1

Step 2 Modify the viewing window and graph again.

- The origin and the x-intercept are displayed in the standard viewing window. But notice that the y-intercept is outside of the viewing window.

Find the y-intercept.

$y = 5x - 14$ Original equation

$ = 5(0) - 14$ Replace x with 0.

$ = -14$ Simplify.

> This window allows the complete graph, including the y-intercept, to be displayed.

[−10, 10] scl: 1 by [−20, 5] scl: 1

Since the y-intercept is -14, choose a viewing window that includes a number less than -14. The window $[-10, 10]$ by $[-20, 5]$ with a scale of 1 on each axis is a good choice.

     KEYSTROKES: [WINDOW] −10 [ENTER] 10 [ENTER] 1 [ENTER] −20 [ENTER] 5 [ENTER] 1 [GRAPH]

Exercises

Work cooperatively. Use a graphing calculator to graph each equation in the standard viewing window. Sketch the result.

1. $y = x + 5$ **2.** $y = 5x + 6$ **3.** $y = 9 - 4x$

4. $3x + y = 5$ **5.** $x + y = -4$ **6.** $x - 3y = 6$

MP SENSE-MAKING Graph each equation in the standard viewing window. Determine whether the graph is complete. If the graph is not complete, adjust the viewing window and graph the equation again.

7. $y = 4x + 7$ **8.** $y = 9x - 5$ **9.** $y = 2x - 11$

10. $4x - y = 16$ **11.** $6x + 2y = 23$ **12.** $x + 4y = -36$

Consider the linear equation $y = 3x + b$.

13. Choose several different positive and negative values for b. Graph each equation in the standard viewing window.

14. For which values of b is the complete graph in the standard viewing window?

15. How is the value of b related to the y-intercept of the graph of $y = 3x + b$?

Algebra Lab
Rate of Change of a Linear Function

In mathematics, you can measure the steepness of a line using a ratio.

Mathematical Practices

MP 8 Look for and express regularity in repeated reasoning.

Set Up the Lab

- Stack three books on your desk.

- Lean a ruler on the books to create a ramp.

- Tape the ruler to the desk.

- Measure the **rise** and the **run**. Record your data in a table like the one at the right.

- Calculate and record the ratio $\frac{rise}{run}$.

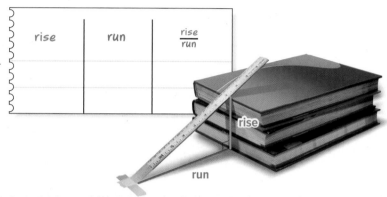

rise	run	$\frac{rise}{run}$

Activity

Step 1

Move the books to make the ramp steeper. Measure and record the **rise** and the **run**. Calculate and record $\frac{rise}{run}$.

Step 2

Add books to the stack to make the ramp even steeper. Measure, calculate, and record your data in the table.

Analyze the Results

1. Examine the ratios you recorded. How did they change as the ramp became steeper?

2. **MAKE A PREDICTION** Suppose you want to construct a skateboard ramp that is not as steep as the one shown at the right. List three different sets of $\frac{rise}{run}$ measurements that will result in a less steep ramp. Verify your predictions by calculating the ratio $\frac{rise}{run}$ for each ramp.

18 in.

24 in.

$m = \frac{18}{24} = \frac{3}{4}$

3. Copy the coordinate graph shown and draw a line through the origin with a $\frac{rise}{run}$ ratio greater than the original line. Then draw a line through the origin with a ratio less than that of the original line. Explain using the words *rise* and *run* why the lines you drew have a ratio greater or less than the original line.

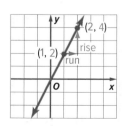

4. We have seen what happens on the graph as the $\frac{rise}{run}$ ratio gets closer to zero. What would you predict will happen when the ratio is zero? Explain your reasoning. Give an example to support your prediction.

Rate of Change and Slope

::Then	::Now	::Why?

• You graphed ordered pairs in the coordinate plane.

1 Use rate of change to solve problems.

2 Find the slope of a line.

• The Daredevil Drop at Wet 'n Wild Emerald Pointe in Greensboro, North Carolina, is a water slide that drops riders 76 feet. *A rate of change* of the ride might describe the distance a rider has fallen over a length of time.

New Vocabulary
rate of change
slope

Mathematical Practices
2 Reason abstractly and quantitatively.

1 Rate of Change **Rate of change** is a ratio that describes, on average, how much one quantity changes with respect to a change in another quantity.

Key Concept Rate of Change

If x is the independent variable and y is the dependent variable, then

$$\text{rate of change} = \frac{\text{change in } y}{\text{change in } x}$$

Real-World Example 1 Find Rate of Change

ENTERTAINMENT Use the table to find the rate of change. Then explain its meaning.

$$\text{rate of change} = \frac{\text{change in } y}{\text{change in } x} \;\;\begin{array}{l}\leftarrow \text{dollars}\\ \leftarrow \text{games}\end{array}$$

$$= \frac{\text{change in cost}}{\text{change in number of games}}$$

$$= \frac{156 - 78}{4 - 2}$$

$$= \frac{78}{2} \text{ or } \frac{39}{1}$$

Number of Computer Games	Total Cost ($)
x	y
2	78
4	156
6	234

The rate of change is $\frac{39}{1}$ or 39. This means that the cost per game is $39.

▶ **Guided Practice**

1. REMODELING The table shows how the tiled surface area changes with the number of floor tiles.

A. Find the rate of change.

B. Explain the meaning of the rate of change.

Number of Floor Tiles	Area of Tiled Surface (in²)
x	y
3	48
6	96
9	144

So far, you have seen rates of change that are *constant*. Many real-world situations involve rates of change that are not constant.

Real-World Example 2   Compare Rates of Change

PETS The graph shows the amount spent on pets in the U.S. in recent years.

a. **Find the rates of change for 2003–2007 and 2011–2015.**

2003–2007:

$$\frac{\text{change in sales}}{\text{change in time}} = \frac{41.2 - 32.4}{2007 - 2003} \frac{\text{dollars}}{\text{years}} \qquad \text{Substitute.}$$
$$= 2.2 \qquad \text{Simplify.}$$

Over this 4-year period, sales increased by 8.8 billion, for a rate of change of 2.2 billion per year.

2011–2015:

$$\frac{\text{change in sales}}{\text{change in time}} = \frac{60.28 - 50.96}{2015 - 2011} \frac{\text{dollars}}{\text{years}} \qquad \text{Substitute.}$$
$$= 2.33 \qquad \text{Simplify.}$$

Over this 4-year period, sales increased by 9.32 billion, for a rate of change of 2.33 billion per year.

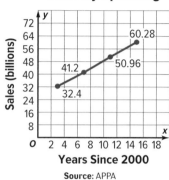

Total U.S. Pet Industry Spending

Sales (billions) vs. Years Since 2000

Source: APPA

b. **Explain the meaning of the rate of change in each case.**

For 2003–2007, on average, 2.2 billion more was spent on pets each year than the previous.

For 2011–2015, on average, 2.33 billion more was spent on pets each year than the previous.

c. **How are the different rates of change shown on the graph?**

There is a greater vertical change for 2011–2015 than for 2003–2007. Therefore, the section of the graph for 2011–2015 is steeper.

▶ **Guided Practice**

2. Refer to the graph above. Without calculating, find the 4-year period that has the greatest rate of change. Then calculate to verify your answer.

Study Tip

MP **Reasoning** A positive rate of change indicates an increase over time. A negative rate of change indicates that a quantity is decreasing.

A rate of change is constant for a function when the rate of change is the same between any pair of points on the graph of the function. Linear functions have a constant rate of change.

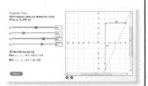
Example 3   Constant Rates of Change

Determine whether each function is linear. Explain.

a.

x	y
1	−6
4	−8
7	−10
10	−12
13	−14

b.

x	y
−3	10
−1	12
1	16
3	18
5	22

Study Tip

Linear or Nonlinear Function? Notice that the changes in x and y are not the same. For the rate of change to be linear, the change in x-values must be constant, and the change in y-values must be constant.

---

x	y	rate of change	
1	−6	$\dfrac{-8-(-6)}{4-1}$	or $-\dfrac{2}{3}$
4	−8	$\dfrac{-10-(-8)}{7-4}$	or $-\dfrac{2}{3}$
7	−10	$\dfrac{-12-(-10)}{10-7}$	or $-\dfrac{2}{3}$
10	−12	$\dfrac{-14-(-12)}{13-10}$	or $-\dfrac{2}{3}$
13	−14		

The rate of change is constant. Thus, the function is linear.

x	y	rate of change	
−3	10	$\dfrac{12-10}{-1-(-3)}$	or 1
−1	12	$\dfrac{16-12}{1-(-1)}$	or 2
1	16	$\dfrac{18-16}{3-1}$	or 1
3	18	$\dfrac{22-18}{5-3}$	or 2
5	22		

This rate of change is not constant. Thus, the function is not linear.

 Guided Pracatice

3A.

x	y
−3	11
−2	15
−1	19
1	23
2	27

3B.

x	y
12	−4
9	1
6	6
3	11
0	16

---

2 Find Slope The **slope** of a nonvertical line is the ratio of the change in the y-coordinates (rise) to the change in the x-coordinates (run) as you move from one point to another.

It can be used to describe a rate of change. Slope describes how steep a line is. The greater the absolute value of the slope, the steeper the line.

The graph shows a line that passes through $(-1, 3)$ and $(2, -2)$.

$$\textbf{slope} = \frac{\text{rise}}{\text{run}}$$

$$= \frac{\text{change in } y\text{-coordinates}}{\text{change in } x\text{-coordiantes}}$$

$$= \frac{-2-3}{2-(-1)} \text{ or } -\frac{5}{3}$$

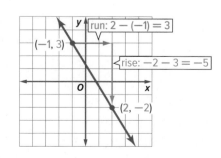

So, the slope of the line is $-\dfrac{5}{3}$.

Because a linear function has a constant rate of change, any two points on a nonvertical line can be used to determine its slope.

Key Concept Slope

Words	The slope of a nonvertical line is the ratio of the rise to the run.	Graph	
Symbols	The slope m of a nonvertical line through any two points, (x_1, y_1) and (x_2, y_2), can be found as follows.		

$$m = \frac{y_2 - y_1}{x_2 - x_1} \longleftarrow \text{change in } y$$
$$\longleftarrow \text{change in } x$$

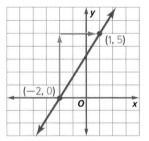

Reading Math

(MP) Structure y_1 is read as *y sub one*, and x_2 is read as *x sub two*. The 1 and 2 are subscripts and refer to the first and second point to which the *x*- and *y*-values correspond.

The slope of a line can be positive, negative, zero, or undefined. If the line is not horizontal or vertical, then the slope is either positive or negative.

Example 4 — Positive, Negative, and Zero Slope

Find the slope of a line that passes through each pair of points.

a. $(-2, 0)$ and $(1, 5)$

$$m = \frac{y_2 - y_1}{x_2 - x_1} \qquad \frac{\text{rise}}{\text{run}}$$

$$= \frac{5 - 0}{1 - (-2)} \qquad (-2, 0) = (x_1, y_1) \text{ and } (1, 5) = (x_2, y_2)$$

$$= \frac{5}{3} \qquad \text{Simplify.}$$

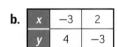

Watch Out!

Order Be careful not to transpose the order of the *x*-values or the *y*-values.

b.

x	−3	2
y	4	−3

$$m = \frac{y_2 - y_1}{x_2 - x_1} \qquad \frac{\text{rise}}{\text{run}}$$

$$= \frac{-3 - 4}{2 - (-3)} \qquad (-3, 4) = (x_1, y_1) \text{ and } (2, -3) = (x_2, y_2)$$

$$= \frac{-7}{5} \text{ or } -\frac{7}{5} \qquad \text{Simplify.}$$

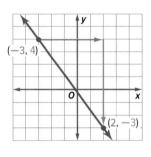

c. $(-3, -1)$ and $(2, -1)$

$$m = \frac{y_2 - y_1}{x_2 - x_1} \qquad \frac{\text{rise}}{\text{run}}$$

$$= \frac{-1 - (-1)}{2 - (-3)} \qquad \text{Substitute.}$$

$$= \frac{0}{5} \text{ or } 0 \qquad \text{Simplify.}$$

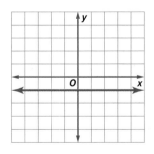

Guided Practice

Find the slope of the line that passes through each pair of points.

4A.

x	3	4
y	6	8

4B. $(-4, -2)$ and $(0, -2)$

4C.

x	−4	−2
y	2	10

Example 5 Undefined Slope

Find the slope of the line that passes through $(-2, 4)$ and $(-2, -3)$.

$$m = \frac{y_2 - y_1}{x_2 - x_1} \qquad \frac{\text{rise}}{\text{run}}$$

$$= \frac{-3 - 4}{-2 - (-2)} \qquad \text{Substitute.}$$

$$= \frac{-7}{0} \text{ or undefined} \qquad \text{Simplify.}$$

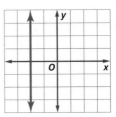

Study Tip

Zero and Undefined Slopes
If the change in y-values is 0, then the graph of the line is horizontal. If the change in x-values is 0, then the slope is undefined. This graph is a vertical line.

▶ **Guided Practice**

Find the slope of the line that passes through each pair of points.

5A. $(6, 3), (6, 7)$ **5B.** $(-3, 2), (-3, -1)$

The graphs of lines with different slopes are summarized below.

> **Key Concept** Slope

positive slope	negative slope	slope of 0	undefined slope
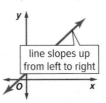	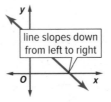	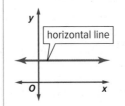	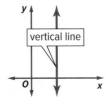
The function values are increasing over the entire domain.	The function values are decreasing over the entire domain.	The function values are constant over the entire domain.	The relation is not a function.

Example 6 Find Coordinates Given the Slope

Find the value of r so that the line through $(1, 4)$ and $(-5, r)$ has a slope of $\frac{1}{3}$.

$$m = \frac{y_2 - y_1}{x_2 - x_1} \qquad \text{Slope Formula}$$

$$\frac{1}{3} = \frac{r - 4}{-5 - 1} \qquad \text{Let } (1, 4) = (x_1, y_1) \text{ and } (-5, r) = (x_2, y_2).$$

$$\frac{1}{3} = \frac{r - 4}{-6} \qquad \text{Subtract.}$$

$$3(r - 4) = 1(-6) \qquad \text{Find the cross products.}$$

$$3r - 12 = -6 \qquad \text{Distributive Property}$$

$$3r = 6 \qquad \text{Add 12 to each side and simplify.}$$

$$r = 2 \qquad \text{Divide each side by 3 and simplify.}$$

So, the line goes through $(-5, 2)$.

▶ **Guided Practice**

Find the value of r so the line that passes through each pair of points has the given slope.

6A. $(-2, 6), (r, -4); m = -5$ **6B.** $(r, -6), (5, -8); m = -8$

Check Your Understanding   ◯ = Step-by-Step Solutions begin on page R11.

✓ **Go Online!** for a
Self-Check Quiz

Example 1

Find the rate of change represented in each table or graph.

1.

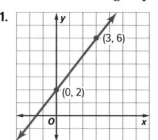

2.

x	y
3	−6
5	2
7	10
9	18
11	26

Example 2

3. **MP** **SENSE-MAKING** Refer to the graph at the right.

  a. Find the rate of change of prices from 2006 to 2008. Explain the meaning of the rate of change.

  b. Without calculating, find a two-year period that had a greater rate of change than 2006–2008. Explain.

  c. Between which years would you guess the new stadium was built? Explain your reasoning.

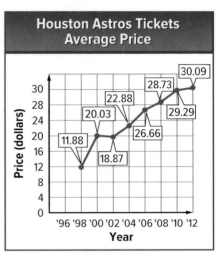

Source: *Team Marketing Report*

Example 3

Determine whether each function is linear. Write *yes* or *no*. Explain.

4.

x	−7	−4	−1	2	5
y	5	4	3	2	1

5.

x	8	12	16	20	24
y	7	5	3	0	−2

Examples 4–5 Find the slope of the line that passes through each pair of points.

6. (5, 3), (6, 9)

7. (−4, 3), (−2, 1)

8. (6, −2), (8, 3)

9. (1, 10), (−8, 3)

10. (−3, 7), (−3, 4)

11. (5, 2), (−6, 2)

Example 6

Find the value of *r* so the line that passes through each pair of points has the given slope.

12. (−4, r), (−8, 3), m = −5

13. (5, 2), (−7, r), m = $\frac{5}{6}$

Practice and Problem Solving   Extra Practice is on page R3.

Example 1

Find the rate of change represented in each table or graph.

14.

x	y
5	2
10	3
15	4
20	5

15

x	y
1	15
2	9
3	3
4	−3

Example 1   Find the rate of change represented in each table or graph.

16.

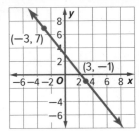

(−3, 7)

(3, −1)

17.

(−4, −6)   (4, −2)

Example 2   **18. SPORTS** What was the annual rate of change from 2011 to 2015 for women participating in collegiate lacrosse? Explain the meaning of the rate of change.

Year	Number of Women
2011	13,532
2015	15,902

19. RETAIL The average retail price in a recent year for a used car is shown in the table at the right.

Age (years)	Value ($)
2	19,116
3	17,772

   a. Write a linear function to model the price of the car with respect to age.

   b. Interpret the meaning of the slope of the line.

   c. Assuming a constant rate of change, predict the average retail price for a 7-year-old car.

Example 3   Determine whether each function is linear. Write *yes* or *no*. Explain.

20.

x	4	2	0	−2	−4
y	−1	1	3	5	7

21.

x	−7	−5	−3	−1	0
y	11	14	17	20	23

22.

x	−0.2	0	0.2	0.4	0.6
y	0.7	0.4	0.1	0.3	0.6

23.

x	$\frac{1}{2}$	$\frac{3}{2}$	$\frac{5}{2}$	$\frac{7}{2}$	$\frac{9}{2}$
y	$\frac{1}{2}$	1	$\frac{3}{2}$	2	$\frac{5}{2}$

Examples 4–5   Find the slope of the line that passes through each pair of points.

24. (4, 3), (−1, 6)   **25** (8, −2), (1, 1)   **26.** (2, 2), (−2, −2)

27. (−3, 5), (3, 6)   **28.** (−3, 2), (7, 2)   **29.** (8, 10), (−4, −6)

30. (−8, 6), (−8, 4)   **31.** (−12, 15), (18, −13)   **32.** (−8, −15), (−2, 5)

Example 6   Find the value of *r* so the line that passes through each pair of points has the given slope.

33. (12, 10), (−2, r), m = −4   **34.** (r, −5), (3, 13), m = 8

35. (3, 5), (−3, r), $m = \frac{3}{4}$ **36.** (−2, 8), (r, 4), $m = -\frac{1}{2}$

Find the rate of change of each line.

37. $y = 3x - 4$ **38.** $y = -\frac{2}{3}x + 5$ **39.** $y = \frac{4}{5}x - \frac{7}{5}$

MP TOOLS Use a ruler to estimate the slope of each object.

40.

41.

42. HIKING When hiking up a certain hill, you rise 15 feet for every 1250 feet you walk forward. What is the slope of the trail?

Find the slope of the line that passes through each pair of points.

43.

x	y
4.5	−1
5.3	2

44.

x	y
0.75	1
0.75	−1

45.

x	y
$2\frac{1}{2}$	$-1\frac{1}{2}$
$-\frac{1}{2}$	$\frac{1}{2}$

46. MULTIPLE REPRESENTATIONS In this problem, you will investigate why the slope of a line through any two points on that line is constant.

   a. Visual Sketch a line ℓ that contains points A, B, A' and B' on a coordinate plane.

   b. Geometric Add segments to form right triangles ABC and $A'B'C'$ with right angles at C and C'. Describe $\overline{AC}$ and $\overline{A'C'}$, and $\overline{BC}$ and $\overline{B'C'}$.

   c. Verbal How are triangles ABC and $A'B'C'$ related? What does that imply for the slope between any two distinct points on line ℓ?

47 BASKETBALL The table shown below shows the average points per game (PPG) Michael Redd has scored in each of his first 9 seasons with the NBA's Milwaukee Bucks.

Season	1	2	3	4	5	6	7	8	9
PPG	2.2	11.4	15.1	21.7	23.0	25.4	26.7	22.7	21.2

   a. Make a graph of the data. Connect each pair of adjacent points with a line.

   b. Use the graph to determine in which period Michael Redd's PPG increased the fastest. Explain your reasoning.

   c. Discuss the difference in the rate of change from season 1 through season 4, from season 4 through season 7, and from season 7 through season 9.

H.O.T. Problems   Use **H**igher-**O**rder **T**hinking Skills

48. MP REASONING Why does the Slope Formula not work for vertical lines? Explain.

49. OPEN ENDED Use what you know about rate of change to describe the function represented by the table.

Time (wk)	Height of Plant (in.)
4	9.0
6	13.5
8	18.0

50. CHALLENGE Find the value of d so the line that passes through (a, b) and (c, d) has a slope of $\frac{1}{2}$.

51. WRITING IN MATH Explain how the rate of change and slope are related and how to find the slope of a line.

52. MP CRITIQUE ARGUMENTS Kyle and Luna are finding the value of a so the line that passes through $(10, a)$ and $(−2, 8)$ has a slope of $\frac{1}{4}$. Is either of them correct? Explain.

Kyle

$$\frac{-2 - 10}{8 - a} = \frac{1}{4}$$
$$1(8 - a) = 4(-12)$$
$$8 - a = -48$$
$$a = 56$$

Luna

$$\frac{8 - a}{-2 - 10} = \frac{1}{4}$$
$$4(8 - a) = 1(-12)$$
$$32 - 4a = -12$$
$$a = 11$$

53. Emily earns extra spending money by babysitting children in her neighborhood. On Tuesday, she earned $16 for babysitting for two hours. On Saturday, she earned $40 for babysitting for 5 hours. Let x represent the number of hours and y represent the amount earned. What is the rate of change? **MP** 2

○ **A** $\frac{24}{7}$

○ **B** $\frac{8}{1}$

○ **C** $-\frac{38}{11}$

○ **D** $\frac{5}{2}$

54. The table shows the number of minutes it takes a bicyclist to travel along a level road. What is the rate of change in meters per minute? **MP** 2

Bicycling	
Time (min)	Distance (m)
3	1125
5	1875
8	3000

○ **A** $-\frac{375}{1}$

○ **B** $-\frac{1}{375}$

○ **C** $\frac{1}{375}$

○ **D** $\frac{375}{1}$

55. The graph shows the temperature in degrees Celsius over time in minutes. **MP** 2

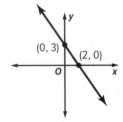

What is the rate of change in temperature in degrees Celsius per minute?

56. The graph shows the relationship between the weight in pounds of chocolate covered almonds and their cost. **MP** 1, 4

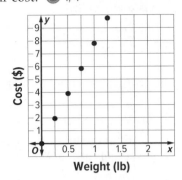

Which is the best estimate for the cost per pound?

○ **A** $1.95

○ **B** $3.90

○ **C** $5.85

○ **D** $7.80

○ **E** $9.75

57. The table shows the perimeter and side length of a regular polygon. **MP** 2, 3

Side length (cm)	4	5.5	7	9.1	10.25
Perimeter (cm)	12	16.5	21	27.3	30.75

Use the rate of change for the data in the table to determine the type of polygon. Explain your reasoning.

58. MULTI-STEP The table shows the amount in Tamina's savings account over several months. **MP** 1, 3

Month	0	1	2	3	4	5	6	7	8
Amount ($)	225	300	375	450	525	550	575	575	575

a. Make a graph of the data. Connect adjacent points with a line segment.

b. Over what interval did Tamina's balance have the greatest rate of change? Explain your reasoning.

c. If the rate of change from month 0 to month 1 had continued, how much money would have been in Tamina's account in month 8? Justify your answer.

MP REASONING Determine whether each equation is a linear equation. Write *yes* or *no*. If yes, write the equation in standard form. (Lesson 3-1)

1. $y = -4x + 3$

2. $x^2 + 3y = 8$

3. $\frac{1}{4}x - \frac{3}{4}y = -1$

Graph each equation using the *x*- and *y*-intercepts. (Lesson 3-1)

4. $y = 3x - 6$

5. $2x + 5y = 10$

Graph each equation by making a table. (Lesson 3-1)

6. $y = -2x$

7. $x = 8 - y$

8. MP MODELING The equation $8x + 18y = 288$ describes the total amount of money collected when selling x paperback books at \$8 per book and y hardback books at \$18 per book. Graph the equation using the *x*- and *y*-intercepts. (Lesson 3-1)

Find the zeros of each function by graphing. (Lesson 3-2)

9. $f(x) = x + 8$

10. $f(x) = 4x - 24$

11. $f(x) = 18 + 8x$

12. $f(x) = \frac{3}{5}x - \frac{1}{2}$

13. $f(x) = -5x + 35$

14. $f(x) = 14x - 84$

15. $f(x) = 121 + 11x$

16. The function $y = -15 + 3x$ represents the outside temperature, in degrees Fahrenheit, in a small Alaskan town where x represents the number of hours after midnight. The function is accurate for x-values representing midnight through 4:00 P.M. Find the zero of this function. (Lesson 3-2)

17. Find the rate of change represented in the table. (Lesson 3-3)

x	y
1	2
4	6
7	10
10	14

Find the slope of the line that passes through each pair of points. (Lesson 3-3)

18. $(2, 6), (4, 12)$

19. $(1, 5), (3, 8)$

20. $(-3, 4), (2, -6)$

21. $\left(\frac{1}{3}, \frac{3}{4}\right), \left(\frac{2}{3}, \frac{1}{4}\right)$

22. MP STRUCTURE Find the value of r so the line that passes through the pair of points has the given slope. (Lesson 3-3)

$$(-4, 8), (r, 12), m = \frac{4}{3}$$

23. Find the slope of the line that passes through the pair of points. (Lesson 3-3)

x	y
2.6	−2
3.1	4

24. MP CONSTRUCT ARGUMENTS The graph shows the population growth in Heckertsville since 2008. (Lesson 3-3)

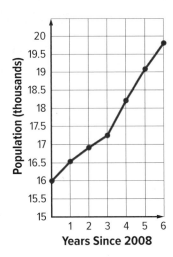

a. For which time period is the rate of change the greatest?

b. Explain the meaning of the slope from 2008 to 2014.

c. MP Which mathematical practice did you use to solve this problem?

25. Explain the relationship between the *x*-intercepts of a graph and the zeros of the related function. (Lessons 3-1, 3-2)

26. A certain linear function has a negative slope. How do the *y*-values change as the *x*-values increase? (Lesson 3-3)

Graphing Technology Lab
Investigating Slope-Intercept Form

Set Up the Lab

- Cut a small hole in a top corner of a plastic sandwich bag. Hang the bag from the end of the force sensor.

- Connect the force sensor to your data collection device.

Mathematical Practices
MP **5** Use appropriate tools strategically.

Activity Collect Data

Work cooperatively.

Step 1 Use the sensor to collect the weight with 0 washers in the bag. Record the data pair in the calculator.

Step 2 Place one washer in the plastic bag. Wait for the bag to stop swinging, then measure and record the weight.

Step 3 Repeat the experiment, adding different numbers of washers to the bag. Each time, record the number of washers and the weight.

Analyze the Results

1. The domain contains values of the independent variable, number of washers. The range contains values of the dependent variable, weight. Use the graphing calculator to create a scatter plot using the ordered pairs (washers, weight).

2. Write a sentence that describes the points on the graph.

3. Describe the position of the point that represents the trial with no washers in the bag.

4. The rate of change can be found by using the formula for slope.

$$\frac{\text{rise}}{\text{run}} = \frac{\text{change in weight}}{\text{change in number of washers}}$$

Find the rate of change in the weight as more washers are added.

5. Explain how the rate of change is shown on the graph.

Make a Conjecture

The graph shows sample data from a washer experiment.
Describe the graph for each situation.

6. a bag that hangs weighs 0.8 N when empty and increases in weight at the rate of the sample

7. a bag that has the same weight when empty as the sample and increases in weight at a faster rate

8. a bag that has the same weight when empty as the sample and increases in weight at a slower rate

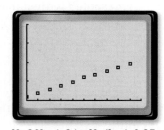

[0, 20] scl: 2 by [0, 1] scl: 0.25

Ed Imaging

Slope-Intercept Form

:: **Then**

- You found rates of change and slopes.

:: **Now**

1. Write and graph linear equations in slope-intercept form.

2. Model real-world data with equations in slope-intercept form.

:: **Why?**

- Jamil has 500 songs on his smartphone. He joins a music club that lets him download 30 songs per month for a monthly fee. The number of songs that Jamil could eventually have if he does not delete any songs is represented by $y = 30x + 500$ for x months.

 New Vocabulary

slope-intercept form
constant function

 Mathematical Practices

2 Reason abstractly and quantitatively.

8 Look for and express regularity in repeated reasoning.

1 **Slope-Intercept Form** An equation of the form $y = mx + b$, where m is the slope and b is the y-intercept, is in **slope-intercept form**. The variables m and b are called *parameters* of the equation. Changing either value changes the equation's graph.

🔁 Key Concept Slope-Intercept Form

Words	The slope-intercept form of a linear equation is $y = mx + b$, where m is the slope and b is the y-intercept.

Example

$$y = mx + b$$
$$y = 2x + 6$$

slope↑ ↑ y-intercept

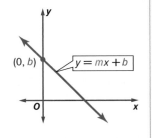

Example 1 Write and Graph an Equation

Write an equation in slope-intercept form for the line with a slope of $\frac{3}{4}$ and a y-intercept of -2. Then graph the equation.

$y = mx + b$ — Slope-intercept form

$y = \frac{3}{4}x + (-2)$ — Replace m with $\frac{3}{4}$ and b with -2.

$y = \frac{3}{4}x - 2$ — Simplify.

Now graph the equation.

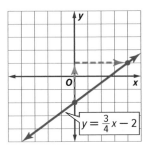

Step 1 Plot the y-intercept $(0, -2)$.

Step 2 The slope is $\frac{\text{rise}}{\text{run}} = \frac{3}{4}$. From $(0, -2)$, move up 3 units and right 4 units. Plot the point.

Step 3 Draw a line through the two points.

▶ Guided Practice

Write an equation of a line in slope-intercept form with the given slope and y-intercept. Then graph the equation.

1A. slope: $-\frac{1}{2}$, y-intercept: 3

1B. slope: -3, y-intercept: -8

Go Online!

Investigate how changing the values of *m* and *b* in the slope-intercept form of a linear equation affects the graph by using the **Virtual Manipulatives**.

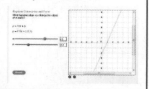

When an equation is not written in slope-intercept form, it may be easier to rewrite it before graphing.

Example 2   Graph Linear Equations

Graph $3x + 2y = 6$. Then state the slope and y-intercept.

Rewrite the equation in slope-intercept form.

$3x + 2y = 6$	Original equation
$3x + 2y - 3x = 6 - 3x$	Subtract 3x from each side.
$2y = 6 - 3x$	Simplify.
$2y = -3x + 6$	$6 - 3x = 6 + (-3x)$ or $-3x + 6$
$\dfrac{2y}{2} = \dfrac{-3x + 6}{2}$	Divide each side by 2.
$y = -\dfrac{3}{2}x + 3$	Slope-intercept form

Now graph the equation. The slope is $-\dfrac{3}{2}$, and the y-intercept is 3.

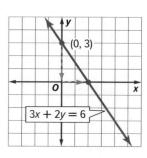

Step 1 Plot the y-intercept (0, 3).

Step 2 The slope is $\dfrac{\text{rise}}{\text{run}} = -\dfrac{3}{2}$. From (0, 3), move down 3 units and right 2 units. Plot the point.

Step 3 Draw a line through the two points.

▶ **Guided Practice**

Graph each equation. Then state the slope and y-intercept.

2A. $3x - 4y = 12$ **2B.** $-2x + 5y = 10$

Study Tip

MP **Structure** When counting rise and run, a negative sign may be associated with the value in the numerator or denominator. If with the numerator, begin by counting down for the rise. If with the denominator, count left when counting the run. The resulting line will be the same.

Except for the graph of $y = 0$, which lies on the x-axis, horizontal lines have a slope of 0. They are graphs of **constant functions**, which can be written in slope-intercept form as $y = 0x + b$ or $y = b$, where b is any number. Constant functions do not cross the x-axis. Their domain is all real numbers, and their range is b.

Example 3   Graph Linear Equations

Graph $y = -3$. Then state the slope and y-intercept.

Step 1 Plot the y-intercept (0, −3).

Step 2 The slope is 0. Draw a line through the points with y-coordinate −3.

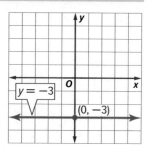

▶ **Guided Practice**

Graph each equation. Then state the slope and y-intercept.

3A. $y = 5$ **3B.** $2y = 1$

Vertical lines have no slope. So, equations of vertical lines cannot be written in slope-intercept form.

There are times when you will need to write an equation when given a graph. To do this, locate the *y*-intercept and use the rise and run to find another point on the graph. Then write the equation in slope-intercept form.

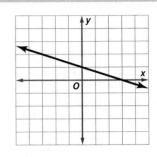

Example 4 — Write an Equation in Slope-Intercept Form

Which of the following is an equation in slope-intercept form for the line shown?

A $y = -3x + 1$

B $y = -3x + 3$

C $y = -\frac{1}{3}x + 1$

D $y = -\frac{1}{3}x + 3$

<aside>
StudyTip

Eliminating Choices Analyze the graph to determine the slope and the *y*-intercept. Then you can save time by eliminating answer choices that do not match the graph.
</aside>

Read the Item

You need to find the slope and *y*-intercept of the line to write the equation.

Solve the Item

Step 1 The line crosses the *y*-axis at (0, 1), so the *y*-intercept is 1. The answer is either A or C.

Step 2 To get from (0, 1) to (3, 0), go down 1 unit and 3 units to the right.
The slope is $-\frac{1}{3}$.

Step 3 Write the equation.

$$y = mx + b$$
$$y = -\frac{1}{3}x + 1$$

CHECK The graph also passes through (−3, 2). If the equation is correct, this should be a solution.

$$y = -\frac{1}{3}x + 1$$
$$2 \overset{?}{=} -\frac{1}{3}(-3) + 1$$
$$2 \overset{?}{=} 1 + 1$$
$$2 = 2 \checkmark \qquad \text{The answer is C.}$$

▶ **Guided Practice**

4. Which of the following is an equation in slope-intercept form for the line shown?

A $y = \frac{1}{4}x - 1$

B $y = \frac{1}{4}x + 4$

C $y = 4x - 1$

D $y = 4x + 4$

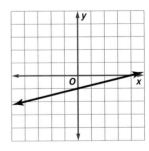

2 Modeling Real-World Data Real-world data can be modeled by a linear equation if there is a constant rate of change. The rate of change represents the slope. The *y*-intercept is the point where the value of the independent variable is 0.

Real-World Example 5   **Write and Graph a Linear Equation**

SPORTS Use the information at the left about high school sports.

a. Write a linear equation to find the number of girls in high school sports after 2002.

Words	Number of girls competing	equals	rate of change	times	number of years	plus	amount at start.
Variables	Let G = number of girls competing.			Let n = number of years since 2002.			
Equation	G	=	0.05	×	n	+	2.8

The equation is $G = 0.05n + 2.8$.

b. Graph the equation. Then state the slope and y-intercept.

The y-intercept is where the data begins. So, the graph passes through $(0, 2.8)$.

The rate of change is the slope, so the slope is 0.05. The y-intercept is 2.8.

c. Estimate the number of girls competing in 2027.

The year 2027 is 25 years after 2002.

$G = 0.05n + 2.8$ Write the equation.

$\quad = 0.05(\mathbf{25}) + 2.8$ Replace n with 25.

$\quad = 4.05$ Simplify.

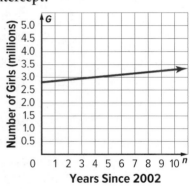

There will be about 4.05 million girls competing in high school sports in 2027.

Real-World Link

In 2002, about 2.8 million girls competed in high school sports. The number of girls competing in high school sports has increased by an average of 0.05 million per year since 2002.

Source: National Federation of High School Associations

> **Guided Practice**

5. FUNDRAISERS The band boosters are selling sandwiches for $5 each. They bought $1160 in ingredients.

    a. Write an equation for the profit P made on n sandwiches.

    b. Graph the equation. Then state the slope and y-intercept.

    c. Find the total profit if 1400 sandwiches are sold.

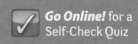

Go Online! for a Self-Check Quiz

Check Your Understanding   = Step-by-Step Solutions begin on page R11.

Example 1   Write an equation of a line in slope-intercept form with the given slope and y-intercept. Then graph the equation.

    (1) slope: 2, y-intercept: 4   **2.** slope: -5, y-intercept: 3

    3. slope: $\frac{3}{4}$, y-intercept: -1 **4.** slope: $-\frac{5}{7}$, y-intercept: $-\frac{2}{3}$

Examples 2–3   Graph each equation. Then state the slope and y-intercept.

    5. $-4x + y = 2$ **6.** $2x + y = -6$

    7. $-3x + 7y = 21$ **8.** $6x - 4y = 16$

    9. $y = -1$ **10.** $15y = 3$

Example 4   Write an equation in slope-intercept form for each graph shown.

11.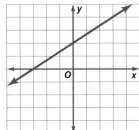

12.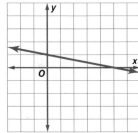

13.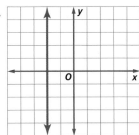

14.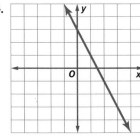

Example 5   **15. FINANCIAL LITERACY** Rondell is buying a new stereo system for his car using Jack's Stereo layaway plan.

Jack's Stereo Layaway Plan
$75 down and
$10 each week

  a. Write an equation for the total amount S that he paid after w weeks.

  b. Graph the equation. Then state the slope and y-intercept.

  c. Find out how much Rondell will have paid after 8 weeks.

16. MP REASONING Ana is driving from her home in Miami, Florida, to her grandmother's house in Dallas. On the first day, she will travel 240 miles to Orlando, Florida, to pick up her cousin. Then they will travel 350 miles each day.

  a. Write an equation that models the total number of miles m Ana has traveled, if d represents the number of days after she picks up her cousin.

  b. Graph the equation.

  c. How long will the drive take if the total length of the trip is 1313 miles?

  d. What assumption(s) did you make?

Practice and Problem Solving

Extra Practice is on page R3.

Example 1   Write an equation of a line in slope-intercept form with the given slope and y-intercept. Then graph the equation.

(17) slope: 5, y-intercept: 8

18. slope: 3, y-intercept: 10

19. slope: -4, y-intercept: 6

20. slope: -2, y-intercept: 8

21. slope: 3, y-intercept: -4

22. slope: 4, y-intercept: -6

Examples 2–3   Graph each equation. Then state the slope and y-intercept.

23. $-3x + y = 6$

24. $-5x + y = 1$

25. $-2x + y = -4$

26. $y = 7x - 7$

27. $5x + 2y = 8$

28. $4x + 9y = 27$

29. $y = 7$

30. $y = -\dfrac{2}{3}$

31. $21 = 7y$

32. $3y - 6 = 2x$

Example 4 **Write an equation in slope-intercept form for each graph shown.**

33.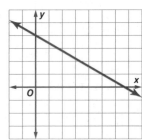

34.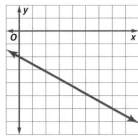

35.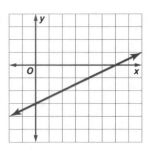

36.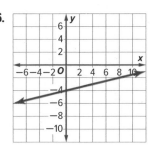

Example 5

37 **MANATEES** In 1991, 1267 manatees inhabited Florida's waters. The manatee population has increased at a rate of 123 manatees per year.

a. Write an equation for the manatee population, P, t years since 1991.

b. Graph this equation.

c. In 2006, the manatee was removed from Florida's endangered species list. What was the manatee population in 2006?

Write an equation of a line in slope-intercept form with the given slope and y-intercept.

38. slope: $\frac{1}{2}$, y-intercept: -3

39. slope: $\frac{2}{3}$, y-intercept: -5

40. slope: $-\frac{5}{6}$, y-intercept: 5

41. slope: $-\frac{3}{7}$, y-intercept: 2

42. slope: 1, y-intercept: 4

43. slope: 0, y-intercept: 5

Graph each equation.

44. $y = \frac{3}{4}x - 2$

45. $y = \frac{5}{3}x + 4$

46. $3x + 8y = 32$

47. $5x - 6y = 36$

48. $-4x + \frac{1}{2}y = -1$

49. $3x - \frac{1}{4}y = 2$

50. TRAVEL A rental company charges $75 per hour for a jet ski plus a $5 fee for a life jacket.

a. Write an equation in slope-intercept form for the total rental cost C for a jet ski and a life jacket for t hours.

b. Graph this equation. Then state the slope and y-intercept.

c. What would the cost be for 2 life jackets and 2 jet skis for 8 hours?

51. **MP** **REASONING** For Illinois residents, the average tuition at Chicago State University is $157 per credit hour. Fees cost $218 per year.

a. Write an equation in slope-intercept form for the tuition T for c credit hours in one year.

b. Find the cost for a student who is taking 32 credit hours in one year.

Write an equation of a line in slope-intercept form with the given slope and y-intercept.

52. slope: -1, y-intercept: 0

53. slope: 0.5, y-intercept: 7.5

54. slope: 0, y-intercept: 7

55. slope: -1.5, y-intercept: -0.25

56. Write an equation of a horizontal line that crosses the y-axis at $(0, -5)$.

57. Write an equation of a line that passes through the origin and has a slope of 3.

58. TEMPERATURE The temperature dropped rapidly overnight. Starting at 80°F, the temperature dropped 3° per hour.

    a. Draw a graph that represents this drop from 0 to 8 hours.

    b. Write an equation that describes this situation. Describe the meaning of each variable as well as the slope and y-intercept.

59. FITNESS Refer to the information at the right.

    a. Write an equation that represents the cost C of a membership for m months.

    b. What does the slope represent?

    c. What does the C-intercept represent?

    d. What is the cost of a two-year membership?

GET FIT GYM
startup fee **$145**
$45 monthly fee

60. MULTI-STEP The rate of change in circulation of a certain magazine has been declining steadily for the last 10 years. In 2013, the circulation was at about 450,000 subscribers.

    a. Describe and explain every characteristic of a graph representing the circulation after 2013.

    b. What assumptions did you make?

    c. Make a prediction about the future of the magazine.

61. SMART PHONES A company sold 3305 apps in the first week of availability. Suppose, on average, they expect to sell 25 apps per hour.

    a. Write an equation for the number of apps A sold t weeks after the first week of availability.

    b. If sales continue at this rate, how many weeks will it take for the company to sell 100,000 apps?

H.O.T. Problems　　　Use **H**igher-**O**rder **T**hinking Skills

62. OPEN ENDED Draw a graph representing a real-world linear function and write an equation for the graph. Describe what the graph represents.

63. (MP) REASONING Determine whether the equation of a vertical line can be written in slope-intercept form. Explain your reasoning.

64. CHALLENGE Summarize the characteristics that the graphs $y = 3x + 4$, $y = 2x + 4$, $y = -x + 4$, and $y = -5x + 4$ have in common.

65. (MP) REGULARITY If given an equation in standard form, explain how to determine the rate of change.

66. WRITING IN MATH Explain how you would use a given y-intercept and the slope to predict a y-value for a given x-value without graphing.

67. Which of the following is an equation of the graph shown in slope-intercept form?
 1

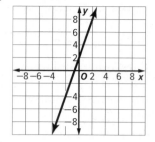

  - A $y = 3x + 2$
  - B $y = x - 1$
  - C $y = -3x + 2$
  - D $y = 2x + 2$

68. Which of the following is an equation of the line in slope-intercept form with a slope of 3 and a y-intercept of -8? MP 7

  - A $y = -3x + 8$
  - B $y = 8x + 3$
  - C $y = 3x - \dfrac{8}{3}$
  - D $y = 3x - 8$

69. Write the equation of a line in slope-intersect form with the given slope and y-intercept. MP 7

  a. slope: 4, y-intercept: -12

  b. slope: 0.25, y-intercept: 1.5

70. Rewrite the equation $6x + 3y = -5$ in slope-intercept form. MP 7

  - A $y = 6x - 5$
  - B $y = 2x - 5$
  - C $y = -2x - \dfrac{5}{3}$
  - D $y = -2x + 5$

71. Ronald graphs the equation of a constant function on the coordinate plane. The graph has a y-intercept of -3. MP 1, 7

  a. What is the slope of the line?

  b. What is an equation of the line?

72. Which of the following is an equation for the line shown? MP 2

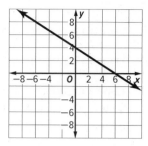

  - A $y = -2x + 6$
  - B $y = -\dfrac{2}{3}x + 4$
  - C $y = -\dfrac{2}{3}x + 6$
  - D $y = \dfrac{2}{3}x + 4$

73. At 5:00 A.M., the temperature was 58°F, and then it climbed steadily at a rate of 2°F per hour throughout the morning. Which of the following is an equation representing the situation, where T is the temperature and x is the number of hours? MP 2

  - A $T = 58x + 2$
  - B $T = 2x + 5$
  - C $T = 2x - 58$
  - D $T = 2x + 58$

74. **MULTI-STEP** Paul joins a gym that has an initial membership fee and a monthly cost. He pays a total of $295 after three months, and $495 after eight months. MP 4

  a. Identify the monthly cost.

  b. What is the initial membership fee?

  c. Write an equation for the total cost C as a function of the number of months, m.

  d. Find how much Paul will have paid after one year.

Algebra Lab
Linear Growth Patterns

If you know the equation of a linear function, you can predict how it will change over any interval. In mathematics, an **interval** is defined as the points between two points or values. So on a number line, the interval between 2 and 5 has a length of 3.

Mathematical Practices
 8 Look for and express regularity in repeated reasoning.

Let points A and B be any two points on the line $y = mx$. Let k be the interval between the x-coordinates of the two points, and ℓ be the interval between the y-coordinates.

Activity 1 $y = x$

Work cooperatively. Use the graph to explore growth over x- and y-intervals.

1A What is the slope of $y = x$?

1B For points A (2, 2) and B (5, 5), what are the lengths of intervals between the x-coordinates, k, and between the y-coordinates, ℓ?

1C What is the relationship between these lengths?

1D Make a prediction about the relationship between k and ℓ for any two points on the line.

1E Choose two other points on the line to check your prediction.

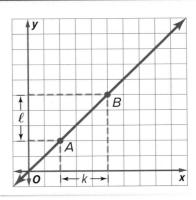

Activity 2 $y = 2x$

Use the graph to explore growth over x- and y-intervals.

2A What is the slope of $y = 2x$?

2B Keep the x-coordinates of points A and B the same, but change the slope to 2. Find k and ℓ.

2C What is the relationship between k and ℓ.

2D Keep $m = 2$ and $k = 3$, but move points A and B. Find ℓ. Does the relationship between k and ℓ change?

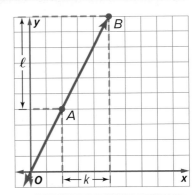

Activity 3 $y = 2x + 1$

Explore growth over x- and y-intervals.

3A The line $y = 2x + 1$ does not go through the origin. If A and B are points on the line with x-coordinates of 2 and 5, what are the y-coordinates of the points?

3B Find k and ℓ.

3C Make and test a prediction about k and ℓ for any two other points.

3D What is the relationship between k and ℓ?

(continued on the next page)

Algebra Lab
Linear Growth Patterns Continued

Analyze the Results

Work cooperatively.

1. Consider any two points along the line $y = mx + b$.

   a. Suppose the x-coordinate of the first point is a. If the length of the interval along the x-axis is k, what is the x-coordinate of the second point?

   b. To find the y-coordinate of the first point, substitute the x-coordinate of the first point into $y = mx + b$. What is the y-coordinate of the first point?

   c. Substitute the x-coordinate of the second point into $y = mx + b$. What is the y-coordinate of the second point?

   d. To find the height of the interval on the y-axis, ℓ, subtract the two y-values. What is the difference?

   e. How does this result prove that linear functions grow by equal differences over equal intervals?

   f. Write the slope m of the line in terms of k and ℓ.

2. **MAKE A PREDICTION** If the slope m is a negative number, how do the y-values change as the x-values increase?

3. Use the graph. When the x-value between two points increases by 1, by how much does the y-value increase?

4. Examine $y = 4x + 3$.

   a. When the x-values between two points increase by 1, by how much do the y-values increase?

   b. If the x-value increases from 5 to 6, how do the corresponding y-values change?

   c. If the x-value increases from 5 to 10, how do the corresponding y-values change?

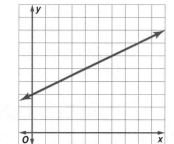

5. Use the graph. How do the function values change over each interval of 1 on the x-axis?

6. Think about the linear function $y = -2x + 7$.

   a. When the x-values between two points increase by 1, how do the y-values change?

   b. If the x-value increases from 2 to 3, how do the corresponding y-values change?

   c. If the x-value increases from 2 to 6, how do the corresponding y-values change?

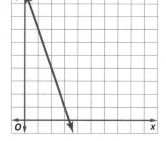

7. A line contains the points (1, 5) and (3, 13). How do the y-values change as the x-values increase?

8. A line contains the points (0, 4) and (1, 1). How do the y-values change as the x-values increase?

9. **MP CONSTRUCT ARGUMENTS** Is the function represented in the table a linear function? Justify your answer by discussing intervals and differences.

x	−3	−2	−1	0	1	2	3
y	15	5	−1	−3	−1	5	15

Transformations of Linear Functions

- You defined and graphed linear functions.

1 Identify the effects on the graphs of linear functions by replacing $f(x)$ with $f(x) + k$ and $f(x - h)$ for positive and negative values.

2 Identify the effects on the graphs of linear functions by replacing $f(x)$ with $af(x)$, $f(ax)$, $-af(x)$ and $f(-ax)$.

- Military pilots fly in formations for protection and defense. The path of each plane can be described as a function that is a transformation of the flight leader's path.

New Vocabulary

transformation
translation
dilation
reflection

MP Mathematical Practices

2 Reason abstractly and quantitatively.

3 Construct viable arguments and critique the reasoning of others.

4 Model with mathematics.

1 Translations The parent function of the family of linear functions is $f(x) = x$. A **transformation** of a linear function moves the graph on the coordinate plane to create a new linear function. One type of transformation is a **translation**. A translation moves a graph up, down, left, right, or in two directions. The graph of the function $f(x) = x + k$ is the graph of $f(x)$ translated vertically. The graph of the function $f(x) = (x - h)$ is the graph of $f(x)$ translated horizontally.

🔑 Key Concept   Translations of Linear Functions

Vertical Translations

For $f(x) + k$, if $k > 0$, the graph of $f(x)$ is translated k units up.

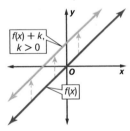

Every point on the graph of $f(x)$ moves k units up.

For $f(x) + k$, if $k < 0$, the graph of $f(x)$ is translated $|k|$ units down.

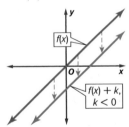

Every point on the graph of $f(x)$ moves $|k|$ units down.

Horizontal Translations

For $f(x - h)$, if $h > 0$, the graph of $f(x)$ is translated h units right.

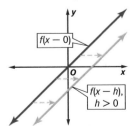

Every point on the graph of $f(x)$ moves h units right.

For $f(x - h)$, if $h < 0$, the graph of $f(x)$ is translated $|h|$ units left.

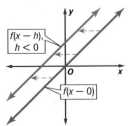

Every point on the graph of $f(x)$ moves $|h|$ units left.

Example 1   Translations of Linear Functions

Describe the translation in each function as it relates to the graph of $f(x) = x.$

a. $g(x) = x - 2$

The graph of $g(x) = x - 2$ is a translation of the graph of $f(x) = x$ down 2 units.

b. $p(x) = (x - 4)$

The graph of $h(x) = (x - 4)$ is a translation of the graph of $f(x) = x$ right 4 units.

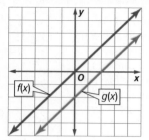

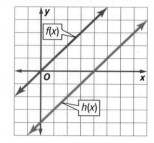

▶ **Guided Practice**

Describe the translation in each function as it relates to the graph of $f(x) = x.$

1A. $g(x) = x + 1$

1B. $p(x) = (x + 2)$

---

2 Dilations and Reflections A **dilation** stretches or compresses the graph of a function. When a linear function is multiplied by a constant a before or after the function is evaluated, the result is a dilation. The graph of $g(x) = a \cdot f(x)$ is the graph of $f(x)$ stretched or compressed vertically. The graph of $g(x) = f(a \cdot x)$ is the graph of $f(x)$ stretched or compressed horizontally.

⬤ Key Concept   Dilations of Linear Functions

Vertical Translations

For $a \cdot f(x)$, if $a > 1$, the graph of $f(x)$ is stretched vertically.

For $a \cdot f(x)$, if $0 < |a| < 1$, the graph of $f(x)$ is compressed vertically.

Study Tip

Slope Notice that the slope of a line is affected by a dilation. A vertical stretch or a horizontal compression makes a graph steeper. A vertical compression or a horizontal stretch makes a graph less steep.

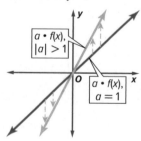

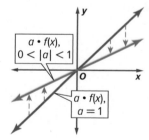

Horizontal Translations

For $f(a \cdot x)$, if $a > 1$, the graph of $f(x)$ is compressed horizontally .

For $f(a \cdot x)$, if $0 < |a| < 1$, the graph of $f(x)$ is stretched horizontally.

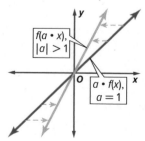

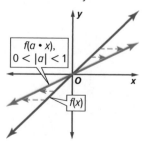

Example 2   Dilations of Linear Functions

Describe the dilation in each function as it relates to the graph of $f(x) = x$.

a. $g(x) = \frac{1}{2}x$

The graph of $g(x) = \frac{1}{2}x$ is a vertical dilation of $f(x) = x$ because the function is multiplied by a positive constant a after it has been evaluated.

Because $0 < |a| < 1$, the graph of $f(x) = x$ is compressed vertically.

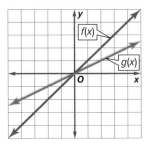

b. $p(x) = (3x)$

The graph of $p(x) = (3x)$ is a horizontal dilation of $f(x) = x$ because x is multiplied by the positive constant a before the function is evaluated.

Because $|a| > 1$, the graph of $f(x) = x$ is compressed horizontally.

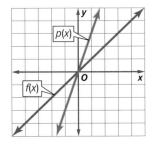

▶ **Guided Practice**

Describe the dilation in each function as it relates to the graph of $f(x) = x$.

2A. $g(x) = (0.8\,x)$ **2B.** $p(x) = 3x$

A **reflection** flips the graph of a function across a line. When a is negative, the graph of $a \cdot f(x)$ is reflected across the x-axis. When a is negative, the graph of $f(a \cdot x)$ is reflected across the y-axis.

🔑 **Key Concept** Reflections of Linear Functions	
Reflection across x-axis	**Reflection across y-axis**
For $a \cdot f(x)$, if $a < 0$, the graph of $f(x)$ is reflected across the x-axis.	For $f(a \cdot x)$, if $a < 0$, the graph of $f(x)$ is reflected across the x-axis.

When a is negative and $a \neq -1$, the graphs of $a \cdot f(x)$ and $f(a \cdot x)$ are both reflections and dilations.

Example 3　　Reflections and Dilations of Linear Functions

Describe the transformations in each function as it relates to the graph of $f(x) = x$.

a. $g(x) = -2x$

Because a is negative in $g(x)$, the graph of $f(x)$ is reflected across the x-axis. Because $|a| > 1$, the graph is also stretched vertically.

b. $p(x) = (-0.25x)$

Because a is negative in $p(x)$, the graph of $f(x)$ is reflected across the y-axis. Because $0 < |a| < 1$, the graph is also stretched horizontally.

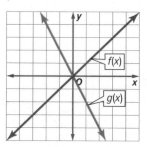

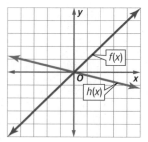

▶ **Guided Practice**

Describe the transformations in each function as it relates to the graph of $f(x) = x$.

3A. $g(x) = (-0.1\ x)$

3B. $p(x) = -2.5x$

Often more than one transformation is applied to a function. Examine each transformation to analyze how the graph is affected.

Example 4　　Analyze Transformations

Describe the transformations in each function as it relates to the graph of $f(x) = x$.

a. $g(x) = -2x + 4$

Because a is negative in $g(x)$, the graph of $f(x)$ is reflected across the x-axis. Because $|a| > 1$, the graph is also stretched vertically. The graph is also translated up 4 units.

b. $h(x) = 0.5\ (x + 1)$

Because $0 < |a| < 1$, the graph is compressed vertically. The graph is also translated left 1 unit.

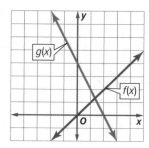

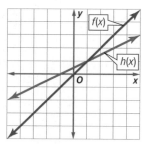

▶ **Guided Practice**

Describe the transformations in each function as it relates to the graph of $f(x) = x$.

4A. $g(x) = (4.1x) - 2$

4B. $h(x) = -2.5(x - 1)$

You can examine a graph and compare it to the graph of the parent function to write a function for the graph.

Example 5   Analyze a Graph

Write a function represented by each graph.

a.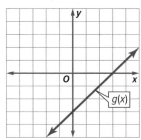

b.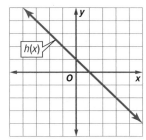

Study Tip

Multiple Functions A graph may be represented by more than one equivalent function. Notice that the graph in part **b** is also correctly represented as a reflection in the y-axis and a translation 1 unit right, $h(x) = -(x - 1)$.

The slope of the graph is the same as the slope of the parent graph so it has not been dilated or reflected. The graph is translated 3 units down. The graph can be represented by $g(x) = x - 3$.

The graph is reflected but not dilated. It is also translated 1 unit up. The graph can be represented by $h(x) = -x + 1$.

▶ **Guided Practice**

5. Write a function represented by the graph.

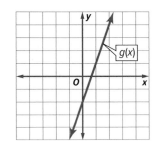

You can perform transformations on functions other than parent functions to represent real-world situations.

Real-World Example 6   Analyze a Function

BUSINESS The prices of servings of frozen yogurt are modeled by $p(x) = 0.39x + 0.60$ where x is the number of ounces of yogurt in the serving.

a. Write a function $n(x)$ that represents the store owner raising the prices by 20%.

b. Find the price of an 8-ounce serving of yogurt before and after the price increase.

a. Raising the prices 20% is represented by a dilation of the function. The increased price will be 120% of the original price.
$$n(x) = 1.2p(x)$$
$$= 1.2(0.39x + 0.60)$$

b. $p(8) = 0.39(8) + 0.60$ $n(8) = 1.2(0.39(8) + 0.60)$
   $= 3.12 + 0.60$ $= 1.2(3.72)$
   $= \$3.72$ $\approx \$4.46$

An 8-ounce serving is $3.72 before the price increase and $4.46 after.

▶ **Guided Practice**

6. Suppose a customer uses a coupon for $1 off any yogurt after the price increase.

   a. Write a function $c(x)$ that represents the price of a serving of yogurt.

   b. Find the customer's price for a 6-ounce serving of yogurt.

Examples 1, 2 **Describe the transformation in each function as it relates to the graph of $f(x) = x$.**

1. $g(x) = (x + 4)$

2. $g(x) = x + 7$

3. $g(x) = 7x$

4. $g(x) = (0.25x)$

Examples 3, 4 **Describe the transformations in each function as it relates to the graph of $f(x) = x$.**

5. $g(x) = -12x$

6. $g(x) = (-0.5x)$

7. $g(x) = 2x - 9$

8. $g(x) = 3.5(x + 1)$

Example 5 **Write a function represented by each graph.**

9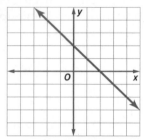

10.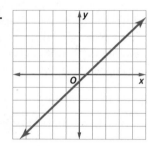

Example 6 **11. FINANCIAL LITERACY** Jerome is buying paint for a mural. The cost of the paints can be modeled by $f(p) = 6.99p$, where p is the number of paints he buys. He has a coupon for 15% off his purchase at the paint store.

a. Write a function $n(p)$ that represents the costs of the paints with the 15% discount.

b. Find the cost of purchasing 5 paints before and after the discount.

Practice and Problem Solving   Extra Practice is found on page R3.

Examples 1–2 **Describe the transformation in each function as it relates to the graph of $f(x) = x$.**

12. $g(x) = x - 5$

13. $g(x) = (x - 3)$

14. $g(x) = \frac{1}{6}x$

15. $g(x) = (2x)$

16. $g(x) = x - 14$

17. $g(x) = (x + 6)$

18. $g(x) = (x - 8)$

19. $g(x) = x + 3$

20. $g(x) = \frac{3}{4}x$

21. $g(x) = (5x)$

22. $g(x) = 15x$

23. $g(x) = (0.4x)$

24. $g(x) = (x + 17)$

25. $g(x) = \frac{7}{16}x$

Examples 3, 4 **Describe the transformations in each function as it relates to the graph of $f(x) = x$.**

26. $g(x) = (-0.75x)$

27. $g(x) = -8x$

28. $g(x) = 0.5(x - 3)$

29. $g(x) = -7x - 5$

30. $g(x) = -11x$

31. $g(x) = (-0.65x)$

32. $g(x) = (-0.8x)$

33 $g(x) = -17x$

34. $g(x) = 0.2(x + 9)$

35. $g(x) = -2x - 6$

36. $g(x) = 3x + 5$

37. $g(x) = 14.4(x - 2)$

Example 5 **Write a function represented by each graph.**

38.
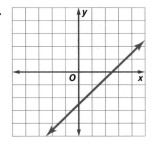

39.
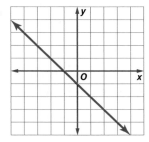

40.
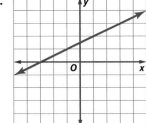

41.
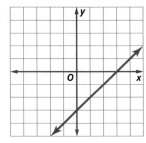

42.
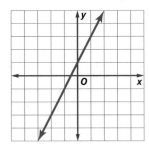

43.
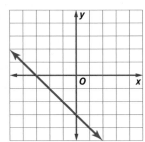

Example 6 **44. FINANCIAL LITERACY** The total price of the tickets to a concert can be modeled by $f(c) = 64.95c$, where c represents the number of tickets purchased. A ticket website charges a $12 service fee per order.

a. Write a function $g(c)$ that represents the cost of buying tickets from the website.

b. Find the cost of purchasing 4 tickets before and after the service fee.

Tell how each translation changes $f(x)$.

45. $f(x) + k$ when $k > 0$ translates the graph k units _____.

46. $f(x) + k$ when $k < 0$ translates the graph $|k|$ units _____.

47. $f(x - h)$ when $h > 0$ translates the graph h units _____.

48. $f(x - h)$ when $h < 0$ translates the graph $|h|$ units _____.

49. $a \cdot f(x)$ when $a = -1$ reflects the graph across the _____.

50. The graph shows $f(x) = x$ and transformations of $f(x)$. Match each function to its graph.

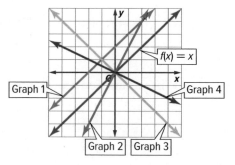

    a. $g(x) = -\frac{1}{2}f(x)$

    b. $p(x) = 2f(x)$

    c. $r(x) = f(x) + 2$

    d. $j(x) = -f(x)$

51. **MULTI-STEP** A weekly subscription meal service delivers recipes and ingredients for a certain number of meals per week.

    a. The initial subscription fee is \$25, and the two-meal plan costs \$70 per week. Write a function $f(x)$ to represent the cost for x weeks.

    b. The initial subscription fee was changed to \$30. Write a function $h(x)$ to represent the cost of the two-meal plan for x weeks.

    c. What type of transformation describes the change from $f(x)$ to $g(x)$?

    d. A promotion is offered for 10% off if you order online. Write a function $p(x)$ to represent the cost of the two-week meal with the discount.

    e. What type of transformation describes the change from $h(x)$ to $p(x)$?

H.O.T. Problems   Use Higher-Order Thinking Skills

52. **MP CRITIQUE ARGUMENTS** Alex thinks that $af(x)$ and $f(ax)$ are equivalent functions. Is he correct? Explain your answer.

53. **CHALLENGE** Start with the linear equation $f(x) = mx + b$. The line can be translated vertically by adding a positive constant k: $g(x) = mx + b + k$. Tell what horizontal translation of $f(x) = mx + b$ would produce the same line. Justify your answer.

54. **MP REASONING** Suppose linear function $f(x)$ is translated 5 units down to get $g(x)$. The same function $f(x)$ translated 5 units right results in the same $g(x)$. What must be true about the slope of $f(x)$? Explain your answer.

55. **OPEN ENDED** Write an equation of a line with a slope of 2 and sketch its graph. Draw the reflection of the line over the y-axis and write the equation of the reflection.

56. **WRITING IN MATH** Describe how the following functions are related to the function $f(x) = x$.

    a. $g(x) = x + 3$

    b. $h(x) = -x$

    c. $k(x) = (x - 1) + 5$

57. The graph of $g(x) = x + 4$ is a translation of the graph of $f(x) = x$. Which of the following describe the translation? **MP** 2

  ◯ **A**   a horizontal translation 4 units to the right

  ◯ **B**   a horizontal translation 4 units to the left

  ◯ **C**   a vertical translation 4 units up

  ◯ **D**   a vertical translation 4 units down

58. The graph of $f(x) = x$ is translated 6 units down. What is the transformed function? **MP** 2,7

  ◯ **A** $g(x) = \frac{1}{6}x$

  ◯ **B** $g(x) = -x + 6$

  ◯ **C** $g(x) = x + 6$

  ◯ **D** $g(x) = x - 6$

59. Which of the following transformations makes the graph of $f(x) = x$ less steep? **MP** 2

  ☐ **A** $g(x) = f(x + 2)$

  ☐ **B** $h(x) = f(x - 2)$

  ☐ **C** $j(x) = f\left(\frac{1}{2}x\right)$

  ☐ **D** $k(x) = f(2x)$

  ☐ **E** $m(x) = f\left(-\frac{1}{2}x\right)$

  ☐ **F** $n(x) = \frac{1}{2}f(x)$

60. The graph of $f(x) = x$ is translated left 8 units. Write the transformed function. **MP** 2,7

  $g(x) = \boxed{}$

61. The graph of $f(x) = x$ is reflected across the x-axis and translated up 3 units. Write the transformed function. **MP** 2,7

  $g(x) = \boxed{}$

62. The graph of $g(x) = a(x - h)$ is a transformation of the graph of $f(x) = x$. If $a < 0$ and $h > 0$, describe the transformations in $g(x)$ as it relates to the graph of $f(x) = x$. Select all that apply. **MP** 2,7

  ◯ **A**   reflection across the x-axis

  ◯ **B**   reflection across the y-axis

  ◯ **C**   translation h units left

  ◯ **D**   translation h units right

63. Which function represents a vertical stretch of the graph of $f(x) = x$? **MP** 2

  ◯ **A** $g(x) = x + 7$

  ◯ **B** $g(x) = (0.7x)$

  ◯ **C** $g(x) = \frac{1}{7}x$

  ◯ **D** $g(x) = 7x$

64. MULTI-STEP The graph of $g(x) = (ax) + k$ shown is a transformation of the graph of $f(x) = x$. **MP** 2,7

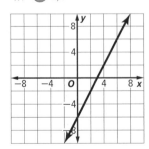

  a.   Describe the transformations across $g(x)$ as it relates to the graph of $f(x) = x$.

  b.   What is the value of k?

  c.   What is the value of a?

  d.   Write the function $g(x)$ to represent the graph.

Arithmetic Sequences as Linear Functions

:: **Then**

- You indentified linear functions.

:: **Now**

1. Recognize arithmetic sequences.
2. Relate arithmetic sequences to linear functions.

:: **Why?**

- During a 2000-meter race, the coach of a women's crew team recorded the team's times at several intervals.

  - At 400 meters, the time was 1 minute 32 seconds.
  - At 800 meters, it was 3 minutes 4 seconds.
  - At 1200 meters, it was 4 minutes 36 seconds.
  - At 1600 meters, it was 6 minutes 8 seconds.

  They completed the race with a time of 7 minutes 40 seconds.

New Vocabulary

sequence
terms of the sequence
arithmetic sequence
common difference

Mathematical Practices

8 Look for and express regularity in repeated reasoning.

1 Recognize Arithmetic Sequences You can relate the pattern of team times to linear functions. A **sequence** is a set of numbers, called the **terms of the sequence**, in a specific order. Look for a pattern in the information given for the women's crew team. Make a table to analyze the data.

Distance (m)	400	800	1200	1600	2000
Time (min : sec)	1:32	3:04	4:36	6:08	7:40

+ 1:32   + 1:32   + 1:32   + 1:32

As the distance increases in regular intervals, the time increases by 1 minute 32 seconds. Since the difference between successive terms is constant, this is an **arithmetic sequence**. The difference between the terms is called the **common difference** d.

Key Concept Arithmetic Sequence

Words	An arithmetic sequence is a numerical pattern that increases or decreases at a constant rate called the *common difference*.

Examples

3, 5, 7, 9, 11, . . .
+2 +2 +2 +2
$d = 2$

33, 29, 25, 21, 17, . . .
−4 −4 −4 −4
$d = -4$

The three dots used with sequences are called an *ellipsis*. The ellipsis indicates that there are more terms in the sequence that are not listed.

Aurora Photos/Alamy

Example 1　Identify Arithmetic Sequences

Determine whether each sequence is an arithmetic sequence. Explain.

a. $-4, -2, 0, 2, \ldots$

$$\underset{\substack{+2\quad +2\quad +2}}{-4 \quad -2 \quad 0 \quad 2}$$

The difference between terms in the sequence is constant. Therefore, this sequence is arithmetic.

b. $\frac{1}{2}, \frac{5}{8}, \frac{3}{4}, \frac{13}{16}, \ldots$

$$\underset{\substack{+\frac{1}{8}\quad +\frac{1}{8}\quad +\frac{1}{16}}}{\frac{1}{2} \quad \frac{5}{8} \quad \frac{3}{4} \quad \frac{13}{16}}$$

This is not an arithmetic sequence. The difference between terms is not constant.

Guided Practice

1A. $-26, -22, -18, -14, \ldots$

1B. $1, 4, 9, 25, \ldots$

You can use the common difference of an arithmetic sequence to find the next term.

Example 2　Find the Next Term

Find the next three terms of the arithmetic sequence $15, 9, 3, -3, \ldots$.

Step 1 Find the common difference by subtracting successive terms.

$$\underset{\substack{-6\quad -6\quad -6}}{15 \quad 9 \quad 3 \quad -3}$$

The common difference is -6.

Step 2 Add -6 to the last term of the sequence to get the next term.

$$\underset{\substack{-6\quad -6\quad -6}}{-3 \quad -9 \quad -15 \quad -21}$$

The next three terms in the sequence are $-9, -15,$ and -21.

Guided Practice

2. Find the next four terms of the arithmetic sequence $9.5, 11.0, 12.5, 14.0, \ldots$.

Math History Link

Mina Rees (1902–1997) Rees received the first award for Distinguished Service to Mathematics from the Mathematical Association of America. She was the first president of the Graduate Center at The City University of New York. Her work in analyzing patterns is still inspiring young women to study mathematics today.

Study Tip

MP Regularity Notice the regularity in the way expressions in terms of a_1 and d change with each term of the sequence.

Each term in an arithmetic sequence can be expressed in terms of the first term a_1 and the common difference d.

Term	Symbol	In Terms of a_1 and d	Numbers
first term	a_1	a_1	8
second term	a_2	$a_1 + 1d$	$8 + 1(3) = 11$
third term	a_3	$a_1 + 2d$	$8 + 2(3) = 14$
fourth term	a_4	$a_1 + 3d$	$8 + 3(3) = 17$
$\vdots$	$\vdots$	$\vdots$	$\vdots$
nth term	a_n	$a_1 + (n-1)d$	$8 + (n-1)(3)$

Key Concept　nth Term of an Arithmetic Sequence

The nth term of an arithmetic sequence with first term a_1 and common difference d is given by $a_n = a_1 + (n-1)d$, where n is a positive integer.

Example 3   Find the nth Term

a. Write an equation for the nth term of the arithmetic sequence −12, −8, −4, 0, ...

Step 1 Find the common difference.

$$-12 \quad -8 \quad -4 \quad 0$$

$+4 \quad +4 \quad +4$ The common difference is 4.

Step 2 Write an equation.

$$a_n = a_1 + (n-1)d \qquad \text{Formula for the nth term}$$
$$= -12 + (n-1)4 \qquad a_1 = -12 \text{ and } d = 4$$
$$= -12 + 4n - 4 \qquad \text{Distributive Property}$$
$$= 4n - 16 \qquad \text{Simplify.}$$

Study Tip

*n*th Terms Since *n* represents the number of the term, the inputs for *n* are the counting numbers.

Go Online!

You can quickly generate terms of arithmetic sequences on a graphing calculator or on the **calculator** in the eToolkit. Enter the first term and press ENTER or =. Then add or subtract the common difference and press ENTER or = repeatedly.

b. Find the 9th term of the sequence.

Substitute 9 for *n* in the formula for the *n*th term.

$$a_n = 4n - 16 \qquad \text{Formula for the nth term}$$
$$a_9 = 4(9) - 16 \qquad n = 9$$
$$a_9 = 36 - 16 \qquad \text{Multiply.}$$
$$a_9 = 20 \qquad \text{Simplify.}$$

c. Graph the first five terms of the sequence.

n	$4n-16$	a_n	(n, a_n)
1	$4(1)-16$	-12	$(1, -12)$
2	$4(2)-16$	-8	$(2, -8)$
3	$4(3)-16$	-4	$(3, -4)$
4	$4(4)-16$	0	$(4, 0)$
5	$4(5)-16$	4	$(5, 4)$

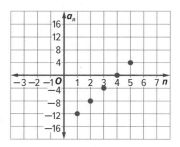

d. Which term of the sequence is 32?

In the formula for the *n*th term, substitute 32 for a_n.

$$a_n = 4n - 16 \qquad \text{Formula for the nth term}$$
$$32 = 4n - 16 \qquad a_n = 32$$
$$32 + 16 = 4n - 16 + 16 \qquad \text{Add 16 to each side.}$$
$$48 = 4n \qquad \text{Simplify.}$$
$$12 = n \qquad \text{Divide each side by 4.}$$

▶ **Guided Practice**

Consider the arithmetic sequence 3, −10, −23, −36, ...

3A. Write an equation for the *n*th term of the sequence.

3B. Find the 15th term in the sequence.

3C. Graph the first five terms of the sequence.

3D. Which term of the sequence is −114?

2 Arithmetic Sequences and Functions

As you can see from Example 3, the graph of the first five terms of the arithmetic sequence lie on a line. An arithmetic sequence is a discrete linear function in which n is the independent variable, a_n is the dependent variable, and d is the slope. The formula can be rewritten as the function $f(n) = (n - 1)d + a_1$, where n is a counting number.

While the domain of most linear functions are all real numbers, in Example 3 the domain of the function is the set of counting numbers and the range of the function is the set of integers on the line.

Real-World Example 4   Arithmetic Sequences as Functions

INVITATIONS Marisol is mailing invitations to her quinceañera. The arithmetic sequence $0.66, $1.32, $1.98, $2.64, ... represents the cost of postage.

a. Write a function to represent this sequence.

The first term, a_1, is 0.66. Find the common difference.

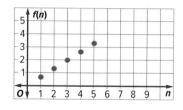

The common difference is 0.66.

$$a_n = a_1 + (n - 1)d \qquad \text{Formula for the } n\text{th term}$$
$$= 0.66 + (n - 1)0.66 \qquad a_1 = 0.66 \text{ and } d = 0.66$$
$$= 0.66 + 0.66n - 0.66 \qquad \text{Distributive Property}$$
$$= 0.66n \qquad \text{Simplify.}$$

The function is $f(n) = 0.66n$.

b. Graph the function, and determine the domain and range.

The rate of change of the function is 0.66. Make a table and plot points.

n	f(n)
1	0.66
2	1.32
3	1.98
4	2.64
5	3.30

The domain of a function is the number of invitations Marisol mails. So, the domain is {1, 2, 3, 4, ...}. The range of the function is the postage cost or {0.66, 1.32, 1.98, 2.64, 3.30 ...}.

Guided Practice

4. TRACK The chart shows the length of Martin's long jumps.

Jump	1	2	3	4
Length (ft)	8	9.5	11	12.5

A. Write a function to represent this arithmetic sequence.

B. Then graph the function.

Check Your Understanding

◯ = Step-by-Step Solutions begin on page R11.

Example 1   Determine whether each sequence is an arithmetic sequence. Write *yes* or *no*. Explain.

1. 18, 16, 15, 13, …   **2.** 4, 9, 14, 19, …

Example 2   Find the next three terms of each arithmetic sequence.

3. 12, 9, 6, 3, …   **4.** −2, 2, 6, 10, …

Example 3   Write an equation for the *n*th term of each arithmetic sequence. Then graph the first five terms of the sequence.

5. 15, 13, 11, 9, …   **6.** −1, −0.5, 0, 0.5, …

Example 4   **7. SAVINGS** Kaia has \$525 in a savings account. After one month she has \$580 in the account. The next month the balance is \$635. The balance after the third month is \$690. Write a function to represent the arithmetic sequence. Then graph the function.

Practice and Problem Solving

Extra Practice is on page R3.

Example 1   Determine whether each sequence is an arithmetic sequence. Write *yes* or *no*. Explain.

8. −3, 1, 5, 9, …   **9.** $\frac{1}{2}, \frac{3}{4}, \frac{5}{8}, \frac{7}{16}, …$

10. −10, −7, −4, 1, …   **11.** −12.3, −9.7, −7.1, −4.5, …

Example 2   Find the next three terms of each arithmetic sequence.

12. 0.02, 1.08, 2.14, 3.2, …   **13.** 6, 12, 18, 24, …

14. 21, 19, 17, 15, …   **⑮** $-\frac{1}{2}, 0, \frac{1}{2}, 1, …$

16. $2\frac{1}{3}, 2\frac{2}{3}, 3, 3\frac{1}{3}, …$ **17.** $\frac{7}{12}, 1\frac{1}{3}, 2\frac{1}{12}, 2\frac{5}{6}, …$

Example 3   Write an equation for the *n*th term of the arithmetic sequence. Then graph the first five terms in the sequence.

18. −3, −8, −13, −18, …   **19.** −2, 3, 8, 13, …

20. −11, −15, −19, −23, …   **21.** −0.75, −0.5, −0.25, 0, …

Example 4   **22. AMUSEMENT PARKS** Shiloh and her friends spent the day at an amusement park. In the first hour, they rode two rides. After 2 hours, they had ridden 4 rides. They had ridden 6 rides after 3 hours.

     a. Write a function to represent the arithmetic sequence.

     b. Graph the function.

     c. Determine the domain and range of the function.

23. **MP MODELING** The table shows how Ryan is paid at his lumber yard job.

Number of 10-ft 2×4 Planks Cut	1	2	3	4	5	6	7
Amount Paid in Commission (\$)	8	16	24	32	40	48	56

     a. Write a function to represent Ryan's commission.

     b. Graph the function.

     c. Determine the domain and range of the function.

24. The graph is a representation of an arithmetic sequence.

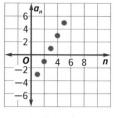

    a. List the first five terms.

    b. Write the formula for the nth term.

    c. Write the function.

    d. Find the domain and range of the function.

25. **NEWSPAPERS** A local newspaper charges by the number of words for advertising. Write a function to represent the advertising costs.

DAILY NEWS ADVERTISING	
10 words $7.50	20 words $10.00
15 words $8.75	25 words $11.25

26. The fourth term of an arithmetic sequence is 8. If the common difference is 2, what is the first term?

27. The common difference of an arithmetic sequence is -5. If a_{12} is 22, what is a_1?

28. The first four terms of an arithmetic sequence are 28, 20, 12, and 4. Which term of the sequence is -36?

29. **CARS** Jamal's odometer of his car reads 24,521. If Jamal drives 45 miles every day, what will the odometer reading be after 25 days?

30. **YEARBOOKS** The yearbook staff is unpacking a box of school yearbooks. The arithmetic sequence 281, 270, 259, 248 ... represents the total number of ounces that the box weighs as each yearbook is taken out of the box.

    a. Write a function to represent this sequence.

    b. Determine the weight of each yearbook.

    c. If the box weighs at least 17 ounces empty and 292 ounces when it is full, how many yearbooks were in the box?

    d. Find the domain and range of the function.

31. **SPORTS** To train for a marathon, Olivia runs 3 miles per day for the first week and then increases the daily distance by a half mile each of the following weeks.

    a. Write an equation to represent the nth term of the sequence.

    b. If the pattern continues, during which week will she run 10 miles per day?

    c. Is it reasonable to think that this pattern will continue indefinitely? Explain.

H.O.T. Problems   Use **H**igher-**O**rder **T**hinking Skills

32. **OPEN ENDED** Create an arithmetic sequence with a common difference of -10.

33. **MP** **PERSEVERANCE** Find the value of x that makes $x + 8$, $4x + 6$, and $3x$ the first three terms of an arithmetic sequence.

34. **MP** **REASONING** Compare and contrast the domain and range of the linear functions described by $Ax + By = C$ and $a_n = a_1 + (n - 1)d$.

35. **CHALLENGE** Determine whether each sequence is an arithmetic sequence. Write *yes* or *no*. Explain. If yes, find the common difference and the next three terms.

    a. $2x + 1, 3x + 1, 4x + 1...$ **b.** $2x, 4x, 8x, ...$

36. **WRITING IN MATH** How are graphs of arithmetic sequences and linear functions similar? different?

37. A sequence is represented by the function $f(n) = 9 + 3n$. What are the first three terms in the sequence? **MP** 8

a. $f(1) =$ ☐

b. $f(2) =$ ☐

c. $f(3) =$ ☐

38. The first four terms of a sequence are 5, −3, −11, −19. Which of the following is an equation for the nth term of the sequence? **MP** 8

○ **A** $a_n = 8n − 3$

○ **B** $a_n = n − 4$

○ **C** $a_n = −8n − 8$

○ **D** $a_n = −8n + 13$

39. The points on the graph represent the terms of an arithmetic sequence. Which of the following is a formula for the nth term of the sequence? **MP** 8

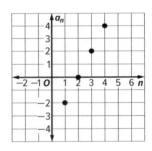

○ **A** $a_n = 2n − 4$

○ **B** $a_n = n$

○ **C** $a_n = −2n$

○ **D** $a_n = −2n − 3$

40. The first four terms in a sequence are 17, 22, 27, 32. What is an equation for the nth term of the arithmetic sequence? **MP** 8

$a_n =$ ☐

41. Madison High School has student identification cards that work like prepaid credit cards. Franklin charges the same amount on his card every day. The function $f(x) = −4.5x + 50$ represents the value on Franklin's card after x number of meals. What are the first three terms in the sequence? **MP** 4, 8

○ **A** −4.5, 9, −13.5

○ **B** 50, 54.5, 59

○ **C** 45.5, 41, 36.5

○ **D** 50, 45.5, 41

42. The first three terms in a sequence are 25, 19, 13. **MP** 8

a. Write an equation for the nth term of the arithmetic sequence.

$a_n =$ ☐

b. Give the ninth term in the sequence.

$a_9 =$ ☐

43. MULTI-STEP The athletic boosters are selling sweatshirts for a fundraiser. The arithmetic sequence 12, 15, 18, 21 represents the number sold at four consecutive games. **MP** 4, 8

a. Write a function to represent this sequence.

$f(n) =$ ☐

b. Find the additional number of sweatshirts sold at each game. ☐

c. Graph the function and determine the domain and range.

domain = ☐

range = ☐

d. The boosters sold 45 sweatshirts at one game. Which game was it? ☐

Piecewise and Step Functions

:: **Then**

- You identified and graphed linear functions.

:: **Now**

1 Identify and graph step functions.

2 Identify and graph piecewise-defined functions.

:: **Why?**

- Kim is ordering books online. The site charges for shipping based on the amount of the order. If the order is less than $10, shipping costs $3. If the order is at least $10 but less than $20, it will cost $5 to ship it.

 New Vocabulary

step function
piecewise-linear function
greatest integer function
piecewise-defined function

 Mathematical Practices

4 Model with mathematics.

---

1 Step Functions The graph of a **step function** is a series of line segments. Because each part of a step function is linear, this type of function is called a **piecewise-linear function**. One example of a step function is the **greatest integer function**, written as $f(x) = [\![x]\!]$, where $f(x)$ is the greatest integer not greater than x. For example, $[\![6.8]\!] = 6$ because 6 is the greatest integer that is not greater than 6.8.

Key Concept Greatest Integer Function

Parent function:	$f(x) = [\![x]\!]$
Type of graph:	disjointed line segments
Domain:	all real numbers
Range:	all integers

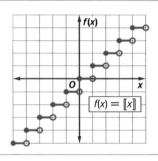

$f(x) = [\![x]\!]$

Example 1 Greatest Integer Function

Graph $f(x) = [\![x + 2]\!]$. State the domain and range.

First, make a table. Select a few values between integers. On the graph, dots represent included points. Circles represent points not included.

x	$x + 2$	$[\![x + 2]\!]$
0	2	2
0.25	2.25	2
0.5	2.5	2
1	3	3
1.25	3.25	3
1.5	3.5	3
2	4	4
2.25	4.25	4

$f(x) = [\![x + 2]\!]$

Note that this is the graph of $f(x) = [\![x]\!]$ shifted 2 units to the left.

Because the dots and circles overlap, the domain is all real numbers. The range is all integers. Notice that the graph has no symmetry and no maximum or minimum values. As x increases, $f(x)$ increases, and as x decreases, $f(x)$ decreases.

▶ **Guided Practice**

1. Graph $g(x) = 2[\![x]\!]$. State the domain and range.

Step functions can be used to represent many real-world situations involving money.

Real-World Example 2   Step Function

SMART PHONE PLANS Data service providers charge for the amount of data used by the megabyte, not by the kilobyte. A provider charges $8 per 100-megabyte or any fraction thereof for exceeding the number of megabytes allotted on each plan. Draw a graph that represents this situation.

The total cost for the extra megabytes will be a multiple of $8, and the graph will be a step function. If the data used is greater than 0 but less than or equal to 100 megabytes, the charge will be $8. If the data used is greater than 200 but is less than or equal to 300 megabytes, you will be charged for 300 megabytes or $24.

Real-World Link

China recently became the country with the most smartphone users, about 246 million. They surpassed the United States who had approximately 230 million smartphone users. About 27% of Chinese smartphone users use more than 21 apps on their smartphones.

Source: Business Insider

x	f(x)
$0 < x \le 100$	8
$100 < x \le 200$	16
$200 < x \le 300$	24
$300 < x \le 400$	32
$400 < x \le 500$	40
$500 < x \le 600$	48
$600 < x \le 700$	56

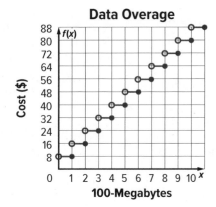

Guided Practice

2. **PARKING** A garage charges $4 for the first hour and $1 for each additional hour. Draw a graph that represents this situation.

2 **Piecewise-Defined Functions** Piecewise-linear functions are examples of *piecewise-defined functions*. The rule for a **piecewise-defined function** includes two or more expressions. Each expression in the rule for a piecewise-defined function applies to a different interval of the function's domain.

Example 3   Piecewise-Defined Function

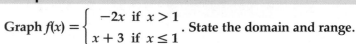

Graph $f(x) = \begin{cases} -2x & \text{if } x > 1 \\ x + 3 & \text{if } x \le 1 \end{cases}$. State the domain and range.

Graph the first expression. Create a table of values for $f(x) = -2x$ using values of $x > 1$, and draw the graph. Since x is not equal to 1, place a circle at $(1, -2)$.

x	$f(x) = -2x$
1.5	−3
2	−4
2.5	−5

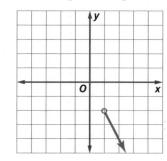

Next, graph the second expression.

Create a table of values for $f(x) = x + 3$ using values of $x \le 1$, and draw the graph. If $x = 1$, $f(x) = 4$; draw a point at $(1, 4)$.

x	f(x) = x + 3
−3	0
−1	2
1	4

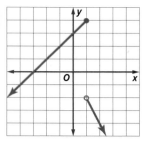

The domain is all real numbers. The range is $y \le 4$.

> **Guided Practice**

3. Graph $f(x) = \begin{cases} 2x + 1 \text{ if } x > 0 \\ 3 \text{ if } x \le 0 \end{cases}$. State the domain and range.

Piecewise-defined functions can be used to model a variety of real-world situations.

Real-World Example 4   Piecewise-Defined Function

COFFEE The cost of purchasing coffee beans can be represented by

$f(x) = \begin{cases} 10x \text{ if } x \le 1 \\ 8x + 2 \text{ if } 1 < x \le 4, \\ 5x + 14 \text{ if } x > 4 \end{cases}$ where x represents the weight of the beans in pounds.

Graph the function, and then interpret the meaning of each piece in the context of the situation.

Make a table of values, and use it to graph each piece.

For each piece of the function, evaluate the expression for values of x in the interval defined for that piece of the function.

Graph each piece of the function using the values in the table. The endpoints may overlap.

x	f(x)	
0	0	f(x) = 10x
1	10	
2	18	
3	26	f(x) = 8x + 2
4	34	
5	39	
6	44	f(x) = 5x + 14
7	49	

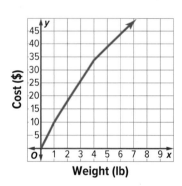

First piece: The slope of 10 shows that the cost of coffee beans is $10 per pound for 1 pound or less.

Second piece: The slope of 8 shows that the cost of coffee beans is $8 per pound, plus $2 for more than 1 pound and up to 4 pounds.

Third piece: The slope of 8 shows that the cost of coffee beans is $5 per pound, plus $14 for over 4 pounds.

> **Guided Practice**

4. JOBS The function $f(x) = \begin{cases} 8x \text{ if } 0 \le x \le 40 \\ 12x - 160 \text{ if } x > 40 \end{cases}$ models Jovan's earnings in dollars in a week when he works x hours. Graph the function, and then interpret each piece of the function.

> **Study Tip**
>
> **MP Structure** To graph a piecewise-defined function, graph each "piece" separately. Each part of the function rule will only be graphed on its given interval.

Study Tip

MP Regularity The greatest integer function and piecewise-defined functions are nonlinear functions.

Concept Summary Piecewise and Step Functions

Step Function	Piecewise-Defined Function
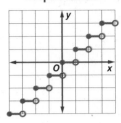	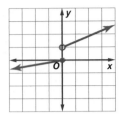

Go Online! for a Self-Check Quiz

Check Your Understanding   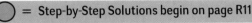 = Step-by-Step Solutions begin on page R11

Example 1   **Graph each function. State the domain and range.**

1. $f(x) = \frac{1}{2}[\![x]\!]$

2. $g(x) = -[\![x]\!]$

3. $h(x) = [\![2x]\!]$

Example 2   **4. EXERCISE** Ilan is hiring a personal trainer to help him exercise. The table shows the rates, in 15-minute intervals, for the cost of a personal trainer.

Time (minutes)	Cost ($)
$0 < x \le 15$	30
$15 < x \le 30$	55
$30 < x \le 45$	75
$45 < x \le 60$	90
$60 < x \le 75$	100
$75 < x \le 90$	105

   a. Graph the step function.

   b. What is the cost for a session that lasts 70 minutes?

Example 3   **Graph each function. State the domain and range.**

5. $f(x) = \begin{cases} 2x - 1 & \text{if } x > -1 \\ -x & \text{if } x \le -1 \end{cases}$

6. $g(x) = \begin{cases} -3x - 2 & \text{if } x > -2 \\ -x + 1 & \text{if } x \le -2 \end{cases}$

Example 4   **7 ELEVATORS** The function $f(x) = \begin{cases} 2x & \text{if } 0 \le x < 15 \\ 30 & \text{if } 15 \le x < 30 \text{ models the height} \\ -2x + 90 & \text{if } 30 \le x \le 45 \end{cases}$

above ground, in meters, of an elevator after x seconds. Graph the function, and then interpret each piece of the function.

Practice and Problem Solving   Extra Practice is on page R3.

Example 1   **Graph each function. State the domain and range.**

8. $f(x) = 3[\![x]\!]$ **9.** $f(x) = [\![-x]\!]$ **10.** $g(x) = -2[\![x]\!]$

11. $g(x) = [\![x]\!] + 3$ **12.** $h(x) = [\![x]\!] - 1$ **13.** $h(x) = \frac{1}{2}[\![x]\!] + 1$

Example 2   **14. CAB FARES** Lauren wants to take a taxi from a hotel to a friend's house. The rate is $3 plus $1.50 per mile after the first mile. Every fraction of a mile is rounded up to the next mile.

   a. Draw a graph to represent the cost of using a taxi cab.

   b. What is the cost if the trip is 8.5 miles long?

15. **(MP) MODELING** The United States Postal Service increases the rate of postage periodically. The table shows the cost to mail a letter weighing 1 ounce or less from 2001 through 2013. Draw a step graph to represent the data.

Year	2001	2002	2006	2007	2008	2009	2012	2013
Cost ($)	0.34	0.37	0.39	0.41	0.42	0.44	0.45	0.46

Example 3

Graph each function. State the domain and range.

16. $f(x) = \begin{cases} \frac{1}{2}x - 1 \text{ if } x > 3 \\ -2x + 3 \text{ if } x \le 3 \end{cases}$

17. $f(x) = \begin{cases} 2x - 5 \text{ if } x > 1 \\ 4x - 3 \text{ if } x \le 1 \end{cases}$

18. $f(x) = \begin{cases} 2x + 3 \text{ if } x \ge -3 \\ -\frac{1}{3}x + 1 \text{ if } x < -3 \end{cases}$

19. $f(x) = \begin{cases} 3x + 4 \text{ if } x \ge 1 \\ x + 3 \text{ if } x < 1 \end{cases}$

20. $f(x) = \begin{cases} 3x + 2 \text{ if } x > -1 \\ -\frac{1}{2}x - 3 \text{ if } x \le -1 \end{cases}$

21. $f(x) = \begin{cases} 2x + 1 \text{ if } x < -2 \\ -3x - 1 \text{ if } x \ge -2 \end{cases}$

Example 4

22. **PUBLIC SAFETY** The function $f(x) = \begin{cases} 10x + 100 \text{ if } 0 < x < 20 \\ 300 \text{ if } x \ge 20 \end{cases}$ models the cost in dollars of a speeding ticket for a person driving x mi/h over the speed limit. Graph the function, and then interpret each piece of the function.

23. **AVIATION** The function $f(x) = \begin{cases} -1000x + 30,000 \text{ if } 0 \le x < 5 \\ 25,000 \text{ if } x \ge 5 \end{cases}$ models the altitude in feet of an airplane after x minutes. Graph the function, and then interpret each piece of the function.

Determine the domain and range of each function.

24.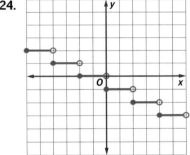

25.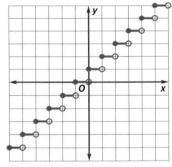

26.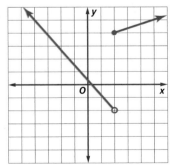

27.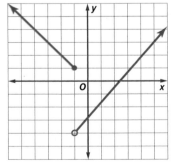

28. **BOATING** According to Boat Minnesota, the maximum number of people that can safely ride in a boat is determined by the boat's length and width. The table shows some guidelines for the length ℓ of a boat that is 6 feet wide. Graph this relation.

Length of Boat (ft)	$18 \le \ell < 20$	$20 \le \ell < 23$	$23 \le \ell < 24$
Number of People	7	8	9

29. **CAR LEASE** As part of Marcus' leasing agreement, he will be charged $0.20 per mile for each mile over 12,000. Any fraction of a mile is rounded up to the next mile. Make a step graph to represent the cost of going over the mileage.

30. **MULTI-STEP** A baseball team is ordering warm-up jackets with the team logo on the front and the players' names on the back. The store charges $10 to set up the artwork. Jackets are $10 each, artwork on the front is $4, and name imprints are $2. The store says that they would be willing to give a percent discount on orders of 11 or more jackets. Additionally, for orders of more than 20 jackets, they will give $15 off the entire order, as well as a percent discount.

   a. How much would it cost for the 9 starters to get warm-up jackets?

   b. The team has raised $455. If all 27 members of the team and 3 of their coaches want jackets, what is the lowest whole percent discount the coach needs to negotiate?

   c. Explain your solution process.

31. **DANCE** A local studio owner will teach up to four students by herself. Her instructors can teach up to 5 students each. Draw a step function graph that best describes the number of instructors needed for the different numbers of students.

32. **THEATERS** A community theater will only perform a show if there are at least 250 pre-sale ticket requests. Additional performances will be added for each 250 requests after that. Draw a step function graph that best describes this situation.

Graph each function.

33. $f(x) = \begin{cases} [\![x]\!] \text{ if } -2 \le x \le 1 \\ x + 2 \text{ if } x > -1 \end{cases}$

34. $f(x) = \begin{cases} 2x - 1 \text{ if } x < 2 \\ -[\![x]\!] \text{ if } x > 2 \end{cases}$

35. **SMOOTHIES** Write a step function for the number of medium smoothies that can be bought with x dollars if a medium smoothie costs $3.79. Then use the function to find the number of smoothies that can bought with $10.

36. **MULTIPLE REPRESENTATIONS** Renting a kayak for 1 hour or less costs $15 per hour. Each additional hour costs $10 per hour, up to a maximum of $35 per day.

   a. Graphical Draw a graph of a piecewise-defined function that best describes this situation.

   b. Symbolic Write the equation of the piecewise-defined function, and explain what each piece of the function rule represents.

H.O.T. Problems   Use Higher-Order Thinking Skills

37. **MP CONSTRUCT ARGUMENTS** Does the piecewise relation $y = \begin{cases} -2x + 4 \text{ if } x \ge 2 \\ -\frac{1}{2}x - 1 \text{ if } x \le 0 \end{cases}$ represent a function? Why or why not?

38. **ERROR ANALYSIS** Dylan claims that the graph of the function $f(x) = \begin{cases} -3x + 4 \text{ if } x < 0 \\ x - 2 \text{ if } x \ge 0 \end{cases}$ includes the point (0, 4).

   a. What error did Dylan make? Explain your reasoning.

   b. Find the correct value of $f(0)$.

39. **OPEN ENDED** Research the cost of tickets to a local zoo or museum. Then graph a step function that models how the cost of a ticket depends on a person's age.

40. **WRITING IN MATH** Compare and contrast the graphs of step and piecewise-defined functions with the graphs of linear functions. Discuss the domains, ranges, maxima, and minima.

41. MULTI-STEP The graph shows how a car's speed changes over time after passing through an intersection.

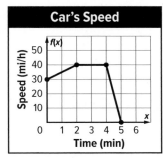

Determine whether each statement about the graph is true or false. **MP** 2

a. The interval for which the car is moving is $0 \leq x < 5$.

b. The interval for which the car maintains a constant speed is $2 \leq x \leq 4$.

c. The y-intercept shows that the car started with a speed of 0 mi/h.

d. The x-intercept shows that the car comes to a stop after 5 minutes.

e. The car came to a stop from 2 to 4 minutes.

42. A store is having a sale that offers discounts based on the amount a customer spends. Customers who spend at least $75 but less than $100 get a 25% discount. Customers who spend at least $100 but less than $125 get a 30% discount, and customers who spend at least $125 get a 40% discount. Draw a graph of a piecewise-defined function that models this situation. **MP** 4

43. The least integer function is written as $f(x) = \lceil x \rceil$, where $f(x)$ is the least integer not less than x. **MP** 8

a. Graph the least integer function.

b. What are the domain and range of the function?

c. Describe at least one pattern you notice in the graph of the least integer function.

d. Why is the least integer function an example of a step function?

44. The graph shows how the cost of having a gift wrapped at a store depends on the length of the box being wrapped.

Which interpretations of the graph are correct? Select all that apply. **MP** 2

☐ **A** Any box longer than 30 in. costs $10 to wrap.

☐ **B** A box with a length of 18 in. costs $8 to wrap.

☐ **C** There are 4 different cost levels for gift wrapping.

☐ **D** Wrapping a box that is 6 inches long costs more than wrapping a box that is 4 inches long.

45. Which function is shown in the graph? **MP** 7

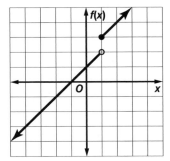

○ **A** $f(x) = \begin{cases} x + 1 \text{ if } x < 1 \\ x + 2 \text{ if } x \geq 1 \end{cases}$

○ **B** $f(x) = \begin{cases} x + 1 \text{ if } x < 2 \\ x + 2 \text{ if } x \geq 3 \end{cases}$

○ **C** $f(x) = \begin{cases} x + 2 \text{ if } x < 1 \\ x + 3 \text{ if } x \geq 1 \end{cases}$

○ **D** $f(x) = \begin{cases} x + 2 \text{ if } x < 2 \\ x + 3 \text{ if } x \geq 3 \end{cases}$

EXTEND 3-7

Graphing Technology Lab
Piecewise-Linear Functions

You can use a graphing calculator to graph and analyze various piecewise functions, including greatest integer functions and absolute value functions.

Mathematical Practices
MP 5 Use appropriate tools strategically.

Activity 1   Greatest Integer Functions

Work cooperatively. Graph $f(x) = [\![x]\!]$ in the standard viewing window.

The calculator may need to be changed to dot mode for the function to graph correctly. Press [MODE] then use the arrow and [ENTER] keys to select **DOT**.

Enter the equation in the **Y=** list. Then graph the equation.

KEYSTROKES: [Y=] [MATH] [▶] 5 [X,T,θ,n] [ )] [ZOOM] 6

1A. How does the graph of $f(x) = [\![x]\!]$ compare to the graph of $f(x) = x$?

1B. What are the domain and range of the function $f(x) = [\![x]\!]$? Explain.

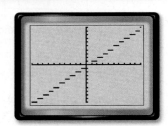

$[-10, 10]$ scl: 1 by $[-10, 10]$ scl: 1

The graphs of piecewise functions are affected by changes in parameters. An **absolute value function** is a type of piecewise linear function. Absolute value functions can be written in the form $f(x) = a|x - h| + k$, where a, h, and k are constants.

Activity 2   Graph a Complete Graph

Work cooperatively. Graph $y = |x| - 3$ and $y = |x| + 1$ in the standard viewing window.

Enter the equations in the **Y=** list. Then graph.

KEYSTROKES: [Y=] [MATH] [▶] 1 [X,T,θ,n] [ )] [−] 3 [ENTER] [MATH] [▶] 1
[X,T,θ,n] [ )] [+] 1 [ZOOM] 6

2A. Compare and contrast the graphs to the graph of $y = |x|$.

2B. How does the value of k affect the graph of $y = |x| + k$?

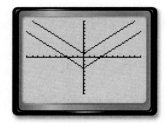

$[-10, 10]$ scl: 1 by $[-10, 10]$ scl: 1

Analyze the Results

Use a graphing calculator to graph each function.

1. $y = |x + 2|$

2. $y = [\![x]\!] - 3$

3. $y = 2[\![x]\!]$

4. $y = |x - 5|$

5. $y = [\![x]\!] + 1$

6. $y = \frac{1}{2}|x|$

7. A parking garage charges \$4 for every hour or fraction of an hour. Is this situation modeled by a *linear* function or a *step* function? Explain your reasoning.

8. A maintenance technician is testing an elevator system. The technician starts the elevator at the fifth floor. It is sent to the ground floor, then back to the fifth floor. Assume the elevator travels at a constant rate. Should the height of the elevator be modeled by a step function or an absolute value function? Explain.

Absolute Value Functions

:: **Then**

- You identified and graphed piecewise and step functions.

:: **Now**

1 Identify and graph translations of absolute value functions.

2 Identify and graph reflections and dilations of absolute value functions.

:: **Why?**

- Cyril is approaching an overlook point on a straight hiking trail at a rate of 3 mi/h. He is currently 1 mi from the overlook. An absolute value function could model his distance from the overlook both before and after he passes it.

 New Vocabulary

absolute value function

vertex

 Mathematical Practices

1 Make sense of problems and persevere in solving them.

7 Look for and make use of structure.

1 **Translations** An **absolute value function** is a type of piecewise linear function that can be written in the form $f(x) = a|x - h| + k$, where a, h, and k are constants. The **vertex** of an absolute value function is the minimum or maximum point of its graph.

Key Concept Absolute Value Function

Parent function: $f(x) = |x|$, defined as

$$f(x) = \begin{cases} x \text{ if } x > 0 \\ 0 \text{ if } x = 0 \\ -x \text{ if } x < 0 \end{cases}$$

Type of graph: V-shaped

Domain: all real numbers

Range: all nonnegative real numbers

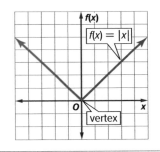

Example 1 Graphing $f(x) + k$

Graph $g(x) = |x| + 2$. State the domain and range, and describe how the graph is related to the graph of $f(x) = |x|$.

Since $|x|$ cannot be negative, the vertex of the graph occurs when $|x| = 0$, or when $x = 0$. Make a table of values. Include values for $x < 0$ and $x > 0$. Then graph.

$g(x) = |x| + 2$	
x	$f(x)$
−4	6
−2	4
0	2
2	4
4	6

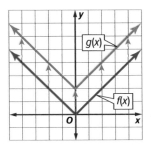

The domain is all real numbers.

The range is $g(x) \geq 2$.

The graph of $g(x)$ is the graph of $f(x)$ translated 2 units up.

▶ **Guided Practice**

1. Graph $g(x) = |x| - 1$. State the domain and range, and describe how the graph is related to the graph of $f(x) = |x|$.

andipantz/E+/Getty Images

Example 1 shows that the graph of $g(x) = |x| + k$ is the graph of $f(x) = |x|$ translated vertically. Now consider horizontal translations of the graph of $f(x) = |x|$.

Example 2   **Graphing f(x − h)**

Graph $g(x) = |x - 4|$. State the domain and range, and describe how the graph is related to the graph of $f(x) = |x|$.

Since $|x - 4|$ cannot be negative, the vertex of the graph occurs when $|x - 4| = 0$, or when $x = 4$. Make a table of values. Include values for $x < 4$ and $x > 4$.

$g(x) = |x - 4|$	
x	$g(x)$
0	4
2	2
4	0
6	2
8	4

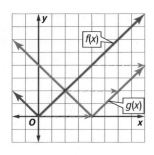

The domain is all real numbers.

The range is $g(x) \geq 0$.

The graph of $g(x)$ is the graph of $f(x)$ translated right 4 units.

> **Study Tip**
>
> **MP Regularity** Subtracting a number from x inside the absolute value symbols shifts the graph to the right. Adding a number to x inside the absolute value symbols shifts the graph to the left.

Guided Practice

2. Graph $g(x) = |x + 2|$. State the domain and range, and describe how the graph is related to the graph of $f(x) = |x|$.

---

2 Reflections and Dilations

The graphs of absolute value functions can also be reflected and dilated. One way is by multiplying the function rule by a nonzero constant.

Example 3   **Graphing a • f(x)**

Graph $g(x) = -2|x|$. State the domain and range, and describe how the graph is related to the graph of $f(x) = |x|$.

Since $|x|$ cannot be negative, the vertex of the graph occurs when $|x| = 0$, or when $x = 0$. Make a table of values. Include values for $x < 0$ and $x > 0$.

$g(x) = -2|x|$	
x	$g(x)$
−4	−8
−2	−4
0	0
2	−4
4	−8

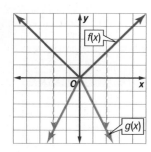

The domain is all real numbers.

The range is $g(x) \leq 0$.

The graph of $g(x)$ is the graph of $f(x)$ reflected across the x-axis and vertically stretched by a factor of 2.

> **Study Tip**
>
> **Direction** For an absolute value function of the form $f(x) = a|x|$, the graph opens upward if $a > 0$ and opens downward if $a < 0$.

Guided Practice

3. Graph $g(x) = \frac{1}{2}|x|$. State the domain and range, and describe how the graph is related to the graph of $f(x) = |x|$.

Absolute value functions can also be dilated by multiplying x in the function rule by a nonzero constant.

Example 4   Graphing *f*(*bx*)

Graph $g(x) = |2x|$. State the domain and range, and describe how the graph is related to the graph of $f(x) = |x|$.

Since $|2x|$ cannot be negative, the vertex of the graph occurs when $|2x| = 0$, or when $x = 0$. Make a table of values. Include values for $x < 0$ and $x > 0$.

$g(x) = |2x|$	
x	$g(x)$
−4	8
−2	4
0	0
2	4
4	8

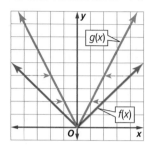

The domain is all real numbers.

The range is $g(x) \geq 0$.

The graph of $g(x)$ is the graph of $f(x)$ horizontally compressed by a factor of $\frac{1}{2}$.

▶ **Guided Practice**

4. Graph $g(x) = \left|\frac{1}{2}x\right|$. State the domain and range, and describe how the graph is related to the graph of $f(x) = |x|$.

Absolute value functions can be used to represent real-world situations.

Real-World Example 5   Absolute Value Function

TRANSPORTATION A commuter train is approaching a sensor on the track at a speed of 100 km/h. The function $f(x) = \frac{4}{3}|x - 30|$ models the train's distance in kilometers from the sensor after x minutes. Graph this function. After how many minutes will the train reach the sensor?

Since $|x - 30|$ cannot be negative, the vertex occurs when $|x - 30| = 0$, or when $x = 30$. Make a table of values. Include values for $x < 30$ and $x > 30$.

x	$f(x)$
0	40
15	20
30	0
45	20
60	40

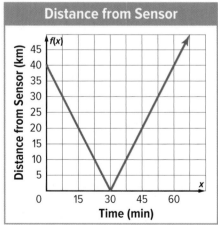

Distance from Sensor

The x-intercept of 30 shows that the train will reach the sensor after 30 min.

▶ **Guided Practice**

5. MARINE SCIENCE The function $f(x) = 3|x - 8| - 24$ models a dolphin's elevation in feet compared to sea level after x seconds. Graph the function. How far below sea level is the dolphin at the deepest point in its dive?

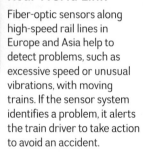

Real-World Link

Fiber-optic sensors along high-speed rail lines in Europe and Asia help to detect problems, such as excessive speed or unusual vibrations, with moving trains. If the sensor system identifies a problem, it alerts the train driver to take action to avoid an accident.

Source: *The Optical Society*

Go Online!

Investigate the family of absolute value functions with the **Geometer's Sketchpad®** activity in ConnectED.

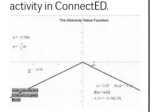

Concept Summary Absolute Value Functions of the Form $g(x) = a|x - h| + k$		
Effect of h	**Effect of k**	**Effect of a**
$h > 0$: translation h units right $h < 0$: translation $|h|$ units left	$k > 0$: translation k units up $k < 0$: translation $|k|$ units down	$a < 0$: reflection across the x-axis $|a| > 1$: vertical stretch by a factor of $|a|$ $0 < |a| < 1$: vertical compression by a factor of $|a|$

Check Your Understanding   = Step-by-Step Solutions begin on page R11.

 **Go Online!** for a Self-Check Quiz

Examples 1–4 Graph each function. State the domain and range, and describe how the graph is related to the graph of $f(x) = |x|$.

1. $g(x) = |x| + 1$ 2. $g(x) = |x| - 2$ 3. $g(x) = |x| + 3$

4. $g(x) = |x + 5|$ 5. $g(x) = |x - 3|$ 6. $g(x) = |x + 1|$

7. $g(x) = -|x|$ 8. $g(x) = 3|x|$ 9. $g(x) = -\frac{1}{2}|x|$

10. $g(x) = |3x|$ 11. $g(x) = |\frac{1}{2}x|$ 12. $g(x) = |-2x|$

Example 5 13. **ARCHITECTURE** The function $f(x) = -\frac{1}{2}|x - 16| + 20$ for $0 \le x \le 32$ models the shape of a house's roof, where $f(x)$ is the height in feet of the roof at a horizontal distance of x feet from the left end of the roof. Graph the function. How high is the peak of the roof above the ground?

Practice and Problem Solving   Extra Practice is found on page R3.

Examples 1–4 Graph each function. State the domain and range, and describe how the graph is related to the graph of $f(x) = |x|$.

14. $g(x) = |x| - 4$ 15. $g(x) = |x| + 6$ 16. $g(x) = |x| - 3$

17. $g(x) = |x + 4|$ 18. $g(x) = |x - 1|$ 19. $g(x) = |x - 2|$

20. $g(x) = 2|x|$ 21. $g(x) = -\frac{1}{4}|x|$ 22. $g(x) = -3|x|$

23. $g(x) = |4x|$ 24. $g(x) = |-3x|$ 25. $g(x) = |\frac{1}{4}x|$

Example 5 26. **ANCIENT EGYPT** The function $f(x) = -\frac{147}{115}|x - 115| + 147$ for $0 \le x \le 230$ models the shape of the Great Pyramid of Giza, viewed from the side, when it was originally built. In this function, $f(x)$ is the height in meters of the surface of the pyramid at a horizontal distance of x meters from the left vertex. Graph the function. How tall was the Great Pyramid when it was originally built?

Determine the domain and range of each function.

27.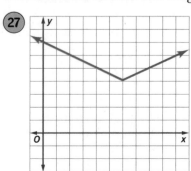

28.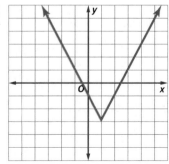

Graph each function. Identify its vertex, and describe how the graph is related to the graph of $f(x) = |x|$.

29. $g(x) = |x + 2| + 3$ **30.** $g(x) = 2|x - 1|$ **31.** $g(x) = -\frac{1}{2}|x| + 1$

Write the equation of each graphed function, and explain how you determined your answer.

32.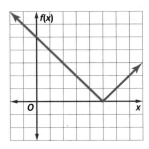

33.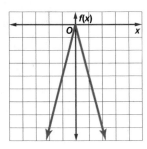

34. **MP** **PERSEVERANCE** A marathon runner is headed along a straight path toward mile marker 8 at a speed of 600 ft/min. The runner is currently 1200 ft from mile marker 8. Write and graph the equation of an absolute value function that models the runner's distance in feet from mile marker 8 after x minutes.

35 Consider the function $f(x) = |2x + 3|$.

     a. Make a table of values where x is all integers from -5 to 5, inclusive.

     b. Plot the points on a coordinate grid.

     c. Graph the function.

36. Consider the function $f(x) = |2x| + 3$.

     a. Make a table of values where x is all integers from -5 to 5, inclusive.

     b. Plot the points on a coordinate grid.

     c. Graph the function.

     d. Describe how this graph is different from the graph in Exercise 35.

Classify each function as *step*, *piecewise*, or *absolute value*.

37. $g(x) = |x - 2| - 5$ **38.** $g(x) = \frac{1}{2}[\![x]\!] + 4$ **39.** $g(x) = \begin{cases} 2x + 1 \text{ if } x \geq 1 \\ -x + 4 \text{ if } x < 1 \end{cases}$

40. **MULTI-STEP** The function $f(x) = |4.5 - 0.9x|$ models a car's distance in miles from a sensor on a highway that measures vehicle speed. In this function, x represents the number of minutes since timing began.

     a. In how many minutes will the car reach the sensor?

     b. The speed limit on the highway is 55 mi/h. Will the sensor indicate that the car is speeding? Explain your reasoning.

41. **MP** **REGULARITY** Answer each question for absolute value functions of the general form $f(x) = a|x - h| + k$ in terms of the parameters a, h, and k. If needed, use a graphing calculator other graphing program to explore how changing the values of the parameters affects the graph of the function.

     a. What are the coordinates of the vertex?

     b. What is the equation of the line of symmetry for the function's graph?

     c. What are the slopes of the rays that form the function's graph?

Write each absolute-value function as a piecewise function.

42. $f(x) = |x|$ **43.** $f(x) = |x - 3|$ **44.** $f(x) = 2|x| + 1$

45. What is the relationship between the slopes of the two parts of an absolute-value function?

Write a function for each transformation.

46. Translate the graph of $f(x) = |x|$ to the left 3 units.

47 Reflect the graph of $f(x) = |2x|$ across the x-axis.

48. Compress the graph of $f(x) = -4|x| - 3$ vertically by a factor of $\frac{1}{4}$.

49. Translate the graph of $f(x) = -|3x| + 5$ down 6 units.

50. MULTIPLE REPRESENTATIONS In this problem, you will explore piecewise linear functions.

    a. Tabular Copy and complete the table of values for $f(x) = |[\![x]\!]|$ and $g(x) = [\![|x|]\!]$.

x	$[\![x]\!]$	$f(x) = |[\![x]\!]|$	$|x|$	$g(x) = [\![|x|]\!]$
-3	-3	3	3	3
-2.5				
-2				
0				
0.5				
1				
1.5				

    b. Graphical Graph each function on a coordinate plane.

    c. Analytical Compare and contrast the graphs of $f(x)$ and $g(x)$.

GEOMETRY Find the area of the triangle that is above the x-axis and below the graph of the function.

51. $f(x) = -|x| + 10$ **52.** $f(x) = -|2x| + 6$ **53.** $f(x) = -4|x| + 4$

H.O.T. Problems   Use Higher-Order Thinking Skills

MP SENSE-MAKING Refer to the graph.

54. Write an absolute value function that represents the graph.

55. Write a piecewise function to represent the graph.

56. What are the domain and range?

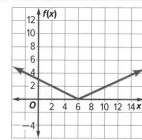

57. MP CRITIQUE ARGUMENTS Tayden argues that the graph of an absolute value function can have more than one x-intercept. Is Caleb correct? Justify your answer.

58. ERROR ANALYSIS Darcy claims that the graph of the function $g(x) = |x + 8|$ is a horizontal translation 8 units to the right of the graph of the parent function $f(x) = |x|$. What error did Darcy make? Explain your reasoning.

59. WRITING IN MATH Explain how to determine whether the graph of an absolute value function has a maximum point or a minimum point.

60. Use the numbers and symbols in the box to complete the equation for the absolute value function shown in the graph. You may use a number or symbol more than once. **MP** 7

2	3	5	+	−

$$f(x) = |x \ \square \ \square| \ \square \ \square$$

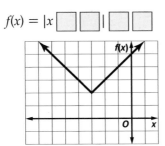

61. Which transformations of the graph of $f(x) = |x|$ are needed to produce the graph of $g(x) = -|x - 1|$? Select all that apply. **MP** 7

☐ **A** reflection across the x-axis

☐ **B** reflection across the y-axis

☐ **C** translation 1 unit right

☐ **D** translation 1 unit left

☐ **E** translation 1 unit up

☐ **F** translation 1 unit down

62. Consider the function $g(x) = \left|\frac{1}{3}x\right|$. **MP** 7

a. Graph the function.

b. Describe the graph of $g(x)$ as a transformation of the graph of $f(x) = |x|$.

63. Complete the sentence using one of the choices in the box. **MP** 7

The graph of $g(x) = |x| + 1.3$ is the same as the graph of the parent function translated 1.3 units _____.

○ **A** left

○ **B** right

○ **C** up

○ **D** down

64. Select an equation of the function shown in the graph? **MP** 7

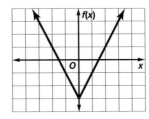

○ **A** $f(x) = 2|x - 3|$

○ **B** $f(x) = 2|x| - 3$

○ **C** $f(x) = 2|x| + 3$

○ **D** $f(x) = -3|x - 2|$

○ **E** $f(x) = -3|x + 2|$

○ **F** $f(x) = -3|x| + 2$

65. Determine whether each statement about $g(x) = -\frac{1}{2}|x| + 4$ is true or false. **MP** 7

a. The graph has a minimum point.

b. The function has two x-intercepts.

c. The vertex of the function is $(0, 4)$.

d. The graph is symmetric about the x-axis.

e. The graph is translated 4 units up from the graph of $f(x) = |x|$.

66. **MULTI-STEP** The function $g(x) = \frac{1}{5}|x - 20| + 2$ models a ship's distance in miles from a lighthouse x minutes after the captain first spots the lighthouse. **MP** 2, 4

a. Graph the function.

b. What is the vertex of the function, and what do its coordinates represent in this situation?

c. How far is the ship from the lighthouse when the captain first spots it?

d. Describe the graph of $g(x)$ as a transformation of the graph of $f(x) = |x|$.

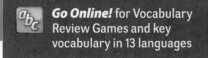

Go Online! for Vocabulary Review Games and key vocabulary in 13 languages

Study Guide

Key Concepts

Graphing Linear Equations (Lesson 3-1)

- The standard form of a linear equation is $Ax + By = C$, where $A \geq 0$, A and B are not both zero, and A, B, and C are integers whose greatest common factor is 1.

Zeros of Linear Functions (Lesson 3-2)

- Values of x for which $f(x) = 0$ are called zeros of the function f. A zero of a function is located at an x-intercept of the graph of the function.

Rate of Change and Slope (Lesson 3-3)

- If x is the independent variable and y is the dependent variable, then rate of change equals

$$\frac{\text{change in } y}{\text{change in } x}.$$

- The slope of a line is the ratio of the rise to the run.

$$m = \frac{y_2 - y_1}{x_2 - x_1}$$

Slope-Intercept Form (Lesson 3-4)

- The slope-intercept form of a linear equation is $y = mx + b$, where m is the slope and b is the y-intercept.

Arithmetic Sequences (Lesson 3-5)

- The nth term a_n of an arithmetic sequence with first term a_1 and common difference d is given by $a_n = a_1 + (n - 1)d$, where n is a positive integer.

Special Functions (Lesson 3-7, 3-8)

- The greatest integer function is written as $f(x) = [\![x]\!]$, where $f(x)$ is the greatest integer not greater than x.
- The absolute value function is written as $f(x) = |x|$, where $f(x)$ is the distance from x to 0 on a number line.

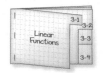

FOLDABLES® Study Organizer

Use your Foldable to review the chapter. Working with a partner can be helpful. Ask for clarification of concepts as needed.

Key Vocabulary

arithmetic sequence (p. 190)	root (p. 151)
common difference (p. 190)	sequence (p. 190)
constant (p. 143)	slope (p. 162)
constant function (p. 172)	slope-intercept form (p. 171)
greatest integer function (p. 197)	standard form (p. 143)
	step function (p. 197)
linear equation (p. 143)	transformations (p. 181)
linear function (p. 151)	x-intercept (p. 144)
piecewise-linear function (p. 197)	y-intercept (p. 144)
rate of change (p. 160)	zero of a function (p. 151)

Vocabulary Check

State whether each sentence is *true* or *false*. If *false*, replace the underlined word or number to make a true sentence.

1. The x-coordinate of the point at which the graph of an equation crosses the x-axis is a(n) <u>x-intercept</u>.

2. Values of x for which $f(x) = 0$ are called <u>zeros</u> of the function f.

3. Any two points on a nonvertical line can be used to determine the <u>slope</u>.

4. The slope of the line $y = 5$ is <u>5</u>.

5. A ratio that describes, on average, how much one quantity changes with respect to a change in another quantity is a <u>rate of change</u>.

6. The <u>y-intercept</u> is the y-coordinate of the point where the graph crosses the y-axis.

7. The <u>range</u> of the greatest integer function is the set of all real numbers.

Concept Check

8. How is an arithmetic sequence like a linear function?

9. Explain how to find the common difference for an arithmetic sequence.

10. Find the slope of $y = 3x - 6$.

Lesson-by-Lesson Review

3-1 Graphing Linear Functions

Find the *x*-intercept and *y*-intercept of the graph of each linear function.

11.

x	y
−8	0
−4	3
0	6
4	9
8	12

12.

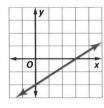

Graph each equation. Then state the domain and range.

13. $y = -x + 2$

14. $x + 5y = 4$

15. $2x - 3y = 6$

16. $5x + 2y = 10$

17. SOUND The distance *d* in kilometers that sound waves travel through water is given by $d = 1.6t$, where *t* is the time in seconds.

   a. Make a table of values and graph the equation.

   b. Use the graph to estimate how far sound can travel through water in 7 seconds.

Example 1

Graph $3x - y = 4$ by using the *x*- and *y*-intercepts.

Find the *x*-intercept.

$3x - y = 4$

$3x - \mathbf{0} = 4$ Let $y = 0$.

$3x = 4$

$x = \dfrac{4}{3}$

Find the *y*-intercept.

$3x - y = 4$

$3(\mathbf{0}) - y = 4$ Let $x = 0$.

$-y = 4$

$y = -4$

x-intercept: $\dfrac{4}{3}$

y-intercept: -4

The graph intersects the *x*-axis at $\left(\dfrac{4}{3}, 0\right)$ and the *y*-axis at $(0, -4)$. Plot these points. Then draw the line through them.

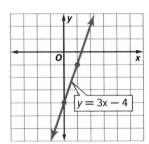

3-2 Zeros of Linear Functions

Find the zeros of each function by graphing.

18. $f(x) = 2x + 8$

19. $f(x) = 4x - 24$

20. $f(x) = 3x - 5$

21. $f(x) = 6x + 3$

22. $f(x) = 16 - 8x$

23. $f(x) = 21 + 3x$

24. $f(x) = -4x - 28$

25. $f(x) = 25x - 225$

26. FUNDRAISING Sean's class is selling boxes of popcorn to raise money for a class trip. Sean's class paid $85 for the popcorn, and they are selling each box for $1. The function $y = x - 85$ represents their profit *y* for the number of boxes sold *x*. Find the zero and describe what it means in this situation.

Example 2

Find the zeros of $f(x) = 3x + 3$ by graphing.

The graph intersects the *x*-axis at −1. So, the solution is −1.

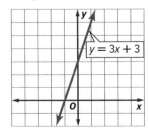

3-3 Rate of Change and Slope

Find the rate of change represented in each table or graph.

27.

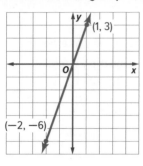

28.

x	y
−2	−3
0	−3
4	−3
12	−3

Find the slope of the line that passes through each pair of points.

29. $(0, 5), (6, 2)$ 30. $(−6, 4), (−6, −2)$

31. **PHOTOS** The average cost of online photos decreased from $0.50 per print to $0.13 per print between 2002 and 2013. Find the average rate of change in the cost. Explain what it means.

Example 3

Find the slope of the line that passes through $(0, −4)$ and $(3, 2)$.

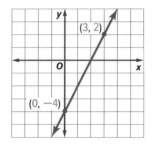

Let $(0, −4) = (x_1, y_1)$ and $(3, 2) = (x_2, y_2)$.

$m = \dfrac{y_2 - y_1}{x_2 - x_1}$ Slope formula

$= \dfrac{2 - (−4)}{3 - 0}$ $x_1 = 0, x_2 = 3, y_1 = −4, y_2 = 2$

$= \dfrac{6}{3}$ or 2   Simplify.

3-4 Slope-Intercept Form

Write an equation of a line in slope-intercept form with the given slope and y-intercept. Then graph the equation.

32. slope: 3, y-intercept: 5

33. slope: −2, y-intercept: −9

34. slope: $\dfrac{2}{3}$, y-intercept: 3

35. slope: $-\dfrac{5}{8}$, y-intercept: −2

Graph each equation. Then state the slope and y-intercept.

36. $y = 4x − 2$ 37. $y = −3x + 5$

38. $y = \dfrac{1}{2}x + 1$ 39. $3x + 4y = 8$

40. **SKI RENTAL** Write an equation in slope-intercept form for the total cost of skiing for *h* hours with one lift ticket and ski rental.

Apache Slope
Ski Lodge

Lift Ticket $50/day
Ski Rental $10/hour

Example 4

Write an equation of a line in slope-intercept form with slope −5 and y-intercept −3. Then graph the equation.

$y = mx + b$ Slope-intercept form

$y = −5x + (−3)$ $m = −5$ and $b = −3$

$y = −5x − 3$ Simplify.

To graph the equation, plot the y-intercept $(0, −3)$.

Then move up 5 units and left 1 unit. Plot the point. Draw a line through the two points.

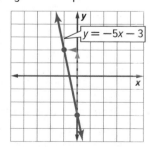

3-5 Transformations of Linear Functions

Describe the transformation in each function as it relates to the graph of $f(x) = x$.

41. $g(x) = x - 7$

42. $g(x) = -3x$

43. $g(x) = 4(x - 2)$

44. $g(x) = 0.25(x + 4)$

45. Rachel earns a weekly base salary of $400 plus a commission on her sales. Her total weekly pay is described by the function $p(x) = 0.25x + 400$. How will the graph of this function change if her base salary increases to $450?

46. Gabe runs the same number of miles x every weekday. The total number of miles he runs in one week is represented by $r(x) = 5x$. If Gabe decides to run 30% more each day, write a function $g(x)$ to represent the increase.

Example 5

The cost of picking your own strawberries at a farm is described by $p(x) = 2.5x + 5$, where x is the number of pounds.

a. Write a function $g(x)$ to represent a 10% discount. Discounting prices by 10% is represented by a dilation of the function. The decrease in price will be 90% of the original price.
$$g(x) = 0.9p(x)$$
$$= 0.9(2.5x + 5)$$

b. Find the price of picking 8 pounds of strawberries before and after the discount.

$$
\begin{aligned}
p(8) &= 2.5(8) + 5 & g(8) &= 0.9[2.5(8) + 5] \\
&= 20 + 5 & &= 0.9(25) \\
&= \$25 & &= \$20
\end{aligned}
$$

3-6 Arithmetic Sequences as Linear Functions

Find the next three terms of each arithmetic sequence.

47. 6, 11, 16, 21, …

48. 1.4, 1.2, 1.0, …

49. 100, 91, 82, 73, …

50. 3, 10, 17, 24, …

51. −14, −10, −6, −2, …

52. 51, 48, 45, 42, …

Write an equation for the nth term of each arithmetic sequence.

53. $a_1 = 6$, $d = 5$

54. 28, 25, 22, 19, …

55. $a_1 = 12$, $d = 3$

56. 14, 19, 24, 29, …

57. SCIENCE The table shows the distance traveled by sound in water. Write an equation for this sequence. Then find the time for sound to travel 72,300 feet.

Time (s)	1	2	3	4
Distance (ft)	4820	9640	14,460	19,280

Example 6

Find the next three terms of the arithmetic sequence 10, 23, 36, 49, … .

Find the common difference.

10 23 36 49
  +13 +13 +13

So, $d = 13$.

Add 13 to the last term of the sequence. Continue adding 13 until the next three terms are found.

49 62 75 88
  +13 +13 +13

The next three terms are 62, 75, and 88.

3-7 Piecewise and Step Functions

Graph each function. State the domain and range.

58. $f(x) = [\![x]\!]$

59. $f(x) = [\![2x]\!]$

60. $f(x) = \begin{cases} x - 2 \text{ if } x < 1 \\ 3x \text{ if } x \geq 1 \end{cases}$

61. $f(x) = \begin{cases} 2x - 3 \text{ if } x \leq 2 \\ x + 1 \text{ if } x > 2 \end{cases}$

Give the domain and range of each function.

62. $f(x) = \begin{cases} 2x - 8 \text{ if } x < 4 \\ -\frac{1}{2}x + 2 \text{ if } x \geq 4 \end{cases}$

63. $f(x) = \begin{cases} 6 \text{ if } x < -2 \\ 3x + 6 \text{ if } x \geq -2 \end{cases}$

64. $f(x) = \begin{cases} -3x - 2 \text{ if } x < -1 \\ 2x + 3 \text{ if } x \geq -1 \end{cases}$

65. $f(x) = \begin{cases} 10 \text{ if } x < 17 \\ 15 \text{ if } x \geq 17 \end{cases}$

66. Amy is going snorkeling with her friends. She can rent snorkeling gear for $4 in 20 minute intervals plus a $25 insurance fee. How much would it cost to rent the gear for 30 minutes? Explain whether Amy would pay more to rent the gear for 40 minutes.

Example 7

Graph $f(x) = \begin{cases} x + 5 \text{ if } x \leq -1 \\ 2x \text{ if } x > -1 \end{cases}$. State the domain and range.

Create a table of values for $f(x) = x + 5$ using values of $x \leq -1$. If $x = -1$, $f(x) = 4$, so place a point at $(-1, 4)$.

Then add values to the table for $f(x) = -2x$ using values of $x > -1$. Since x is not equal to -1 in this expression, place a circle at $(-1, -2)$.

x	f(x)
−3	2
−2	3
−1	4
0	0
1	2
2	4

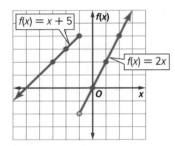

The domain is all real numbers, and the range is all real numbers.

3-8 Absolute Value Functions

Graph each function. State the domain and range.

67. $f(x) = |x|$

68. $f(x) = |2x - 2|$

69. $f(x) = |x + 1|$

70. $f(x) = |2x|$

71. $f(x) = |3x - 1|$

72. $f(x) = |x + 2|$

Describe how the graph of each function is related to the graph of $f(x) = |x|$.

73. $g(x) = |x| - 4$

74. $g(x) = 3|x|$

75. $g(x) = -|x|$

76. $g(x) = |x| + 7$

77. $g(x) = \frac{1}{2}|x|$

78. John is jogging on a straight jogging trail toward a flagpole at a speed of 500 ft/min. He is currently 1500 feet from the flagpole. Write an absolute value function that models John's distance from the flagpole after x minutes. How many minutes will it take John to reach the flagpole?

Example 8

Graph $f(x) = |x + 3|$. State the domain and range.

x	f(x)
−5	2
−4	1
−3	0
−2	1
−1	2

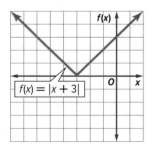

The domain is all real numbers, and the range is $f(x) \geq 0$.

1. **TEMPERATURE** The equation to convert Celsius temperature C to Kelvin temperature K is shown.

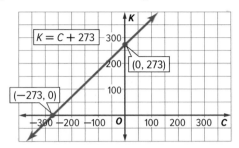

$K = C + 273$

$(0, 273)$

$(-273, 0)$

   a. State the independent and dependent variables. Explain.

   b. Determine the C- and K-intercepts and describe what the intercepts mean in this situation.

Graph each equation.

2. $y = x + 2$

3. $y = 4x$

4. $x + 2y = -1$

5. $-3x = 5 - y$

Find the zeros of each function by graphing.

6. $f(x) = 4x + 2$

7. $f(x) = 6 - 3x$

8. $f(x) = 5x + 2$

9. $f(x) = 16 - 8x$

Find the slope of the line that passes through each pair of points.

10. $(5, 8), (-3, 7)$

11. $(5, -2), (3, -2)$

12. $(-4, 7), (8, -1)$

13. $(6, -3), (6, 4)$

14. **MULTIPLE CHOICE** Which is the slope-intercept form of the equation for the linear function shown in the graph?

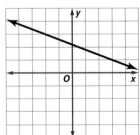

A $y = -\frac{2}{5}x + \frac{1}{5}$

B $y = -\frac{2}{5}x + \frac{11}{5}$

C $y = \frac{2}{5}x - \frac{1}{5}$

D $y = -\frac{5}{2}x + \frac{17}{2}$

Describe the transformation(s) in each function as it relates to the graph of $f(x) = x$.

15. $g(x) = -4x$

16. $g(x) = -(x + 2)$

17. $g(x) = x - 1$

18. $g(x) = 3x$

Determine whether each sequence is an arithmetic sequence. If it is, write an equation for the nth term in the sequence.

19. $-40, -32, -24, -16, \ldots$

20. $0.75, 1.5, 3, 6, 12, \ldots$

21. $5, 17, 29, 41, \ldots$

22. **MULTIPLE CHOICE** In each figure, only one side of each regular pentagon is shared with another pentagon. The length of each side is 1 centimeter. If the pattern continues, what is the perimeter of a figure that has 6 pentagons?

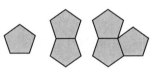

A 30 cm

B 25 cm

C 20 cm

D 15 cm

Graph each function.

23. $f(x) = |x - 1|$

24. $f(x) = -|2x|$

25. $f(x) = [\![x]\!]$

26. $f(x) = \begin{cases} 2x - 1 \text{ if } x < 2 \\ x - 3 \text{ if } x \geq 2 \end{cases}$

27. Determine the domain and range of the function.

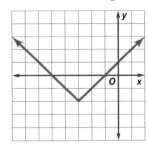

Performance Task

Provide a clear solution to each part of the task. Be sure to show all of your work, include all relevant drawings, and justify your answers.

FINANCIAL LITERACY Adrith runs a lawn-mowing business for his neighbors. He has different plans homeowners can purchase based on their needs.

Part A

With the Basic plan, homeowners pay each time Adrith mows their lawn. The table shows the total amount paid for number of service visits.

Number of visits	1	2	3	4
Amount paid ($)	15	30	45	60

1. What is the common difference?
2. Write an equation to represent the sequence.

Part B

Homeowners pay a $50 fee to sign up for the Premium plan, which includes both lawn mowing and trimming shrubs. The graph for the equation representing this plan goes through the points (1, 65) and (3, 95), where x represents the number of service visits and y represents the total amount paid.

3. What is the slope for the Premium plan?
4. Write an equation in slope-intercept form to represent the total amount paid for any number of service visits.

Part C

Compare the common difference for the sequence representing the Basic plan and the slope for the Premium plan.

5. What does the common difference for the Basic plan represent?
6. What does the slope for the Premium plan represent?
7. Explain how the costs per service visit are related.

Part D

8. Graph the solutions of the equations for the Basic plan and Premium plan. Describe the change from the Basic plan to the Premium plan as a transformation of a linear function.

Part E

9. **Construct an Argument** For homeowners on the Basic plan, Adrith will trim shrubs for an additional charge of $5 each time. Describe an advantage of the Premium plan.

Part F

10. **Structure** Describe a transformation that would make the graph of the Premium plan steeper. Explain how the transformation affects the cost per service visit.

Test-Taking Strategy

Example

Test-Taking Tip

Reading Math Problems
The first step to solving any math problem is to read the problem. When reading a math problem to get the information you need to solve, it is helpful to use special reading strategies.

Read the problem. Identify what you need to know. Then use the information in the problem to solve.

Jamal, Gina, Lisa, and Renaldo are renting a car for a road trip. The cost of renting the car is given by the function $C = 22.5 + 25d$, where C is the total cost for renting the car for d days. What does the slope of the function represent?

A number of people

B cost per day

C number of days

D miles per gallon

Step 1 **What do you need to find?**
The question asks what the slope represents.

Step 2 **Is there enough information given to solve the problem?**
Yes. You know that C represents the total cost and d represents the number of days.

Step 3 **What information, if any, is not needed to solve the problem?**
You do not need to know the number of people on the trip.

Step 4 **Are there any obvious wrong answers? If so, which one(s)? Explain.**
A and C; Slope is a ratio. Choices A and C are not ratios, so you can eliminate them.

Step 5 **What is the correct answer?** B

Apply the Strategy

Read the problem. Identify what you need to know. Then use the information in the problem to solve.

What does the x-intercept mean in the context of the situation given in the graph?

A   amount of time needed to drain the bathtub

B   number of gallons in the tub when the drain plug is pulled

C   number of gallons in the tub after x minutes

D   amount of water drained each minute

Draining a Bathtub

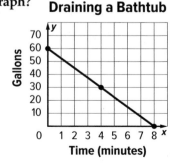

Answer the questions below.

a.   What do you need to find?

b.   Is there enough information given to solve the problem?

c.   What information, if any, is not needed to solve the problem?

d.   Are there any obvious wrong answers? If so, which one(s)? Explain.

e.   What is the correct answer?

Read each question. Then fill in the correct answer on the answer document provided by your teacher or on a sheet of paper.

1. Carlos delivered 16 newspapers in 10 minutes and 24 newspapers in 15 minutes. Let x represent the time and y represent the number of newspapers delivered. What is the rate of change?

   ○ A $\frac{8}{5}$

   ○ B $\frac{5}{8}$

   ○ C $-\frac{a + m}{2}$

   ○ D $\frac{2}{3}$

2. Use the graph shown.

   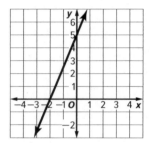

   Select all the equations that represent the graph.

   ☐ A $y = \frac{5}{2}x + 5$

   ☐ B $5x + 2y = 10$

   ☐ C $-5x + 2y = 10$

   ☐ D $-5x + 2y = -10$

   ☐ E $2x - 5y = 10$

3. Which of the following is equivalent to $5(4 - m) - 2(m + 3)$?

   ○ A $14 - 7m$

   ○ B $26 - 7m$

   ○ C $23 - 7m$

   ○ D $23 - 3m$

4. The first four terms of a sequence are $-12, -5, 2, 9$. Which of the following is an equation for the nth term of the sequence?

   ○ A $a_n = 7n - 12$

   ○ B $a_n = -7n - 5$

   ○ C $a_n = 7n - 5$

   ○ D $a_n = 7n - 19$

5. What is the solution of $\frac{4a + 7}{5} = -18$?

6. To earn money for a contest, the math team pays $50 for a box of spirit banners. They sell each banner for $2.50. The function $y = 2.5x - 50$ indicates the profit y the team makes after selling x number of banners.

   a. What is the zero of the function?

   ☐

   b. What does the zero represent?

   ○ A the price of one banner

   ○ B the number of banners to be sold

   ○ C the number of banners that must be sold before making a profit

   ○ D how much the team owes for the banners

7. A line has an x-intercept of -2 and a y-intercept of 8. What is the slope of the line?

   ○ A -2

   ○ B $-\frac{1}{4}$

   ○ C $\frac{1}{4}$

   ○ D 4

8. The scale on a map is 1.5 inches = 6 miles. If two cities are 4 inches apart on the map, what is the actual distance between the cities?

   ☐ miles

9. Which of the following relations is a function?

- ○ **A** {(3, 1), (−2, −1), (5, 1)}
- ○ **B** {(2, 4), (−1, 2), (2, −4)}
- ○ **C** {(3, 1), (3, 4), (3, 7)}
- ○ **D** {(3, 1), (2, 1), (3, −1)}

10. Which expression can be used to find the slope of the line that passes through the pair of points shown in the graph?

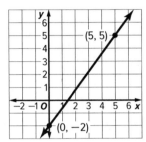

- ○ **A** $\dfrac{0-5}{-2-5}$
- ○ **B** $\dfrac{5-0}{5-2}$
- ○ **C** $\dfrac{5-(-2)}{5-0}$
- ○ **D** $\dfrac{5-5}{-2-0}$

11. A sequence is represented by $f(n) = 7 - 4n$. What are the first three terms of the sequence?

- ○ **A** 3, −1, −5
- ○ **B** 11, 15, 19
- ○ **C** 7, 11, 15
- ○ **D** −4, −8, −12

12. The x-coordinates of the endpoints of a horizontal line are a and b. The x-coordinate of the midpoint m of the line can be found using the formula $m = \dfrac{a+b}{2}$. Which equation shows b in terms of a and m?

- ○ **A** $b = 2m + a$
- ○ **B** $b = 2m - a$
- ○ **C** $2m = a + b$
- ○ **D** $b = \dfrac{a+m}{2}$

13. A hot-air balloon was at a height of 60 feet above the ground when it began to ascend. The balloon climbed at a rate of 15 feet per minute.

- **a.** Find the height of the hot-air balloon after climbing for 1, 2, 3, and 4 minutes.

- **b.** Let t represent the time in minutes since the balloon began climbing. Write an algebraic equation for a sequence that can be used to find the height, h, of the balloon after t minutes.

- **c.** Use your equation from part **b** to find the height, in feet, of the hot-air balloon after climbing for 8 minutes.

  [   ] ft

- **d.** **MP** Which mathematical practice did you use to solve this problem?

14. Consider the absolute value function $g(x) = 3|x - 2| + 1$.

- **a.** Find the vertex of the graph.

- **b.** Describe the graph of the function as a transformation of the graph of $f(x) = |x|$.

- **c.** Write the function as a piecewise function.

Need Extra Help?

If you missed Question...	1	2	3	4	5	6	7	8	9	10	11	12	13	14
Go to Lesson...	3-3	3-4	1-4	3-6	2-3	3-2	3-3	2-6	1-7	3-3	3-6	2-7	3-6	3-8

CHAPTER 4
Equations of Linear Functions

THEN
You graphed linear functions.

NOW
In this chapter, you will:

- Write and graph linear equations in various forms.
- Use scatter plots and lines of fit, and write equations of best-fit lines using linear regression.
- Find inverse linear functions.
- Explore causation and correlation.

MP WHY

TRAVEL The travel industry tracks how many, where, and when people travel. From the yearly data, patterns emerge. Rate of change can be applied to these data to determine a linear model which can be used to predict travel trends in future years.

Use the Mathematical Practices to complete the activity.

1. Use Tools Use the Internet to determine current trends in leisure travel. Obtain data from past years as well as current data. Choose specific information such as number of people traveling or money spent.

2. Make Sense of Problems Make a table showing the values for travel trends that you obtained.

3. Graph Your Data Use the Line Graph tool in ConnectED to graph your data. What trends can you see in the data after you graph it?

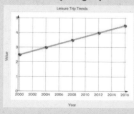

4. Discuss Compare your graph with that of other students. Are there any common trends in leisure travel?

Go Online to Guide Your Learning

Explore & Explain	Organize

Graphing Tools

Use the **Graphing Tools** to enhance your understanding of equations of linear functions. Graphing slope-intercept form is discussed in Lesson 4-1.

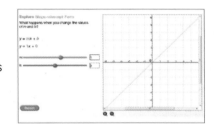

Foldables

Equations of Linear Functions Make this to help you organize your notes about linear functions. Begin with one sheet of 11" by 17" paper.

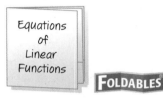

The Geometer's Sketchpad

Use **The Geometer's Sketchpad** to evaluate predictions using lines of fit, to illustrate how to write equations in slope-intercept form when given a graph, to find parallel and perpendicular lines, and to write equations of best-fit and median-fit lines.

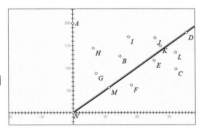

Collaborate

Chapter Project

In the **Statistically Speaking** project, you'll use what you have learned about linear regression to complete a civic literacy project.

eBook

Interactive Student Guide

Before starting the chapter, answer the **Chapter Focus** preview questions. Check your answers as you complete each lesson. At the end of the chapter, try the **Performance Task**.

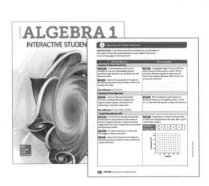

Focus

 ## LEARNSMART

Need help studying? Complete the **Descriptive Statistics** and **Relationships Between Quantities and Reasoning with Equations** domains in LearnSmart to review for the chapter test.

ALEKS

You can use the **Functions and Lines** topic in ALEKS to find out what you know about linear functions and what you are ready to learn.*

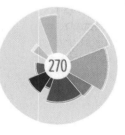

* Ask your teacher if this is part of your program.

Get Ready for the Chapter

Go Online! for Vocabulary Review Games and key vocabulary in 13 languages.

Connecting Concepts

Concept Check

Review the concepts used in this chapter by answering the questions below.

1. Explain the steps to determine the ordered pair for point *A*.

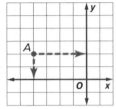

2. Given $2(m - n)^2 + 3p$ for $m = 5$, $n = 2$, and $p = -3$, what would be the first step to solve this equation?

3. Based on the order of operations, which part of the equation given in question **2** should be solved first?

4. Given $= 2(3)^2 + 3(3)$, what should be evaluated first?

5. Explain how to solve the equation $5x + 15y = 9$ for *x*.

6. Point *A* has coordinates (7, 9). Name another point that has the same *y*-coordinate.

7. Find the slope of a line that passes through (2, 1) and (4, 7).

8. What is the *y*-intercept of the line $y = 5x + 8$?

9. Describe the slope of horizontal and vertical lines.

New Vocabulary

English		Español
linear extrapolation	p. 227	extrapolación lineal
parallel lines	p. 239	rectas paralelas
perpendicular lines	p. 240	rectas perpendiculares
scatter plot	p. 247	gráfica de dispersión
line of fit	p. 248	recta de ajuste
linear interpolation	p. 249	interpolación lineal
best-fit line	p. 259	recta de ajuste óptimo
linear regression	p. 259	retroceso lineal
correlation coefficient	p. 259	coeficiente de correlación
median-fit line	p. 262	línea de mediana-ataque
inverse relation	p. 267	relación inversa
inverse function	p. 268	función inversa

Performance Task Preview

Understanding equations of linear functions will help you finish the Performance Task at the end of the chapter. You will use the concepts and skills in this chapter to help a team of scientists conduct a study on temperature trends around the world using equations.

MP **In this Performance Task you will:**

• look for and make use of structure

• attend to precision

Review Vocabulary

coefficient coeficiente the numerical factor of a term

function función a relation in which each element of the domain is paired with exactly one element of the range

ratio razón a comparison of two numbers by division

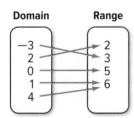

Writing Equations in Slope-Intercept Form

:: **Then**

- You graphed lines given the slope and the *y*-intercept.

:: **Now**

1 Write an equation of a line in slope-intercept form given the slope and one point.

2 Write an equation of a line in slope-intercept form given two points.

:: **Why?**

- The 2014 attendance at the Columbus Zoo and Aquarium was about 1.1 million. The zoo's attendance in 2016 was about 1.3 million. You can find the average rate of change for this data. Then you can write an equation that would model the average attendance at the zoo for a given year.

 New Vocabulary

constraint

linear extrapolation

 Mathematical Practices

3 Construct viable arguments and critique the reasoning of others.

6 Attend to precision.

1 **Write an Equation Given the Slope and a Point** The next example shows how to write an equation of a line if you are given a slope and a point other than the *y*-intercept.

Example 1 Write an Equation Given the Slope and a Point

Write an equation in slope-intercept form for the line that passes through (2, 1) with a slope of 3.

You are given the slope but not the *y*-intercept.

Step 1 Find the *y*-intercept.

$y = mx + b$	Slope-intercept form
$1 = 3(2) + b$	Replace *m* with 3, *y* with 1, and *x* with 2.
$1 = 6 + b$	Simplify.
$1 - 6 = 6 + b - 6$	Subtract 6 from each side.
$-5 = b$	Simplify.

Step 2 Write the equation in slope-intercept form.

$y = mx + b$	Slope-intercept form
$y = 3x - 5$	Replace *m* with 3 and *b* with −5.

Therefore, the equation of the line is $y = 3x - 5$.

▶ **Guided Practice**

Write an equation in slope-intercept form for a line that passes through the given point and has the given slope.

1A. (−2, 5), slope 3

1B. (4, −7), slope −1

2 **Write an Equation Given Two Points** If you are given two points through which a line passes, you can use them to find the slope first. Then follow the steps in Example 1 to write the equation.

Example 2　　**Write an Equation Given Two Points**

Write an equation in slope-intercept form for the line that passes through each pair of points.

a. **(3, 1) and (2, 4)**

> **Step 1** Find the slope of the line containing the given points.
>
> $$m = \frac{y_2 - y_1}{x_2 - x_1}$$　　Slope Formula
>
> $$= \frac{4 - 1}{2 - 3}$$　　$(x_1, y_1) = (3, 1)$ and $(x_2, y_2) = (2, 4)$
>
> $$= \frac{3}{-1} \text{ or } -3$$　　Simplify.

> **Step 2** Use either point to find the y-intercept.
>
> $$y = mx + b$$　　Slope-intercept form
>
> $$4 = (-3)(2) + b$$　　Replace m with -3, x with 2, and y with 4.
>
> $$4 = -6 + b$$　　Simplify.
>
> $$4 - (-6) = -6 + b - (-6)$$　　Subtract -6 from each side.
>
> $$10 = b$$　　Simplify.

> **Step 3** Write the equation in slope-intercept form.
>
> $$y = mx + b$$　　Slope-intercept form
>
> $$y = -3x + 10$$　　Replace m with -3 and b with 10.

Therefore, the equation is $y = -3x + 10$.

b. **(−4, −2) and (−5, −6)**

> **Step 1** Find the slope of the line containing the given points.
>
> $$m = \frac{y_2 - y_1}{x_2 - x_1}$$　　Slope Formula
>
> $$= \frac{-6 - (-2)}{-5 - (-4)}$$　　$(x_1, y_1) = (-4, -2)$ and $(x_2, y_2) = (-5, -6)$
>
> $$= \frac{-4}{-1} \text{ or } 4$$　　Simplify.

> **Step 2** Use either point to find the y-intercept.
>
> $$y = mx + b$$　　Slope-intercept form
>
> $$-2 = 4(-4) + b$$　　Replace m with 4, x with -4, and y with -2.
>
> $$-2 = -16 + b$$　　Simplify.
>
> $$-2 - (-16) = -16 + b - (-16)$$　　Subtract -16 from each side.
>
> $$14 = b$$　　Simplify.

> **Step 3** Write the equation in slope-intercept form.
>
> $$y = mx + b$$　　Slope-intercept form
>
> $$y = 4x + 14$$　　Replace m with 4 and b with 14.

Therefore, the equation is $y = 4x + 14$.

▶ **Guided Practice**

Write an equation in slope-intercept form for the line that passes through each pair of points.

2A. $(-1, 12), (4, -8)$　　　　　　**2B.** $(5, -8), (-7, 0)$

Study Tip

MP **Precision**　Given two points on a line, you may select either point to be (x_1, y_1). Be sure to remain consistent throughout the problem.

Study Tip

Slope　If the (x_1, y_1) coordinates are negative, be sure to account for both the negative signs and the subtraction symbols in the Slope Formula.

In mathematics, a **constraint** is a condition that a solution must satisfy. Equations can be viewed as constraints that the solutions of the equations meet.

Real-World Example 3   Use Slope-Intercept Form

FLIGHTS The table shows the number of domestic flights in the U.S. from 2008 to 2012. Write an equation in slope-intercept form that could be used to predict the number of flights if the number continues to decrease at the same rate.

Year	Flights (millions)
2008	9.39
2009	8.77
2010	8.70
2011	8.65
2012	8.44

Understand You know the number of flights for 2008–2012.

Plan Let x represent the number of years since 2005, and let y represent the number of flights. Write an equation of the line that passes through $(3, 9.39)$ and $(7, 8.44)$.

Solve Find the slope.

$$m = \frac{y_2 - y_1}{x_2 - x_1}$$ Slope Formula

$$= \frac{8.44 - 9.39}{7 - 3}$$ Let $(x_1, y_1) = (3, 9.39)$ and $(x_2, y_2) = (7, 8.44)$.

$$= -\frac{0.95}{4} \text{ or } -0.2375$$ Simplify.

Use $(7, 8.44)$ to find the y-intercept of the line.

$y = mx + b$	Slope-intercept form
$8.44 = -0.2375(7) + b$	Replace y with 8.44, m with -0.2375, and x with 7.
$8.44 = -1.6625 + b$	Simplify.
$10.1025 = b$	Add 1.6625 to each side.

Write the equation in slope-interept form.

$y = -0.2375x + 10.1025$ Replace m with -0.2375 and b with 10.1025.

Check Check your result by using the coordinates of the other point.

$y = -0.2375x + 10.1025$	Original equation
$9.39 \stackrel{?}{=} -0.2375(3) + 10.1025$	Replace y with 9.39 and x with 3.
$9.39 = 9.39 \checkmark$	Simplify.

The solution method assumes the decrease in flights is linear. Assumptions are common in modeling and result in valid solutions as long as they do not contradict the constraints.

▸ **Guided Practice**

3. FINANCIAL LITERACY In addition to his weekly salary, Ethan is paid $16 per delivery. Last week, he made 5 deliveries, and his total pay was $215. Write a linear equation in slope-intercept form to find Ethan's total weekly pay T if he makes d deliveries.

You can use a linear equation to make predictions about values that are beyond the range of the data. This process is called **linear extrapolation**.

Problem-Solving Tip

MP Precision Deciding whether an answer is reasonable is useful when an exact answer is not necessary.

Real-World Example 4   Predict from Slope-Intercept Form

FLIGHTS Use the equation in Example 3 to estimate the number of domestic flights in 2025.

$y = -0.2375x + 10.1025$ Original equation

$ = -0.2375(20) + 10.1025 \text{ or } 5.3525 \text{ million}$ Replace x with 20.

▸ **Guided Practice**

4. MONEY Use the equation in Guided Practice 3 to predict how much money Ethan will earn in a week if he makes 8 deliveries.

Check Your Understanding   ⃝ = Step-by-Step Solutions begin on page R11.

Go Online! for a Self-Check Quiz

Example 1   Write an equation in slope-intercept form for the line that passes through the given point and has the given slope.

1. (3, −3), slope 3
2. (2, 4), slope 2
3. (1, 5), slope −1
4. (−4, 6), slope −2

Example 2   Write an equation in slope-intercept form for the line that passes through each pair of points.

5. (4, −3), (2, 3)
6. (−7, −3), (−3, 5)
7. (−1, 3), (0, 8)
8. (−2, 6), (0, 0)

Examples 3, 4   **9. WHITEWATER RAFTING** Ten people from a local youth group went to RioGrande Whitewater Rafting Tour Company for a one-day rafting trip. The group paid $525.

Guide's FEE *plus* **$45.00** per person for **1-day** trip

  a. Write an equation in slope-intercept form to find the total cost C for p people.

  b. How much would it cost for 15 people?

Practice and Problem Solving   Extra Practice is on page R4.

Example 1   Write an equation in slope-intercept form for the line that passes through the given point and has the given slope.

10. (3, 1), slope 2
11 (−1, 4), slope −1
12. (1, 0), slope 1
13. (7, 1), slope 8
14. (2, 5), slope −2
15. (2, 6), slope 2

Example 2   Write an equation in slope-intercept form for the line that passes through each pair of points.

16. (9, −2), (4, 3)
17. (−2, 5), (5, −2)
18. (−5, 3), (0, −7)
19. (3, 5), (2, −2)
20. (−1, −3), (−2, 3)
21. (−2, −4), (−2, 4)

Examples 3, 4   **22.** Ⓜ **MODELING** Greg is driving a remote control car at a constant speed. He starts the timer when the car is 5 feet away. After 2 seconds the car is 35 feet away.

  a. Write a linear equation in slope-intercept form to find the distance d of the car from Greg.

  b. Estimate the distance the car has traveled after 10 seconds.

23. ZOOS Refer to the beginning of the lesson.

  a. Write a linear equation to find the attendance y after x years. Let x be the number of years since 2010.

  b. Estimate the zoo's attendance in 2024.

24. BOOKS In 1904, a dictionary cost 30¢. Since then the cost of a dictionary has risen an average of 6¢ per year.

  a. Write a linear equation in slope-intercept form to find the cost C of a dictionary y years after 1904.

  b. If this trend continues, what will the cost of a dictionary be in 2025?

Write an equation in slope-intercept form for the line that passes through the given point and has the given slope.

25. (4, 2), slope $\frac{1}{2}$
26. (3, −2), slope $\frac{1}{3}$
27. (6, 4), slope $-\frac{3}{4}$
28. (2, −3), slope $\frac{2}{3}$
29. (2, −2), slope 0
30. (−4, −2), slope $-\frac{3}{5}$

31. DOGS In 2001, there were about 56.1 thousand golden retrievers registered in the United States. In 2002, the number was 62.5 thousand.

    a. Write a linear equation in slope-intercept form to find the number of thousands of golden retrievers G that will be registered in year t, where $t = 0$ is the year 2000.

    b. Estimate the number of golden retrievers that will be registered in 2027.

32. GYM MEMBERSHIPS A local recreation center offers a yearly membership and charges an additional fee for aerobics classes as shown in the table.

Number of classes (x)	1	5	10	15
Total cost (y)	270	290	315	340

    a. Write an equation in slope-intercept form that represents the total cost of the membership.

    b. Carly spent $500 one year. How many aerobics classes did she take?

33. SUBSCRIPTION An online magazine allows you to view up to 25 articles free. To view 30 articles, you pay $49.15. To view 33 articles, you pay $57.40.

    a. What is the cost of each article for which you pay a fee?

    b. What is the cost of the magazine subscription?

Write an equation in slope-intercept form for the line that passes through the given points.

34. $(5, -2), (7, 1)$ **35** $(5, -3), (2, 5)$ **36.** $\left(\dfrac{5}{4}, 1\right), \left(-\dfrac{1}{4}, \dfrac{3}{4}\right)$ **37.** $\left(\dfrac{5}{12}, -1\right), \left(-\dfrac{3}{4}, \dfrac{1}{6}\right)$

Determine whether the given point is on the line. Explain why or why not.

38. $(3, -1); y = \dfrac{1}{3}x + 5$ **39.** $(6, -2); y = \dfrac{1}{2}x - 5$

For Exercises 40–42, determine which equation best represents each situation. Explain the meaning of each variable.

A $\quad y = -\dfrac{1}{3}x + 72$	**B** $\quad y = 2x + 225$	**C** $\quad y = 12x + 4$

40. CONCERTS Tickets to a concert cost $12 each plus a processing fee of $4 per order.

41. FUNDRAISING The freshman class has $225. They sell raffle tickets at $2 each.

42. POOLS The current water level of a swimming pool is 6 feet. The rate of evaporation is $\dfrac{1}{3}$ inch per day.

43. **MP** **SENSE-MAKING** A manufacturer implemented a program to reduce waste. The table shows the progress they have made since 2004.

Years since 2004	0	5	10	15
Waste (tons)	946	804	662	520

    a. How many tons of waste were sent to the landfill in 2016?

    b. In what year will it become impossible for this trend to continue? Explain.

44. COMBINING FUNCTIONS Amy opens an account with a deposit of $5000, and she deposits $100 to the account every week.

    a. Write a function $d(t)$ to express the amount of money in the account t weeks after the initial deposit.

    b. Amy spends $600 the first week and $250 in each of the following weeks. Write a function $w(t)$ to express the amount of money taken out of the account each week.

    c. Find $B(t) = d(t) - w(t)$. What does this new function represent?

    d. Will Amy run out of money? If so, when?

45 **CONCERT TICKETS** Jackson is ordering tickets for a concert online. There is a processing fee for each order, and the tickets are $76 each. Jackson ordered 5 tickets and the cost was $398.

   a. Determine the processing fee. Write a linear equation to represent the total cost C for t tickets.

   b. Make a table of values for at least three other numbers of tickets.

   c. Graph this equation. Predict the cost of 8 tickets.

46. **MULTI-STEP** Ricky is saving money to buy a TV listed at $936. He currently has $40. He charges $20 for every lawn he mows, and he spends about $6 in gas for every three lawns. He also has a paper route, which earns him $45 per month.

   a. In how many weeks will he have enough money if he mows three lawns per week?

   b. Explain your solution process.

   c. What assumptions did you make?

H.O.T. Problems   Use Higher-Order Thinking Skills

47. **ERROR ANALYSIS** Tess and Jacinta are writing an equation of the line through $(3, -2)$ and $(6, 4)$. Is either of them correct? Explain your reasoning.

Tess
$m = \dfrac{4 - (-2)}{6} - 3 = \dfrac{6}{3}$ or 2
$y = mx + b$
$6 = 2(4) + b$
$6 = 8 + b$
$-2 = b$
$y = 2x - 2$

Jacinta
$m = \dfrac{4 - (-2)}{6 - 3} = \dfrac{6}{3}$ or 2
$y = mx + b$
$-2 = 2(3) + b$
$-2 = 6 + b$
$-8 = b$
$y = 2x - 8$

48. **MP** **SENSE-MAKING** Consider three points, $(3, 7)$, $(-6, 1)$, and $(9, p)$, on the same line. Find the value of p and explain your steps.

49. **MP** **STRUCTURE** Consider the standard form of a linear equation, $Ax + By = C$.

   a. Rewrite the equation in slope-intercept form.

   b. What is the slope?

   c. What is the y-intercept?

   d. Is this true for all real values of A, B, and C?

50. **OPEN ENDED** Create a real-world situation that fits the graph at the right. Define the two quantities and describe the functional relationship between them. Write an equation to represent this relationship and describe what the slope and y-intercept mean.

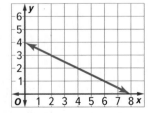

51. **WRITING IN MATH** Linear equations are useful in predicting future events. Describe some factors in real-world situations that might affect the reliability of the graph in making any predictions.

52. **MP** **ARGUMENTS** What information is needed to write the equation of a line? Explain.

53. Which of the following equations in slope-intercept form best represents the line that passes through the points $(-2, 4)$ and $(3, -1)$? **MP** 2

- ○ **A** $y = 5x + 14$
- ○ **B** $y = -x + 2$
- ○ **C** $y = x + 6$
- ○ **D** $y = -x - 4$

54. The temperature at 6:00 P.M. is $70°$F. The temperature will drop $6°$F each hour. Write an equation in slope-intercept form that models the temperature in y degrees Fahrenheit after x hours. **MP** 4

- ○ **A** $y = 70x - 6$
- ○ **B** $y = 6x - 70$
- ○ **C** $y = -6x - 70$
- ○ **D** $y = -6x + 70$

55. What is the y-intercept for the equation of a line that passes through the point $(-1, 5)$ and has a slope of 5? **MP** 2

- ○ **A** -5
- ○ **B** 0
- ○ **C** 5
- ○ **D** 10

56. The table shows the cost of purchasing various amounts of grapes.

Cost of Grapes			
Grapes, g (lb)	1.8	2	4
Cost, C (dollars)	2.25	2.50	5.00

Which of the following equations best represents this situation? **MP** 4

- ○ **A** $C = 0.8g$
- ○ **B** $C = 4g + 5$
- ○ **C** $C = 1.25g$
- ○ **D** $C = g + 0.50$

57. The population of California was about 37.3 million in 2010 and 39 million in 2015. **MP** 2, 4

a. Write an equation to find the population of California y after x years. Let x be the number of years since 2010.

b. Which of the following statements about the population of California are true?

- ☐ **A** The population is increasing by about 0.34 million people per year.
- ☐ **B** The population is decreasing.
- ☐ **C** In 2030, the population will be about 44.1 million.
- ☐ **D** The population is increasing by about 1.7 million people per year.
- ☐ **E** In 2025, the population will be about 42.4 million.

58. MULTI-STEP An educational company offers a math summer camp that costs $250 to attend. At the camp, tutoring sessions can be purchased. The table shows the total cost given the number of tutoring sessions purchased. **MP** 2, 4

Cost of Math Help			
Number Tutoring Sessions	5	10	15
Total Cost (Including Math Summer Camp)	$400	$550	$700

a. Write an equation for the total cost of the camp c given the number of tutoring sessions purchased n.

b. What would it cost if a student went to summer camp and got 20 tutoring sessions in one year?

c. Explain the meaning of the slope of the line.

d. If a customer spent $1450 in one year (including Math Summer Camp), how many tutoring sessions did they receive in one year?

Writing Equations in Standard and Point-Slope Forms

::Then	::Now	::Why?

::Then
- You wrote equations in slope-intercept form.

::Now
1 Write equations of lines in standard form and point-slope form.

2 Write linear equations in different forms.

::Why?
Most humane societies have foster homes for newborn puppies, kittens, and injured or sick animals. During the spring and summer, a large shelter can place 3000 animals in homes each month.

If a shelter had 200 animals in foster homes at the beginning of spring, $y = 3000x + 200$ could be used to represent the number of animals placed in foster homes y after x months.

New Vocabulary
standard form
point-slope form

Mathematical Practices
1 Make sense of problems.
4 Model with mathematics.
7 Make use of structure.

1 **Standard and Point-Slope Forms** An equation of a line can be written in **standard form** when given the intercepts or other coordinates of points of the line.

⬢ Key Concept Standard Form

Words The linear equation $Ax + By = C$ is written in standard form, where $A \geq 0$, A and B are not both zero, and A, B, and C, are integers with a greatest common factor of 1.

Symbols $Ax + By = C$

Example $3x + 2y = 12$

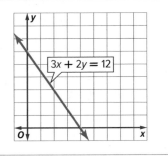
$3x + 2y = 12$

Real-World Example 1   Write and Graph an Equation in Standard Form

FIELD TRIP A group of 48 people is traveling to a nature preserve in small and/or large vans. A small van holds 8 people and a large van holds 12 people.

a. Write an equation in standard form that models the number of small and large vans that your class could use.

Let $x =$ the number of small vans and $y =$ the number of large vans the group uses.

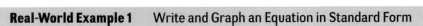

words	8 ·	number of small vans	+	12 ·	number of large vans	holds	48 people.
equation	8 ·	x	+	12 ·	y	=	48

The equation is $8x + 12y = 48$.

> **Study Tip**
> **MP** **Reason Abstractly** Read the problem carefully and write a verbal model that can be represented by an equation in standard form.

b. Use the intercepts to graph the equation.

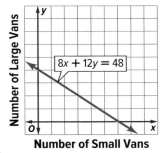

x-intercept	*y*-intercept
$8x + 12y = 48$	$8x + 12y = 48$
$8x + 12(0) = 48$	$8(0) + 12y = 48$
$8x = 48$	$12y = 48$
$x = 6$	$y = 4$
The *x*-intercept is 6.	The *y*-intercept is 4.

Graph the intercepts and then draw a line through them.

c. If the group takes 2 large vans, how many small vans will they need?

$8x + 12y = 48$	Original equation.
$8x + 12(2) = 48$	Substitute 2 for *y*
$8x + 24 = 48$	Simplify.
$8x = 24$	Subtract 24 from each side.
$x = 3$	Divide each side by 8.

The group uses 3 small vans and 2 large vans.

> **Guided Practice**

1. In Example 1, suppose 96 people decided to go to the nature preserve. Write an equation that can represent the new situation.

An equation of a line can be written in **point-slope form** when given the coordinates of one known point and the slope of that line.

Key Concept Point-Slope Form

Words   The linear equation $y - y_1 = m(x - x_1)$ is written in point-slope form, where (x_1, y_1) is a given point on a nonvertical line and *m* is the slope of the line.

Symbols $y - y_1 = m(x - x_1)$

Example $y - 2 = 3(x - 4)$

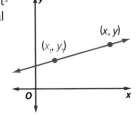

Example 2   Write and Graph an Equation in Point-Slope Form

Write an equation in point-slope form for the line that passes through $(3, -2)$ with a slope of $\frac{1}{4}$. Then graph the equation.

$y - y_1 = m(x - x_1)$ Point-slope form

$y - (-2) = \frac{1}{4}(x - 3)$ $(x_1, y_1) = (3, -2), m = \frac{1}{4}$

$y + 2 = \frac{1}{4}(x - 3)$ Simplify.

Plot the point at $(3, -2)$ and use the slope to find another point on the line. Draw a line through the two points.

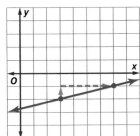

> **Guided Practice**

2. Write an equation in point-slope form for the line that passes through $(-2, 2)$ with a slope of -3. Then graph the equation.

2 Forms of Linear Equations
If you are given the slope and the coordinates of one or two points, you can write a linear equation in the following ways.

Concept Summary Writing Equations

Given the Slope and One Point

Step 1 Substitute the value of m and let the x and y coordinates be (x_1, y_1). Or, substitute the value of m, x, and y into the slope-intercept form and solve for b.

Step 2 Rewrite the equation in the needed form.

Given Two Points

Step 1 Find the slope.

Step 2 Choose one of the two points to use.

Step 3 Follow the steps for writing an equation given the slope and one point.

Example 3 Find an Equation of a Line

Write an equation in point-slope form, slope-intercept form, and standard form for each line.

a. line through point (5, 1) with a slope of $-\dfrac{2}{3}$

Point-Slope Form:	$y - y_1 = m(x - x_1)$	Point-slope form
	$y - 1 = -\dfrac{2}{3}(x - 5)$	Substitution
Slope-Intercept Form:	$y - 1 = -\dfrac{2}{3}x + \dfrac{10}{3}$	Distributive Property
	$y = -\dfrac{2}{3}x + \dfrac{10}{3} + \dfrac{3}{3}$	Add 1 to each side.
	$y = -\dfrac{2}{3}x + \dfrac{13}{3}$	Simplify.
Standard Form:	$\dfrac{2}{3}x + y = \dfrac{13}{3}$	Add $\dfrac{2}{3}x$ to each side.
	$2x + 3y = 13$	Multiply each side by 3.

b. line through points (−1, −3) and (5, 6)

Find the slope.	$m = \dfrac{6 - (-3)}{5 - (-1)} = \dfrac{9}{6}$ or $\dfrac{3}{2}$	
Point-Slope Form:	$y - 6 = \dfrac{3}{2}(x - 5)$	Substitution
Slope-Intercept Form:	$y - 6 = \dfrac{3}{2}x - \dfrac{15}{2}$	Distributive Property
	$y = \dfrac{3}{2}x - \dfrac{3}{2}$	Add 6 to each side.
Standard Form:	$-\dfrac{3}{2}x + y = -\dfrac{3}{2}$	Subtract $\dfrac{3}{2}x$ from each side.
	$3x - 2y = 3$	Multiply each side by −2.

▶ Guided Practice

3A. line through point (−5, 1) with a slope of 7
3B. line through points (4, −6) and (1, 3)

Being able to use a variety of forms of linear equations may be useful in other subjects.

Study Tip

Slopes in squares Nonvertical opposite sides of a square have equal slopes. If the coordinates for one of the vertices are unavailable, use the slope of the opposite side.

Example 4   Write Equations in Point-Slope and Standard Forms

GEOMETRY The figure shows square $RSTU$. Write an equation in point-slope form and standard form for the line containing side $\overline{TU}$.

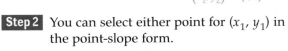

Step 1   Find the slope of $\overline{TU}$.

$$m = \frac{y_2 - y_1}{x_2 - x_1}$$ Slope Formula

$$= \frac{5 - 2}{7 - 4} \text{ or } 1$$ $(x_1, y_1) = (4, 2)$ and $(x_2, y_2) = (7, 5)$

Step 2   You can select either point for (x_1, y_1) in the point-slope form.

$$y - y_1 = m(x - x_1)$$ Point-slope form

$$y - 2 = 1(x - 4)$$ $(x_1, y_1) = (4, 2)$

$$y - 5 = 1(x - 7)$$ $(x_1, y_1) = (7, 5)$

Step 3   Write in standard form

$$y - 5 = 1(x - 7)$$ Original equation

$$y - 5 = 1x - 7$$ Distributive Property

$$y = 1x - 2$$ Add 5 to each side.

$$-1x + y = -2$$ Subtract 1x from each side.

$$x - y = 2$$ Multiply each side by −1.

Guided Practice

Write an equation in point-slope form and standard form for each line.

4A. the line containing side $\overline{ST}$

4B. the line containing $\overline{SR}$

Check Your Understanding   ◯ = Step-by-Step Solutions begin on page R11

Go Online! for a Self-Check Quiz

Example 1

1. ANIMAL CARE You plan to spend $200 at a dog kennel that charges $40 per night for boarding and $5 for walking your dog.

   a. Write an equation in standard form that models the number of nights you could board and the number of walks you could buy for your dog.

   b. Graph the equation.

   c. If you board your dog for 4 nights, how many walks could you buy?

Example 2

Write an equation in point-slope form for the line that passes through the given point with the slope provided. Then graph the equation.

   2. $(-2, -8)$, slope $\frac{5}{6}$ $(-2, 5)$, slope $= -6$ **4.** $(4, 3)$, vertical line

Example 3 Write an equation in point-slope form, slope-intercept form, and standard form for each line.

5. $m = \frac{7}{8}$, $(3, -2)$

6. $m = -5$, $(-3, -7)$

7. $m = \frac{5}{3}$, $(-6, -2)$

8. $(-6, 10)$, $(-4, 18)$

9. $(-5, 7)$, $(3, 1)$

10. $(-4, 9)$, $(1, 14)$

Example 4 **11** **GEOMETRY** Use right triangle FGH.

a. Write an equation in point-slope form for the line containing $\overline{GH}$.

b. Write the standard form of the line containing $\overline{GH}$.

c. Write the slope-intercept form of the line containing $\overline{GH}$.

d. Write an equation in point-slope form for the line containing $\overline{FH}$.

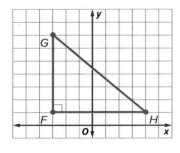

12. **MP** **REASONING** Describe how to use the point-slope form of a line, $y - y_1 = m(x - x_1)$, to graph the line.

Practice and Problem Solving

Extra Practice begins on page R4.

Example 2 Write an equation in point-slope form for the line that passes through each point with the given slope. Then graph the equation.

13. $(-6, -3)$, $m = -1$

14. $(-7, 6)$, $m = 0$

15. $(-2, 11)$, vertical line

16. $(-6, -8)$, $m = -\frac{5}{8}$

17. $(-2, -9)$, $m = -\frac{7}{5}$

18. $(-6, 0)$, horizontal line

Example 3 Write an equation in point-slope form, slope-intercept form, and standard form for each line.

19. $m = -6$, $(-9, 9)$

20. $m = -3$, $(-2, 6)$

21. $m = \frac{9}{10}$, $(-3, -7)$

22. $m = \frac{2}{3}$, $(-7, -4)$

23. $m = -\frac{1}{3}$, $(2, -3)$

24. $m = \frac{17}{10}$, $(1, 2)$

25. $(7, 6)$, $(14, -8)$

26. $(-4, 11)$, $(-8, -1)$

27. $(-7, -5)$, $(-10, 13)$

28. $(-5, 1)$, $(5, 9)$

29. $(4, -2)$, $(-8, -4)$

30. $(-8, -6)$, $(4, -15)$

Example 1 **31.** **BOATS** A rental company charges \$100 per hour to rent a paddle boat and \$150 per hour to rent a pontoon boat.

a. Define the variables and write an equation in standard form that models the number of hours you could rent a paddle boat or pontoon boat if you have \$1200 to spend.

b. Graph the equation from Part **a**.

c. If you rent a paddle boat for 6 hours, for how many hours can you rent a pontoon boat?

32. **MP** **MODELING** A bus ride costs \$1.75 and a train ride costs \$2.00. A monthly transportation pass including unlimited bus and train rides costs \$90.

a. Define the variables and write an equation in standard form that models the number of bus and train rides you must take to cover the cost of a transportation pass.

b. If you ride the bus 24 times per month, how many times must you ride the train to cover the cost of the transportation pass?

Write an equation for the line described in standard form.

33. through $(-1, 7)$ and $(8, -2)$

34. through $(-4, 3)$ with y-intercept 0

35. with x-intercept 4 and y-intercept 5

36. **REGULARITY** Explain how to write an equation for the line that passes through $(-4, 8)$ and $(3, -7)$.

Example 4 **Write an equation in point-slope form and standard form for each line.**

37.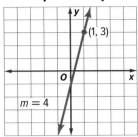

38.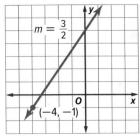

39.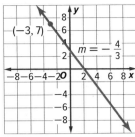

40. **REASONING** Rita's Rentals rents wave runners for $50 per hour plus a one-time service fee. The total cost for service and 6 rental hours is $325.

    a. Write an equation in point-slope form to find the total price y for any number of hours x. (*Hint:* The point $(6, 325)$ is a solution to the equation.)

    b. Write the equation in slope-intercept form.

    c. What is the service fee?

41 **WEATHER** The barometric pressure is 598 millimeters of mercury (mmHg) at an altitude of 1.8 kilometers and 577 millimeters of mercury at 2.1 kilometers.

    a. Write a formula for the barometric pressure as a function of the altitude.

    b. What is the altitude if the pressure is 657 millimeters of mercury?

H.O.T. Problems   Use Higher-Order Thinking Skills

42. **WHICH ONE DOESN'T BELONG?** Identify the equation that does not belong. Explain your reasoning.

| $y - 5 = 3(x - 1)$ | $y + 1 = 3(x + 1)$ | $y + 4 = 3(x + 1)$ | $y - 8 = 3(x - 2)$ |

43. **ERROR ANALYSIS** Juana thinks that $f(x)$ and $g(x)$ have the same slope but different intercepts. Sabrina thinks that $f(x)$ and $g(x)$ describe the same line. Is either of them correct? Explain your reasoning.

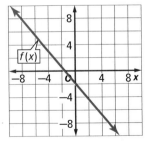

The graph of $g(x)$ is the line that passes through $(3, -7)$ and $(-6, 4)$.

44. **OPEN-ENDED** Describe a real-life scenario that has a constant rate of change and a value of y for a particular value of x. Represent this situation using an equation in point-slope form, an equation in standard form, and an equation in slope-intercept form.

45. **REASONING** Write an equation in point-slope form for the line that passes through the points (f, g) and (h, j).

46. **WRITING IN MATH** Explain how to write an equation in standard form that models all the possible combinations of lengths and widths for a rectangle having a perimeter of 100 meters. Make a table that shows five possible lengths and widths.

47. **ERROR ANALYSIS** Henry says that the slope of a line in standard form $Ax + By = C$ is $-\frac{A}{B}$, while Stephanie says that the slope is $\frac{C}{B}$. Is either correct? Explain.

48. A nursery sells Irish moss groundcover for $5.50 per plant and Creeping Thyme groundcover for $6.00 per plant. You have $330 to spend. **MP** 1, 4

a. Write an equation in standard form that models the number of moss plants x and thyme plants y that you can buy.

⬚

b. Which combinations of plants will NOT cost exactly $330, excluding tax? Select all that apply.

☐ **A** 0 moss, 55 thyme

☐ **B** 8 moss, 62 thyme

☐ **C** 12 moss, 66 thyme

☐ **D** 24 moss, 33 thyme

c. Graph the equation.

d. Can you buy an equal number of plants? Justify your argument.

49. Which of the following is an equation of the line that passes through (2, 4) and (−3, −6)? **MP** 7

○ **A** $y - 4 = 2(x - 2)$

○ **B** $y - 4 = \frac{1}{2}(x - 2)$

○ **C** $y + 4 = 2(x + 2)$

○ **D** $y + 4 = \frac{1}{2}(x + 2)$

50. Kellie earns $12.00 per hour working at the coffee shop. Using the table below, write an equation that models Kellie's wages y for the number of hours worked x in point-slope form. **MP** 4, 7

Hours (x)	Wages (y)
1	$12.00
3	$36.00
5	$60.00
8	$96.00

51. Write an equation in point-slope form, slope-intercept form, and standard form for the line with $m = -4$ and point (−6, 8). **MP** 7

52. Write an equation in point-slope form, slope-intercept form, and standard form for the line containing points (−3, −3) and (3, −7). **MP** 7

53. Which equation is represented by the graph? **MP** 1

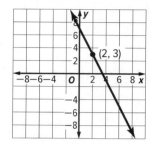

○ **A** $x - 2 = -2(y - 3)$

○ **B** $y - 3 = -2(x - 2)$

○ **C** $x - 2 = y - 3$

○ **D** $y - 2 = -2(x - 3)$

54. GEOMETRY Use the right triangle EFG. **MP** 1, 7

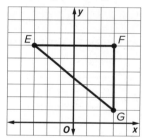

a. Write an equation in point-slope form for the line containing $\overline{EG}$.

⬚

b. Write the standard form of the line containing $\overline{EG}$.

⬚

55. a. If you have $2.25, write an equation in standard form that models the number of nickels x and dimes y that you could have. **MP** 1, 7

⬚

b. Graph the equation.

c. If you have 25 nickels, how many dimes do you have?

LESSON 3

Parallel and Perpendicular Lines

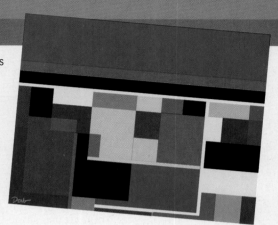

∴Then

- You wrote equations in point-slope form.

∴Now

1 Write an equation of the line that passes through a given point, parallel to a given line.

2 Write an equation of the line that passes through a given point, perpendicular to a given line.

∴Why?

- Notice the squares, rectangles and lines in the piece of art shown at the right. Some of the lines intersect forming right angles. Other lines do not intersect at all.

 New Vocabulary
parallel lines
perpendicular lines

MP Mathematical Practices

5 Use appropriate tools strategically.

1 **Parallel Lines** Lines in the same plane that do not intersect are called **parallel lines**. Nonvertical parallel lines have the same slope.

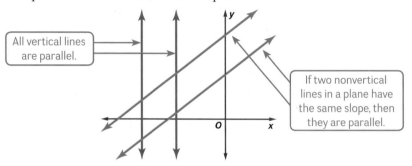

All vertical lines are parallel.

If two nonvertical lines in a plane have the same slope, then they are parallel.

You can write an equation of a line parallel to a given line if you know a point on the line and an equation of the given line. First find the slope of the given line. Then, substitute the point provided and the slope from the given line into the point-slope form.

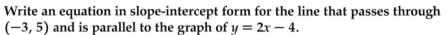

Example 1   Parallel Line Through a Given Point

Write an equation in slope-intercept form for the line that passes through $(-3, 5)$ and is parallel to the graph of $y = 2x - 4$.

Step 1 The slope of the line with equation $y = 2x - 4$ is 2. The line parallel to $y = 2x - 4$ has the same slope, 2.

Step 2 Find the equation in slope-intercept form.

$$y - y_1 = m(x - x_1)$$ Point-slope form
$$y - 5 = 2[x - (-3)]$$ Replace m with 2 and (x_1, y_1) with $(-3, 5)$.
$$y - 5 = 2(x + 3)$$ Simplify.
$$y - 5 = 2x + 6$$ Distributive Property
$$y - 5 + 5 = 2x + 6 + 5$$ Add 5 to each side.
$$y = 2x + 11$$ Write the equation in slope-intercept form.

▶ **Guided Practice**

1. Write an equation in slope-intercept form for the line that passes through $(4, -1)$ and is parallel to the x-axis.

Purestock/ Getty Images

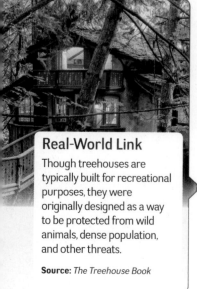

<table>
<tr><td>

Review Vocabulary

opposite reciprocals

The opposite reciprocal of $\frac{a}{b}$ is $-\frac{b}{a}$. Their product is −1.

</td></tr>
</table>

2 Perpendicular Lines

Perpendicular Lines Lines that intersect at right angles are called **perpendicular lines**. The slopes of nonvertical perpendicular lines are opposite reciprocals. That is, if the slope of a line is 4, the slope of the line perpendicular to it is $-\frac{1}{4}$.

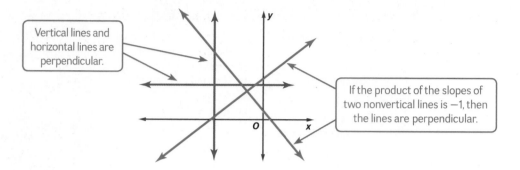

> Vertical lines and horizontal lines are perpendicular.

> If the product of the slopes of two nonvertical lines is −1, then the lines are perpendicular.

You can use slope to determine whether two lines are perpendicular.

Real-World Example 2   Slopes of Perpendicular Lines

DESIGN The outline of a company's new logo is shown on a coordinate plane.

a. Is ∠DFE a right angle in the logo?

If $\overline{BE}$ and $\overline{AD}$ are perpendicular, then ∠DFE is a right angle. Find the slopes of $\overline{BE}$ and $\overline{AD}$.

slope of $\overline{BE}$: $m = \frac{1-3}{7-2}$ or $-\frac{2}{5}$

slope of $\overline{AD}$: $m = \frac{6-1}{4-2}$ or $\frac{5}{2}$

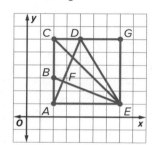

The line segments are perpendicular because $-\frac{2}{5} \times \frac{5}{2} = -1$. Therefore, ∠DFE is a right angle.

b. Is each pair of opposite sides parallel?

If a pair of opposite sides are parallel, then they have the same slope.

slope of $\overline{AC}$: $m = \frac{6-1}{2-2}$ or undefined

Since $\overline{AC}$ and $\overline{GE}$ are both parallel to the *y*-axis, they are vertical and are therefore parallel.

slope of $\overline{CG}$: $m = \frac{6-6}{7-2}$ or 0

Since $\overline{CG}$ and $\overline{AE}$ are both parallel to the *x*-axis, they are horizontal and are therefore parallel.

Guided Practice

2. **CONSTRUCTION** On the plans for a treehouse, a beam represented by $\overline{QR}$ has endpoints $Q(-6, 2)$ and $R(-1, 8)$. A connecting beam represented by $\overline{ST}$ has endpoints $S(-3, 6)$ and $T(-8, 5)$. Are the beams perpendicular? Explain.

<table>
<tr><td>

Real-World Link

Though treehouses are typically built for recreational purposes, they were originally designed as a way to be protected from wild animals, dense population, and other threats.

Source: *The Treehouse Book*

</td></tr>
</table>

You can determine whether the graphs of two linear equations are parallel or perpendicular by comparing the slopes of the lines.

Example 3   Parallel or Perpendicular Lines

Determine whether the graphs of $y = 5$, $x = 3$, and $y = -2x + 1$ are *parallel* or *perpendicular*. Explain.

Graph each line on a coordinate plane.

From the graph, you can see that $y = 5$ is parallel to the x-axis and $x = 3$ is parallel to the y-axis. Therefore, they are perpendicular. None of the lines are parallel.

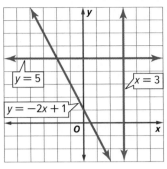

▶ **Guided Practice**

3. Determine whether the graphs of $6x - 2y = -2$, $y = 3x - 4$, and $y = 4$ are *parallel* or *perpendicular*. Explain.

You can write the equation of a line perpendicular to a given line if you know a point on the line and the equation of the given line.

Example 4   Perpendicular Line Through a Given Point

Write an equation in slope-intercept form for the line that passes through $(-4, 6)$ and is perpendicular to the graph of $2x + 3y = 12$.

Step 1 Find the slope of the given line by solving the equation for y.

$$2x + 3y = 12 \qquad \text{Original equation}$$
$$2x - 2x + 3y = -2x + 12 \qquad \text{Subtract } 2x \text{ from each side.}$$
$$3y = -2x + 12 \qquad \text{Simplify.}$$
$$\frac{3y}{3} = \frac{-2x + 12}{3} \qquad \text{Divide each side by 3.}$$
$$y = -\frac{2}{3}x + 4 \qquad \text{Simplify.}$$

The slope is $-\frac{2}{3}$.

Step 2 The slope of the perpendicular line is the opposite reciprocal of $-\frac{2}{3}$ or $\frac{3}{2}$. Find the equation of the perpendicular line.

$$y - y_1 = m(x - x_1) \qquad \text{Point-slope form}$$
$$y - 6 = \frac{3}{2}[x - (-4)] \qquad (x_1, y_1) = (-4, 6) \text{ and } m = \frac{3}{2}$$
$$y - 6 = \frac{3}{2}(x + 4) \qquad \text{Simplify.}$$
$$y - 6 = \frac{3}{2}x + 6 \qquad \text{Distributive Property}$$
$$y - 6 + 6 = \frac{3}{2}x + 6 + 6 \qquad \text{Add 6 to each side.}$$
$$y = \frac{3}{2}x + 12 \qquad \text{Simplify.}$$

▶ **Guided Practice**

4. Write an equation in slope-intercept form for the line that passes through $(4, 7)$ and is perpendicular to the y-axis.

Verify the special relationships in the slopes of parallel and perpendicular lines by completing the **Algebra Lab** in the Resources.

Concept Summary Parallel and Perpendicular Lines

	Parallel Lines	Perpendicular Lines
Words	Two nonvertical lines are parallel if they have the same slope.	Two nonvertical lines are perpendicular if the product of their slopes is −1.
Symbols	$\overleftrightarrow{AB} \parallel \overleftrightarrow{CD}$	$\overleftrightarrow{EF} \perp \overleftrightarrow{GH}$
Models	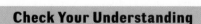	

Check Your Understanding   ◯ = Step-by-Step Solutions begin on page R11

Go Online! for a Self-Check Quiz

Example 1   Write an equation in slope-intercept form for the line that passes through the given point and is parallel to the graph of the given line.

1. $(-1, 2)$, $y = \frac{1}{2}x - 3$

2. $(2, 4)$, x-axis

Example 2

3. GARDENS A garden is in the shape of a quadrilateral with vertices $A(-2, 1)$, $B(3, -3)$, $C(5, 7)$, and $D(-3, 4)$. Two paths represented by $\overline{AC}$ and $\overline{BD}$ cut across the garden. Are the paths perpendicular? Explain.

4. ⓂⓅ **PRECISION** A square is a quadrilateral that has opposite sides parallel, consecutive sides that are perpendicular, and diagonals that are perpendicular. Determine whether the quadrilateral is a square. Explain.

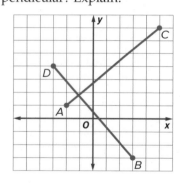

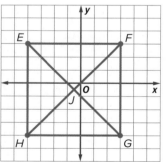

Example 3   Determine whether the graphs of the following equations are *parallel* or *perpendicular*. Explain.

⑤ $y = -2x$, $2y = x$, $4y = 2x + 4$

6. $y = \frac{1}{2}x$, $3y = x$, $y = -\frac{1}{2}x$

Example 4   Write an equation in slope-intercept form for the line that passes through the given point and is perpendicular to the graph of the line.

7. $(-2, 3)$, $y = -\frac{1}{2}x - 4$

8. $(-1, 4)$, $y = 3x + 5$

9. $(2, 3)$, x-axis

10. $(3, 6)$, $3x - 4y = -2$

Example 1   Write an equation in slope-intercept form for the line that passes through the given point and is parallel to the graph of the given equation.

11. $(3, -2)$, $y = x + 4$ **12.** $(4, -3)$, $y = 3x - 5$ **13.** $(0, 2)$, $y = -5x + 8$

14. $(-4, 2)$, $y = -\frac{1}{2}x + 6$ **15.** $(-2, 3)$, $y = -\frac{3}{4}x + 4$ **16.** $(9, 12)$, $y = 13x - 4$

Example 2   **17. GEOMETRY** A trapezoid is a quadrilateral that has exactly one pair of parallel opposite sides. Is *ABCD* a trapezoid? Explain your reasoning.

18. GEOMETRY *CDEF* is a kite. Are the diagonals of the kite perpendicular? Explain your reasoning.

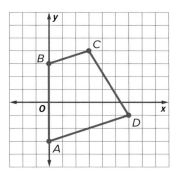

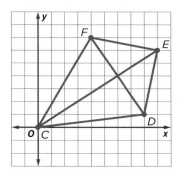

19. Determine whether the graphs of $y = -6x + 4$ and $y = \frac{1}{6}x$ are perpendicular. Explain.

20. MAPS On a map, Dogwood Drive passes through $R(0, -9)$ and $S(4, -11)$, and Lewis Road passes through $J(4, -5)$ and $K(6, -2)$. If they are straight lines, are the two streets perpendicular? Explain.

Example 3   **MP PERSERVERANCE** Determine whether the graphs of the following equations are *parallel* or *perpendicular*. Explain.

21. $2x - 8y = -24$, $4x + y = -2$, $x - 4y = 4$

22. $3x - 9y = 9$, $3y = x + 12$, $2x - 6y = 12$

Example 4   Write an equation in slope-intercept form for the line that passes through the given point and is perpendicular to the graph of the equation.

23 $(-3, -2)$, $y = -2x + 4$ **24.** $(-5, 2)$, $y = \frac{1}{2}x - 3$ **25.** $(-4, 5)$, $y = \frac{1}{3}x + 6$

26. $(2, 6)$, $y = -\frac{1}{4}x + 3$ **27.** $(3, 8)$, $y = 5x - 3$ **28.** $(4, -2)$, $y = 3x + 5$

Write an equation in slope-intercept form for a line perpendicular to the graph of the equation that passes through the *x*-intercept of that line.

29. $y = -\frac{1}{2}x - 4$ **30.** $y = \frac{2}{3}x - 6$ **31.** $y = 5x + 3$

32. Write an equation in slope-intercept form for the line that is perpendicular to the graph of $3x + 2y = 8$ and passes through the *y*-intercept of that line.

Determine whether the graphs of each pair of equations are *parallel*, *perpendicular*, or *neither*.

33. $y = 4x + 3$
$4x + y = 3$

34. $y = -2x$
$2x + y = 3$

35. $3x + 5y = 10$
$5x - 3y = -6$

36. $-3x + 4y = 8$
$-4x + 3y = -6$

37. $2x + 5y = 15$
$3x + 5y = 15$

38. $2x + 7y = -35$
$4x + 14y = -42$

39. Write an equation of the line that is parallel to the graph of $y = 7x - 3$ and passes through the origin.

 40. EXCAVATION Scientists excavating a dinosaur mapped the site on a coordinate plane. If one bone lies from $(-5, 8)$ to $(10, -1)$ and a second bone lies from $(-10, -3)$ to $(-5, -6)$, are the bones parallel? Explain.

41 ARCHAEOLOGY In the ruins of an ancient civilization, an archaeologist found pottery at $(2, 6)$ and hair accessories at $(4, -1)$. A pole is found with one end at $(7, 10)$ and the other end at $(14, 12)$. Is the pole perpendicular to the line through the pottery and the hair accessories? Explain.

42. GRAPHICS To create a design on a computer, Andeana must enter the coordinates for points on the design. One line segment she drew has endpoints of $(-2, 1)$ and $(4, 3)$. The other coordinates that Andeana entered are $(2, -7)$ and $(8, -3)$. Could these points be the vertices of a rectangle? Explain.

43. MULTIPLE REPRESENTATIONS In this problem, you will explore parallel and perpendicular lines.

    a. Graphical Graph the points $A(-3, 3)$, $B(3, 5)$, and $C(-4, 0)$ on a coordinate plane.

    b. Analytical Determine the coordinates of a fourth point D that would form a parallelogram. Explain your reasoning.

    c. Analytical What is the minimum number of points that could be moved to make the parallelogram a rectangle? Describe which points should be moved, and explain why.

H.O.T. Problems   Use **H**igher-**O**rder **T**hinking Skills

44. MP SENSE-MAKING If the line through $(-2, 4)$ and $(5, d)$ is parallel to the graph of $y = 3x + 4$, what is the value of d?

45. MP STRUCTURE Which key features of the graphs of two parallel lines are the same, and which are different? Which key features of the graphs of two perpendicular lines are the same, and which are different?

46. MP SENSE-MAKING Graph a line that is parallel and a line that is perpendicular to $y = 2x - 1$.

47. ERROR ANALYSIS Carmen and Chase are finding an equation of the line that is perpendicular to the graph of $y = \frac{1}{3}x + 2$ and passes through the point $(-3, 5)$. Is either of them correct? Explain your reasoning.

Carmen	Chase
$y - 5 = -3[x - (-3)]$	$y - 5 = 3[x - (-3)]$
$y - 5 = -3(x + 3)$	$y - 5 = 3(x + 3)$
$y = -3x - 9 + 5$	$y = 3x + 9 + 5$
$y = -3x - 4$	$y = 3x + 14$

48. WRITING IN MATH Illustrate how you can determine whether two lines are parallel or perpendicular. Write an equation for the graph that is parallel and an equation for the graph that is perpendicular to the line shown. Explain your reasoning.

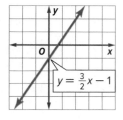

49. The city is planning to create a new street which will be parallel to Vine Street. Which of the following is an equation that could represent the new street? **MP** 4

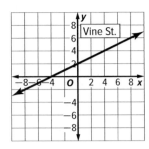

○ **A** $y = 2x - 4$

○ **B** $y = \frac{1}{2}x + 2$

○ **C** $y = -\frac{1}{2}x + 2$

○ **D** $y = \frac{1}{2}x + 4$

50. Which of the following equations represents a line parallel to $y = -2x + 4$? **MP** 1

I. $y + 1 = -2x + 3$

II. $2x + y = 5$

III. $y = 2(3 - x)$

○ **A** I only

○ **B** I and III

○ **C** II and III

○ **D** I, II, and III

51. A point and a line are shown below.

$$(1, 3); y = -\frac{1}{3}x + 5$$

Which of the following equations passes through the point and is perpendicular to the line? **MP** 2

○ **A** $y = -\frac{1}{3}x + 3\frac{1}{3}$

○ **B** $y = 3x + 6$

○ **C** $y = -\frac{1}{3}x + 2\frac{2}{3}$

○ **D** $y = 3x$

52. A line through which two points would be parallel to a line with a slope of $\frac{3}{4}$? **MP** 2

○ **A** (0, 5) and (−4, 2)

○ **B** (0, 2) and (−4, 1)

○ **C** (0, 0) and (0, −2)

○ **D** (0, −2) and (−4, −2)

53. Which of the following is the equation of a line in slope-intercept form that passes through (3, −5) and is parallel to the graph of $2x - y = 8$? **MP** 2

○ **A** $y = 2x + 13$

○ **B** $y = -\frac{1}{2}x - \frac{7}{2}$

○ **C** $y = 2x - 11$

○ **D** $y = -\frac{1}{2}x$

54. MULTI-STEP The line segment represents one side of a square **MP** 1

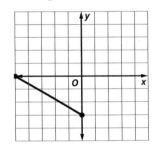

a. Write an equation for the line containing the segment shown.

b. What must be true about the sides adjacent and opposite the segment to form a square?

c. Write equations for the lines containing the sides of the square adjacent to the segment.

d. If (3, 2) is another vertex of the square, write an equation for the line containing the opposite side of the square.

e. What are the four vertices of the square?

f. The diagonals of a square must be perpendicular. Is the figure with the vertices you found a square? Explain.

1. **BOATS** Write an equation in slope-intercept form for the total rental cost C for a pontoon boat used for t hours. (Lesson 4-1)

Pontoon Boat Rentals
$105 for 1 hr
$180 for 2 hr

Write an equation in slope-intercept form for the line with the given conditions. (Lesson 4-1)

2. $(2, 5)$; slope 3

3. $(-3, -1)$, slope $\frac{1}{2}$

4. $(-3, 4)$, $(1, 12)$

5. $(-1, 6)$, $(2, 4)$

6. $(2, 1)$, slope 0

7. **MULTIPLE CHOICE** Write an equation of the line that passes through the point $(0, 0)$ and has slope -4. (Lesson 4-1)

   A $y = x - 4$ **C** $y = -4x$

   B $y = x + 4$ **D** $y = 4 - x$

8. **PHONES** The number of students at Jefferson High School who have a smart phone can be represented by the equation $y = 0.482x + 148$, where x is the number of students.

   a. Estimate the number of smart phones if there are 1450 students. (Lesson 4-1)

   b. **MP** Which mathematical practice did you use to solve this problem?

Write an equation in point-slope form for the line that passes through each point with the given slope. (Lesson 4-2)

9. $(1, 4)$, $m = 6$ 10. $(-2, -1)$, $m = -3$

11. Write an equation in point-slope form for the line that passes through the point $(8, 3)$, $m = -2$. (Lesson 4-2)

12. Write $y + 3 = \frac{1}{2}(x - 5)$ in standard form. (Lesson 4-2)

13. Write $y + 4 = -7(x - 3)$ in slope-intercept form. (Lesson 4-2)

Write each equation in standard form. (Lesson 4-2)

14. $y - 5 = -2(x - 3)$ 15. $y + 4 = \frac{2}{3}(x - 3)$

16. **MULTIPLE CHOICE** Determine whether the graphs of the pair of equations are *parallel, perpendicular,* or *neither.* (Lesson 4-3)

$$y = -6x + 8$$
$$3x + \frac{1}{2}y = -3$$

   A parallel

   B perpendicular

   C neither

   D not enough information

Write an equation in slope-intercept form for the line that passes through the given point and is parallel to the graph of the equation. (Lesson 4-3)

17. $(2, -1)$; $y = \frac{1}{4}x + 8$

18. $(-8, -4)$; $3x - 4y = 12$

19. **DESIGN** A graphic artist is designing a company logo using drawing software.

   a. Part of the logo includes line segments from $(-6, 10)$ to $(6, 16)$ and from $(0, 6)$ to $(12, 12)$. Are the segments parallel? Explain. (Lesson 4-3)

   b. Another part of the logo includes a triangle with vertices $(0,13)$, $(2, 7)$, and $(6, 9)$. Could these be the vertices of a right triangle? Explain.

Write an equation in slope-intercept form for the line that passes through the given point and is perpendicular to the graph of the equation. (Lesson 4-3)

20. $(3, -4)$; $y = -\frac{1}{3}x - 5$

21. $(0, -3)$; $y = -2x + 4$

22. $(-4, -5)$; $-4x + 5y = -6$

23. $(-1, -4)$; $-x - 2y = 0$

Scatter Plots and Lines of Fit

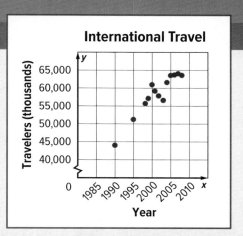

International Travel

:: **Then**	:: **Now**	:: **Why?**

Then
- You wrote linear equations given a point and the slope.

Now
1. Investigate relationships between quantities by using points on scatter plots.
2. Use lines of fit to make and evaluate predictions.

Why?
The graph shows the number of people from the United States who travel to other countries. The points do not all lie on the same line; however, you may be able to draw a line that is close to all the points. That line would show a linear relationship between the year x and the number of travelers each year y. Generally, international travel has increased.

New Vocabulary
bivariate data
scatter plot
correlation
association
line of fit
linear interpolation

MP **Mathematical Practices**

1 Make sense of problems and persevere in solving them.

4 Model with mathematics.

1 **Investigate Relationships Using Scatter Plots** Data with two variables are called **bivariate data**. A **scatter plot** shows the relationship between a set of data with two variables, graphed as ordered pairs on a coordinate plane. That relationship is known as a **correlation** or an **association**.

🔖 Key Concept Scatter Plots

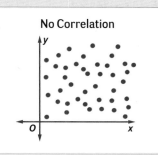

Positive Correlation	Negative Correlation	No Correlation
As *x* increases, *y* increases.	As *x* increases, *y* decreases.	*x* and *y* are not related.

Real-World Example 1   Evaluate a Correlation

WAGES Determine whether the graph shows a *positive*, *negative*, or *no* correlation. If there is a positive or negative correlation, describe its meaning in the situation.

The graph shows a positive correlation. As the number of hours worked increases, the wages usually increase.

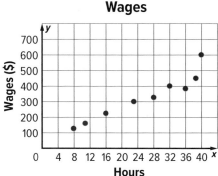

Wages

▶ **Guided Practice**

1. Refer to the graph on annual attendance at the Houston Livestock Show and Rodeo in Example 3. Determine whether the graph shows a *positive*, *negative*, or *no* correlation. If there is a positive or negative correlation, describe its meaning.

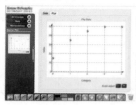
2 **Use Lines of Fit** Scatter plots can show whether there is a trend in a set of data. When the data points all lie close to a line, a **line of fit** or *trend line* can model the trend.

Key Concept Using a Linear Function to Model Data

Step 1 Make a scatter plot. Determine whether any relationship exists in the data.

Step 2 Draw a line that seems to pass close to most of the data points.

Step 3 Use two points on the line of fit to write an equation for the line.

Step 4 Use the line of fit to make predictions.

Real-World Example 2   Write a Line of Fit

INVASIVE SPECIES The table below shows the population growth of the zebra mussels in a local recreational area. Identify the independent and dependent variables. Is there a relationship in the data? If so, predict the number of zebra mussels in 2039.

Years since 2009	1	2	3	4	5	6	7
Number of mussels (thousands)	15	125	175	225	325	435	483

Real-World Link

Zebra mussels are an invasive species that cause damage in many places in the United States. Boaters and anglers spread mussels from one body of water to another. Zebra mussels clog public water intake pipes, ruin boats, and are dangerous to swimmers.

Step 1 Make a scatter plot.

The independent variable is the year, and the dependent variable is the number of zebra mussels. As the year increases, the number of zebra mussels increases. There is a positive correlation between the two variables.

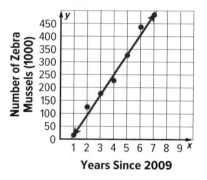

Invasion of Zebra Mussels

Step 2 Draw a line of fit.

No one line will pass through all the data points. Draw a line that passes close to the points. A line of fit is shown.

Step 3 Write an equation in slope-intercept form for the line of fit.

The line of fit passes close to the data points (1, 15) and (7, 483).

Find the slope.

$$m = \frac{y_2 - y_1}{x_2 - x_1}$$

$$= \frac{483 - 15}{7 - 1} \qquad (x_1, y_1) = (1, 15),$$
$$\qquad\qquad\qquad (x_2, y_2) = (7, 483)$$

$$= \frac{468}{6} \text{ or } 78$$

Use $m = 78$ and either the point-slope form or the slope-intercept form to write an equation of the line of fit.

$$y - y_1 = m(x - x_1)$$

$$y - 15 = 78(x - 1)$$

$$y - 15 = 78x - 78$$

$$y = 78x - 63$$

A slope of 78 means that the number of zebra mussels increased by an average of 78,000 per year. To predict the number of mussels in 2039, substitute 30 for x in the equation. The number of mussels is $78(30) - 63 = 2277$ thousand or 2,277,000 zebra mussels in 2039.

Amy Benson, US Geological Survey

> **Guided Practice**

2. **RESTAURANTS** The table shows revenue in billions for fast food restaurants by year. Make a scatter plot and determine what relationship exists, if any.

Year	2004	2005	2006	2007	2008	2009	2010	2011	2012
Sales	172.10	180.58	189.54	186.5	184.77	178.59	183.99	190.25	195.19

Source: IBIS World

Recall that linear extrapolation is used to predict values *outside* the range of the data. You can also use a linear equation to predict values *inside* the range of the data. This is called **linear interpolation**.

Real-World Example 3   Use Interpolation or Extrapolation

ENTERTAINMENT Use a scatter plot to find the approximate attendance at the 2010 Houston Livestock Show and Rodeo.

Step 1 Draw a line of fit. The line should be as close to as many points as possible.

Step 2 Write the slope-intercept form of the equation. The line of fit passes through (5, 1070) and (30, 2144).

Houston Livestock Show and Rodeo

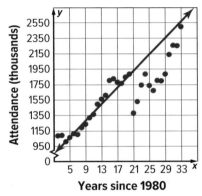

Find the slope.

$$m = \frac{y_2 - y_1}{x_2 - x_1}$$ Slope Formula

$$= \frac{2144 - 1070}{30 - 5}$$ $(x_1, y_1) = (5, 1070),$ $(x_2, y_2) = (30, 2144)$

$$= \frac{1074}{25} \text{ or} \approx 43$$ Simplify.

Use $m = \frac{1074}{25}$ and either the point-slope form or the slope-intercept form to write the equation of the line of fit.

$$y - y_1 = m(x - x_1)$$

$$y - 1070 = \frac{1074}{25}(x - 5)$$

$$y - 1070 = \frac{1074}{25}x - \frac{1074}{5}$$

$$y = \frac{1074}{25}x + \frac{4276}{5}$$

Step 3 Evaluate the function for $x = 2010 - 1980$ or 30.

$$y = \frac{1074}{25}x + \frac{4276}{5}$$ Equation of best-fit line

$$= \frac{1074}{25}(30) + \frac{4276}{5}$$ $x = 30$

$$= \frac{6444}{5} + \frac{4276}{5} \text{ or } 2144$$ Add.

In 2010, there were 2144 thousand or 2,144,000 people who attended the Houston Livestock Show and Rodeo.

> **Guided Practice**

3. **RESTAURANTS** Use the equation for the line of fit for the data in Guided Practice 2 to estimate revenue in 2030.

Check Your Understanding ◯ = Step-by-Step Solutions begin on page R11.

✓ **Go Online!** for a Self-Check Quiz

Example 1   Determine whether each graph shows a *positive*, *negative*, or *no* correlation.
If there is a positive or negative correlation, describe its meaning in the situation.

1. **Free Throws**

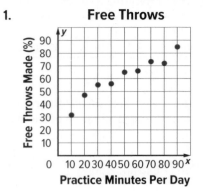

2. **Lemonade Sales**

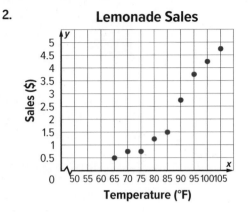

Example 2   3. Ⓜ️ **SENSE-MAKING** The table shows the median age of
females when they were first married.

a. Make a scatter plot and determine what relationship
exists, if any, in the data. Identify the independent and
the dependent variables.

b. Draw a line of fit for the scatter plot.

c. Write an equation in slope-intercept form for the
line of fit.

Example 3   d. Predict what the median age of females when they are
first married will be in 2026.

e. Do you think the equation can give a reasonable
estimate for the year 2056? Explain.

Year	Age
2001	25.1
2002	25.3
2003	25.3
2004	25.3
2005	25.5
2006	25.9
2007	26
2008	26.2
2009	26.5
2010	26.7
2011	26.9

Source: U.S. Bureau of Census

Practice and Problem Solving   Extra Practice is on page R4.

Example 1   Determine whether each graph shows a *positive*, *negative*, or *no* correlation. If there is a
positive or negative correlation, describe its meaning in the situation.

4. **Game Tickets
at the Fair**

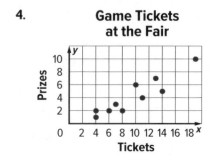

5 **NBA 3-Point
Percentage**

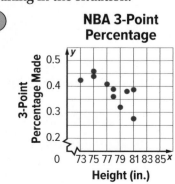

6.

Salaries

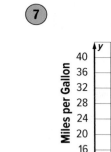

7

Gas Mileage of Various Vehicles

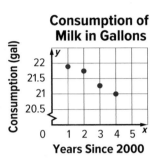

Examples 2–3   **8. MILK** Refer to the scatter plot of gallons of milk consumption per person for selected years.

    a. Use the points (2, 21.75) and (4, 21) to write the slope-intercept form of an equation for the line of fit.

    b. Predict the milk consumption in 2025.

    c. Predict in what year milk consumption will be 10 gallons.

    d. Is it reasonable to use the equation to estimate the consumption of milk for any year? Explain.

Consumption of Milk in Gallons

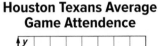

9. FOOTBALL Use the scatter plot.

    a. Use the points (6, 70,183) and (10, 71,080) to write the slope-intercept form of an equation for the line of fit shown in the scatter plot.

    b. Predict the average attendance at a game in 2025.

    c. Can you use the equation to make a decision about the average attendance in any given year in the future? Explain.

Houston Texans Average Game Attendence

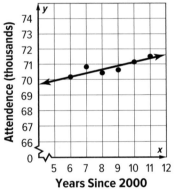

10. MP SENSE-MAKING The table at the right shows the amount of milk, in ounces, that a baby goat needs based on its weight, in pounds.

    a. Make a scatter plot comparing the weight in pounds to the amount of milk needed in ounces.

    b. Draw a line of fit for the data.

    c. Write the slope-intercept form of an equation for the line of fit.

    d. Predict the amount of milk needed for a baby goat that weighs 55 pounds.

    e. Use the equation of the line of fit to predict the weight of a baby goat that needs 38 ounces of milk.

Weight (lb)	Milk (oz)
5	12
7	16
10	20
15	28
20	32
25	40
30	48
40	64
50	80

11 **GEYSERS** The time to the next eruption of Old Faithful can be predicted by using the duration of the current eruption.

Duration (min)	1.5	2	2.5	3	3.5	4	4.5	5
Interval (min)	48	55	70	72	74	82	93	100

a. Identify the independent and the dependent variables. Make a scatter plot and determine what relationship, if any, exists in the data. Draw a line of fit for the scatter plot.

b. Let x represent the duration of the previous interval. Let y represent the time between eruptions. Write the slope-intercept form of the equation for the line of fit. Predict the interval after a 7.5-minute eruption.

c. Make a critical judgment about using the equation to predict the duration of the next eruption. Would the equation be a useful model?

12. **COLLECT DATA** Use a tape measure to measure both the foot size and the height in inches of ten individuals.

a. Record your data in a table.

b. Make a scatter plot and draw a line of fit for the data.

c. Write an equation for the line of fit.

d. Make a conjecture about the relationship between foot size and height.

H.O.T. Problems Use Higher-Order Thinking Skills

13. **MP SENSE-MAKING** Describe a real-life situation that can be modeled using a scatter plot. Decide whether there is a *positive*, *negative*, or *no* correlation. Explain what this correlation means.

14. **WHICH ONE DOESN'T BELONG?** Analyze the following situations and determine which one does not belong.

hours worked and amount of money earned	height of an athlete and favorite color

seedlings that grow an average of 2 centimeters each week	number of photos stored on a camera and capacity of camera

15. **MP ARGUMENTS** Determine which line of fit is better for the scatter plot at the right. Explain your reasoning.

16. **CHALLENGE** What can make a scatter plot and line of fit more useful for accurate predictions? Does an accurate line of fit always predict what will happen in the future? Explain.

17. **WRITING IN MATH** Make a scatter plot that shows the height of a person and age up to 16 years. Explain how you could use the scatter plot to predict the age of a person given his or her height. How can the information from a scatter plot be used to identify trends and make decisions?

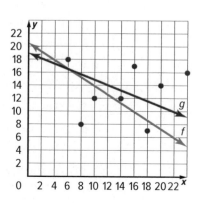

18. Which term best describes the correlation shown in the graph? **MP** 1

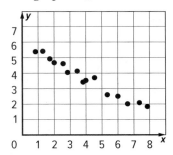

- **A** Positive
- **B** Negative
- **C** Zero
- **D** No correlation

19. Which equation best represents the data in the table? **MP** 4

Study Hours, h	0	1	2	3	4	5	6
Exam Score, S	55	70	83	76	87	92	90

- **A** $S = 40 + 15h$
- **B** $S = 60 + 6h$
- **C** $S = 55 + 10h$
- **D** $S = 80 + 3h$

20. The table shows the height of a skyscraper during its construction, where month 0 is when the base was completed. Use the heights at 9 and 18 months to write the equation for the line of fit and approximate the height of the building at 6 months. **MP** 1, 4

Month	0	9	12	18	20
Height (ft)	26	252	345	495	553

- **A** 113 ft
- **B** 171 ft
- **C** 184 ft
- **D** 711 ft

21. Use the equation for the line of fit in exercise 20 to predict in which month the skyscraper will reach its full height of 820 feet. **MP** 3

- **A** Month 10
- **B** Month 30
- **C** Month 39
- **D** Month 50

22. MULTI-STEP The table below shows the change in sea level over a 50-year period. **MP** 1

Year	Change in Sea Level (in.)
1960	5.7
1965	4.8
1970	5.6
1975	6.2
1980	6.0
1985	6.2
1990	6.8
1995	7.3
2000	7.8
2005	8.2
2010	8.4

a. Make a scatter plot of the data. Draw a line of fit as close to as many points as possible.

b. Determine whether the scatter plot shows a positive, negative, or no correlation. Explain your answer in terms of the situation.

c. Use the points (1970, 5.6) and (1995, 7.3) to write an equation for the line of best fit.

d. Which of the following is the best prediction of the change in sea level in 2030? Show your work.

- **A** 5.7 inches
- **B** 8.4 inches
- **C** 9.7 inches
- **D** 11.4 inches

LESSON 5

Correlation and Causation

:: Then

- Investigate relationships between variables by using points on scatter plots.

:: Now

1. Distinguish between correlation and causation.

:: Why?

- A study found a negative correlation between understanding HTML and liking licorice. But does understanding HTML cause a person to dislike licorice?

 New Vocabulary

correlation
causation

 Mathematical Practices

2 Reason abstractly and quantitatively.

3 Construct viable arguments and critique the reasoning of others.

8 Look for and express regularity in repeated reasoning.

1 Correlation and Causation Sets of data can show a relationship, or **correlation**, between two variables even when the variables are not related or dependent on each other. Because of this, correlation does not prove **causation**, where a change in one variable causes a change in another variable. It is important to determine whether one variable is causing the other to change or if the data just show a correlation.

🎵 Key Concept Correlation and Causation

Follow the steps to analyze data.

Step 1 Graph ordered pairs to create a scatter plot.

Step 2 Determine whether the scatter plot shows a positive or negative correlation.

Step 3 Determine whether the data show causation. Does one variable *cause* the other? Could other factors be influencing the data results?

Real-World Example 1   Correlation and Causation by Graphing

PETS The local humane society tracks the number of dogs and cats that are adopted each month. The table shows their records for five months.

Month	January	February	March	April	May
Number of Dogs	90	54	62	39	85
Number of Cats	124	72	95	53	121

a. **Graph the ordered pairs to create a scatter plot (number of dogs, number of cats).**

b. **Determine whether the scatter plot shows a *positive* or *negative* correlation.**

The graph shows a positive correlation. As the number of adopted dogs increases, the number of adopted cats increases.

c. **Determine whether the data show causation. Does one variable *cause* the other? Could other factors be influencing the data results?**

No, the data show a correlation only. Adopting a dog does not cause anyone to adopt a cat. Other factors, such as the human society holding adoption drives or reducing fees may cause an increase in the adoptions of both animals.

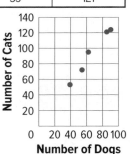

Graph with y-axis "Number of Cats" (0 to 140) and x-axis "Number of Dogs" (0 to 100).

Ronnie Kaufman/Larry Hirshowitz/Blend Images/Getty Images

Guided Practice

JOBS Lane works as a cook at a pizza parlor. The table shows the number of hours he worked and the amount he was paid each month.

Month	May	June	July	August	September	October
Hours Worked	17	51	49	42.5	29.5	31
Pay Earned	149.60	448.80	431.20	374	259.60	272.80

1A. Graph the ordered pairs as a scatter plot (hours worked, pay earned).

1B. Does the scatter plot show a *positive* or *negative* correlation? Explain.

1C. Determine whether the data show causation. Does one variable *cause* the other? Could other factors be influencing the data results?

While a positive or negative correlation can be observed between many variables, this relationship may be purely coincidental. That is, the change in one variable may not cause the change in the other variable.

Once a correlation between variables has been determined, it is important to analyze a situation to determine whether the data also show causation.

Real-World Example 2   Correlation and Causation by Situation

Determine whether each situation illustrates *correlation but not causation*, or *causation*.

a. Rebecca records the pounds of each type of vegetable her garden produces each year. She discovers a negative correlation between the pounds of sweet potatoes and the pounds of spinach it produces.

This situation models correlation, but not causation. One vegetable growing well is unlikely to cause the other to grow poorly. Other factors could be affecting the plants, such as how each type of vegetable responds to various weather and soil conditions

b. A local police report states that there is a positive correlation between the number of days it snows and the number of car accidents annually.

This situation models causation. Snow can cause car accidents for several reasons, including loss of visibility and slick roads.

Guided Practice

Determine whether each situation illustrates *correlation but not causation*, or *causation*.

2A. When mixing orange juice from concentrate, there is a negative correlation between the amount of water added and the concentration of the orange juice.

2B. Teresa sells crafts at the local flea market every weekend. She noticed a positive correlation between the number of necklaces and the number of toy puppets she sells each day.

Check Your Understanding   ◯ = Step-by-Step Solutions begin on page R11.

Go Online! for a
Self-Check Quiz

Example 1

1. **GRADES** A teacher tracks her students' test scores in math class. She wonders whether the class test scores are related to the final exam scores. The table shows class test scores and the final exam scores for six students.

Student	1	2	3	4	5	6
Class Test	50	68	89	61	74	84
Final Exam	42	60	90	38	70	79

   a. Graph the ordered pairs as a scatter plot (class test, final exam).

   b. Is the correlation *negative* or *positive*? Explain.

   c. Is the relationship also a causation? Explain.

Example 2   **Determine whether each situation illustrates *correlation but not causation*, or *causation*.**

2. During their annual summer vacation, the Brown family found a positive correlation between the number of gallons of gas they used and the number of miles they drove each day.

3. A study found a negative correlation between the average number of hours people spend exercising and the average number of hours they spend on social media each week.

Practice and Problem Solving   Extra Practice is found on page R4.

Example 1

4. **FUNDRAISING** Laura is selling scones for a school fundraiser. She record the number of scones she sells and the amount she is paid for each transaction. The table shows her first six transactions.

Transaction	1	2	3	4	5	6
Number of Scones	12	6	1	36	24	10
Payment Amount	5	3	1	12	11	4

   a. Graph the ordered pairs as a scatter plot (number of scones, payment amount).

   b. Is the correlation *negative* or *positive*? Explain.

   c. Does the data illustrate causation? Explain.

5. **DESIGN** Mateo designs and sells graphic T-shirts. He records his inventory and the number of T-shirts he sells for each batch. The table shows the number of T-shirts he has left in his batch at the end of each week and the total number of T-shirts he has sold so far.

Week	1	2	3	4	5	6
Total Number Sold	5	8	12	15	20	22
Current Inventory	25	22	18	15	10	8

   a. Graph the ordered pairs as a scatter plot (total number sold, current inventory).

   b. Is the correlation *negative* or *positive*? Explain.

   c. Does the data illustrate causation? Explain.

6. **ANALYSIS** Darren decides to track the number of hours he spends doing chores each week. He wants to compare them to the number of text messages he sends. The table shows the number of hours he does chores and the number of text messages he sends each week for six weeks.

Week	1	2	3	4	5	6
Hours Doing Chores	2	4	2.5	4.5	1.5	1
Text Messages Sent	123	152	130	165	125	120

   a. Graph the ordered pairs as a scatter plot (hours doing chores, text messages sent).

   b. Is the correlation *negative* or *positive*? Explain.

   c. Does the data illustrate causation? Explain.

7. **ANALYSIS** The data show the per capita consumption of mozzarella cheese and the number of civil engineering doctoral degrees awarded in the United States over several years.

Year	2006	2008	2010	2012	2014	2016
Mozzarella Consumed (pounds)	10.5	10.6	11.3	11.5	11.9	12.3
Civil Engineering Doctorates	655	712	645	714	767	805

a. Graph the ordered pairs as a scatter plot (mozzarella consumed, civil engineering doctorates).

b. Is the correlation *negative* or *positive*? Explain.

c. Does the data illustrate causation? Explain.

Example 2
Determine whether each situation illustrates *correlation but not causation*, or *causation*.

8. The local hardware store notices a positive correlation between the number of snow shovels and the pounds of salt purchased each day.

9. George read online that the amount of time spent exercising is positively correlated with the amount of water a person consumes.

10. When doing his taxes, Josef found that the amount of his commission was positively correlated with the number of appliances he sold each week.

11. Jane compared the places to eat in her town, and found a negative correlation between the number of fast food places and upscale restaurants.

12. **FINANCIAL LITERACY** What type of correlation would you expect to find between the amount of electricity a family uses each month and the amount of their monthly electric bill? Does this situation also illustrate causation? Why or why not?

13. **DEPRECIATION** Marie researched the value of her car, and found that its value is negatively correlated with the number of years since she bought it. Does this situation illustrate causation? If so, would you predict her car to be worth more or less a year from now?

14. **SMALL BUSINESS** Isabelle owns a small hair salon. She discovered a positive correlation between the number of clients she has and the amount of money she makes each week. Does this situation illustrate causation? If so, what should she do to make more money?

15. **SURVEY** Stella surveyed her friends at school to find how many video games and construction sets they have. She found a positive correlation between the number of video games and the number of construction sets the students had. Does this situation illustrate causation? If not, what other factors might be contributing to the correlation?

H.O.T. Problems   Use Higher-Order Thinking Skills

16. **CONSTRUCT ARGUMENTS** Explain the difference between correlation and causation.

17. **SENSE-MAKING** Can variables show a correlation even when they do not illustrate causation? Justify your answer.

18. **CONSTRUCT ARGUMENTS** Why does correlation not prove causation?

19. **OPEN ENDED** Describe a situation that illustrates causation.

20. **OPEN ENDED** Describe a situation that illustrates positive correlation, but not causation, and another situation that illustrates negative correlation, but not causation.

21. Which situation shows a positive correlation and causation? **MP** 1, 3

   ○ **A**   the number of withdrawals and the amount in a bank account

   ○ **B**   the number of checks deposited and the amount in a bank account

   ○ **C**   the time of day and the amount in a bank account

   ○ **D**   the amount spent and the amount in a bank account

22. Which situation does not illustrate causation? **MP** 1, 3

   ○ **A**   the amount of snow and number of school closings

   ○ **B**   the amount of sugar consumed and risk of health issues

   ○ **C**   the age of an actor and number of awards won

   ○ **D**   the amount of time exercising and the number of calories burned

23. Which of the following describe the relationship between the number of loaves of bread baked and the amount of flour used? **MP** 1, 3

   ☐ **A**   positive correlation

   ☐ **B**   negative correlation

   ☐ **C**   causation

   ☐ **D**   no causation

24. Which of the following describe the relationship between the number of hours training employees on safety and the number of employee injuries? **MP** 1, 3

   ☐ **A**   positive correlation

   ☐ **B**   negative correlation

   ☐ **C**   causation

   ☐ **D**   no causation

25. **MULTI-STEP** The table shows the number of minutes played and the number of points scored by a basketball player. **MP** 1, 3, 8

Game	1	2	3	4	5	6
Minutes	22	25	23	27	29	31
Points	16	22	18	24	28	35

   a.   Graph the ordered pairs as a scatter plot (minutes, points).

   b.   Is the correlation *negative* or *positive*? Explain.

   c.   Does the data illustrate causation? Explain.

26. **MULTI-STEP** The following table shows meal portion size and the number of Calories. **MP** 1, 3, 8

Item	1	2	3	4	5	6
Portion Size (ounces)	3	1	2	1.5	2.5	3.5
Calories	275	50	150	75	200	300

   a.   Graph the ordered pairs as a scatter plot (portion size, Calories).

   b.   Is the correlation *negative* or *positive*? Explain.

   c.   Does the data illustrate causation? Explain.

27. Which graph shows negative correlation and causation? **MP** 1

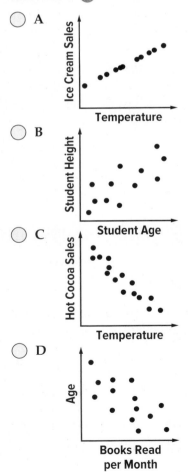

Regression and Median-Fit Lines

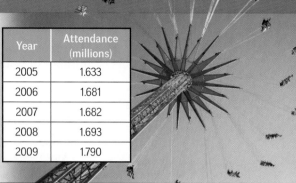

	:: Then	:: Now	:: Why?

:: **Then**

● You used lines of fit and scatter plots to evaluate trends and make predictions.

:: **Now**

1 Write equations of best-fit lines using linear regression.

2 Write equations of median-fit lines.

:: **Why?**

● The table shows the total attendance, in millions of people, at the Minnesota State Fair from 2005 to 2009. You can use a graphing calculator to find the equation of a *best-fit line* and use it to make predictions about future attendance at the fair.

Year	Attendance (millions)
2005	1.633
2006	1.681
2007	1.682
2008	1.693
2009	1.790

New Vocabulary
best-fit line
linear regression
correlation coefficient
residual
median-fit line

Mathematical Practices
4 Model with mathematics.

1 Best-Fit Lines You have learned how to find and write equations for lines of fit by hand. Many calculators use complex algorithms that find a more precise line of fit called the **best-fit line**. One algorithm is called **linear regression**.

Your calculator may also compute a number called the **correlation coefficient**. This number will tell you if your correlation is positive or negative and how closely the equation is modeling the data. The closer the correlation coefficient is to 1 or −1, the more closely the equation models the data.

Real-World Example 1   Best-Fit Line

MOVIES The table shows the amount of money made by movies in the United States. Use a graphing calculator to write an equation for the best-fit line for that data. Find and interpret the correlation coefficient.

Year	1995	1997	1999	2001	2003	2005	2007	2009	2011	2013
Revenue ($ billion)	$5.29	$6.51	$7.30	$8.13	$9.35	$8.93	$9.63	$10.65	$10.20	$9.28

Before you begin, make sure that your Diagnostic setting is on. You can find this under the **CATALOG** menu. Press **D** and then scroll down and click **DiagnosticOn**. Then press ENTER .

Step 1 Enter the data by pressing STAT and selecting the **Edit** option. Let the year 1995 be represented by 0. Enter the years since 1995 into List 1 (**L1**). These will represent the *x*-values. Enter the income ($ billion) into List 2 (**L2**). These will represent the *y*-values.

Step 2 Perform the regression by pressing STAT and selecting the **CALC** option. Scroll down to **LinReg (ax+b)** and press ENTER twice.

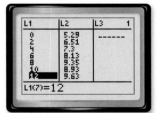

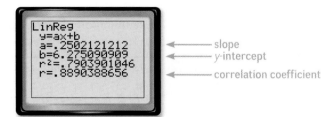

```
LinReg
y=ax+b
a=.2502121212    ←——— slope
b=6.275090909    ←——— y-intercept
r²=.7903901046
r=.8890388656    ←——— correlation coefficient
```

Step 3 Write the equation of the regression line by rounding the a and b values on the screen. The form that we chose for the regression was $ax + b$, so the equation is $y = 0.25x + 6.28$. The correlation coefficient is about 0.8890, which means that the equation models the data fairly well.

> **Guided Practice**

Write an equation of the best-fit line for the data in each table. Find and interpret the correlation coefficient. Round to the nearest ten-thousandth. Let x be the number of years since 2007.

1A. HOCKEY The table shows the number of goals of leading scorers for the Mustang Girls Hockey Team.

Year	2007	2008	2009	2010	2011	2012	2013	2014
Goals	30	23	41	35	31	43	33	45

1B. HOCKEY The table gives the number of goals scored by the team each season.

Year	2007	2008	2009	2010	2011	2012	2013	2014
Goals	63	44	55	63	81	85	93	84

We know that not all of the points will lie on the best-fit line. The difference between an observed y-value and its predicted y-value (found on the best-fit line) is called a **residual**. Residuals measure how much the data deviate from the regression line. When residuals are plotted on a scatter plot they can help to assess how well the best-fit line describes the data. If the best-fit line is a good fit, there is no pattern in the residual plot.

Real-World Example 2   Graph and Analyze a Residual Plot

HOCKEY Graph and analyze the residual plot for the data for Guided Practice 1A. Determine if the best-fit line models the data well.

After calculating the best-fit line in Guided Practice 1A, you can obtain the residual plot of the data. Turn on **Plot2** under the **STAT PLOT** menu and choose ⌊∴⌋. Use **L1** for the **Xlist** and **RESID** for the **Ylist**. You can obtain **RESID** by pressing 2nd [STAT] and selecting **RESID** from the list of names. Graph the scatter plot of the residuals by pressing ZOOM and choosing **ZoomStat**.

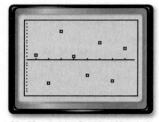

[0, 8] scl: 1 by [−10, 10] scl: 2

The residuals appear to be randomly scattered and centered about the line $y = 0$. Thus, the best-fit line seems to model the data well.

> **Guided Practice**

2. UNEMPLOYMENT Graph and analyze the residual plot for the following data comparing graduation rates and unemployment rates.

Graduation Rate	73	85	64	81	68	82
Unemployment Rate	6.9	4.1	3.2	5.5	4.3	5.1

A residual is positive when the observed value is above the line, negative when the observed value is below the line, and zero when it is on the line. One common measure of how well a line fits is the sum of squared vertical distances from the points to the line. The best-fit line, which is also called the *least-squares regression line*, minimizes the sum of the squares of those distances.

We can use points on the best-fit line to estimate values that are not in the data. Recall that when we estimate values that are between known values, this is called *linear interpolation*. When we estimate a number outside of the range of the data, it is called *linear extrapolation*.

Real-World Example 3   Use Interpolation and Extrapolation

PAINTBALL The table shows the points received by the top ten paintball teams at a tournament. **Estimate how many points the 20th-ranked team received.**

Rank	1	2	3	4	5	6	7	8	9	10
Score	100	89	96	99	97	98	78	70	64	80

Write an equation of the best-fit line for the data. Then extrapolate to find the missing value.

Step 1 Enter the data from the table into the lists. Let the ranks be the *x*-values and the scores be the *y*-values. Then graph the scatter plot.

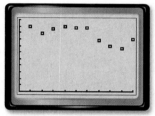

[0, 10] scl: 1 by [0, 110] scl: 10

Step 2 Perform the linear regression using the data in the lists. Find the equation of the best-fit line.

The equation is about $y = -3.32x + 105.3$.

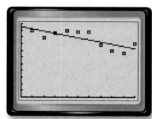

[0, 10] scl: 1 by [0, 110] scl: 10

Step 3 Graph the best-fit line. Press Y= VARS and choose **Statistics**. From the **EQ** menu, choose **RegEQ**. Then press GRAPH.

Step 4 Use the graph to predict the points that the 20th-ranked team received. Change the viewing window to include the *x*-value to be evaluated. Press 2nd [CALC] ENTER 20 ENTER to find that when $x = 20$, $y \approx 39$. It is estimated that the 20th ranked team received 39 points.

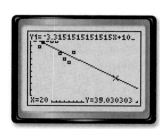

[0, 25] scl: 1 by [0, 110] scl: 1

SWIMMING Predict the total attendance at a community pool for each of the following temperatures.

Temperature	92	102	97	104	108	110	111	111	107	105
Attendance	395	309	332	307	251	178	164	149	273	299

3A. 90 degrees   **3B.** 115 degrees

2 Median-Fit Lines A second type of fit line that can be found using a graphing calculator is a **median-fit line**. The equation of a median-fit line is calculated using the medians of the coordinates of the data points.

Example 4   Median-Fit Line

PAINTBALL Find and graph the equation of a median-fit line for the data in Example 3. Then predict the score of the 15th ranked team.

Step 1 Reenter the data if it is not in the lists. Clear the **Y=** list and graph the scatter plot.

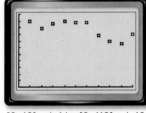

[0, 10] scl: 1 by [0, 110] scl: 10

Step 2 To find the median-fit equation, press the STAT key and select the **CALC** option. Scroll down to the **Med-Med** option and press ENTER . The value of a is the slope, and the value of b is the y-intercept.

The equation for the median-fit line is about $y = -3.71x + 108.26$.

Step 3 Copy the equation to the **Y=** list and graph. Use the **value** option to find the value of y when $x = 15$.

The 15th ranked team scored about 53 points.

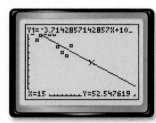

[0, 25] scl: 1 by [0, 110] scl: 1

Notice that the equations for the regression line and the median-fit line are very similar.

Real-World Link

Paintball is more popular with 12- to 17-year-olds than any other age group. In a recent year, 3,649,000 teens participated in paintball while 2,195,000 18- to 24-year-olds participated.

Source: *Statistical Abstract of the United States*

Guided Practice

4. Use the data from Guided Practice 3 and a median-fit line to estimate the attendance for temperatures of 90 and 115 degrees. Compare these values with the answers from the regression line.

Check Your Understanding   ◯ = Step-by-Step Solutions begin on page R11.

Go Online! for a
Self-Check Quiz

Examples 1, 2   **1. POTTERY** A local university is keeping track of the number of art students who use the pottery studio each day.

Day	1	2	3	4	5	6	7
Students	10	15	18	15	13	19	20

  a. Write an equation of the regression line and find and interpret the correlation coefficient.

  b. Graph the residual plot and determine if the regression line models the data well.

Example 3   **2. BASEBALL** The table below shows the number of stolen bases (SB) and times caught stealing (CS) for a few major league baseball teams during the 2013 season. Use linear interpolation to estimate the number of times caught stealing by a team that had 35 stolen bases.

SB	23	28	29	31	45	48	51	59	60	67	72	76
CS	13	23	18	14	22	20	22	26	24	24	16	18

Example 4   **3. VACATION** The Smiths want to rent a house on the lake that sleeps eight people. The cost of the house per night is based on how close it is to the water.

Distance from Lake (mi)	0.0 (houseboat)	0.3	0.5	1.0	1.25	1.5	2.0
Price/Night ($)	785	325	250	200	150	140	100

  a. Find and graph an equation for the median-fit line.

  b. What would you estimate is the cost of a rental 1.75 miles from the lake?

Practice and Problem Solving   Extra Practice is on page R4.

Example 1   **Write an equation of the regression line for the data in each table. Then find and interpret the correlation coefficient.**

  4. SKYSCRAPERS The table ranks the ten tallest buildings in which Shai has been.

Rank	1	2	3	4	5	6	7	8	9	10
Stories	101	88	110	88	88	80	69	102	78	70

  ⑤ **MUSIC** The table gives the number of annual violin auditions held by a youth symphony each year since 2009. Let x be the number of years since 2009.

Year	2009	2010	2011	2012	2013	2014	2015
Auditions	22	19	25	37	32	35	42

Example 2   **6. RETAIL** The table gives the sales at a clothing chain since 2008. Let x be the number of years since 2008.

Year	2008	2009	2010	2011	2012	2013	2014
Sales (Millions of Dollars)	6.84	7.6	10.9	15.4	17.6	21.2	26.5

  a. Write an equation of the regression line.

  b. Graph and analyze the residual plot.

7 MARATHON The number of entrants in the Boston Marathon every five years since 1975 is shown. Let x be the number of years since 1975.

Year	1975	1980	1985	1990	1995	2000	2005	2010
Entrants	2395	5417	5594	9412	9416	17,813	20,453	26,735

a. Find an equation for the median-fit line.

b. According to the equation, how many entrants were there in 2003?

8. **CAMPING** A campground keeps a record of the number of campsites rented the week of July 4 for several years. Let x be the number of years since 2007.

Year	2009	2010	2011	2012	2013	2014	2015	2016	2017
Sites Rented	34	45	42	53	58	47	57	65	59

a. Find an equation for the regression line.

b. Predict the number of campsites that will be rented in 2019.

c. Predict the number of campsites that will be rented in 2027.

9. **ICE CREAM** An ice cream company keeps a count of the tubs of chocolate ice cream delivered to each of their stores in a particular area.

a. Find an equation for the median-fit line.

b. Graph the points and the median-fit line.

c. How many tubs would be delivered to a 1500-square-foot store? a 5000-square-foot store?

Store Size (ft²)	2100	2225	3135	3569	4587
Tubs (hundreds)	110	102	215	312	265

10. **MP REASONING** The prices of the eight top-selling brands of jeans at Jeanie's Jeans are given in the table below.

Sales Rank	1	2	3	4	5	6	7	8
Price ($)	43	44	50	61	64	135	108	78

a. Find the equation for the regression line.

b. According to the equation, what would be the price of a pair of the 12th best-selling brand?

c. Is this a reasonable prediction? Explain.

11. **STATE FAIR** Refer to the beginning of the lesson.

a. Graph a scatter plot of the data, where $x = 1$ represents 2005. Then find and graph the equation for the best-fit line.

b. Graph and analyze the residual plot.

c. Predict the total attendance in 2025. Is your answer reasonable?

12. **FIREFIGHTERS** The table shows statistics from the U.S. Fire Administration.

   a. Find an equation for the median-fit line.

   b. Graph the points and the median-fit line using a graphing calculator.

   c. Does the median-fit line give you an accurate model of the number of firefighters? Explain.

Age	Number of Firefighters
18	40,919
25	245,516
35	330,516
45	296,665
55	167,087
65	54,559

13. **ATHLETICS** The table shows the number of participants in high school athletics.

Year Since 1980	0	10	20	25	30
Athletes	5,356,913	5,298,671	6,705,223	7,159,904	7,667,955

Source: The National Federation of State High School Associations

   a. Find an equation for the regression line.

   b. According to the equation, how many participated in 2008?

14. **ART** A count was kept on the number of paintings sold at an auction each year. Let x be the number of years since 1990.

Years Painted	1990	1995	2000	2005	2010	2015
Paintings Sold	8	5	25	21	9	22

   a. Find the equation for the linear regression line.

   b. How many paintings were sold in 2001?

   c. Is the linear regression equation an accurate model of the data? Explain why or why not.

H.O.T. Problems   Use **H**igher-**O**rder **T**hinking Skills

15. **MP** **CRITIQUE** Below are the results of the World Superpipe Championships in 2013.

Men	Score	Rank	Women	Score
Shaun White	98.00	1	Kelly Clark	90.33
Ayumu Hirano	92.33	2	Elena Hight	90.00
Markus Malin	91.33	3	Arielle Gold	85.00
Scotty Lago	90.00	4	Torah Bright	82.00
Greg Bretz	84.66	5	Hannah Teter	81.00

   Find an equation of the regression line for each, and graph them on the same coordinate plane. Compare and contrast the men's and women's graphs.

16. **CHALLENGE** For a class project, the scores that 10 randomly selected students earned on the first eight tests of the school year are given. Explain how to find a line of best fit. Could it be used to predict the scores of other students? Explain your reasoning.

17. **OPEN ENDED** For 10 different people, measure their heights and the lengths of their heads from chin to top. Use these data to generate a linear regression equation and a median-fit equation. Make a prediction using both of the equations.

18. **WRITING IN MATH** How are lines of fit and linear regression similar? different?

19. The table shows yo-yo sales for a company over a 4-year period.

Year	2014	2015	2016	2017
Yo-Yos Sold (thousands)	9.7	7.1	6.2	5.8

Use a graphing calculator to find an equation for the best-fit line that models the data where x represents the number of years since 2014 and y represents the number of yo-yos sold in thousands. **MP** 4

○ **A** $y = -1.26x + 9.09$

○ **B** $y = -1.26x + 10.35$

○ **C** $y = -0.68x + 7.42$

○ **D** $y = -1.26x + 25.47$

20. Atma used his graphing calculator to find a best-fit equation for some data. In the equation $w = 0.32r + 4.8$, w represents the weight of a cantaloupe in pounds and r represents its radius in inches. Predict the weight of a cantaloupe with a radius of 5 inches. **MP** 4

○ **A** 0.625 pound

○ **B** 1.6 pounds

○ **C** 4.8 pounds

○ **D** 6.4 pounds

21. The table shows the canceled check fee for a bank from 2013 to 2016.

Canceled check fee, years since 2013				
Years	0	1	2	3
Fee ($)	5	10	10	15

Which of the following best predicts the fee in the year 2020? **MP** 4

○ **A** $15

○ **B** $20

○ **C** $26.50

○ **D** $30

22. The table shows the average cost of a 10-person Thanksgiving dinner. **MP** 4, 6

Year	2006	2008	2010	2012	2014
Cost ($)	38.10	44.61	43.47	49.48	49.41

a. Find an equation of the regression line. Let x be the number of years since 2006.

b. Which of the following statements are true?

☐ **A** The correlation coefficient indicates that the best-fit line does not model the data well.

☐ **B** The average cost in 2009 was about $43.64.

☐ **C** The average cost in 2026 will be about $75.25.

☐ **D** The residual plot indicates that the best-fit line models the data well.

23. MULTI-STEP The table shows U.S. online sales, in millions of dollars, on Cyber Monday since 2010. **MP** 2, 3, 4

Year	2010	2011	2012	2013	2014
Sales	1028	1251	1465	1735	2038

a. Find an equation of the regression line. Let x be the number of years since 2010.

b. Predict the Cyber Monday sales in 2025.

c. Find the correlation coefficient.

d. A classmate says that the correlation coefficient is not close to 0, so the equation in part **a** does not model the data well. Do you agree? Explain.

Inverses of Linear Functions

:: Then	:: Now	:: Why?
• You represented relations as tables, graphs, and mappings.	**1** Find the inverse of a relation. **2** Find the inverse of a linear function.	• Randall is writing a report on Santiago, Chile, and he wants to include a brief climate analysis. He found a table of temperatures recorded in degrees Celsius. He knows that a formula for converting degrees Fahrenheit to degrees Celsius is $C(x) = \frac{5}{9}(x - 32)$. He will need to find the *inverse* function to convert from degrees Celsius to degrees Fahrenheit.

Average Temp (°C)		
Month	Min	Max
Jan.	12	29
March	9	27
May	5	18
July	3	15
Sept.	6	29
Nov.	9	26

New Vocabulary
inverse relation
inverse function

Mathematical Practices
7 Look for and make use of structure.

1 **Inverse Relations** An **inverse relation** is the set of ordered pairs obtained by exchanging the x-coordinates with the y-coordinates of each ordered pair in a relation. If (5, 3) is an ordered pair of a relation, then (3, 5) is an ordered pair of the inverse relation.

Key Concept Inverse Relations

Words	If one relation contains the element (a, b), then the inverse relation will contain the element (b, a).

Example A and B are inverse relations.

A		B
$(-3, -16)$	⟶	$(-16, -3)$
$(-1, 4)$	⟶	$(4, -1)$
$(2, 14)$	⟶	$(14, 2)$
$(5, 32)$	⟶	$(32, 5)$

Notice that the domain of a relation becomes the range of its inverse, and the range of the relation becomes the domain of its inverse.

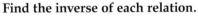

Example 1   Inverse Relations

Find the inverse of each relation.

a. $\{(4, -10), (7, -19), (-5, 17), (-3, 11)\}$

To find the inverse, exchange the coordinates of the ordered pairs.

$(4, -10) \rightarrow (-10, 4)$ $(-5, 17) \rightarrow (17, -5)$

$(7, -19) \rightarrow (-19, 7)$ $(-3, 11) \rightarrow (11, -3)$

The inverse is $\{(-10, 4), (-19, 7), (17, -5), (11, -3)\}$.

b.

x	−4	−1	5	9
y	−13	−8.5	0.5	6.5

Write the coordinates as ordered pairs. Then exchange the coordinates of each pair.

$(-4, -13) \rightarrow (-13, -4)$ $(5, 0.5) \rightarrow (0.5, 5)$

$(-1, -8.5) \rightarrow (-8.5, -1)$ $(9, 6.5) \rightarrow (6.5, 9)$

The inverse is $\{(-13, -4), (-8.5, -1), (0.5, 5), (6.5, 9)\}$.

1A. {(−6, 8), (−15, 11), (9, 3), (0, 6)}

1B.

x	−10	−4	−3	0
y	5	11	12	15

The graphs of relations can be used to find and graph inverse relations.

Example 2   Graph Inverse Relations

Graph the inverse of the relation.

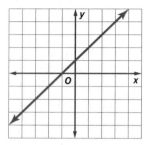

The graph of the relation passes through the points at (−4, −3), (−2, −1), (0, 1), (2, 3), and (3, 4). To find points through which the graph of the inverse passes, exchange the coordinates of the ordered pairs. The graph of the inverse passes through the points at (−3, −4), (−1, −2), (1, 0), (3, 2), and (4, 3). Graph these points and then draw the line that passes through them.

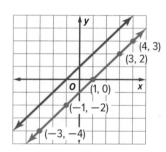

▶ **Guided Practice**

Graph the inverse of each relation.

2A.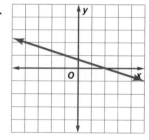

2B.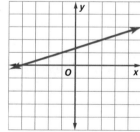

The graphs from Example 2 are graphed on the right with the line $y = x$. Notice that the graph of an inverse is the graph of the original relation reflected in the line $y = x$. For every point (x, y) on the graph of the original relation, the graph of the inverse will include the point (y, x).

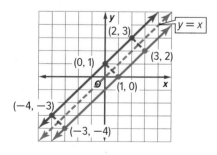

2 **Inverse Functions** A linear relation that is described by a function has an **inverse function** that can generate ordered pairs of the inverse relation. The inverse of the linear function $f(x)$ can be written as $f^{-1}(x)$ and is read *f of x inverse* or *the inverse of f of x*.

(I)McGraw-Hill Education

🔁 Key Concept Finding Inverse Functions

To find the inverse function $f^{-1}(x)$ of the linear function $f(x)$, complete the following steps.

Step 1 Replace $f(x)$ with y in the equation for $f(x)$.

Step 2 Interchange y and x in the equation.

Step 3 Solve the equation for y.

Step 4 Replace y with $f^{-1}(x)$ in the new equation.

Example 3 Find Inverse Linear Functions

Find the inverse of each function.

a. $f(x) = 4x - 8$

Step 1	$f(x) = 4x - 8$	Original equation
	$y = 4x - 8$	Replace $f(x)$ with y.
Step 2	$x = 4y - 8$	Interchange y and x.
Step 3	$x + 8 = 4y$	Add 8 to each side.
	$\dfrac{x + 8}{4} = y$	Divide each side by 4.
Step 4	$\dfrac{x + 8}{4} = f^{-1}(x)$	Replace y with $f^{-1}(x)$.

The inverse of $f(x) = 4x - 8$ is $f^{-1}(x) = \dfrac{x + 8}{4}$ or $f^{-1}(x) = \dfrac{1}{4}x + 2$.

Watch Out!

Notation The -1 in $f^{-1}(x)$ is *not* an exponent.

CHECK Graph both functions and the line $y = x$ on the same coordinate plane. $f^{-1}(x)$ appears to be the reflection of $f(x)$ in the line $y = x$. ✓

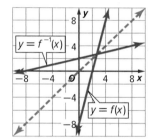

b. $f(x) = -\dfrac{1}{2}x + 11$

Step 1	$f(x) = -\dfrac{1}{2}x + 11$	Original equation
	$y = -\dfrac{1}{2}x + 11$	Replace $f(x)$ with y.
Step 2	$x = -\dfrac{1}{2}y + 11$	Interchange y and x.
Step 3	$x - 11 = -\dfrac{1}{2}y$	Subtract 11 from each side.
	$-2(x - 11) = y$	Multiply each side by -2.
	$-2x + 22 = y$	Distributive Property
Step 4	$-2x + 22 = f^{-1}(x)$	Replace y with $f^{-1}(x)$.

The inverse of $f(x) = -\dfrac{1}{2}x + 11$ is $f^{-1}(x) = -2x + 22$.

▶ **Guided Practice**

3A. $f(x) = 4x - 12$

3B. $f(x) = \dfrac{1}{3}x + 7$

Real-World Example 4   Use an Inverse Function

TEMPERATURE Refer to the beginning of the lesson. Randall wants to convert the temperatures from degrees Celsius to degrees Fahrenheit.

a. Find the inverse function $C^{-1}(x)$.

Step 1	$C(x) = \frac{5}{9}(x - 32)$	Original equation
	$y = \frac{5}{9}(x - 32)$	Replace $C(x)$ with y.
Step 2	$x = \frac{5}{9}(y - 32)$	Interchange y and x.
Step 3	$\frac{9}{5}x = y - 32$	Multiply each side by $\frac{9}{5}$.
	$\frac{9}{5}x + 32 = y$	Add 32 to each side.
Step 4	$\frac{9}{5}x + 32 = C^{-1}(x)$	Replace y with $C^{-1}(x)$.

The inverse function of $C(x)$ is $C^{-1}(x) = \frac{9}{5}x + 32$.

b. What do x and $C^{-1}(x)$ represent in the context of the inverse function?

x represents the temperature in degrees Celsius. $C^{-1}(x)$ represents the temperature in degrees Fahrenheit.

c. Find the average temperatures for July in degrees Fahrenheit.

The average minimum and maximum temperatures for July are 3° C and 15° C, respectively. To find the average minimum temperature, find $C^{-1}(3)$.

$C^{-1}(x) = \frac{9}{5}x + 32$ Original equation

$C^{-1}(3) = \frac{9}{5}(3) + 32$ Substitute 3 for x.

$= 37.4$ Simplify.

To find the average maximum temperature, find $C^{-1}(15)$.

$C^{-1}(x) = \frac{9}{5}x + 32$ Original equation

$C^{-1}(15) = \frac{9}{5}(15) + 32$ Substitute 15 for x.

$= 59$ Simplify.

The average minimum and maximum temperatures for July are 37.4° F and 59° F, respectively.

▶ **Guided Practice**

4. RENTAL CAR Peggy rents a car for the day. The total cost $C(x)$ in dollars is given by $C(x) = 24.99 + 0.3x$, where x is the number of miles she drives.

A. Find the inverse function $C^{-1}(x)$.

B. What do x and $C^{-1}(x)$ represent in the context of the inverse function?

C. How many miles did Peggy drive if her total cost was $39.99?

Real-World Career

The winter months in Chile occur during the summer months in the U.S. due to Chile's location in the southern hemisphere. The average daily high temperature of Santiago during its winter months is about 60° F.

Source: World Weather Information Service

Check Your Understanding ◯ = Step-by-Step Solutions begin on page R11.

Go Online! for a
Self-Check Quiz

Example 1   **Find the inverse of each relation.**

1. $\{(4, -15), (-8, -18), (-2, -16.5), (3, -15.25)\}$

2.

x	−3	0	1	6
y	11.8	3.7	1	−12.5

Example 2   **Graph the inverse of each relation.**

3.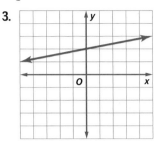

4.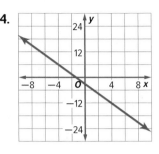

Example 3   **Find the inverse of each function.**

5. $f(x) = -2x + 7$ **6.** $f(x) = \frac{2}{3}x + 6$

Example 4

7. **MP** **REASONING** Dwayne and his brother purchase season tickets to the Dallas Stars games. The ticket package requires a one-time purchase of a personal seat license costing $1200 for two seats. A ticket to each game costs $70. The cost $C(x)$ in dollars for Dwayne for the first season is $C(x) = 600 + 70x$, where x is the number of games Dwayne attends.

   a. Find the inverse function.

   b. What do x and $C^{-1}(x)$ represent in the context of the inverse function?

   c. How many games did Dwayne attend if his total cost for the season was $950?

Practice and Problem Solving
Extra Practice is on page R4.

Example 1   **Find the inverse of each relation.**

8. $\{(-5, 13), (6, 10.8), (3, 11.4), (-10, 14)\}$

9 $\{(-4, -49), (8, 35), (-1, -28), (4, 7)\}$

10.

x	y
−8	−36.4
−2	−15.4
1	−4.9
5	9.1
11	30.1

11.

x	y
−3	7.4
−1	4
1	0.6
3	−2.8
5	−6.2

Example 2   **Graph the inverse of each relation.**

12.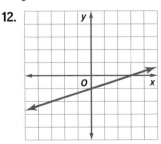

13.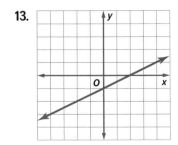

Example 3   Find the inverse of each function.

14. $f(x) = 25 + 4x$

15 $f(x) = 17 - \frac{1}{3}x$

16. $f(x) = 4(x + 17)$

17. $f(x) = 12 - 6x$

18. $f(x) = \frac{2}{5}x + 10$

19. $f(x) = -16 - \frac{4}{3}x$

Example 4   **20. RIDING** An equestrian school charges students a fee of $12 per hour to ride after paying a monthly charge of $50. The total monthly cost $C(x)$ of the service in dollars is $C(x) = 50 + 12x$, where x is the number of hours spent riding.

  a. Find the inverse function.

  b. What do x and $C^{-1}(x)$ represent in the context of the inverse function?

  c. How many hours did the student ride if the monthly bill is $338?

21. LANDSCAPING At the start of the mowing season, Chuck collects a one-time maintenance fee of $25 from his customers. He charges the Fosters $45 for each cut. The total amount collected from the Fosters in dollars for the season is $C(x) = 25 + 45x$, where x is the number of times Chuck mows the Fosters' lawn.

  a. Find the inverse function.

  b. What do x and $C^{-1}(x)$ represent in the context of the inverse function?

  c. How many times did Chuck mow the Fosters' lawn if he collected a total of $1015 from them?

Write the inverse of each equation in $f^{-1}(x)$ notation.

22. $3y - 12x = -72$

23. $x + 5y = 15$

24. $-42 + 6y = x$

25. $3y + 24 = 2x$

26. $-7y + 2x = -28$

27. $3y - x = 3$

MP **TOOLS** Match each function with the graph of its inverse.

A.
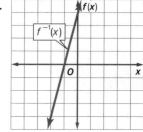

B.
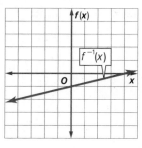

C.
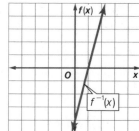

D.
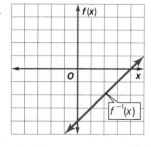

28. $f(x) = x + 4$

29. $f(x) = 4x + 4$

30. $f(x) = \frac{1}{4}x + 1$

31. $f(x) = \frac{1}{4}x - 1$

Write an equation for the inverse function $f^{-1}(x)$ that satisfies the given conditions.

32. slope of $f(x)$ is 7; graph of $f^{-1}(x)$ contains the point (13, 1)

33 graph of $f(x)$ contains the points (−3, 6) and (6, 12)

34. graph of $f(x)$ contains the point (10, 16); graph of $f^{-1}(x)$ contains the point (3, −16)

35. slope of $f(x)$ is 4; $f^{-1}(5) = 2$

36. TV MOVIES Mary Ann pays a monthly fee for her cable package which includes one TV movie a month. She gets billed an additional charge for every additional movie. During her first month, Mary Ann watched 5 additional movies and her bill was $135.50. During her second month, Mary Ann watched 9 additional movies and her bill was $157.50.

  a. Write a function that represents the total monthly cost $C(x)$ of Mary Ann's cable package, where x is the number of additional movies watched.

  b. Find the inverse function.

  c. What do x and $C^{-1}(x)$ represent in the context of the inverse function?

  d. How many additional movies did Mary Ann watch if her bill for her third month was $174?

37. MULTIPLE REPRESENTATIONS In this problem, you will explore the domain and range of inverse functions.

  a. Algebraic Write a function for the area $A(x)$ of the rectangle shown.

  b. Graphical Graph $A(x)$. Describe the domain and range of $A(x)$ in the context of the situation.

  c. Algebraic Write the inverse of $A(x)$. What do x and $A^{-1}(x)$ represent in the context of the situation?

  d. Graphical Graph $A^{-1}(x)$. Describe the domain and range of $A^{-1}(x)$ in the context of the situation.

  e. Logical Determine the relationship between the domains and ranges of $A(x)$ and $A^{-1}(x)$.

8   Area = $A(x)$

$(x - 3)$

H.O.T. Problems   Use **H**igher-**O**rder **T**hinking Skills

38. **MP** **PERSEVERANCE** If $f(x) = 5x + a$ and $f^{-1}(10) = -1$, find a.

39. **MP** **PERSEVERANCE** If $f(x) = \frac{1}{a}x + 7$ and $f^{-1}(x) = 2x - b$, find a and b.

MP **ARGUMENTS** Determine whether the following statements are *sometimes*, *always*, or *never* true. Explain your reasoning.

40. If $f(x)$ and $g(x)$ are inverse functions, then $f(a) = b$ and $g(b) = a$.

41. If $f(a) = b$ and $g(b) = a$, then $f(x)$ and $g(x)$ are inverse functions.

42. OPEN ENDED Give an example of a function and its inverse. Verify that the two functions are inverses by graphing the functions and the line $y = x$ on the same coordinate plane.

43. WRITING IN MATH Explain why it may be helpful to find the inverse of a function.

44. For the linear function $f(x) = \frac{1}{2}x + 5$, what domain value corresponds to a range value of 3? **MP** 7

- **A** -4
- **B** 5
- **C** $6\frac{1}{2}$
- **D** 16

45. The table shows some values of a linear function.

x	−2	0	3	7
y	0	1	2.5	?

What is the missing value in the table? **MP** 7

- **A** 1
- **B** 4
- **C** 4.5
- **D** 5.5

46. For the function $f(x) = -\frac{1}{3}x - 3$, what is the value of x when $f(x) = -6$? **MP** 7

- **A** -5
- **B** -1
- **C** 3
- **D** 9

47. For what value of the domain does $x = f(x)$ if $f(x) = \frac{1}{2}x + 5$? **MP** 1

48. Suppose $f(x) = \frac{2}{3}x - 3$. What is the value of $f^{-1}(-15)$? **MP** 7

49. Suppose $f(x) = 2x + 8$. Through which point does the graph of $f^{-1}(x)$ pass? **MP** 7

- **A** $4, -2$
- **B** $(-4, 2)$
- **C** $(-2, 4)$
- **D** $(0, 8)$

50. The graph shows the adjusted price y for an item with an original price of x after a certain percent increase.

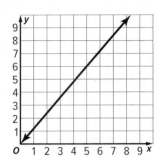

What is the slope of the line that takes the increased price as an input x and returns the original price as the output y? **MP** 4

- **A** $\frac{1}{5}$
- **B** $\frac{5}{6}$
- **C** $\frac{6}{5}$
- **D** 5

51. Which is the inverse of $2x + 3y = 9$ in $f^{-1}(x)$ notation? **MP** 7

- **A** $3x + 2y = 9$
- **B** $f^{-1}(x) = -\frac{2}{3}x + 3$
- **C** $f^{-1}(x) = \frac{3}{2}x + 3$
- **D** $f^{-1}(x) = -\frac{3}{2}x + 4\frac{1}{2}$

52. Daya's new job pays $12 per hour. For the job, she must buy a uniform for $36, which will be taken out of her first paycheck. **MP** 4

a. Write a function to find the first paycheck.

b. What do x and $f(x)$ represent in the context of this function?

c. Write the inverse function.

d. What do x and $f^{-1}(x)$ represent in the context of of the inverse function?

e. How much money does Daya have in her paycheck if she works 24 hours?

f. How many hours would Daya have worked if the paycheck were for $132.00?

Algebra Lab
Drawing Inverses

You can use patty paper to draw the graph of an inverse relation by reflecting the original graph in the line $y = x$.

Mathematical Practices

MP 5 Use appropriate tools strategically.

Activity Draw an Inverse

Work cooperatively. Consider the graphs shown.

Step 1 Trace the graphs onto a square of patty paper, waxed paper, or tracing paper.

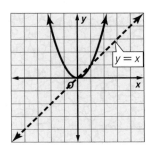

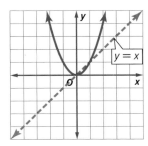

Step 2 Flip the patty paper over and lay it on the original graph so that the traced $y = x$ is on the original $y = x$.

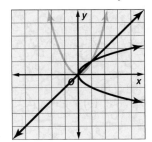

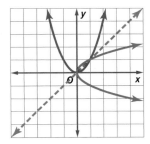

Notice that the result is the reflection of the graph in the line $y = x$ or the inverse of the graph.

Analyze the Results

Work cooperatively.

1. Is the graph of the original relation a function? Explain.

2. Is the graph of the inverse relation a function? Explain.

3. What are the domain and range of the original relation? of the inverse relation?

4. If the domain of the original relation is restricted to $D = \{x \mid x \geq 0\}$, is the inverse relation a function? Explain.

5. If the graph of a relation is a function, what can you conclude about the graph of its inverse?

6. **CHALLENGE** The vertical line test can be used to determine whether a relation is a function. Write a rule that can be used to determine whether a function has an inverse that is also a function.

Go Online! for Vocabulary Review Games and key vocabulary in 13 languages

Study Guide

Key Concepts

Slope-Intercept Form (Lesson 4-1)

- The slope-intercept form of a linear equation is $y = mx + b$, where m is the slope and b is the y-intercept.

Standard and Point-Slope Forms (Lesson 4-2)

- The linear equation $y - y_1 = m(x - x_1)$ is written in point-slope form, where (x_1, y_1) is a given point on a nonvertical line and m is the slope of the line.

Parallel and Perpendicular Lines (Lesson 4-3)

- Nonvertical parallel lines have the same slope.
- Lines that intersect at right angles are called perpendicular lines. The slopes of perpendicular lines are opposite reciprocals.

Scatter Plots and Lines of Fit (Lesson 4-4)

- Data with two variables are called bivariate data.
- A scatter plot is a graph in which two sets of data are plotted as ordered pairs in a coordinate plane.

Correlation and Causation (Lesson 4-5)

- Correlation has a positive, negative, or no relationship between paired data.
- Causation is a relationship that occurs when a change in one variable produces a change in another variable.

Regression and Median-Fit Lines (Lesson 4-6)

- A graphing calculator can be used to find regression lines and median-fit lines.

Inverses of Linear Functions (Lesson 4-7)

- An inverse relation is the set of ordered pairs obtained by exchanging the x-coordinates with the y-coordinates of each ordered pair of a relation.
- A linear function $f(x)$ has an inverse function that can be written as $f^{-1}(x)$ and is read *f of x inverse* or *the inverse of f of x*.

FOLDABLES® Study Organizer

Use your foldable to review the chapter. Working with a partner can be helpful. Ask for clarification of concepts as needed.

Equations of Linear Functions

Key Vocabulary

association (p. 247)	linear interpolation (p. 249)
best-fit line (p. 259)	linear regression (p. 259)
bivariate data (p. 247)	line of fit (p. 248)
causation (p. 254)	median-fit line (p. 262)
constraint (p. 227)	parallel lines (p. 239)
correlation (p. 247)	perpendicular lines (p. 240)
correlation coefficient (p. 259)	point-slope form (p. 233)
inverse function (p. 268)	residual (p. 260)
inverse relation (p. 267)	scatter plot (p. 247)
linear extrapolation (p. 227)	standard form (p. 232)

Vocabulary Check

State whether each sentence is *true* or *false*. If *false*, replace the underlined term to make a true sentence.

1. A <u>constraint</u> is a condition that a solution must satisfy.

2. The process of using a linear equation to make predictions about values that are beyond the range of the data is called <u>linear regression</u>.

3. An <u>inverse relation</u> is the set of ordered pairs obtained by exchanging the x-coordinates with the y-coordinates of each ordered pair of a relation.

4. The <u>correlation coefficient</u> describes whether the correlation between the variables is positive or negative and how closely the regression equation is modeling the data.

5. Lines in the same plane that do not intersect are called <u>parallel</u> lines.

6. A <u>residual</u> is the vertical distance between an observed data value and an estimated value on a line of best fit.

Concept Check

7. Are any two lines graphed in the plane always parallel or perpendicular? (Lesson 4-3)

8. For a given set of data, why does correlation not prove causation? (Lesson 4-5)

9. How are points on the best-fit line used with data? (Lesson 4-6)

Lesson-by-Lesson Review

4-1 Writing Equations in Slope-Intercept Form

Write an equation in slope-intercept form for the line that passes through the given point and has the given slope.

10. $(1, 2)$, slope 3

11. $(2, -6)$, slope -4

12. $(-3, -1)$, slope $\frac{2}{5}$

13. $(5, -2)$, slope $-\frac{1}{3}$

Write an equation in slope-intercept form for the line that passes through the given points.

14. $(2, -1)$, $(5, 2)$

15. $(-4, 3)$, $(1, 13)$

16. $(3, 5)$, $(5, 6)$

17. $(2, 4)$, $(7, 2)$

18. **CAMP** In 2010, a camp had 450 campers. Five years later, the number of campers rose to 750. Write a linear equation in slope-intercept form for the the number of campers that attend camp x years after 2010.

Example 1

Write an equation in slope-intercept form for the line that passes through $(3, 2)$ with a slope of 5.

Step 1 Find the y-intercept.

$y = mx + b$	Slope-intercept form
$2 = 5(3) + b$	$m = 5$, $y = 2$, and $x = 3$
$2 = 15 + b$	Simplify.
$-13 = b$	Subtract 15 from each side.

Step 2 Write the equation in slope-intercept form.

$y = mx + b$	Slope-intercept form
$y = 5x - 13$	$m = 5$ and $b = -13$

4-2 Writing Equations in Standard and Point-Slope Forms

Write an equation in point-slope form for the line that passes through the given point with the slope provided.

19. $(6, 3)$, slope 5

20. $(-2, 1)$, slope -3

21. $(-4, 2)$, slope 0

Write each equation in standard form.

22. $y - 3 = 5(x - 2)$

23. $y - 7 = -3(x + 1)$

24. $y + 4 = \frac{1}{2}(x - 3)$

25. $y - 9 = -\frac{4}{5}(x + 2)$

Write each equation in slope-intercept form.

26. $y - 2 = 3(x - 5)$

27. $y - 12 = -2(x - 3)$

28. $y + 3 = 5(x + 1)$

29. $y - 4 = \frac{1}{2}(x + 2)$

Example 2

Write an equation in point-slope form for the line that passes through $(3, 4)$ with a slope of -2.

$y - y_1 = m(x - x_1)$	Point-slope form
$y - 4 = -2(x - 3)$	Replace m with -2 and (x_1, y_1) with $(3, 4)$.

Example 3

Write $y + 6 = -4(x - 3)$ in standard form.

$y + 6 = -4(x - 3)$	Original equation
$y + 6 = -4x + 12$	Distributive Property
$4x + y + 6 = 12$	Add $4x$ to each side.
$4x + y = 6$	Subtract 6 from each side.

4-3 Parallel and Perpendicular Lines

Write an equation in slope-intercept form for the line that passes through the given point and is parallel to the graph of each equation.

30. $(2, 5)$, $y = x - 3$

31. $(0, 3)$, $y = 3x + 5$

32. $(-4, 1)$, $y = -2x - 6$

33. $(-5, -2)$, $y = -\frac{1}{2}x + 4$

Write an equation in slope-intercept form for the line that passes through the given point and is perpendicular to the graph of the given equation.

34. $(2, 4)$, $y = 3x + 1$

35. $(1, 3)$, $y = -2x - 4$

36. $(-5, 2)$, $y = \frac{1}{3}x + 4$

37. $(3, 0)$, $y = -\frac{1}{2}x$

Example 4

Write an equation in slope-intercept form for the line that passes through $(-2, 4)$ and is parallel to the graph of $y = 6x - 3$.

The slope of the line with equation $y = 6x - 3$ is 6. The line parallel to $y = 6x - 3$ has the same slope, 6.

$y - y_1 = m(x - x_1)$	Point-slope form
$y - 4 = 6[x - (-2)]$	Substitute.
$y - 4 = 6(x + 2)$	Simplify.
$y - 4 = 6x + 12$	Distributive Property
$y = 6x + 16$	Add 4 to each side.

4-4 Scatter Plots and Lines of Fit

38. Determine whether the graph shows a *positive*, *negative*, or *no* correlation. If there is a positive or negative correlation, describe its meaning.

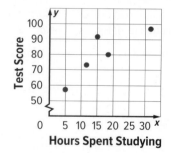

39. BUSINESS A scatter plot of data compares the number of years since a business has opened and its annual number of sales. It contains the ordered pairs $(2, 650)$ and $(5, 1280)$. Write an equation in slope-intercept form for the line of fit for this situation.

Example 5

The scatter plot displays the number of texts and the number of calls made daily. Write an equation for the line of fit.

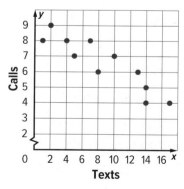

First, find the slope using $(2, 9)$ and $(17, 4)$.

$m = \dfrac{4 - 9}{17 - 2} = \dfrac{-5}{15}$ or $-\dfrac{1}{3}$ Substitute and simplify.

Then find the y-intercept.

$9 = -\dfrac{1}{3}(2) + b$ Substitute.

$9\frac{2}{3} = b$ Add $\frac{2}{3}$ to each side.

Write the equation. $y = -\dfrac{1}{3}x + 9\frac{2}{3}$

4-5 Correlation and Causation

OUTDOORS The table shows the average high temperature, in degrees Fahrenheit, and the number of hours Aiko spent outside each month.

Month	Aug.	Sept.	Oct.	Nov.	Dec.
Temperature (°F)	78	70	64	48	36
Time Outside (hours)	94	75	60	34	20

40. Graph the ordered pairs to create a scatter plot.

41. Does the graph show a *positive*, *negative*, or *no* correlation? Explain.

42. Do you think the change in the average high temperature caused the number of hours Aiko spent outside to change? Explain.

Example 6

ELECTRONICS The table shows the projected number of smartphones and tablets, in millions, in the U.S. from 2014–2018.

Year	2014	2015	2016	2017	2018
Phones	171	191	207	220	229
Tablets	149	159	166	172	177

a. Graph the ordered pairs to create a scatter plot.

b. Describe the correlation, if any, shown by the graph. As the number of phones increase, the number of tablets also increases. So, there is a positive correlation.

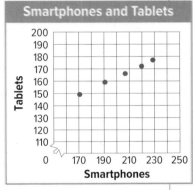

c. Do the data illustrate causation? Explain. No, the number of smartphones increasing does not cause the number of tablets to increase. Other factors, such as increased production or technological advances may increase the number of both.

4-6 Regression and Median-Fit Lines

43. **SALE** The table shows the number of purchases made at an outerwear store during a sale. Write an equation of the regression line. Then estimate the daily purchases on day 10 of the sale.

Days Since Sale Began	1	2	3	4	5	6	7
Daily Purchases	15	21	32	30	40	38	51

44. **MOVIES** The table shows ticket sales at a certain theater during the first week after a movie opened. Write an equation of the regression line. Then estimate the daily ticket sales on the 15th day.

Days Since Movie Opened	1	2	3	4	5	6	7
Daily Ticket Sales ($)	185	192	189	178	165	168	155

Example 7

ATTENDANCE The table shows the annual attendance at an amusement park. Write an equation of the regression line for the data.

Years Since 2008	0	1	2	3	4	5	6
Attendance (thousands)	75	80	72	68	65	60	53

Step 1 Enter the data by pressing ☐ STAT ☐ and selecting the **Edit** option.

Step 2 Perform the regression by pressing ☐ STAT ☐ and selecting the **CALC** option. Scroll down to **LinReg (ax + b)** and press ☐ ENTER ☐.

Step 3 Write the equation of the regression line by rounding the a- and b-values on the screen.
$y = -4.04x + 79.68$

4-7 Inverses of Linear Functions

Find the inverse of each relation.

45. $\{(7, 3.5), (6.2, 8), (-4, 2.7), (-12, 1.4)\}$

46. $\{(1, 9), (13, 26), (-3, 4), (-11, -2)\}$

47.

X	Y
−4	2.7
−1	3.8
0	4.1
3	7.2

48.

X	Y
−12	4
−8	0
−4	−4
0	−8

Find the inverse of each function.

49. $f(x) = \frac{5}{11}x + 10$

50. $f(x) = 3x + 8$

51. $f(x) = -4x - 12$

52. $f(x) = \frac{1}{4}x - 7$

53. $f(x) = -\frac{2}{3}x + \frac{1}{4}$

54. $f(x) = -3x + 3$

55. $f(x) = 3x$

56. $f(x) = -x$

57. $f(x) = 4x - 7$

58. $f(x) = \frac{3x + 4}{7}$

59. $f(x) = x$

60. $f(x) = -5x - \frac{6}{7}$

61. $f(x) = \frac{2}{5}x - \frac{1}{7}$

62. $f(x) = 2 - 3x$

63. $f(x) = -3x$

64. $f(x) = 2 + \frac{3x}{5}$

65. $f(x) = \frac{-3x - 1}{8}$

66. $f(x) = -\frac{3}{5}x$

67. $f(x) = x - 1$

Example 8

Find the inverse of the relation.

$$\{(5, -3), (11, 2), (-6, 12), (4, -2)\}$$

To find the inverse, exchange the coordinates of the ordered pairs.

$(5, -3) \rightarrow (-3, 5)$ $(-6, 12) \rightarrow (12, -6)$

$(11, 2) \rightarrow (2, 11)$ $(4, -2) \rightarrow (-2, 4)$

The inverse is $\{(-3, 5), (2, 11), (12, -6), (-2, 4)\}$.

Example 9

Find the inverse of $f(x) = \frac{1}{4}x + 9$.

$f(x) = \frac{1}{4}x + 9$	Original equation
$y = \frac{1}{4}x + 9$	Replace $f(x)$ with y.
$x = \frac{1}{4}y + 9$	Interchange y and x.
$x - 9 = \frac{1}{4}y$	Subtract 9 from each side.
$4(x - 9) = y$	Multiply each side by 4.
$4x - 36 = y$	Distributive Property
$4x - 36 = f^{-1}(x)$	Replace y with $f^{-1}(x)$.

Example 10

Find the inverse of $f(x) = \frac{2x + 15}{5}$.

$f(x) = \frac{9x + 15}{5}$	Original equation
$y = \frac{2x + 15}{5}$	Replace $f(x)$ with y.
$x = \frac{2y + 15}{5}$	Interchange y and x.
$5x = 2y + 15$	Multiply each side by 5.
$2y = 9x - 15$	Subtract 15 from each side.
$y = \frac{5x - 15}{2}$	Divide each side by 2.
$f^{-1}(x) = \frac{5x - 15}{2}$	Replace y with $f^{-1}(x)$.

Write an equation of a line in slope-intercept form that passes through the given point and has the given slope.

1. $(-4, 2)$; slope -3 **2.** $(3, -5)$; slope $\frac{2}{3}$

Write an equation of the line in slope-intercept form that passes through the given points.

3. $(1, 4), (3, 10)$ **4.** $(2, 5), (-2, 8)$

5. $(0, 4), (-3, 0)$ **6.** $(7, -1), (9, -4)$

The figure shows trapezoid EFGH.

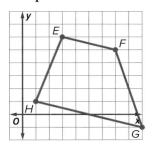

7. Write an equation in point-slope form for the line containing $\overline{EF}$.

8. Write an equation in standard form for the line containing $\overline{FG}$.

9. PAINTING The data in the table show the size of a room in square feet and the time it takes to paint the room in minutes.

Room Size	100	150	200	400	500
Painting Time	160	220	270	500	680

  a. Use the points $(100, 160)$ and $(500, 680)$ to write an equation in slope-intercept form.

  b. Predict the amount of time required to paint a room measuring 750 square feet.

10. SALARY The table shows the relationship between years of experience and teacher salary.

Years Experience	1	5	10	15	20
Salary (thousands of dollars)	28	31	42	49	64

  a. Write an equation for the best-fit line.

  b. Find the correlation coefficient and explain what it tells us about the relationship between experience and salary.

Write an equation in slope-intercept form for the line that passes through the given point and is parallel to the graph of each equation.

11. $(2, -3), y = 4x - 9$

12. $(-5, 1), y = -3x + 2$

Write an equation in slope-intercept form for the line that passes through the given point and is perpendicular to the graph of the equation.

13. $(1, 4), y = -2x + 5$ **14.** $(-3, 6), y = \frac{1}{4}x + 2$

15. MULTIPLE CHOICE The graph shows the relationship between outside temperature and daily ice cream cone sales. What type of correlation is shown?

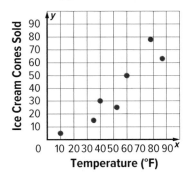

  A positive correlation

  B negative correlation

  C no correlation

  D not enough information

16. ADOPTION The table shows the number of children from Ethiopia adopted by U.S. citizens.

Years Since 2000	5	7	9	10	11
Number of Children	442	1254	2275	2511	1732

  a. Write the slope-intercept form of the equation for the line of fit.

  b. Predict the number of children from Ethiopia who will be adopted in 2025.

Find the inverse of each function.

17. $f(x) = -5x - 30$

18. $f(x) = 4x + 10$

19. $f(x) = \frac{1}{6}x - 2$

20. $f(x) = \frac{3}{4}x + 12$

Performance Task

Provide a clear solution to each part of the task. Be sure to show all work, include all relevant drawings, and justify your answers.

TEMPERATURE CHANGES Scientists often work with models to analyze current temperature trends or predict future changes.

Part A

A scientist in China found that the average annual temperature has risen over the last 100 years and calculated the following equation to model the change: $y = 0.08x + 4$.

1. **Structure** Graph this equation on a coordinate plane. Let x be the number of years and let y be the temperature, in degrees Celsius, on the y-axis.

2. Write the equation in point-slope form.

Part B

A scientist in Siberia also calculated the average annual temperature in her region over the last 100 years. The graph shows her findings.

3. Find the slope and y-intercept of the graph.

4. Write the equation of the line shown in the graph.

5. **Precision** According to the graph and the equation you wrote, what would you predict the average annual temperature to be after 150 years? Explain your reasoning.

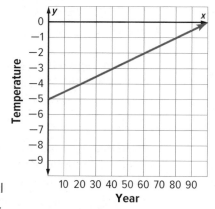

Part C

Three scientists came up with the following three equations to represent the average annual temperatures of their respective regions: $y = 0.04x + 18$, $y = -25x + 65$, and $y + 1.96 = 0.04(x - 1)$.

6. Determine whether the equations represent lines that are parallel, perpendicular, or intersect with one another.

Part D

A scientist in the United States records the following average annual temperatures in (x, y) form, where x represents the number of years after the annual temperatures began to be recorded and y represents the average annual temperature in degrees Fahrenheit: (100, 56.0), (90, 54.75), (80, 54.5), (70, 53.25), (60, 53.0), (50, 52.75), (40, 51.75), (30, 50.25), (20, 49.75), and (10, 49.25).

7. Use a graphing calculator to find the equation of the regression line for the data. Round all values to the nearest hundredth.

8. What is the correlation coefficient of the data? Round your answer to the nearest ten-thousandth.

9. Use the regression line you calculated above to predict the average annual temperature in the United States 180 years after the start of the study. Round your answer to the nearest hundredth.

10. Find the inverse of the function represented by the regression line you found above. Round all values to the nearest hundredth.

Test-Taking Strategy

Example

Read the problem. Identify what you need to know. Then use the information in the problem to solve.

The table shows production costs for building different numbers of skateboards. Which of the following best models the information given in the table?

Skateboards Built	Production Costs
14	$325
18	$375
22	$425
28	$500

A $y = 24x + 100$

C $y = 18x + 150$

B $y = 12.5x + 150$

D $y = 12x + 100$

Test-Taking Tip

Look for a Pattern One of the most common problem-solving strategies is to look for a pattern. The ability to recognize patterns, model them algebraically, and extend them is a valuable problem-solving tool.

Step 1 **What do you notice about the values in the table?**
As the number of skateboards built increases, so do the production costs.

Step 2 **How are the terms of the pattern related?**
The rate of change between each pair of points is the same. The terms form a linear relationship. For each additional skateboard built, the production costs increase $12.50.

Step 3 **How can you write a rule to describe the pattern?**
The data can be modeled by a linear equation, $y = mx + b$, where x is the number of skateboards built and y is the production costs. The slope is 12.5. To find the y-intercept, substitute any point from the table and solve for b. The equation is $y = 12.5x + 150$.

Step 4 **What is the correct answer?**
The correct answer is B.

Step 5 **How can you check that your answer is correct?**
I can substitute any of the points in the table into the equation to see if the statement is true. I can also graph the points to see if the line that passes through them has the same slope and y-intercept in the equation.

Apply the Strategy

Read the problem. Identify what you need to know. Then use the information in the problem to solve. Show your work.

What is the equation of the line shown in the graph?

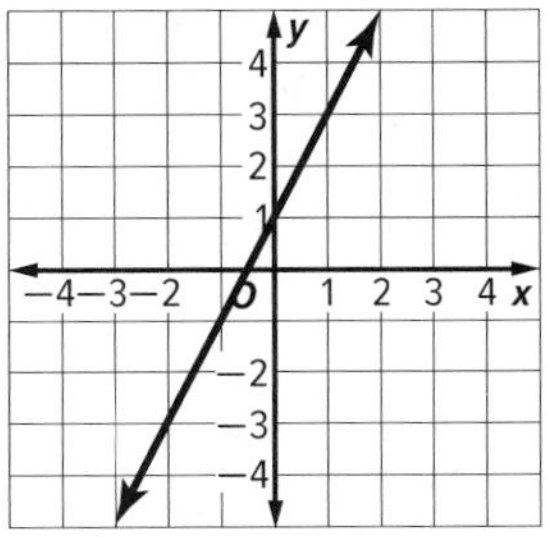

A $y = -2x + 1$

C $y = 2x - \frac{1}{2}$

B $y = \frac{1}{2}x + 1$

D $y = 2x + 1$

Answer the questions below.

a. What do you notice about the x- and y-coordinates of the graph?

b. How are the coordinates related?

c. How can you write a rule to describe the pattern between points?

d. What is the correct answer?

e. How can you check that your answer is correct?

Read each question. Then fill in the correct answer on the answer document provided by your teacher or on a sheet of paper.

1. 10-packs of gum are on sale for $4, but there is a store limit of 5 packs per customer. The function $C(x)$ gives the cost C for x 10-packs. Which of the following best represents the range of this function?

    A {4}

    B {5}

    C {0, 1, 2, 3, 4, 5}

    D {0, 4, 8, 12, 16, 20}

2. What is the y-intercept for a line with a slope of -2 that passes through the point (3.6, 7.4)?

    [ ]

3. Which of the following best expresses the equation $3x + 2y = 18$ in slope-intercept form?

    A $3x = -2y + 18$

    B $y - 2 = 3(x + 18)$

    C $y = -\frac{3}{2}x + 9$

    D $y = \frac{3}{2}x + 9$

4. Which of the following best represents the line graphed below?

    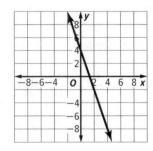

    A $y = -x + 4$

    B $y = 4x + 1$

    C $y = \frac{1}{3}x + 4$

    D $y = -3x + 4$

5. Which of the situations illustrates causation?

    A the number of pages in a book and the cost of the book

    B the number of banks in a town and the number of post offices in the town

    C the number of full tennis ball canisters and the total number of tennis balls

    D the number of miles driven in a car and the gallons of gas consumed by the car

    E the number of toasters sold in a store and the number of shoes sold in the store each year

6. Find the inverse of the function $f(x) = 8 - 3x$.

    [ ]

7. Which of the following best represents the line that passes through (2, 10) and is parallel to the graph of $y = -2x + 3$?

    A $y = \frac{1}{2}x + 9$

    B $y = -2x$

    C $y = -2x + 14$

    D $y = -2x - 6$

8. Zhao writes the arithmetic sequence below.

    $$4\frac{1}{2}, 3, 1\frac{1}{2}, 0, -1\frac{1}{2}, \ldots$$

    Which expression represents the nth term?

    A $\frac{10 - n}{2}$

    B $6 - \frac{3n}{2}$

    C $-0.5n$

    D $4.5 - 1.5n$

Test Taking Tip

Question 8 Remember that the first term of the sequence corresponds to $n = 1$.

9. Which statement best describes the graphs of the two linear functions below?

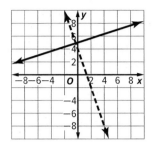

- ○ A The lines are perpendicular, because their slopes are opposites.

- ○ B The lines are perpendicular, because their slopes are negative reciprocals.

- ○ C The lines are perpendicular, because the product of their slopes is 1.

- ○ D The lines intersect but are not perpendicular.

10. Aimee finds the best-fit line and correlation coefficient r for some data. If $r \approx -0.97$, which statements are true?

- ☐ A The data show a nonlinear correlation.

- ☐ B The data show a weak linear correlation.

- ☐ C The data show a strong linear correlation.

- ☐ D The data show a positive correlation.

- ☐ E The data show a negative correlation.

11. Which of the following lines is parallel to the y-axis and passes through $(-3, 5)$?

- ○ A $y = -3$

- ○ B $y = 5$

- ○ C $x = -3$

- ○ D $x = 5$

12. Solve $\dfrac{n - 1}{4} = 21$

$n = $ ⬚

13. Part of the graph of the linear function $f(x)$ is shown below.

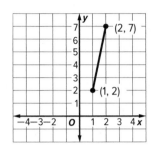

What is the x-intercept of the function?

- ○ A -3

- ○ B 0

- ○ C $\dfrac{3}{5}$

- ○ D 1

14. George cuts a plastic strip into two pieces, one of which is 3 inches longer than the other. He then cuts the longer piece in half. Two of the three pieces are each 1 inch longer than the third piece.

a. How long was the original plastic strip?

⬚ inches

b. **MP** What mathematical practice did you use to solve this problem?

15. What is the slope of a line that is parallel to the line with equation $3x - 4y = 5$?

Need Extra Help?

If you missed Question...	1	2	3	4	5	6	7	8	9	10	11	12	13	14	15
Go to Lesson...	1-7	4-1	4-1	3-4	4-5	4-7	4-3	3-6	4-3	4-6	4-3	2-3	4-2	2-1	4-3

THEN

You solved linear equations.

NOW

In this chapter, you will:

- Solve one-step and multi-step inequalities.

- Solve compound inequalities and inequalities involving absolute value.

- Graph inequalities in two variables.

 WHY

PETS In the United States, about 78.2 million dogs are kept as pets. Approximately 23% of these were adopted from animal shelters. About 44% of all households own at least one dog.

Use the Mathematical Practices to complete the activity.

1. **Attend to Precision** Survey your class to determine how many people have pets, how many of these pets are dogs, and how many have pets other than dogs.

2. **Reasoning** How could you use the information from your survey to determine the percentage or students who have pets and the percentage who have dogs? Explain.

3. **Model with Mathematics** Pick one element of the data you collected and use the Fraction Models tool in ConnectED to show this data.

4. **Discuss** Compare the different data collected. How might this data be represented in one equation? Can you think of other ways to represent this data?

Go Online to Guide Your Learning

| Explore & Explain | Organize |

 Graphing Tools

Investigate the graphs of linear inequalities, as discussed in Lesson 5-6, by using the **Graphing Tools** in ConnectED. These are outstanding tools to enhance your understanding.

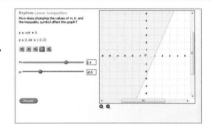

 Foldables

Linear Inequalities Make this Foldable to help you organize your Chapter 5 notes about linear inequalities.

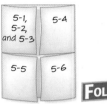

eToolkit

You can use the number line **Virtual Manipulative** on the eToolkit to graph inequalities. Create the number line, then use the pen and segment tools to graph the inequality.

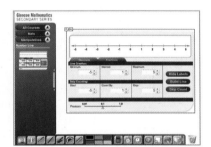

| Collaborate |

 **Chapter Project**

In the **Caring For Our BFFs** project, you will use what you have learned about inequalities to complete an entrepreneurial literacy project.

eBook

Interactive Student Guide

Before starting the chapter, answer the **Chapter Focus** preview questions. Check your answers as you complete each lesson. At the end of the chapter, try the **Performance Task**.

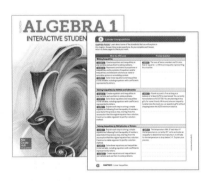

| Focus |

 LEARNSMART

Need help studying? Complete the **Linear and Exponential Relationships** and **Relationships Between Quantities and Reasoning with Equations** domains in LearnSmart to review for the chapter test.

ALEKS

You can use the **Linear Inequalities** topic in ALEKS to find out what you know about linear functions and what you are ready to learn.*

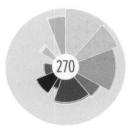

* Ask your teacher if this is part of your program.

Get Ready for the Chapter

Go Online! for Vocabulary Review Games and key vocabulary in 13 languages.

Connecting Concepts

Concept Check

Review the concepts used in this chapter by answering the questions below.

1. Claudia opened a savings account with $325. She saves $100 per month. Write an equation to explain how many dollars d she has after m months.

2. Which property can be used to justify that $7 < m$ can be written as $m > 7$?

3. How is solving inequalities like solving equations?

4. A survey shows 85% of students at a university have cars. If the results were reported within 2% accuracy, what are the maximum and minimum percent of students who have cars?

5. The expression mi/gal represents the gas mileage of a car. Find the gas mileage of a car that goes 295 miles on 12 gallons of gasoline. Round to the nearest tenth.

6. Jenny's cabin is at least 9 miles away. Write an inequality to find the time it will take Jenny to get to the cabin if she walks at a rate of $\frac{3}{4}$ mile per hour.

New Vocabulary

English		Español
inequality	p. 289	desigualdad
compound inequality	p. 309	desigualdad compuesta
intersection	p. 309	intersección
union	p. 310	unión
boundary	p. 321	frontera
half-plane	p. 321	semiplano
closed half-plane	p. 321	semiplano cerrada
open half-plane	p. 321	semiplano abierto

Performance Task Preview

Understanding linear equalities will help you finish the Performance Task at the end of the chapter. You will use the concepts and skills in this chapter to calculate taxi cab fare and to help a taxi driver address his budgetary concerns.

MP In this Performance Task you will:

- reason abstractly and quantitatively
- construct an argument
- make sense of problems

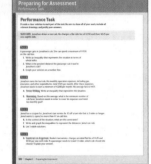

Review Vocabulary

equivalent equations ecuaciones equivalentes equations that have the same solution

linear equation ecuación lineal an equation in the form $Ax + By = C$, with a graph consisting of points on a straight line

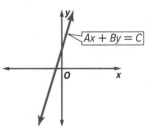

solution set conjunto solución the set of elements from the replacement set that makes an open sentence true

Solving Inequalities by Addition and Subtraction

Then	**Now**	**Why?**

- You solved equations by using addition and subtraction.

- **1** Solve linear inequalities by using addition.
- **2** Solve linear inequalities by using subtraction.

- Ashley is having lunch at a bistro. She has selected an appetizer from the menu that costs $6.50 and now needs to decide which sandwiches she could order so that she stays within the $15 she budgeted for lunch.

$$6.5 + w \leq 15$$
$$\underline{-6.5 \qquad -6.5}$$
$$w \leq 8.5$$

Ashley needs to order a sandwich that is $8.50 or less.

Sandwich	Price
BLT	$8.75
Club	$9.25
Turkey	$7.50
Rueben	$10.00
Veggie	$7.00

 New Vocabulary
inequality
set-builder notation

 Mathematical Practices
2 Reason abstractly and quantitatively.
4 Model with mathematics.

1 **Solve Inequalities by Addition** An open sentence that contains $<$, $>$, $\leq$, or $\geq$ is an **inequality**. The example above illustrates the Addition Property of Inequalities.

> **Key Concept** Addition Property of Inequalities
>
> **Words** If the same number is added to each side of a true inequality, the resulting inequality is also true.
>
> **Symbols** For all numbers a, b, and c, the following are true.
> **1.** If $a > b$, then $a + c > b + c$.
> **2.** If $a < b$, then $a + c < b + c$.

This property is also true for $\geq$ and $\leq$.

Example 1 Solve by Adding

Solve $x - 12 \geq 8$. Check your solution.

$x - 12 \geq 8$	Original inequality
$x - 12 + 12 \geq 8 + 12$	Add 12 to each side.
$x \geq 20$	Simplify.

The solution is the set {all numbers greater than or equal to 20}.

CHECK To check, substitute three different values into the original inequality: 20, a number less than 20, and a number greater than 20.

> **Guided Practice**

Solve each inequality. Check your solution.

1A. $22 > m - 8$

1B. $d - 14 \geq -19$

Recall that a more concise way of writing a solution set is to use set-builder notation. In set-builder notation, the solution set in Example 1 is $\{x \mid x \geq 20\}$.

This solution set can be graphed on a number line. Be sure to check if the endpoint of the graph of an inequality should be a circle or a dot. If the endpoint is not included in the graph, use a circle; otherwise use a dot.

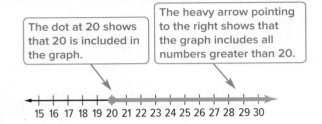

The dot at 20 shows that 20 is included in the graph.

The heavy arrow pointing to the right shows that the graph includes all numbers greater than 20.

15 16 17 18 19 20 21 22 23 24 25 26 27 28 29 30

2 Solve Inequalities by Subtraction
Subtraction can also be used to solve inequalities.

🔑 Key Concept Subtraction Property of Inequalities

Words If the same number is subtracted from each side of a true inequality, the resulting inequality is also true.

Symbols For all numbers a, b, and c, the following are true.

1. If $a > b$, then $a - c > b - c$.

2. If $a < b$, then $a - c < b - c$.

This property is also true for $\geq$ and $\leq$.

Example 2   Solve by Subtracting

Solve $m + 19 > 56$.

A $\{m \mid m < 75\}$ **B** $\{m \mid m < 37\}$ **C** $\{m \mid m > 37\}$ **D** $\{m \mid m > 75\}$

Read the Item

You need to find the solution set for the inequality.

Solve the Item

Step 1 Solve the inequality.

$$m + 19 > 56 \qquad \text{Original inequality}$$

$$m + 19 - 19 > 56 - 19 \qquad \text{Subtract 19 from each side.}$$

$$m > 37 \qquad \text{Simplify.}$$

Step 2 Write in set-builder notation: $\{m \mid m > 37\}$. The answer is C.

CHECK You can verify the solution by substituting 37 back into the original inequality and then testing values to confirm the direction of the inequality symbol.

▶ **Guided Practice**

2. Solve $p + 8 \leq 18$.

   A $\{p \mid p \geq 10\}$ **B** $\{p \mid p \leq 10\}$ **C** $\{p \mid p \leq 26\}$ **D** $\{p \mid p \geq 126\}$

Terms that are constants are not the only terms that can be subtracted. Terms with variables can also be subtracted from each side to solve inequalities.

Example 3   Variables on Each Side

Solve $3a + 6 \leq 4a$. Then graph the solution set on a number line.

$$3a + 6 \leq 4a \qquad \text{Original inequality}$$
$$3a - 3a + 6 \leq 4a - 3a \qquad \text{Subtract } 3a \text{ from each side.}$$
$$6 \leq a \qquad \text{Simplify.}$$

Since $6 \leq a$ is the same as $a \geq 6$, the solution set is $\{a \mid a \geq 6\}$.

Guided Practice

Solve each inequality. Then graph the solution set on a number line.

3A. $9n - 1 < 10n$ **3B.** $5h \leq 12 + 4h$

Verbal problems containing phrases like *greater than* or *less than* can be solved by using inequalities. The chart shows some other phrases that indicate inequalities.

Concept Summary   Phrases for Inequalities

<	>	≤	≥
less than fewer than	greater than more than	at most, no more than, less than or equal to	at least, no less than, greater than or equal to

Real-World Example 4   Use an Inequality to Solve a Problem

PETS Felipe needs for the temperature of his leopard gecko's basking spot to be at least 82°F. Currently the basking spot is 62.5°F. How much warmer does the basking spot need to be?

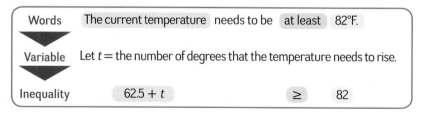

$$62.5 + t \geq 82 \qquad \text{Original inequality}$$
$$62.5 + t - 62.5 \geq 82 - 62.5 \qquad \text{Subtract 62.5 from each side.}$$
$$t \geq 19.5 \qquad \text{Simplify.}$$

Felipe needs to raise the temperature of the basking spot 19.5°F or more.

Guided Practice

4. SHOPPING Sanjay has $65 to spend at the mall. He bought a T-shirt for $18 and a belt for $14. If Sanjay wants a pair of jeans, how much can he spend?

Check Your Understanding ◯ = Step-by-Step Solutions begin on page R11.

✓ **Go Online!** for a Self-Check Quiz

Examples 1–3 Solve each inequality. Then graph the solution set on a number line.

1. $x - 3 > 7$
2. $5 \geq 7 + y$
3. $g + 6 < 2$
4. $11 \leq p + 4$
5. $10 > n - 1$
6. $k + 24 > -5$
7. $8r + 6 < 9r$
8. $8n \geq 7n - 3$

Example 4 Define a variable, write an inequality, and solve each problem. Check your solution.

9. Twice a number increased by 4 is at least 10 more than the number.

10. Three more than a number is less than twice the number.

11. **AMUSEMENT** A thrill ride swings passengers back and forth, a little higher each time up to 137 feet. Suppose the height of the swing after 30 seconds is 45 feet. How much higher will the ride swing?

Practice and Problem Solving   Extra Practice is on page R5.

Examples 1–3 Solve each inequality. Then graph the solution set on a number line.

12. $m - 4 < 3$
13. $p - 6 \geq 3$
14. $r - 8 \leq 7$
15. $t - 3 > -8$
16. $b + 2 \geq 4$
17. $13 > 18 + r$
18. $5 + c \leq 1$
19. $-23 \geq q - 30$
20. $11 + m \geq 15$
21. $h - 26 < 4$
22. $8 \leq r - 14$
23. $-7 > 20 + c$
24. $2a \leq -4 + a$
25. $z + 4 \geq 2z$
26. $w - 5 \leq 2w$
27. $3y + 6 \leq 2y$
28. $6x + 5 \geq 7x$
29. $-9 + 2a < 3a$

Example 4 Define a variable, write an inequality, and solve each problem. Check your solution.

30. Twice a number is more than the sum of that number and 9.

31. The sum of twice a number and 5 is at most 3 less than the number.

32. The sum of three times a number and −4 is at least twice the number plus 8.

33. Six times a number decreased by 8 is less than five times the number plus 21.

MP MODELING Define a variable, write an inequality, and solve each problem. Then interpret your solution.

34. **FINANCIAL LITERACY** Keisha is babysitting at $8 per hour to earn money for a car. So far she has saved $1300. The car that Keisha wants to buy costs at least $5440. How much money does Keisha still need to earn to buy the car?

35. **SURVEY** A recent survey at Seth's school found that more than 210 people in the junior class like the Indiana Hoosiers basketball team. Of those, 50 will attend a game this season. How many of the juniors like the Indiana Hoosiers basketball team, but will not be attending a game this season?

36. **TABLET** Stephanie added 20 more apps to her tablet, making the total more than 61. How many apps were originally on the tablet?

37. TEMPERATURE The water temperature in a swimming pool increased 4°F this morning. The temperature is now less than 81°F. What was the water temperature this morning?

38. BASKETBALL A player's goal was to score at least 150 points this season. So far, she has scored 123 points. If there is one game left, how many points must she score to reach her goal?

39 SPAS Samantha received a $75 gift card for a local day spa for her birthday. She plans to get a haircut and a manicure. How much money will be left on her gift card after her visit?

Service	Cost ($)
haircut	at least 32
manicure	at least 26

40. VOLUNTEER Kono knows that he can only volunteer up to 25 hours per week. If he has volunteered for the times recorded at the right, how much more time can Kono volunteer this week?

Center	Time (h)
Shelter	3 h 15 min
Kitchen	2 h 20 min

Solve each inequality. Check your solution, and then graph it on a number line.

41. $c + (-1.4) \geq 2.3$

42. $9.1g + 4.5 < 10.1g$

43. $k + \frac{3}{4} > \frac{1}{3}$

44. $\frac{3}{2}p - \frac{2}{3} \leq \frac{4}{9} + \frac{1}{2}p$

45. MULTIPLE REPRESENTATIONS In this problem, you will explore multiplication and division in inequalities.

   a. Geometric Suppose a balance has 12 pounds on the left side and 18 pounds on the right side. Draw a picture to represent this situation.

   b. Numerical Write an inequality to represent the situation.

   c. Tabular Create a table showing the result of doubling, tripling, or quadrupling the weight on each side of the balance. Create a second table showing the result of reducing the weight on each side of the balance by a factor of $\frac{1}{2}$, $\frac{1}{3}$, or $\frac{1}{4}$. Include a column in each table for the inequality representing each situation.

   d. Verbal Describe the effect multiplying or dividing each side of an inequality by the same positive value has on the inequality.

MP REASONING If $m + 7 \geq 24$, then complete each inequality.

46. $m \geq \underline{\ ?\ }$

47. $m + \underline{\ ?\ } \geq 27$

48. $m - 5 \geq \underline{\ ?\ }$

49. $m - \underline{\ ?\ } \geq 14$

50. $m - 19 \geq \underline{\ ?\ }$

51. $m + \underline{\ ?\ } \geq 43$

H.O.T. Problems   Use Higher-Order Thinking Skills

52. MP REASONING Compare and contrast the graphs of $a < 4$ and $a \leq 4$.

53. CHALLENGE Suppose $b > d + \frac{1}{3}$, $c + 1 < a - 4$, and $d + \frac{5}{8} > a + 2$. Order a, b, c, and d from least to greatest.

54. OPEN ENDED Write three linear inequalities that are equivalent to $y < -3$.

55. WRITING IN MATH Summarize the process of solving and graphing linear inequalities.

56. WRITING IN MATH Explain why $x - 2 > 5$ has the same solution set as $x > 7$.

57. At the school store, Raul bought 4 notebooks and 2 pens. Helen bought 5 notebooks. If Helen spent at least as much as Raul, which statement about the costs of notebooks and pens is true? **MP** 4

- ○ **A** If pens cost $1.50, then notebooks must cost at least $5.

- ○ **B** If pens cost $1.50, then notebooks must cost more than $5.

- ○ **C** If pens cost $2.50, then notebooks must cost at least $5.

- ○ **D** If pens cost $2.50, then notebooks must cost no more than $5.

58. A florist sells bouquets of carnations and roses. Carnations cost $2 each, and roses cost $5 each. The total cost of each bouquet is no more than $25. Which best describes the quantities of carnations and roses in each bouquet? **MP** 4

- ○ **A** 3 carnations, at least 5 roses

- ○ **B** 3 carnations, no more than 5 roses

- ○ **C** 5 carnations, at least 3 roses

- ○ **D** 5 carnations, no more than 3 roses

59. Which inequality does not have a solution set of $\{x \mid x < -1\}$? **MP** 2

- ○ **A** $9x - 1 > 10x$

- ○ **B** $9x + 1 > 10x - 2$

- ○ **C** $10x + 1 < 9x$

- ○ **D** $10x - 1 < 9x - 2$

60. Marta scores 85, 70, and 92 on three math tests. What does she need to score on her fourth math test to have an average of 85 or better for all four tests? **MP** 4

- ○ **A** 85 or better   ○ **C** 93 or better

- ○ **B** better than 93   ○ **D** 100

61. Blaine solved an inequality and graphed its solution set on the number line below.

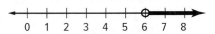

Which inequality did Blaine solve? **MP** 2, 3

- ○ **A** $5(x - 3) > 15$

- ○ **B** $5(x + 3) > 15$

- ○ **C** $5(x - 3) < 15$

- ○ **D** $5(x + 3) < 15$

62. **MULTI-STEP** On Monday, Marlene withdraws $50 from an ATM, writes a check for her cable bill for $160, and uses her debit card to buy $42.25 worth of groceries. At the end of the day, her account balance is at least $400. **MP** 2, 4

a. Which of the inequalities can be used to model Marlene's account activity for Monday?

- ☐ **A** $x - 202.25 \leq 350$

- ☐ **B** $x - 252.25 \geq 400$

- ☐ **C** $x \geq 752.25$

- ☐ **D** $x - 202.25 > 350$

- ☐ **E** $x - 252.25 \leq 400$

- ☐ **F** $x - 92.25 \geq 240$

b. What is the least amount that Marlene has in her account on Monday before her first transaction?

| |

c. Write an inequality that models Marlene's balance before her first transaction.

| |

d. Use set-builder notation to write your answer to part **c**.

| |

e. Graph your answer to part c on a number line.

You can use algebra tiles to solve inequalities.

Mathematical Practices
MP 7 Look for and make use of Structure

Activity Solve Inequalities

Work cooperatively. Solve $-2x \le 4$.

Step 1 Use a self-adhesive note to cover the equals sign on the equation mat. Then write a $\le$ symbol on the note. Model the inequality.

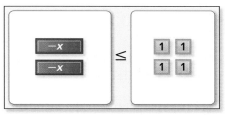

$-2x \le 4$

Step 2 Since you do not want to solve for a negative x-tile, eliminate the negative x-tiles by adding 2 positive x-tiles to each side. Remove the zero pairs.

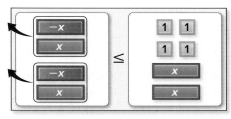

$-2x + 2x \le 4 + 2x$

Step 3 Add 4 negative 1-tiles to each side to isolate the x-tiles. Remove the zero pairs.

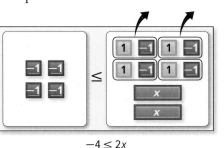

$-4 \le 2x$

Step 4 Separate the tiles into 2 groups.

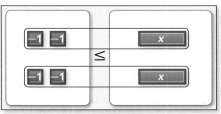

$-2 \le x$ or $x \ge -2$

Model and Analyze Work cooperatively. Use algebra tiles to solve each inequality.

1. $-3x < 9$
2. $-4x > -4$
3. $-5x \ge 15$
4. $-6x \le -12$

5. In Exercises 1–4, is the coefficient of x in each inequality positive or negative?

6. Compare the inequality symbols and locations of the variable in Exercises 1–4 with those in their solutions. What do you find?

7. Model the solution for $3x \le 12$. How is this different from solving $-3x \le 12$?

8. Write a rule for solving inequalities involving multiplication and division. (*Hint:* Remember that dividing by a number is the same as multiplying by its reciprocal.)

Solving Inequalities by Multiplication and Division

··Then	··Now	··Why?
• You solved equations by using multiplication and division.	**1** Solve linear inequalities by using multiplication. **2** Solve linear inequalities by using division.	• Terry received a gift card for $20 to her favorite frozen yogurt shop. If each ounce of frozen yogurt costs $0.55 per ounce, the number of ounces of frozen yogurt she can purchase can be represented by the inequality $0.55y \le 20$.

 Mathematical Practices

1 Make sense of problems and persevere in solving them.

6 Attend to precision.

1 **Solve Inequalities by Multiplication** If you multiply each side of an inequality by a positive number, then the inequality remains true.

$$4 > 2$$ Original inequality

$$4(3) \underset{\text{?}}{=} 2(3)$$ Multiply each side by 3.

$$12 > 6$$ Simplify.

Notice that the direction of the inequality remains the same.

If you multiply each side of an inequality by a negative number, the inequality symbol changes direction.

$$7 < 9$$ Original inequality

$$7(-2) \underset{\text{?}}{=} 9(-2)$$ Multiply each side by −2.

$$-14 > -18$$ Simplify.

These examples demonstrate the Multiplication Property of Inequalities.

Key Concept Multiplication Property of Inequalities

Words	Symbols	Examples
If both sides of an inequality that is true are multiplied by a positive number, the resulting inequality is also true.	For any real numbers a and b and any positive real number c, if a > b, then $ac > bc$. And, if a < b, then $ac < bc$.	$6 > 3.5$ $6(2) > 3.5(2)$ $12 > 7$ and $2.1 < 5$ $2.1(0.5) < 5(0.5)$ $1.05 < 2.5$
If both sides of an inequality that is true are multiplied by a negative number, the direction of the inequality sign is reversed to make the resulting inequality also true.	For any real numbers a and b and any negative real number c, if a > b, then $ac < bc$. And, if a < b, then $ac > bc$.	$7 > 4.5$ $7(-3) < 4.5(-3)$ $-21 < -13.5$ and $3.1 < 5.2$ $3.1(-4) > 5.2(-4)$ $-12.4 > -20.8$

This property also holds for inequalities involving $\le$ and $\ge$.

Real-World Example 1　Write and Solve an Inequality

SURVEYS Of the students surveyed at Madison High School, fewer than 84 said they have never purchased an item online. This is about one eighth of those surveyed. How many students were surveyed?

Understand You know the number of students who have never purchased an item online and the portion this is of the number of students surveyed.

Plan Let n = the number of students surveyed. Write an open sentence that represents this situation.

One eighth of the number of students surveyed is less than 84 can be written as $\frac{1}{8} \cdot n < 84$.

Solve Solve for n.

$$\frac{1}{8}n < 84 \qquad \text{Original inequality}$$

$$(8)\frac{1}{8}n < (8)84 \qquad \text{Multiply each side by 8.}$$

$$n < 672 \qquad \text{Simplify.}$$

Fewer than 672 students were surveyed.

Check Check the endpoint with 672 and the direction of the inequality with a value less than 672.

$$\frac{1}{8}(672) \stackrel{?}{=} 84 \qquad \text{Check endpoint.} \qquad \frac{1}{8}(0) \stackrel{?}{<} 84 \qquad \text{Check direction.}$$

$$84 = 84 \checkmark \qquad\qquad\qquad\qquad 0 < 84 \checkmark$$

You could also check the solution by substituting a number greater than 672 and verifying that the resulting inequality is false. 672 is a reasonable number for a high school survey.

▶ **Guided Practice**

1. **BIOLOGY** Mount Kinabalu in Malaysia has the greatest concentration of wild orchids on Earth. It contains more than 750 species, or about one fourth of all orchid species in Malaysia. How many orchid species are there in Malaysia?

You can also use multiplicative inverses with the Multiplication Property of Inequalities to solve an inequality.

Example 2　Solve by Multiplying

Solve $-\frac{3}{7}r < 21$. Graph the solution on a number line.

$$-\frac{3}{7}r < 21 \qquad \text{Original inequality}$$

$$\left(-\frac{7}{3}\right)\left(-\frac{3}{7}r\right) > \left(-\frac{7}{3}\right)21 \qquad \text{Multiply each side by } -\frac{7}{3}. \text{ Reverse the inequality symbol.}$$

$$r > -49 \qquad \text{Simplify. Check by substituting values.}$$

The solution set is $\{r \mid r > -49\}$.

$$\overset{\longleftarrow}{\underset{-51\ -50\ -49\ -47\ -45\ -43\ -41}{|+|+|+|\circ+|+|+|+|+|\longrightarrow}}$$

▶ **Guided Practice**

Solve each inequality. Graph the solution.

2A. $-\frac{n}{6} \leq 8$ 　　**2B.** $-\frac{4}{3}p > -10$ 　　**2C.** $\frac{1}{5}m \geq -3$ 　　**2D.** $\frac{3}{8}t < 5$

2 Solve Inequalities by Division

If you divide each side of an inequality by a positive number, then the inequality remains true. Notice that the direction of the inequality remains the same.

$$-10 < -5 \quad \text{Original inequality}$$
$$\frac{-10}{5} \; \underset{?}{} \; \frac{-5}{5} \quad \text{Divide each side by } -5.$$
$$-2 < -1 \quad \text{Simplify.}$$

If you divide each side of an inequality by a negative number, the inequality symbol changes direction.

$$15 < 18 \quad \text{Original inequality}$$
$$\frac{15}{-3} \; \underset{?}{} \; \frac{18}{-3} \quad \text{Divide each side by } -3.$$
$$-5 > -6 \quad \text{Simplify.}$$

These examples demonstrate the Division Property of Inequalities.

Watch Out!

Negatives A negative sign in an inequality does not necessarily mean that the direction of the inequality should change. For example, when solving $\frac{x}{6} > -3$, do not change the direction of the inequality.

Math History Link

Thomas Harriot
(1560–1621) Harriot was a prolific astronomer. He was the first to map the Moon's surface and to see sunspots. Harriot is best known for his work in algebra.

🔑 Key Concept Division Property of Inequalities

Words	Symbols	Examples	
If both sides of a true inequality are divided by a positive number, the resulting inequality is also true.	For any real numbers a and b and any positive real number a, if $a > b$, then $\frac{a}{c} > \frac{b}{c}$. And, if $a < b$, then $\frac{a}{c} < \frac{b}{c}$.	$4.5 > 2.1$ $\frac{4.5}{3} > \frac{2.1}{3}$ $1.5 > 0.7$	and	$1.5 < 5$ $\frac{1.5}{0.5} < \frac{5}{0.5}$ $3 < 10$
If both sides of a true inequality are divided by a negative number, the direction of the inequality sign is reversed to make the resulting inequality also true.	For any real numbers a and b, and any negative real number c, if $a > b$, then $\frac{a}{c} < \frac{b}{c}$. And, if $a < b$, then $\frac{a}{c} < \frac{b}{c}$.	$6 > 2.4$ $\frac{6}{-6} < \frac{2.4}{-6}$ $-1 < -0.4$	and	$-1.8 < 3.6$ $\frac{-1.8}{-9} < \frac{3.6}{-9}$ $0.2 > -0.4$

This property also holds true for inequalities involving $\leq$ and $\geq$.

Example 3 Divide to Solve an Inequality

Solve each inequality. Graph the solution on a number line.

a. $60t > 8$

$$60t > 8 \quad \text{Original inequality}$$
$$\frac{60t}{60} > \frac{8}{60} \quad \text{Divide each side by 60.}$$
$$t > \frac{2}{15} \quad \text{Simplify.}$$
$$\left\{ t \mid t > \frac{2}{15} \right\}$$

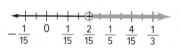

b. $-7d \leq 147$

$$-7d \leq 147 \quad \text{Original inequality}$$
$$\frac{-7d}{-7} \geq \frac{147}{-7} \quad \text{Divide each side by } -7. \text{ Reverse the inequality symbol.}$$
$$d \geq -21 \quad \text{Simplify.}$$
$$\{ d \mid d \geq -21 \}$$

▶ **Guided Practice**

3A. $8p \leq 58$

3C. $-12h > 15$

3B. $-42 \geq 6r$

3D. $-\frac{1}{2}n \leq 6$

Check Your Understanding ◯ = Step-by-Step Solutions begin on page R11.

✓ **Go Online!** for a Self-Check Quiz

Example 1

1. **FUNDRAISING** The Jefferson High School Band Boosters raised more than $5500 from sales of $12 T-shirts. Define a variable, and write an inequality to represent the number of T-shirts they sold. Solve the inequality and interpret the results.

Examples 2–3 Solve each inequality. Graph the solution on a number line.

2. $30 > \frac{1}{2}n$

3. $-\frac{3}{4}r \le -6$

4. $-\frac{c}{6} \ge 7$

5. $\frac{h}{2} < -5$

6. $9t > 108$

7. $-84 < 7v$

8. $-28 \le -6x$

9. $40 \ge -5z$

Practice and Problem Solving

Extra Practice is on page R5.

Example 1

Define a variable, write an inequality, and solve each problem. Then interpret your solution.

10. **BUSINESS** Mario is the manager at a local store. He has budgeted $1600 for the wages of the cashiers for the week. If each cashier makes $8.75 an hour, how many total hours can the cashiers work that week?

11. **FINANCIAL LITERACY** Rodrigo needs at least $560 to pay for his spring break expenses, and he is saving $25 from each of his weekly paychecks. How long will it be before he can pay for his trip?

Examples 2–3 Solve each inequality. Graph the solution on a number line.

12. $\frac{1}{4}m \le -17$

⑬ $\frac{1}{2}a < 20$

14. $-11 > -\frac{c}{11}$

15. $-2 \ge -\frac{d}{34}$

16. $-10 \le \frac{x}{-2}$

17. $-72 < \frac{f}{-6}$

18. $\frac{2}{3}h > 14$

19. $-\frac{3}{4}j \ge 12$

20. $-\frac{1}{6}n \le -18$

21. $6p \le 96$

22. $4r < 64$

23. $32 > -2y$

24. $-26 < 26t$

25. $-6v > -72$

26. $-33 \ge -3z$

27. $4b \le -3$

28. $-2d < 5$

29. $-7f > 5$

30. **CHEERLEADING** To remain on the cheerleading squad, Lakita must attend at least $\frac{3}{5}$ of the study table sessions offered. She attends 15 sessions. If Lakita met the requirements, what is the maximum number of study table sessions?

31. **BRACELETS** How many bracelets can Caitlin buy for herself and her friends if she wants to spend no more than $22?

$4.75

32. **MP** **PRECISION** The National Honor Society at Pleasantville High School wants to raise at least $500 for a local charity. Each student earns $0.50 for every quarter of a mile walked in a walk-a-thon. How many miles will the students need to walk?

33. **MUSEUM** The American history classes are planning a trip to a local museum. Admission is $8 per person. Determine how many people can go for $260.

34. **GASOLINE** Suppose gasoline costs $2.69 per gallon. To the nearest tenth, how many gallons of gasoline can Jan buy for $24.

Match each inequality to the graph of its solution.

35. $-\frac{2}{3}h \le 9$ **36.** $25j \ge 8$ **37.** $3.6p < -4.5$ **38.** $2.3 < -5t$

a.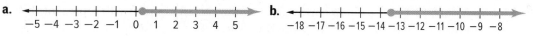

b.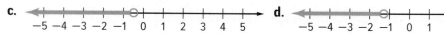

c.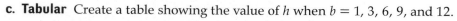

d.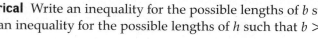

39 **CANDY** Fewer than 42 employees at a factory stated that they preferred fudge over fruit candy. This is about two thirds of the employees. How many employees are there?

40. EMPLOYMENT A certain bank employs more than 275 people at all of its branches. Approximately three fifths of all the people are employed at the west branch. How many people work at the west branch?

41. MULTIPLE REPRESENTATIONS The equation for the volume of a pyramid is $\frac{1}{3}$ the area of the base times the height.

   a. Geometric Draw a pyramid with a square base b cm long and a height of h cm.

   b. Numerical Suppose the pyramid has a volume of 72 cm^3. Write an equation to find the height.

   c. Tabular Create a table showing the value of h when $b = 1, 3, 6, 9,$ and 12.

   d. Numerical Write an inequality for the possible lengths of b such that $b < h$. Write an inequality for the possible lengths of h such that $b > h$.

H.O.T. Problems   Use **H**igher-**O**rder **T**hinking Skills

42. ERROR ANALYSIS Taro and Jamie are solving $6d \ge -84$. Is either of them correct? Explain your reasoning.

Taro	Jamie
$6d \ge -84$	$6d \ge -84$
$\dfrac{6d}{6} \ge \dfrac{-84}{6}$	$\dfrac{6d}{6} \le \dfrac{-84}{6}$
$d \ge -14$	$d \le -14$

43. CHALLENGE Solve each inequality for x. Assume that $a > 0$.

   a. $-ax < 5$ **b.** $\frac{1}{a}x \ge 8$ **c.** $-6 \ge ax$

44. (MP) **STRUCTURE** Determine whether $x^2 > 1$ and $x > 1$ are equivalent. Explain.

45. (MP) **REASONING** Explain whether the statement *If $a > b$, then $\frac{1}{a} > \frac{1}{b}$* is *sometimes, always,* or *never* true.

46. OPEN ENDED Create a real-world situation to represent the inequality $-\frac{5}{8} \ge x$.

47. (e) **WRITING IN MATH** How are solving linear inequalities and linear equations similar? different?

48. Norman must make 20 party grab bags for a total cost of less than $75. Each grab bag contains an identical fruit bar and an identical prize. Which inequality can be used to model the situation? Let f be fruit bars and p be prizes. **MP** 4

- ○ **A** $f + p \geq 3.75$
- ○ **B** $f + p < 75$
- ○ **C** $20(f + p) < 75$
- ○ **D** $20f + 20p \leq 75$

49. Fyodor solved an inequality that has a solution set of $\{n \mid n \geq -5\}$. Which inequality did he solve? **MP** 3

- ○ **A** $-\frac{2}{5}n \leq 2$
- ○ **B** $-\frac{2}{5}n \geq 2$
- ○ **C** $-\frac{5}{2}n \leq 2$
- ○ **D** $-\frac{5}{2}n \geq 2$

50. Which inequality does not have the solution set graphed below? **MP** 6

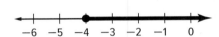

- ○ **A** $5x \geq -20$
- ○ **B** $5x \leq -20$
- ○ **C** $-5x \leq 20$
- ○ **D** $20 \geq -5x$

51. Each week, Donna earns b dollars babysitting and t dollars tutoring. She saves 10% of what she earns. She wants to save at least $25 in one week. If she earns a total of $100 babysitting in that week, what is the least amount of money, in dollars, she will have to earn tutoring in that week? **MP** 1, 4

[   ]

52. Which of the following is the solution to the inequality $-0.25y \leq -4$ **MP** 6

- ○ **A** $\{y \mid y \leq 16\}$
- ○ **B** $\{y \mid y \geq 16\}$
- ○ **C** $\{y \mid y \leq 1\}$
- ○ **D** $\{y \mid y \geq 1\}$

53. A jar contains at least 600 marbles. Two-thirds of the marbles are red. The rest are blue. How many marbles are blue? **MP** 1

- ○ **A** at least 200 marbles
- ○ **B** no more than 200 marbles
- ○ **C** at least 400 marbles
- ○ **D** no more than 400 marbles

54. MULTI-STEP Brianna is saving $15 each week to purchase a new tablet for $389.99. **MP** 1, 4

a. Write an inequality that models the minimum number of weeks n Brianna needs to save.

[   ]

b. What is the least number of weeks Brianna needs to save?

[   ]

c. Suppose Brianna has saved $75 so far. She starts working a part-time job and discovers that she can now save $25 each week. Which of the following is true?

- ☐ **A** $75 + 25n \geq 389.99$
- ☐ **B** $25n \geq 464.99$
- ☐ **C** $25n \geq 314.99$
- ☐ **D** $n \geq 13$
- ☐ **E** $n \geq 19$
- ☐ **F** $n \leq 13$

d. How many weeks will Brianna need to save after starting her new job?

[   ]

e. Suppose 10 weeks after starting her job, the tablet goes on sale for 20% off the original price. Will Brianna be able to buy the tablet? Explain.

[   ]

55. Maya has 85 geodes in her collection, which is over 4 times more than she had a year ago. $85 \geq 4x$. At most how many geodes did Maya have a year ago? **MP** 4, 6

- ○ **A** 20
- ○ **B** 21
- ○ **C** 22
- ○ **D** 81

∷Then

- You solved multi-step equations.

∷Now

1. Solve linear inequalities involving more than one operation.

2. Solve linear inequalities involving the Distributive Property.

∷Why?

- A salesperson may make a base monthly salary and earn a commission on each of her sales. To find the number of sales she needs to make to pay her monthly bills, you can use a multi-step inequality.

 Mathematical Practices

7 Look for and make use of structure.

1 Solve Multi-Step Inequalities Multi-step inequalities can be solved by undoing the operations in the same way you would solve a multi-step equation.

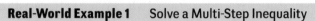

Real-World Example 1 Solve a Multi-Step Inequality

SALES Write and solve an inequality to find the sales Mrs. Jones needs if she earns a monthly salary of $2000 plus a 10% commission on her sales. Her goal is to make at least $4000 per month. What sales does she need to meet her goal?

base salary + (commission × sales) ≥ income needed

$$2000 + 0.10x \geq 4000 \qquad \text{Substitution}$$
$$0.10x \geq 2000 \qquad \text{Subtract 2000 from each side.}$$
$$x \geq 20{,}000 \qquad \text{Divide each side by 0.10.}$$

She must make at least $20,000 in sales to meet her monthly goal.

▸ **Guided Practice**

1. **FINANCIAL LITERACY** The Print Shop advertises a special to print 400 flyers for less than the competition. The price includes a $3.50 set-up fee. If the competition charges $35.50, what does the Print Shop charge for each flyer?

When multiplying or dividing by a negative number, the direction of the inequality symbol changes. This holds true for multi-step inequalities.

Example 2 Inequality Involving a Negative Coefficient

Solve $-11y - 13 > 42$. **Graph the solution on a number line.**

$$-11y - 13 > 42 \qquad \text{Original inequality}$$
$$-11y > 55 \qquad \text{Add 13 to each side and simplify.}$$
$$\frac{-11y}{-11} < \frac{55}{-11} \qquad \text{Divide each side by } -11, \text{ and reverse the inequality.}$$
$$y < -5 \qquad \text{Simplify.}$$

The solution set is $\{y \mid y < -5\}$.

-10 -8 -6 -4 -2 0 2

▸ **Guided Practice** **Solve each inequality. Graph the solution on a number line.**

2A. $23 \geq 10 - 2w$

2B. $43 > -4y + 11$

You can translate sentences into multi-step inequalities and then solve them using the Properties of Inequalities.

Example 3   Write and Solve an Inequality

Define a variable, write an inequality, and solve the problem.

Three minus eleven times a number is more than twenty-one minus five times the number.

Let n be the number.

Three	minus	eleven times a number	is more than	twenty-one	minus	five times the number.
3	$-$	$11n$	$>$	21	$-$	$5n$

$3 - 6n > 21$ Add $5n$ to each side and simplify.

$-6n > 18$ Subtract 8 from each side and simplify.

$\dfrac{-6n}{-6} < \dfrac{18}{-6}$ Divide each side by -6, and reverse the inequality.

$n < -3$ Simplify.

The solution set is $\{n \mid n < -3\}$.

▶ **Guided Practice**

3. *Two more than half of a number is greater than twenty-seven.*

---

2 **Solve Inequalities Involving the Distributive Property** When solving inequalities that contain grouping symbols, use the Distributive Property to remove the grouping symbols first. Then use the order of operations to simplify the resulting inequality.

Example 4   Distributive Property

Solve $4(3t - 5) + 7 \geq 8t + 3$. Graph the solution on a number line.

$4(3t - 5) + 7 \geq 8t + 3$ Original inequality

$12t - 20 + 7 \geq 8t + 3$ Distributive Property

$12t - 13 \geq 8t + 3$ Combine like terms.

$4t - 13 \geq 3$ Subtract $8t$ from each side and simplify.

$4t \geq 16$ Add 13 to each side.

$\dfrac{4t}{4} \geq \dfrac{16}{4}$ Divide each side by 4.

$t \geq 4$ Simplify.

The solution set is $\{t \mid t \geq 4\}$.

   −2   0   2   4   6   8   10   12

▶ **Guided Practice**

Solve each inequality. Graph the solution on a number line.

4A. $6(5z - 3) \leq 36z$ **4B.** $2(h + 6) > -3(8 - h)$

Go Online!

You can use the number line **Virtual Manipulative** on the eToolkit to graph inequalities. Create the number line, then use the pen and segment tools to graph the inequality.

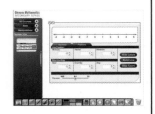

If solving an inequality results in a statement that is always true, the solution set is the set of all real numbers. This solution set is written as $\{x \mid x \text{ is a real number.}\}$. If solving an inequality results in a statement that is never true, the solution set is the empty set, which is written as the symbol $\varnothing$. The empty set has no members.

Example 5 Empty Set and All Real Numbers

Solve each inequality. Check your solution.

a. $9t - 5(t - 5) \leq 4(t - 3)$

$9t - 5(t - 5) \leq 4(t - 3)$	Original inequality
$9t - 5t + 25 \leq 4t - 12$	Distributive Property
$4t + 25 \leq 4t - 12$	Combine like terms.
$4t + 25 - 4t \leq 4t - 12 - 4t$	Subtract $4t$ from each side.
$25 \leq -12$	Simplify.

Since the inequality results in a false statement, the solution set is the empty set, $\varnothing$.

b. $3(4m + 6) \leq 42 + 6(2m - 4)$

$3(4m + 6) \leq 42 + 6(2m - 4)$	Original inequality
$12m + 18 \leq 42 + 12m - 24$	Distributive Property
$12m + 18 \leq 12m + 18$	Combine like terms.
$12m + 18 - 12m \leq 12m + 18 - 12m$	Subtract $12m$ from each side.
$18 \leq 18$	Simplify.

All values of m make the inequality true. All real numbers are solutions.

> **Guided Practice**

Solve each inequality. Check your solution.

5A. $18 - 3(8c + 4) \geq -6(4c - 1)$ **5B.** $46 \leq 8m - 4(2m + 5)$

Study Tip

MP Structure Notice that the inequality $4t + 25 < 4t - 12$ means *some number $4t$ plus 25 is less than or equal to that number minus 12*. No real number makes that inequality true. Observing the meaning of the expressions in each step in this way can lead you to solutions more quickly.

Check Your Understanding   = Step-by-Step Solutions begin on page R11.

Go Online! for a Self-Check Quiz

Example 1

1. **ELEVATORS** If 15 people with 225 total pounds of luggage plan to ride an elevator with a 3000-pound weight limit, write and solve an inequality to find the allowable average weight per person.

2. **PARTY** Rita is ordering a pizza for $14.49 and a few 2-liter bottles of soda. She has $25 to spend. The cost of delivery and tip is $5. If each 2-liter bottle costs $1.99, write and solve an inequality to find the greatest number of bottles she can buy.

Example 2

MP STRUCTURE Solve each inequality. Graph the solution on a number line.

3 $6h - 10 \geq 32$ **4.** $-3 \leq \frac{2}{3}r + 9$

5. $-3x + 7 > 43$ **6.** $4m - 17 < 6m + 25$

Example 3

Define a variable, write an inequality, and solve each problem. Then check your solution.

7. Four times a number minus 6 is greater than eight plus two times the number.

8. Negative three times a number plus 4 is less than five times the number plus 8.

Examples 4–5 Solve each inequality. Graph the solution on a number line.

9. $-6 \leq 3(5v - 2)$ **10.** $-5(g + 4) > 3(g - 4)$ **11.** $3 - 8x \geq 9 + 2(1 - 4x)$

Examples 1–2   **MP** **STRUCTURE** Solve each inequality. Graph the solution on a number line.

12. $5b - 1 \geq -11$ **13.** $21 > 15 + 2a$

14. $-9 \geq \frac{2}{5}m + 7$ **15.** $\frac{w}{8} - 13 > -6$

16. $-a + 6 \leq 5$ **17.** $37 < 7 - 10w$

18. $8 - \frac{z}{3} \geq 11$ **19.** $-\frac{5}{4}p + 6 < 12$

20. $3b - 6 \geq 15 + 24b$ **21.** $15h + 30 < 10h - 45$

Example 3   **Define a variable, write an inequality, and solve each problem. Check your solution.**

22. Three fourths of a number decreased by nine is at least forty-two.

23. Two thirds of a number added to six is at least twenty-two.

24. Seven tenths of a number plus 14 is less than forty-nine.

25. Eight times a number minus twenty-seven is no more than the negative of that number plus eighteen.

26. Ten is no more than 4 times the sum of twice a number and three.

27. Three times the sum of a number and seven is greater than five times the number less thirteen.

28. The sum of nine times a number and fifteen is less than or equal to the sum of twenty-four and ten times the number.

Examples 4–5   **MP** **STRUCTURE** Solve each inequality. Graph the solution on a number line.

29. $-3(7n + 3) < 6n$ **30.** $21 \geq 3(a - 7) + 9$

31. $2y + 4 > 2(3 + y)$ **32.** $3(2 - b) < 10 - 3(b - 6)$

33. $7 + t \leq 2(t + 3) + 2$ **34.** $8a + 2(1 - 5a) \leq 20$

Define a variable, write an inequality, and solve each problem. Then interpret your solution.

35. **CARS** A car salesperson is paid a base salary of $35,000 a year plus 8% of sales. What are the sales needed to have an annual income greater than $65,000?

36. **CARGO** Keith has 40-pound bags of mulch in his truck that weigh a total of 3600 pounds. His Owner's Manual lists the truck's capacity as at most 3000 pounds. How many bags does Keith need to remove in order to meet the weight requirements?

37. Solve $6(m - 3) > 5(2m + 4)$. Show each step and justify your work.

38. Solve $8(a - 2) \leq 10(a + 2)$. Show each step and justify your work.

39. **MUSICAL** A high school drama club is performing a musical to benefit a local charity. Tickets are $5 each. They also received donations of $565. They want to raise at least $1500.

     a. Write an inequality that describes this situation. Then solve the inequality.

     b. Graph the solution.

40. **ICE CREAM** Benito has $6 to spend. A sundae costs $3.25 plus $0.65 per topping. Write and solve an inequality to find how many toppings he can order.

41 SCIENCE The normal body temperature of a camel is 97.7°F in the morning. If it has had no water by noon, its body temperature can be greater than 104°F.

   a. Write an inequality that represents a camel's body temperature at noon if the camel had no water.

   b. If C represents degrees Celsius, then $F = \frac{9}{5}C + 32$. Write and solve an inequality to find the camel's body temperature at noon in degrees Celsius.

42. NUMBER THEORY Find all sets of three consecutive positive even integers with a sum no greater than 36.

43. NUMBER THEORY Find all sets of four consecutive positive odd integers with a sum that is less than 42.

Solve each inequality. Check your solution.

44. $2(x - 4) \leq 2 + 3(x - 6)$

45. $\frac{2x - 4}{6} \geq -5x + 2$

46. $5.6z + 1.5 < 2.5z - 4.7$

47. $0.7(2m - 5) \geq 21.7$

GRAPHING CALCULATOR Use a graphing calculator to solve each inequality.

48. $3x + 7 > 4x + 9$

49. $13x - 11 \leq 7x + 37$

50. $2(x - 3) < 3(2x + 2)$

51. $\frac{1}{2}x - 9 < 2x$

52. $2x - \frac{2}{3} \geq x - 22$

53. $\frac{1}{3}(4x + 3) \geq \frac{2}{3}x + 2$

54. MULTIPLE REPRESENTATIONS In this problem, you will solve compound inequalities. A number x is greater than 4, and the same number is less than 9.

   a. Numerical Write two separate inequalities for the statement.

   b. Graphical Graph the solution set for the first inequality in red. Graph the solution set for the second inequality in blue. Highlight where they overlap.

   c. Tabular Make a table using ten points from your number line, including points from each section. Use one column for each inequality and a third column titled "Both are True." Complete the table by writing true or false.

   d. Verbal Describe the relationship between the colored regions of the graph and the chart.

   e. Logical Make a prediction of what the graph of $4 < x < 9$ looks like.

H.O.T. Problems   Use **H**igher-**O**rder **T**hinking Skills

55. MP REASONING Explain how you could solve $-3p + 7 \geq -2$ without multiplying or dividing each side by a negative number.

56. CHALLENGE If $ax + b < ax + c$ is true for all real values of x, what will be the solution of $ax + b > ax + c$? Explain how you know.

57. CHALLENGE Solve each inequality for x. Assume that $a > 0$.
   a. $ax + 4 \geq -ax - 5$ **b.** $2 - ax < x$ **c.** $-\frac{2}{a}x + 3 > -9$

58. WHICH ONE DOESN'T BELONG? Name the inequality that does not belong. Explain.

$4y + 9 > -3$	$3y - 4 > 5$	$-2y + 1 < -5$	$-5y + 2 < -13$

59. WRITING IN MATH Explain when the solution set of an inequality will be the empty set or the set of all real numbers. Show an example of each.

60. MULTI-STEP Jed's online music club allows him to download 25 songs per month for $14.99. Additional songs cost $1.29 each. **MP** 4,7

a. Write an inequality to represent this situation. Let t be his monthly spending limit and m represent the total number of songs downloaded.

- A $1.29m \leq t + 10.01$
- B $1.29m \leq t + 17.26$
- C $1.29t \leq m + 10.01$
- D $1.29t \leq m + 17.26$

b. Which of the following statements best describes the number of songs Jed can download each month without going over his spending limit?

- A Jed will not go over his spending limit of $30 if he downloads 37 songs.
- B Jed will not go over his spending limit of $40 if he downloads 45 songs.
- C Jed can download a maximum of 36 songs if his budget is $30.
- D Jed can download a maximum of 36 songs if his budget is $40.

c. Suppose that the music club changes its plan so that 50 songs can be downloaded a month. How will the inequality representing this inequality change?

- A The coefficient of t will decrease.
- B The coefficient of t will increase.
- C The coefficient of m will increase.
- D The constant will increase.

61. Sonia's doctor said to call if her temperature is over 38°C. Her thermometer shows temperature in degrees Fahrenheit. To convert, she uses the formula

$$C = \frac{5}{9}(F - 32),$$

where C is degrees Celsius and F is degrees Fahrenheit. To the nearest tenth, what is the least Fahrenheit temperature for which Sonia should call her doctor? **MP** 7

62. Which of the inequalities does not have $\{x | x > 2\}$ as its solution set? **MP** 1

- A $3x + 1 > 2x + 3$
- B $5 > -x + 7$
- C $-2x + 1 > 5$
- D $7x - 1 > 6x + 1$

63. Tyrone solved the inequality below and graphed its solution set on a number line.

$$\frac{2x}{3} > 4x + 10$$

Which is Tyrone's graph? **MP** 1

- A
- B
- C
- D

64. MULTI-STEP Use the following inequality to answer the following questions.
$-5(x - 2) < 2x + 8 + 3x$ **MP** 7

a. Which statements are true?

- ☐ A The solution set is the empty set.
- ☐ B The solution set is all real numbers.
- ☐ C $-5x + 10 < 5x + 8$ is an equivalent inequality.
- ☐ D $5x - 10 < 5x + 8$ is an equivalent inequality.
- ☐ E The number line is partially shaded.

b. Suppose the solution is restricted to negative numbers. What is the solution?

c. Suppose the solution is restricted to integers only. Graph the solution on a number line.

65. It costs $30 to rent a moving van plus $0.75 per mile. What is the maximum distance that can be driven to keep the total cost below $50? **MP** 4,7

- A 20
- C 26
- B 24
- D 27

Solve each inequality. Then graph it on a number line. (Lesson 5-1)

1. $x - 8 > 4$

2. $m + 2 \geq 6$

3. $p - 4 < -7$

4. $12 \leq t - 9$

5. CONCERTS Lupe's allowance for the month is $60. She wants to go to a concert for which a ticket costs $45. (Lesson 5-1)

    a. Write and solve an inequality that shows how much money she can spend that month after buying a concert ticket.

    b. She spends $9.99 on music downloads and $2 on lunch in the cafeteria. Write and solve an inequality that shows how much she can spend after these purchases and the concert ticket.

    c. **MP** What mathematical practice did you use to solve this problem?

Define a variable, write an inequality, and solve each problem. Check your solution. (Lesson 5-1)

6. The sum of a number and -2 is no more than 6.

7. A number decreased by 4 is more than -1.

8. Twice a number increased by 3 is less than the number decreased by 4.

9. MULTIPLE CHOICE Jane is saving money to buy a new smartphone that costs no more than $150. So far, she has saved $82. How much more money does Jane need to save? (Lesson 5-1)

    A $68

    B more than $68

    C no more than $68

    D at least $68

Solve each inequality. Check your solution. (Lesson 5-2)

10. $\frac{1}{3}y \geq 5$

11. $4 < \frac{c}{5}$

12. $-8x > 24$

13. $2m \leq -10$

14. $\frac{x}{2} < \frac{5}{8}$

15. $-9a \geq -45$

16. $\frac{w}{6} > -3$

17. $\frac{k}{7} < -2$

18. ANIMALS Black-tailed prairie dogs are commonly found throughout the Great Plains of North America. Adults can be up to 21 inches long, including their tails, and weigh up to 3 pounds. (Lesson 5-2)

    a. Write inequalities to describe the lengths and weights of black-tailed prairie dogs.

    b. If a black-tailed prairie dog's tail is one seventh of its total length, write and solve an inequality that describes the lengths of black-tailed prairie dogs' tails.

19. GARDENING Bill is building a fence around a square garden to keep deer out. He has 60 feet of fencing. Find the maximum length of a side of the garden. (Lesson 5-2)

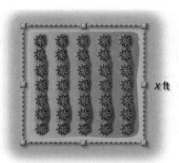

x ft

Solve each inequality. Check your solution. (Lesson 5-3)

20. $4a - 2 > 14$

21. $2x + 11 \leq 5x - 10$

22. $-p + 4 < -9$

23. $\frac{d}{4} + 1 \geq -3$

24. $-2(4b + 1) < -3b + 8$

Define a variable, write an inequality, and solve each problem. Check your solution. (Lesson 5-3)

25. Three times a number increased by 8 is no more than the number decreased by 4.

26. Two thirds of a number plus 5 is greater than 17.

27. MULTIPLE CHOICE Shoe rental costs $2, and each game bowled costs $3. How many games can Kyle bowl without spending more than $15? (Lesson 5-3)

    A 2

    B 3

    C 4

    D 5

You hear and use the words *and* as well as *or* every day. Those words have special use in math. A compound statement is made up of two simple statements connected by the word *and* or *or*. Before you can determine whether a compound statement is true or false, you must understand what the words *and* and *or* mean.

Mathematical Practices
MP **2** Reason abstractly and quantitatively

A spider has eight legs, *and* a dog has five legs.
For a compound statement connected by the word *and* to be true, both simple statements must be true.

A spider has eight legs. ⟶ true

A dog has five legs. ⟶ false

Since one of the statements is false, the compound statement is false.

A compound statement connected by the word *or* may be *exclusive* or *inclusive*. For example, the statement "With your lunch, you may have milk *or* juice," is exclusive. In everyday language, *or* means one or the other, but not both. However, in mathematics, *or* is inclusive. It means one or the other or both.

A spider has eight legs, *or* a dog has five legs.
For a compound statement connected by the word *or* to be true, at least one of the simple statements must be true. Since it is true that a spider has eight legs, the compound statement is true.

Exercises

Work cooperatively. Is each compound statement *true* or *false*? Explain.

1. Most top 20 movies in 2010 were rated PG, *or* most top 20 movies in 2008 were rated G.

2. In 2011, more top 20 movies were rated PG than were rated G, *and* more were rated PG than rated PG-13.

3. For the years shown most top 20 movies are rated PG-13, *and* the least top 20 movies are rated G.

4. $11 < 5$ or $9 < 7$

5. $-2 > 0$ and $3 < 7$

6. $5 > 0$ and $-3 < 0$

7. $-2 > -3$ or $0 = 0$

8. $8 \neq 8$ or $-2 > -5$

9. $5 > 10$ and $4 > -2$

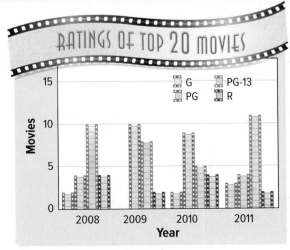

RATINGS OF TOP 20 MOVIES

G   PG-13
PG   R

Source: National Association of Theater Owners

Solving Compound Inequalities

:::Then

- You solved absolute value equations with two cases.

:::Now

1. Solve compound inequalities containing the word *and*, and graph their solution set.

2. Solve compound inequalities containing the word *or*, and graph their solution set.

:::Why?

- In 2014, the National Museum of Natural History in Washington, D.C., displayed its first Tyrannosaurus rex skeleton in the museum's history. The fossils were discovered by a hiker in Montana and are one of the most complete T. rex skeletons ever found. The fossils are estimated to be at least 65 million years old but less than 68 million years old. This can be written as a compound inequality $65 \leq t < 68$, where t is the age of the fossil in millions.

New Vocabulary

compound inequality
intersection
union

Mathematical Practices

1 Make sense of problems and persevere in solving them.

8 Look for and express regularity in repeated reasoning.

1 **Inequalities Containing *and*** When considered together, two inequalities such as $v \geq 4$ and $v < 18$ form a **compound inequality**. A compound inequality containing *and* is only true if both inequalities are true. Its graph is where the graphs of the two inequalities overlap. This is called the **intersection** of the two graphs.

The intersection can be found by graphing each inequality and then determining where the graphs intersect.

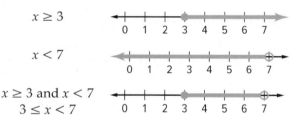

$x \geq 3$

$x < 7$

$x \geq 3$ and $x < 7$
$3 \leq x < 7$

The statement $3 \leq x < 7$ can be read as *x is greater than or equal to 3 and less than 7* or *x is between 3 and 7 including 3*.

Example 1   Solve and Graph an Intersection

Solve $-2 \leq x - 3 < 4$. Then graph the solution set.

First, express $-2 \leq x - 3 < 4$ using *and*. Then solve each inequality.

$-2 \leq x - 3$	**and**	$x - 3 < 4$	Write the inequalities.
$-2 + 3 \leq x - 3 + 3$		$x - 3 + 3 < 4 + 3$	Add 3 to each side.
$1 \leq x$		$x < 7$	Simplify.

The solution set is $\{x \mid 1 \leq x < 7\}$. Now graph the solution set.

Graph $1 \leq x$ or $x \geq 1$.

Graph $x < 7$.

Find the intersection of the graphs.

Solve each compound inequality. Then graph the solution set.

1A. $y - 3 \geq -11$ and $y - 3 \leq -8$ **1B.** $6 \leq r + 7 < 10$

2 **Inequalities Containing *or*** Another type of compound inequality contains the word *or*. A compound inequality containing *or* is true if at least one of the inequalities is true. Its graph is the **union** of the graphs of two inequalities.

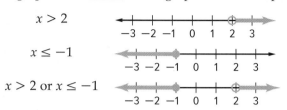

When solving problems involving inequalities, *within* is meant to be inclusive, so use $\geq$ or $\leq$. *Between* is meant to be exclusive, so use $<$ or $>$.

Real-World Example 2 Write and Graph a Compound Inequality

SOUND **The human ear can only detect sounds between the frequencies 20 Hertz and 20,000 Hertz. Write and graph a compound inequality that describes the frequency of sounds humans cannot hear.**

The problem states that humans can hear the frequencies between 20 Hz and 20,000 Hz. We are asked to find the frequencies humans cannot hear.

Words	The frequency	is at most	20 Hertz	or	The frequency	is at least	20,000 Hertz.
Variable	Let f be the frequency.						
Inequality	f	$\leq$	20	or	f	$\geq$	20,000

Now, graph the solution set.

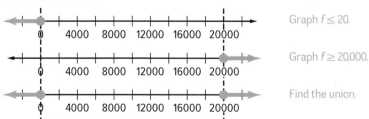

Graph $f \leq 20$.

Graph $f \geq 20{,}000$.

Find the union.

Notice that the graphs do not intersect. Humans cannot hear sounds at a frequency less than or equal to 20 Hertz or greater than or equal to 20,000 Hertz. The compound inequality is
$\{f \mid f \leq 20 \text{ or } f \geq 20{,}000\}$.

2. MANUFACTURING A company is manufacturing an action figure that must be at least 11.2 centimeters and at most 11.4 centimeters tall. Write and graph a compound inequality that describes how tall the action figure can be.

Go Online!

The phrase *at most* in Example 2 indicates $\leq$. It could have been phrased as *no more than* or *less than or equal to*. Work with a partner to list as many phrases for each type of inequality as you can. Add your list to your **eStudent Edition** notes to reference anytime.

(l)McGraw-Hill Education

Example 3   Solve and Graph a Union

Solve $-2m + 7 \leq 13$ or $5m + 12 > 37$. Then graph the solution set.

$-2m + 7 \leq 13$		**or**	$5m + 12 > 37$
$-2m + 7 - 7 \leq 13 - 7$	Subtract.		$5m + 12 - 12 > 37 - 12$
$-2m \leq 6$	Simplify.		$5m > 25$
$\dfrac{-2m}{-2} \geq \dfrac{6}{-2}$	Divide.		$\dfrac{5m}{5} > \dfrac{25}{5}$
$m \geq -3$	Simplify.		$m > 5$

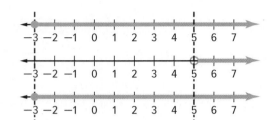

Graph $m \geq -3$.

Graph $m > 5$.

Find the union.

Notice that the graph of $m \geq -3$ contains every point in the graph of $m > 5$. So, the union is the graph of $m \geq -3$. The solution set is $\{m \,|\, m \geq -3\}$.

> **Guided Practice**

Solve each compound inequality. Then graph the solution set.

3A. $a + 1 < 4$ or $a - 1 \geq 3$ **3B.** $x \leq 9$ or $2 + 4x < 10$

Study Tip

Intersections and Unions The graph of a compound inequality containing *and* will be an intersection. The graph of a compound inequality containing *or* will be a union.

Go Online! for a Self-Check Quiz

Check Your Understanding

◯ = Step-by-Step Solutions begin on page R11.

Examples 1–3   Solve each compound inequality. Then graph the solution set.

1. $4 \leq p - 8$ and $p - 14 \leq 2$ **2.** $r + 6 < -8$ or $r - 3 > -10$

3. $4a + 7 \geq 31$ or $a > 5$ **4.** $2 \leq g + 4 < 7$

Example 2   **5.** (MP) **SENSE-MAKING** The recommended weight for an adult beagle is at least 18 pounds, but no more than 30 pounds. If a beagle puppy is currently 8 pounds, what is the recommended number of pounds the puppy should gain by the time it is an adult?

Practice and Problem Solving

Extra Practice is on page R5.

Examples 1–3   Solve each compound inequality. Then graph the solution set.

6. $f - 6 < 5$ and $f - 4 \geq 2$ **7** $n + 2 \leq -5$ and $n + 6 \geq -6$

8. $y - 1 \geq 7$ or $y + 3 < -1$ **9.** $t + 14 \geq 15$ or $t - 9 < -10$

10. $-5 < 3p + 7 \leq 22$ **11.** $-3 \leq 7c + 4 < 18$

12. $5h - 4 \geq 6$ and $7h + 11 < 32$ **13.** $22 \geq 4m - 2$ or $5 - 3m \leq -13$

14. $-4a + 13 \geq 29$ and $10 < 6a - 14$ **15.** $-y + 5 \geq 9$ or $3y + 4 < -5$

Example 2

16. **SPEED** The posted speed limit on an interstate highway is shown. Write an inequality that represents the sign. Graph the inequality.

17. **NUMBER THEORY** Find all sets of two consecutive positive odd integers with a sum that is at least 8 and less than or equal to 24.

Write a compound inequality for each graph.

18. ![number line]
−2 −1 0 1 2 3 4

19. ![number line]
−4 −3 −2 −1 0 1 2

20. ![number line]
−1 0 1 2 3 4

21. ![number line]
−6 −5 −4 −3 −2 −1 0

22. ![number line]
1 2 3 4 5 6 7

23. ![number line]
−4 −3 −2 −1 0 1 2

Solve each compound inequality. Then graph the solution set.

24. $3b + 2 < 5b − 6 \leq 2b + 9$

25. $−2a + 3 \geq 6a − 1 > 3a − 10$

26. $10m − 7 < 17m$ or $−6m > 36$

27. $5n − 1 < −16$ or $−3n − 1 < 8$

28. **ZOO** Groups of 20 or more receive a 20% discount to the Memphis Zoo. Tickets range in price from $10.00 to $15.00, depending on your age.

   a. What is the range of ticket prices after the 20% discount is used?

   b. If a group purchases 25 tickets, find the range in prices of the total cost.

Define a variable, write an inequality, and solve each problem. Then check your solution.

29. Eight less than a number is no more than 14 and no less than 5.

30. The sum of 3 times a number and 4 is between −8 and 10.

31. The product of −5 and a number is greater than 35 or less than 10.

32. One half a number is greater than 0 and less than or equal to 1.

33. **SNAKES** Most snakes live where the temperature ranges from 75°F to 90°F, inclusive. Write an inequality for temperatures where snakes will *not* thrive.

34. **FUNDRAISING** Yumas is selling gift cards to raise money for a class trip. He can earn prizes depending on how many cards he sells. So far, he has sold 34 cards. How many more does he need to sell to earn a prize in category 4?

Cards	Prize
1–15	1
16–30	2
31–45	3
46–60	4
+61	5

35. **TURTLES** Atlantic sea turtle eggs that incubate below 23°C or above 33°C rarely hatch. Write the temperature requirements in two ways: as a pair of simple inequalities, and as a compound inequality.

36. **MP STRUCTURE** The *Triangle Inequality Theorem* states that the sum of the measures of any two sides of a triangle is greater than the measure of the third side.

   a. Write and solve three inequalities to express the relationships among the measures of the sides of the triangle shown at the right.

   b. What are four possible lengths for the third side of the triangle?

   c. Write a compound inequality for the possible values of x.

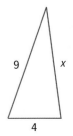

37. HURRICANES The Saffir-Simpson Hurricane Scale rates hurricanes on a scale from 1 to 5 based on their wind speed.

Category	Wind Speed (mph)	Example (year)
1	74–95	Isaac (2012)
2	96–110	Ernesto (2012)
3	111–129	Sandy (2012)
4	130–156	Ophelia (2011)
5	> 157	Felix (2007)

   a. Write compound inequalities for the wind speeds of category 3 and category 4 hurricanes.

   b. What is the intersection of the two graphs of the inequalities you found in part **a**?

38. MULTIPLE REPRESENTATIONS In this problem, you will investigate measurements. The **absolute error** of a measurement is equal to one half the unit of measure. The **relative error** of a measure is the ratio of the absolute error to the expected measure.

   a. Tabular Copy and complete the table.

Measure	Absolute Error	Relative Error
14.3 cm	$\frac{1}{2}(0.1) = 0.05$ cm	$\frac{\text{absolute error}}{\text{expected measure}} = \frac{0.05 \text{ cm}}{14.3 \text{ cm}}$ ≈ 0.0035 or 0.4%
1.85 cm		
61.2 cm		
237 cm		

   b. Analytical You measured a length of 12.8 centimeters. Compute the absolute error and then write the range of possible measures.

   c. Logical To what precision would you have to measure a length in centimeters to have an absolute error of less than 0.05 centimeter?

   d. Analytical To find the relative error of an area or volume calculation, add the relative errors of each linear measure. If the measures of the sides of a rectangular box are 6.5 centimeters, 7.2 centimeters, and 10.25 centimeters, what is the relative error of the volume of the box?

H.O.T. Problems   Use **H**igher-**O**rder **T**hinking Skills

39. ERROR ANALYSIS Chloe and Jonas are solving $3 < 2x - 5 < 7$. Is either of them correct? Explain your reasoning.

Chloe	Jonas
$3 < 2x - 5 < 7$	$3 < 2x - 5 < 7$
$3 < 2x < 12$	$8 < 2x < 7$
$\frac{3}{2} < x < 6$	$4 < x < \frac{7}{2}$

40. (MP) PERSEVERANCE Solve each inequality for x. Assume a is constant and $a > 0$.

   a. $-3 < ax + 1 \le 5$ **b.** $-\frac{1}{a}x + 6 < 1$ or $2 - ax > 8$

41. OPEN ENDED Create an example of a compound inequality containing *or* that has infinitely many solutions.

42. CHALLENGE Determine whether the following statement is *always, sometimes,* or *never* true. Explain. *The graph of a compound inequality that involves an* or *statement is bounded on the left and right by two values of x.*

43. WRITING IN MATH Give an example of a compound inequality you might encounter at an amusement park. Does the example represent an intersection or a union?

44. Which of the following is not a correct description of the graph of the solution set of $-5n + 2 \geq 17$ or $2n + 1 > 5$? **MP** 1

○ **A** The graph has a dot on -3 and a circle on 2.

○ **B** The graph has a ray pointing to the right starting at -3.

○ **C** The graph has a ray pointing to the left starting at -3.

○ **D** The graph has a ray pointing to the right starting at 2.

45. Which of the following compound inequalities does not have the same solution set as the sum of a number and 5 is between 2 and 10? **MP** 2

○ **A** $-3 < x < 5$

○ **B** $2 < x + 5 < 10$

○ **C** $7 < x < 15$

○ **D** $-6 < 2x < 10$

46. Which compound inequality has a graph that includes the entire number line? **MP** 1

○ **A** $-4p > 8$ or $2p < 16$

○ **B** $-2q > 6$ or $2q < 8$

○ **C** $-2r < 4$ or $4r < -8$

○ **D** $-s > 2$ or $2s > -16$

47. Gianni solved a compound inequality and graphed its solution set below.

Which inequality did Gianni solve? **MP** 3

○ **A** $12 \geq 3a + 4 > -3$

○ **B** $12 \geq 3a + 4 > 1$

○ **C** $16 \geq 3a + 4 > 1$

○ **D** $16 \geq 3a + 4 > -3$

48. MULTI-STEP According to 2015 Bureau of Labor Statistics data, the annual salary for a full-time truck driver is within $26,240 and $62,010.

a. Use set-builder notation to write a compound inequality for the weekly salary w of a truck driver, rounded to the nearest cent.

b. Which graph represents the weekly salary for truck drivers?

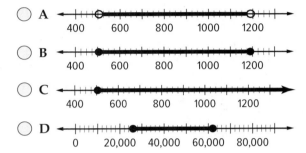

c. Determine the how many weeks it would take for a truck driver to earn $5000.

49. A certain type of glue will not work if the temperature is less than 45°F or if the temperature is 78°F or above. Which compound inequality represents the temperatures t when the glue will work? **MP** 4

○ **A** $45 < t < 78$

○ **B** $45 \leq t < 78$

○ **C** $45 < t \leq 78$

○ **D** $45 \leq t \leq 78$

50. Celeste wrote the compound inequality shown below.

$$-5 \leq 2x + b < 7$$

For what value of b will the inequality have a solution set of $\{x \mid -7 \leq x < -1\}$? **MP** 7

Solving Inequalities Involving Absolute Value

::Then	::Now	::Why?
• You solved equations involving absolute value.	**1** Solve and graph absolute value inequalities (<).	• Some companies use absolute value inequalities to control the quality of their product. To make baby carrots, long carrots are sliced into 3-inch sections
	2 Solve and graph absolute value inequalities (>).	and peeled. If the machine is accurate to within $\frac{1}{8}$ of an inch, the length ranges from $2\frac{7}{8}$ inches to $3\frac{1}{8}$ inches.

 Mathematical Practices

3 Construct viable arguments and critique the reasoning of others.

7 Look for and make use of structure.

1 Absolute Value Inequalities (<) The inequality $|x| < 3$ means that the distance between x and 0 is less than 3.

So, $x > -3$ and $x < 3$. The solution set is $\{x \mid -3 < x < 3\}$.

When solving absolute value inequalities, there are two cases to consider.

Case 1 The expression inside the absolute value symbols is nonnegative.

Case 2 The expression inside the absolute value symbols is negative.

The solution is the intersection of the solutions of these two cases.

Example 1   Solve Absolute Value Inequalities (<)

Solve each inequality. Then graph the solution set.

a. $|m + 2| < 11$

Rewrite $|m + 2| < 11$ for Case 1 *and* Case 2.

Case 1 $m + 2$ is nonnegative.	**and**	**Case 2** $m + 2$ is negative.
$m + 2 < 11$		$-(m + 2) < 11$
$m + 2 - 2 < 11 - 2$		$m + 2 > -11$
$m < 9$		$m + 2 - 2 > -11 - 2$
		$m > -13$

So, $m < 9$ and $m > -13$. The solution set is $\{m \mid -13 < m < 9\}$.

```
 +—⊖———————————⊖—+
-14 -12 -10 -8 -6 -4 -2  0  2  4  6  8  10
```

b. $|y - 1| < -2$

$|y - 1|$ cannot be negative since the results inside an absolute value is always positive. So it is not possible for $|y - 1|$ to be less than -2. Therefore, there is no solution, and the solution set is the empty set, $\varnothing$.

▶ **Guided Practice**

1A. $|n - 8| \leq 2$ **1B.** $|2c - 5| < -3$

Real-World Link

Financial Future Almost two thirds of teens, ages 14 to 18, believe that they will be as well-off or better off than their parents or guardians.

Source: Junior Achievement

Real-World Example 2   Apply Absolute Value Inequalities

COLLEGE A recent survey showed that 52% of teens, ages 14 to 18, believe that students are borrowing too much for college. The margin of error was within 3.5 percentage points. Find the range of teens who believe students are borrowing too much for college.

The difference between the actual percent of teens and the percent from the survey is less than or equal to 3.5. Let x be the actual percent. Then $|x - 52| \leq 3.5$.

Solve each case of the inequality.

Case 1 $x - 52$ is nonnegative.   **and**   **Case 2** $x - 52$ is negative.

$$x - 52 \leq 3.5$$
$$x - 52 + 52 \leq 3.5 + 52$$
$$x \leq 55.5$$

$$-(x - 52) \leq 3.5$$
$$x - 52 \geq -3.5$$
$$x \geq 48.5$$

The percent of teens who believe students are borrowing too much for college is $\{x \mid 48.5 \leq x \leq 55.5\}$.

Guided Practice

2. CHEMISTRY The melting point of ice is 0°C. During a chemistry experiment, Jill observed ice melting within 2°C of this measurement. Write the range of temperatures that Jill observed.

2 Absolute Value Inequalities (>) The inequality $|x| > 3$ means that the distance between x and 0 is greater than 3.

So, $x < -3$ or $x > 3$. The solution set is $\{x \mid x < -3 \text{ or } x > 3\}$.

As in the previous example, we must consider both cases.

Case 1 The expression inside the absolute value symbols is nonnegative.

Case 2 The expression inside the absolute value symbols is negative.

Example 3   Solve Absolute Value Inequalities (>)

Solve $|3n + 6| \geq 12$. Then graph the solution set.

Rewrite $|3n + 6| \geq 12$ for Case 1 *or* Case 2.

Case 1 $3n + 6$ is nonnegative.   **or**   **Case 2** $3n + 6$ is negative.

$$3n + 6 \geq 12$$
$$3n + 6 - 6 \geq 12 - 6$$
$$3n \geq 6$$
$$n \geq 2$$

$$-(3n + 6) \geq 12$$
$$3n + 6 \leq -12$$
$$3n \leq -18$$
$$n \leq -6$$

So, $n \geq 2$ or $n \leq -6$. The solution set is $\{n \mid n \geq 2 \text{ or } n \leq -6\}$.

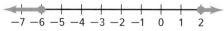

Study Tip

MP **Structure** For $|a| \geq b$, where a is any linear expression in one variable and b is a negative number, the solution set will always be the set of all real numbers. Since $|a|$ is always greater than or equal to zero, $|a|$ is always greater than or equal to b.

Guided Practice

Solve each inequality. Then graph the solution set.

3A. $|2k + 1| > 7$ **3B.** $|r - 6| \geq -5$

Go Online! for a
Self-Check Quiz

Examples 1–3 Solve each inequality. Then graph the solution set.

1. $|a - 5| < 3$ **2.** $|u + 3| < 7$ **3.** $|t + 4| \leq -2$

4. $|c + 2| > -2$ **5.** $|n + 5| \geq 3$ **6.** $|p - 2| \geq 8$

Example 2   **7. FINANCIAL LITERACY** Jerome bought stock in his favorite restaurant at $70.85. It has fluctuated up to $0.75 in July. Find the range of prices for which the stock could trade in July.

Practice and Problem Solving   Extra Practice is on page R5.

Examples 1–3 Solve each inequality. Then graph the solution set.

8. $|x + 8| < 16$ **9** $|r + 1| \leq 2$ **10.** $|2c - 1| \leq 7$

11. $|3h - 3| < 12$ **12.** $|m + 4| < -2$ **13.** $|w + 5| < -8$

14. $|r + 2| > 6$ **15.** $|k - 4| > 3$ **16.** $|2h - 3| \geq 9$

17. $|4p + 2| \geq 10$ **18.** $|5v + 3| > -9$ **19.** $|-2c - 3| > -4$

Example 4   **20. FISH** A fish tank should be within 2°F of the recommended temperature of 78°F. Write a range for the ideal temperatures of the fish tank.

Solve each inequality. Then graph the solution set.

21. $|4n + 3| \geq 18$ **22.** $|5t - 2| \leq 6$ **23.** $\left|\dfrac{3h + 1}{2}\right| < 8$

24. $\left|\dfrac{2p - 8}{4}\right| \geq 9$ **25.** $\left|\dfrac{7c + 3}{2}\right| \leq -5$ **26.** $\left|\dfrac{2g + 3}{2}\right| > -7$

27. $|-6r - 4| < 8$ **28.** $|-3p - 7| > 5$ **29.** $|-h + 1.5| < 3$

30. MULTI-STEP Each year Kareem receives a $150 gift card to download music and apps onto his phone. The card expires in one year.

   a. Suppose Kareem gives himself a monthly allowance to use from his card. If he spent within $3 of his monthly allowance for each of the first 11 months, find the range Kareem has left to spend on music and apps in the last month of the year.

   b. Describe your solution process.

   c. What assumptions did you make?

   31. CHEMISTRY Water can be present in our atmosphere as a solid, liquid, or gas. Water freezes at 32°F and vaporizes at 212°F.

   a. Write and graph the range of temperatures in which water is not a liquid.

   b. Write the absolute value inequality that describes this situation.

(MP) STRUCTURE Write an open sentence involving absolute value for each graph.

32.
```
  ←+——+——+——⊕——+——+——+——⊕——+——+——+→
   −5 −4 −3 −2 −1  0  1  2  3  4  5
```

33.
```
  ←+——+——●——+——+——+——+——+——●——+——+→
   −6 −5 −4 −3 −2 −1  0  1  2  3  4
```

34.
```
  ←——+——+——●——+——+——+——+——●——+——+——→
   −6 −5 −4 −3 −2 −1  0  1  2  3  4
```

35.
```
  ←+——⊕——+——+——+——+——+——+——+——⊕——+→
    0  1  2  3  4  5  6  7  8  9  10 11
```

36. TEMPERATURE Normal body temperature for humans when taken orally is 36.8°C. However, this number varies from person to person and can be different at different times of the day. It is normal for temperatures to be 0.5°C higher or lower. What is the range of normal body temperatures in humans?

37 MINIATURE GOLF Ginger's score was within 5 strokes of her average score of 52. Determine the range of scores for Ginger's game.

Express each statement using an inequality involving absolute value. Do *not* solve.

38. The pH of a swimming pool must be within 0.3 of a pH of 7.5.

39. The temperature inside a refrigerator should be within 1.5 degrees of 38°F.

40. Ramona's bowling score was within 6 points of her average score of 98.

41. The cruise control of a car should keep the speed within 3 miles per hour of 55.

42. MULTIPLE REPRESENTATIONS In this problem, you will investigate the graphs of linear inequalities on a coordinate plane.

a. Tabular Copy and complete the table. Substitute the x and $f(x)$ values for each point into each inequality. Mark whether the resulting statement is *true* or *false*.

Point	$f(x) \geq x - 1$	true/false	$f(x) \leq x - 1$	true/false
$(-4, 2)$				
$(-2, 2)$				
$(0, 2)$				
$(2, 2)$				
$(4, 2)$				

b. Graphical Graph $f(x) = x - 1$.

c. Graphical Plot each point from the table that made $f(x) \geq x - 1$ a true statement on the graph in red. Plot each point that made $f(x) \leq x - 1$ a true statement in blue.

d. Logical Make a conjecture about what the graphs of $f(x) \geq x - 1$ and $f(x) \leq x - 1$ look like. Complete the table with other points to verify your conjecture.

e. Logical Use what you discovered to describe the graph of a linear inequality.

H.O.T. Problems   Use **H**igher-**O**rder **T**hinking Skills

43. ERROR ANALYSIS Lucita sketched a graph of her solution to $|2a - 3| > 1$. Is she correct? Explain your reasoning.

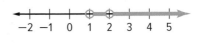

44. MP REASONING The graph of an absolute value inequality is *sometimes, always,* or *never* the union of two graphs. Explain.

45. MP CONSTRUCT ARGUMENTS Demonstrate why the solution of $|t| > 0$ is not all real numbers. Explain your reasoning.

46. e WRITING IN MATH How are symbols used to represent mathematical ideas? Use an example to justify your reasoning.

47. WRITING IN MATH Explain how to determine whether an absolute value inequality uses a compound inequality with *and* or a compound inequality with *or*. Then summarize how to solve absolute value inequalities.

48. Reggie solved the inequality $|2n + 4| > 2$. Then he graphed its solution set. Which graph is Reggie's graph? **MP** 1

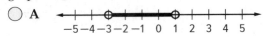

- A
- B
- C
- D

49. Which inequality has the same solution set as $|2x - 12| \geq 6$? **MP** 2

- **A** $|12 - 2x| \leq 6$
- **B** $|x - 6| \geq 3$
- **C** $|12 - 2x| \geq -6$
- **D** $|x - 6| \geq 9$

50. Natalie solved the inequality $|5 - b| < 2$ and graphed its solution set. Which statement best describes the graph? **MP** 1

- **A** circles on −3 and 7 with shading in between
- **B** circles on 3 and 7 with shading in between
- **C** circle on 3 with a ray pointing to the right
- **D** circle on 7 with a ray pointing to the left

51. Which inequality has the solution set graphed below? **MP** 1

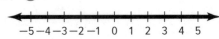

- **A** $|2x - 3| \leq -7$
- **B** $|2x + 3| \geq 7$
- **C** $|2x + 3| \leq -7$
- **D** $|2x + 3| \geq -7$

52. MULTI-STEP Jon wants to save \$500. He earns b dollars each week over a period of x weeks. **MP** 4

a. Which equation can be used to find how much Jon can save each week to be within \$20 of his goal?

- **A** $|bx - 20| \geq 500$
- **B** $|bx - 500| \geq 20$
- **C** $|bx - 20| \leq 500$
- **D** $|bx - 500| \leq 20$

b. How many weeks should Jon plan to save if he would save the minimum amount for the summer with \$48 per week?

53. For quality control, a package of 8 hamburger patties must be within fixed range from the weight printed on the package. The inequality $|8x - 2| < 0.1$ represents the acceptable range. How close must the actual weight be to the printed amount? **MP** 4, 7

- **A** 0.1 unit
- **B** 0.5 unit
- **C** 0.8 unit
- **D** 2 units

54. A survey shows that 36% of voters are in favor of building a new library. The inequality $|x - 36| \leq 2$ represents the range of voters who are actually in favor. Which statement about the survey is not true? **MP** 4, 7

- **A** The margin of error for the survey was within 2 percentage points.
- **B** The actual percentage p of voters who are in favor is $\{p \mid 35 \leq p \leq 37\}$.
- **C** The actual percentage p of voters who are in favor is $\{p \mid 34 \leq p \leq 38\}$.
- **D** It is possible that 38% of the voters surveyed are in favor.

55. For what value of y will the absolute value inequality $|x + y| \leq 5$ have a solution set of $\{x \mid -8 \leq x \leq 2\}$? **MP** 2

Graphing Inequalities in Two Variables

:: Then	:: Now	:: Why?
• You graphed linear equations.	**1** Graph linear inequalities on the coordinate plane. **2** Solve inequalities by graphing.	• Hannah has budgeted $70 every three months for car maintenance. From this she must buy oil costing $8 and filters that cost $12 each. How much oil and how many filters can Hannah buy and stay within her budget?

 New Vocabulary
boundary
half-plane
closed half-plane
open half-plane

 Mathematical Practices
5 Use appropriate tools strategically.

1 **Graph Linear Inequalities** The graph of a linear inequality is the set of points that represent all of the possible solutions of that inequality. An equation defines a **boundary**, which divides the coordinate plane into two **half-planes**.

The boundary may or may not be included in the solution. When it is included, the solution is a **closed half-plane**. When not included, the solution is an **open half-plane**.

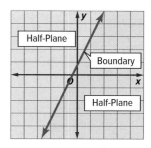

Key Concept Graphing Linear Inequalities

Step 1 Graph the boundary. Use a solid line when the inequality contains ≤ or ≥. Use a dashed line when the inequality contains < or >.

Step 2 Use a test point to determine which half-plane should be shaded.

Step 3 Shade the half-plane that contains the solution.

Example 1 Graph an Inequality (< or >)

Graph $3x - y < 2$.

Step 1 First, solve for y in terms of x.
$$3x - y < 2$$
$$-y < -3x + 2$$
$$y > 3x - 2$$

Then, graph $y = 3x - 2$. Because the inequality involves >, graph the boundary with a dashed line.

Step 2 Select (0, 0) as a test point.

$$3x - y < 2 \quad \text{Original inequality}$$
$$3(0) - 0 < 2 \quad x = 0 \text{ and } y = 0$$
$$0 < 2 \quad \text{true}$$

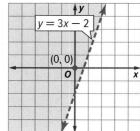

Step 3 So, the half-plane containing the origin is the solution. Shade this half-plane.

▶ **Guided Practice** Graph each inequality.

1A. $y > \frac{1}{2}x + 3$

1B. $x - 1 > y$

Example 2   Graph an Inequality (≤ or ≥)

Graph $x + 5y \leq 10$.

Step 1   Solve for y in terms of x.

$$x + 5y \leq 10 \qquad \text{Original inequality}$$
$$5y \leq -x + 10 \qquad \text{Subtract } x \text{ from each side and simplify.}$$
$$y \leq -\frac{1}{5}x + 2 \qquad \text{Divide each side by 5.}$$

Graph $y = -\frac{1}{5}x + 2$. Because the inequality symbol is ≤, graph the boundary with a solid line.

Step 2   Select a test point. Let's use (3, 3). Substitute the values into the original inequality.

$$x + 5y \leq 10 \qquad \text{Original inequality}$$
$$3 + 5(3) \leq 10 \qquad x = 3 \text{ and } y = 3$$
$$18 \not\leq 10 \qquad \text{Simplify.}$$

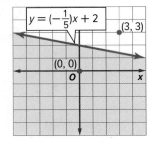

Step 3   Since this statement is false, shade the other half-plane.

▶ **Guided Practice**

Graph each inequality.

2A. $x - y \leq 3$ **2B.** $2x + 3y \geq 18$

---

2 Solve Linear Inequalities We can use a coordinate plane to solve inequalities with two variables.

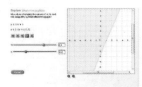
Example 3   Write Inequalities From Graphs

Write an inequality to represent the graph.

Step 1   First find the equation of the boundary.

The boundary intersects the y-axis at $(0, -3)$, so $b = -3$.

$$m = \frac{3 - 1}{3 - 2} = 2 \qquad (x_1, y_1) = (2, 1), (x_2, y_2) = (3, 3)$$

The equation of the boundary is $y = 2x - 3$.

Step 2   The boundary is solid, so the inequality contains a ≤ or ≥ sign.

Step 3   (0, 0) is in the shaded region, so it must make the inequality true.

$$y \geq 2x - 3 \qquad y \leq 2x - 3$$
$$0 \geq -3 \text{ true} \qquad 0 \leq -3 \text{ false}$$

So the inequality is $y \geq 2x - 3$.

▶ **Guided Practice**

Write an inequality to represent the graph.

3.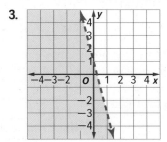

An inequality can be viewed as a constraint in a problem situation. Each solution of the inequality represents a combination that meets the constraint. In real-world problems, the domain and range are often restricted to nonnegative or whole numbers. If a solution satisfies the constraints and makes sense in the context of the situation, then it is considered a viable solution.

Real-World Link

As a supplement to traditional yearbooks, many schools are producing digital versions. They include features that allow you to click on a picture and see a short video clip.

Source: eSchool News

Real-World Example 4   Write and Solve an Inequality

CLASS PICNIC A yearbook company promises to give the junior class a picnic if they spend at least $72,000 on yearbooks and class rings. Each yearbook costs $75, and each class ring costs $375. How many yearbooks and/or class rings must the junior class buy to get their picnic?

Understand You know the cost of each item and the minimum amount the class needs to spend. You need to determine how many of each item needs to be purchased.

Plan Let x = the number of yearbooks and y = the number of class rings the class must purchase. Write an inequality.

$75	times	the number of yearbooks	plus	$375	times	the number of rings	is at least	$72,000.
75	$\cdot$	x	+	375	$\cdot$	y	$\geq$	72,000

Solve Solve for y in terms of x.

$$75x + 375y - 75x \geq 72{,}000 - 75x \qquad \text{Subtract } 75x \text{ from each side.}$$

$$375y \geq -75x + 72{,}000 \qquad \text{Simplify.}$$

$$\frac{375y}{375} \geq \frac{-75x}{375} + \frac{72000}{375} \qquad \text{Divide each side by 375.}$$

$$y \geq -0.2x + 192 \qquad \text{Simplify.}$$

Because the company cannot sell a negative number of items, the domain and range must be nonnegative numbers. Graph the boundary with a solid line. If we test (0, 0), the result is $0 \geq 72{,}000$, which is false. Shade the closed half-plane that does not include (0, 0).

One solution is (500, 100), or 500 yearbooks and 100 class rings.

Graph: Number of Class Rings (y-axis, 20 to 180) vs. Number of Yearbooks (x-axis, 200 to 1000); line $y = -0.2x + 192$; point (500, 100).

Check Test (500, 100). The result is $100 \geq 92$, which is true.

The graph shows all possible solutions based on the constraints. There are often other constraints, not explicitly stated, that determine the reasonableness of a solution. For example, the size of the junior class: if the junior class has only 50 students, they would not be expected to earn a picnic. The solution is reasonable for the given information.

Guided Practice

4. **CONCERT** Neil is practicing for a concert. Write and graph an inequality for the hours y he will practice in x weeks.

Number of Weeks	1	2	3	4	5	6
Minimum Number of Hours of Practice	6	12	18	24	30	36

Check Your Understanding = Step-by-Step Solutions begin on page R11.

Go Online! for a
Self-Check Quiz

Examples 1–2 **Graph each inequality.**

1. $y > x + 3$

2. $y \geq -8$

3. $x + y > 1$

4. $y \leq x - 6$

5. $y < 2x - 4$

6. $x - y \leq 4$

Example 3 **Write an inequality to represent each graph.**

7.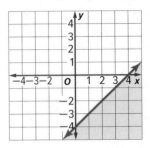

8.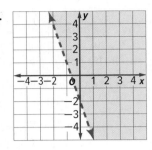

Example 4 **9. FINANCIAL LITERACY** The surf shop has a weekly overhead of $2300.

   a. Write an inequality to represent the number of skimboards and longboards the shop sells each week to make a profit.

   b. How many skimboards and longboards must the shop sell each week to make a profit?

KOWABUNGA
SURF SHOP

Skimboards $115
Longboards $685

Practice and Problem Solving

Extra Practice is on page R5.

Examples 1–2 **Graph each inequality.**

10. $y < x - 3$

11. $y > x + 12$

12. $y \geq 3x - 1$

13. $y \leq -4x + 12$

14. $6x + 3y > 12$

15. $2x + 2y < 18$

16. $-2x + y \geq -4$

17. $8x + y \leq 6$

18. $10x + 2y \leq 14$

Examples 3–4 **19. MP MODELING** The girls' soccer team wants to raise at least $2000 to buy new goals. If they make $1.00 from each hot dog and $1.25 from each soda, how many of each item do they need to sell?

   a. Write an inequality to represent the situation.

   b. Graph an inequality to represent the situation.

   c. Plot at least 5 possible solutions on your graph.

Write an inequality to represent each graph.

20.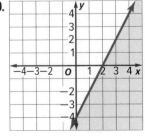

21.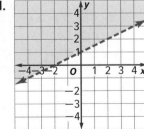

22.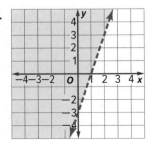

Use a graph to solve each inequality.

23. $3x + 2 < 0$

24. $4x - 1 > 3$

25. $-6x - 8 \geq -4$

26. $-5x + 1 < 3$

27. $-7x + 13 < 10$

28. $-4x - 4 \leq -6$

29. MULTI-STEP Sybrina wants to paint her bedroom, which has a wall area of 640 square feet, her daughter's bedroom, which has a wall area of 470 square feet, and her two sons' bedrooms that have wall areas of 320 square feet each. She wants to paint some of the rooms in satin, at $40 a gallon, and the others in flat, at $20 per gallon. She has a budget of $350.

  a. If each gallon of paint covers 350 square feet, how many rooms can she paint in satin and still be able to paint all the rooms with two coats? Explain your solution process.

  b. What assumptions did you make?

Graph each inequality. Determine which of the ordered pairs are part of the solution set for each inequality.

30. $y \geq 6$; {(0, 4), (−2, 7), (4, 8), (−4, −8), (1, 6)}

31 $x < −4$; {(2, 1), (−3, 0), (0, −3), (−5, −5), (−4, 2)}

32. $2x − 3y \leq 1$; {(2, 3), (3, 1), (0, 0), (0, −1), (5, 3)}

Write an inequality to represent each table.

33.

x	y
−1	−6
1	2
10	38

34.

x	y
−4	−3
0	−1
4	1

35.

x	y
−4	−14
4	10
8	22

36. RECYCLING Mr. Jones would like to spend no more than $37.50 per week on recycling. A recycling service will remove up to 50 pounds of plastic and paper products per week. They charge $0.25 per pound of plastic and $0.75 per pound of paper products.

  a. Write an inequality that describes Mr. Jones' weekly cost for the service if he stays within his budget.

  b. Graph an inequality for the weekly costs for the service.

37. MULTIPLE REPRESENTATIONS Use inequalities A and B to investigate graphing compound inequalities on a coordinate plane.

  A. $7(y + 6) \leq 21x + 14$ **B.** $−3y \leq 3x − 12$

  a. Numerical Solve each inequality for y.

  b. Graphical Graph both inequalities on one graph. Shade the half-plane that makes A true in red. Shade the half-plane that makes B true in blue.

  c. Verbal What does the overlapping region represent?

H.O.T. Problems   Use **H**igher-**O**rder **T**hinking Skills

38. (MP) **REASONING** Explain why a point on the boundary should not be used as a test point.

39. (MP) **TOOLS** Write a linear inequality for which (−1, 2), (0, 1), and (3, −4) are solutions, but (1, 1) is not.

40. WRITING IN MATH Summarize the steps to graph an inequality in two variables.

41. OPEN ENDED Write a two-variable inequality with a restricted domain and range to represent a real-world situation. Give the domain and range, and explain why they are restricted.

42. Lola can spend up to $20 on snacks for a study session. Muffins cost $4 each, and bagels cost $2 each. To see how many of each item she can buy, Lola graphs the inequality $4x + 2y \leq 20$, where x is the number of muffins and y is the number of bagels. Which of the following is not a correct description of Lola's graph? **MP** 4, 5

- **A** The boundary is included in the solution.

- **B** The boundary has a slope of -2 and a y-intercept of 10.

- **C** The half-plane containing the point for 5 muffins and 2 bagels is the solution.

- **D** The half-plane containing the point for 2 muffins and 5 bagels is the solution.

43. Which situation could be modeled by the inequality $10x + 12y \geq 225$? **MP** 4

- **A** It takes Ann 10 minutes to walk a kilometer. Jaxon takes 12 minutes to walk a kilometer. Find their total distance after 225 minutes.

- **B** A grocery store sells 10-pound and 12-pound bags of flour. The flour display shelf has a 225-pound limit.

- **C** Children's tickets to the movies cost $10 each and adult tickets cost $12 each. Jay can spend no more than $225.

- **D** Shane earns $10 per hour at one job and $12 per hour at another job. He wants to earn at least $225.

44. The graph of a linear inequality in two variables is shown below.

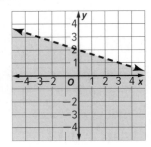

What is the linear inequality? **MP** 1

- **A** $-2x - 6y > -12$

- **B** $-2x - 6y < -12$

- **C** $-2x + 6y > -12$

- **D** $-2x + 6y < -12$

45. MULTI-STEP Consider the inequality $3x - y > 1$. **MP** 8

- **a.** Which inequality has the same set of solutions?

  - **A** $y > 3x - 1$

  - **B** $y < 3x - 1$

  - **C** $y > -3x + 1$

  - **D** $y < -3x + 1$

- **b.** If $(0, n)$ is a solution to the inequality, what is the greatest possible value of n, where n is an integer?

46. Which inequality includes $(3, 3)$ in its solution set? **MP** 8

- **A** $x > -2$

- **B** $x < 2$

- **C** $y > 3$

- **D** $y < -3$

47. The graph below represents the hours x and the miles y that Noah plans to walk in a walkathon.

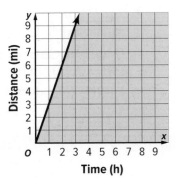

Which best describes Noah's planned walking pace? **MP** 4, 5

- **A** at least 3 miles per hour

- **B** no more than 3 miles per hour

- **C** less than 3 miles per hour

- **D** greater than 3 miles per hour

You can use a graphing calculator to investigate the graphs of inequalities.

Mathematical Practices
MP 5 Use appropriate tools strategically.

Activity 1   Graph a Linear Inequality (< or ≤)

Work cooperatively. Graph $y \leq 2x + 5$.

Clear all functions from the **Y=** list.

KEYSTROKES: [Y=] [CLEAR]

Graph $y \leq 2x + 5$ in a standard viewing window.

KEYSTROKES: 2 [X,T,θ,n] [+] 5 [◄] [◄] [◄] [◄] [◄] [◄] [ENTER] [ENTER] [ENTER] [ZOOM] 6

All ordered pairs for which y is *less than or equal to* $2x + 5$ lie *below or on* the line and are solutions.

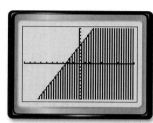

[−10, 10] scl: 1 by [−10, 10] scl: 1

Activity 2   Graph a Linear Inequality (> or ≥)

Work cooperatively. Graph $y - 2x \geq 5$.

Clear the graph that is currently displayed.

KEYSTROKES: [Y=] [CLEAR]

Rewrite $y - 2x \geq 5$ as $y \geq 2x + 5$ and graph it.

KEYSTROKES: 2 [X,T,θ,n] [+] 5 [◄] [◄] [◄] [◄] [◄] [◄] [ENTER] [ENTER] [ZOOM] 6

All ordered pairs for which y is *greater than or equal to* $2x + 5$ lie *above or on* the line and are solutions.

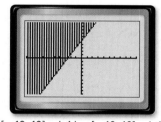

[−10, 10] scl: 1 by [−10, 10] scl: 1

Exercises

Work cooperatively.

1. Compare and contrast the two graphs shown above.

2. Graph $y \geq -3x + 1$ in the standard viewing window. Using your graph, name four solutions of the inequality.

3. Suppose student water park tickets cost $16 and adult water park tickets cost $20. You would like to buy at least 10 tickets but spend no more than $200.
   a. Let x = number of student tickets and y = number of adult tickets. Write two inequalities, one representing the number of tickets and the other representing the cost.

   b. Graph the inequalities. Use the viewing window [0, 20] scl: 1 by [0, 20] scl: 1.

   c. Name four possible combinations of student and adult tickets.

Go Online! for Vocabulary Review Games and key vocabulary in 13 languages

Study Guide

Key Concepts

Solving One-Step Inequalities (Lessons 5-1 and 5-2)

For all numbers a, b, and c, the following are true.

- If $a > b$, then $a + c > b + c$.
- If $a < b$, then $a + c < b + c$.
- If $a > b$ and c is positive, $ac > bc$.
- If $a > b$ and c is negative, $ac < bc$.

Multi-Step and Compound Inequalities (Lessons 5-3 and 5-4)

- Multi-step inequalities can be solved by undoing the operations in the same way you would solve a multi-step equation.
- A compound inequality containing *and* is only true if both inequalities are true.
- A compound inequality containing *or* is true if at least one of the inequalities is true.

Absolute Value Inequalities (Lesson 5-5)

- The absolute value of any number x is its distance from zero on a number line and is written as $|x|$. If $x \geq 0$, then $|x| = x$. If $x < 0$, then $|x| = -x$.
- If $|x| < n$ and $n > 0$, then $-n < x < n$.
- If $|x| > n$ and $n > 0$, then $x > n$ or $x < -n$.

Inequalities in Two Variables (Lesson 5-6)

To graph an inequality:

Step 1 Graph the boundary. Use a solid line when the inequality contains $\leq$ or $\geq$. Use a dashed line when the inequality contains $<$ or $>$.

Step 2 Use a test point to determine which half-plane should be shaded.

Step 3 Shade the half-plane.

FOLDABLES® Study Organizer

Use your Foldable to review the chapter. Working with a partner can be helpful. Ask for clarification of concepts as needed.

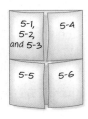

Key Vocabulary

boundary (p. 321)

closed half-plane (p. 321)

compound inequality (p. 309)

half-plane (p. 321)

inequality (p. 289)

intersection (p. 309)

open half-plane (p. 321)

union (p. 310)

Vocabulary Check

State whether each sentence is *true* or *false*. If *false*, replace the underlined term to make a true sentence.

1. The graph of a compound inequality containing *and* shows the <u>union</u> of the individual graphs.

2. The graph of an inequality of the form $y < ax + b$ is a region on the coordinate plane called a <u>half-plane</u>.

3. A <u>point</u> defines the boundary of an open half-plane.

4. The <u>boundary</u> is the graph of the equation of the line that defines the edge of each half-plane.

5. The solution set to the inequality $y \geq x$ includes the <u>boundary</u>.

6. When solving an inequality, <u>multiplying</u> each side by a negative number reverses the inequality symbol.

7. The graph of a compound inequality that contains <u>*and*</u> is the intersection of the graphs of the two inequalities.

Concept Check

8. Explain why set-builder notation is a more concise way of writing a solution set.

9. Explain why a compound inequality containing *or* is true if one or both of the inequalities is true.

Lesson-by-Lesson Review

5-1 Solving Inequalities by Addition and Subtraction

Solve each inequality. Then graph it on a number line.

10. $w - 4 > 9$

11. $x + 8 \leq 3$

12. $6 + h < 1$

13. $-5 < a + 2$

14. $13 - p \geq 15$

15. $y + 1 \leq 8$

16. FIELD TRIP A bus can hold 44 people. If there are 35 students in Samantha's class, how many more people can ride on the bus?

Example 1

Solve $x - 9 < -4$. Then graph it on a number line.

$x - 9 < -4$	Original inequality
$x - 9 + 9 < -4 + 9$	Add 9 to each side.
$x < 5$	Simplify.

The solution set is $\{x \mid x < 5\}$.

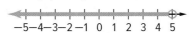

5-2 Solving Inequalities by Multiplication and Division

Solve each inequality. Graph the solution on a number line.

17. $\frac{1}{3}x > 6$

18. $\frac{1}{5}g \geq -4$

19. $4p < 32$

20. $-55 \leq -5w$

21. $-2m > 100$

22. $\frac{2}{3}t < -48$

23. PROM The prom committee has no more than $600 to spend on centerpieces for the prom. If each centerpiece costs $25, what is the maximum number of centerpieces the committee can buy?

Example 2

Solve $-14h < 56$. Check your solution.

$-14h < 56$	Original inequality
$\dfrac{-14h}{-14} > \dfrac{56}{-14}$	Divide each side by -14.
$h > -4$	Simplify.
$\{h \mid h > -4\}$	

CHECK To check, substitute three different values into the original inequality: -4, a number less than -4, and a number greater than -4.

5-3 Solving Multi-Step Inequalities

Solve each inequality. Graph the solution on a number line.

24. $3h - 7 < 14$

25. $4 + 5b > 34$

26. $18 \leq -2x + 8$

27. $\frac{t}{3} - 6 > -4$

28. Four times a number decreased by 6 is less than -2. Define a variable, write an inequality, and solve for the number.

29. TICKET SALES The drama club collected $160 from ticket sales for the spring play. They need to collect at least $400 to pay for new lighting for the stage. If tickets sell for $3 each, how many more tickets need to be sold?

Example 3

Solve $-6y - 13 > 29$. Check your solution.

$-6y - 13 > 29$	Original inequality
$-6y - 13 + 13 > 29 + 13$	Add 13 to each side.
$-6y > 42$	Simplify.
$\dfrac{-6y}{-6} < \dfrac{42}{-6}$	Divide each side by -6 and change $>$ to $<$.
$y < -7$	Simplify.

The solution set is $\{y \mid y < -7\}$.

CHECK $-6y - 13 > 29$	Original inequality
$-6(-10) - 13 \overset{?}{>} 29$	Substitute -10 for y.
$47 > 29$ ✓	Simplify.

5-4 Solving Compound Inequalities

Solve each compound inequality. Then graph the solution set.

30. $m - 3 < 6$ and $m + 2 > 4$

31. $-4 < 2t - 6 < 8$

32. $3x + 2 \le 11$ or $5x - 8 > 22$

33. KITES A kite can be flown in wind speeds no less than 7 miles per hour and no more than 16 miles per hour. Write an inequality for the wind speeds at which the kite can fly.

Example 4

Solve $-3w + 4 > -8$ and $2w - 11 > -19$. Then graph the solution set.

$$-3w + 4 > -8 \qquad \text{and} \qquad 2w - 11 > -19$$
$$w < 4 \qquad\qquad\qquad w > -4$$

To graph the solution set, graph $w < 4$ and graph $w > -4$. Then find the intersection.

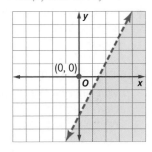

5-5 Solving Inequalities Involving Absolute Value

Solve each inequality. Then graph the solution set.

34. $|x - 4| < 9$

35. $|p + 2| > 7$

36. $|2c + 3| \le 11$

37. $|f - 9| \ge 2$

38. $|3d - 1| \le 8$

39. $\left|\dfrac{4b - 2}{3}\right| < 12$

40. $\left|\dfrac{2t + 6}{2}\right| > 10$

41. $|-4y - 3| < 13$

42. $|m + 19| \le 1$

43. $|-k - 7| \ge 4$

Example 5

Solve $|x - 6| < 9$. Then graph the solution set.

Case 1 $x - 6$ is nonnegative.	**Case 2** $x - 6$ is negative.
$x - 6 < 9$	$-(x - 6) < 9$
$x < 15$	$x > -3$

The solution set is $\{x \mid -3 < x < 15\}$.

5-6 Graphing Inequalities in Two Variables

Graph each inequality.

44. $y > x - 3$

45. $y < 2x + 1$

46. $3x - y \le 4$

47. $y \ge -2x + 6$

48. $5x - 2y < 10$

49. $x + y \ge 1$

Graph each inequality. Determine which of the ordered pairs are part of the solution set for each inequality.

50. $y \le 4$; $\{(3, 6), (1, 2), (-4, 8), (3, -2), (1, 7)\}$

51. $-2x + 3y \ge 12$; $\{(-2, 2), (-1, 1), (0, 4), (2, 2)\}$

52. BAKERY Ben has $24 to spend on cookies and cupcakes. Write and graph an inequality that represents what Ben can buy.

$2 $3

Example 6

Graph $2x - y > 3$.

Solve for y in terms of x.

$2x - y > 3$	Original inequality
$-y > -2x + 3$	Subtract $2x$ from each side.
$y < 2x - 3$	Multiply each side by -1.

Graph the boundary using a dashed line. Choose $(0, 0)$ as a test point.

$2(0) - 0 \overset{?}{>} 3$

$0 \not> 3$

Since 0 is not greater than 3, shade the plane that does not contain $(0, 0)$.

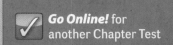

Go Online! for another Chapter Test

Solve each inequality. Then graph it on a number line.

1. $x - 9 < -4$

2. $6p \geq 5p - 3$

3. MULTIPLE CHOICE Drew currently has 31 comic books in his collection. His friend Connor has 58 comic books. How many more comic books does Drew need to add to his collection in order to have a larger collection than Connor?

  A no more than 21

  B 27

  C at least 28

  D more than 30

Solve each inequality. Graph the solution on a number line.

4. $\frac{1}{5}h > 3$

5. $7w \leq -42$

6. $-\frac{2}{3}t \geq 24$

7. $-9m < -36$

8. $3c - 7 < 11$

9. $\frac{g}{4} + 3 \leq -9$

10. $-2(x - 4) > 5x - 13$

11. ZOO The 8th grade science class is going to the zoo. The class can spend up to $300 on admission.

Zoo Admission	
Visitor	**Cost**
student	$8
adult	$10

  a. Write an inequality for this situation.

  b. If there are 32 students in the class and 1 adult will attend for every 8 students, can the entire class go to the zoo?

Solve each compound inequality. Then graph the solution set.

12. $y - 8 < -3$ or $y + 5 > 19$

13. $-11 \leq 2h - 5 \leq 13$

14. $3z - 2 > -5$ and $7z + 4 < -17$

Define a variable, write an inequality, and solve the problem. Check your solution.

15. The difference of a number and 4 is no more than 8.

16. Nine times a number decreased by four is at least twenty-three.

17. MULTIPLE CHOICE Write a compound inequality for the graph shown below.

  A $-2 \leq x < 3$ **C** $x < -2$ or $x \geq 3$

  B $x \leq -2$ or $x \geq 3$ **D** $-2 < x \leq 3$

Solve each inequality. Then graph the solution set.

18. $|p - 5| < 3$ **19.** $|2f + 7| \geq 21$

20. $|-4m + 3| \leq 15$ **21.** $\left|\frac{x - 3}{4}\right| > 5$

22. RETAIL A sporting goods store is offering a $15 coupon on any pair of shoes.

  a. The most and least expensive pairs of shoes are $149.95 and $24.95. What is the range of costs for customers with coupons?

  b. When buying a pair of $109.95 shoes, you can use a coupon or a 15% discount. Which option is best?

Graph each inequality.

23. $y < 4x - 1$ **24.** $2x + 3y \geq 12$

25. Graph $y > -2x + 5$. Then determine which of the ordered pairs in $\{(-2, 0), (-1, 5), (2, 3), (7, 3)\}$ are in the solution set.

26. PRESCHOOL Mrs. Jones is buying new books and puzzles for her preschool classroom. Each book costs $6, and each puzzle costs $4. Write and graph an inequality to determine how many books and puzzles she can buy for $96.

Performance Task

Provide a clear solution to each part of the task. Be sure to show all of your work, include all relevant drawings, and justify your answers.

TAXI CABS Jonathon drives a taxi cab. He charges a flat ride fee of $2.50 and then $0.25 per one-eighth mile.

Part A

A passenger gets in Jonathon's cab. She can spend a maximum of $11.00 on the cab fare.

1. Write an inequality that represents the situation in terms of whole miles.

2. What is the greatest distance the passenger can travel in Jonathon's cab?

3. Graph your solution on a number line.

Part B

Jonathon owns his taxi cab. His monthly operation expenses, including gas, insurance, and other expenditures, total $1125 per month. After these expenses, Jonathon wants to earn a minimum of $2200 per month. His average fare is $4.75.

4. **Sense-Making** Write an inequality that represents the situation.

5. **Reasoning** Based on this average, what is the minimum number of cab fares Jonathon needs in order to cover his expenses and meet his monthly goal?

Part C

Jamal has a coupon for Jonathon's taxi service for $3 off a taxi ride that is 3 miles or longer. Jamal wants to spend no more than $7 on cab fare.

6. In the context of the situation, what are the constraints?

7. Write and graph the inequalities to represent the distances Jamal can ride.

8. List 3 viable solutions.

Part D

9. **Construct an Argument** Rosita's taxi service charges an initial flat fee of $1.25 and $0.50 per one sixth mile. If a passenger needs to travel 1.5 miles, which cab should she choose? Explain your answer.

Test-Taking Strategy

Example

Read the problem. Identify what you need to know. Then use the information in the problem to solve. Show your work.

Pedro's test scores so far are 89, 74, 79, 85, and 88. He needs a test average of at least 85 to earn an A for the semester. What score must he have on the final test to earn an A for the semester?

A 33 **B** 85 **C** 95 **D** 100

Step 1 **What do you need to find?**

The lowest score Pedro can get and still earn an A for the semester.

Step 2 **Is there enough information given to solve the problem?**

Yes, you know Pedro's current test scores and the average test score he needs to get an A.

Step 3 **How can you solve the problem?**

Write and solve an inequality. His test average must be greater than or equal to 85.

Step 4 **What is the correct answer?**

$$\frac{89 + 74 + 79 + 85 + 88 + t}{6} \geq 85$$

$$\frac{415 + t}{6} \geq 85$$

$$415 + t \geq 510$$

$$t \geq 95$$

Pedro's final test score must be 95 or better in order for him to earn an A. The answer is C.

Apply the Strategy

Read the problem. Identify what you need to know. Then use the information in the problem to solve.

Rosa earns $200 a month delivering newspapers, plus an average of $11 per hour babysitting. If her goal is to earn at least $295 this month, how many hours will she have to babysit?

A at least 8 hours **C** at least 20 hours

B at least 9 hours **D** at least 26 hours

Answer the questions below.

a. What do you need to find?

b. Is there enough information given to solve the problem?

c. How can you solve the problem?

d. What is the correct answer? $200 + 11h \geq 295$

Read each question. Then fill in the correct answer on the answer document provided by your teacher or on a sheet of paper.

1. Which inequality has the solution set graphed below?

- ⚪ A $5 - 2x < 9$ or $3x + 4 \leq 1$
- ⚪ B $5 - 2x > 9$ or $3x + 4 \leq 1$
- ⚪ C $5 - 2x > 9$ or $3x + 4 \geq 1$
- ⚪ D $5 - 2x < 9$ or $3x + 4 \geq 1$

2. Five times the greater of two consecutive even integers is 14 greater than three times the lesser number. What is the greater number?

3. At the carnival, Jamie bought 4 plain balloons and 3 picture balloons. Paola bought 4 picture balloons. If Paola spent no more than Jamie, which statement is true?

- ⚪ A If plain balloons cost $0.75, then picture balloons must cost $3.00 or less.
- ⚪ B If plain balloons cost $0.75, then picture balloons must cost $3.00 or more.
- ⚪ C If plain balloons cost $1.50, then picture balloons must cost $3.00 or less.
- ⚪ D If plain balloons cost $1.50, then picture balloons must cost $3.00 or more.

Test-Taking Tip

Question 3 To model the problem, write an inequality using two variables, one for the cost of plain balloons and one for the cost of picture balloons. Then substitute a value for one variable and solve the inequality.

4. Bianca solved the inequality below and graphed its solution set on a number line.

$$-\frac{3x}{4} > -2x - 5$$

Which is Bianca's number line?

- ⚪ A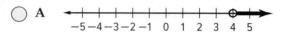
- ⚪ B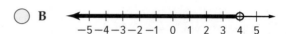
- ⚪ C
- ⚪ D

5. An election poll reports that 52% of voters are in favor of the incumbent candidate. The margin of error is within 3 percentage points, which means that the range of voters who are actually in favor of the incumbent is within 3 percentage points above or below the reported percent. Which inequality represents the range of voters who are actually in favor of the incumbent candidate?

- ⚪ A $|x + 52| > 3$
- ⚪ B $|x - 52| \geq 3$
- ⚪ C $|x + 52| \leq 3$
- ⚪ D $|x - 52| \leq 3$

6. Which compound inequality has a solution set of $\{x \mid -3 < x \leq 1\}$?

- ⚪ A $-9 < 3x + 5 \leq 3$
- ⚪ B $-4 < 3x + 5 \leq 8$
- ⚪ C $-9 < 3x - 5 \leq 3$
- ⚪ D $-4 < 3x - 5 \leq 8$

7. The graph of a linear inequality in two variables is shown below.

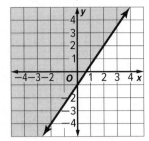

What is the linear inequality?

○ **A** $2y + 3x \geq 2$ ○ **C** $2y + 3x \geq -2$

○ **B** $2y - 3x \geq 2$ ○ **D** $2y - 3x \geq -2$

8. Which inequality does not have the solution set graphed below?

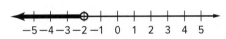

○ **A** $-3x > 6$ ○ **C** $-3x < 6$

○ **B** $3x < -6$ ○ **D** $-6 > 3x$

9. Select all pairs of equations that represent a pair of perpendicular lines.

☐ **A** $y = \frac{1}{3}x + 5,\ y - 4 = \frac{1}{3}(x + 3)$

☐ **B** $x + 4y = 10,\ y = 4x + 7$

☐ **C** $x - \frac{1}{2}y = 6,\ y = 2x - 5$

☐ **D** $y = -5x + 2,\ x + 5y = 4$

☐ **E** $y = \frac{1}{6}x - 10,\ 6x + y = 4$

10. Cindy's online music club allows up to 25 song downloads per month for $15. If she downloads more than 25 songs, she pays an additional $1.25 per song. Cindy writes the inequality $1.25(n - 25) + 15 \leq t$, where n represents the number of songs downloaded and t represents her monthly spending limit for the music club.

a. If $t = 25$, what is the greatest value of n for which Cindy will stay within her monthly spending limit?

        ⬚

b. **MP** What mathematical practice did you use to solve this problem?

11. Which of the following is not a correct statement about the function $f(x) = 3x - 7$?

○ **A** The domain of the function is all real numbers.

○ **B** The range of the function is all real numbers.

○ **C** The function has zeros of 7 and −7.

○ **D** The graph of the function intersects the y-axis at −7.

12. The inequality $85 \leq t \leq 94$ represents the acceptable temperature range for water in the pool at a local health club, where t is the temperature in degrees Fahrenheit. Which statement is true?

○ **A** Temperatures of 94°F and above or 85°F and below are unacceptable.

○ **B** Temperatures of 94°F and below or 85°F and above are unacceptable.

○ **C** Temperatures greater than 94°F or less than 85°F are unacceptable.

○ **D** Temperatures less than 94°F or greater than 85°F are unacceptable.

Need Extra Help?

If you missed Question...	1	2	3	4	5	6	7	8	9	10	11	12
Go to Lesson...	5-4	2-4	5-1	5-3	5-5	5-4	5-6	5-2	4-3	5-3	3-2	5-4

THEN

You solved linear equations in one variable.

NOW

In this chapter you will:

- Solve systems of linear equations by graphing, substitution, and elimination.
- Solve systems of linear inequalities by graphing.

 WHY

MUSIC Marching band competitions are a source of funds for the bands that host them. They use these funds to buy new uniforms, instruments, and to help with travel expenses.

Use the Mathematical Practices to complete the activity.

1. Use Tools Use the Internet to find information about a local band competition or find out if your school hosts a competition. What are the admission fees for students and adults for this event?

2. Reasoning If you knew how many adult and student tickets were sold, what could you determine? Could you use an equation to determine how much money the band made?

3. Apply Math Using the admission fees you found, write an equation showing the total amount of band income if 200 adults and 100 students attend the competition.

4. Model with Mathematics Use the Equation Chart tool in ConnectED to represent the necessary equation.

5. Discuss Could you calculate how many student tickets and how many adult tickets were sold if you knew only the price of each and the total amount of income?

Go Online to Guide Your Learning

Explore & Explain	Organize

 ### The Geometer's Sketchpad

Visualize and model slope using the Slope of a Line sketch in Lesson 6-1.

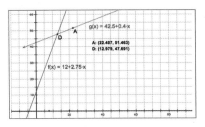

 ### eToolkit

Investigate using the **Equation Chart Mat** and the **Algebra Tiles** tool to model substitution.

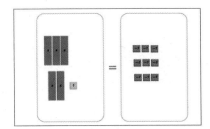

eBook

Interactive Student Guide

Before starting the chapter, answer the **Chapter Focus** preview questions. Check your answers as you complete each lesson. At the end of the chapter, try the **Performance Task**.

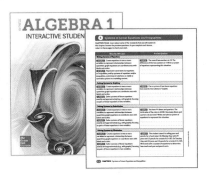

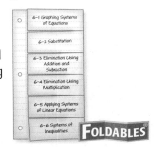 ### Foldables

Systems of Linear Equations and Inequalities Make this Foldable to help you organize your notes about solving systems of equations and inequalities. Begin with a sheet of notebook paper folded lengthwise. Cut and label six tabs using the lesson titles.

Collaborate

 ### Chapter Project

In the Save the Date project, you will use what you have learned about systems of equations to complete a project.

Focus

 ### LEARNSMART

Need help studying? Complete the **Linear and Exponential Relationships** domain in LearnSmart to review for the chapter test.

 ### ALEKS

You can use the **Systems** and **Linear Inequalities** topics in ALEKS to find out what you know about linear functions and what you are ready to learn.*

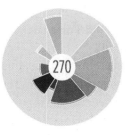

* Ask your teacher if this is part of your program.

Get Ready for the Chapter

Go Online! for Vocabulary Review Games and key vocabulary in 13 languages.

Connecting Concepts

Concept Check

Review the concepts used in this chapter by answering each question below.

1. Is the expression 6^2 the same as $(6)^2$?

2. Is the expression $3 + 6^2$ the same as $(3 + 6)^2$?

3. What would be the first step in evaluating the expression $\left(\frac{7}{11}\right)^2$?

4. The formula for the area of a triangle is $A = \frac{1}{2}bh$, where A represents the area, b is the base, and h is the height. Solve the formula for the base of the triangle.

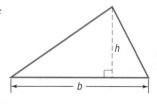

5. Explain what step you would do first to solve $2x + 4y = 12$ for x.

6. Given $3x = 36 - 12y$, what would you do to simplify the equation?

7. Given $5x - 10y = 40$, what would the equation be if you solve for y?

8. What do you call two or more inequalities with the same variables?

New Vocabulary

English		Español
system of equations	p. 339	sistema de ecuaciones
consistent	p. 339	consistente
independent	p. 339	independiente
dependent	p. 339	dependiente
inconsistent	p. 339	inconsistente
substitution	p. 348	sustitución
elimination	p. 354	eliminación
matrix	p. 374	matriz
element	p. 374	elemento
dimension	p. 374	dimensión
augmented matrix	p. 374	matriz ampliada
row reduction	p. 375	reducción de fila
identity matrix	p. 375	matriz
system of inequalities	p. 376	sistema de desigualdades

Performance Task Preview

You can use the concepts and skills in the chapter to solve problems in a real-world setting. Understanding systems of equations will help you finish the Performance Task at the end of the chapter.

MP **In this Performance Task you will:**

- make sense of problems
- reason abstractly and quantitatively
- construct viable arguments
- model with mathematics
- look for and make use of structure

Review Vocabulary

domain dominio the set of the first numbers of the ordered pairs in a relation

intersection intersección the graph of a compound inequality containing *and*; the solution is the set of elements common to both graphs

proportion proporción an equation stating that two ratios are equal

Proportion

$$\frac{24}{30} = \frac{4}{5}$$

Graphing Systems of Equations

∷Then	∷Now	∷Why?

• You graphed linear equations.

1 Determine the number of solutions a system of linear equations has.

2 Solve systems of linear equations by graphing.

A volleyball team is selling T-shirts. There is a $600 set-up fee and then each T-shirt costs $3 to print. They plan to sell the T-shirts for $15 each. The volleyball team wants to know how many shirts they will need to sell in order to make a profit.

Graphing a system can show when a profit is made. The cost of producing the T-shirts can be modeled by the equation $y = 3x + 600$, where y represents the cost of production and x is the number of T-shirts produced.

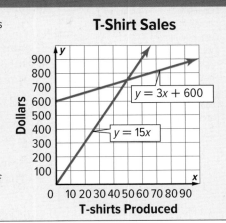

T-Shirt Sales

$y = 3x + 600$

$y = 15x$

Dollars

T-shirts Produced

 New Vocabulary

system of equations
consistent
independent
dependent
inconsistent

 Mathematical Practices

3 Construct viable arguments and critique the reasoning of others.

8 Look for and express regularity in repeated reasoning.

1 Possible Number of Solutions The income from the T-shirts sold can be modeled by the equation $y = 15x$, where y represents the total income of selling the T-shirts, and x is the number of T-shirts sold.

If we graph these equations, we can see at which point the volleyball team begins making a profit. The point where the two graphs intersect is where the volleyball team breaks even. At this point, the income equals the cost. This happens when the volleyball team sells 50 T-shirts. If the volleyball team sells more than 50 T-shirts, they will make a profit.

The two equations, $y = 3x + 600$ and $y = 15x$, form a **system of equations**. The ordered pair that is a solution of both equations is the solution of the system. A system of two linear equations can have one solution, an infinite number of solutions, or no solution.

• If a system has at least one solution, it is said to be **consistent**. The graphs intersect at one point or are the same line.

• If a consistent system has exactly one solution, it is said to be **independent**. If it has an infinite number of solutions, it is **dependent**. This means that there are unlimited solutions that satisfy both equations.

• If a system has no solution, it is said to be **inconsistent**. The graphs are parallel.

Concept Summary Possible Solutions			
Number of Solutions	exactly one	infinite	no solution
Terminology	consistent and independent	consistent and dependent	inconsistent
Graph			

Study Tip

MP **Reasoning** When both equations are of the form $y = mx + b$, the values of m and b can determine the number of solutions.

Compare m and b	Number of Solutions
different m values	one
same m value, but different b values	none
same m value, and same b value	infinite

Example 1   Number of Solutions

Use the graph at the right to determine whether each system is *consistent* or *inconsistent* and if it is *independent* or *dependent*.

a. $y = -2x + 3$
$y = x - 5$

Since the graphs of these two lines intersect at one point, there is exactly one solution. Therefore, the system is consistent and independent.

b. $y = -2x - 5$
$y = -2x + 3$

Since the graphs of these two lines are parallel, there is no solution of the system. Therefore, the system is inconsistent.

> **Guided Practice**

1A. $y = 2x + 3$
$y = -2x - 5$

1B. $y = x - 5$
$y = -2x - 5$

2 **Solve by Graphing**   One method of solving a system of equations is to graph the equations carefully on the same coordinate grid and find their point of intersection. This point is the solution of the system.

Example 2   Solve by Graphing

Graph each system and determine the number of solutions that it has. If it has one solution, name it.

a. $y = -3x + 10$
$y = x - 2$

The graphs appear to intersect at the point (3, 1). You can check this by substituting 3 for x and 1 for y.

CHECK $y = -3x + 10$ Original equation

 $1 \stackrel{?}{=} -3(3) + 10$ Substitution

 $1 \stackrel{?}{=} -9 + 10$ Multiply.

 $1 = 1 \checkmark$

 $y = x - 2$ Original equation

 $1 \stackrel{?}{=} 3 - 2$ Substitution

 $1 = 1 \checkmark$ Multiply.

The solution is (3, 1).

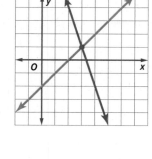

b. $2x - y = -1$
$4x - 2y = 6$

The lines have the same slope but different y-intercepts, so the lines are parallel. Since they do not intersect, there is no solution of this system. The system is inconsistent.

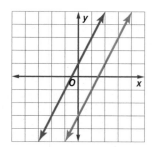

Review Vocabulary

parallel lines never intersect and have the same slope

Graph each system and determine the number of solutions that it has. If it has one solution, name it.

2A. $x - y = 2$
$3y + 2x = 9$

2B. $y = -2x - 3$
$6x + 3y = -9$

We can use what we know about systems of equations to solve many real-world problems involving constraints that are modeled by two or more different functions.

Real-World Example 3   Write and Solve a System of Equations

The number of girls participating in high school soccer and track and field has steadily increased over the past few years. Use the information in the table to predict the approximate year when the number of girls participating in these two sports will be the same.

High School Sport	Number of Girls Participating in 2007 (thousands)	Number of Girls Participating in 2012 (thousands)
soccer	345	371
track and field	458	467

Source: National Federation of State High School Associations

Step 1   Find the average rate of increase for both sports.

Soccer: $\dfrac{371 - 345}{2012 - 2007} = 5.2$ Track and field: $\dfrac{467 - 458}{2012 - 2007} = 1.8$

Step 2   Write equations describing the number of girls participating in each sport if the average rate of increase stays the same.

Let y = the number of girls participating. Let x = the number of years after 2012.

Soccer: $y = 5.2x + 371$ Track and field: $y = 1.8x + 467$

Step 3   Graph both equations. The graph appears to intersect at (28, 517).

Use substitution to check this answer.

$y = 5.2x + 371$ $y = 1.8x + 467$

$517 \stackrel{?}{=} 5.2(28) + 371$ $517 \stackrel{?}{=} 1.8(28) + 467$

$517 \approx 516.6$ ✔ $517 \approx 517.4$ ✔

The solution means that approximately 28 years after 2012—or in 2040—the number of girls participating in high school soccer and track and field will be the same, about 517,000.

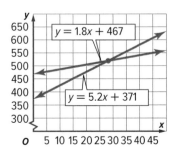

Go Online!

Interact with the graphs of systems of linear equations by using the Graphing Tools in ConnectED.

▶ **Guided Practice**

3. VIDEO GAMES Joe and Josh each want to buy a video game. Joe has $14 and saves $10 a week. Josh has $26 and saves $7 a week. In how many weeks will they have the same amount?

Check Your Understanding = Step-by-Step Solutions begin on page R11.

Example 1 Use the graph at the right to determine whether each system is *consistent* or *inconsistent* and whether it is *independent* or *dependent*.

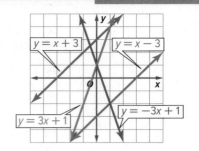

1. $y = -3x + 1$
$y = 3x + 1$

2. $y = 3x + 1$
$y = x - 3$

3. $y = x - 3$
$y = x + 3$

4. $y = x + 3$
$x - y = -3$

5. $x - y = -3$
$y = -3x + 1$

6. $y = -3x + 1$
$y = x - 3$

Example 2 Graph each system and determine the number of solutions that it has. If it has one solution, name it.

7. $y = x + 4$
$y = -x - 4$

8. $y = x + 3$
$y = 2x + 4$

Example 3 **9.** **MP MODELING** Alberto and Ashanti are reading a graphic novel.

 a. Write an equation to represent the pages each boy has read.

 b. Graph each equation.

 c. How long will it be before Alberto has read more pages than Ashanti? Check and interpret your solution.

Alberto
35 pages read;
20 pages each day

Ashanti
85 pages read;
10 pages each day

Practice and Problem Solving Extra Practice is on page R6.

Example 1 Use the graph at the right to determine whether each system is *consistent* or *inconsistent* and whether it is *independent* or *dependent*.

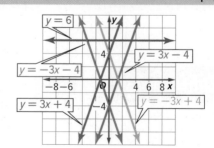

10. $y = 6$
$y = 3x + 4$

11. $y = 3x + 4$
$y = -3x + 4$

12. $y = -3x + 4$
$y = -3x - 4$

13 $y = -3x - 4$
$y = 3x - 4$

14. $3x - y = -4$
$y = 3x + 4$

15. $3x - y = 4$
$3x + y = 4$

Example 2 Graph each system and determine the number of solutions that it has. If it has one solution, name it.

16. $y = -3$
$y = x - 3$

17. $y = 4x + 2$
$y = -2x - 4$

18. $y = x - 6$
$y = x + 2$

19. $x + y = 4$
$3x + 3y = 12$

20. $x - y = -2$
$-x + y = 2$

21. $x + 2y = 3$
$x = 5$

22. $2x + 3y = 12$
$2x - y = 4$

23. $2x + y = -4$
$y + 2x = 3$

24. $2x + 2y = 6$
$5y + 5x = 15$

Example 3

25. SCHOOL DANCE Akira and Jen are competing to see who can sell the most tickets for the Winter Dance. On Monday, Akira sold 22 and then sold 30 per day after that. Jen sold 53 on Monday and then sold 20 per day after that.

a. Write equations for the number of tickets each person has sold.

b. Graph each equation.

c. Solve the system of equations. Check and interpret your solution.

26. (MP) **MODELING** If x is the number of years since 2013 and y is the percent of people paying bills, the following equations represent the percent of people writing checks to pay their bills and the percent of people paying their bills online.

Writing checks: $y = -2x + 24$ Online: $y = 6x + 55$

a. Graph the system of equations.

b. Estimate the year check writing and the online payments were used equally.

Graph each system and determine the number of solutions that it has. If it has one solution, name it.

27. $y = \frac{1}{2}x$

$y = x + 2$

28. $y = 6x + 6$

$y = 3x + 6$

29. $y = 2x - 17$

$y = x - 10$

30. $8x - 4y = 16$

$-5x - 5y = 5$

31. $3x + 5y = 30$

$3x + y = 18$

32. $-3x + 4y = 24$

$4x - y = 7$

33. $2x - 8y = 6$

$x - 4y = 3$

34. $4x - 6y = 12$

$-2x + 3y = -6$

35. $2x + 3y = 10$

$4x + 6y = 12$

36. $3x + 2y = 10$

$2x + 3y = 10$

37. $3y - x = -2$

$y - \frac{1}{3}x = 2$

38. $\frac{8}{5}y = \frac{2}{5}x + 1$

$\frac{2}{5}y = \frac{1}{10}x + \frac{1}{4}$

39. MULTI-STEP In 2012, website A had 512 million visitors, and website B had 131 million visitors. In 2016, website A had 476 million visitors, and website B had 251 million visitors.

a. What conclusions can you make?

b. Describe your solution process in making your conclusions.

c. What assumptions did you make?

40. ELECTRONICS Suppose x represents the number of years since 2012 and y represents the number of units sold. Then the number of tablets sold each year since 2012, in millions, can be modeled by the equation $y = 12.5x + 10.9$. The number of laptops sold each year since 2012, in millions, can be modeled by the equation $y = -9.1x + 78.8$.

a. Graph each equation and find its domain and range.

b. In which year did tablet sales surpass laptop sales?

Graph each system and determine the number of solutions that it has. If it has one solution, name it.

41. $2y = 1.2x - 10$

$4y = 2.4x$

42. $x = 6 - \frac{3}{8}y$

$4 = \frac{2}{3}x + \frac{1}{4}y$

Write a system of linear equations for each graph given. Then name the solution, if it has one.

43.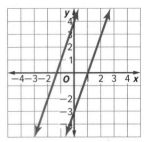

44.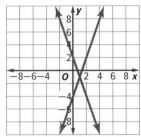

45.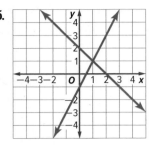

46. **MULTIPLE REPRESENTATIONS** In this problem, you will explore different methods for finding the intersection of the graphs of two linear equations.

   a. **Algebraic** Use algebra to solve the equation $\frac{1}{2}x + 3 = -x + 12$.

   b. **Graphical** Use a graph to solve $y = \frac{1}{2}x + 3$ and $y = -x + 12$.

   c. **Analytical** How is the equation in part **a** related to the system in part **b**?

   d. **Verbal** Explain how to use the graph in part **b** to solve the equation in part **a**.

H.O.T. Problems   Use **H**igher-**O**rder **T**hinking Skills

47. **CHALLENGE** Use graphing to find the solution of the system of equations $2x + 3y = 5$, $3x + 4y = 6$, and $4x + 5y = 7$.

48. **ERROR ANALYSIS** Store A is offering a 10% discount on the purchase of all electronics in their store. Store B is offering $10 off all the electronics in their store. Francisca and Alan are deciding which offer will save them more money. Is either of them correct? Explain your reasoning.

Francisca	Alan
You can't determine which store has the better offer unless you know the price of the items you want to buy.	Store A has the better offer because 10% of the sale price is a greater discount than $10.

49. **CONSTRUCT ARGUMENTS** Determine whether a system of two linear equations with (0, 0) and (2, 2) as solutions *sometimes, always,* or *never* has other solutions. Explain.

50. **WHICH ONE DOESN'T BELONG?** Which one of the following systems of equations doesn't belong with the other three? Explain your reasoning.

$4x - y = 5$	$-x + 4y = 8$	$4x + 2y = 14$	$3x - 2y = 1$
$-2x + y = -1$	$3x - 6y = 6$	$12x + 6y = 18$	$2x + 3y = 18$

51. **OPEN-ENDED** Write three equations such that they form three systems of equations with $y = 5x - 3$. The three systems should be inconsistent, consistent and independent, and consistent and dependent, respectively.

52. **WRITING IN MATH** Describe the advantages and disadvantages to solving systems of equations by graphing.

53. Jenny is selling T-shirts and sweatshirts to raise money for the pep squad. She sold 12 shirts total. The number of T-shirts Jenny sold was 2 more than 4 times the number of sweatshirts she sold. How many of each type of shirt did Jenny sell? MP 4

   A 2 sweatshirts; 10 T-shirts

   B 10 sweatshirts; 2 T-shirts

   C 10 sweatshirts; 50 T-shirts

   D 50 sweatshirts; 10 T-shirts

54. Which system of equations could have (1, 5) as a solution, if $b < 0$ and $c > 0$? MP 2

   A $y = -x + 6$
      $y = -x + b$

   B $y = -x + 6$
      $y = cx + 6$

   C $y = 2x + 3$
      $y = bx + c$

   D $y = 2x + 3$
      $y = 2x + b$

55. Some values for two linear equations are shown in the tables below.

Equation 1			Equation 2	
x	**y**		**x**	**y**
−2	8		−4	−8
0	6		−1	−2
5	1		2	4
10	−4		5	10

What is the solution to the system of equations represented by these tables? MP 8

   A (0, 6)

   B (2, 4)

   C (4, 2)

   D (6, 0)

56. A quarter, some dimes, and some nickels are worth $1. If there are 12 coins altogether, how many dimes are there? MP 2, 4

57. MULTI-STEP In 2017, the population of the town of Smithfield was 7200 and growing by 100 residents per year. The population of the town of Plymouth was 8850 in 2017 and decreasing by 50 residents per year. MP 4

   a. Write equations to represent the population of each town.

   b. In what year will the towns be expected to have the same population?

   c. What is the population of both cities when they have the same population?

   d. Describe your solution process.

   e. What assumptions did you make?

58. What system of equations is given by the lines in the graph? MP 1

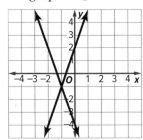

   A $y = 3x + 4$
      $y = -3x + 2$

   B $y = -x + 1$
      $y = -2x - 2$

   C $y = 2x + 3$
      $y = -4x - 3$

   D $y = 3x + 2$
      $y = -3x - 4$

59. On a computer game, Fermin's score was 12 points less than twice Lisa's score. The total of both scores was 18 points. How many points did each person score? MP 4

   A Lisa: 2 points, Fermin: 16 points

   B Lisa: 8 points, Fermin: 10 points

   C Lisa: 10 points, Fermin: 8 points

   D Lisa: 15 points, Fermin: 3 points

You can use a graphing calculator to graph and solve a system of equations.

Mathematical Practices

5 Use appropriate tools strategically.

Activity 1   Solve a System of Equations

Work cooperatively. Solve the system of equations. Round to the nearest hundredth.

$5.23x + y = 7.48$
$6.42x - y = 2.11$

Step 1 Solve each equation for y to enter them into the calculator.

$5.23x + y = 7.48$	First equation
$5.23x + y - 5.23x = 7.48 - 5.23x$	Subtract 5.23x from each side.
$y = 7.48 - 5.23x$	Simplify.
$6.42x - y = 2.11$	Second equation
$6.42x - y - 6.42x = 2.11 - 6.42x$	Subtract 6.42x from each side.
$-y = 2.11 - 6.42x$	Simplify.
$(-1)(-y) = (-1)(2.11 - 6.42x)$	Multiply each side by −1.
$y = -2.11 + 6.42x$	Simplify.

Step 2 Enter these equations in the **Y=** list and graph in the standard viewing window.

KEYSTROKES: $\boxed{\text{Y=}}$ 7.48 $\boxed{-}$ 5.23 $\boxed{\text{X,T,θ,}n}$
$\boxed{\text{ENTER}}$ $\boxed{(-)}$ 2.11 $\boxed{+}$
6.42 $\boxed{\text{X,T,θ,}n}$ $\boxed{\text{ZOOM}}$ 6

Step 3 Use the **CALC** menu to find the point of intersection.

KEYSTROKES: $\boxed{\text{2nd}}$ [CALC] 5 $\boxed{\text{ENTER}}$ $\boxed{\text{ENTER}}$
$\boxed{\text{ENTER}}$

[−10, 10] scl: 1 by [−10, 10] scl: 1

The solution is approximately (0.82, 3.17).

When you solve a system of equations with $y = f(x)$ and $y = g(x)$, the solution is an ordered pair that satisfies both equations. The solution always occurs when $f(x) = g(x)$. Thus, the x-coordinate of the solution is the value of x where $f(x) = g(x)$.

One method you can use to solve an equation with one variable is by graphing and solving a system of equations based on the equation. To do this, write a system using both sides of the equation. Then use a graphing calculator to solve the system.

Activity 2   Use a System to Solve a Linear Equation

Work cooperatively. Use a system of equations to solve $5x + 6 = -4$.

Step 1 Write a system of equations. Set each side of the equation equal to y.

$y = 5x + 6$ First equation
$y = -4$ Second equation

Step 2 Enter these equations in the **Y=** list and graph.

Step 3 Use the **CALC** menu to find the point of intersection.

[−10, 10] scl: 1 by [−10, 10] scl: 1

The solution is −2.

Exercises

Work cooperatively. Use a graphing calculator to solve each system of equations. Write decimal solutions to the nearest hundredth.

1. $y = 2x - 3$
$y = -0.4x + 5$

2. $y = 6x + 1$
$y = -3.2x - 4$

3. $x + y = 9.35$
$5x - y = 8.75$

4. $2.32x - y = 6.12$
$4.5x + y = -6.05$

5. $5.2x - y = 4.1$
$1.5x + y = 6.7$

6. $1.8 = 5.4x - y$
$y = -3.8 - 6.2x$

7. $7x - 2y = 16$
$11x + 6y = 32.3$

8. $3x + 2y = 16$
$5x + y = 9$

9. $0.62x + 0.35y = 1.60$
$-1.38x + y = 8.24$

10. $75x - 100y = 400$
$33x - 10y = 70$

Use a graphing calculator to solve each equation. Write decimal solutions to the nearest hundredth.

11. $4x - 2 = -6$

12. $3 = 1 + \frac{x}{2}$

13. $\frac{x + 4}{-2} = -1$

14. $\frac{3}{2}x + \frac{1}{2} = 2x - 3$

15. $4x - 9 = 7 + 7x$

16. $-2 + 10x = 8x - 1$

17. WRITING IN MATH Explain why you can solve an equation like $r = ax + b$ by solving the system of equations $y = r$ and $y = ax + b$.

LESSON 2

Substitution

:·Then

- You solved systems of equations by graphing.

:·Now

1. Solve systems of equations by using substitution.

2. Solve real-world problems involving systems of equations by using substitution.

:·Why?

- Two movies were released at the same time. Movie A earned $31 million in its opening week, but fell to $15 million the following week. Movie B opened earning $21 million and fell to $11 million the following week. If the earnings for each movie continue to decrease at the same rate, when will they earn the same amount?

New Vocabulary

substitution

Mathematical Practices

2 Reason abstractly and quantitatively.

---

1 Solve by Substitution You can use a system of equations to find when the movie earnings are the same. One method of finding an exact solution of a system of equations is called **substitution**.

> #### Key Concept Solving by Substitution
>
> **Step 1** When necessary, solve at least one equation for one variable.
>
> **Step 2** Substitute the resulting expression from Step 1 into the other equation to replace the variable. Then solve the equation.
>
> **Step 3** Substitute the value from Step 2 into either equation, and solve for the other variable. Write the solution as an ordered pair.

Example 1   Solve a System by Substitution

Use substitution to solve the system of equations.

$y = 2x + 1$ ⟵   **Step 1** The first equation is already solved for y.
$3x + y = -9$

Step 2 Substitute $2x + 1$ for y in the second equation.

$3x + y = -9$	Second equation
$3x + 2x + 1 = -9$	Substitute $2x + 1$ for y.
$5x + 1 = -9$	Combine like terms.
$5x = -10$	Subtract 1 from each side.
$x = -2$	Divide each side by 5.

Step 3 Substitute -2 for x in either equation to find y.

$y = 2x + 1$	First equation
$= 2(-2) + 1$	Substitute -2 for x.
$= -3$	Simplify.

The solution is $(-2, -3)$.

CHECK You can check your solution by graphing.

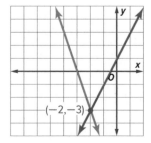

$(-2, -3)$

▶ Guided Practice

1A. $y = 4x - 6$
   $5x + 3y = -1$

1B. $2x + 5y = -1$
   $y = 3x + 10$

Study Tip

MP Perseverance If both equations are in the form $y = mx + b$, they can simply be set equal to each other and then solved for x. The solution for x can then be used to find the value of y.

If a variable is not isolated in one of the equations in a system, solve an equation for a variable first. Then you can use substitution to solve the system.

Example 2 Solve and then Substitute

Use substitution to solve the system of equations.

$$x + 2y = 6$$
$$3x - 4y = 28$$

Step 1 Solve the first equation for x since the coefficient is 1.

$x + 2y = 6$	First equation
$x + 2y - 2y = 6 - 2y$	Subtract $2y$ from each side.
$x = 6 - 2y$	Simplify.

Step 2 Substitute $6 - 2y$ for x in the second equation to find the value of y.

$3x - 4y = 28$	Second equation
$3(6 - 2y) - 4y = 28$	Substitute $6 - 2y$ for x.
$18 - 6y - 4y = 28$	Distributive Property
$18 - 10y = 28$	Combine like terms.
$18 - 10y - 18 = 28 - 18$	Subtract 18 from each side.
$-10y = 10$	Simplify.
$y = -1$	Divide each side by -10.

Step 3 Find the value of x.

$x + 2y = 6$	First equation
$x + 2(-1) = 6$	Substitute -1 for y.
$x - 2 = 6$	Simplify.
$x = 8$	Add 2 to each side.

Guided Practice

2A. $4x + 5y = 11$
    $y - 3x = -13$

2B. $x - 3y = -9$
    $5x - 2y = 7$

Generally, if you solve a system of equations and the result is a false statement such as $3 = -2$, there is no solution. If the result is an identity, such as $3 = 3$, then there are an infinite number of solutions.

Example 3 No Solution or Infinitely Many Solutions

Use substitution to solve the system of equations.

$$y = 2x - 4$$
$$-6x + 3y = -12$$

Substitute $2x - 4$ for y in the second equation.

$-6x + 3y = -12$	Second equation
$-6x + 3(2x - 4) = -12$	Substitute $2x - 4$ for y.
$-6x + 6x - 12 = -12$	Distributive Property
$-12 = -12$	Combine like terms.

Study Tip

Dependent Systems There are infinitely many solutions of the system in Example 3 because the equations in slope-intercept form are equivalent, and they have the same graph.

This statement is an identity. Thus, there are an infinite number of solutions.

Guided Practice** Use substitution to solve each system of equations.

3A. $2x - y = 8$
$\quad y = 2x - 3$

3B. $4x - 3y = 1$
$\quad 6y - 8x = -2$

---

2 Solve Real-World Problems

You can use substitution to find the solution of a real-world problem involving constraints modeled by a system of equations.

Real-World Example 4 — Write and Solve a System of Equations

MUSIC A store sold a total of **125** car stereo systems and speakers in one week. The stereo systems sold for **\$104.95**, and the speakers sold for **\$18.95**. The sales from these two items totaled **\$6926.75**. How many of each item were sold?

Number of Units Sold	c	t	125
Sales (\$)	$104.95c$	$18.95t$	6926.75

Let c = the number of car stereo systems sold, and let t = the number of speakers sold.

So, the two equations are $c + t = 125$ and $104.95c + 18.95t = 6926.75$.

Notice that $c + t = 125$ represents combinations of car stereo systems and speakers with a sum of 125. The equation $104.95c + 18.95t = 6926.75$ represents the combinations of car stereo systems and speakers with a sales of \$6926.75. The solution of the system of equations represents the option that meets both of the constraints.

Step 1 Solve the first equation for c.

$$c + t = 125 \qquad \text{First equation}$$
$$c + t - t = 125 - t \qquad \text{Subtract } t \text{ from each side.}$$
$$c = 125 - t \qquad \text{Simplify.}$$

Step 2 Substitute $125 - t$ for c in the second equation.

$$104.95c + 18.95t = 6926.75 \qquad \text{Second equation}$$
$$104.95(125 - t) + 18.95t = 6926.75 \qquad \text{Substitute } 125 - t \text{ for } c.$$
$$13{,}118.75 - 104.95t + 18.95t = 6926.75 \qquad \text{Distributive Property}$$
$$13{,}118.75 - 86t = 6926.75 \qquad \text{Combine like terms.}$$
$$-86t = -6192 \qquad \text{Subtract 13,118.75 from each side.}$$
$$t = 72 \qquad \text{Divide each side by } -86.$$

Step 3 Substitute 72 for t in either equation to find the value of c.

$$c + t = 125 \qquad \text{First equation}$$
$$c + 72 = 125 \qquad \text{Substitute 72 for } t.$$
$$c = 53 \qquad \text{Subtract 72 from each side.}$$

The store sold 53 car stereo systems and 72 speakers.

Guided Practice

4. BASEBALL As of 2016, the New York Yankees and the Cincinnati Reds together had won a total of 32 World Series. The Yankees had won 5.4 times as many as the Reds. How many World Series had each team won?

Examples 1–3 **Use substitution to solve each system of equations.**

1. $y = x + 5$
   $3x + y = 25$

2. $x = y - 2$
   $4x + y = 2$

3. $3x + y = 6$
   $4x + 2y = 8$

4. $2x + 3y = 4$
   $4x + 6y = 9$

5. $x - y = 1$
   $3x = 3y + 3$

6. $2x - y = 6$
   $-3y = -6x + 18$

Example 4

7. **GEOMETRY** The sum of the measures of angles X and Y is 180°. The measure of angle X is 24° greater than the measure of angle Y.

   a. Define the variables, and write equations for this situation.

   b. Find the measure of each angle.

Practice and Problem Solving   Extra Practice is on page R6.

Examples 1–3 **Use substitution to solve each system of equations.**

8. $y = 5x + 1$
   $4x + y = 10$

⑨ $y = 4x + 5$
   $2x + y = 17$

10. $y = 3x - 34$
    $y = 2x - 5$

11. $y = 3x - 2$
    $y = 2x - 5$

12. $2x + y = 3$
    $4x + 4y = 8$

13. $3x + 4y = -3$
    $x + 2y = -1$

14. $y = -3x + 4$
    $-6x - 2y = -8$

15. $-1 = 2x - y$
    $8x - 4y = -4$

16. $x = y - 1$
    $-x + y = -1$

17. $y = -4x + 11$
    $3x + y = 9$

18. $y = -3x + 1$
    $2x + y = 1$

19. $3x + y = -5$
    $6x + 2y = 10$

20. $5x - y = 5$
    $-x + 3y = 13$

21. $2x + y = 4$
    $-2x + y = -4$

22. $-5x + 4y = 20$
    $10x - 8y = -40$

Example 4

23. **ECONOMICS** In 2000, the demand for nurses was 2,000,000, while the supply was only 1,890,000. The projected demand for nurses in 2020 is 2,810,414, while the supply is only projected to be 2,001,998.

    a. Define the variables, and write equations to represent these situations.

    b. Use substitution to determine during which year the supply of nurses was equal to the demand.

24. (MP) **REASONING** The table shows the approximate number of tourists in two areas of the world during a recent year and the average rates of change in tourism.

Destination	Number of Tourists	Average Rates of Change in Tourists (millions per year)
South America and the Caribbean	40.3 million	increase of 0.8
Middle East	17.0 million	increase of 1.8

    a. Define the variables, and write an equation for each region's tourism rate.

    b. If the trends continue, in how many years would you expect the number of tourists in the regions to be equal?

25 SPORTS The table shows the winning times for the Ironman World Championship.

Year	Men's	Women's
2000	8:21.00	9:26.16
2012	8:18.37	9:15.54

a. The times are in hours, minutes, and seconds. Rewrite the times rounded to the nearest minute.

b. Let the year 2000 be 0. Assume that the rate of change remains the same for years after 2000. Write an equation to represent each of the men's and women's winning times y in any year x.

c. If the trend continues, when would you expect the men's and women's winning times to be the same? Explain your reasoning.

26. CONCERT TICKETS Booker is buying tickets online for a concert. He finds tickets for himself and his friends for $65 each plus a one-time fee of $10. Paula is looking for tickets to the same concert. She finds them at another website for $69 and a one-time fee of $13.60.

a. Define the variables, and write equations to represent this situation.

b. Create a table of values for 1 to 5 tickets for each person's purchase.

c. Graph each of these equations.

d. Use the graph to determine who received the better deal. Explain why.

H.O.T. Problems   Use Higher-Order Thinking Skills

27. ERROR ANALYSIS In the system $a + b = 7$ and $1.29a + 0.49b = 6.63$, a represents pounds of apples and b represents pounds of bananas. Guillermo and Cara are finding and interpreting the solution. Is either of them correct? Explain.

Guillermo	Cara
$1.29a + 0.49b = 6.63$	$1.29a + 0.49b = 6.63$
$1.29a + 0.49(a + 7) = 6.63$	$1.29(7 - b) + 0.49b = 6.63$
$1.29 + 0.49a + 3.43 = 6.63$	$9.03 - 1.29b + 0.49b = 6.63$
$0.49a = 3.2$	$-0.8b = -2.4$
$a = 1.9$	$b = 3$
$a + b = 7$, so $b = 5$. The solution (2, 5) means that 2 pounds of apples and 5 pounds of bananas were bought.	The solution $b = 3$ means that 3 pounds of apples and 3 pounds of bananas were bought.

28. MP SENSE-MAKING A local charity has 60 volunteers. The ratio of boys to girls is 7:5. Find the number of boy volunteers and the number of girl volunteers.

29. MP REASONING Compare and contrast the solution of a system found by graphing and the solution of the same system found by substitution.

30. OPEN-ENDED Create a system of equations that has one solution. Illustrate how the system could represent a real-world situation and describe the significance of the solution in the context of the situation.

31. WRITING IN MATH Explain how to determine what to substitute when using the substitution method of solving systems of equations.

32. The perimeter of a rectangle is 72 inches. Its length is 6 inches greater than twice its width. What is the length of the rectangle? **MP** 8

- ◯ **A** 10 inches
- ◯ **B** 11 inches
- ◯ **C** 26 inches
- ◯ **D** 28 inches

33. Solve the system of equations. **MP** 8

$$3x + 2y = -8$$
$$-6x - 4y = 12$$

- ◯ **A** $(0, -2)$
- ◯ **B** $(-2, 0)$
- ◯ **C** no solution
- ◯ **D** infinitely many solutions

34. For which value of c does the system of equations have infinitely many solutions? **MP** 2

$$9x = -3y + c$$
$$y + 3x = 5$$

- ◯ **A** 3
- ◯ **B** 5
- ◯ **C** 10
- ◯ **D** 15

35. A collection of nickels and dimes has a value of $1.85. The value of the dimes is $0.10 less than twice the value of the nickels. Which system of equations can be used to find the number of nickels and the number of dimes? **MP** 4

- ◯ **A** $n + d = 185; n - d = 1$
- ◯ **B** $n + d = 185; 2n - d = 1$
- ◯ **C** $5n + 10d = 185; d = 2n - 1$
- ◯ **D** $5n + 10d = 185; n - d = 1$

36. Given a temperature in degrees Celsius C, you can find the temperature in degrees Fahrenheit using the following formula.

$$F = \frac{9}{5}C + 32$$

Which of these equations could be used to create a system of equations to find the temperature in degrees Fahrenheit that is double the temperature in degrees Celsius? **MP** 4

- ◯ **A** $2F = C$
- ◯ **C** $F = 2 + C$
- ◯ **B** $2C = F$
- ◯ **D** $C = 2 + F$

37. MULTI-STEP The sum of two numbers is three times the difference of the two numbers. The greater number is five more than the lesser number. **MP** 2

a. Write a system of equations to represent this situation.

b. What are the values of the lesser and greater numbers?

c. Could you have substituted for x or y to solve the system of equations? Explain.

38. How many solutions are possible for the following system of equations? **MP** 2

$$x + 2y = 0$$
$$2x = y$$

- ◯ **A** 0
- ◯ **B** 1
- ◯ **C** 2
- ◯ **D** infinitely many

39. A school paid $248 for new basketball and soccer jerseys. The number of basketball jerseys purchased was 6 less than 3 times the number of soccer ball jerseys purchased. Basketball jerseys cost $28 each and soccer ball jerseys cost $20 each. How many total jerseys did the school purchase? **MP** 4

- ◯ **A** 4 jerseys
- ◯ **C** 10 jerseys
- ◯ **B** 6 jerseys
- ◯ **D** 12 jerseys

LESSON 3
Elimination Using Addition and Subtraction

::**Then**

- You solved systems of equations by using substitution.

::**Now**

1. Solve systems of equations by using elimination with addition.

2. Solve systems of equations by using elimination with subtraction.

::**Why?**

- In Chicago, Illinois, there are two more months a when the mean high temperature is below 70°F than there are months b when it is above 70°F. The system of equations $a + b = 12$ and $a - b = 2$ represents this situation.

 New Vocabulary

elimination

 Mathematical Practices

7 Look for and make use of structure.

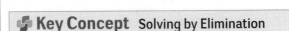

1 **Elimination Using Addition** If you add these equations, the variable y will be eliminated. Using addition or subtraction to solve a system is called **elimination**.

🔧 Key Concept Solving by Elimination

Step 1 Write the system so like terms with the same or opposite coefficients are aligned.

Step 2 Add or subtract the equations, eliminating one variable. Then solve the equation.

Step 3 Substitute the value from Step 2 into one of the equations and solve for the other variable. Write the solution as an ordered pair.

Example 1 Elimination Using Addition

Use elimination to solve the system of equations.

$4x + 6y = 32$
$3x - 6y = 3$ ⟵   **Step 1** $6y$ and $-6y$ have opposite coefficients.

Step 2 Add the equations.

$$4x + 6y = 32$$
$$\underline{(+)\ 3x - 6y = 3}$$
$$7x = 35 \qquad \text{The variable } y \text{ is eliminated.}$$
$$\frac{7x}{7} = \frac{35}{7} \qquad \text{Divide each side by 7.}$$
$$x = 5 \qquad \text{Simplify.}$$

Step 3 Substitute 5 for x in either equation to find the value of y.

$$4x + 6y = 32 \qquad \text{First equation}$$
$$4(5) + 6y = 32 \qquad \text{Replace } x \text{ with 5.}$$
$$20 + 6y = 32 \qquad \text{Multiply.}$$
$$20 + 6y - 20 = 32 - 20 \qquad \text{Subtract 20 from each side.}$$
$$6y = 12 \qquad \text{Simplify.}$$
$$\frac{6y}{6} = \frac{12}{6} \qquad \text{Divide each side by 6.}$$
$$y = 2 \qquad \text{Simplify.}$$

The solution is (5, 2).

Tetra Images/Alamy

> **Guided Practice**

1A. $-4x + 3y = -3$
$4x - 5y = 5$

1B. $4y + 3x = 22$
$3x - 4y = 14$

We can use elimination to find specific numbers that are described as being related to each other.

Example 2   Write and Solve a System of Equations

Negative three times one number plus five times another number is −11. Three times the first number plus seven times the other number is −1. Find the numbers.

Negative three times one number	plus	five times another number	is	−11.
$-3x$	$+$	$5y$	$=$	-11

Three times the first number	plus	seven times the other number	is	−1.
$3x$	$+$	$7y$	$=$	-1

Steps 1 and 2   Write the equations vertically and add.

$$-3x + 5y = -11$$
$$\underline{(+)\ 3x + 7y = \ -1}$$
$$12y = -12 \qquad \text{The variable } x \text{ is eliminated.}$$
$$\frac{12y}{12} = \frac{-12}{12} \qquad \text{Divide each side by 12.}$$
$$y = -1 \qquad \text{Simplify.}$$

Step 3   Substitute −1 for y in either equation to find the value of x.

$3x + 7y = -1$	Second equation
$3x + 7(-1) = -1$	Replace y with −1.
$3x + (-7) = -1$	Simplify.
$3x + (-7) + 7 = -1 + 7$	Add 7 to each side.
$3x = 6$	Simplify.
$\dfrac{3x}{3} = \dfrac{6}{3}$	Divide each side by 3.
$x = 2$	Simplify.

The numbers are 2 and −1.

CHECK		
	$-3x + 5y = -11$	First equation
	$-3(2) + 5(-1) \stackrel{?}{=} -11$	Substitute 2 for x and −1 for y.
	$-11 = -11$ ✔	Simplify.
	$3x + 7y = -1$	Second equation
	$3(2) + 7(-1) \stackrel{?}{=} -1$	Substitute 2 for x and −1 for y.
	$-1 = -1$ ✔	Simplify.

> **Guided Practice**

2. The sum of two numbers is −10. Negative three times the first number minus the second number equals 2. Find the numbers.

Study Tip

Coefficients When the coefficients of a variable are the same, subtracting the equations will eliminate the variable. When the coefficients are opposites, adding the equations will eliminate the variable.

Problem-Solving Tip

 Perseverance
Checking your answers in both equations of a system helps ensure there are no calculation errors.

2 Elimination Using Subtraction
Sometimes we can eliminate a variable by subtracting one equation from another.

Example 3   Solve a System of Equations

Solve the system of equations. $2t + 5r = 6$
                                         $9r + 2t = 22$

A $(-7, 15)$ **B** $\left(7, \frac{8}{9}\right)$ **C** $(4, -7)$ **D** $\left(4, -\frac{2}{5}\right)$

Read the Item

Since both equations contain $2t$, use elimination by subtraction.

Solve the Item

Step 1 Subtract the equations.

$$
\begin{array}{rrl}
5r + 2t = & 6 & \text{Write the system so like terms are aligned.} \\
(-)\ 9r + 2t = & 22 & \\
\hline
-4r \quad\quad = & -16 & \text{The variable } t \text{ is eliminated.} \\
r = & 4 & \text{Simplify.}
\end{array}
$$

Step 2 Substitute 4 for r in either equation to find the value of t.

$$
\begin{array}{rl}
5r + 2t = 6 & \text{First equation} \\
5(4) + 2t = 6 & r = 4 \\
20 + 2t = 6 & \text{Simplify.} \\
20 + 2t - 20 = 6 - 20 & \text{Subtract 20 from each side.} \\
2t = -14 & \text{Simplify.} \\
t = -7 & \text{Simplify.}
\end{array}
$$

The solution is $(4, -7)$. The correct answer is C.

▶ **Guided Practice**

3. Solve the system of equations. $8b + 3c = 11$
                                                $8b + 7c = 7$

  F $(1.5, -1)$ **G** $(1.75, -1)$ **H** $(1.75, 1)$ **J** $(1.5, 1)$

Real-World Example 4   Write and Solve a System of Equations

JOBS Cheryl and Jackie work at an ice cream shop. Cheryl earns $9.25 per hour and Jackie earns $8.75 per hour. During a typical week, Cheryl and Jackie earn $334.75 together. One week, Jackie doubles her work hours, and the girls earn $466. How many hours does each girl work during a typical week?

Understand You know how much Cheryl and Jackie each earn per hour and how much they earned together.

Plan Let c = Cheryl's hours and j = Jackie's hours.

Cheryl's pay	plus	Jackie's pay	equals	$334.75
$9.25c$	$+$	$8.75j$	$=$	334.75
Cheryl's pay	plus	Jackie's pay	equals	$466
$9.25c$	$+$	$8.75(2)j$	$=$	466

Real-World Link

The five most dangerous jobs for teens are: agriculture, construction, traveling sales, landscaping, and driving.

Source: National Consumers League

Realistic Reflections

Solve Subtract the equations to eliminate one of the variables. Then solve for the other variable.

$$9.25c + 8.75j = 334.75 \qquad \text{Write the equations vertically.}$$
$$(-)\ 9.25c + 8.75(2)j = 466$$
$$\overline{-8.75j = -131.25} \qquad \text{Subtract. The variable } c \text{ is eliminated.}$$
$$\frac{-8.75j}{-8.75} = \frac{-131.25}{-8.75} \qquad \text{Divide each side by } -8.75.$$
$$j = 15 \qquad \text{Simplify.}$$

Now substitute 15 for j in either equation to find the value of c.

$$9.25c + 8.75j = 334.75 \qquad \text{First equation}$$
$$9.25c + 8.75(\mathbf{15}) = 334.75 \qquad \text{Substitute 15 for } j.$$
$$9.25c + 131.25 = 334.75 \qquad \text{Simplify.}$$
$$9.25c = 203.50 \qquad \text{Subtract 131.25 from each side.}$$
$$c = 22 \qquad \text{Divide each side by 9.25.}$$

Check Substitute both values into the other equation to see if the equation holds true. If $c = 22$ and $j = 15$, then $9.25(22) + 17.50(15)$ or 466.

Cheryl works 22 hours, while Jackie works 15 hours during a typical week.

Systems of equations can be solved using any of the solution methods you have learned. By thinking about the system before beginning the solution process, you reduce the work required to solve the problem.

> **Guided Practice**

4. PARTIES Tamera and Adelina are throwing a birthday party for their friend. Tamera invited 5 fewer friends than Adelina. Together they invited 47 guests. How many guests did each girl invite?

Check Your Understanding 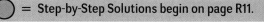 = Step-by-Step Solutions begin on page R11.

 *Go Online!* for a Self-Check Quiz

Examples 1, 3 — **Use elimination to solve each system of equations.**

1. $5m - p = 7$
$7m - p = 11$

2. $8x + 5y = 38$
$-8x + 2y = 4$

3 $7f + 3g = -6$
$7f - 2g = -31$

4. $6a - 3b = 27$
$2a - 3b = 11$

Example 2 — **5.** **MP REASONING** The sum of two numbers is 24. Five times the first number minus the second number is 12. What are the two numbers?

Example 4 — **6. RECYCLING** The recycling and reuse industry employs approximately 1,025,000 more workers than the waste management industry. Together they provide 1,275,000 jobs. How many jobs does each industry provide?

Examples 1, 3 Use elimination to solve each system of equations.

7. $-v + w = 7$
$v + w = 1$

8. $y + z = 4$
$y - z = 8$

9. $-4x + 5y = 17$
$4x + 6y = -6$

10. $5m - 2p = 24$
$3m + 2p = 24$

11. $a + 4b = -4$
$a + 10b = -16$

12. $6r - 6t = 6$
$3r - 6t = 15$

13. $6c - 9d = 111$
$5c - 9d = 103$

14. $11f + 14g = 13$
$11f + 10g = 25$

15. $9x + 6y = 78$
$3x - 6y = -30$

16. $3j + 4k = 23.5$
$8j - 4k = 4$

17. $-3x - 8y = -24$
$3x - 5y = 4.5$

18. $6x - 2y = 1$
$10x - 2y = 5$

Example 2   **19.** The sum of two numbers is 22, and their difference is 12. What are the numbers?

20. Find the two numbers with a sum of 41 and a difference of 9.

21 Three times a number minus another number is -3. The sum of the numbers is 11. Find the numbers.

22. A number minus twice another number is 4. Three times the first number plus two times the second number is 12. What are the numbers?

Example 4   **23. MOVIES** The Blackwells and Joneses are going to a 3D movie. Find the adult price and the children's price of admission.

Family	Number of Adults	Number of Children	Total Cost
Blackwell	2	5	$105.65
Jones	2	3	$77.75

Use elimination to solve each system of equations.

24. $4(x + 2y) = 8$
$4x + 4y = 12$

25. $3x - 5y = 11$
$5(x + y) = 5$

26. $4x + 3y = 6$
$3x + 3y = 7$

27. $6x - 7y = -26$

$6x + 5y = 10$

28. $\frac{1}{2}x + \frac{2}{3}y = 2\frac{3}{4}$

$\frac{1}{4}x - \frac{2}{3}y = 6\frac{1}{4}$

29. $\frac{3}{5}x + \frac{1}{2}y = 8\frac{1}{3}$

$-\frac{3}{5}x + \frac{3}{4}y = 8\frac{1}{3}$

30. **MP** **SENSE-MAKING** The total height of an office building b and the granite statue that stands on top of it g is 326.6 feet. The difference in heights between the building and the statue is 295.4 feet.

a. How tall is the statue?

b. How tall is the building?

31. BIKE RACING Professional Mountain Bike Racing currently has 66 teams. The number of non-U.S. teams is 30 more than the number of U.S. teams.

a. Let x represent the number of non-U.S. teams and y represent the number of U.S. teams. Write a system of equations that represents the number of U.S. teams and non-U.S. teams.

b. Use elimination to find the solution of the system of equations.

c. Interpret the solution in the context of the situation.

d. Graph the system of equations to check your solution.

32. BUSINESS David is analyzing the costs associated with each factory that he uses to ship his product. Assume that units shipped and total cost represent a linear relationship.

Month	Factory A		Factory B	
	Units Shipped	Total Cost	Units Shipped	Total Cost
January	386	$39,160	415	$45,200
February	421	$41,260	502	$52,160

a. Write a system of equations to represent the situation.

b. Use elimination to find the solution to the system.

c. What does the solution represent?

33 MULTIPLE REPRESENTATIONS Collect 9 pennies and 9 paper clips. For this game, you use objects to score points. Each paper clip is worth 1 point and each penny is worth 3 points. Let p represent the number of pennies and c represent the number of paper clips.

$$9 \text{ points} = \text{(pennies)} + \text{(paper clips)} = 3p + c = 3(2) + 3$$

a. **Concrete** Choose a combination of 9 objects and find your score.

b. **Analytical** Write and solve a system of equations to find the number of paper clips and pennies used for 15 points, if 9 total objects are used.

c. **Tabular** If 9 total objects are used, make a table showing the number of paper clips used and the total number of points when the number of pennies is 0, 1, 2, 3, 4, or 5.

d. **Verbal** Does the result in the table match the results in part **b**? Explain.

H.O.T. Problems   Use **H**igher-**O**rder **T**hinking Skills

34. MP REASONING Describe the solution of a system of equations if after you added two equations the result was $0 = 0$.

35. MP REASONING What is the solution of a system of equations if the sum of the equations is $0 = 2$?

36. OPEN-ENDED Create a system of equations that can be solved by using addition to eliminate one variable. Formulate a general rule for creating such systems.

37. MP STRUCTURE The solution of a system of equations is $(-3, 2)$. One equation in the system is $x + 4y = 5$. Find a second equation for the system. Explain how you derived this equation.

38. CHALLENGE The sum of the digits of a two-digit number is 8. The result of subtracting the units digit from the tens digit is -4. Define the variables and write the system of equations that you would use to find the number. Then solve the system and find the number.

39. WRITING IN MATH Describe when it would be most beneficial to use elimination to solve a system of equations.

40. In June, Kayla spent all of the $80 she had earned. The difference between her allowance in June and the money she had left over from May was $20. How much allowance money did she earn in June? **MP** 4, 7

○ **A** $30

○ **B** $50

○ **C** $60

○ **D** $100

41. What is the solution of the system of equations? **MP** 7

$$-2x + y = 4$$
$$2x + y = 10$$

○ **A** $(0, 4)$

○ **B** $(5, -10)$

○ **C** $(1, 6)$

○ **D** $(1.5, 7)$

○ **E** $(6, -2)$

42. The sum of two numbers is 120. Three times their difference is 48. What are the numbers? **MP** 2, 7

○ **A** 16, 120

○ **B** 36, 52

○ **C** 52, 68

○ **D** 68, 84

43. Find the values of x and y that make both equations true. **MP** 7

$$3x + y = 4$$
$$4x + y = 10$$

○ **A** $x = 2$ and $y = -2$

○ **B** $x = -1$ and $y = 7$

○ **C** $x = 1$ and $y = 6$

○ **D** $x = 12$ and $y = -32$

○ **E** $x = 6$ and $y = -14$

44. The perimeter of the parallelogram shown is 244 inches. The difference between the lengths of the given sides is 20 inches.

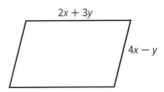

Assuming $2x + 3y$ is the longest side, find the length of each side. **MP** 2

○ **A** 16 inches, 13 inches

○ **B** 66 inches, 56 inches

○ **C** 71 inches, 51 inches

○ **D** 74 inches, 54 inches

45. Swati won a computer game that she played with Matt. The sum of their scores was 754. The difference of their scores was 176. How many points did each person score? **MP** 4, 7

○ **A** Matt 176; Swati 578

○ **B** Matt 289; Swati 465

○ **C** Matt 465; Swati 289

○ **D** Matt 754; Swati 578

46. MULTI-STEP There are two fractions with one fraction that is greater than the other. The sum of the two fractions is $\frac{11}{12}$. If you subtract the lesser fraction from the greater fraction, the difference is $\frac{5}{12}$. Find both fractions. **MP** 2, 7

a. Define the variables.

b. Write a system of equations to represent the situation.

c. Use elimination to find the lesser and greater fractions.

d. Explain how you used elimination to solve the system.

LESSON 4
Elimination Using Multiplication

::Then

- You used elimination with addition and subtraction to solve systems of equations.

::Now

1. Solve systems of equations by using elimination with multiplication.
2. Solve real-world problems involving systems of equations.

::Why?

The table shows the number of cars at Scott's Auto Repair Shop for each type of service.

The manager has allotted 1110 minutes for body work and 570 minutes for engine work. The system $3r + 4m = 1110$ and $2r + 2m = 570$ can be used to find the average time for each service.

Item	Repairs	Maintenance
body	3	4
engine	2	2

Mathematical Practices
1 Make sense of problems and persevere in solving them.

---

1 Elimination Using Multiplication In the system above, neither variable can be eliminated by adding or subtracting. You can use multiplication to solve.

🔑 Key Concept   Solving by Elimination Using Multiplication

Step 1 Multiply at least one equation by a constant to get two equations that contain opposite terms.

Step 2 Add the equations, eliminating one variable. Then solve the equation.

Step 3 Substitute the value from Step 2 into one of the equations and solve for the other variable. Write the solution as an ordered pair.

Example 1   Multiply One Equation to Eliminate a Variable

Use elimination to solve the system of equations.
$$5x + 6y = -8$$
$$2x + 3y = -5$$

Steps 1 and 2

$$5x + 6y = -8$$
$$2x + 3y = -5 \quad \text{Multiply each term by } -2.$$

$$5x + 6y = -8$$
$$(+) -4x - 6y = 10 \quad \text{Add.}$$
$$\overline{x = 2} \quad y \text{ is eliminated.}$$

Step 3
$$2x + 3y = -5 \quad \text{Second equation}$$
$$2(2) + 3y = -5 \quad \text{Substitution, } x = 2$$
$$4 + 3y = -5 \quad \text{Simplify.}$$
$$3y = -9 \quad \text{Subtract 4 from each side and simplify.}$$
$$y = -3 \quad \text{Divide each side by 3 and simplify.}$$

The solution is $(2, -3)$.

▶ Guided Practice

1A. $6x - 2y = 10$
$3x - 7y = -19$

1B. $9r + q = 13$
$3r + 2q = -4$

Sometimes you have to multiply each equation by a different number in order to solve the system.

Example 2 Multiply Both Equations to Eliminate a Variable

Use elimination to solve the system of equations.
$$4x + 2y = 8$$
$$3x + 3y = 9$$

Method 1 Eliminate x.

$4x + 2y = 8$ Multiply by 3.
$3x + 3y = 9$ Multiply by -4.

$$12x + 6y = 24$$
$$(+) \ -12x - 12y = -36 \qquad \text{Add equations.}$$
$$\overline{\ -6y = -12} \qquad x \text{ is eliminated.}$$
$$\frac{-6y}{-6} = \frac{-12}{-6} \qquad \text{Divide each side by } -6.$$
$$y = 2 \qquad \text{Simplify.}$$

Now substitute 2 for y in either equation to find the value of x.

$$3x + 3y = 9 \qquad \text{Second equation}$$
$$3x + 3(2) = 9 \qquad \text{Substitute 2 for } y.$$
$$3x + 6 = 9 \qquad \text{Simplify.}$$
$$3x = 3 \qquad \text{Subtract 6 from each side and simplify.}$$
$$\frac{3x}{3} = \frac{3}{3} \qquad \text{Divide each side by 3.}$$
$$x = 1 \qquad \text{The solution is } (1, 2).$$

Method 2 Eliminate y.

$4x + 2y = 8$ Multiply by 3.
$3x + 3y = 9$ Multiply by -2.

$$12x + 6y = 24$$
$$(+) \ -6x - 6y = -18 \qquad \text{Add equations.}$$
$$\overline{\ 6x = 6} \qquad y \text{ is eliminated.}$$
$$\frac{6x}{6} = \frac{6}{6} \qquad \text{Divide each side by 6.}$$
$$x = 1 \qquad \text{Simplify.}$$

Now substitute 1 for x in either equation to find the value of y.

$$3x + 3y = 9 \qquad \text{Second equation}$$
$$3(1) + 3y = 9 \qquad \text{Substitute 1 for } x.$$
$$3 + 3y = 9 \qquad \text{Simplify.}$$
$$3y = 6 \qquad \text{Subtract 3 from each side and simplify.}$$
$$\frac{3y}{3} = \frac{6}{3} \qquad \text{Divide each side by 3.}$$
$$y = 2 \qquad \text{Simplify.}$$

The solution is (1, 2), which matches the result obtained with Method 1.

CHECK Substitute 1 for x and 2 for y in the first equation.

$$4x + 2y = 8 \qquad \text{Original equation}$$
$$4(1) + 2(2) \stackrel{?}{=} 8 \qquad \text{Substitute (1, 2) for } (x, y).$$
$$4 + 4 \stackrel{?}{=} 8 \qquad \text{Multiply.}$$
$$8 = 8 \ \checkmark \qquad \text{Add.}$$

Guided Practice

2A. $5x - 3y = 6$
$2x + 5y = -10$

2B. $6a + 2b = 2$
$4a + 3b = 8$

2 Solve Real-World Problems
Sometimes it is necessary to use multiplication before elimination in real-world problem solving too.

Real-World Example 3   Solve a System of Equations

FLIGHT A personal aircraft traveling with the wind flies 520 miles in 4 hours. On the return trip, the airplane takes 5 hours to travel the same distance. Find the speed of the airplane if the air is still.

You are asked to find the speed of the airplane in still air.

Let a = the rate of the airplane if the air is still.
Let w = the rate of the wind.

	r	t	d	$r \cdot t = d$
With the Wind	$a + w$	4	520	$(a + w)4 = 520$
Against the Wind	$a - w$	5	520	$(a - w)5 = 520$

So, our two equations are $4a + 4w = 520$ and $5a - 5w = 520$.

$4a + 4w = 520$ Multiply by 5.   → $20a + 20w = 2600$
$5a - 5w = 520$ Multiply by 4. $(+) \ 20a - 20w = 2080$
$\overline{\qquad\qquad\qquad\quad 40a \qquad\quad = 4680}$ w is eliminated.

$$\frac{40a}{40} = \frac{4680}{40}$$ Divide each side by 40.

$$a = 117$$ Simplify.

The rate of the airplane in still air is 117 miles per hour.

> **Guided Practice**

3. CANOEING A canoeist travels 4 miles downstream in 1 hour. The return trip takes the canoeist 1.5 hours. Find the rate of the boat in still water.

Check Your Understanding

 = Step-by-Step Solutions begin on page R11.

Go Online! for a Self-Check Quiz

Examples 1–2   Use elimination to solve each system of equations.

1. $2x - y = 4$
$\quad 7x + 3y = 27$

2. $2x + 7y = 1$
$\quad x + 5y = 2$

3 $4x + 2y = -14$
$\quad 5x + 3y = -17$

4. $9a - 2b = -8$
$\quad -7a + 3b = 12$

Example 3

5.  **SENSE-MAKING** A kayaking group with a guide travels 16 miles downstream, stops for a meal, and then travels 16 miles upstream. The speed of the current remains constant throughout the trip. Find the speed of the kayak in still water.

Leave	10:00 A.M.
Stop for meal	12:00 noon
Return	1:00 P.M.
Finish	5:00 P.M.

6. PODCASTS Steve subscribed to 10 podcasts for a total of 340 minutes. He used his two favorite tags, Hobbies and Recreation, and Soliloquies. Each of the Hobbies and Recreation episodes lasted about 32 minutes. Each Soliloquies episode lasted 42 minutes. To how many of each tag did Steve subscribe?

Examples 1–2 Use elimination to solve each system of equations.

7. $x + y = 2$
$-3x + 4y = 15$

8. $x - y = -8$
$7x + 5y = 16$

9. $x + 5y = 17$
$-4x + 3y = 24$

10. $6x + y = -39$
$3x + 2y = -15$

11. $2x + 5y = 11$
$4x + 3y = 1$

12. $3x - 3y = -6$
$-5x + 6y = 12$

13. $3x + 4y = 29$
$6x + 5y = 43$

14. $8x + 3y = 4$
$-7x + 5y = -34$

15. $8x + 3y = -7$
$7x + 2y = -3$

16. $4x + 7y = -80$
$3x + 5y = -58$

17. $12x - 3y = -3$
$6x + y = 1$

18. $-4x + 2y = 0$
$10x + 3y = 8$

Example 3   **19** **NUMBER THEORY** Four times a number minus 5 times another number is equal to 21. Three times the sum of the two numbers is 36. What are the two numbers?

20. FOOTBALL A field goal is 3 points and the extra point after a touchdown is 1 point. In the 2016 season, Adam Vinatieri of the Indianapolis Colts made a total of 71 field goals and extra points for 125 points. Find the number of field goals and extra points that he made.

Use elimination to solve each system of equations.

21. $2.2x + 3y = 15.25$
$4.6x + 2.1y = 18.325$

22. $-0.4x + 0.25y = -2.175$
$2x + y = 7.5$

23. $\frac{1}{4}x + 4y = 2\frac{3}{4}$
$3x + \frac{1}{2}y = 9\frac{1}{4}$

24. $\frac{2}{5}x + 6y = 24\frac{1}{5}$
$3x + \frac{1}{2}y = 3\frac{1}{2}$

25. MULTI-STEP Michelle and Julie work at a catering company. They need to bake 264 cookies for a birthday party that starts in a little over an hour and a half. Each tube of cookie dough claims to make 36 cookies, but Michelle eats about $\frac{1}{5}$ of every tube and Julie makes cookies that are 1.5 times as large as the recommended cookie size. It takes about 8 minutes to bake a container of cookies, but since Julie's cookies are larger, they take 12 minutes to bake.

   a. How many tubes should each girl plan to bake? How long does each girl use the oven?

   b. Explain your solution process.

   c. What assumptions did you make?

26. GEOMETRY The graphs of $x + 2y = 6$ and $2x + y = 9$ contain two of the sides of a triangle. A vertex of the triangle is at the intersection of the graphs.

   a. What are the coordinates of the vertex?

   b. Draw the graph of the two lines. Identify the vertex of the triangle.

   c. The line that forms the third side of the triangle is the line $x - y = -3$. Draw this line on the previous graph.

   d. Name the other two vertices of the triangle.

27 ENTERTAINMENT At an entertainment center, two groups of people bought batting tokens and miniature golf games, as shown in the table.

Group	Number of Trips to the Batting Cage	Number of Miniature Golf Games	Total Cost
A	16	3	$44.81
B	22	5	$67.73

a. Define the variables, and write a system of linear equations for this situation.

b. Solve the system of equations, and explain what the solution represents.

28. **TESTS** Mrs. Henderson discovered that she had accidentally reversed the digits of a test score and did not give a student 36 points. Mrs. Henderson told the student that the sum of the digits was 14 and agreed to give the student his correct score plus extra credit if he could determine his actual score. What was his correct score?

H.O.T. Problems   Use Higher-Order Thinking Skills

29. **MP REASONING** Explain how you could recognize a system of linear equations with infinitely many solutions.

30. **MP CRITIQUE ARGUMENTS** Jason and Daniela are solving a system of equations. Is either of them correct? Explain your reasoning.

Jason	Daniela
$2r + 7t = 11$	$2r + 7t = 11$
$r - 9t = -7$	$(-)\ r - 9t = -7$
	$r = 18$
$2r + 7t = 11$	
$(-)\ 2r - 18t = -14$	$2r + 7t = 11$
$25t = 25$	$2(18) + 7t = 11$
$t = 1$	$36 + 7t = 11$
$2r + 7t = 11$	$7t = -25$
$2r + 7(1) = 11$	$\dfrac{7t}{7} = -\dfrac{25}{7}$
$2r + 7 = 11$	$t = -3.6$
$2r = 4$	The solution is $(18, -3.6)$.
$\dfrac{2r}{2} = \dfrac{4}{2}$	
$r = 2$	
The solution is $(2, 1)$.	

31. **OPEN-ENDED** Write a system of equations that can be solved by multiplying one equation by -3 and then adding the two equations together.

32. **CHALLENGE** The solution of the system $4x + 5y = 2$ and $6x - 2y = b$ is $(3, a)$. Find the values of a and b. Discuss the steps that you used.

33. **Q WRITING IN MATH** Why is substitution sometimes more helpful than elimination, and vice versa?

34. Seana and Mikayla sold tickets to the school play. Seana sold 3 adult tickets and 2 student tickets for $84. Mikayla sold 2 adult tickets and 5 student tickets for $100. How much did a student ticket cost? **MP** 2, 4

○ **A** $12

○ **B** $20

○ **C** $36

○ **D** $44

35. The perimeter of the first rectangle below is 120 inches. The perimeter of the second rectangle is 180 inches.

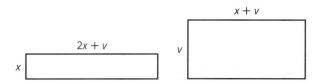

What are the values of x and v? **MP** 2, 7

○ **A** $x = 6; v = 42$

○ **B** $x = 13; v = 42$

○ **C** $x = 42; v = 6$

○ **D** $x = 48; v = 28$

36. Nick and Cal are playing an electronic game. The table shows the number of each kind of play made by each player and each player's score at the end of the game. How many points does a player earn for a double play? **MP** 4, 7

Number of Plays Made			
Player	Single Plays	Double Plays	Total Points
Nick	16	5	197
Cal	9	6	165

○ **A** 6 points

○ **B** 7 points

○ **C** 15 points

○ **D** 17 points

37. FALL FUNDRAISER A fundraiser by the football and track teams was held during the fall. Both teams sold boxes of cookies and candy bars, which were delivered in cartons. The total number of cartons of cookies and candy bars sold by each team is listed in the table below along with the total amount of money that was collected. **MP** 2, 4, 7

Team	Cartons of Cookies	Cartons of Candy Bars	Money Collected
Football	9	12	$1,845
Track	8	15	$2,127.50

a. Define the variables and write a system of linear equations for this situation.

b. Solve the system of equations to determine the amount of money collected for each carton of cookies and candy bars sold.

c. If a carton of cookies contained 10 boxes and a carton of candy bars contains 50 candy bars, what was the selling price for one box of cookies and for one candy bar?

38. MULTI-STEP Two cousins, Jay and Carolyn, planned to take flights and meet at a connecting city A and then fly together to a destination city B where they would meet their other cousin Diane for a vacation. **MP** 1, 4

a. Jay's flight path followed the linear equation $y = 2x + 3$ and Carolyn's flight path followed the linear equation $y = -3x - 2$. Solve the system of equations to find the coordinates of city A.

b. Suppose Jay and Carolyn fly from city A to city B on a flight path of slope -1. What is the equation of Jay and Carolyn's flight path to city B?

c. Suppose Diane is traveling from the north on a flight path following the linear equation $y = 2x + 24$. Find the coordinates of the destination city B where all three cousins would enjoy a vacation together.

Use the graph to determine whether each system is *consistent* **or** *inconsistent* **and whether it is** *independent* **or** *dependent.* (Lesson 6-1)

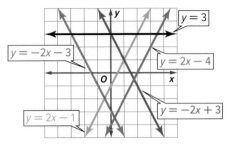

1. $y = 2x - 1$
$y = -2x + 3$

2. $y = -2x + 3$
$y = -2x - 3$

Graph each system and determine the number of solutions that it has. If it has one solution, name it. (Lesson 6-1)

3. $y = 2x - 3$
$y = x + 4$

4. $x + y = 6$
$x - y = 4$

5. $x + y = 8$
$3x + 3y = 24$

6. $x - 4y = -6$
$y = -1$

7. $3x + 2y = 12$
$3x + 2y = 6$

8. $2x + y = -4$
$5x + 3y = -6$

Use substitution to solve each system of equations. (Lesson 6-2)

9. $y = x + 4$
$2x + y = 16$

10. $y = -2x - 3$
$x + y = 9$

11. $x + y = 6$
$x - y = 8$

12. $y = -4x$
$6x - y = 30$

13. **MP** **PERSEVERANCE** The cost of two orders at a restaurant is shown in the table. (Lesson 6-2)

Order	Total Cost
3 tacos, 2 burritos	$23.35
4 tacos, 1 burrito	$19.80

a. Define variables to represent the cost of a taco and the cost of a burrito.

b. Write a system of equations to find the cost of a single taco and a single burrito.

c. Solve the systems of equations, and explain what the solution means.

d. How much would a customer pay for 2 tacos and 2 burritos?

14. **AMUSEMENT PARKS** The cost of two groups who recently visited an aquarium is shown in the table. (Lesson 6-3)

Group	Total Cost
4 adults, 2 children	$110
4 adults, 3 children	$123

a. Define variables to represent the cost of an adult ticket and the cost of a child ticket.

b. Write a system of equations to find the cost of an adult ticket and a child ticket.

c. Solve the system of equations, and explain what the solution means.

d. **MP** Which mathematical practice did you use to solve this problem?

e. How much will a group of 3 adults and 5 children be charged for admission?

15. **MULTIPLE CHOICE** Angelina spent $16 for 12 pieces of candy to take to a meeting. Each chocolate bar costs $2, and each lollipop costs $1. Determine how many of each she bought. (Lesson 6-3)

A 6 chocolate bars, 6 lollipops

B 4 chocolate bars, 8 lollipops

C 7 chocolate bars, 5 lollipops

D 3 chocolate bars, 9 lollipops

Use elimination to solve each system of equations. (Lessons 6-3 and 6-4)

16. $x + y = 9$
$x - y = -3$

17. $x + 3y = 11$
$x + 7y = 19$

18. $9x - 24y = -6$
$3x + 4y = 10$

19. $-5x + 2y = -11$
$5x - 7y = 1$

20. **MULTIPLE CHOICE** The Blue Mountain High School Drama Club is selling tickets to their spring musical. Adult tickets are $4 and student tickets are $1. A total of 285 tickets are sold for $765. How many of each type of ticket are sold? (Lesson 6-4)

A 145 adult, 140 student

B 120 adult, 165 student

C 180 adult, 105 student

D 160 adult, 125 student

Applying Systems of Linear Equations

- You solved systems of equations by using substitution and elimination.

1 Determine the best method for solving systems of equations.

2 Apply systems of equations.

In speed skating, competitors race two at a time on a double track. Indoor speed skating rinks have two track sizes for race events: an official track and a short track.

Speed Skating Tracks	
official track	x
short track	y

The total length of the two tracks is 511 meters. The official track is 44 meters less than four times the short track. The total length is represented by $x + y = 511$. The length of the official track is represented by $x = 4y - 44$.

You can solve the system of equations to find the length of each track.

 Mathematical Practices

2 Reason abstractly and quantitatively.

4 Model with mathematics.

1 Determine the Best Method You have learned five methods for solving systems of linear equations. The table summarizes the methods and the types of systems for which each method works best.

Concept Summary Solving Systems of Equations	
Method	**The Best Time to Use**
Graphing	To estimate solutions, since graphing usually does not give an exact solution.
Substitution	If one of the variables in either equation has a coefficient of 1 or −1.
Elimination Using Addition	If one of the variables has opposite coefficients in the two equations.
Elimination Using Subtraction	If one of the variables has the same coefficient in the two equations.
Elimination Using Multiplication	If none of the coefficients are 1 or −1 and neither of the variables can be eliminated by simply adding or subtracting the equations.

Substitution and elimination are algebraic methods for solving systems of equations. An algebraic method is best for an exact solution. Graphing, with or without technology, is a good way to estimate a solution.

A system of equations can be solved using each method. To determine the best approach, analyze the coefficients of each term in each equation.

Example 1   Choose the Best Method

Determine the best method to solve the system of equations. Then solve the system.

$$4x - 4y = 8$$
$$-8x + y = 19$$

Understand   To determine the best method to solve the system of equations, look closely at the coefficients of each term.

Plan   Neither the coefficients of x nor y are the same or additive inverses, so you cannot add or subtract to eliminate a variable. Since the coefficient of y in the second equation is 1, you can use substitution.

Solve   First, solve the second equation for y.

$-8x + y = 19$	Second equation
$-8x + y + 8x = 19 + 8x$	Add $8x$ to each side.
$y = 19 + 8x$	Simplify.

Next, substitute $19 + 8x$ for y in the first equation.

$4x - 4y = 8$	First equation
$4x - 4(19 + 8x) = 8$	Substitution
$4x - 76 - 32x = 8$	Distributive Property
$-28x - 76 = 8$	Simplify.
$-28x - 76 + 76 = 8 + 76$	Add 76 to each side.
$-28x = 84$	Simplify.
$\dfrac{-28x}{-28} = \dfrac{84}{-28}$	Divide each side by -28.
$x = -3$	Simplify.

Last, substitute -3 for x in the second equation.

$-8x + y = 19$	Second equation
$-8(-3) + y = 19$	$x = -3$
$y = -5$	Simplify.

The solution of the system of equations is $(-3, -5)$.

Check   Use a graphing calculator to check your solution. If your algebraic solution is correct, then the graphs will intersect at $(-3, -5)$.

The system of equations can also be solved by using elimination with multiplication. You can multiply the first equation by 2 and then add to eliminate the x-term.

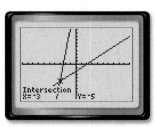

$[-10, 10]$ scl: 1 $[-10, 10]$ scl: 1

Math History Link

Leonardo Pisano
(1170–1250) Leonardo Pisano is better known by his nickname *Fibonacci*. His book introduced the Hindu-Arabic place-value decimal system. Systems of linear equations are studied in this work.

▶ **Guided Practice**

1A. $5x + 7y = 2$
      $-2x + 7y = 9$

1B. $3x - 4y = -10$
      $5x + 8y = -2$

1C. $5x - y = 17$
      $3x + 2y = 5$

2 Apply Systems of Linear Equations
When applying systems of linear equations to problems, it is important to analyze each solution in the context of the situation.

Example 2   Apply Systems of Linear Equations

PRACTICE Carter has been practicing for a band concert. The blue line represents the total time he will spend practicing. Avery just joined the band today. The red line represents the time she will spend practicing for the concert.

a. Write a system of linear equations based on the information in the graph.

b. Interpret the meaning of each equation.

c. Solve the system and describe its meaning in the context of the situation.

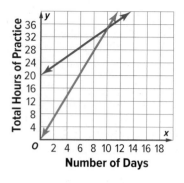

a. First, write an equation for the blue line.

Choose two points on the graph to find the slope of the line.

$m = \dfrac{y_2 - y_1}{x_2 - x_1}$ Slope-intercept formula

$m = \dfrac{32 - 20}{8 - 0}$ $(x_1, y_1) = (0, 20), (x_2, y_2) = (8, 32)$

$m = \dfrac{12}{8}$ or 1.5 Simplify.

The y-intercept is $(0, 20)$, so $b = 20$.

The equation for the blue line is $y = 1.5x + 20$.

Now, write an equation for the red line.

$m = \dfrac{y_2 - y_1}{x_2 - x_1}$ Slope-intercept formula

$m = \dfrac{28 - 0}{8 - 0}$ $(x_1, y_1) = (0, 0), (x_2, y_2) = (8, 28)$

$m = \dfrac{28}{8}$ or 3.5 Simplify.

The y-intercept is $(0, 0)$, so $b = 0$.

The equation for the red line is $y = 3.5x$.

b. Carter has already spent 20 hours practicing and he will spend an additional 1.5 hours each day practicing. Avery hasn't spent any time practicing yet, but she will spend 3.5 hours each day practicing.

c. The solution of this system of equations is the point where the two lines intersect, $(10, 35)$. This means that after 10 days, Carter and Avery will have spent an equal amount of time practicing, 35 hours.

Guided Practice

2. EMPLOYMENT Jared has worked 70 hours this summer. He plans to work 2.5 hours in each of the coming days. Clementine just started working and plans to work 5 hours a day in the coming days. Write and solve a system of equations to find how long it will be before they will have worked the same number of hours.

Check Your Understanding ⬤ = Step-by-Step Solutions begin on page R11.

Go Online! for a
Self-Check Quiz

Example 1

Determine the best method to solve each system of equations. Then solve the system.

1. $2x + 3y = -11$
$-8x - 5y = 9$

2. $3x + 4y = 11$
$2x + y = -1$

3. $3x - 4y = -5$
$-3x + 2y = 3$

4. $3x + 7y = 4$
$5x - 7y = -12$

Example 2

5. FINANCIAL LITERACY The debate team is selling pizzas and subs. The blue line represents the total profit the debate team can earn and the red line represents the total number of items the debate team can sell.

a. Write a system of linear equations based on the information in the graph.

b. Interpret the meaning of each equation.

c. Solve the system and describe its meaning in the context of the situation.

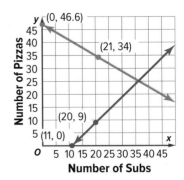

Practice and Problem Solving

Extra Practice is on page R6.

Example 1

Determine the best method to solve each system of equations. Then solve the system.

6. $-3x + y = -3$
$4x + 2y = 14$

7. $2x + 6y = -8$
$x - 3y = 8$

8. $3x - 4y = -5$
$-3x - 6y = -5$

9. $5x + 8y = 1$
$-2x + 8y = -6$

10. $y + 4x = 3$
$y = -4x - 1$

11 $-5x + 4y = 7$
$-5x - 3y = -14$

Example 2

12. FINANCIAL LITERACY The lacrosse team is hosting a dance as a fundraiser. They sold 38 more individual tickets than couples' tickets and earned a total of $730. Write and solve a system of equations to represent this situation.

Ticket Type	Selling Price
individual	$3
couple	$5

13. BAND There are 40 students in band third period. The number of girls is 4 less than 3 times the number of boys. Write and solve a system of equations to find the number of girls and boys in third period band.

14. CAVES The Caverns of Sonora have two different tours: the Crystal Palace tour and the Horseshoe Lake tour. The total length of both tours is 3.25 miles. The Crystal Palace tour is a half-mile less than twice the length of the Horseshoe Lake tour. Determine the length of each tour.

15. **MP** **MODELING** The *break-even point* is the point at which income equals expenses. Ridgemont High School is paying $13,200 for the writing and research of their yearbook plus a printing fee of $25 per book. If they sell the books for $40 each, how many will they have to sell to break even? Explain.

Write a system of linear equations given each table. Then name the solution.

16.

x	y_1	y_2
−6	−10	−12
0	−6	4
6	−2	20

17.

x	y_1	y_2
−4	11	−7
4	−5	1
8	−13	5

18. PAINTBALL Clara and her friends are planning a trip to a paintball park. Find the cost of lunch and the cost of each paintball. What would be the cost for 400 paintballs and lunch?

Paintball in the Park
- $25 for 500 paintballs
- $16 for 200 paintballs

Lunch is included

19 RECYCLING Mara and Ling each recycled aluminum cans and newspaper, as shown in the table. Mara earned $3.77, and Ling earned $4.65.

   a. Define variables and write a system of linear equations from this situation.

   b. What was the price per pound of aluminum? Determine the reasonableness of your solution.

Materials	Pounds Recycled	
	Mara	Ling
aluminum cans	9	9
newspaper	26	114

20. BAKE SALE The tennis team is having a bake sale. Cookies sell for $1 each, and cupcakes are $3 each. If Connie spends $15 for 7 items, how many cupcakes did she buy?

21. MUSIC An online music club offers individual songs for one price or entire albums for another. Kendrick pays $14.90 to download 5 individual songs and 1 album. Geoffrey pays $21.75 to download 3 individual songs and 2 albums.

   a. How much does the music club charge to download a song?

   b. How much does the music club charge to download an entire album?

22. CANOEING Malik canoed against the current for 2 hours and then with the current for 1 hour before resting. Julio traveled against the current for 2.5 hours and then with the current for 1.5 hours before resting. If they traveled a total of 9.5 miles against the current, 20.5 miles with the current, and the current is 3 miles per hour, how fast do Malik and Julio travel in still water?

H.O.T. Problems   Use Higher-Order Thinking Skills

23. OPEN-ENDED Formulate a system of equations that represents a situation in your school. Describe the method that you would use to solve the system. Then solve the system and explain what the solution means.

24. MP REASONING In a system of equations, x represents the time spent riding a bike, and y represents the distance traveled. You determine the solution to be $(-1, 7)$. Use this problem to discuss the importance of analyzing solutions in the context of real-world problems.

25. CHALLENGE Solve $4x + y = 13$ and $6x - y = 7$ by using three different methods. Show your work.

26. WRITE A QUESTION A classmate says that elimination is the best way to solve a system of equations. Write a question to challenge his conjecture.

27. WHICH ONE DOESN'T BELONG? Which system is different? Explain.

$x - y = 3$ $x + \frac{1}{2}y = 1$	$-x + y = 0$ $5x = 2y$	$y = x - 4$ $y = \frac{2}{x}$	$y = x + 1$ $y = 3x$

28. ⓔ WRITING IN MATH How do you know what method to use when solving a system of equations?

29. Erin has dimes and nickels in her pocket. She has a total of 30 coins. The total value of the change in her pocket is $2.10. How many dimes does Erin have in her pocket? MP 2, 4

[   ]

30. In right triangle ABC, angle B measures 90°. The measure of angle A is 6 less than twice the measure of angle C. What is the measure of angle A? MP 2, 4

- ○ **A** 32°
- ○ **B** 58°
- ○ **C** 90°
- ○ **D** 180°

31. Terence and his father went on a driving trip. Terence drove 32 miles more than his father. Together, they drove 260 miles. How many miles did Terence drive? MP 2, 4

- ○ **A** 82
- ○ **B** 114
- ○ **C** 146
- ○ **D** 151

32. Kate bought 3 pounds of rice and 1 pound of beans for $4.50. Elise bought 4 pounds of rice and 2 pounds of beans for $7.00. Select all of the TRUE statements. MP 2, 4

- ☐ **A** Beans cost more than rice.
- ☐ **B** The cost of 1 pound of rice is $1.50.
- ☐ **C** The cost of 1 pound of beans is $1.00
- ☐ **D** 1 pound of rice and 1 pound beans would cost $2.50.
- ☐ **E** The cost of 3 pounds of rice is $3.00.
- ☐ **F** The equations used to solve the problem are $3x + y = 4.5$ and $4x + 2y = 7$.

33. For a graph of $x + y = 10$ and $y = 5x$, select all the statements that are TRUE for this system of equations. MP 7

- ☐ **A** Both linear equations have y-intercepts through the origin.
- ☐ **B** x and y equal to 5 is one of the solutions.
- ☐ **C** The solution is at x equal to $\frac{5}{3}$.
- ☐ **D** The solution is $(\frac{5}{3}, \frac{25}{3})$.
- ☐ **E** The solution is (5, 25).
- ☐ **F** The solution is $(\frac{5}{3}, 25)$.
- ☐ **G** The solution is the intersection point of the lines $x + y = 10$ and $y = 5x$.

34. MULTI-STEP There were 114 people that attended a particular seminar. There were twice as many women as men at the seminar. Use the variables m for men and w for women. MP 2, 4

a. Write a system of equations to represent this situation.

[   ]

b. Which equation shows the substitution method being used to solve the system of equations?

- ○ **A** $m + 2w = 114$
- ○ **B** $m + 2m = 114$
- ○ **C** $2m + w = 114$
- ○ **D** $2m + 2w = 228$

c. Solve the system of equations.

35. The sum of two numbers is 29. The difference of the same two numbers is 5. What are the numbers? MP 2

[   ]

Algebra Lab
Using Matrices to Solve Systems of Equations

A **matrix** is a rectangular arrangement of numbers, called **elements**, in rows and columns enclosed in brackets. Usually named using an uppercase letter, a matrix can be described by its **dimensions** or by the number of rows and columns in the matrix. A matrix with m rows and n columns is an $m \times n$ matrix (read "m by n").

Mathematical Practices
MP.7 Look for and make use of structure.

$$A = \begin{bmatrix} 7 & -9 & 5 & 3 \\ -1 & 3 & -3 & 6 \\ 0 & -4 & 8 & 2 \end{bmatrix}$$

3 rows

A is a 3×4 matrix.

4 columns

The element 2 is in Row 3, Column 4.

You can use an augmented matrix to solve a system of equations. An **augmented matrix** consists of the coefficients and the constant terms of a system of equations. Make sure that the coefficients of the x-terms are listed in one column, the coefficients of the y-terms are in another column, and the constant terms are in a third column. The coefficients and constant terms are usually separated by a dashed line.

Linear System

$$x - 3y = 8$$
$$-9x + 2y = -4$$

Augmented Matrix

$$\begin{bmatrix} 1 & -3 & | & 8 \\ -9 & 2 & | & -4 \end{bmatrix}$$

Activity 1 Write an Augmented Matrix

Work cooperatively. Write an augmented matrix for each system of equations.

a. $-2x + 7y = 11$
$6x - 4y = 2$

Place the coefficients of the equations and the constant terms into a matrix.

$-2x + 7y = 11$
$6x - 4y = 2$ $\longrightarrow$ $\begin{bmatrix} -2 & 7 & | & 11 \\ 6 & -4 & | & 2 \end{bmatrix}$

b. $x - 2y = 5$
$y = -4$

$x - 2y = 5$
$y = -4$ $\longrightarrow$ $\begin{bmatrix} 1 & -2 & | & 5 \\ 0 & 1 & | & -4 \end{bmatrix}$

You can solve a system of equations by using an augmented matrix. By performing row operations, you can change the form of the matrix. The operations are the same as the ones used when working with equations.

🔑 Key Concept Elementary Row Operations

The following operations can be performed on an augmented matrix.

• Interchange any two rows.

• Multiply all entries in a row by a nonzero constant.

• Replace one row with the sum of that row and a multiple of another row.

(continued on the next page)

Row operations produce a matrix equivalent to the original system. **Row reduction** is the process of performing elementary row operations on an augmented matrix to solve a system. The goal is to get the coefficients portion of the matrix to have the form $\begin{bmatrix} 1 & 0 \\ 0 & 1 \end{bmatrix}$, which is called the **identity matrix**. The first row will give you the solution for x, because the coefficient of y is 0. The second row will give you the solution for y, because the coefficient of x is 0.

Activity 2   Use Row Operations to Solve a System

Work cooperatively. Use an augmented matrix to solve the system of equations.

$-5x + 3y = 6$
$x - y = 4$

Step 1 Write the augmented matrix: $\begin{bmatrix} -5 & 3 & \vdots & 6 \\ 1 & -1 & \vdots & 4 \end{bmatrix}$

Step 2 Notice that the first element in the second row is 1. Interchange the rows so 1 can be in the upper left-hand corner.

$\begin{bmatrix} -5 & 3 & \vdots & 6 \\ 1 & -1 & \vdots & 4 \end{bmatrix}$ → Interchange R₁ and R₂. → $\begin{bmatrix} 1 & -1 & \vdots & 4 \\ -5 & 3 & \vdots & 6 \end{bmatrix}$

Step 3 To make the first element in the second row a 0, multiply the first row by 5 and add the result to row 2.

$\begin{bmatrix} 1 & -1 & \vdots & 4 \\ -5 & 3 & \vdots & 6 \end{bmatrix}$ → $5R_1 + R_2$ → $\begin{bmatrix} 1 & -1 & \vdots & 4 \\ 0 & -2 & \vdots & 26 \end{bmatrix}$ $1(5) + (-5) = 0;\ -1(5) + 3 = -2;$ $4(5) + 6 = 26$

Step 4 To make the second element in the second row a 1, multiply the second row by $-\dfrac{1}{2}$.

$\begin{bmatrix} 1 & -1 & \vdots & 4 \\ 0 & -2 & \vdots & 26 \end{bmatrix}$ → $-\dfrac{1}{2}R_2$ → $\begin{bmatrix} 1 & -1 & \vdots & 4 \\ 0 & 1 & \vdots & -13 \end{bmatrix}$ $0\left(-\dfrac{1}{2}\right) = 0;\ -2\left(-\dfrac{1}{2}\right) = 1;$ $26\left(-\dfrac{1}{2}\right) = -13$

Step 5 To make the second element in the second row a 0, add the rows together.

$\begin{bmatrix} 1 & -1 & \vdots & 4 \\ 0 & 1 & \vdots & -13 \end{bmatrix}$ → $R_2 + R_1$ → $\begin{bmatrix} 1 & 0 & \vdots & -9 \\ 0 & 1 & \vdots & -13 \end{bmatrix}$ $1 + 0 = 1;\ -1 + 1 = 0;$ $4 + (-13) = -9$

The solution is $(-9, -13)$.

Model and Analyze

Work cooperatively. Write an augmented matrix for each system of equations. Then solve the system.

1. $x + y = -3$
$x - y = 1$

2. $x - y = -2$
$2x + 2y = 12$

3. $3x - 4y = -27$
$x + 2y = 11$

4. $x + 4y = -6$
$2x - 5y = 1$

5. $x - 3y = -2$
$4x + y = 31$

6. $x + 2y = 3$
$-3x + 3y = 27$

Systems of Inequalities

● You graphed and solved linear inequalities.

1 Solve systems of linear inequalities by graphing.

2 Apply systems of linear inequalities.

● Jacui is beginning an exercise program that involves an intense cardiovascular workout. Her trainer recommends that for a person her age, her heart rate should stay within the following range as she exercises.

- It should be higher than 102 beats per minute.
- It should not exceed 174 beats per minute.

The graph shows the maximum and minimum target heart rate for people ages 0 to 30 as they exercise. If the preferred range is in light green, how old do you think Jacui is?

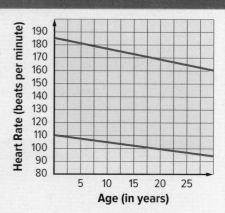

New Vocabulary

system of inequalities

Mathematical Practices

6 Attend to precision.

7 Look for and make use of structure.

1 Systems of Inequalities The graph above is a graph of two inequalities. A set of two or more inequalities with the same variables is called a **system of inequalities**.

The solution of a system of inequalities with two variables is the set of ordered pairs that satisfy all of the inequalities in the system. The solution set is represented by the overlap, or intersection, of the graphs of the inequalities.

Example 1   Solve by Graphing

Solve the system of inequalities by graphing.

$y > -2x + 1$
$y \leq x + 3$

The graph of $y = -2x + 1$ is dashed and is not included in the graph of the solution. The graph of $y = x + 3$ is solid and is included in the graph of the solution.

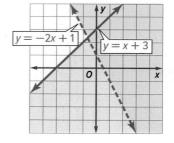

The solution of the system is the set of ordered pairs in the intersection of the graphs of $y > -2x + 1$ and $y \leq x + 3$. This region is shaded in green.

When graphing more than one region, it is helpful to use two different colored pencils or two different patterns for each region. This will make it easier to see where the regions intersect and find possible solutions.

▶ **Guided Practice**

1A. $y \leq 3$
     $x + y \geq 1$

1B. $2x + y \geq 2$
     $2x + y < 4$

1C. $y \geq -4$
     $3x + y \leq 2$

1D. $x + y > 2$
     $-4x + 2y < 8$

Sometimes the regions never intersect. When this happens, there is no solution because there are no points in common.

Study Tip

 Reasoning A system of equations represented by parallel lines does not have a solution. However, a system of inequalities with parallel boundaries can have a solution. For example:

Example 2   No Solution

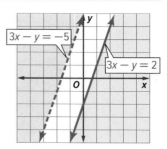

Solve the system of inequalities by graphing.

$3x - y \geq 2$
$3x - y < -5$

The graphs of $3x - y = 2$ and $3x - y = -5$ are parallel lines. The two regions do not intersect at any point, so the system has no solution.

> **Guided Practice**

2A. $y > 3$
 $y < 1$

2B. $x + 6y \leq 2$
 $y \geq -\dfrac{1}{6}x + 7$

---

2 **Apply Systems of Inequalities** When using a system of inequalities to describe constraints on the possible combinations in a real-world problem, sometimes only whole-number solutions will make sense.

Real-World Example 3   Whole-Number Solutions

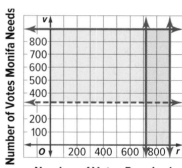

ELECTIONS Monifa is running for student council. The election rules say that for the election to be valid, at least 80% of the 900 students must vote. Monifa knows that she needs more than 330 votes to win.

a. Define the variables, and write a system of inequalities to represent this situation. Then graph the system.

Let r = the number of votes required by the election rules; 80% of 900 students is 720 students. So $r \geq 720$.

Let v = the number of votes that Monifa needs to win. So $v > 330$.

There are 900 students, so $r \leq 900$ and $v \leq 900$. The system of inequalities is $720 \leq r \leq 900$ and $330 < v \leq 900$.

b. Name one viable option.

Only whole-number solutions make sense in this problem. One possible solution is (800, 400); 800 students voted and Monifa received 400 votes.

Real-World Link

Student government might be a good activity for you if you like to bring about change, plan events, and work with others.

> **Guided Practice**

3. FUNDRAISING The Theater Club is selling shirts. They have only enough supplies to print 120 shirts. They will sell sweatshirts for $22 and T-shirts for $15, with a goal of at least $2000 in sales.

A. Define the variables, and write a system of inequalities to represent this situation.

B. Then graph the system.

C. Name one possible solution.

D. Is (45, 30) a solution? Explain.

Examples 1–2 **Solve each system of inequalities by graphing.**

1. $x \geq 4$
$y \leq x - 3$

2. $y > -2$
$y \leq x + 9$

3. $y < 3x + 8$
$y \geq 4x$

4. $3x - y \geq -1$
$2x + y \geq 5$

5. $y \leq 2x - 7$
$y \geq 2x + 7$

6. $y > -2x + 5$
$y \geq -2x + 10$

7. $2x + y \leq 5$
$2x + y \leq 7$

8. $5x - y < -2$
$5x - y > 6$

Example 3

9. **AUTO RACING** At a racecar driving school there are safety requirements.

  a. Define the variables, and write a system of inequalities to represent the height and weight requirements in this situation. Then graph the system.

  b. Name one possible solution.

  c. Is (50, 180) a solution? Explain.

UR **FAST** DRIVING SCHOOL
RULES TO QUALIFY
18 years of age or older
Good physical condition
Under 6 ft 7 in. tall
Under 295 lb

Practice and Problem Solving

Extra Practice is on page R6.

Examples 1–2 **Solve each system of inequalities by graphing.**

10. $y < 6$
$y > x + 3$

11 $y \geq 0$
$y \leq x - 5$

12. $y \leq x + 10$
$y > 6x + 2$

13. $y < 5x - 2$
$y > -6x + 2$

14. $2x - y \leq 6$
$x - y \geq -1$

15. $3x - y > -5$
$5x - y < 9$

16. $y \geq x + 10$
$y \leq x - 3$

17. $y < 5x - 5$
$y > 5x + 9$

18. $y \geq 3x - 5$
$3x - y > -4$

19. $4x + y > -1$
$y < -4x + 1$

20. $3x - y \geq -2$
$y < 3x + 4$

21. $y > 2x - 3$
$2x - y \geq 1$

22. $5x - y < -6$
$3x - y \geq 4$

23. $x - y \leq 8$
$y < 3x$

24. $4x + y < -2$
$y > -4x$

Example 3

25. **ICE RINKS** Ice resurfacers are used for rinks of at least 1000 square feet and up to 17,000 square feet. The price ranges from as little as $10,000 to as much as $150,000.

  a. Define the variables, and write a system of inequalities to represent this situation. Then graph the system.

  b. Name one possible solution.

  c. Is (15,000, 30,000) a solution? Explain.

26. **MP MODELING** Josefina works between 10 and 30 hours per week at a pizzeria. She earns $8.50 an hour, but can earn tips when she delivers pizzas.

  a. Write a system of inequalities to represent the dollars d she could earn for working h hours in a week.

  b. Graph this system.

  c. Josefina earned $195.50 last week. What range of hours could she have worked?

Solve each system of inequalities by graphing.

27. $x + y \geq 1$
$x + y \leq 2$

28. $3x - y < -2$
$3x - y < 1$

29. $2x - y \leq -11$
$3x - y \geq 12$

30. $y < 4x + 13$
$4x - y \geq 1$

31. $4x - y < -3$
$y \geq 4x - 6$

32. $y \leq 2x + 7$
$y < 2x - 3$

33. $y > -12x + 1$
$y \leq 9x + 2$

34. $2y \geq x$
$x - 3y > -6$

35. $x - 5y > -15$
$5y \geq x - 5$

36. CLASS PROJECT An economics class formed a company to sell school supplies. They would like to sell at least 20 notebooks and 50 pens per week, with a goal of earning at least $150 per week.

a. Define the variables, and write a system of inequalities to represent this situation.

b. Graph the system.

c. Name one possible solution.

37. FINANCIAL LITERACY Opal makes $15 per hour working for a photographer. She also coaches a competitive soccer team for $10 per hour. Opal needs to earn at least $90 per week, but she does not want to work more than 20 hours per week.

a. Define the variables, and write a system of inequalities to represent this situation.

b. Graph this system.

c. Give two possible solutions to describe how Opal can meet her goals.

d. Is (2, 2) a solution? Explain.

H.O.T. Problems   Use Higher-Order Thinking Skills

38. CHALLENGE Create a system of inequalities equivalent to $|x| \leq 4$.

39. MP REASONING State whether the following statement is *sometimes*, *always*, or *never* true. Explain your answer with an example or counterexample.

Systems of inequalities with parallel boundaries have no solutions.

40. MP REASONING Describe the graph of the solution of this system without graphing.
$6x - 3y \leq -5$
$6x - 3y \geq -5$

41. OPEN-ENDED One inequality in a system is $3x - y > 4$. Write a second inequality so that the system will have no solution.

42. MP PRECISION Graph the system of inequalities. Estimate the area of the solution.
$y \geq 1$
$y \leq x + 4$
$y \leq -x + 4$

43. WRITING IN MATH Refer to the beginning of the lesson. Explain what each colored region of the graph represents. Explain how shading in various colors can help to clearly show the solution set of a system of inequalities.

44. Fernando wants to start an exercise regimen where each workout will include swimming and walking on a treadmill. He is going to spend at least 15 minutes swimming and at least 5 minutes walking on the treadmill. Each workout will be less than 25 minutes. He graphs a system of inequalities as shown to represent this situation, where x represents the time spent swimming and y represents the time spent walking. Which of the solutions could describe a workout, where x is the number of minutes swimming and y is the number of minutes walking? **MP** 4

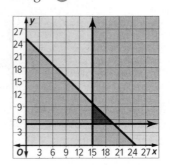

○ **A** $(11, 7)$

○ **B** $(13, 10)$

○ **C** $(15, 8)$

○ **D** $(18, 9)$

45. Which system of inequalities is shown on the graph? **MP** 7

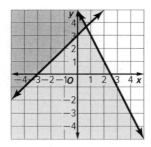

○ **A** $2x + y \leq 5$
   $-x + y \geq 3$

○ **B** $2x + y \geq 5$
   $-x + y \geq 3$

○ **C** $2x + y \geq 5$
   $-x + y \leq 3$

○ **D** $2x + y \leq 5$
   $-x + y \leq 3$

46. A system of inequalities is shown below.
MP 7

$$y \leq x + 5$$
$$y \leq -3x + 3$$
$$y \geq 0$$

What is the area, in square units, of the triangular region described by the system?

47. A system of inequalities is shown below. **MP** 7

$$-6x + 2y \geq 5$$
$$y < 3x + 2$$

Which of the following is not a correct description of the graph of the system?

○ **A** The graphs are parallel.

○ **B** The point $(-3, 6)$ is included in the graph $-6x + 2y > 5$.

○ **C** The solution to the system is the area between the two lines.

○ **D** The point $(2, 5)$ is included in the graph $y < 3x + 2$.

48. MULTI-STEP A contractor was given constraints before creating a rectangular playground area at a city park. She was told that the width of one side of the fencing had to be at least 100 yards and the total fencing around the playground had to be no more than 350 yards. **MP** 2, 4

a. Define variables, and write a system of inequalities to represent this situation.

b. Select all possible solutions.
   ☐ **A** Fence with length of 70 yards and width of 100 yards
   ☐ **B** Fence with length of 75 yards and width of 105 yards
   ☐ **C** Fence with length of 80 yards and width of 100 yards
   ☐ **D** Fence with length of 85 yards and width of 85 yards
   ☐ **E** $(25, 150)$
   ☐ **F** $(50, 125)$
   ☐ **G** $(75, 100)$

c. Is $(-50, 225)$ a solution? Explain.

You can use TI-Nspire technology to explore systems of inequalities. To prepare your calculator, add a new Graphs page from the Home screen.

Mathematical Practices

  5 Use appropriate tools strategically.

Activity 1 Graph Systems of Inequalities

Mr. Jackson owns a car washing and detailing business. It takes 20 minutes to wash a car and 60 minutes to detail a car. He works at most 8 hours per day and does at most 4 details per day. Work cooperatively to write and graph a system of linear inequalities to represent this situation.

First, write a linear inequality that represents the time it takes for car washing and car detailing. Let x represent the number of car washes, and let y represent the number of car details. Then $20x + 60y \leq 480$.

To graph this using a graphing calculator, solve for y.

$20x + 60y \leq 480$ Original inequality

$\quad 60y \leq -20x + 480$ Subtract 20x from each side and simplify.

$\quad\quad y \leq -\dfrac{1}{3}x + 8$ Divide each side by 60 and simplify.

Mr. Jackson does at most 4 details per day. This means that $y \leq 4$.

Step 1 Adjust the viewing window and then graph $y \leq 4$. Use the **Window Settings** option from the **Window/Zoom** menu to adjust the window to -4 to 30 for x and -2 to 10 for y. Keep the scales as **Auto**. Then enter **del** ≤ 4 **enter**.

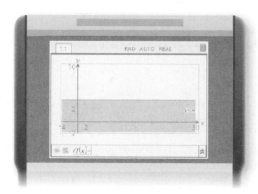

Step 2 Graph $y \leq -\dfrac{1}{3}x + 8$. Press **tab del** $\leq$ and then enter $-\dfrac{1}{3}x + 8$.

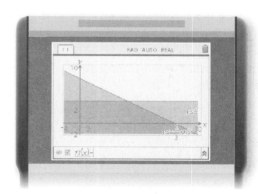

The darkest shaded region of the graph represents the solutions.

Analyze the Results

Work cooperatively.

1. If Mr. Jackson charges $75 for each car he details and $25 for each car wash, what is the maximum amount of money he could earn in one day?

2. What is the greatest number of washes that Mr. Jackson could do in a day? Explain.

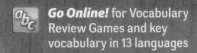
Go Online! for Vocabulary Review Games and key vocabulary in 13 languages

Study Guide

Key Concepts

Systems of Equations (Lessons 6-1 through 6-5)

- A system with a graph of two intersecting lines has one solution and is *consistent* and *independent.*

- Graphing a system of equations can only provide approximate solutions. For exact solutions, you must use algebraic methods.

- In the substitution method, one equation is solved for a variable and the expression substituted into the second equation to find the value of another variable.

- In the elimination method, one variable is eliminated by adding or subtracting the equations.

- Sometimes multiplying one or both equations by a constant makes it easier to use the elimination method.

- The best method for solving a system of equations depends on the coefficients of the variables.

Systems of Inequalities (Lesson 6-6)

- A system of inequalities is a set of two or more inequalities with the same variables.

- The solution of a system of inequalities is the intersection of the graphs.

 **Study Organizer**

Use your Foldable to review the chapter. Working with a partner can be helpful. Ask for clarification of concepts as needed.

Key Vocabulary

augmented matrix (p. 374)	**inconsistent** (p. 339)
consistent (p. 339)	**independent** (p. 339)
dependent (p. 339)	**matrix** (p. 374)
dimension (p. 374)	**substitution** (p. 348)
element (p. 374)	**system of equations** (p. 339)
elimination (p. 354)	**system of inequalities** (p. 376)

Vocabulary Check

State whether each sentence is *true* or *false*. If *false*, replace the underlined term to make a true sentence.

1. If a system has at least one solution, it is said to be <u>consistent</u>.

2. If a consistent system has exactly <u>two</u> solution(s), it is said to be independent.

3. If a consistent system has an infinite number of solutions, it is said to be <u>inconsistent</u>.

4. If a system has no solution, it is said to be <u>inconsistent</u>.

5. <u>Substitution</u> involves substituting an expression from one equation for a variable in the other.

6. In some cases, <u>dividing</u> two equations in a system together will eliminate one of the variables. This process is called elimination.

7. A set of two or more inequalities with the same variables is called a <u>system of equations</u>.

8. When the graphs of the inequalities in a system of inequalities <u>do not intersect</u>, there are no solutions to the system.

Concept Check

9. Is only addition or subtraction used in the elimination method?

10. Does a consistent and independent system of equations have an infinite number of solutions?

Lesson-by-Lesson Review

6-1 Graphing Systems of Equations

Graph each system and determine the number of solutions that it has. If it has one solution, name it.

11. $x - y = 1$
$x + y = 5$

12. $y = 2x - 4$
$4x + y = 2$

13. $2x - 3y = -6$
$y = -3x + 2$

14. $-3x + y = -3$
$y = x - 3$

15. $x + 2y = 6$
$3x + 6y = 8$

16. $3x + y = 5$
$6x = 10 - 2y$

17. MAGIC NUMBERS Sean is trying to find two numbers with a sum of 14 and a difference of 4. Define two variables, write a system of equations, and solve by graphing.

Example 1

Graph the system and determine the number of solutions it has. If it has one solution, name it.

$y = 2x + 2$
$y = -3x - 3$

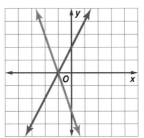

The lines appear to intersect at the point $(-1, 0)$. You can check this by substituting -1 for x and 0 for y.

CHECK $y = 2x + 2$ Original equation
$0 \stackrel{?}{=} 2(-1) + 2$ Substitution
$0 \stackrel{?}{=} -2 + 2$ Multiply.
$0 = 0$ ✔

$y = -3x - 3$ Original equation
$0 \stackrel{?}{=} -3(-1) - 3$ Substitution
$0 \stackrel{?}{=} 3 - 3$ Multiply.
$0 = 0$ ✔

The solution is $(-1, 0)$.

6-2 Substitution

Use substitution to solve each system of equations.

18. $x + y = 3$
$x = 2y$

19. $x + 3y = -28$
$y = -5x$

20. $3x + 2y = 16$
$x = 3y - 2$

21. $x - y = 8$
$y = -3x$

22. $y = 5x - 3$
$x + 2y = 27$

23. $x + 3y = 9$
$x + y = 1$

24. GEOMETRY The perimeter of a rectangle is 48 inches. The length is 6 inches greater than the width. Define the variables, and write equations to represent this situation. Solve the system by using substitution.

Example 2

Use substitution to solve the system.

$3x - y = 18$
$y = x - 4$

$3x - y = 18$ First equation
$3x - (x - 4) = 18$ Substitute $x - 4$ for y.
$2x + 4 = 18$ Simplify.
$2x = 14$ Subtract 4 from each side.
$x = 7$ Divide each side by 2.

Use the value of x and either equation to find the value for y.

$y = x - 4$ Second equation
$= 7 - 4$ or 3   Substitute and simplify.

The solution is $(7, 3)$.

6-3 Elimination Using Addition and Subtraction

Use elimination to solve each system of equations.

25. $x + y = 13$
$x - y = 5$

26. $-3x + 4y = 21$
$3x + 3y = 14$

27. $x + 4y = -4$
$x + 10y = -16$

28. $2x + y = -5$
$x - y = 2$

29. $6x + y = 9$
$-6x + 3y = 15$

30. $x - 4y = 2$
$3x + 4y = 38$

31. $2x + 2y = 4$
$2x - 8y = -46$

32. $3x + 2y = 8$
$x + 2y = 2$

33. BASEBALL CARDS Cristiano bought 24 baseball cards for $50. One type cost $1 per card, and the other cost $3 per card. Define the variables, and write equations to find the number of each type of card he bought. Solve by using elimination.

Example 3

Use elimination to solve the system of equations.

$3x - 5y = 11$
$x + 5y = -3$

$$\begin{array}{r} 3x - 5y = 11 \\ (+)\quad x + 5y = -3 \\ \hline 4x \quad\quad = 8 \\ x = 2 \end{array}$$

The variable y is eliminated.
Divide each side by 4.

Now, substitute 2 for x in either equation to find the value of y.

$3x - 5y = 11$ — First equation
$3(2) - 5y = 11$ — Substitute.
$6 - 5y = 11$ — Multiply.
$-5y = 5$ — Subtract 6 from each side.
$y = -1$ — Divide each side by -5.

The solution is $(2, -1)$.

6-4 Elimination Using Multiplication

Use elimination to solve each system of equations.

34. $x + y = 4$
$-2x + 3y = 7$

35. $x - y = -2$
$2x + 4y = 38$

36. $3x + 4y = 1$
$5x + 2y = 11$

37. $-9x + 3y = -3$
$3x - 2y = -4$

38. $8x - 3y = -35$
$3x + 4y = 33$

39. $2x + 9y = 3$
$5x + 4y = 26$

40. $-7x + 3y = 12$
$2x - 8y = -32$

41. $8x - 5y = 18$
$6x + 6y = -6$

42. BAKE SALE On the first day, a total of 40 items were sold for $356. Define the variables, and write a system of equations to find the number of cakes and pies sold. Solve by using elimination.

MONARCH MIDDLE SCHOOL

Bake Sale

Pies $10

Cakes $8

Example 4

Use elimination to solve the system of equations.

$3x + 6y = 6$
$2x + 3y = 5$

Notice that if you multiply the second equation by -2, the coefficients of the y-terms are additive inverses.

$3x + 6y = 6$ $3x + 6y = 6$
$2x + 3y = 5$ Multiply by -2. $(+) -4x - 6y = -10$
                                      $-x\quad\quad = -4$
                                      $x = 4$

Now, substitute 4 for x in either equation to find the value of y.

$2x + 3y = 5$ — Second equation
$2(4) + 3y = 5$ — Substitution
$8 + 3y = 5$ — Multiply.
$3y = -3$ — Subtract 8 from both sides.
$y = -1$ — Divide each side by 3.

The solution is $(4, -1)$.

6-5 Applying Systems of Linear Equations

Determine the best method to solve each system of equations. Then solve the system.

43. $y = x - 8$
   $y = -3x$

44. $y = -x$
   $y = 2x$

45. $x + 3y = 12$
   $x = -6y$

46. $x + y = 10$
   $x - y = 18$

47. $3x + 2y = -4$
   $5x + 2y = -8$

48. $6x + 5y = 9$
   $-2x + 4y = 14$

49. $3x + 4y = 26$
   $2x + 3y = 19$

50. $11x - 6y = 3$
   $5x - 8y = -25$

51. COINS Tionna has saved dimes and quarters in her piggy bank. Define the variables, and write a system of equations to determine the number of dimes and quarters. Then solve the system using the best method for the situation.

$4.00
25 coins

52. FAIR At a county fair, the cost for 4 slices of pizza and 2 orders of French fries is $21.00. The cost of 2 slices of pizza and 3 orders of French fries is $16.50. To find out how much a single slice of pizza and an order of French fries costs, define the variables and write a system of equations to represent the situation. Determine the best method to solve the system of equations. Then solve the system. (Lesson 6-5)

Example 5

Determine the best method to solve the system of equations. Then solve the system.

$3x + 5y = 4$
$4x + y = -6$

The coefficient of y is 1 in the second equation. So solving by substitution is a good method. Solve the second equation for y.

$4x + y = -6$	Second equation
$y = -6 - 4x$	Subtract $4x$ from each side.

Substitute $-6 - 4x$ for y in the first equation.

$3x + 5(-6 - 4x) = 4$	Substitute.
$3x - 30 - 20x = 4$	Distributive Property
$-17x - 30 = 4$	Simplify.
$-17x = 34$	Add 30 to each side.
$x = -2$	Divide by -17.

Last, substitute -2 for x in either equation to find y.

$4x + y = -6$	Second equation
$4(-2) + y = -6$	Substitute.
$-8 + y = -6$	Multiply.
$y = 2$	Add 8 to each side.

The solution is $(-2, 2)$.

6-6 Systems of Inequalities

Solve each system of inequalities by graphing.

53. $x > 3$
$y < x + 2$

54. $y \leq 5$
$y > x - 4$

55. $y < 3x - 1$
$y \geq -2x + 4$

56. $y \leq -x - 3$
$y \geq 3x - 2$

57. JOBS Kishi makes $9 an hour working at the grocery store and $12 an hour delivering newspapers. She cannot work more than 20 hours per week. Graph two inequalities that Kishi can use to determine how many hours she needs to work at each job if she wants to earn at least $150 per week.

Example 6

Solve the system of inequalities by graphing.

$y < 3x + 1$
$y \geq -2x + 3$

The solution set of the system is the set of ordered pairs in the intersection of the two graphs. This portion is shaded in the graph below.

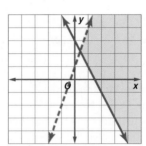

Go Online! for another Chapter Test

Graph each system and determine the number of solutions that it has. If it has one solution, name it.

1. $y = 2x$
$y = 6 - x$

2. $y = x - 3$
$y = -2x + 9$

3. $x - y = 4$
$x + y = 10$

4. $2x + 3y = 4$
$2x + 3y = -1$

Use substitution to solve each system of equations.

5. $y = x + 8$
$2x + y = -10$

6. $x = -4y - 3$
$3x - 2y = 5$

7. GARDENING Corey has 42 feet of fencing around his garden. The garden is rectangular in shape, and its length is equal to twice the width minus 3 feet. Define the variables, and write a system of equations to find the length and width of the garden. Solve the system by using substitution.

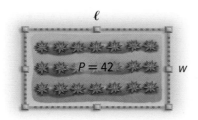

8. MULTIPLE CHOICE Use elimination to solve the system.
$$6x - 4y = 6$$
$$-6x + 3y = 0$$

A $(5, 6)$

B $(-3, -6)$

C $(1, 0)$

D $(4, -8)$

9. Shelly has $300 to shop for jeans and sweaters. Each pair of jeans cost $65, each sweater costs $34, and she buys 7 items. Determine the number of pairs of jeans and sweaters Shelly bought.

Use elimination to solve each system of equations.

10. $x + y = 13$
$x - y = 5$

11. $3x + 7y = 2$
$3x - 4y = 13$

12. $x + y = 8$
$x - 3y = -4$

13. $2x + 6y = 18$
$3x + 2y = 13$

14. MAGAZINES Julie subscribes to a sports magazine and a fashion magazine. She received 24 issues this year. The number of fashion issues is 6 less than twice the number of sports issues. Define the variables, and write a system of equations to find the number of issues of each magazine.

Determine the best method to solve each system of equations. Then solve the system.

15. $y = 3x$
$x + 2y = 21$

16. $x + y = 12$
$y = x - 4$

17. $x + y = 15$
$x - y = 9$

18. $3x + 5y = 7$
$2x - 3y = 11$

19. OFFICE SUPPLIES At a sale, Ricardo bought 24 reams of paper and 4 inkjet cartridges for $320. Britney bought 2 reams of paper and 1 inkjet cartridge for $50. The reams of paper were all the same price and the inkjet cartridges were all the same price. Write a system of equations to represent this situation. Determine the best method to solve the system of equations. Then solve the system.

Solve each system of inequalities by graphing.

20. $x > 2$
$y < 4$

21. $x + y \leq 5$
$y \geq x + 2$

22. $3x - y > 9$
$y > -2x$

23. $y \geq 2x + 3$
$-4x - 3y > 12$

Preparing for Assessment

Performance Task

Provide a clear solution to each part of the task. Be sure to show all of your work, include all relevant drawings, and justify your answers.

SMARTPHONES Smartphones contain gold and silver. A smartphone has 10 times as much silver as it has gold for a total weight of 0.385 gram. How much gold and silver is in that smartphone?

Part A

1. **Sense Making** Define the variables, and write a system of linear equations to represent the situation.

Part B

2. Interpret the meaning of each equation.

Part C

3. Determine the best method to solve the system of equations by looking closely at the coefficients of each term. Explain your reasoning.

Part D

4. **Tools** Solve the system and describe its meaning in the context of the situation. Show your work.

Part E

5. Determine the number of solutions that the system has. If it has one solution, name it.

Part F

6. Use substitution to check the answer. Is the solution reasonable? Explain your reasoning.

Test-Taking Strategy

Example

Read the problem. Identify what you need to know. Then use the information in the problem to solve.

Solve $\begin{cases} 4x - 8y = 20 \\ -3x + 5y = -14 \end{cases}$.

A $(5, 0)$ **C** $(3, -1)$

B $(4, -2)$ **D** $(-6, -5)$

Step 1 **What do you need to find?** The solution to the system of equations.

Step 2 **Is there enough information given to solve the problem?** Yes.

Step 3 **What information, if any, is not needed to solve the problem?** None that one can discern.

Step 4 **Are there any obvious wrong answers?** No.

Step 5 **What is the correct answer?**
Find the answer choice that satisfies both equations of the system.

	First Equation	Second Equation
Guess: (5, 0)	$4x - 8y = 20$ $4(5) - 8(0) = 20$ ☐	$-3x + 5y = -14$ $-3(5) + 5(0) \neq -14$ ☐

	First Equation	Second Equation
Guess: (4, −2)	$4x - 8y = 20$ $4(4) - 8(-2) \neq 20$ ☐	$-3x + 5y = -14$ $-3(4) + 5(-2) \neq -14$ ☐

	First Equation	Second Equation
Guess: (3, −1)	$4x - 8y = 20$ $4(3) - 8(-1) = 20$ ☐	$-3x + 5y = -14$ $-3(3) + 5(-1) = -14$ ☐

The correct answer is C.

Test-Taking Tip

Guess and Check It is very important to pace yourself and keep track of how much time you have left. If time is running short, or if you are unsure how to solve a problem, the guess and check strategy may help you determine the correct answer quickly.

Apply the Strategy

Gina bought 5 hot dogs and 3 soft drinks at the ball game for \$32.25. Renaldo bought 4 hot dogs and 2 soft drinks for \$24.00. How much do a single hot dog and a single drink cost?

A   hot dog: \$4.50 soft drink: \$3.75   **C**   hot dog: \$4.50 soft drink: \$4.25

B   hot dog: \$3.75 soft drink: \$4.50   **D**   hot dog: \$3.75 soft drink: \$4.25

Answer the questions below.

a. What do you need to find?

b. Is there enough information given to solve the problem?

c. What information, if any, is not needed to solve the problem?

d. Are there any obvious wrong answers?

e. What is the correct answer?

Read each question. Then fill in the correct answer on the answer document provided by your teacher or on a sheet of paper.

1. What is the slope of a line that is perpendicular to $x + 2y = -3$?

2. Which of the following statements is true of the system below?

$$3x + y = -5$$
$$y + 3x = 4$$

- A The system has one solution at $\left(\frac{7}{6}, -\frac{1}{2}\right)$.
- B The system has no solution.
- C The system has infinitely many solutions.
- D The system has one solution at $\left(-\frac{5}{3}, 9\right)$.

3. Nadia's vegetable garden has a perimeter of 48 feet. Expressions for the dimensions of her garden are shown in the diagram below.

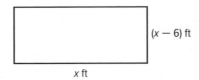

$(x - 6)$ ft

x ft

   a. What are the dimensions of Nadia's vegetable garden?
   - A length = 27 feet, width = 21 feet
   - B length = 15 feet, width = 8 feet
   - C length = 9 feet, width = 3 feet
   - D length = 15 feet, width = 9 feet

   b. What is the area of her garden?

Test-Taking Tip

Question 5 To find 30% more than Ryan weighs, first let r represent Ryan's weight. Then $0.3r$ is 30% of r, and $r + 0.3r$ or $1.3r$ is 30% more than what Ryan weighs.

4. If $(-3h, 2h)$ and $(2h, 6h)$ are two points on the graph of a line and h is not equal to 0, what is the slope of the line?
   - A -8
   - B -4
   - C $\frac{3}{5}$
   - D $\frac{4}{5}$

5. Ryan's brother, Joel, weighs 30% more than Ryan. Together, the boys weigh 276 pounds. In pounds, how much does Joel weigh?

6. The sum of two numbers is -3. Three times the first number minus the second number is 27. What are the two numbers?
   - A 6 and -3
   - B 6 and -9
   - C 12 and -15
   - D $\frac{7}{15}$ and -18

7. Which of the following are accurate descriptions of the graph of $3x + y = -5$?
   - A The graph contains the points $(-3, 4)$, $(-1, -2)$, and $(3, -14)$.
   - B The graph is a line with a y-intercept of $(0, -5)$ and a slope of -3.
   - C The graph has an x-intercept of $\left(-\frac{5}{3}, 0\right)$.
   - D The graph passes through the points $(1, -8)$ and $(2, 1)$.

8. Two friends own a total of 25 books. One owns four times as many books as the other. Find the number of books each friend owns.

9. The band at Jason's high school had a concert for charity. They raised $1500. Adult tickets were $5 each and student tickets were $3 each. The members of the band sold a total of 400 tickets. How many student tickets did they sell?

   ○ **A** 150

   ○ **B** 175

   ○ **C** 250

   ○ **D** 750

10. The graph of a linear inequality is shown below.

    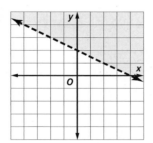

    What is the inequality that describes the graph?

    [   ]

11. Which expressions are equivalent to $-3(-x + 5) + 6(x - 4) - 2x$?

    ☐ **A** $3(x - 5) + 4(x - 6)$

    ☐ **B** $9x - 39$

    ☐ **C** $7x - 39$

    ☐ **D** $x - 39$

12. At a garage sale, Julia sold a total of 32 T-shirts for $114. The prices she sold the shirts for are shown in the table.

T-shirt Prices	
Sports T-shirts	$5
Concert T-shirts	$3

How many concert T-shirts did Julia sell?

13. Which system of inequalities describes all points in Quadrant III?

    ○ **A** $x > 0, y < 0$

    ○ **B** $x < 0, y > 0$

    ○ **C** $x < 0, y < 0$

    ○ **D** $x < y, y < 0$

14. There are 28 students in Lisa's Algebra 1 class. There are 12 more girls than boys. How many girls are in the class?

    ○ **A** 20   ○ **C** 16

    ○ **B** 18   ○ **D** 8

15. The sum of the perimeters of two different equilateral triangles is 48 inches. The difference between the perimeters of the two triangles is 6 inches.

    a. Write the system of equations to find the side lengths for each triangle.

    b. What are the perimeters of the two triangles?

Need Extra Help?

If you missed Question...	1	2	3	4	5	6	7	8	9	10	11	12	13	14	15
Go to Lesson...	4-4	6-1	2-3	3-3	6-2	6-3	4-1	6-2	6-5	5-6	1-4	6-5	6-6	6-2	6-3

CHAPTER 7
Exponents and Exponential Functions

THEN
You evaluated expressions involving exponents.

NOW
In this chapter, you will:

- Simplify and perform operations on expressions involving exponents.
- Extend the properties of integer exponents to rational exponents.
- Write and transform exponential functions.
- Graph and use exponential functions.

 WHY

SPACE NASA specializes in space exploration. NASA employees use math to, among other things, calculate travel time and to determine the solar power available to spacecraft as they travel away from the sun.

Use the Mathematical Practices to complete the activity.

1. Using Tools Use the Internet to find out how NASA calculated the solar energy available for the Juno spacecraft or another spacecraft and to learn the amount of solar power available on Earth (in w/m ^{2}).

2. Apply Math Create a table to record distances from the sun (in AU) and then calculate the energy available to solar panels on the spacecraft at different points.

3. Model with Mathematics Use the Line Graph tool to graph the results of your research and calculations.

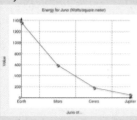

Go Online to Guide Your Learning

Explore & Explain		Organize

The Geometer's Sketchpad

With The Geometer's Sketchpad, you can multiply monomials using properties of exponents. It can also be used to simplify expressions containing negative and zero exponents, to simplify expressions containing negative and zero exponents, and to graph exponential functions.

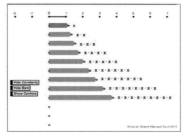

Graphing Tools

Explore **Exponential Functions** using the graphing tools to enhance your understanding in Lessons 7-5 and 7-7.

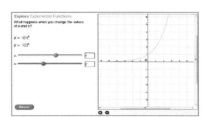

eBook

Interactive Student Guide

Before starting the chapter, answer the **Chapter Focus** preview questions. Check your answers as you complete each lesson. At the end of the chapter, try the **Performance Task**.

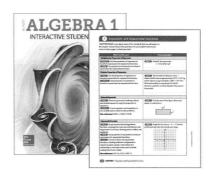

Foldables

Exponents and Exponential Functions Make this Foldable to help you organize your notes about exponents and exponential functions. Begin with eleven sheets of notebook paper arranged in a stack and stapled. Starting with the second sheet of paper, cut along the right side to form tabs and label each tab with a lesson **FOLDABLES** number.

Collaborate

Chapter Project

In the **Interesting Thing About Credit Cards** project, you will use what you have learned about exponential functions to complete a project.

Focus

 LEARNSMART

Need help studying? Complete the **Linear and Exponential Relationships** domain in LearnSmart to review for the chapter test.

ALEKS

You can use the **Functions and Lines** and **Exponents** topics in ALEKS to find out what you know about linear functions and what you are ready to learn.*

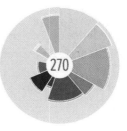

* Ask your teacher if this is part of your program.

Get Ready for the Chapter

Go Online! for Vocabulary Review Games and key vocabulary in 13 languages.

Connecting Concepts

Concept Check

Review the concepts used in this chapter by answering the questions below.

1. What is a benefit of writing $4 \cdot 4 \cdot 4 \cdot 4 \cdot 4$ using exponents?

2. If you know a photo is 4 inches on one side and 6 inches on the other, how would you determine the area of the photo?

3. What would the units be for the area of the picture in question 2?

4. A cube is 5 feet on each side. How would you determine the volume of the cube?

5. What would the units be for question 4?

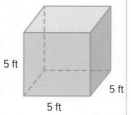
5 ft
5 ft
5 ft

6. Is the expression -5^2 different from $(-5)^2$? Why?

7. Is the expression $2 + 5^2$ different from $(2 + 5)^2$?

8. How would you evaluate the expression $\frac{5}{7} \times \frac{5}{7}$? How else could you write the expression?

Performance Task Preview

You can use the concepts and skills in the chapter to solve problems in a real-world setting. Understanding exponents and exponential functions will help you finish the Performance Task at the end of the chapter.

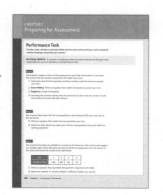

MP **In this Performance Task you will:**

• make sense of problems
• look for and express regularity in reasoning

New Vocabulary

English		Español
monomial	p. 395	monomio
constant	p. 395	constante
zero exponent	p. 403	cero exponente
order of magnitude	p. 405	orden de magnitud
rational exponent	p. 410	exponent racional
cube root	p. 411	raíz cúbica
nth root	p. 411	raíz enésima
exponential equation	p. 413	ecuación exponencial
radical expression	p. 419	expresión radical
rationalizing the denominator	p. 421	racionalizar el denominador
conjugate	p. 421	conjugado
exponential function	p. 430	función exponencial
exponential growth	p. 430	crecimiento exponencial
exponential decay	p. 430	desintegración exponencial
compound interest	p. 451	interés es compuesta
geometric sequence	p. 462	secuencia geométrica
common ratio	p. 462	proporción común
recursive formula	p. 469	fórmula recursiva

Review Vocabulary

base base In an expression of the form x^n, the base is x.

Distributive Property Propiedad distributiva
For any numbers a, b, and c, $a(b + c) = ab + ac$ and $a(b - c) = ab - ac$.

exponent exponente
In an expression of the form x^n, the exponent is n. It indicates the number of times x is used as a factor.

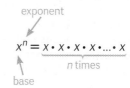

$$x^n = \underbrace{x \cdot x \cdot x \cdot x \cdot \ldots \cdot x}_{n \text{ times}}$$
exponent
base

Multiplication Properties of Exponents

- You evaluated expressions with exponents.

1 Multiply monomials using the properties of exponents.

2 Simplify expressions using the multiplication properties of exponents.

Many formulas contain *monomials*. For example, the formula for the horsepower of a car is $H = w\left(\frac{v}{234}\right)^3$. *H* represents the horsepower produced by the engine, *w* equals the weight of the car with passengers, and *v* is the velocity of the car at the end of a quarter of a mile. As the velocity increases, the horsepower increases.

 New Vocabulary
monomial
constant

 Mathematical Practices
8 Look for and express regularity in repeated reasoning.

1 Multiply Monomials A **monomial** is a number, a variable, or the product of a number and one or more variables with nonnegative integer exponents. It has only one term. In the formula to calculate the horsepower of a car, the term $w\left(\frac{v}{234}\right)^3$ is a monomial.

An expression that involves division by a variable, like $\frac{ab}{c}$, is not a monomial.

A **constant** is a monomial that is a real number. The monomial $3x$ is an example of a *linear expression* since the exponent of x is 1. The monomial $2x^2$ is a *nonlinear expression* since the exponent is a positive number other than 1.

Example 1   Identify Monomials

Determine whether each expression is a monomial. Write *yes* or *no*. Explain your reasoning.

a. 10   Yes; this is a constant, so it is a monomial.

b. $f + 24$ No; this expression has addition, so it has more than one term.

c. h^2 Yes; this expression is a product of variables.

d. j Yes; single variables are monomials.

▶ **Guided Practice**

1A. $-x + 5$ **1B.** $23abcd^2$

1C. $\dfrac{xyz^2}{2}$ **1D.** $\dfrac{mp}{n}$

Recall that an expression of the form x^n is called a *power* and represents the result of multiplying x by itself n times. x is the *base*, and n is the *exponent*. The word *power* is also used sometimes to refer to the exponent.

$$\underset{\text{base}}{\overset{\text{exponent}}{3^4}} = \overset{\text{4 factors}}{\overbrace{3 \cdot 3 \cdot 3 \cdot 3}} = 81$$

By applying the definition of a power, you can find the product of powers. Look for a pattern in the exponents.

$$2^2 \cdot 2^4 = \overbrace{2 \cdot 2}^{2\,\text{factors}} \cdot \overbrace{2 \cdot 2 \cdot 2 \cdot 2}^{4\,\text{factors}} \qquad 4^3 \cdot 4^2 = \overbrace{4 \cdot 4 \cdot 4}^{3\,\text{factors}} \cdot \overbrace{4 \cdot 4}^{2\,\text{factors}}$$
$$\underset{2+4=6\,\text{factors}}{} \qquad\qquad\qquad \underset{3+2=5\,\text{factors}}{}$$

These examples demonstrate the property for the product of powers.

🔩 Key Concept Product of Powers

Words	To multiply two powers that have the same base, add their exponents.
Symbols	For any real number a and any integers m and p, $a^m \cdot a^p = a^{m+p}$.
Examples	$b^3 \cdot b^5 = b^{3+5}$ or b^8 $\qquad$ $g^4 \cdot g^6 = g^{4+6}$ or g^{10}

Example 2 Product of Powers

Simplify each expression.

a. $(6n^3)(2n^7)$

$\begin{aligned}
(6n^3)(2n^7) &= (6 \cdot 2)(n^3 \cdot n^7) &&\text{Group the coefficients and the variables.}\\
&= (6 \cdot 2)(n^{3+7}) &&\text{Product of Powers}\\
&= 12n^{10} &&\text{Simplify.}
\end{aligned}$

> **Study Tip**
>
> **Coefficients and Powers of 1** A variable with no exponent or coefficient shown can be assumed to have an exponent and coefficient of 1. For example, $x = 1x^1$.

b. $(3pt^3)(p^3t^4)$

$\begin{aligned}
(3pt^3)(p^3t^4) &= (3 \cdot 1)(p \cdot p^3)(t^3 \cdot t^4) &&\text{Group the coefficients and the variables.}\\
&= (3 \cdot 1)(p^{1+3})(t^{3+4}) &&\text{Product of Powers}\\
&= 3p^4t^7 &&\text{Simplify.}
\end{aligned}$

▶ **Guided Practice**

2A. $(3y^4)(7y^5)$ $\qquad\qquad\qquad$ **2B.** $(-4rx^2t^3)(-6r^5x^2t)$

We can use the Product of Powers Property to find the power of a power. In the following examples, look for a pattern in the exponents.

$$(3^2)^4 = \overbrace{(3^2)(3^2)(3^2)(3^2)}^{4\,\text{factors}} \qquad (r^4)^3 = \overbrace{(r^4)(r^4)(r^4)}^{3\,\text{factors}}$$
$$= 3^{2+2+2+2} \qquad\qquad\qquad = r^{4+4+4}$$
$$= 3^8 \qquad\qquad\qquad\qquad = r^{12}$$

These examples demonstrate the property for the power of a power.

🔩 Key Concept Product of Powers

Words	To find the power of a power, multiply the exponents.
Symbols	For any real number a and any integers m and p, $(a^m)^p = a^{m \cdot p}$.
Examples	$(b^3)^5 = b^{3 \cdot 5}$ or b^{15} $\qquad$ $(g^6)^7 = g^{6 \cdot 7}$ or g^{42}

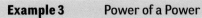

Study Tip

MP **Regularity** The power rules are general methods. If you are unsure about when to multiply the exponents and when to add the exponents, write the expression in expanded form.

Example 3   Power of a Power

Simplify $\left[(2^3)^2\right]^4$.

A 2^{24} **B** 2^{12} **C** 2^{10} **D** 2^9

Read the Item

You need to apply the power of a power rule.

Solve the Item

$$\left[(2^3)^2\right]^4 = (2^{3 \cdot 2})^4 \qquad \text{Power of a Power}$$
$$= (2^6)^4 \qquad \text{Simplify.}$$
$$= 2^{6 \cdot 4} \text{ or } 2^{24} \qquad \text{Power of a Power}$$

The correct choice is A.

> **Guided Practice**

3. Simplify $\left[(2^2)^2\right]^4$.

    A 2^8 **B** 2^{10} **C** 2^{16} **D** 2^{24}

We can use the Product of Powers Property and the Power of a Power Property to find the power of a product. Look for a pattern in the exponents below.

$$\overbrace{(tw)^3 = (tw)(tw)(tw)}^{\text{3 factors}} \qquad \overbrace{(2yz^2)^3 = (2yz^2)(2yz^2)(2yz^2)}^{\text{3 factors}}$$
$$= (t \cdot t \cdot t)(w \cdot w \cdot w) \qquad\qquad = (2 \cdot 2 \cdot 2)(y \cdot y \cdot y)(z^2 \cdot z^2 \cdot z^2)$$
$$= t^3 w^3 \qquad\qquad\qquad\qquad = 2^3 y^3 z^6 \text{ or } 8y^3 z^6$$

These examples demonstrate the property for the power of a product.

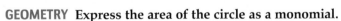

Key Concept   Power of a Product

Words	To find the power of a product, find the power of each factor and multiply.
Symbols	For any real numbers a and b and any integer m, $(ab)^m = a^m b^m$.
Example	$(-2xy^3)^5 = (-2)^5 x^5 y^{15}$ or $-32x^5 y^{15}$

Math-History Link

Albert Einstein
(1879–1955) Albert Einstein is perhaps the most well-known scientist of the 20th century. His formula, $E = mc^2$, where E represents the energy, m is the mass of the material, and c is the speed of light, shows that if mass is accelerated enough, it could be converted into usable energy.

Example 4   Power of a Product

GEOMETRY Express the area of the circle as a monomial.

$$\text{Area} = \pi r^2 \qquad \text{Formula for the area of a circle}$$
$$= \pi (2xy^2)^2 \qquad \text{Replace } r \text{ with } 2xy^2.$$
$$= \pi (2^2 x^2 y^4) \qquad \text{Power of a Product}$$
$$= 4x^2 y^4 \pi \qquad \text{Simplify.}$$

$2xy^2$

The area of the circle is $4x^2 y^4 \pi$ square units.

> **Guided Practice**

4A. Express the area of a square with sides of length $3xy^2$ as a monomial.

4B. Express the area of a triangle with height $4a$ and base $5ab^2$ as a monomial.

2 Simplify Expressions

We can combine and use these properties to simplify expressions involving monomials.

🔖 Key Concept   Simplifying Monomial Expressions

To simplify a monomial expression, write an equivalent expression in which:

- each variable base appears exactly once,
- there are no powers of powers, and
- all fractions are in simplest form.

Example 5   Simplify Expressions

Simplify $\left(3xy^4\right)^2\left[(-2y)^2\right]^3$.

$$
\begin{aligned}
\left(3xy^4\right)^2\left[(-2y)^2\right]^3 &= \left(3xy^4\right)^2(-2y)^6 && \text{Power of a Power} \\
&= (3)^2 x^2 \left(y^4\right)^2 (-2)^6 y^6 && \text{Power of a Product} \\
&= 9x^2 y^8 (64) y^6 && \text{Power of a Power} \\
&= 9(64) x^2 \cdot y^8 \cdot y^6 && \text{Commutative} \\
&= 576 x^2 y^{14} && \text{Product of Powers}
\end{aligned}
$$

▶ **Guided Practice**

5. Simplify $\left(\frac{1}{2}a^2 b^2\right)^3 \left[(-4b)^2\right]^2$.

Check Your Understanding

◯ = Step-by-Step Solutions begin on page R11.

Example 1   Determine whether each expression is a monomial. Write *yes* or *no*. Explain your reasoning.

1. 15
2. $2 - 3a$
3. $\dfrac{5c}{d}$
4. $-15g^2$
5. $\dfrac{r}{2}$
6. $7b + 9$

Examples 2–3   Simplify each expression.

7. $k(k^3)$
8. $m^4(m^2)$
9. $2q^2(9q^4)$
10. $\left(5u^4 v\right)\left(7u^4 v^3\right)$
11. $\left[(3^2)^2\right]^2$
12. $\left(xy^4\right)^6$
13. $\left(4a^4 b^9 c\right)^2$
14. $\left(-2f^2 g^3 h^2\right)^3$
15. $\left(-3p^5 t^6\right)^4$

Example 4   16. **GEOMETRY** The formula for the surface area of a cube is $SA = 6s^2$, where SA is the surface area and s is the length of any side.

    a. Express the surface area of the cube as a monomial.

    b. What is the surface area of the cube if $a = 3$ and $b = 4$?

$a^3 b$

Example 5   Simplify each expression.

17. $\left(5x^2 y\right)^2 \left(2xy^3 z\right)^3 (4xyz)$
18. $\left(-3d^2 f^3 g\right)^2 \left[\left(-3d^2 f\right)^3\right]^2$
19. $\left(-2g^3 h\right)\left(-3gj^4\right)^2 \left(-ghj\right)^2$
20. $\left(-7ab^4 c\right)^3 \left[\left(2a^2 c\right)^2\right]^3$

Example 1　Determine whether each expression is a monomial. Write *yes* or *no*. Explain your reasoning.

21. 122

22. $3a^4$

23. $2c + 2$

24. $\dfrac{-2g}{4h}$

25. $\dfrac{5k}{10}$

26. $6m + 3n$

Examples 2–3　Simplify each expression.

㉗ $(q^2)(2q^4)$

28. $(-2u^2)(6u^6)$

29. $(9w^2x^8)(w^6x^4)$

30. $(y^6z^9)(6y^4z^2)$

31. $(b^8c^6d^5)(7b^6c^2d)$

32. $(14fg^2h^2)(-3f^4g^2h^2)$

33. $(j^5k^7)^4$

34. $(n^3p)^4$

35. $[(2^2)^2]^2$

36. $[(3^2)^2]^4$

37. $[(4r^2t)^3]^2$

38. $[(-2xy^2)^3]^2$

Example 4　**GEOMETRY** Express the area of each triangle as a monomial.

39.

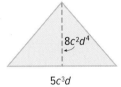

$8c^2d^4$

$5c^3d$

40.

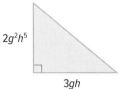

$2g^2h^5$

$3gh$

Example 5　Simplify each expression.

41. $(2a^3)^4(a^3)^3$

42. $(c^3)^2(-3c^5)^2$

43. $(2gh^4)^3[(-2g^4h)^3]^2$

44. $(5k^2m)^3[(4km^4)^2]^2$

45. $(p^5r^2)^4(-7p^3r^4)^2(6pr^3)$

46. $(5x^2y)^2(2xy^3z)^3(4xyz)$

47. $(5a^2b^3c^4)(6a^3b^4c^2)$

48. $(10xy^5z^3)(3x^4y^6z^3)$

49. $(0.5x^3)^2$

50. $(0.4h^5)^3$

51. $\left(-\dfrac{3}{4}c\right)^3$

52. $\left(\dfrac{4}{5}a^2\right)^2$

53. $(8y^3)(-3x^2y^2)\left(\dfrac{3}{8}xy^4\right)$

54. $\left(\dfrac{4}{7}m\right)^2(49m)(17p)\left(\dfrac{1}{34}p^5\right)$

55. $(-3r^3w^4)^3(2rw)^2(-3r^2)^3(4rw^2)^3(2r^2w^3)^4$

56. $(3ab^2c)^2(-2a^2b^4)^2(a^4c^2)^3(a^2b^4c^5)^2(2a^3b^2c^4)^3$

57. **FINANCIAL LITERACY** Cleavon has money in an account that earns 3% simple interest. The formula for computing simple interest is $I = Prt$, where I is the interest earned, P represents the principal that he put into the account, r is the interest rate (in decimal form), and t represents time in years.

　a. Cleavon makes a deposit of \$2c and leaves it for 2 years. Write a monomial that represents the interest earned.

　b. If c represents a birthday gift of \$250, how much will Cleavon have in this account after 2 years?

MP **TOOLS** Express the volume of each solid as a monomial.

58.

$2x$

$3x^2$

59.

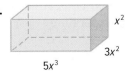

x^2

$3x^2$

$5x^3$

60.

$4x^4$

$2x^3$

$2x^2$

61 **PACKAGING** For a commercial art class, Aiko must design a new container for individually wrapped pieces of candy. The shape that she chose is a cylinder. The formula for the volume of a cylinder is $V = \pi r^2 h$.

   a. The radius that Aiko would like to use is $2p^3$, and the height is $4p^3$. Write a monomial that represents the volume of her container.

   b. Make a table for five possible measures for the radius and height of a cylinder having the same volume.

   c. What is the volume of Aiko's container if the height is doubled?

62. ENERGY Albert Einstein's formula $E = mc^2$ shows that if mass is accelerated enough, it could be converted into usable energy. Energy E is measured in joules, mass m in kilograms, and the speed c of light is about 300 million meters per second.

   a. Complete the calculations to convert 3 kilograms of gasoline completely into energy.

   b. What happens to the energy if the amount of gasoline is doubled?

63. MULTIPLE REPRESENTATIONS In this problem, you will explore exponents.

   a. Tabular Copy and use a calculator to complete the table.

Power	3^4	3^3	3^2	3^1	3^0	3^{-1}	3^{-2}	3^{-3}	3^{-4}
Value						$\frac{1}{3}$	$\frac{1}{9}$	$\frac{1}{27}$	$\frac{1}{81}$

   b. Analytical What do you think the values of 5^0 and 5^{-1} are? Verify your conjecture using a calculator.

   c. Analytical Complete: For any nonzero number a and any integer n, $a^{-n} = $ _____.

   d. Verbal Describe the value of a nonzero number raised to the zero power.

H.O.T. Problems   Use **H**igher-**O**rder **T**hinking Skills

64. **MP PERSEVERANCE** For any nonzero real numbers a and b and any integers m and t, simplify the expression $\left(-\dfrac{a^m}{b^t}\right)^{2t}$ and describe each step.

65. **MP REGULARITY** Copy the table below.

Equation	Related Expression	Power of x	Linear or Nonlinear
$y = x$			
$y = x^2$			
$y = x^3$			

   a. For each equation, write the related expression and record the power of x.

   b. Graph each equation using a graphing calculator.

   c. Classify each graph as *linear* or *nonlinear*.

   d. Explain how to determine whether an equation, or its related expression, is linear or nonlinear without graphing.

66. OPEN-ENDED Write three different expressions that can be simplified to x^6.

67. WRITING IN MATH Write two formulas that have monomial expressions in them. Explain how each is used in a real-world situation.

68. Which expression represents the area of the trapezoid? **MP** 7

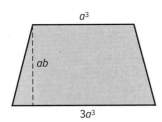

- **A** $1.5a^4b$
- **B** $2a^4b$
- **C** $3a^7b$
- **D** $5a^3b$

69. Which expression is equivalent to $(3m^2np^4)(-5m^3n^2p)^2$? **MP** 8

- **A** $-15m^5n^3p^6$
- **B** $-15m^8n^5p^6$
- **C** $75m^5n^3p^5$
- **D** $75m^8n^5p^6$

70. $WXYZ$ is a parallelogram.

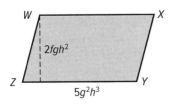

What is the area of $WXYZ$, expressed as a monomial? **MP** 7

- **A** $7fg^3h^5$
- **B** $7fg^2h^6$
- **C** $10fg^3h^5$
- **D** $10fg^2h^6$

71. Which of the following expressions simplifies to the monomial $12h^6j^8$? **MP** 8

- **A** $(4h^2j^4)(3h^3j^2)$
- **B** $(2h^3j^2)(6h^3j^4)$
- **C** $(3h^3j^6)(4h^3j^2)$
- **D** $\frac{(8h^2j^4)(3h^3j^2)}{2}$

72. Identify which expressions are monomials. Select all that apply. **MP** 8

- **A** 12
- **B** $\frac{3x + 6}{9}$
- **C** $\frac{5x^2y}{22}$
- **D** $23a^3b^2c$
- **E** $\frac{mn^4}{v}$

73. MULTI-STEP Consider the rectangular prism shown. **MP** 7, 8

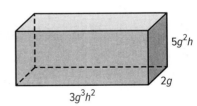

a. Which expression represents the area of the face with a length of $3g^3h^2$ and a width of $5g^2h$?

- **A** $15g^5h^2$
- **B** $15g^5h^3$
- **C** $15g^6h^2$
- **D** $15g^9h^2$

b. What is the volume of the prism?

74. Paul rewrote the expression $[(13^2)^5]^8$ as a power with a single exponent. He kept the base 13. What was the new exponent? **MP** 4, 8

Division Properties of Exponents

::**Then**

- You multiplied monomials using the properties of exponents.

::**Now**

1. Divide monomials using the properties of exponents.

2. Simplify expressions containing negative and zero exponents.

::**Why?**

- The tallest redwood tree is 112 meters or about 10^2 meters tall. The average height of a redwood tree is 15 meters. The closest power of ten to 15 is 10^1, so an average redwood is about 10^1 meters tall. The ratio of the tallest tree's height to the average tree's height is $\frac{10^2}{10^1}$ or 10^1. This means the tallest redwood tree is approximately 10 times as tall as the average redwood tree.

 New Vocabulary

order of magnitude

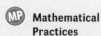 **Mathematical Practices**

2 Reason abstractly and quantitively.

1 Divide Monomials We can use the principles for reducing fractions to find quotients of monomials like $\frac{10^2}{10^1}$. In the following examples, look for a pattern in the exponents.

$$\frac{2^7}{2^4} = \frac{\overbrace{\cancel{2} \cdot \cancel{2} \cdot \cancel{2} \cdot \cancel{2} \cdot 2 \cdot 2 \cdot 2}^{7\,\text{factors}}}{\underbrace{\cancel{2} \cdot \cancel{2} \cdot \cancel{2} \cdot \cancel{2}}_{4\,\text{factors}}} = 2 \cdot 2 \cdot 2 \text{ or } 2^3 \qquad \frac{t^4}{t^3} = \frac{\overbrace{\cancel{t} \cdot \cancel{t} \cdot \cancel{t} \cdot t}^{4\,\text{factors}}}{\underbrace{\cancel{t} \cdot \cancel{t} \cdot \cancel{t}}_{3\,\text{factors}}} = t$$

These examples demonstrate the Quotient of Powers Rule.

Key Concept Quotient of Powers

Words	To divide two powers with the same base, subtract the exponents.
Symbols	For any nonzero number a, and any integers m and p, $\frac{a^m}{a^p} = a^{m-p}$.
Examples	$\frac{c^{11}}{c^8} = c^{11-8}$ or c^3 $\qquad$ $\frac{r^5}{r^2} = r^{5-2} = r^3$

Example 1 Quotient of Powers

Simplify $\frac{g^3 h^5}{g h^2}$. Assume that no denominator equals zero.

$$\frac{g^3 h^5}{g h^2} = \left(\frac{g^3}{g}\right)\left(\frac{h^5}{h^2}\right) \qquad \text{Group powers with the same base.}$$

$$= \left(g^{3-1}\right)\left(h^{5-2}\right) \qquad \text{Quotient of Powers}$$

$$= g^2 h^3 \qquad \text{Simplify.}$$

▸ **Guided Practice**

Simplify each expression. Assume that no denominator equals zero.

1A. $\dfrac{x^3 y^4}{x^2 y}$

1B. $\dfrac{k^7 m^{10} p}{k^5 m^3 p}$

We can use the Product of Powers Property to find the powers of quotients for monomials. In the following example, look for a pattern in the exponents.

$$\left(\frac{3}{4}\right)^3 = \overbrace{\left(\frac{3}{4}\right)\left(\frac{3}{4}\right)\left(\frac{3}{4}\right)}^{3 \text{ factors}} = \frac{\overbrace{3 \cdot 3 \cdot 3}^{3 \text{ factors}}}{\underbrace{4 \cdot 4 \cdot 4}_{3 \text{ factors}}} = \frac{3^3}{4^3}$$

$$\left(\frac{c}{d}\right)^2 = \overbrace{\left(\frac{c}{d}\right)\left(\frac{c}{d}\right)}^{2 \text{ factors}} = \frac{\overbrace{c \cdot c}^{2 \text{ factors}}}{\underbrace{d \cdot d}_{2 \text{ factors}}} = \frac{c^2}{d^2}$$

Key Concept Power of a Quotient

Words To find the power of a quotient, find the power of the numerator and the power of the denominator.

Symbols For any real numbers a and $b \neq 0$, and any integer m, $\left(\frac{a}{b}\right)^m = \frac{a^m}{b^m}$.

Examples $\left(\frac{3}{5}\right)^4 = \frac{3^4}{5^4}$ $\left(\frac{r}{t}\right)^5 = \frac{r^5}{t^5}$

Real-World Career

Astronomer An astronomer studies the universe and analyzes space travel and satellite communications. To be a technician or research assistant, a bachelor's degree is required.

Example 2   **Power of a Quotient**

Simplify $\left(\frac{3p^3}{7}\right)^2$.

$\left(\frac{3p^3}{7}\right)^2 = \frac{(3p^3)^2}{7^2}$ Power of a Quotient

$= \frac{3^2(p^3)^2}{7^2}$ Power of a Product

$= \frac{9p^6}{49}$ Power of a Power

Guided Practice

Simplify each expression.

2A. $\left(\frac{3x^4}{4}\right)^3$ **2B.** $\left(\frac{5x^5y}{6}\right)^2$ **2C.** $\left(\frac{2y^2}{3z^3}\right)^2$ **2D.** $\left(\frac{4x^3}{5y^4}\right)^3$

A calculator can be used to explore expressions with 0 as the exponent. There are two methods to explain why a calculator gives a value of 1 for 3^0.

Method 1

$\frac{3^5}{3^5} = 3^{5-5}$ Quotient of Powers

$= 3^0$ Simplify.

Method 2

$\frac{3^5}{3^5} = \frac{\cancel{3} \cdot \cancel{3} \cdot \cancel{3} \cdot \cancel{3} \cdot \cancel{3}}{\cancel{3} \cdot \cancel{3} \cdot \cancel{3} \cdot \cancel{3} \cdot \cancel{3}}$ Definition of Powers

$= 1$ Simplify.

Since $\frac{3^5}{3^5}$ can only have one value, we can conclude that $3^0 = 1$, which leads to the Zero Exponent Property. A **zero exponent** is any nonzero number raised to the zero power.

Go Online!

You have worked with exponents that are positive integers. Explore the concept of zero and negative exponents using the Geometer's Sketchpad® activity in ConnectED.

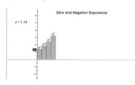

Key Concept Zero Exponent Property

Words   Any nonzero number raised to the zero power is equal to 1.

Symbols   For any nonzero number a, $a^0 = 1$.

Examples $15^0 = 1$ $\left(\dfrac{b}{c}\right)^0 = 1$ $\left(\dfrac{2}{7}\right)^0 = 1$

Example 3   Zero Exponent

Simplify each expression. Assume that no denominator equals zero.

a. $\left(\dfrac{4n^2q^5r^2}{9n^3q^2r}\right)^0$

$\left(\dfrac{4n^2q^5r^2}{9n^3q^2r}\right)^0 = 1$ $a^0 = 1$

b. $\dfrac{x^5y^0}{x^3}$

$\dfrac{x^5y^0}{x^3} = \dfrac{x^5(1)}{x^3}$ $a^0 = 1$

$= x^2$ Quotient of Powers

Guided Practice

3A. $\dfrac{b^4c^2d^0}{b^2c}$

3B. $\left(\dfrac{2f^4g^7h^3}{15f^3g^9h^6}\right)^0$

Study Tip

Zero Exponent Be careful of parentheses. The expression $(5x)^0$ is 1, but $5x^0 = 5$.

---

2 Negative Exponents Any nonzero real number raised to a negative power is a **negative exponent**. To investigate the meaning of a negative exponent, we can simplify expressions like $\dfrac{c^2}{c^5}$ using two methods.

Method 1

$\dfrac{c^2}{c^5} = c^{2-5}$ Quotient of Powers

$= c^{-3}$ Simplify.

Method 2

$\dfrac{c^2}{c^5} = \dfrac{\cancel{c} \cdot \cancel{c}}{\cancel{c} \cdot \cancel{c} \cdot c \cdot c \cdot c}$ Definition of Powers

$= \dfrac{1}{c^3}$ Simplify.

Since $\dfrac{c^2}{c^5}$ can only have one value, we can conclude that $c^{-3} = \dfrac{1}{c^3}$.

Key Concept Negative Exponent Property

Words   For any nonzero number a and any integer n, a^{-n} is the reciprocal of a^n. Also, the reciprocal of a^{-n} is a^n.

Symbols   For any nonzero number a and any integer n, $a^{-n} = \dfrac{1}{a^n}$.

Examples $2^{-4} = \dfrac{1}{2^4} = \dfrac{1}{16}$ $\dfrac{1}{j^{-4}} = j^4$

An expression is considered simplified when it contains only positive exponents, each base appears exactly once, there are no powers of powers, and all fractions are in simplest form.

Example 4   Negative Exponents

Simplify each expression. Assume that no denominator equals zero.

a. $\dfrac{n^{-5}p^4}{r^{-2}}$

$\dfrac{n^{-5}p^4}{r^{-2}} = \left(\dfrac{n^{-5}}{1}\right)\left(\dfrac{p^4}{1}\right)\left(\dfrac{1}{r^{-2}}\right)$ Write as a product of fractions.

$= \left(\dfrac{1}{n^5}\right)\left(\dfrac{p^4}{1}\right)\left(\dfrac{r^2}{1}\right)$ $a^{-n} = \dfrac{1}{a^n}$ and $\dfrac{1}{a^{-n}} = a^n$

$= \dfrac{p^4 r^2}{n^5}$ Multiply.

Study Tip

Negative Signs Be aware of where a negative sign is placed.
$5^{-1} = \dfrac{1}{5}$, while $-5^1 \neq \dfrac{1}{5}$.

b. $\dfrac{5r^{-3}t^4}{-20r^2t^7u^{-5}}$

$\dfrac{5r^{-3}t^4}{-20r^2t^7u^{-5}} = \left(\dfrac{5}{-20}\right)\left(\dfrac{r^{-3}}{r^2}\right)\left(\dfrac{t^4}{t^7}\right)\left(\dfrac{1}{u^{-5}}\right)$ Group powers with the same base.

$= \left(-\dfrac{1}{4}\right)(r^{-3-2})(t^{4-7})(u^5)$ Quotient of Powers and Negative Exponents Property

$= -\dfrac{1}{4}r^{-5}t^{-3}u^5$ Simplify.

$= -\dfrac{1}{4}\left(\dfrac{1}{r^5}\right)\left(\dfrac{1}{t^3}\right)(u^5)$ Negative Exponent Property

$= -\dfrac{u^5}{4r^5t^3}$ Multiply.

c. $\dfrac{2a^2b^3c^{-5}}{10a^{-3}b^{-1}c^{-4}}$

$\dfrac{2a^2b^3c^{-5}}{10a^{-3}b^{-1}c^{-4}} = \left(\dfrac{2}{10}\right)\left(\dfrac{a^2}{a^{-3}}\right)\left(\dfrac{b^3}{b^{-1}}\right)\left(\dfrac{c^{-5}}{c^{-4}}\right)$ Group powers with the same base.

$= \left(\dfrac{1}{5}\right)(a^{2-(-3)})(b^{3-(-1)})(c^{-5-(-4)})$ Quotient of Powers and Negative Exponents Property

$= \dfrac{1}{5}a^5b^4c^{-1}$ Simplify.

$= \dfrac{1}{5}(a^5)(b^4)\left(\dfrac{1}{c}\right)$ Negative Exponent Property

$= \dfrac{a^5b^4}{5c}$ Multiply.

Guided Practice

Simplify each expression. Assume that no denominator equals zero.

4A. $\dfrac{v^{-3}wx^2}{wy^{-6}}$ **4B.** $\dfrac{32a^{-8}b^3c^{-4}}{4a^3b^5c^{-2}}$ **4C.** $\dfrac{5j^{-3}k^2m^{-6}}{25k^{-4}m^{-2}}$

Real-World Link

An adult human weighs about 70 kilograms and an adult dairy cow weighs about 700 kilograms. Their weights differ by 1 order of magnitude.

Order of magnitude is used to compare measures and to estimate and perform rough calculations. The **order of magnitude** of a quantity is the number rounded to the nearest power of 10. For example, the power of 10 closest to 95,000,000,000 is 10^{11}, or 100,000,000,000. So the order of magnitude of 95,000,000,000 is 10^{11}.

Real-World Example 5   **Apply Properties of Exponents**

HEIGHT Suppose the average height of a man is about 1.7 meters, and the average height of an ant is 0.0008 meter. How many orders of magnitude as tall as an ant is a man?

Understand We must find the order of magnitude of the heights of the man and ant. Then find the ratio of the orders of magnitude of the man's height to that of the ant's height.

Plan Round each height to the nearest power of ten. Then find the ratio.

Solve The average height of a man is close to 1 meter. So, the order of magnitude is 10^0 meter. The average height of an ant is about 0.001 meter. So, the order of magnitude is 10^{-3} meters.

The ratio of the height of a man to the height of an ant is about $\frac{10^0}{10^{-3}}$.

$\frac{10^0}{10^{-3}} = 10^{0-(-3)}$ Quotient of Powers

$= 10^3$ or 1000 $0 - (-3) = 0 + 3$ or 3

So, a man is approximately 1000 times as tall as an ant, or a man is 3 orders of magnitude as tall as an ant.

Check The ratio of the man's height to the ant's height is $\frac{1.7}{0.0008} = 2125$. The order of magnitude of 2125 is 10^3. ✔

Rounding and analyzing the orders of magnitude of a solution is an effective tool to determine the reasonableness of a solution. Our solution is reasonable because the order of magnitude matches the actual height ratio.

Real-World Link

There are over 14,000 species of ants living all over the world. Some ants can carry objects that are 50 times their own weight.

Source: Maine Animal Coalition

> **Guided Practice**

5. ASTRONOMY The order of magnitude of the mass of Earth is about 10^{27}. The order of magnitude of the Milky Way galaxy is about 10^{44}. How many orders of magnitude as big is the Milky Way galaxy as Earth?

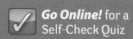

Go Online! for a Self-Check Quiz

Check Your Understanding   ◯ = Step-by-Step Solutions begin on page R11.

Examples 1–4 Simplify each expression. Assume that no denominator equals zero.

1. $\dfrac{t^5u^4}{t^2u}$

2. $\dfrac{a^6b^4c^{10}}{a^3b^2c}$

 $\dfrac{m^6r^5p^3}{m^5r^2p^3}$

4. $\dfrac{b^4c^6f^8}{b^4c^3f^5}$

5. $\dfrac{g^8h^2m}{hg^7}$

6. $\dfrac{r^4t^7v^2}{t^7v^2}$

7. $\dfrac{x^3y^2z^6}{z^5x^2y}$

8. $\dfrac{n^4q^4w^6}{q^2n^3w}$

9. $\left(\dfrac{2a^3b^5}{3}\right)^2$

10. $\dfrac{r^3v^{-2}}{t^{-7}}$

11. $\left(\dfrac{2c^3d^5}{5g^2}\right)^5$

12. $\left(-\dfrac{3xy^4z^2}{x^3yz^4}\right)^0$

13. $\left(\dfrac{3f^4gh^4}{32f^3g^4h}\right)^0$

14. $\dfrac{4r^2v^0t^5}{2rt^3}$

15. $\dfrac{f^{-3}g^2}{h^{-4}}$

16. $\dfrac{-8x^2y^8z^{-5}}{12x^4y^{-7}z^7}$

17. $\dfrac{2a^2b^{-7}c^{10}}{6a^{-3}b^2c^{-3}}$

Example 5 **18. POPULATION** The 2013 population of Nevada was 2,790,136 with an average of 2.70 persons living in each household. Use order of magnitude to approximate the number of households in Nevada in 2013.

Examples 1–4
A.SSE.2

Simplify each expression. Assume that no denominator equals zero.

19. $\dfrac{m^4p^2}{m^2p}$

20. $\dfrac{p^{12}t^3r}{p^2tr}$

21. $\dfrac{3m^{-3}r^4p^2}{12t^4}$

22. $\dfrac{c^4d^4f^3}{c^2d^4f^3}$

23. $\left(\dfrac{3xy^4}{5z^2}\right)^2$

24. $\left(\dfrac{3t^6u^2v^5}{9tuv^{21}}\right)^0$

25. $\left(\dfrac{p^2t^7}{10}\right)^3$

26. $\dfrac{x^{-4}y^9}{z^{-2}}$

27. $\dfrac{a^7b^8c^8}{a^5bc^7}$

28. $\left(\dfrac{3np^3}{7q^2}\right)^2$

29. $\left(\dfrac{2r^3t^6}{5u^9}\right)^4$

30. $\left(\dfrac{3m^5r^3}{4p^8}\right)^4$

31. $\left(-\dfrac{5f^9g^4h^2}{fg^2h^3}\right)^0$

32. $\dfrac{p^{12}t^7r^2}{p^2t^7r}$

33. $\dfrac{p^4t^{-3}}{r^{-2}}$

34. $-\dfrac{5c^2d^5}{8cd^5f^0}$

35. $\dfrac{-2f^3g^2h^0}{8f^2g^2}$

36. $\dfrac{12m^{-4}p^2}{-15m^3p^{-9}}$

37. $\dfrac{k^4m^3p^2}{k^2m^2}$

38. $\dfrac{14f^{-3}g^2h^{-7}}{21k^3}$

39. $\dfrac{39t^4uv^{-2}}{13t^{-3}u^7}$

40. $\left(\dfrac{a^{-2}b^4c^5}{a^{-4}b^{-4}c^3}\right)^2$

41. $\dfrac{r^3t^{-1}x^{-5}}{tx^5}$

42. $\dfrac{g^0h^7j^{-2}}{g^{-5}h^0j^{-2}}$

Example 5
A.SSE.2,
F.IF.8b

43. SOCIAL NETWORKING In a recent year, a social networking site had about 750,000 servers. Suppose there were 1.15 billion active users on the site. Determine the order of magnitude for the servers and active users. Using the orders of magnitude, how many active users were there compared to servers?

44. PROBABILITY The probability of rolling a die and getting an even number is $\frac{1}{2}$. If you roll the die twice, the probability of getting an even number both times is $\left(\frac{1}{2}\right)\left(\frac{1}{2}\right)$ or $\left(\frac{1}{2}\right)^2$.
a. What does $\left(\frac{1}{2}\right)^4$ represent?
b. Write an expression to represent the probability of rolling a die d times and getting an even number every time. Write the expression as a power of 2.

Simplify each expression. Assume that no denominator equals zero.

45. $\dfrac{-4w^{12}}{12w^3}$

46. $\dfrac{13r^7}{39r^4}$

47. $\dfrac{(4k^3m^2)^3}{(5k^2m^{-3})^{-2}}$

48. $\dfrac{3wy^{-2}}{(w^{-1}y)^3}$

49. $\dfrac{20qr^{-2}t^{-5}}{4q^0r^4t^{-2}}$

50. $\dfrac{-12c^3d^0f^{-2}}{6c^5d^{-3}f^4}$

51. $\dfrac{(2g^3h^{-2})^2}{(g^2h^0)^{-3}}$

52. $\dfrac{(5pr^{-2})^{-2}}{(3p^{-1}r)^3}$

53. $\left(\dfrac{-3x^{-6}y^{-1}z^{-2}}{6x^{-2}yz^{-5}}\right)^{-2}$

54. $\left(\dfrac{2a^{-2}b^4c^2}{-4a^{-2}b^{-5}c^{-7}}\right)^{-1}$

55. $\dfrac{(16x^2y^{-1})^0}{(4x^0y^{-4}z)^{-2}}$

56. $\left(\dfrac{4^0c^2d^3f}{2c^{-4}d^{-5}}\right)^{-3}$

57. MP SENSE-MAKING The processing speed of an older desktop computer is about 10^8 instructions per second. A new computer can process about 10^{11} instructions per second. The newer computer is how many times as fast as the older one?

58. ASTRONOMY The brightness of a star is measured in magnitudes. The lower the magnitude, the brighter the star. A magnitude 9 star is 2.51 times as bright as a magnitude 10 star. A magnitude 8 star is $2.51 \cdot 2.51$ or 2.51^2 times as bright as a magnitude 10 star.

a. How many times as bright is a magnitude 3 star as a magnitude 10 star?

b. Write an expression to compare a magnitude m star to a magnitude 10 star.

c. A full moon is considered to be magnitude -13, approximately. Does your expression make sense for this magnitude? Explain.

59 PROBABILITY The probability of rolling a die and getting a 3 is $\frac{1}{6}$. If you roll the die twice, the probability of getting a 3 both times is $\frac{1}{6} \cdot \frac{1}{6}$ or $\left(\frac{1}{6}\right)^2$.

a. Write an expression to represent the probability of rolling a die d times and getting a 3 each time.

b. Write the expression as a power of 6.

60. MULTIPLE REPRESENTATIONS To find the area of a circle, use $A = \pi r^2$. The formula for the area of a square is $A = s^2$.

a. **Algebraic** Find the ratio of the area of the circle to the area of the square.

b. **Algebraic** If the radius of the circle and the length of each side of the square are doubled, find the ratio of the area of the circle to the square.

c. **Tabular** Copy and complete the table.

Radius	Area of Circle	Area of Square	Ratio
r			
$2r$			
$3r$			
$4r$			
$5r$			
$6r$			

d. **Analytical** What conclusion can be drawn from this?

H.O.T. Problems   Use Higher-Order Thinking Skills

61. **MP REASONING** Is $x^y \cdot x^z = x^{yz}$ sometimes, always, or never true? Explain.

62. OPEN-ENDED Name two monomials with a quotient of $24a^2b^3$.

63. CHALLENGE Use the Quotient of Powers Property to explain why $x^{-n} = \frac{1}{x^n}$.

64. **MP REGULARITY** Write a convincing argument to show why $3^0 = 1$.

65. WRITING IN MATH Explain how to use the Quotient of Powers Property and the Power of a Quotient Property.

66. Which of the following is a simplified form of the given expression? Assume the denominator is not 0. **MP** 7, 8

$$\left(\frac{15n^5m^3}{5m^4n^2}\right)^2$$

- **A** $\frac{3n^6}{m^2}$
- **B** $\frac{9n^6}{m^2}$
- **C** $\frac{9n^5}{m^3}$
- **D** $9m^2n^2$

67. The value of the expression is 1. What is the value of m? **MP** 7

$$\left(\frac{x^2y^3}{y^2z}\right)^m$$

- **A** -2
- **B** -1
- **C** 0
- **D** 1

68. Simplify $(4^{-2} \cdot 5^0 \cdot 64)^3$. **MP** 7

- **A** $\frac{1}{64}$
- **B** 64
- **C** 1024
- **D** 8000

69. Simplify the expression. Assume the denominator does not equal 0. **MP** 7, 8

$$\frac{3n^2v^{-4}w}{18n^{-3}v^5w^{-2}}$$

- **A** $\frac{1}{6nvw}$
- **B** $\frac{n^5w^2}{6v^9}$
- **C** $\frac{v}{6nw}$
- **D** $\frac{n^5w^3}{6v^9}$

70. In the simplified form of the expression $\left(\frac{8b^5c^4d^9}{2b^7c^2d^5}\right)^3$, what is the exponent of the variable d? 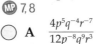 7, 8

- **A** 4
- **B** 12
- **C** 14
- **D** 64

71. Which expression simplifies to $\frac{q^{13}r^{10}}{3p^{13}}$? **MP** 7, 8

- **A** $\frac{4p^5q^{-4}r^{-7}}{12p^{-8}q^9r^3}$
- **B** $\frac{2p^{-5}q^4r^{-7}}{6p^8q^{-9}r^3}$
- **C** $\frac{3p^{-5}q^4r^7}{9p^8q^{-9}r^{-3}}$
- **D** $\frac{5p^5q^{-4}r^7}{15p^{-8}q^9r^{-3}}$

72. MULTI-STEP Find the product $(v^{-2}w^3x^2y^{-6})(vw^{-2}x^{-3}y^4)$ by following the steps below. 7, 8

a. Express the product as a fraction with all the current factors having positive exponents.

b. Which is the simplified form of the expression in part **a**?

- **A** $\frac{xy^2}{vw}$
- **B** $\frac{wy^2}{vx}$
- **C** $\frac{x}{vwy^2}$
- **D** $\frac{w}{vxy^2}$

c. Examine the expression in part **b**. Which variables cannot equal 0? Select all that apply.

- **A** v
- **B** w
- **C** x
- **D** y

::Then
- You used the laws of exponents to find products and quotients of monomials.

::Now
1. Evaluate and rewrite expressions involving rational exponents.
2. Solve equations involving expressions with rational exponents.

::Why?
- It's important to protect your skin with sunscreen to prevent damage. The sun protection factor (SPF) of a sunscreen indicates how well it protects you. Sunscreen with an SPF of f absorbs about p percent of the UV-B rays, where $p = 50f^{0.2}$.

 New Vocabulary
rational exponent
cube root
nth root
exponential equation

 Mathematical Practices
5 Use appropriate tools strategically.

1 Rational Exponents You know that an exponent represents the number of times that the base is used as a factor. But how do you evaluate an expression with an exponent that is not an integer like the one above? Let's investigate **rational exponents** by assuming that they behave like integer exponents.

$$\left(b^{\frac{1}{2}}\right)^2 = b^{\frac{1}{2}} \cdot b^{\frac{1}{2}} \qquad \text{Write as a multiplication expression.}$$

$$= b^{\frac{1}{2} + \frac{1}{2}} \qquad \text{Product of Powers}$$

$$= b^1 \text{ or } b \qquad \text{Simplify.}$$

Thus, $b^{\frac{1}{2}}$ is a number with a square equal to b. So $b^{\frac{1}{2}} = \sqrt{b}$.

Key Concept $b^{\frac{1}{2}}$

Words For any nonnegative real number b, $b^{\frac{1}{2}} = \sqrt{b}$.

Examples $16^{\frac{1}{2}} = \sqrt{16}$ or 4 $\qquad\qquad$ $38^{\frac{1}{2}} = \sqrt{38}$

Example 1 Radical and Exponential Forms

Write each expression in radical form, or write each radical in exponential form.

a. $25^{\frac{1}{2}}$

$\qquad 25^{\frac{1}{2}} = \sqrt{25} \qquad$ Definition of $b^{\frac{1}{2}}$

$\qquad\qquad = 5 \qquad$ Simplify.

b. $\sqrt{18}$

$\qquad \sqrt{18} = 18^{\frac{1}{2}} \qquad$ Definition of $b^{\frac{1}{2}}$

c. $5x^{\frac{1}{2}}$

$\qquad 5x^{\frac{1}{2}} = 5\sqrt{x} \qquad$ Definition of $b^{\frac{1}{2}}$

d. $\sqrt{8p}$

$\qquad \sqrt{8p} = (8p)^{\frac{1}{2}} \qquad$ Definition of $b^{\frac{1}{2}}$

▶ **Guided Practice**

1A. $a^{\frac{1}{2}}$ $\qquad$ **1B.** $\sqrt{22}$ $\qquad$ **1C.** $(7w)^{\frac{1}{2}}$ $\qquad$ **1D.** $2\sqrt{x}$

You know that to find the square root of a number a you find a number with a square of a. In the same way, you can find other roots of numbers. If $a^3 = b$, then a is the **cube root** of b, and if $a^n = b$ for a positive integer n, then a is an ***n*th root** of b.

🔑 Key Concept *n*th Root

Words	For any real numbers a and b and any positive integer n, if $a^n = b$, then a is an nth root of b.
Symbols	If $a^n = b$, then $\sqrt[n]{b} = a$.
Example	Because $2^4 = 16$, 2 is a fourth root of 16; $\sqrt[4]{16} = 2$.

Study Tip

 Tools You can use a graphing calculator to find nth roots. Enter n, then press [MATH] and choose $\sqrt[x]{}$.

Since $3^2 = 9$ and $(-3)^2 = 9$, both 3 and -3 are square roots of 9. Similarly, since $2^4 = 16$ and $(-2)^4 = 16$, both 2 and -2 are fourth roots of 16. The positive roots are called *principal roots*. Radical symbols indicate principal roots, so $\sqrt{16} = 2$.

Example 2 *n*th Roots

Simplify.

a. $\sqrt[3]{27}$

$\sqrt[3]{27} = \sqrt[3]{3 \cdot 3 \cdot 3}$

$\quad\quad = 3$

b. $\sqrt[5]{32}$

$\sqrt[5]{32} = \sqrt[5]{2 \cdot 2 \cdot 2 \cdot 2 \cdot 2}$

$\quad\quad = 2$

▶ **Guided Practice**

2A. $\sqrt[3]{64}$

2B. $\sqrt[4]{10,000}$

Like square roots, nth roots can be represented by rational exponents.

$$\left(b^{\frac{1}{n}}\right)^n = \underbrace{b^{\frac{1}{n}} \cdot b^{\frac{1}{n}} \cdot \,\cdots\, \cdot b^{\frac{1}{n}}}_{n \text{ factors}} \quad\quad \text{Write as a multiplication expression.}$$

$$= b^{\frac{1}{n} + \frac{1}{n} + \,\cdots\, + \frac{1}{n}} \quad\quad \text{Product of Powers}$$

$$= b^1 \text{ or } b \quad\quad \text{Simplify.}$$

Thus, $b^{\frac{1}{n}}$ is a number with an nth power equal to b. So $b^{\frac{1}{n}} = \sqrt[n]{b}$.

🔑 Key Concept $b^{\frac{1}{n}}$

Words	For any positive real number b and any integer $n > 1$, $b^{\frac{1}{n}} = \sqrt[n]{b}$.
Example	$8^{\frac{1}{3}} = \sqrt[3]{8} = \sqrt[3]{2 \cdot 2 \cdot 2}$ or 2

Example 3   Evaluate $b^{\frac{1}{n}}$ Expressions

Simplify.

a. $125^{\frac{1}{3}}$

$$125^{\frac{1}{3}} = \sqrt[3]{125} \qquad {\scriptstyle b^{\frac{1}{n}} = \sqrt[n]{b}}$$

$$= \sqrt[3]{5 \cdot 5 \cdot 5} \qquad {\scriptstyle 125 = 5^3}$$

$$= 5 \qquad {\scriptstyle \text{Simplify.}}$$

b. $1296^{\frac{1}{4}}$

$$1296^{\frac{1}{4}} = \sqrt[4]{1296} \qquad {\scriptstyle b^{\frac{1}{n}} = \sqrt[n]{b}}$$

$$= \sqrt[4]{6 \cdot 6 \cdot 6 \cdot 6} \qquad {\scriptstyle 1296 = 6^4}$$

$$= 6 \qquad {\scriptstyle \text{Simplify.}}$$

▶ **Guided Practice**

3A. $27^{\frac{1}{3}}$ **3B.** $256^{\frac{1}{4}}$

The Power of a Power Property allows us to extend the definition of $b^{\frac{1}{n}}$ to $b^{\frac{m}{n}}$.

$$b^{\frac{m}{n}} = \left(b^{\frac{1}{n}}\right)^m \qquad {\scriptstyle \text{Power of a Power}}$$

$$= \left(\sqrt[n]{b}\right)^m \text{ or } \sqrt[n]{b^m} \qquad {\scriptstyle b^{\frac{1}{n}} = \sqrt[n]{b}}$$

🔁 Key Concept $b^{\frac{m}{n}}$

Words   For any positive real number b and any integers m and $n > 1$,
$$b^{\frac{m}{n}} = \left(\sqrt[n]{b}\right)^m \text{ or } \sqrt[n]{b^m}.$$

Example $8^{\frac{2}{3}} = \left(\sqrt[3]{8}\right)^2 = 2^2$ or 4

Example 4   Evaluate $b^{\frac{m}{n}}$ Expressions

Simplify.

a. $64^{\frac{2}{3}}$

$$64^{\frac{2}{3}} = \left(\sqrt[3]{64}\right)^2 \qquad {\scriptstyle b^{\frac{m}{n}} = \left(\sqrt[n]{b}\right)^m}$$

$$= \left(\sqrt[3]{4 \cdot 4 \cdot 4}\right)^2 \qquad {\scriptstyle 64 = 4^3}$$

$$= 4^2 \text{ or } 16 \qquad {\scriptstyle \text{Simplify.}}$$

b. $36^{\frac{3}{2}}$

$$36^{\frac{3}{2}} = \left(\sqrt[2]{36}\right)^3 \qquad {\scriptstyle b^{\frac{m}{n}} = \left(\sqrt[n]{b}\right)^m}$$

$$= 6^3 \qquad {\scriptstyle \sqrt{36} = 6}$$

$$= 216 \qquad {\scriptstyle \text{Simplify.}}$$

▶ **Guided Practice**

4A. $27^{\frac{2}{3}}$ **4B.** $256^{\frac{5}{4}}$

2 Solve Exponential Equations

In an **exponential equation**, variables occur as exponents. The Power Property of Equality and the other properties of exponents can be used to solve exponential equations.

> **Key Concept** Power Property of Equality
>
> Words   For any real number $b > 0$ and $b \neq 1$, $b^x = b^y$ if and only if $x = y$.
>
> Examples   If $5^x = 5^3$, then $x = 3$. If $n = \frac{1}{2}$, then $4^n = 4^{\frac{1}{2}}$.

Example 5 Solve Exponential Equations

Solve each equation.

a. $6^x = 216$

$6^x = 216$	Original equation
$6^x = 6^3$	Rewrite 216 as 6^3.
$x = 3$	Property of Equality

CHECK $6^x = 216$

$6^3 \overset{?}{=} 216$

$216 = 216$ ✓

b. $25^{x-1} = 5$

$25^{x-1} = 5$	Original equation
$(5^2)^{x-1} = 5$	Rewrite 25 as 5^2.
$5^{2x-2} = 5^1$	Power of a Power, Distributive Property
$2x - 2 = 1$	Power Property of Equality
$2x = 3$	Add 2 to each side.
$x = \frac{3}{2}$	Divide each side by 2.

CHECK $25^{x-1} = 5$

$25^{\frac{3}{2} - 1} \overset{?}{=} 5$

$25^{\frac{1}{2}} = 5$ ✓

Guided Practice

5A. $5^x = 125$

5B. $12^{2x+3} = 144$

Real-World Example 6 Solve Exponential Equations

SUNSCREEN Refer to the beginning of the lesson. Find the SPF that absorbs 100% of UV-B rays.

$p = 50f^{0.2}$	Original equation
$100 = 50f^{0.2}$	$p = 100$
$2 = f^{0.2}$	Divide each side by 50.
$2 = f^{\frac{1}{5}}$	$0.2 = \frac{1}{5}$
$(2^5)^{\frac{1}{5}} = f^{\frac{1}{5}}$	$2 = 2^1 = (2^5)^{\frac{1}{5}}$
$2^5 = f$	Power Property of Equality
$32 = f$	Simplify.

Real-World Link

Use extra caution near snow, water, and sand because they reflect the damaging rays of the Sun, which can increase your chance of sunburn.

Source: American Academy of Dermatology

Guided Practice

6. CHEMISTRY The radius r of the nucleus of an atom of mass number A is $r = 1.2A^{\frac{1}{3}}$ femtometers. Find A if $r = 3.6$ femtometers.

Check Your Understanding ⬭ = Step-by-Step Solutions begin on page R11.

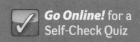
Go Online! for a
Self-Check Quiz

Example 1 **Write each expression in radical form, or write each radical in exponential form.**

1. $12^{\frac{1}{2}}$ **2.** $3x^{\frac{1}{2}}$ **3.** $\sqrt{33}$ **4.** $\sqrt{8n}$

Examples 2–4 Simplify.

5. $\sqrt[3]{512}$ **6.** $\sqrt[5]{243}$ **7.** $343^{\frac{1}{3}}$ **8.** $\left(\frac{1}{16}\right)^{\frac{1}{4}}$

9. $343^{\frac{2}{3}}$ **10.** $81^{\frac{3}{4}}$ **⑪** $216^{\frac{4}{3}}$ **12.** $\left(\frac{1}{49}\right)^{\frac{3}{2}}$

Example 5 **Solve each equation.**

13. $8^x = 4096$ **14.** $3^{3x+1} = 81$ **15.** $4^{x-3} = 32$

Example 6 **16.** **MP TOOLS** A weir is used to measure water flow in a channel. For a rectangular broad-crested weir, the flow Q in cubic feet per second is related to the weir length L in feet and height H of the water by $Q = 1.6LH^{\frac{3}{2}}$. Find the water height for a weir that is 3 feet long and has flow of 38.4 cubic feet per second.

Practice and Problem Solving   Extra Practice is on page R7.

Example 1 **Write each expression in radical form, or write each radical in exponential form.**

17. $15^{\frac{1}{2}}$ **18.** $24^{\frac{1}{2}}$ **19.** $4k^{\frac{1}{2}}$ **20.** $(12y)^{\frac{1}{2}}$

21. $\sqrt{26}$ **22.** $\sqrt{44}$ **23.** $2\sqrt{ab}$ **24.** $\sqrt{3xyz}$

Examples 2–4 Simplify.

25. $\sqrt[3]{8}$ **26.** $\sqrt[5]{1024}$ **27.** $\sqrt[3]{216}$ **28.** $\sqrt[4]{10{,}000}$

29. $\sqrt[3]{0.001}$ **30.** $\sqrt[4]{\frac{16}{81}}$ **31.** $1331^{\frac{1}{3}}$ **32.** $64^{\frac{1}{6}}$

33. $3375^{\frac{1}{3}}$ **34.** $512^{\frac{1}{9}}$ **35.** $\left(\frac{1}{81}\right)^{\frac{1}{4}}$ **36.** $\left(\frac{3125}{32}\right)^{\frac{1}{5}}$

37. $8^{\frac{2}{3}}$ **38.** $625^{\frac{3}{4}}$ **39.** $729^{\frac{5}{6}}$ **40.** $256^{\frac{3}{8}}$

41. $125^{\frac{4}{3}}$ **42.** $49^{\frac{5}{2}}$ **43.** $\left(\frac{9}{100}\right)^{\frac{3}{2}}$ **44.** $\left(\frac{8}{125}\right)^{\frac{4}{3}}$

Example 5

Solve each equation.

45. $3^x = 243$

46. $12^x = 144$

47. $16^x = 4$

48. $27^x = 3$

49. $9^x = 27$

50. $32^x = 4$

51. $2^{x-1} = 128$

52. $4^{2x+1} = 1024$

53. $6^{x-4} = 1296$

54. $9^{2x+3} = 2187$

55 $4^{3x} = 512$

56. $128^{3x} = 8$

Example 6

57. **CONSERVATION** Water collected in a rain barrel can be used to water plants and reduce city water use. Water flowing from an open rain barrel has velocity $v = 8h^{\frac{1}{2}}$, where v is in feet per second and h is the height of the water in feet. Find the height of the water if it is flowing at 16 feet per second.

58. **ELECTRICITY** The radius r in millimeters of a platinum wire L centimeters long with resistance 0.1 ohm is $r = 0.059L^{\frac{1}{2}}$. How long is a wire with radius 0.236 millimeter?

Write each expression in radical form, or write each radical in exponential form.

59. $17^{\frac{1}{3}}$

60. $q^{\frac{1}{4}}$

61. $7b^{\frac{1}{3}}$

62. $m^{\frac{2}{3}}$

63. $\sqrt[3]{29}$

64. $\sqrt[5]{h}$

65. $2\sqrt[3]{a}$

66. $\sqrt[3]{xy^2}$

Simplify.

67. $\sqrt[3]{0.027}$

68. $\sqrt[4]{\dfrac{n^4}{16}}$

69. $a^{\frac{1}{3}} \cdot a^{\frac{2}{3}}$

70. $c^{\frac{1}{2}} \cdot c^{\frac{3}{2}}$

71. $(8^2)^{\frac{2}{3}}$

72. $\left(y^{\frac{3}{4}}\right)^{\frac{1}{2}}$

73. $9^{-\frac{1}{2}}$

74. $16^{-\frac{3}{2}}$

75. $(3^2)^{-\frac{3}{2}}$

76. $\left(81^{\frac{1}{4}}\right)^{-2}$

77. $k^{-\frac{1}{2}}$

78. $\left(d^{\frac{4}{3}}\right)^0$

Solve each equation.

79. $2^{5x} = 8^{2x-4}$

80. $81^{2x-3} = 9^{x+3}$

81. $2^{4x} = 32^{x+1}$

82. $16^x = \dfrac{1}{2}$

83. $25^x = \dfrac{1}{125}$

84. $6^{8-x} = \dfrac{1}{216}$

85. **MP MODELING** The frequency f in hertz of the nth key on a piano is $f = 440\left(2^{\frac{1}{12}}\right)^{n-49}$.

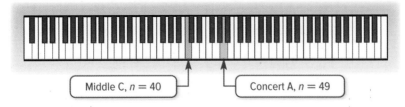

Middle C, $n = 40$ Concert A, $n = 49$

a. What is the frequency of Concert A?

b. Which note has a frequency of 220 Hz?

86. RANDOM WALKS Suppose you go on a walk where you choose the direction of each step at random. The path of a molecule in a liquid or a gas, the path of a foraging animal, and a fluctuating stock price are all modeled as random walks. The number of possible random walks w of n steps where you choose one of d directions at each step is $w = d^n$.

a. How many steps have been taken in a 2-direction random walk if there are 4096 possible walks?

b. How many steps have been taken in a 4-direction random walk if there are 65,536 possible walks?

c. If a walk of 7 steps has 2187 possible walks, how many directions could be taken at each step?

87 SOCCER The radius r of a ball that holds V cubic units of air is modeled by $r = 0.62V^{\frac{1}{3}}$. What are the possible volumes of each size soccer ball?

Soccer Ball Dimensions	
Size	Diameter (in.)
3	7.3–7.6
4	8.0–8.3
5	8.6–9.0

88. MULTIPLE REPRESENTATIONS In this problem, you will explore the graph of an exponential function.

a. TABULAR Copy and complete the table below.

x	-2	$-\frac{3}{2}$	-1	$-\frac{1}{2}$	0	$\frac{1}{2}$	1	$\frac{3}{2}$	2
$f(x) = 4^x$									

b. GRAPHICAL Graph $f(x)$ by plotting the points and connecting them with a smooth curve.

c. VERBAL Describe the shape of the graph of $f(x)$. What are its key features? Is it linear?

H.O.T. Problems   Use Higher-Order Thinking Skills

89. OPEN-ENDED Write two different expressions with rational exponents equal to $\sqrt{2}$.

90. MP CONSTRUCT ARGUMENTS Determine whether each statement is *always, sometimes,* or *never* true. Assume that x is a nonnegative real number. Explain your reasoning.

a. $x^2 = x^{\frac{1}{2}}$

b. $x^{-2} = x^{\frac{1}{2}}$

c. $x^{\frac{1}{3}} = x^{\frac{1}{2}}$

d. $\sqrt{x} = x^{\frac{1}{2}}$

e. $\left(x^{\frac{1}{2}}\right)^2 = x$

f. $x^{\frac{1}{2}} \cdot x^2 = x$

91. CHALLENGE For what values of x is $x = x^{\frac{1}{3}}$?

92. ERROR ANALYSIS Anna and Jamal are solving $128^x = 4$. Is either of them correct? Explain your reasoning.

Anna	Jamal
$128^x = 4$	$128^x = 4$
$(2^7)^x = 2^2$	$(2^7)^x = 4$
$2^{7x} = 2^2$	$2^{7x} = 4^1$
$7x = 2$	$7x = 1$
$x = \frac{2}{7}$	$x = \frac{1}{7}$

93. WRITING IN MATH Explain why 2 is the principal fourth root of 16.

94. What is the solution of the equation $64^{2x+1} = 1024$?
MP 8

- ○ **A** $\frac{1}{3}$
- ○ **B** $\frac{5}{3}$
- ○ **C** 2
- ○ **D** $\frac{15}{2}$

95. What is $16\sqrt[4]{c^5}$ written in exponential form?
MP 7

- ○ **A** $2c^{\frac{5}{4}}$
- ○ **B** $2c^{\frac{4}{5}}$
- ○ **C** $16c^{\frac{5}{4}}$
- ○ **D** $16c^{\frac{4}{5}}$

96. If $j = \sqrt[5]{32 \times 32 \times 32 \times 32 \times 32}$, what is the value of j? Select all that apply. **MP** 7, 8

- ☐ **A** 2
- ☐ **B** 2^5
- ☐ **C** 32
- ☐ **D** 32^5

97. **MULTI-STEP** To simulate gravity, space crafts rotate to create artificial gravity through centripetal force.

The formula $t = \dfrac{2\pi r^{\frac{1}{2}}}{g^{\frac{1}{2}}}$ is used to find the time in seconds to complete one rotation in order to create a specific gravity, g, for a space craft of radius r meters. **MP** 2, 4

- **a.** How long does it take a space craft with a radius of 80 meters to complete one rotation in order to simulate Earth's gravity of 9.8 m/s²?

  _____

- **b.** What is the gravity created by a space craft completing one rotation every 20 seconds with a radius of 90 meters?

  _____

98. What is the solution of the equation $9^5 = 3^{x+2}$? **MP** 8

- ○ **A** 3
- ○ **B** 5
- ○ **C** 8
- ○ **D** 10

99. Which expression is equivalent to $3x^{\frac{1}{2}}$?
MP 7

- ○ **A** $\sqrt{3x}$
- ○ **B** $3\sqrt{x}$
- ○ **C** $\frac{1}{3x}$
- ○ **D** $(3x)^2$

100. Which number is equivalent to $64^{\frac{3}{2}}$?
MP 7

- ○ **A** 4
- ○ **B** 8
- ○ **C** 16
- ○ **D** 512

101. What is the solution of the equation $16^{x+2} = 2^{12}$?
MP 8

- ○ **A** 1
- ○ **B** 3
- ○ **C** 4
- ○ **D** 10

102. Which expression is equivalent to $\sqrt{24n}$?
MP 7

- ○ **A** $(24n)^2$
- ○ **B** $24n^2$
- ○ **C** $24n^{\frac{1}{2}}$
- ○ **D** $(24n)^{\frac{1}{2}}$

Simplify each expression. (Lesson 7-1)

1. $(x^3)(4x^5)$

2. $(m^2p^5)^3$

3. $\left[(2xy^3)^2\right]^3$

4. $(6ab^3c^4)(-3a^2b^3c)$

5. MULTIPLE CHOICE Express the volume of the solid as a monomial. (Lesson 7-1)

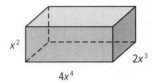

x^2 $2x^3$ $4x^4$

A $6x^9$

B $8x^9$

C $8x^{24}$

D $7x^{24}$

Simplify each expression. Assume that no denominator equals 0. (Lesson 7-2)

6. $\left(\dfrac{2a^4b^3}{c^6}\right)^3$

7. $\dfrac{2xy^0}{6x}$

8. $\dfrac{m^7n^4p}{m^3n^3p}$

9. $\dfrac{p^4t^{-2}}{r^{-5}}$

10. ASTRONOMY Physicists estimate that the number of stars in the universe has an order of magnitude of 10^{21}. The number of stars in the Milky Way galaxy is around 100 billion. (Lesson 7-2)

   a. Using orders of magnitude, how many times as many stars are there in the universe as the Milky Way?

   b. **MP** Which mathematical practice did you use to solve this problem?

Write each expression in radical form, or write each radical in exponential form. (Lesson 7-3)

11. $42^{\frac{1}{2}}$

12. $11x^{\frac{1}{2}}$

13. $(11g)^{\frac{1}{2}}$

14. $\sqrt{55}$

15. $\sqrt{5k}$

16. $4\sqrt{p}$

Simplify. (Lesson 7-3)

17. $\sqrt[3]{729}$

18. $\sqrt[4]{625}$

19. $1331^{\frac{1}{3}}$

20. $\left(\dfrac{16}{81}\right)^{\frac{1}{4}}$

21. $8^{\frac{2}{3}}$

22. $625^{\frac{3}{4}}$

23. $216^{\frac{5}{3}}$

24. $\left(\dfrac{1}{4}\right)^{\frac{3}{2}}$

Solve each equation. (Lesson 7-3)

25. $4^x = 4096$

26. $5^{2x+1} = 125$

27. $4^{x-3} = 128$

Radical Expressions

∴Then	∴Now	∴Why?

● You evaluated expressions involving rational exponents.

1 Simplify square roots by applying the Product and Quotient Properties of Square Roots.

2 Add, subtract, and multiply radical expressions.

● Ximena is going to run in her neighborhood to get ready for the soccer season. She plans to run the course that she has laid out three times each day.

How far does Ximena have to run to complete the course that she laid out?

How far does she run every day?

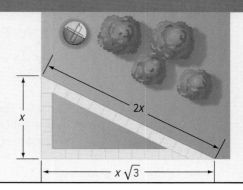

 New Vocabulary

radical expression
rationalizing the denominator
conjugate

 Mathematical Practices

6 Attend to precision.
7 Look for and make use of structure.
8 Look for and express regularity in repeated reasoning.

1 **Properties of Square Roots** A **radical expression** contains a radical, such as a square root. The expression under the radical sign is called the radicand. A radicand under a square root symbol is in simplest form if the following three conditions are true.

• No radicands have perfect square factors other than 1.

• No radicands contain fractions.

• No radicals appear in the denominator of a fraction.

The following property can be used to simplify square roots.

◆ Key Concept **Product Property of Square Roots**

Words	For any nonnegative real numbers a and b, the square root of ab is equal to the square root of a times the square root of b.
Symbols	$\sqrt{ab} = \sqrt{a} \cdot \sqrt{b}$, if $a \geq 0$ and $b \geq 0$
Examples	$\sqrt{4 \cdot 9} = \sqrt{36}$ or 6 $\qquad$ $\sqrt{4 \cdot 9} = \sqrt{4} \cdot \sqrt{9} = 2 \cdot 3$ or 6

Example 1 **Simplify Square Roots**

Simplify $\sqrt{80}$.

$$\sqrt{80} = \sqrt{2 \cdot 2 \cdot 2 \cdot 2 \cdot 5} \qquad \text{Prime factorization of 80}$$

$$= \sqrt{2^2} \cdot \sqrt{2^2} \cdot \sqrt{5} \qquad \text{Product Property of Square Roots}$$

$$= 2 \cdot 2 \cdot \sqrt{5} \text{ or } 4\sqrt{5} \qquad \text{Simplify.}$$

▶ **Guided Practice**

1A. $\sqrt{54}$ 　　　　　　　　　　　　**1B.** $\sqrt{180}$

Example 2   Multiply Square Roots

Simplify $\sqrt{2} \cdot \sqrt{14}$.

$$\sqrt{2} \cdot \sqrt{14} = \sqrt{2} \cdot \sqrt{2} \cdot \sqrt{7} \qquad \text{Product Property of Square Roots}$$

$$= \sqrt{2^2} \cdot \sqrt{7} \text{ or } 2\sqrt{7} \qquad \text{Product Property of Square Roots}$$

▶ **Guided Practice**

2A. $\sqrt{5} \cdot \sqrt{10}$ **2B.** $\sqrt{6} \cdot \sqrt{8}$

Consider the expression $\sqrt{x^2}$. It may seem that $x = \sqrt{x^2}$, but when finding the principal square root of an expression containing variables, you have to be sure that the result is not negative. Consider $x = -3$.

$$\sqrt{x^2} \overset{?}{=} x$$

$$\sqrt{(-3)^2} \overset{?}{=} -3 \qquad \text{Replace } x \text{ with } -3.$$

$$\sqrt{9} \overset{?}{=} -3 \qquad (-3)^2 = 9$$

$$3 \neq -3 \qquad \sqrt{9} = 3$$

Notice in this case, if the right hand side of the equation were $|x|$, the equation would be true. For expressions where the exponent of the variable inside a radical is even and the simplified exponent is odd, you must use absolute value.

$$\sqrt{x^2} = |x| \qquad \sqrt{x^3} = x\sqrt{x} \qquad \sqrt{x^4} = x^2 \qquad \sqrt{x^6} = |x^3|$$

Example 3   Simplify a Square Root with Variables

Simplify $\sqrt{90x^3y^4z^5}$.

$$\sqrt{90x^3y^4z^5} = \sqrt{2 \cdot 3^2 \cdot 5 \cdot x^3 \cdot y^4 \cdot z^5} \qquad \text{Prime factorization}$$

$$= \sqrt{2} \cdot \sqrt{3^2} \cdot \sqrt{5} \cdot \sqrt{x^2} \cdot \sqrt{x} \cdot \sqrt{y^4} \cdot \sqrt{z^4} \cdot \sqrt{z} \qquad \text{Product Property}$$

$$= \sqrt{2} \cdot 3 \cdot \sqrt{5} \cdot x \cdot \sqrt{x} \cdot y^2 \cdot z^2 \cdot \sqrt{z} \qquad \text{Simplify.}$$

$$= 3y^2z^2x\sqrt{10xz} \qquad \text{Simplify.}$$

▶ **Guided Practice**

3A. $\sqrt{32r^2k^4t^5}$ **3B.** $\sqrt{56xy^{10}z^5}$

To divide square roots and simplify radical expressions, you can use the Quotient Property of Square Roots.

Reading Math

Fractions in the Radicand
The expression $\sqrt{\frac{a}{b}}$ is read *the square root of a over b,* or *the square root of the quantity of a over b.*

◆ Key Concept   Quotient Property of Square Roots

Words   For any real numbers a and b, where $a \geq 0$ and $b > 0$, the square root of $\frac{a}{b}$ is equal to the square root of a divided by the square root of b.

Symbols $\sqrt{\dfrac{a}{b}} = \dfrac{\sqrt{a}}{\sqrt{b}}$

You can use the properties of square roots to **rationalize the denominator** of a fraction with a radical. This involves multiplying the numerator and denominator by a factor that eliminates radicals in the denominator.

Example 4   Rationalize a Denominator

Which expression is equivalent to $\sqrt{\dfrac{35}{15}}$?

A $\dfrac{5\sqrt{21}}{15}$ **B** $\dfrac{\sqrt{21}}{3}$ **C** $\dfrac{\sqrt{525}}{15}$ **D** $\dfrac{\sqrt{35}}{15}$

Read the Item The radical expression needs to be simplified.

Solve the Item

$$\sqrt{\frac{35}{15}} = \sqrt{\frac{7}{3}} \qquad \text{Reduce } \frac{35}{15} \text{ to } \frac{7}{3}.$$

$$= \frac{\sqrt{7}}{\sqrt{3}} \qquad \text{Quotient Property of Square Roots}$$

$$= \frac{\sqrt{7}}{\sqrt{3}} \cdot \frac{\sqrt{3}}{\sqrt{3}} \qquad \text{Multiply by } \frac{\sqrt{3}}{\sqrt{3}}.$$

$$= \frac{\sqrt{21}}{3} \qquad \text{Product Property of Square Roots}$$

The correct choice is B.

▶ **Guided Practice**

4. Simplify $\dfrac{\sqrt{6y}}{\sqrt{12}}$.

A $\dfrac{\sqrt{y}}{2}$ **B** $\dfrac{\sqrt{y}}{4}$ **C** $\dfrac{\sqrt{2y}}{2}$ **D** $\dfrac{\sqrt{2y}}{4}$

Binomials of the form $a\sqrt{b} + c\sqrt{d}$ and $a\sqrt{b} - c\sqrt{d}$, where a, b, c, and d are rational numbers, are called **conjugates**. For example, $2 + \sqrt{7}$ and $2 - \sqrt{7}$ are conjugates. The product of two conjugates is a rational number and can be found using the pattern for the difference of squares.

Example 5   Use Conjugates to Rationalize a Denominator

Simplify $\dfrac{3}{5 + \sqrt{2}}$.

$$\frac{3}{5 + \sqrt{2}} = \frac{3}{5 + \sqrt{2}} \cdot \frac{5 - \sqrt{2}}{5 - \sqrt{2}} \qquad \text{The conjugate of } 5 + \sqrt{2} \text{ is } 5 - \sqrt{2}.$$

$$= \frac{3(5 - \sqrt{2})}{5^2 - (\sqrt{2})^2} \qquad (a - b)(a + b) = a^2 - b^2$$

$$= \frac{15 - 3\sqrt{2}}{25 - 2} \text{ or } \frac{15 - 3\sqrt{2}}{23} \qquad (\sqrt{2})^2 = 2$$

▶ **Guided Practice**   Simplify each expression.

5A. $\dfrac{3}{2 + \sqrt{2}}$ **5B.** $\dfrac{7}{3 - \sqrt{7}}$

2 Operations with Radical Expressions

2 Operations with Radical Expressions To add or subtract radical expressions, the radicands must be alike in the same way that monomial terms must be alike to add or subtract.

Monomials   Radical Expressions

$$4a + 2a = (4 + 2)a \qquad\qquad 4\sqrt{5} + 2\sqrt{5} = (4 + 2)\sqrt{5}$$
$$= 6a \qquad\qquad\qquad\qquad = 6\sqrt{5}$$

$$9b - 2b = (9 - 2)b \qquad\qquad 9\sqrt{3} - 2\sqrt{3} = (9 - 2)\sqrt{3}$$
$$= 7b \qquad\qquad\qquad\qquad = 7\sqrt{3}$$

Notice that when adding and subtracting radical expressions, the radicand does not change. This is the same as when adding or subtracting monomials.

Example 6   Add and Subtract Expressions with Like Radicands

Simplify each expression.

a. $5\sqrt{2} + 7\sqrt{2} - 6\sqrt{2}$

$\qquad 5\sqrt{2} + 7\sqrt{2} - 6\sqrt{2} = (5 + 7 - 6)\sqrt{2}$ Distributive Property

$\qquad\qquad\qquad\qquad\quad = 6\sqrt{2}$ Simplify.

b. $10\sqrt{7} + 5\sqrt{11} + 4\sqrt{7} - 6\sqrt{11}$

$\qquad 10\sqrt{7} + 5\sqrt{11} + 4\sqrt{7} - 6\sqrt{11} = (10 + 4)\sqrt{7} + (5 - 6)\sqrt{11}$ Distributive Property

$\qquad\qquad\qquad\qquad\qquad\qquad\quad = 14\sqrt{7} - \sqrt{11}$ Simplify.

▶ **Guided Practice**

6A. $3\sqrt{2} - 5\sqrt{2} + 4\sqrt{2}$ **6B.** $6\sqrt{11} + 2\sqrt{11} - 9\sqrt{11}$

6C. $15\sqrt{3} - 14\sqrt{5} + 6\sqrt{5} - 11\sqrt{3}$ **6D.** $4\sqrt{3} + 3\sqrt{7} - 6\sqrt{3} + 3\sqrt{7}$

Not all radical expressions have like radicands. Simplifying the expressions may make it possible to have like radicands so that they can be added or subtracted.

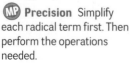

Study Tip

MP Precision Simplify each radical term first. Then perform the operations needed.

Example 7   Add and Subtract Expressions with Unlike Radicands

Simplify $2\sqrt{18} + 2\sqrt{32} + \sqrt{72}.$

$2\sqrt{18} + 2\sqrt{32} + \sqrt{72} = 2(\sqrt{3^2} \cdot \sqrt{2}) + 2(\sqrt{4^2} \cdot \sqrt{2}) + (\sqrt{6^2} \cdot \sqrt{2})$ Product Property

$\qquad\qquad\qquad\qquad = 2(3\sqrt{2}) + 2(4\sqrt{2}) + (6\sqrt{2})$ Simplify.

$\qquad\qquad\qquad\qquad = 6\sqrt{2} + 8\sqrt{2} + 6\sqrt{2}$ Multiply.

$\qquad\qquad\qquad\qquad = 20\sqrt{2}$ Simplify.

▶ **Guided Practice**

7A. $4\sqrt{54} + 2\sqrt{24}$ **7B.** $4\sqrt{12} - 6\sqrt{48}$

7C. $3\sqrt{45} + \sqrt{20} - \sqrt{245}$ **7D.** $\sqrt{24} - \sqrt{54} + \sqrt{96}$

Multiplying radical expressions is similar to multiplying monomial algebraic expressions. Let $x \geq 0$.

<table>
<tr><td align="center">Monomials</td><td align="center">Radical Expressions</td></tr>
<tr><td align="center">$(2x)(3x) = 2 \cdot 3 \cdot x \cdot x$</td><td align="center">$(2\sqrt{x})(3\sqrt{x}) = 2 \cdot 3 \cdot \sqrt{x} \cdot \sqrt{x}$</td></tr>
<tr><td align="center">$= 6x^2$</td><td align="center">$= 6x$</td></tr>
</table>

You can also apply the Distributive Property to radical expressions.

Example 8   Multiply Radical Expressions

Simplify each expression.

a. $3\sqrt{2} \cdot 2\sqrt{6}$

$$
\begin{aligned}
3\sqrt{2} \cdot 2\sqrt{6} &= (3 \cdot 2)(\sqrt{2} \cdot \sqrt{6}) &&\text{Associative Property} \\
&= 6(\sqrt{12}) &&\text{Multiply.} \\
&= 6(2\sqrt{3}) &&\text{Simplify.} \\
&= 12\sqrt{3} &&\text{Multiply.}
\end{aligned}
$$

Watch Out!

Multiplying Radicands Make sure that you multiply the radicands when multiplying radical expressions. A common mistake is to add the radicands rather than multiply.

b. $3\sqrt{5}(2\sqrt{5} + 5\sqrt{3})$

$$
\begin{aligned}
3\sqrt{5}(2\sqrt{5} + 5\sqrt{3}) &= (3\sqrt{5} \cdot 2\sqrt{5}) + (3\sqrt{5} \cdot 5\sqrt{3}) &&\text{Distributive Property} \\
&= [(3 \cdot 2)(\sqrt{5} \cdot \sqrt{5})] + [(3 \cdot 5)(\sqrt{5} \cdot \sqrt{3})] &&\text{Associative Property} \\
&= [6(\sqrt{25})] + [15(\sqrt{15})] &&\text{Multiply.} \\
&= [6(5)] + [15(\sqrt{15})] &&\text{Simplify.} \\
&= 30 + 15\sqrt{15} &&\text{Multiply.}
\end{aligned}
$$

Guided Practice

8A. $2\sqrt{6} \cdot 7\sqrt{3}$

8B. $9\sqrt{5} \cdot 11\sqrt{15}$

8C. $3\sqrt{2}(4\sqrt{3} + 6\sqrt{2})$

8D. $5\sqrt{3}(3\sqrt{2} - \sqrt{3})$

You can also multiply radical expressions with more than one term in each factor. This is similar to multiplying two algebraic binomials with variables.

Real-World Example 9   Multiply Radical Expressions

GEOMETRY Find the area of the rectangle in simplest form.

$A = (5\sqrt{2} - \sqrt{3})(\sqrt{5} + 4\sqrt{3})$ $A = \ell \cdot w$

$$
\begin{aligned}
&= \overbrace{(5\sqrt{2})(\sqrt{5})}^{\textbf{F}\text{irst Terms}} + \overbrace{(5\sqrt{2})(4\sqrt{3})}^{\textbf{O}\text{uter Terms}} + \overbrace{(-\sqrt{3})(\sqrt{5})}^{\textbf{I}\text{nner Terms}} + \overbrace{(-\sqrt{3})(4\sqrt{3})}^{\textbf{L}\text{ast Terms}} \\
&= 5\sqrt{10} + 20\sqrt{6} - \sqrt{15} - 4\sqrt{9} &&\text{Multiply.} \\
&= 5\sqrt{10} + 20\sqrt{6} - \sqrt{15} - 12 &&\text{Simplify.}
\end{aligned}
$$

$\sqrt{5} + 4\sqrt{3}$

$5\sqrt{2} - \sqrt{3}$

Review Vocabulary

FOIL Method Multiply two binomials by finding the sum of the products of the First terms, the Outer terms, the Inner terms, and the Last terms.

Guided Practice

9. GEOMETRY The area A of a rhombus can be found using the equation $A = \frac{1}{2}d_1d_2$, where d_1 and d_2 are the lengths of the diagonals. What is the area of the rhombus at the right?

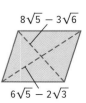

$8\sqrt{5} - 3\sqrt{6}$

$6\sqrt{5} - 2\sqrt{3}$

Concept Summary Operations with Radical Expressions

Operation	Symbols	Example
addition, $b \geq 0$	$a\sqrt{b} + c\sqrt{b} = (a + c)\sqrt{b}$ like radicands	$4\sqrt{3} + 6\sqrt{3} = (4 + 6)\sqrt{3}$ $= 10\sqrt{3}$
subtraction, $b \geq 0$	$a\sqrt{b} - c\sqrt{b} = (a - c)\sqrt{b}$ like radicands	$12\sqrt{5} - 8\sqrt{5} = (12 - 8)\sqrt{5}$ $= 4\sqrt{5}$
multiplication, $b \geq 0, g \geq 0$	$a\sqrt{b}(f\sqrt{g}) = af\sqrt{bg}$ Radicands do not have to be like radicands.	$3\sqrt{2}(5\sqrt{7}) = (3 \cdot 5)(\sqrt{2 \cdot 7})$ $= 15\sqrt{14}$

 **Go Online!** for a Self—Check Quiz

Check Your Understanding 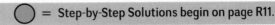 ◯ = Step-by-Step Solutions begin on page R11.

Examples 1–3 **Simplify each expression.**

1. $\sqrt{24}$

2. $3\sqrt{16}$

3. $2\sqrt{25}$

4. $\sqrt{10} \cdot \sqrt{14}$

5. $\sqrt{3} \cdot \sqrt{18}$

6. $3\sqrt{10} \cdot 4\sqrt{10}$

7. $\sqrt{60x^4y^7}$

8. $\sqrt{88m^3p^2r^5}$

9. $\sqrt{99ab^5c^2}$

Example 4 10. Which expression is equivalent to $\sqrt{\dfrac{45}{10}}$?

   A $\dfrac{5\sqrt{2}}{10}$ **B** $\dfrac{\sqrt{45}}{10}$ **C** $\dfrac{\sqrt{50}}{10}$ **D** $\dfrac{3\sqrt{2}}{2}$

Example 5 **Simplify each expression.**

11. $\dfrac{3}{3 + \sqrt{5}}$

12. $\dfrac{5}{2 - \sqrt{6}}$

13. $\dfrac{2}{1 - \sqrt{10}}$

14. $\dfrac{1}{4 + \sqrt{12}}$

15. $\dfrac{4}{6 - \sqrt{7}}$

16. $\dfrac{6}{5 + \sqrt{11}}$

Examples 6–8 **Simplify each expression.**

17. $3\sqrt{5} + 6\sqrt{5}$

18. $8\sqrt{3} + 5\sqrt{3}$

19. $\sqrt{7} - 6\sqrt{7}$

20. $10\sqrt{2} - 6\sqrt{2}$

21. $4\sqrt{5} + 2\sqrt{20}$

22. $\sqrt{12} - \sqrt{3}$

23. $\sqrt{8} + \sqrt{12} + \sqrt{18}$

24. $\sqrt{27} + 2\sqrt{3} - \sqrt{12}$

25. $9\sqrt{2}(4\sqrt{6})$

26. $4\sqrt{3}(8\sqrt{3})$

27. $\sqrt{3}(\sqrt{7} + 3\sqrt{2})$

28. $\sqrt{5}(\sqrt{2} + 4\sqrt{2})$

Example 9 29. **GEOMETRY** The area A of a triangle can be found by using the formula $A = \dfrac{1}{2}bh$, where b represents the base and h is the height. What is the area of the triangle at the right?

$4\sqrt{3} + \sqrt{5}$

$2\sqrt{3} + \sqrt{5}$

Examples 1–3 **Simplify each expression.**

30. $\sqrt{52}$ **31.** $\sqrt{56}$ **32.** $3\sqrt{18}$

33. $4\sqrt{2} \cdot 5\sqrt{8}$ **34.** $\sqrt{10} \cdot \sqrt{20}$ **35.** $5\sqrt{81q^5}$

36. $\sqrt{28a^2b^3}$ **37.** $\sqrt{75qr^3}$ **38.** $7\sqrt{63m^3p}$

39. $4\sqrt{66g^2h^4}$ **40.** $\sqrt{2ab^2} \cdot \sqrt{10a^5b}$ **41.** $\sqrt{4c^3d^3} \cdot \sqrt{8c^3d}$

Examples 4–5 **Simplify each expression.**

42. $\sqrt{\dfrac{27}{m^5}}$ **43** $\sqrt{\dfrac{32}{t^4}}$ **44.** $\sqrt{\dfrac{3}{16}} \cdot \sqrt{\dfrac{9}{5}}$

45. $\dfrac{7}{5 + \sqrt{3}}$ **46.** $\dfrac{9}{6 - \sqrt{8}}$ **47.** $\dfrac{3\sqrt{3}}{-2 + \sqrt{6}}$

48. $\dfrac{3}{\sqrt{7} - \sqrt{2}}$ **49.** $\dfrac{5}{\sqrt{6} + \sqrt{3}}$ **50.** $\dfrac{2\sqrt{5}}{2\sqrt{7} + 3\sqrt{3}}$

51 **ROLLER COASTER** Starting from a stationary position, the velocity v of a roller coaster in feet per second at the bottom of a hill can be approximated by $v = \sqrt{64h}$, where h is the height of the hill in feet.

   a. Simplify the equation.

   b. Determine the velocity of a roller coaster at the bottom of a 134-foot hill.

52. **Evaluate each expression for the given values. Write in simplest form. Assume all square roots are positive.**

   a. $\sqrt{z}$ if $z = 36$ **b.** $\sqrt{wx}$ if $w = 8$ and $x = 16$

   c. $2\sqrt[3]{w}$ if $w = 8$ **d.** $\sqrt[3]{32x}$ if $x = 16$

   e. $\sqrt{w} \cdot \sqrt{y}$ if $w = 8$ and $y = 20$ **f.** $\sqrt[3]{6z}$ if $z = 36$

   g. a^2 if $a = \sqrt{48}$ **h.** ab if $a = \sqrt{8}$ and $b = 2\sqrt{2}$

   i. $a + c$ if $a = \sqrt{49}$ and $c = \sqrt[3]{8}$ **j.** $f - d$ if $f = \sqrt[3]{1}$ and $d = \sqrt[3]{27}$

   k. $\sqrt[3]{ab}$ if $a = \sqrt{8}$ and $b = 2\sqrt{2}$ **l.** $\sqrt[3]{b^2}$ if $b = 3\sqrt{3}$

Examples 6–8 **Simplify each expression.**

   53. $7\sqrt{5} + 4\sqrt{5}$ **54.** $3\sqrt{50} - 3\sqrt{32}$

   55. $\sqrt{5}(\sqrt{2} + 4\sqrt{2})$ **56.** $7\sqrt{3} - 2\sqrt{2} + 3\sqrt{2} + 5\sqrt{3}$

   57. $(\sqrt{3} - \sqrt{2})(\sqrt{15} + \sqrt{12})$ **58.** $5\sqrt{3}(6\sqrt{10} - 6\sqrt{3})$

   59. $(5\sqrt{2} + 3\sqrt{5})(2\sqrt{10} - 5)$ **60.** $(3\sqrt{11} + 3\sqrt{15})(3\sqrt{3} - 2\sqrt{2})$

Example 9   **61.** **GEOMETRY** Find the perimeter and area of the rectangle.

$3\sqrt{7} + 3\sqrt{5}$

$2\sqrt{7} + 2\sqrt{5}$

62. **ELECTRICITY** The amount of current in amperes I that an appliance uses can be calculated using calculated using the formula $I = \sqrt{\frac{P}{R}}$, where P is the power in watts and R is the resistance in ohms.

   a. Simplify the formula.

   b. How much current does an appliance use if the power used is 75 watts and the resistance is 5 ohms?

63. **KINETIC ENERGY** The speed v of a ball can be determined by the equation $v = \sqrt{\frac{2k}{m}}$, where k is the kinetic energy and m is the mass of the ball.

   a. Simplify the formula if the mass of the ball is 3 kilograms.

   b. If the ball is traveling 7 meters per second, what is the kinetic energy of the ball in Joules?

Simplify each expression.

64. $\sqrt{\frac{1}{9}} - \sqrt{2}$

65. $\sqrt{2}\left(\sqrt{8} + \sqrt{6}\right)$

66. $2\sqrt{3}\left(\sqrt{12} - 5\sqrt{3}\right)$

67. $\sqrt{\frac{1}{5}} - \sqrt{5}$

68. $\sqrt{\frac{2}{3}} + \sqrt{6}$

69. $2\sqrt{\frac{1}{2}} + 2\sqrt{2} - \sqrt{8}$

70. $8\sqrt{\frac{5}{4}} + 3\sqrt{20} - 10\sqrt{\frac{1}{5}}$

71. $\left(3 - \sqrt{5}\right)^2$

72. $\left(\sqrt{2} + \sqrt{3}\right)^2$

73. **ROLLER COASTERS** The velocity v in feet per second of a roller coaster at the bottom of a hill is related to the vertical drop h in feet and the velocity v_0 of the coaster at the top of the hill by the formula $v_0 = \sqrt{v^2 - 64h}$.

   a. What velocity must a coaster have at the top of a 225-foot hill to achieve a velocity of 120 feet per second at the bottom?

   b. Explain why $v_0 = v - 8\sqrt{h}$ is not equivalent to the formula given.

74. **FINANCIAL LITERACY** Tadi invests $225 in a savings account. In two years, Tadi has $232 in his account. You can use the formula $r = \sqrt{\frac{v_2}{v_0}} - 1$ to find the average annual interest rate r that the account has earned. The initial investment is v_0, and v_2 is the amount in two years. What was the average annual interest rate that Tadi's account earned?

H.O.T. Problems   Use Higher-Order Thinking Skills

75. **MP STRUCTURE** Explain how to solve $\frac{\sqrt{3} + 2}{x} = \frac{\sqrt{3} - 1}{\sqrt{3}}$.

76. **MP REASONING** Marge takes a number, subtracts 4, multiplies by 4, takes the square root, and takes the reciprocal to get $\frac{1}{2}$. What number did she start with? Write a formula to describe the process.

77. **OPEN-ENDED** Write two binomials of the form $a\sqrt{b} + c\sqrt{f}$ and $a\sqrt{b} - c\sqrt{f}$. Then find their product.

78. **CHALLENGE** Determine whether the following statement is *true* or *false*. Provide a proof or counterexample to support your answer.

$$x + y > \sqrt{x^2 + y^2} \text{ when } x > 0 \text{ and } y > 0$$

79. **MP CONSTRUCT ARGUMENTS** Make a conjecture about the sum of a rational number and an irrational number. Is the sum of a rational number and an irrational number *rational* or *irrational*? Is the product of a nonzero rational number and an irrational number *rational* or *irrational*? Explain your reasoning.

80. **OPEN-ENDED** Write an equation that shows a sum of two radicals with different radicands. Explain how you could combine these terms.

81. **WRITING IN MATH** Describe step by step how to multiply two radical expressions, each with two terms. Write an example to demonstrate your description.

82. Jason correctly simplified the product $\sqrt{2} \cdot \sqrt{22}$. Which of the following is most likely to be a step in the process he followed? **MP** 8

- ○ **A** $\sqrt{2} \cdot \sqrt{22} = \sqrt{2} \cdot \sqrt{2} \cdot \sqrt{11}$
- ○ **B** $\sqrt{2} \cdot \sqrt{22} = \sqrt{2} \cdot \sqrt{2} + \sqrt{11}$
- ○ **C** $\sqrt{2} \cdot \sqrt{22} = (\sqrt{2 \cdot 2})^2 \cdot \sqrt{11}$
- ○ **D** $\sqrt{2} \cdot \sqrt{22} = 4\sqrt{11}$

83. Which expression is equivalent to $\sqrt{\dfrac{40}{12}}$? **MP** 1

- ○ **A** 10
- ○ **B** $\dfrac{\sqrt{30}}{3}$
- ○ **C** $\dfrac{\sqrt{30}}{9}$
- ○ **D** $\dfrac{\sqrt{30}}{\sqrt{3}}$

84. Which expression is equivalent to 4? **MP** 1, 2

- ○ **A** $\dfrac{\sqrt{8}}{2}$
- ○ **B** $\dfrac{\sqrt{8}}{\sqrt{2}}$
- ○ **C** $2\sqrt{8}$
- ○ **D** $\sqrt{2} \cdot \sqrt{8}$

85. An object is dropped off a cliff into a deep canyon. The velocity v of the object at the bottom of the canyon is given by the function $v = \sqrt{64h}$, where h is the height of the cliff, in feet. Which equation is not equivalent to $v = \sqrt{64h}$? **MP** 4

- ○ **A** $v = 2\sqrt{4h}$
- ○ **B** $v = 2\sqrt{16h}$
- ○ **C** $v = 2 \cdot 2\sqrt{4h}$
- ○ **D** $v = 8\sqrt{h}$

86. Which expression is equivalent to $4\sqrt{27} + 2\sqrt{12} + \sqrt{75}$? **MP** 1

- ○ **A** $12 + 9\sqrt{3}$
- ○ **B** $10\sqrt{3}$
- ○ **C** $21\sqrt{3}$
- ○ **D** $25\sqrt{3}$

87. What is the area of the square in simplest form? **MP** 1, 2

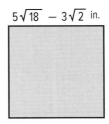

$5\sqrt{18} - 3\sqrt{2}$ in.

- ○ **A** 252 in^2
- ○ **B** 270 in^2
- ○ **C** 288 in^2
- ○ **D** 648 in^2

88. For what value of a does the expression $a\sqrt{5} \cdot 4\sqrt{20}$ equal 200? **MP** 1

[   ]

89. MULTI-STEP The annual interest rate r earned by an investment can be found using the formula $r = \sqrt[t]{\dfrac{A}{P}} - 1$, where P is the initial investment and A is the value of the investment after t years.

a. What is the average interest rate for an initial investment of \$5750 that has a value of \$6080 after two years? Round to the nearest tenth of a percent.

[   ]

b. Solve the function for P.

[   ]

c. Find the initial investment if its value is \$4000 after 12 years at an interest rate of 3.4%.

[   ]

Algebra Lab: Sums and Products of Rational and Irrational Numbers

Irrational numbers are the set of numbers that cannot be expressed as a terminating or repeating decimal. Irrational numbers cannot be written as a fraction $\frac{a}{b}$, where a and b are integers and $b \neq 0$.

Mathematical Practices

MP **2** Reason abstractly and quantitatively.

Real Numbers

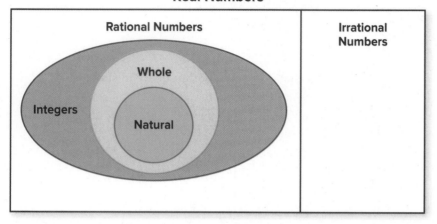

Activity 1   Sum and Product Rational Numbers and Irrational Numbers

Make a conjecture about whether the sum and product of a nonzero rational number and an irrational number by examining specific examples.

Step 1   Make a table to examine the sum of a and b, where a is a nonzero rational number and b is an irrational number.

a	b	$a + b$	Rational or irrational?
-2	$\sqrt{2}$	$-2 + \sqrt{2}$	irrational
0.4	$2\sqrt{3}$	$0.4 + 2\sqrt{3}$	irrational
$\frac{1}{2}$	π	$\frac{1}{2} + \pi$	irrational
-2	$\sqrt{2}$	$-2 + \sqrt{2}$	irrational

Step 2   Make a table to examine the product of a and b.

a	b	ab	Rational or irrational?
-2	$\sqrt{2}$	$-2\sqrt{2}$	irrational
0.4	$2\sqrt{3}$	$0.8\sqrt{3}$	irrational
$\frac{1}{2}$	π	$\frac{\pi}{2}$	irrational

Based on the examples, a conjecture can be made that the sum or product of a nonzero rational number and an irrational number is an irrational number.

It appears from the table that the sum and product of a rational and irrational number are always irrational. Listing examples is not sufficient to prove this conjecture. To prove this is true, it must be true for all rational and irrational numbers. That is, for any rational number a and any irrational number b, $a + b$ and ab are irrational.

Prove that the product of a nonzero rational number and an irrational number is irrational.

Given: *a* is a nonzero rational number and *b* is an irrational number.

Prove: *ab* is irrational.

Assume *ab* is a rational number.

$ab = \dfrac{c}{d}$, where c and d are integers and $d \neq 0$	Definition of a rational number
$b = \dfrac{c}{ad}$	Divide each side by a.
$b \neq \dfrac{c}{ad}$	b is an irrational number.

Because it is given that *b* is irrational and an irrational number cannot be written as a fraction using integer values, the product *ab* is not a rational number. Therefore, *ab* is irrational.

Prove that the sum of a nonzero rational number and an irrational number is irrational.

Given: *a* is a nonzero rational number and *b* is an irrational number.

Prove: *a* + *b* is irrational.

Assume *a* + *b* is a rational number.

$a + b = \dfrac{c}{d}$, where c and d are integers and $d \neq 0$	Definition of a rational number
$b = \dfrac{c}{d} - a$	Subtract each side by a.
$b = \dfrac{c - ad}{d}$	Multiply a by $\dfrac{d}{d}$ and simplify.
$b \neq \dfrac{c - ad}{d}$	b is an irrational number.

Because it is given that *b* is irrational and an irrational number cannot be written as a fraction using integer values, the sum *a* + *b* is not a rational number. Therefore, *a* + *b* is irrational.

Exercises

Give an example of each.

1. Two irrational numbers whose sum is rational

2. Two irrational numbers whose sum is irrational

3. Two irrational numbers whose product is rational

4. Two irrational numbers whose product is irrational

5. Recall that a set is closed under an operation if for any numbers in the set, the result of the operation is also in the set. Explain why the set of irrational numbers is not closed under addition or multiplication.

:: Then	:: Now	:: Why?

Then
- and simplified radical expressions.

Now
1. Graph exponential functions.
2. Identify data that display exponential behavior.

Why?
Tarantulas can appear scary with their large hairy bodies and legs, but they are harmless to humans. The graph shows a tarantula spider population that increases over time. Notice that the graph is not linear.

The graph represents the function $y = 3(2)^x$. This is an example of an *exponential* function.

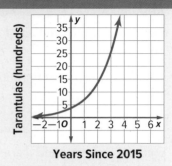

Years Since 2015

 New Vocabulary
exponential function
asymptote
exponential growth function
exponential decay function

 Mathematical Processes

1 Make sense of problems and persevere in solving them.

1 **Graph Exponential Functions** An **exponential function** is a function of the form $y = ab^x$, where $a \neq 0$, $b > 0$, and $b \neq 1$. Notice that the base is a constant and the exponent is a variable. Exponential functions are nonlinear.

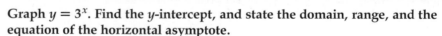

> **Key Concept** Exponential Function
>
> **Words** An exponential function is a function that can be described by an equation of the form $y = ab^x$, where $a \neq 0$, $b > 0$, and $b \neq 1$.
>
> **Examples** $y = 2(3)^x$ $\qquad\qquad$ $y = 4^x$ $\qquad\qquad$ $y = \left(\dfrac{1}{2}\right)^x$

The graphs of exponential functions have a horizontal asymptote. An **asymptote** is a line that a graph approaches.

> **Example 1** Graph with $a > 0$ and $b > 1$
>
> **Graph $y = 3^x$. Find the y-intercept, and state the domain, range, and the equation of the horizontal asymptote.**
>
> The graph crosses the y-axis at 1, so the y-intercept is 1.
> $\mathbf{D} = \{-\infty < x < \infty\}$
> $\mathbf{R} = \{0 < y < \infty\}$
> As x decreases, the graph approaches the x-axis. So, the graph has a horizontal asymptote of $y = 0$.
>
> Notice that the graph approaches the x-axis but there is no x-intercept. The graph is increasing on the entire domain.

x	3^x	y
-2	3^{-2}	$\dfrac{1}{9}$
-1	3^{-1}	$\dfrac{1}{3}$
0	3^0	1
$\dfrac{1}{2}$	$3^{\frac{1}{2}}$	≈ 1.73
1	3^1	3
2	3^2	9

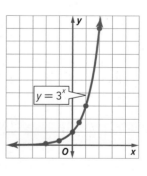

▶ **Guided Practice**

1. Graph $y = 7^x$. Find the y-intercept, and state the domain, range, and the equation of the horizontal asymptote.

Functions of the form $y = ab^x$, where $a > 0$ and $b > 1$, are called **exponential growth functions** and all have the same shape as the graph in Example 1. Functions of the form $y = ab^x$, where $a > 0$ and $0 < b < 1$ are called **exponential decay functions** and also have the same general shape.

Go Online!

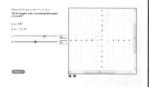

Investigate how changing the values in the equation of an exponential function affects the graph by using the **Graphing Tools** in ConnectED.

Example 2 Graph with $a > 0$ and $0 < b < 1$

Graph $y = \left(\frac{1}{3}\right)^x$. Find the y-intercept, and state the domain, range, and the equation of the asymptote.

The y-intercept is 1.
$D = \{-\infty < x < \infty\}$
$R = \{0 < y < \infty\}$
Notice that as x increases, the y-values decrease less rapidly, but never touch the x-axis. The equation of the asymptote is $y = 0$.

x	$\left(\frac{1}{3}\right)^x$	y
-2	$\left(\frac{1}{3}\right)^{-2}$	9
0	$\left(\frac{1}{3}\right)^{0}$	1
2	$\left(\frac{1}{3}\right)^{2}$	$\frac{1}{9}$

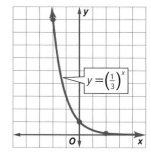

$y = \left(\frac{1}{3}\right)^x$

▶ **Guided Practice**

2. Graph $y = \left(\frac{1}{2}\right)^x - 1$. Find the y-intercept, and state the domain, range, and the equation of the asymptote.

The key features of the graphs of exponential functions can be summarized as follows.

Key Concept Graphs of Exponential Functions

Exponential Growth Functions

Equation: $f(x) = ab^x, a > 0, b > 1$
Domain, Range: $\{-\infty < x < \infty\}; \{y > 0\}$
Intercepts: one y-intercept, no x-intercepts
Horizontal Asymptote: $y = 0$

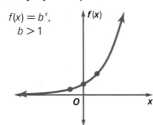

$f(x) = b^x,$
$b > 1$

Exponential Decay Functions

Equation: $f(x) = ab^x, a > 0, 0 < b < 1$
Domain, Range: $\{-\infty < x < \infty\}; \{y > 0\}$
Intercepts: one y-intercept, no x-intercepts
Horizontal Asymptote: $y = 0$

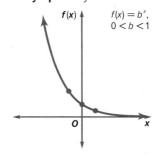

$f(x) = b^x,$
$0 < b < 1$

Exponential functions occur in many real world situations.

Real-World Example 3 Use Exponential Functions to Solve Problems

BOTTLED WATER The function $W = 27.92(1.052)^t$ models the amount of bottled water consumed in the U.S. per capita, where W is gallons of water and t is the number of years since 2010.

a. Graph the function. What values of W and t are meaningful in the context of the problem?

Since t represents time, $t > 0$. At $t = 0$, the consumption is 27.92 gallons. Therefore, in the context of this problem, $W > 27.92$ is meaningful.

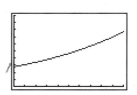

[0, 20] scl: 2 by [0, 100] scl: 10

b. Find and interpret the y-intercept and the equation of the asymptote.

The y-intercept, 27.92, represents that 27.92 gallons of bottled water was consumed in 2010, per capita. The asymptote $y = 0$ indicates that annual bottled water consumption can never be less than or equal to 0 gallons.

c. Predict the per capita consumption of bottled water in 2020.

$$W = 27.92(1.052)^t \qquad \text{Original equation}$$
$$= 27.92(1.052)^{10} \qquad t = 10$$
$$\approx 46.35 \qquad \text{Use a calculator.}$$

The predicted consumption of bottled water in 2020 is about 46.35 gallons in the U.S.

▶ **Guided Practice**

3. **BIOLOGY** Beginning with 10 cells in a bacteria culture, the population can be represented by the function $B = 10(2)^t$, where B is the number of cells and t is time in 20 minute intervals. Graph the function. Then find and interpret the y-intercept and the equation of the asymptote. How many cells will there be after 2 hours?

2 **Identify Exponential Behavior** Recall from Lesson 3-3 that linear functions have a constant rate of change. Exponential functions do not have constant rates of change, but they do have constant ratios.

> **Problem-Solving Tip**
>
> **Make an Organized List**
> Making an organized list of x-values and corresponding y-values is helpful in graphing the function. It can also help you identify patterns in the data.

Example 4 Identify Exponential Behavior

Determine whether the set of data shown at right displays exponential behavior. Write *yes* or *no*. Explain why or why not.

x	0	5	10	15	20	25
y	64	32	16	8	4	2

Method 1 Look for a pattern.

The domain values are at regular intervals of 5. Look for a common factor among the range values. The range values differ by the common factor of $\frac{1}{2}$.

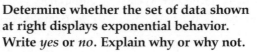

Since the domain values are at regular intervals and the range values differ by a positive common factor, the data are probably exponential. Its equation may involve $\left(\frac{1}{2}\right)^x$.

Method 2 Graph the data.

Plot the points and connect them with a smooth curve. The graph shows a rapidly decreasing value of y as x increases. This is a characteristic of exponential behavior in which the base is between 0 and 1.

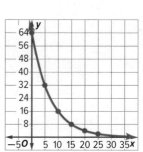

▶ **Guided Practice**

4. Determine whether the set of data shown below displays exponential behavior. Write *yes* or *no*. Explain why or why not.

x	0	3	6	9	12	15
y	12	16	20	24	28	32

Check Your Understanding   ○ = Step-by-Step Solutions begin on page R11.

✓ **Go Online!** for a Self-Check Quiz

Examples 1-2 Graph each function. Find the y-intercept and state the domain, range, and the equation of the asymptote.

1. $y = 2^x$

2. $y = -5^x$

3. $y = -\left(\frac{1}{5}\right)^x$

4. $y = 3\left(\frac{1}{4}\right)^x$

5. $f(x) = 6^x + 3$

6. $f(x) = 2 - 2^x$

Example 3 **7. BIOLOGY** The function $f(t) = 100(1.05)^t$ models the growth of a fruit fly population, where $f(t)$ is the number of flies and t is time in days.

  a. What values for the domain and range are reasonable in the context of this situation? Explain.

  b. After two weeks, approximately how many flies are in this population?

Example 4 Determine whether the set of data shown below displays exponential behavior. Write *yes* or *no*. Explain why or why not.

8.

x	1	2	3	4	5	6
y	−4	−2	0	2	4	6

9.

x	2	4	6	8	10	12
y	1	4	16	64	256	1024

Practice and Problem Solving   Extra Practice is on page R7.

Examples 1–2 Graph each function. Find the y-intercept and state the domain, range, and the equation of the asymptote.

10. $y = 2 \cdot 8^x$

11. $y = 2 \cdot \left(\frac{1}{6}\right)^x$

12. $y = \left(\frac{1}{12}\right)^x$

13. $y = -3 \cdot 9^x$

14. $y = -4 \cdot 10^x$

15. $y = 3 \cdot 11^x$

16. $y = 4^x + 3$

17. $y = \frac{1}{2}(2^x - 8)$

18. $y = 5(3^x) + 1$

19. $y = -2(3^x) + 5$

Example 3 **20. ⓂⓅ MODELING** A population of rabbits increases according to the model $p = 200(2.5)^{0.02t}$, where t is the number of months and $t = 0$ corresponds to January.

  a. Use this model to estimate the number of rabbits in March.

  b. Graph the function. Find the equation of the asymptote, the y-intercept, and describe what they represent in the context of the situation. State the domain and range given the context of the situation.

Example 4 Determine whether the set of data shown below displays exponential behavior. Write *yes* or *no*. Explain why or why not.

㉑

x	−4	0	4	8	12
y	2	−4	8	−16	32

22.

x	−6	−3	0	3
y	5	10	15	20

23.

x	−8	−6	−4	−2
y	0.25	0.5	1	2

24.

x	20	30	40	50	60
y	1	0.4	0.16	0.064	0.0256

25. PHOTOGRAPHY Jameka is enlarging a photograph to make a poster for school. She will enlarge the picture repeatedly at 150%. The function $P = 1.5^x$ models the new size of the picture being enlarged, where x is the number of enlargements. How many times as big is the picture after 4 enlargements?

26. MULTI-STEP Daniel deposited $500 into a savings account that earns interest monthly. After 8 years, his investment is worth $807.07. Daniel wants to save a total of $1500 dollars in the next 5 years.

   a. How much should Daniel add to his savings account to ensure that he has $1500 in 5 years?

   b. Explain your solution process.

   c. What assumptions did you make?

Identify each function as *linear,* *exponential,* **or** *neither.*

27.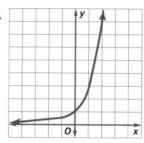

28.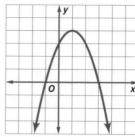

29.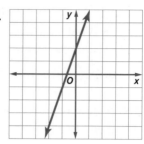

30. $y = 4^x$

31. $y = 2x(x - 1)$

32. $5x + y = 8$

33. BUSINESS The function $N = 380(0.85)^t$ models the number of desktop computers sold per year at a local technology store, where t is the number of years since 2015 and N is the number of desktop computers sold. How many desktop computers will be sold in 2025?

Write the equation of the horizontal asymptote for each function.

34. $y = 2^x + 6$

35. $y = 5\left(\dfrac{2}{3}\right)^x$

36. $y = -\dfrac{1}{4}(2)^x$

37. $y = -3 + 5^x$

38. $y = \left(\dfrac{2}{3}\right)^{x-1}$

39. $y = -5(3)^x + 1$

40. BACTERIA The bacteria in a culture doubles every hour. At $t = 0$, there were 25 bacteria in the culture. The function $N = 25(2)^t$ models the number of bacteria N in the culture t hours after 7 A.M. Estimate the number of bacteria at 10 P.M.

H.O.T. Problems   Use Higher-Order Thinking Skills

41. (MP) PERSEVERANCE Write an exponential function for which the graph passes through the points at $(0, 3)$ and $(1, 6)$.

42. (MP) REASONING Determine whether the graph of $y = ab^x$, where $a \neq 0$, $b > 0$, and $b \neq 1$, *sometimes,* *always,* or *never* has an x-intercept. Explain your reasoning.

43. OPEN-ENDED Find an exponential function that represents a real-world situation, and graph the function. Analyze the graph, and explain why the situation is modeled by an exponential function rather than a linear function.

44. (MP) REASONING Use tables and graphs to compare and contrast an exponential function $f(x) = ab^x + c$, where $a \neq 0$, $b > 0$, and $b \neq 1$, and a linear function $g(x) = ax + c$. Include intercepts, intervals where the functions are increasing, decreasing, positive, or negative, relative maxima and minima, symmetry, and end behavior.

45. WRITING IN MATH Explain how to determine whether a set of data displays exponential behavior.

46. Which of the following functions represents the data in the table? MP 1

x	0	1	2	3
y	1	0.25	0.0625	0.015625

- ○ **A** $y = 0.25x + 1$
- ○ **B** $y = x - 0.75$
- ○ **C** $y = 0.25^x$
- ○ **D** $y = -0.25^x$

47. Which of following gives the equation and y-intercept of the graph shown? MP 1

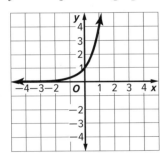

- ○ **A** $y = \left(\frac{1}{4}\right)^x;\ 0$
- ○ **B** $y = \left(\frac{1}{4}\right)^x;\ 1$
- ○ **C** $y = 4^x;\ 0$
- ○ **D** $y = 4^x;\ 1$

48. The function $f(t) = 2(2.25)^t$ models the growth of bacteria cells, where $f(t)$ is the number of bacteria cells and t is time in days. After 10 days, approximately how many bacteria cells are there? MP 4

- ○ **A** 45
- ○ **B** 1139
- ○ **C** 3325^x
- ○ **D** 6651

49. Given the function $y = \left(\frac{2}{3}\right)^x - 6$, MP 1

a. What is the domain?
- ○ **A** $-\infty < x < \infty$
- ○ **B** $-6 < x < 0$
- ○ **C** $-\infty < x < -6$
- ○ **D** $-6 < x < \infty$

b. What is the range?
- ○ **A** $-\infty < y < \infty$
- ○ **B** $0 < y < \infty$
- ○ **C** $-\infty < y < -6$
- ○ **D** $-6 < y < \infty$

50. MULTI-STEP The concentration of a certain medicine in a patient's body is modeled by $f(x) = 500(0.85)^x$, where $f(x)$ is the medicine in milligrams in the patient's body after x hours. MP 1, 4

a. Find the y-intercept and describe its meaning in the context of the situation.

b. In the context of the situation, what is the domain?
- ○ **A** $x \geq 0$
- ○ **B** $x \geq 0.85$
- ○ **C** $0 < x \leq 500$
- ○ **D** all real numbers

c. In the context of the situation, what is the range?
- ○ **A** $f(x) \geq 0$
- ○ **B** $f(x) \leq 500$
- ○ **C** $0 < f(x) \leq 500$
- ○ **D** all real numbers

d. Sketch the graph of the function with the appropriate labels and scales for the graph.

Algebra Lab
Exponential Growth Patterns

IAn exponential function grows in a predictable way. Given the equation of an exponential function, you can predict how the dependent variable will change over an interval of the independent variable.

Mathematical Processes

 8 Look for and express regularity in repeated reasoning.

Let A, B, and C be any three points on the curve $y = ab^x$.
You will explore the pattern for each of the intervals.

Activity 1 $y = 2^x$

Work cooperatively. Explore the growth of $y = 2^x$ over x- and y-intervals.

1A. Complete the table below.

Point	x	y
A	2	
B	3	
C	4	

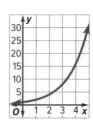

1B. What is the relationship between the x- and y-values in the table?

1C. Make a prediction about the y-values of points on the graph for x-values of 5, 6, and 7.

1D. Check your prediction. Are you correct?

Activity 2 $y = 2(2)^x$

Work cooperatively. Explore the growth of $y = 2(2)^x$ over x- and y-intervals.

2A. Complete the table below.

Point	x	y
A	2	
B	3	
C	4	

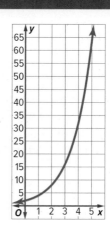

2B. What is the relationship between the intervals for the x- and y-values?

2C. Compare and constrast the graphs from Activity 1 and Activity 2.

Work cooperatively. Explore the growth of $y = 0.5(2)^x$ over x- and y-intervals.

3A. Complete the table below.

Point	x	y
A	2	
B	3	
C	4	

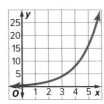

3B. What is the relationship between these intervals?

3C. Compare the graph from Activity 1 to the graph from Activity 3. What is a similarity? What is a difference?

Analyze the Results

Work cooperatively.

1. Consider any points for the graph $y = b^x$.

   a. Suppose the first x-coordinate is n. If the length of the interval along the x-axis is 1, what is the x-coordinate of the second point?

   b. To find the first y-coordinate, substitute the x-coordinate of the first point into $y = b^x$. What is the y-coordinate?

   c. To find the second y-coordinate, substitute the x-coordinate of the second point into $y = b^x$. What is the y-coordinate?

   d. To find the ratio of the y-values, divide the two y-coordinates. What is the result?

   e. Repeat this procedure for consecutive values of x such as $n + 2$, $n + 3$, and so on. What is your conclusion?

   f. Repeat this procedure for a first x-coordinate of n and intervals of length m. How does the result prove that exponential functions grow by equal factors over equal intervals?

2. Examine the function $y = 3^x$.

   a. For an increase of 1 in the x-values, describe the change in the y-value.

   b. If the x-value increases from 3 to 4, how do the corresponding y-values change?

   c. If the x-value increases from 3 to 6, how do the corresponding y-values change?

3. Examine the function $y = \left(\frac{1}{3}\right)^x$.

   a. For an increase of 1 in the x-values, describe the change in the y-value.

   b. If the x-value increases from 3 to 4, how do the corresponding y-values change?

   c. If the x-value increases from 3 to 6, how do the corresponding y-values change?

4. **MP CONSTRUCT ARGUMENTS** Is the function represented in the table an exponential function? Justify your answer by discussing intervals and differences.

x	-3	-2	-1	0	1	2	3
y	-8	-5	-2	1	4	7	10

Transformations of Exponential Functions

Then	Now	Why?

Then

You defined, identified, and graphed exponential functions.

Now

1 Identify the effects on the graphs of exponential functions by replacing $f(x)$ with $f(x) + k$ and $f(x - h)$ for positive and negative values of h and k.

2 Identify the effects on the graphs of exponential functions by replacing $f(x)$ with $af(x)$ and $f(ax)$ with positive and negative values of a.

Why?

In some extreme sports, such as skating, skateboarding, snowboarding, and skiing, athletes use a half-pipe to gain speed and perform tricks. A half-pipe is essentially two ramps, or quarter-pipes, facing each other with an extended flat bottom. Each quarter-pipe can be modeled by an exponential function. When placed together to form a half-pipe, one ramp can be described as a reflection of the other.

 Mathematical Practices

2 Reason abstractly and quantitatively.

4 Model with mathematics.

1 **Translations of Exponential Functions** The general form of an exponential function is $g(x) = ab^{x-h} + k$, where a, h, and k are parameters that dilate, reflect, and translate a parent function $f(x) = b^x$. Recall that a translation moves a graph up, down, left, right, or in two directions.

The graph of the function $g(x) = b^x + k$ is the graph of $f(x)$ translated vertically. The graph of the function $g(x) = b^{x-h}$ is the graph of $f(x)$ translated horizontally.

⬙ Key Concept Translations of Exponential Functions

Vertical Translations

For $f(x) + k$:

- if $k > 0$, the graph of $f(x)$ is translated k units up.

- if $k < 0$, the graph of $f(x)$ is translated $|k|$ units down.

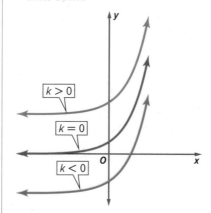

Every point on the graph of $f(x)$ moves vertically k units.

Horizontal Translations

For $f(x - h)$:

- if $h > 0$, the graph of $f(x)$ is translated h units right.

- if $h < 0$, the graph of $f(x)$ is translated $|h|$ units left.

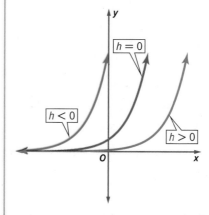

Every point on the graph of $f(x)$ moves horizontally h units.

---

tonsound/iStock/Getty Images

Example 1   Translations of Exponential Functions

Describe the translation in each function as it relates to the graph of $f(x) = 2^x$.

a. $g(x) = 2^x - 4$

The graph of $g(x) = 2^x - 4$ is a translation of the graph of $f(x) = 2^x$ down 4 units.

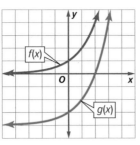

b. $q(x) = 2^{x+3}$

The graph of $q(x) = 2^{x+3}$ is a translation of the graph of $f(x) = 2^x$ left 3 units.

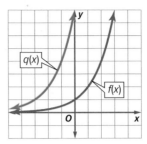

Guided Practice

Describe the translation in each function as it relates to the graph of $f(x) = 4^x$.

1A. $g(x) = 4^x + 3$

2B. $q(x) = 4^{x-5}$

You can compare a graph to the parent function to write a function for the graph.

Example 2   Analyze a Graph

The graph $g(x)$ is a translation of the parent function $f(x) = \left(\dfrac{1}{2}\right)^x$. Use the graph of the function to write an equation for $g(x)$.

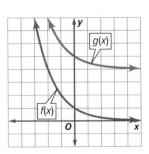

Notice that the horizontal asymptote of $g(x)$ is different than the horizontal asymptote of $f(x)$, implying a vertical translation of the form $g(x) = \left(\dfrac{1}{2}\right)^x + k$.

The parent graph has a y-intercept at $(0, 1)$ and the translated graph has a y-intercept at $(0, 5)$. The y-intercept is shifted 4 units up, so $k = 4$.

The graph is represented by $g(x) = \left(\dfrac{1}{2}\right)^x + 4$.

Guided Practice

2. The graph $g(x)$ is a translation of the parent function $f(x) = 1.5^x$. Use the graph of the function to write an equation for $g(x)$.

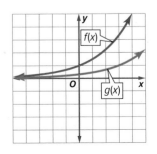

2 Dilations and Reflections Recall that a dilation stretches or compresses the graph of a function. A vertical dilation occurs when the parent function $f(x)$ is multiplied by a positive constant a after the function is evaluated. The graph of $g(x) = a \cdot f(x)$ is the graph of $f(x)$ stretched or compressed vertically by a factor of $|a|$.

A horizontal dilation occurs when the parent function is multiplied by a constant a before the function is evaluated. The graph of $g(x) = f(a \cdot x)$ is the graph of $f(x)$ stretched or compressed horizontally by a factor of $\frac{1}{|a|}$.

Key Concept Dilations of Exponential Functions

Vertical Dilations

For $a \cdot f(x)$, if $|a| > 1$, the graph of $f(x)$ is stretched vertically.

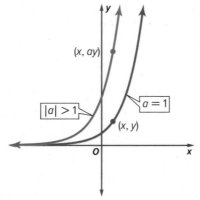

Every point on the graph of $f(x)$ is farther from the x-axis.

For $a \cdot f(x)$, if $0 < |a| < 1$, the graph of $f(x)$ is compressed vertically.

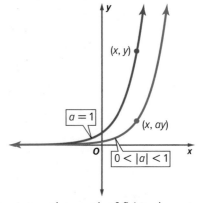

Every point on the graph of $f(x)$ is closer to the x-axis.

Horizontal Dilations

For $f(a \cdot x)$, if $|a| > 1$, the graph of $f(x)$ is compressed horizontally.

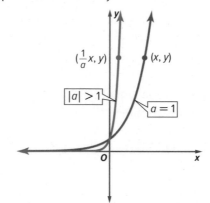

Every point on the graph of $f(x)$ is closer to the y-axis.

For $f(a \cdot x)$, if $0 < |a| < 1$, the graph of $f(x)$ is stretched horizontally.

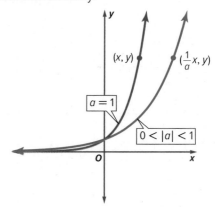

Every point on the graph of $f(x)$ is farther from the y-axis.

Study Tip

MP Reasoning Stretch and compression dilations are related. The graph of a horizontal stretch may appear the same as a vertical compression.

Real-World Example 3   Analyze a Function

SOLAR ENGERGY Photovoltaic cells, or PV cells, convert solar energy into electricity. Since 2000, global solar PV capacity has been growing at a rate that can be approximated by the function $c(x) = 0.897(1.46)^x$, where $c(x)$ is the solar PV capacity in gigawatts, x is the number of years since 2000, and 0.897 is the initial capacity. Describe the dilation in $c(x) = 0.897(1.46)^x$ as it is related to the parent function $f(x) = 1.46^x$.

The function is multiplied by the positive constant a after it has been evaluated and $|a|$ is between 0 and 1, so the graph of $f(x) = 1.46^x$ is compressed vertically by a factor of $|a|$, or 0.897.

Guided Practice

3. **FISH** Suppose the population of a certain fish in a new pond can be modeled by $p(x) = 50(1.5)^x$, where x is the number of generations after an initial stock of 50 fish were added to the pond. Describe the dilation in $p(x) = 50(1.5)^x$ as it relates to the parent function $p(x) = 1.5^x$.

When a is negative, the graph of $a \cdot f(x)$ is reflected across the x-axis, while the graph of $f(a \cdot x)$ is reflected across the y-axis.

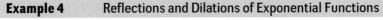

Key Concept Reflections of Linear Functions

Reflection across x-axis	Reflection across y-axis
For $a \cdot f(x)$, if $a < 0$, the graph of $f(x)$ is reflected across the x-axis.	For $f(a \cdot x)$, if $a < 0$, the graph of $f(x)$ is reflected across the y-axis.

When a is negative and $a \neq -1$, the graphs of $a \cdot f(x)$ and $f(a \cdot x)$ are both reflections and dilations.

Example 4 Reflections and Dilations of Exponential Functions

Describe the transformations in each function as it relates to the graph of $f(x) = 3^x$.

a. $g(x) = 3^{-4x}$

Because a is negative in $g(x)$, the graph of $f(x)$ is reflected across the y-axis. Because $|a| > 1$, the graph is also compressed horizontally.

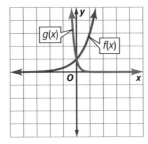

b. $q(x) = -\frac{1}{5}(3)^x$

Because a is negative in $q(x)$, the graph of $f(x)$ is reflected across the x-axis. Because $0 < |a| < 1$, the graph is also compressed vertically.

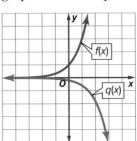

Study Tip

Reflections Determine whether multiplying by −1 occurs before or after the parent function is evaluated. Multiplying by −1 before the parent function is evaluated results in a reflection across the y-axis.

Describe the transformations in each function as it relates to the graph of $f(x) = 2^x$.

4A. $g(x) = 2^{-0.25x}$

4B. $q(x) = -1.75(2)^x$

Often more than one transformation is applied to a function. Examine the transformation to analyze how the graph is affected.

Example 5   Analyze Transformations

Describe the transformations in $g(x) = -3(2)^{x-1} + 5$ as it relates to the graph of $f(x) = 2^x$.

Because $f(x) = 2^x$, $g(x) = af(x - h) + k$, where $a = -3$, $h = 1$, and $k = 5$.

The graph of $f(x)$ is reflected across the x-axis because a is negative. Because $|a| > 1$, the graph is stretched vertically. The graph is translated right 1 unit and up 5 units.

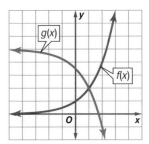

▶ **Guided Practice**

5. Describe the transformations in $g(x) = -\left(\dfrac{3}{4}\right)^{x+5} - 4$ as it relates to the graph of $f(x) = \left(\dfrac{3}{4}\right)^x$.

You can examine a graph and compare it to the graph of the parent function to write a function for the graph.

Example 6   Compare Graphs

Compare the graphs to the parent function $f(x) = 2.5^x$. Then, verify using technology.

$g(x) = 0.5(2.5)^x + 1$, $p(x) = (2.5)^{2x} - 3$, $q(x) = -3(2.5)^{x+2} - 1$

$g(x)$ is compressed vertically by a factor of 0.5 and translated up 1 unit.

$p(x)$ is compressed horizontally by a factor of $\dfrac{1}{2}$ and translated down 3 units.

$q(x)$ is reflected across the x-axis, stretched vertically by a factor of 3, and translated left 2 units and down 1 unit.

To verify using a graphing calculator, enter each equation in the $\boxed{Y=}$ list and graph. Use the features of the calculator to investigate the transformations. Note the change in asymptotes, y-intercepts, and relation to the axes.

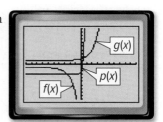

$[-10, 10]$ scl: 1 by $[-10, 10]$ scl: 1

> **Watch Out!**
>
> **Negative Values** Remember that the general form of an exponential function is $g(x) = a(b)^{x-h} + k$, so an exponent of $x + h$ indicates a negative value of h, which is a translation left.

▶ **Guided Practice**

6. Compare the graphs to the parent function $f(x) = 3^x$. Then, verify using technology.

$g(x) = -4(3)^x$, $p(x) = 0.2(3)^{x+3} - 5$, $q(x) = (3)^{-x} + 2$

Check Your Understanding   ◯ = Step-by-Step Solutions begin on page R11.

Go Online! for a
Self-Check Quiz

Example 1   Describe the translation as it relates to the graph of $f(x) = 3^x$.

1. $g(x) = 3^x - 1$

2. $g(x) = 3^{x-5}$

3. $g(x) = 3^{x+1}$

4. $g(x) = 3^{x+2} - 4$

Example 2   The graph $g(x)$ is a translation of the parent function $f(x) = 2^x$. Use the graph of the function to write an equation for $g(x)$.

5.

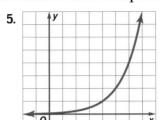

6.

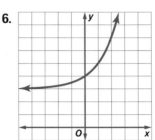

Example 3   **7. FINANCIAL LITERACY** Antonio invests $2200 in a savings account that earns 1.5% interest per year compounded annually. The amount of money in his bank account after t years can be modeled by $g(x) = 2200(1.015)^t$. Describe the dilation in $g(x) = 2200(1.015)^t$ as it relates to the parent function $f(x) = 1.015^t$.

Example 4   Describe the transformations in each function as it relates to the graph of $f(x) = \left(\frac{1}{4}\right)^x$.

8. $g(x) = -\left(\frac{1}{4}\right)^x$

9. $g(x) = \left(\frac{1}{4}\right)^{-3x}$

10. $g(x) = -\frac{1}{8}\left(\frac{1}{4}\right)^x$

11. $g(x) = -9\left(\frac{1}{4}\right)^x$

Examples 5, 6   Describe the transformations in each function as it relates to its parent function.

12. $g(x) = -2.7(3)^{x-8}$

13. $g(x) = \frac{2}{3}(5)^{-x} + 1$

14. $g(x) = -\left(\frac{4}{5}\right)^{x-2} - 6$

15 $g(x) = \left(\frac{5}{4}\right)^{4x} - 2$

Practice and Problem Solving

Extra Practice is found on page R7.

Examples 1, 4   Describe the translation in each function as it relates to the parent function.

16. $g(x) = 4^x + 6$

17. $g(x) = \left(\frac{9}{5}\right)^{x-2}$

18. $g(x) = \left(\frac{2}{5}\right)^x - 1$

19. $g(x) = 1.4^{0.3x}$

20. $g(x) = 1.5^{4x}$

21. $g(x) = 5(3)^x$

22. $g(x) = 0.9^{3x}$

23. $g(x) = \frac{1}{3}\left(\frac{1}{5}\right)^x$

24. $g(x) = 3^{-x}$

25. $g(x) = -7^x$

Examples 5, 6   Describe the transformations in each function as it relates to the parent function.

26. $g(x) = 11^{x+2} + 8$

27. $g(x) = 0.5(2)^{x-1}$

28. $g(x) = -3\left(\frac{1}{6}\right)^x + 2$

29. $g(x) = \left(\frac{2}{7}\right)^{-6x}$

30. $g(x) = -5^{x-8} + 4$

31. $g(x) = -\frac{1}{3}(2)^{x+3}$

32. $g(x) = 2(6.5)^{-x}$

33. $g(x) = 3^{5x} - 1$

34. $g(x) = -8\left(\frac{2}{3}\right)^{x+3} - 4$

35. $g(x) = -0.7(1.26)^{x+4} - 5$

Example 2   The graph of $g(x)$ is a transformation of the parent function $f(x) = \left(\frac{1}{2}\right)^x$. Find the value of n to write an equation for $g(x)$.

36. $g(x) = f(x) + n$

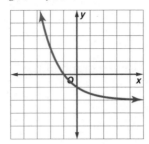

37 $g(x) = f(x + n)$

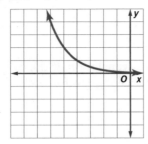

38. $g(x) = f(x) + n$

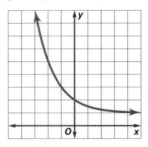

39. $g(x) = n \cdot f(x)$

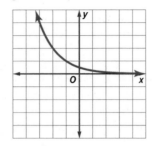

40. $g(x) = f(nx)$

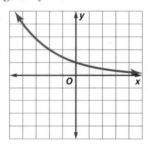

41. $g(x) = n \cdot f(x)$

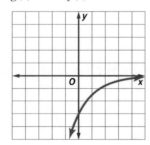

42. $g(x) = n \cdot f(x)$

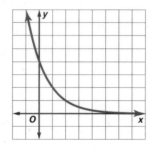

43. $g(x) = f(x) + n$

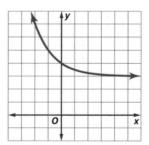

Examples 3, 6 **44. TRAINING** Tara can bench press 130 pounds. Each week of training, she wants to increase the weight by 3%. The function models the weight Tara can lift $w(x) = 130(1.03)^x$ where x represents the number of weeks of training.

    a. Describe the transformation in $w(x)$ as it relates to the parent function $f(x) = 1.03x$.

    b. Robert is also increasing the weight he lifts by 3% each week. The function $v(x) = 95(1.03)^x$ represents the weight he can lift after x weeks. What is his starting weight?

    c. Compare $w(x)$ and $v(x)$.

Use transformations to graph each function.

45. $y = 2^x - 5$

46. $y = -3^{x+1}$

47. $y = \frac{1}{4}^{x+2} - 2$

48. $y = 4(2)^{-x}$

49. $y = -0.5(4)^{x-3} + 1$

50. $y = 2^{0.5x} + 3$

51. $y = \left(\frac{2}{3}\right)^{4x}$

52 $y = -2(3)^x + 5$

Given the parent function $f(x) = \left(\frac{3}{2}\right)^x$ match each function to its graph.

$g(x) = -f(x) + 2$ $j(x) = -2f(x + 2)$ $m(x) = f(x) - 2$

$n(x) = 5f(x)$ $p(x) = f(5x)$ $q(x) = \frac{1}{5}f(x)$

53.

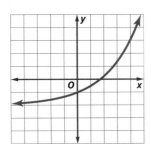

54.

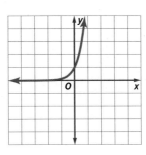

55.

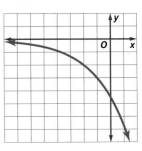

56.

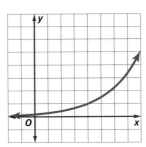

57.

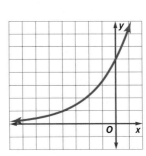

58.

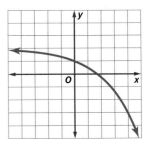

59. **MP REGULARITY** List the functions in order from the most vertically stretched to the least vertically stretched: $g(x) = 0.4(2)^x$, $j(x) = 4(2)^x$, $p(x) = \frac{1}{4}(2)^x$

60. **MP TOOLS** Use a graphing calculator to compare the functions.

$y = 3^x$ $y = 5(3)^x$ $y = \frac{1}{5}(3)^x$

  a. Compare the graphs of $y = 3^x$ and $y = 5(3)^x$. Describe similarities and differences of the key features of the graphs.

  b. Compare the graphs of $y = 3^x$ and $y = \frac{1}{5}(3)^x$. Describe similarities and differences of the key features of the graphs.

61 **MP TOOLS** Use a graphing calculator to compare the functions.

$y = \left(\frac{2}{5}\right)^x$ $y = -\left(\frac{2}{5}\right)^x$ $y = \left(\frac{2}{5}\right)^{-x}$

  a. Compare the graphs of $y = \left(\frac{2}{5}\right)^x$ and $y = -\left(\frac{2}{5}\right)^x$. Describe any similarities and differences you notice about the key features of the graphs.

  b. Compare the graphs of $y = \left(\frac{2}{5}\right)^x$ and $y = \left(\frac{2}{5}\right)^{-x}$. Describe any similarities and differences you notice about the key features of the graphs.

62. **FINANCIAL LITERACY** The balance of Laura's savings over time can be modeled by $g(x) = P(1.002)^t$, where P is the initial amount in her bank account and t is time in years.

  a. Describe the graph of $g(x)$ when she invests $1000 and when she invests $7500 in her bank account.

  b. Suppose Laura keeps $250 cash in a safe in her apartment, which she also considers part of her savings. How would this affect the graph of $g(x)$?

63. **MULTI-STEP** Consider transformations of the function $f(x) = b^x$.

  a. What happens if the function is multiplied by a constant between 0 and 1? Describe the new graph compared to the parent function.

  b. Describe the graph as the constant approaches 0.

  c. What happens is the function is multiplied by a constant greater than 1? Describe the new graph compared to the parent function.

  d. Describe the graph as the constant approaches ∞. .

H.O.T. Problems   Use Higher-Order Thinking Skills

64. **MP** **REASONING** What point is on the graph of every functon of the form $f(x) = b^x$, where $b \neq 0$? Explain your answer.

65. **MP** **REASONING** Explain how a reflection across the x-axis affects each points on the graph of an exponential function.

66. **OPEN ENDED** Using $f(x) = 5^x$ as the parent function, write functions for the following transformations:

  a. reflected across the y-axis and vertically compressed

  b. horizontally stretched and translated left and up

  c. reflected across x-axis, vertically stretched, and translated right

  d. horizontally compressed and translated left and down

67. **MP** **SENSE-MAKING** Examine the following cases and describe the effect a reflection across the x-axis would have on the end behavior of the parent function $f(x) = ab^x$.

  Case 1: $g(x) = -ab^x$ where $b > 1$

  Case 2: $g(x) = -ab^x$ where $0 < b < 1$

68. **MP** **SENSE-MAKING** Examine the following cases and describe the effect a reflection across the y-axis would have on the end behavior of the parent function $f(x) = ab^x$.

  Case 1: $g(x) = ab^{-x}$ where $b > 1$

  Case 2: $g(x) = ab^{-x}$ where $0 < b < 1$

69. **OPEN ENDED** Write an exponential decay function that is reflected across the x-axis and has an asymptote of $y = -2$.

70. **MP** **REASONING** Are the following statements *always*, *sometimes*, or *never* true? Explain.

  a. The graph of $g(x) = b^x + k$ has an asymptote of $y = k$.

  b. The graph of $j(x) = b^{x-h}$ passes through $(h, 1)$.

  c. The graph of $p(x) = -b^x$ decreases as $x \to \infty$.

71. The graph of $g(x) = 5^{x-2} + 8$ is a translation of the graph $f(x) = 5^x$. Which of the following describes the translation? **MP** 7

- ◯ **A** translated left 2 units and down 8 units
- ◯ **B** translated left 2 units and up 8 units
- ◯ **C** translated right 2 units and up 8 units
- ◯ **D** translated right 8 units and down 2 units

72. Which function is a translation of the parent function $f(x) = 3^x$ right 4 units? **MP** 7

- ◯ **A** $g(x) = 3^x + 4$
- ◯ **B** $g(x) = 3^x - 4$
- ◯ **C** $g(x) = 3^{x+4}$
- ◯ **D** $g(x) = 3^{x-4}$

73. Which of the following transformations represents a vertical compression of the parent function $f(x) = 2^x$? **MP** 7

- ☐ **A** $g(x) = -5(2)^x$
- ☐ **B** $g(x) = -0.4(2)^x$
- ☐ **C** $g(x) = \frac{1}{6}(2)^x$
- ☐ **D** $g(x) = 2^{-7x}$
- ☐ **E** $g(x) = 2^{0.1x}$
- ☐ **F** $g(x) = 2^{6x}$

74. The graph of $f(x) = \frac{1}{2}^x$ is reflected across the y-axis and translated up 5 units. Write the transformed function $g(x)$. **MP** 2, 7

75. The graph of $f(x) = \frac{1}{2}^x$ is reflected across the x-axis, stretched vertically by a factor of 7, and translated left 1 unit. Write the transformed function $g(x)$. **MP** 2, 7

76. The graph of $g(x)$ is a translation of the parent function $f(x) = 1.5^x$. Which is the equation for the graph? **MP** 1, 2

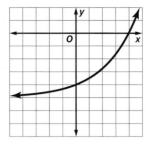

- **A** $g(x) = 1.5^x + 5$
- **B** $g(x) = 1.5^x - 5$
- **C** $g(x) = 1.5^{x+5}$
- **D** $g(x) = 1.5^{x-5}$

77. **MULTI-STEP** Consider the function $g(x) = -2(3)^{x+1} + 4$ as it relates to the parent function $f(x) = 3^x$. **MP** 1, 7

- **a.** Identify the value of a and its affect on the parent function.
- **b.** Identify the value of h and its affect on the parent function.
- **c.** Identify the value of k and its affect on the parent function.
- **d.** Graph $g(x)$.

78. If $f(x) = b^x$, which function represents $g(x)$? **MP** 2, 7

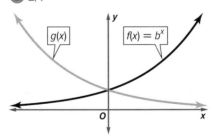

- ◯ **A** $g(x) = -b^x$
- ◯ **B** $g(x) = b^x - 1$
- ◯ **C** $g(x) = b^{-x}$
- ◯ **D** $g(x) = -b^{-x}$

:: **Then**

● You analyzed exponential functions.

:: **Now**

1 Write exponential functions by using a graph, a description, or two points.

2 Create equations and solve problems involving exponential growth and decay.

:: **Why?**

● The first podcast premiered in 2003 and the number of podcasts continues to grow. Since 2010, the number of active podcasts has been increasing by about 17.3% each year. The growth of podcasts can be modeled by $y = 26.5(1 + 0.173)^t$ or $y = 26.5(1.173)^t$, where y represents the number of podcasts in thousands and t is the number of years since 2010.

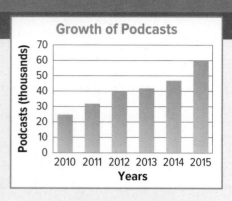

Growth of Podcasts

 New Vocabulary

compound interest

MP **Mathematical Practices**

1 Make sense of problems and persevere in solving them.

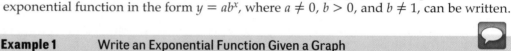

1 **Write Exponential Functions** Given two points, a graph, or description, an exponential function in the form $y = ab^x$, where $a \neq 0$, $b > 0$, and $b \neq 1$, can be written.

Example 1   Write an Exponential Function Given a Graph

Write an exponential function for the graph.

Step 1 Select two points on the graph.

Any two points on the graph can be used to write a system of two equations using $y = ab^x$. Selecting the y-intercept will make writing the equation of the function easier because $y = ab^0 = a$.

Use $(0, 0.5)$ and $(4, 8)$ to write the function.

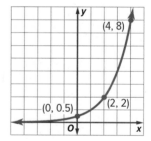

Step 2 Solve for a.

$y = ab^x$	Exponential equation
$0.5 = ab^0$	$(x, y) = (0, 0.5)$
$0.5 = a \cdot 1$	$b^0 = 1$
$0.5 = a$	Simplify.

Step 3 Solve for b.

Use the value of a from Step 2 and the second point to solve for b.

$y = ab^x$	Exponential equation
$8 = (0.5)b^4$	$(x, y) = (4, 8)$, $a = 0.5$
$16 = b^4$	Divide each side by 0.5.
$2 = b$	Simplify.

Step 4 Write the function.

$y = ab^x$	Exponential equation
$y = 0.5(2)^x$	$a = 0.5$, $b = 2$

The graph is represented by the function $y = 0.5(2)^x$.

Guided Practice

1. Write an exponential function for the graph.

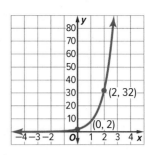

(2, 32)

(0, 2)

Example 2   Write an Exponential Function Given Two Points

Write an exponential function for a graph that passes through (1, 6) and (3, 24).

Substitute the x- and y-coordinates into $y = ab^x$ to get a system of two equations.

Equation 1: $6 = ab^1$	$(x_1, y_1) = (1, 6)$
Equation 2: $24 = ab^3$	$(x_2, y_2) = (3, 24)$

Solve the system for a and b using substitution.

Step 1 Solve for a variable.

$$6 = ab^1 \qquad \text{Equation 1}$$
$$\frac{6}{b} = a \qquad \text{Divide each side by } b.$$

Step 2 Substitute.

Substitute $\frac{6}{b}$ for a in Equation 2 and solve for b.

$$24 = ab^3 \qquad \text{Equation 2}$$
$$24 = \left(\frac{6}{b}\right)b^3 \qquad \text{Substitute } \frac{6}{b} \text{ for } a.$$
$$24 = \frac{6b^3}{b} \qquad \text{Multiply.}$$
$$24 = 6b^2 \qquad \text{Quotient of Powers}$$
$$4 = b^2 \qquad \text{Divide each side by 6.}$$
$$2 = b \qquad \text{Simplify.}$$

Step 3 Solve for a.

Substitute 2 for b in either equation to find a.

$$6 = ab^1 \qquad \text{Equation 1}$$
$$6 = a(2)^1 \qquad b = 2$$
$$3 = a \qquad \text{Divide each side by 2.}$$

$y = 3 \cdot 2^x$ is an exponential function that passes through (1, 6) and (3, 24).

Study Tip

MP **Sense-Making** You can check the equation by making sure that the two given points are solutions of the equation.

Guided Practice

2. Write an exponential function for a graph that passes through (1, 8) and (4, 512).

---

2 **Exponential Growth and Decay** The equation for the number of podcasts over time at the beginning of this lesson is in the form $y = a(1 + r)^t$. This is the general equation for exponential growth.

Key Concept General Equation for Exponential Growth

a is the initial amount.   *t* is time.

$$y = a(1 + r)^t$$

y is the final amount.

r is the rate of change expressed as a decimal, $r > 0$.

Example 3 Interpret the Parameters of an Equation

SOCIAL NETWORKS The number of users, in millions, that belong to a social networking site can be represented by $y = 1.5(1.40)^t$, where *t* is the number of years since 2013. Interpret the parameters of the equation. Make a prediction about when the number of users will exceed 40 million.

The initial number of users *a* is 1.5 million.

The rate of change *r* is $1.40 - 1$, or 0.40. Therefore, the number of users is growing at a rate of 40% per year.

If this growth rate continues, then the number of users that belong to the networking site will rapidly increase. By 2023, more than 40 million users are expected because $1.5(1.40)^{10} \approx 43.4$.

Guided Practice

3. **SMARTPHONE APPS** The number of apps, in thousands, that are on the market for a smartphone can be represented by $y = 320(1.22)^t$, where *t* is the number of years since 2012. Interpret the parameters of the equation. Make a prediction about when the future number of apps on the market will exceed 1 million.

Real-World Example 4 Exponential Growth

CONTEST The prize for a radio station contest begins with a $100 cash prize. Once a day, a name is announced. The person has 15 minutes to call or the prize increases by 2.5% for the next day.

a. Write an equation to represent the amount of the cash prize in dollars after *t* days with no winners.

$y = a(1 + r)^t$ Equation for exponential growth

$y = 100(1 + 0.025)^t$ $a = 100$ and $r = 2.5\%$ or 0.025

$y = 100(1.025)^t$ Simplify.

In the equation $y = 100(1.025)^t$, *y* is the amount of the prize and *t* is the number of days since the contest began.

b. How much will the prize be worth if no one wins after 10 days?

$y = 100(1.025)^t$ Equation for amount of gift card

$= 100(1.025)^{10}$ $t = 10$

≈ 128.01 Use a calculator.

In 10 days, the cash prize will be worth $128.01.

Guided Practice

4. **TUITION** A college's tuition has risen 5% each year since 2010. If the tuition in 2010 was $10,850, write an equation for the amount of the tuition *t* years after 2010. Predict the cost of tuition for this college in 2025.

> **Watch Out!**
>
> **Growth Factor** Be careful when you convert the percent to a decimal. The base of the exponential equation is 1.025, not 1.25.

Key Concept · Equation for Exponential Decay

a is the initial amount.   **t** is time.

$$y = a(1 - r)^t$$

y is the final amount.

r is the rate of decay expressed as a decimal, $0 < r < 1$.

Real-World Example 5 · Exponential Decay

SWIMMING A fully inflated child's raft for a pool is losing 6.6% of its air every day. The raft originally contained 4500 cubic inches of air.

a. Write an equation to represent the loss of air.

$$y = a(1 - r)^t$$ Equation for exponential decay

$$= 4500(1 - 0.066)^t$$ $a = 4500$ and $r = 6.6\%$ or 0.066

$$= 4500(0.934)^t$$ Simplify.

b. Estimate the amount of air in the raft after 7 days.

$$y = 4500(0.934)^t$$ Equation for air loss

$$= 4500(0.934)^7$$ $t = 7$

$$\approx 2790$$ Use a calculator.

The amount of air in the raft after 7 days will be about 2790 cubic inches.

Study Tip

MP **Reasoning** Since r is added to 1, the value inside the parentheses will be greater than 1 for exponential growth functions. For exponential decay functions, this value will be less than 1 since r is subtracted from 1.

▸ **Guided Practice**

5. POPULATION The population of Campbell County, Kentucky, has been decreasing at an average rate of about 0.3% per year. In 2010, its population was 88,647. Write an equation to represent the population since 2010. If the trend continues, predict the population in 2020.

Compound interest is interest earned or paid on both the initial investment and previously earned interest. It is an application of exponential growth.

Key Concept · Equation for Compound Interest

A is the current amount.

n is the number of times the interest is compounded each year, and **t** is time in years.

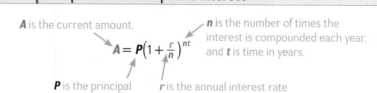

$$A = P\left(1 + \frac{r}{n}\right)^{nt}$$

P is the principal or initial amount.

r is the annual interest rate expressed as a decimal, $r > 0$.

Real-World Career

Financial Advisor Financial advisors help people plan their financial futures. A good financial advisor has mathematical, problem-solving, and communication skills. A bachelor's degree is strongly preferred but not required.

Real-World Example 6 · Compound Interest

FINANCE Maria's parents invested $14,000 at 6% per year compounded monthly. How much money will there be in the account after 10 years?

$$A = P\left(1 + \frac{r}{n}\right)^{nt}$$ Compound interest equation

$$= 14,000\left(1 + \frac{0.06}{12}\right)^{12(10)}$$ $P = 14,000$, $r = 6\%$ or 0.06, $n = 12$, and $t = 10$

$$= 14,000(1.005)^{120}$$ Simplify.

$$\approx 25,471.55$$ Use a calculator.

There will be 25,471.55 in 10 years.

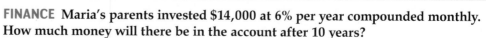

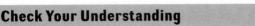

Guided Practice

6. **FINANCE** Determine the amount of an investment if $300 is invested at an interest rate of 3.5% compounded monthly for 22 years.

Check Your Understanding ◯ = Step-by-Step Solutions begin on page R11.

Example 1   Use the information in each graph to construct an exponential function to model the graph.

1.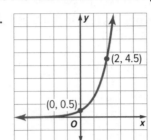

2.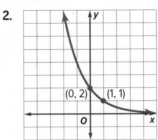

Example 2   Write an exponential function for a graph that passes through each pair of points.

3. (1, 12) and (3, 108)

4. (1, 1.5) and (3, 54)

Example 3   5. **ADVERTISING** The number of people that have "liked" Mindy's Candy Store website can be represented by $y = 270(1.65)^t$, where t is the number of weeks after a review in a national magazine. Interpret the parameters of the equation. Make a prediction about the future number of likes on the website.

Example 4   6. **SALARY** Ms. Acosta received a job as a teacher with a starting salary of $34,000. According to her contract, she will receive a 1.5% increase in her salary every year. How much will Ms. Acosta earn in 7 years?

Example 5   7. **ENROLLMENT** In 2015, 2200 students attended Polaris High School. The enrollment has been declining 2% annually.

   a. Write an equation for the enrollment of Polaris High School t years after 2015.

   b. If this trend continues, how many students will be enrolled in 2030?

Example 6   8. **MONEY** Paul invested $400 into an account with a 5.5% interest rate compounded monthly. How much will Paul's investment be worth in 8 years?

Practice and Problem Solving   Extra Practice is on page R7.

Example 1   Use the information in each graph to construct an exponential function to model the graph.

9.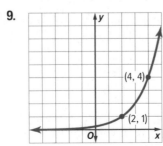

10.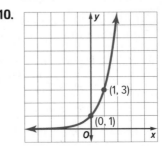

11.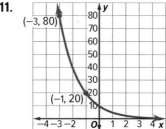

12.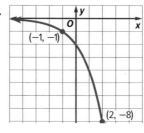

13.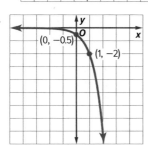

14.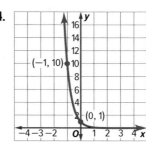

Examples 2 **Write an exponential function for a graph that passes through each pair of points.**

15. (1, 10) and (5, 160)

16. (1, 1) and (4, 64)

Example 3 **17. ANTS** A colony of ants are looting a food source. The amount of food at the source, in grams, can be represented by $y = 82(0.65)^t$, where t is the number of minutes after the looting began. Interpret the parameters of the equation. Make a prediction about the future amount of food left at the source.

18. MUSIC DOWNLOADS The total number of songs, in millions, that have been downloaded from a music sharing site can be represented by $y = 6.7(1.08)^t$, where t is the number of months after the site began. Interpret the parameters of the equation. Make a prediction about the future number of songs downloaded from the site.

Example 4 **19 COINS** Camilo purchased a rare coin from a dealer for $300. The value of the coin increases 5% each year. Determine the value of the coin in 5 years.

20. GAME SHOWS The jackpot on a game show starts with a value of $1000. For every question that a contestant answers correctly, the jackpot's value increases by 25%. Determine the value of the jackpot if the contestant correctly answers 12 questions in a row.

Example 5 **21. INVESTMENTS** Jin's investment of $4500 has been losing its value at a rate of 2.5% each year. What will his investment be worth in 5 years?

22. MUSEUMS The director of a science museum begins a membership drive with the goal of signing up 850 new members. The number of new members that remain to be signed up decreases by 11.6% every month. Determine the number of new members that still need to be signed up at the end of one year.

Example 6 **23. MP PRECISION** Brooke is saving money for a trip to the Bahamas that costs $295.99. She puts $150 into a savings account that pays 7.25% interest compounded quarterly. Will she have enough money in the account after 4 years? Explain.

24. FINANCE Determine the value of an investment of $650 that is invested at an annual interest rate of 1.5% compounded monthly for 16 years.

25. CARS Leonardo purchases a car for $18,995. The car depreciates at a rate of 18% annually. After 6 years, Manuel offers to buy the car for $4500. Should Leonardo sell the car? Explain.

26. POPULATION In the years from 2010 to 2015, the population of the District of Columbia decreased an average of 0.9% annually. In 2010, the population was about 530,000. What is the population of the District of Columbia expected to be in 2030?

27 HOUSING The median house price in the United States decreased an average of 4.25% each year starting in 2010 and continuing through 2012. Assume that this pattern continues.

   a. Write an equation for the median house price for t years after 2010.

   b. Predict the median house price in 2030.

28. ELEMENTS A radioactive element's half-life is the time it takes for one half of the element's quantity to decay. The half-life of Plutonium-241 is 14.4 years. The number of grams A of Plutonium-241 left after t years can be modeled by $A = p(0.5)^{\frac{t}{14.4}}$, where the variable p is the original amount of the element.

   a. How much of a 0.2-gram sample remains after 72 years?

   b. How much of a 5.4-gram sample remains after 1095 days?

Median House Price	
2010	$194,375
2011	$180,046
2012	$178,005

Source: Real Estate Journal

29. COMBINING FUNCTIONS A swimming pool holds a maximum volume of 20,500 gallons of water. It evaporates at a rate of 0.5% per hour. The pool currently contains a volume of 19,000 gallons of water.

   a. Write an exponential function $w(t)$ to express the amount of water remaining in the pool after time t. The variable t represents the number of hours after the pool has reached the volume of 19,000 gallons.

   b. At this same time, a hose is turned on to refill the pool at a net rate of 300 gallons per hour. Write a function $p(t)$ where t is time in hours the hose is running to express the amount of water that is pumped into the pool.

   c. Find $C(t) = p(t) + w(t)$. What does this new function represent?

   d. Use the graph of $C(t)$ to determine how long the hose must run to fill the pool to its maximun capacity.

H.O.T. Problems   Use **H**igher-**O**rder **T**hinking Skills

30. MP REASONING Determine the growth rate (as a percent) of a population that quadruples every year. Explain.

31. MP PRECISION Santos invested $1200 into an account with an interest rate of 8% compounded monthly. Use a calculator to approximate how long it will take for Santos's investment to reach $2500.

32. MP REASONING The amount of water in a container doubles every minute. After 8 minutes, the container is full. After how many minutes was the container half full? Explain.

33. WRITING IN MATH What should you consider when using exponential models to make decisions?

34. WRITING IN MATH Compare and contrast the exponential growth formula and the exponential decay formula.

35. A small town currently has a population of 15,000. Each year the population decreases by 2.25%. What will the population be in 8 years? **MP** 2, 4

- **A** 12,503
- **B** 14,343
- **C** 14,982
- **D** 17,922

36. Robert invested $25,000 in an account that pays 4.5% interest compounded weekly. After 3 years, what is the balance of his account? **MP** 2, 4

- **A** $3611.75
- **B** $28,611.75
- **C** $70,877.18
- **D** $95,877.18

37. The value m of a baseball card is given by the equation $m = 240(1.25)^t$, where t is the number of years after 2012. What is the value of the card m in dollars in the year 2012? **MP** 2, 4

38. Which exponential function passes through (0, 2) and (2, 12.5)? **MP** 1, 2

- **A** $y = 2(2.5)^x$
- **B** $y = 2(12.5)^x$
- **C** $y = 2(3.125)^x$
- **D** $y = 2.5(2)^x$

39. The population of a small town, in thousands, can be represented by $y = 2.6(1.02)^t$, where t is the number of years since 2010. Based on this model, which statements are true? **MP** 1, 4

- **A** The population in 2010 was 2600.
- **B** The population of the town is increasing.
- **C** The population increases by 20% each year.
- **D** The population in 2020 will be greater than 3100.
- **E** The population in 2015 was about 3500.
- **F** From 2010 to 2011, the population increased by 102%.

40. Gina just purchased an antique desk worth $8000. The equation $v = 8000(1.03)^t$ models the value of the desk after t years. Which of the following best describes the meaning of 1.03 in the equation? **MP** 2, 4

- **A** The value of the desk will decrease by 1.03% each year.
- **B** The value of the desk will increase by 1.03% each year.
- **C** The value of the desk will increase by 3% each year.
- **D** The value of the desk will decrease by 3% each year.

41. Which exponential function represents the graph shown? **MP** 1, 2

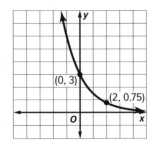

- **A** $y = 3(0.5)^x$
- **B** $y = 3(1.5)^x$
- **C** $y = 0.5(3)^x$
- **D** $y = 1.5(3)^x$

42. **MULTI-STEP** Miguel filled a ball with 1.6 pounds of air. Every 2 months, the ball loses half its air. **MP** 4, 8

a. Which equation represents the pounds of air p that remain in the ball after 9 months?

- **A** $p = 0.8(0.5)^9$
- **B** $p = 1.6(0.5)^{\frac{9}{2}}$
- **C** $p = 1.6(0.5)^9$
- **D** $p = 1.6(1.5)^{\frac{9}{2}}$

b. Explain how you determined the exponent in the equation.

c. Explain how you determined the base of the exponential expression in the equation.

Graphing Technology Lab
Solving Exponential Equations and Inequalities

You can use TI-Nspire Technology to solve exponential equations and inequalities by graphing and by using tables.

Mathematical Practices

 5 Use appropriate tools strategically.

Activity 1   Graph an Exponential Equation

Graph $y = 3^x + 4$ using a graphing calculator.

Step 1 Add a new **Graphs** page.

Step 2 Enter $3^x + 4$ as **f1(x)**.

Step 3 Use the **Window Settings** option from the **Window/Zoom** menu to adjust the window so that x is from -10 to 10 and y is from -100 to 100. Keep the scales as **Auto**.

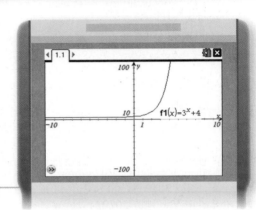

To solve an equation by graphing, graph both sides of the equation and locate the point(s) of intersection.

Activity 2   Solve an Exponential Equation by Graphing

Solve $2^{x-2} = \frac{3}{4}$.

Step 1 Add a new **Graphs** page.

Step 2 Enter 2^{x-2} as **f1(x)** and $\frac{3}{4}$ as **f2(x)**.

Step 3 Use the **Intersection Point(s)** tool from the **Points & Lines** menu to find the intersection of the two graphs. Select the graph of **f1(x) enter** and then the graph of **f2(x) enter**.

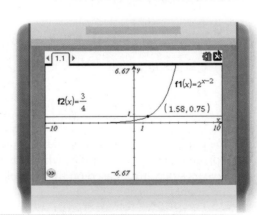

The graphs intersect at about $(1.58, 0.75)$.

Therefore, the solution of $2^{x-2} = \frac{3}{4}$ is 1.58.

Exercises

MP TOOLS Use a graphing calculator to solve each equation.

1. $\left(\dfrac{1}{3}\right)^{x-1} = \dfrac{3}{4}$

2. $2^{2x-1} = 2x$

3. $\left(\dfrac{1}{2}\right)^{2x} = 2^{2x}$

4. $5^{\frac{1}{3}x+2} = -x$

5. $\left(\dfrac{1}{8}\right)^{2x} = -2x + 1$

6. $2^{\frac{1}{4}x-1} = 3^{x+1}$

7. $2^{3x-1} = 4^x$

8. $4^{2x-3} = 5^{-x+1}$

9. $3^{2x-4} = 2^x + 1$

Activity 3 Solve an Exponential Equation by Using a Table

Solve $2\left(\dfrac{1}{2}\right)^{x+2} = \dfrac{1}{4}$ **using a table.**

Step 1 Add a new **Lists & Spreadsheet** page.

Step 2 Label column A as x. Enter values from -4 to 4 in cells A1 to A9.

Step 3 In column B in the formula row, enter the left side of the rational equation. In column C of the formula row, enter $= \dfrac{1}{4}$. Specify **Variable Reference** when prompted.

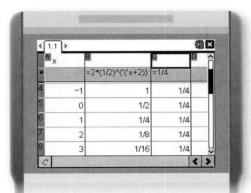

Scroll until you see where the values in Columns B and C are equal. This occurs at $x = 1$. Therefore, the solution of $2\left(\dfrac{1}{2}\right)^{x+2} = \dfrac{1}{4}$ is 1.

You can also use a graphing calculator to solve exponential inequalities.

Activity 4 Solve an Exponential Inequality

Solve $4^{x-3} \le \left(\dfrac{1}{4}\right)^{2x}$.

Step 1 Add a new **Graphs page**.

Step 2 Enter the left side of the inequality into **f1(x)**. Press **del**, select ≥, and enter 4^{x-3}. Enter the right side of the inequality into **f2(x)**. Press **tab del** ≤, and enter $\left(\dfrac{1}{4}\right)^{2x}$.

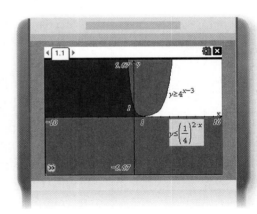

The x-values of the points in the region where the shading overlap is the solution set of the original inequality.

Therefore, the solution of $4^{x-3} \le \left(\dfrac{1}{4}\right)^{2x}$ is $x \le 1$.

Exercises

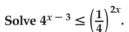 **TOOLS Use a graphing calculator to solve each equation or inequality.**

10. $\left(\dfrac{1}{3}\right)^{3x} = 3^x$

11. $\left(\dfrac{1}{6}\right)^{2x} = 4^x$

12. $3^{1-x} \le 4^x$

13. $4^{3x} \le 2x + 1$

14. $\left(\dfrac{1}{4}\right)^x > 2^{x+4}$

15. $\left(\dfrac{1}{3}\right)^{x-1} \ge 2^x$

Transforming Exponential Expressions

- You wrote and graphed exponential functions.

1 Transform and interpret expressions of exponential functions by applying the properties of exponents.

- When you invest money at the bank, it earns compound interest. To compare two different investment plans, you need to be able to compare interest rates and compounding frequencies to make the best investment with your money.

 Mathematical Practices

4 Model with mathematics.

7 Look for and make use of structure.

1 Transform and Interpret Exponential Expressions You can use the properties of exponents to transform exponential functions into other forms in order to solve real-world problems. Compound interest plans that have different numbers of compoundings per year can be compared by writing their exponential growth functions with the same exponent. For instance, to compare a plan compounded annually with another compounded monthly, write the function for the annual plan in terms of the exponent for the monthly plan. The *effective* monthly interest rate of the annual plan can then be determined from the resulting function.

Example 1   Compare Rates

Monique is trying to decide between two savings account plans. Plan A offers a monthly compounded interest rate of 0.25%, while Plan B offers 2.5% interest compounded annually. Which is the better plan? Explain.

Write a function to represent the amount $A(t)$ that Monique would earn after t years with Plan B. For convenience, let the initial amount of Monique's investment be $1.

$$y = a(1 + r)^t \qquad \text{Equation for exponential growth}$$
$$A(t) = 1(1 + 0.025)^t \qquad y = A(t), a = 1, r = 2.5\% \text{ or } 0.025$$
$$= 1.025^t \qquad \text{Simplify.}$$

Now write a function equivalent to $A(t)$ that represents 12 compoundings per year.

Use the properties of exponents to write an equivalent expression with a power of $12t$, which represents 12 compoundings per year, instead of a power of $1t$, which represents one compounding per year.

$$A(t) = 1.025^{1t} \qquad \text{Original function}$$
$$= 1.025^{\left(\frac{1}{12} \cdot 12\right)t} \qquad 1 = \frac{1}{12} \cdot 12$$
$$= \left(1.025^{\frac{1}{12}}\right)^{12t} \qquad \text{Power of a Power Property}$$
$$\approx 1.0021^{12t} \qquad (1.025)^{\frac{1}{12}} = \sqrt[12]{1.025} \text{ or about } 1.0021$$

From this equivalent function, you can determine that the effective monthly interest rate for Plan B is about 0.0021 or approximately 0.21% per month. This rate is less than the monthly interest rate of 0.25% offered by Plan A.

Plan A is the better plan.

▶ **Guided Practice**

1. Let Me Grow Your Money offers two investment programs. The first plan has a 5% rate of interest compounded annually while the second plan has a quarterly compounded rate of 1.3%. Write a function to represent the amount earned by the first plan after t years. Then, write an equivalent function that represents the first plan with 4 compoundings per year. What is the *effective* quarterly rate of interest of the first plan? Which is the better plan?

You can use the properties of exponents and the compound interest equation to compare rates of exponential growth in a variety of real-world situations.

Real-World Example 2   Compare Rates by Using Compound Interest Equation

BACTERIA *Lactobacillus acidophilus* **is a bacteria used in the production of yogurt and some cheeses. It has a growth rate of 0.92% per minute. The bacteria** *bacillus megaterium* **is synthesized to create penicillin and has a growth rate of 0.046% per second. Determine the effective rate of growth per minute for the bacillus megaterium. Which bacteria grows at a faster rate? Explain.**

Study Tip

MP **Sense-Making** In one equation, t is a variable representing minutes and in the other equation, t is a variable representing seconds. Therefore, t is not equivalent in the two equations and must be converted to represent the same unit.

The compound interest equation can be used to express a growth rate in a greater unit of time. Use the compound interest equation to write a function to represent the amount $A(t)$ of *bacillus megaterium* after t minutes. Let the initial amount of bacteria equal x and the number of compoundings equal 60.

$$A(t) = P\left(1 + \frac{r}{n}\right)^{nt} \qquad \text{Compound interest equation}$$
$$A(t) = x(1 + 0.00046)^{60t} \qquad P = x, \frac{r}{n} = 0.046\% \text{ or } 0.00046, n = 60$$
$$= x(1.00046)^{60t} \qquad \text{Simplify.}$$

The amount of *bacillus megaterium* after t minutes is given by $A(t) = 1.00046^{60t}$.

To determine the per minute rate, write an equivalent equation representing only one compounding per minute by using the Product of Powers Property to change the exponent from $60t$ to $1t$.

$$A(t) = x(1.00046)^{60t} \qquad \text{Original function}$$
$$= x(1.00046^{60})^{t} \qquad \text{Product of Powers Property}$$
$$\approx x(1.028)^{t} \qquad 1.00046^{60} \approx 1.028$$

The equivalent equation representing only one compounding per minute is $A(t) = x(1.028)^{t}$. The effective rate of growth per minute for the *bacillus megaterium* is 0.028 or about 2.8%. *Bacillus megaterium* grows faster because its effective growth rate is 2.8% per minute and the growth rate of *lacatobacillus acidophilus* is 0.92% per minute.

▶ **Guided Practice**

2. SOCIAL MEDIA A business analyst is comparing the growth rates of two social media apps. The number of subscribers to App A is growing by 1.5% per month. The number of subscribers to App B is growing by 16% per year. Write a function to represent the number of subscribers to App A after t years. Then, write an equivalent function that represents the subscribers to App A with only 1 compounding per year. What is the *effective* yearly rate of growth of the subscribers to App A? Which app is growing at a faster rate?

Go Online! for a Self-Check Quiz

Check Your Understanding   ◯ = Step-by-Step Solutions begin on page R11.

Example 1

1 Tareq is planning to invest money in a savings account. Oak Hills Financial offers 3.1% interest compounded annually. First City Bank has a savings account with a quarterly compounded interest rate of 0.7%.

    a. Write a function to represent the amount $A(t)$ that Tareq would earn after t years through Oak Hills Financial, assuming an initial investment of $1. Then write an equivalent function with an exponent of $4t$.

    b. What is the effective quarterly interest rate at Oak Hills Financial?

    c. Which is the better plan? Explain.

Example 2   2. **COLLECTIBLES** Jemma is comparing the growth rates in the value of two items in her baseball collection. The value of a rare baseball card increases by 8.1% per year. The value of a mitt increases by 0.57% each month. Write a function to represent the value of the mitt after t years with an initial value of $650. Then, write an equivalent function that represents the value of the mitt with only one compounding per year. What is the *effective* yearly rate of growth of the mitt? Which item in Jemma's collection is increasing in value at a faster rate?

Practice and Problem Solving
Extra Practice is on page R7.

Example 1   3. **CREDIT** Adam is choosing between two credit cards. World Mutual offers a credit card with 15.3% interest compounded annually. Super City Card has a monthly compounded interest rate of 1.4%. Which card is the better offer? Explain.

Example 2   4. **TECHNOLOGY** According to one source, the value of a desktop computer decreases at a rate of about 6% per month. The value of a laptop computer decreases at a rate of about 66% per year. Write a function to represent the value of the desktop computer after t years with an initial value of $450. Then, write an equivalent function that represents the value of the desktop computer with only 1 compounding per year. What is the effective yearly rate of the decrease in value of the desktop computer? Which type of computer is decreasing in value at a faster rate?

5 **POPULATION** The table shows the population of a small town that experiences a rapid increase in its population. The function $P(t) = 10,200(1.08)^t$ represents the growth of the population using an annual growth rate. Find the approximate effective monthly increase in the town's population.

Year	Population
2012	10,200
2013	11,016
2014	11,897
2015	12,849
2016	13,877

6. **MP** **TOOLS** Using the compound interest equation $A = P\left(1 + \frac{r}{n}\right)^{nt}$ choose fixed and reasonable values for P, n, and r, and use a graphing calculator to graph $A(t)$. Using the same graphing window and the same values for P, n, and r, graph $A = P\left[(1 + r)^{\frac{1}{n}}\right]^{nt}$. What do you notice? What conclusions can you make?

7. **HYPERINFLATION** Hyperinflation is extremely rapid inflation within the economy of a country. In Austria in 1922, the inflation rate for one year was 1426%.

   a. What was the effective monthly inflation rate in Austria in 1922?

   b. During this period, Austria used the crown as their unit of currency. Suppose an item cost 10 crowns at the beginning of May, 1922. What would have been the expected cost of the item at the beginning of August, 1922?

H.O.T. Problems   Use Higher-Order Thinking Skills

8. **OPEN-ENDED** Give an example of two interest rates—one that is compounded monthly and one that is compounded annually—such that the effective monthly interest rates are approximately equal. Justify your answer.

9. **MP** **CRITIQUE ARGUMENTS** Celia said that an annual decrease of 10% is equivalent to an effective monthly decrease of about 98.7% because $(1 - r)^t = (1 - 0.1)^t = 0.9^t = \left(0.9^{\frac{1}{12}}\right)^{12t} \approx 0.987^{12t}$. Do you agree? Explain.

10. **ERROR ANALYSIS** A savings account offers an annual interest rate of 2.1%. Yoshio was asked to estimate the effective quarterly interest rate. His work is shown here. Based on his work, he concluded that the effective quarterly rate is about 8.7%. Is Yoshio correct? If not, explain his error and show how to solve the problem correctly.

$$A(t) = (1 + 0.021)^t$$
$$= 1.021^t$$
$$= (1.021^4)^{\frac{t}{4}}$$
$$= (1.087)^{\frac{t}{4}}$$

11. **WRITING IN MATH** In Example 1, why does it make sense to assume that the initial investment is $1? Would a different initial investment change the result of the problem? Explain.

460 | Lesson 7-8 | Transforming Exponential Expressions

12. The number of players of an online video game is growing according to the model $V = 2000(1.09)^t$, where t is the time in years. Which equation can be used to model the monthly growth in the number of players? **MP** 4, 7

   ○ **A** $V = 2000(2.8167)^{12t}$

   ○ **B** $V = 166.7(2.8167)^{12t}$

   ○ **C** $V = 166.7(1.0072)^{12t}$

   ○ **D** $V = 2000(1.0072)^{12t}$

13. The population of a certain bacteria is given by $A(t) = 200(1.4)^t$, where t is time in hours.

Time (h)	Population
0	200
1	280
2	392
3	549

   Which statements about the growth of the bacteria are true? **MP** 4, 7

   ☐ **A** After 20 minutes, the predicted population of bacteria is 167,337.

   ☐ **B** The number of bacteria grows by 40% per hour.

   ☐ **C** The number of bacteria is modeled by $A(t) = 200(1.0056)^{60t}$, where t is the time in minutes.

   ☐ **D** The number of bacteria grows by about 5.6% per second.

   ☐ **E** The initial population of bacteria was 200.

14. The total number of followers of a musician on a social network is growing at a rate of 24% per year. Which is the best estimate of the effective weekly growth rate in the number of followers? **MP** 4, 7

   ○ **A** 0.045%

   ○ **B** 0.41%

   ○ **C** 11.1%

   ○ **D** 46%

15. The population of a town is increasing at a rate of 0.14% per month. Which equation best models the growth of the population, where P_0 is the initial population and t is the time in years? **MP** 4, 7

   ○ **A** $P(t) = P_0(1.0001)^t$

   ○ **B** $P(t) = P_0(1.017)^t$

   ○ **C** $P(t) = P_0(1.68)^t$

   ○ **D** $P(t) = P_0(4.82)^t$

16. The population of an invasive species of mussels in a river is increasing at a rate of about 1.63% each month. Which is the effective annual growth rate of the mussel population in the river? **MP** 4, 7

   ○ **A** 0.13%

   ○ **B** 6.12%

   ○ **C** 19.56%

   ○ **D** 21.41%

17. The value of a tablet computer decreases by 14% per year. Determine the approximate effective monthly rate of decrease in the value. Write your answer rounded to the nearest tenth. **MP** 4, 7

   ☐

18. **MULTI-STEP** Sharonda is comparing the saving accounts that are offered at the two banks. **MP** 4, 7

Bank	Interest Rate
XYZ Savings	1.8% compounded annually
Statewide Financial	0.12% compounded monthly

   a. Determine the effective monthly interest rate for the savings account at XYZ Savings.

   b. Which of the two banks offers a better deal? Explain.

   c. What is the effective annual interest rate for the savings account at Statewide Financial?

   d. Sharonda learns that Capital Street Bank offers a savings account with 0.55% interest compounded quarterly. Which of the three banks should she choose? Explain.

Geometric Sequences as Exponential Functions

::**Then**

- You related arithmetic sequences to linear functions.

::**Now**

1 Identify and generate geometric sequences.

2 Relate geometric sequences to exponential functions.

::**Why?**

- Genevieve posts a fundraising link on her social media page. Five of her friends repost the link. Each of those five friends have five friends who repost the link. The number of reposts generated forms a geometric sequence.

 New Vocabulary
geometric sequence
common ratio

 Mathematical Practices
7 Look for and make use of structure.

1 Recognize Geometric Sequences The first person generates 5 reposts. If each of these people repost 5 times, 25 links are generated. If each of the 25 people repost 5 times, 125 links are generated. The sequence of reposts generated, 1, 5, 25, 125, ... is an example of a **geometric sequence**.

In a geometric sequence, the first term is nonzero and each term after the first is found by multiplying the previous term by a nonzero constant r called the **common ratio**. The common ratio can be found by dividing any term by its previous term.

Example 1   Identify Geometric Sequences

Determine whether each sequence is *arithmetic*, *geometric*, or *neither*. Explain.

a. 256, 128, 64, 32, ...

Find the ratios of consecutive terms.

256   128   64   32

$$\frac{128}{256} = \frac{1}{2} \qquad \frac{64}{128} = \frac{1}{2} \qquad \frac{32}{64} = \frac{1}{2}$$

Since the ratios are constant, the sequence is geometric. The common ratio is $\frac{1}{2}$.

b. 4, 9, 12, 18, ...

Find the ratios of consecutive terms.

4   9   12   18

$$\frac{9}{4} = 2\frac{1}{4} \qquad \frac{12}{9} = 1\frac{1}{3} \qquad \frac{18}{12} = 1\frac{1}{2}$$

The ratios are not constant, so the sequence is not geometric.

Find the differences of consecutive terms.

4   9   12   18

$$9 - 4 = 5 \qquad 12 - 9 = 3 \qquad 18 - 12 = 6$$

There is no common difference, so the sequence is not arithmetic.
Thus, the sequence is neither geometric nor arithmetic.

▸ **Guided Practice**

1A. 1, 3, 9, 27, ...   **1B.** −20, −15, −10, −5, ...   **1C.** 2, 8, 14, 22, ...

Once the common ratio is known, more terms of a sequence can be generated.

Example 2 Find Terms of Geometric Sequences

Find the next three terms in each geometric sequence.

a. $1, -4, 16, -64, \ldots$

Step 1 Find the common ratio.

$$1 \qquad -4 \qquad 16 \qquad -64$$

$$\frac{-4}{1} = -4 \qquad \frac{16}{-4} = -4 \qquad \frac{-64}{16} = -4$$

Step 2 Multiply each term by the common ratio to find the next three terms.

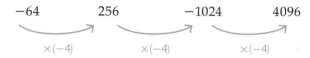

$$-64 \qquad 256 \qquad -1024 \qquad 4096$$

$$\times(-4) \qquad \times(-4) \qquad \times(-4)$$

The next three terms are 256, -1024, and 4096.

b. $9, 3, 1, \frac{1}{3} \ldots$

Step 1 Find the common ratio.

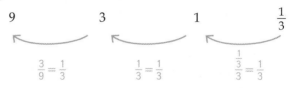

$$9 \qquad 3 \qquad 1 \qquad \frac{1}{3}$$

$$\frac{3}{9} = \frac{1}{3} \qquad \frac{1}{3} = \frac{1}{3} \qquad \frac{\frac{1}{3}}{3} = \frac{1}{3}$$

The value of r is $\frac{1}{3}$.

Step 2 Multiply each term by the common ratio to find the next three terms.

$$\frac{1}{3} \qquad \frac{1}{9} \qquad \frac{1}{27} \qquad \frac{1}{81}$$

$$\times \frac{1}{3} \qquad \times \frac{1}{3} \qquad \times \frac{1}{3}$$

The next three terms are $\frac{1}{9}, \frac{1}{27}$, and $\frac{1}{81}$.

▶ **Guided Practice**

2A. $-3, 15, -75, 375, \ldots$ **2B.** $24, 36, 54, 81, \ldots$

2 Geometric Sequences and Functions Finding the nth term of a geometric sequence would be tedious if we used the above method. The table below shows a rule for finding the nth term of a geometric sequence.

Position, n	1	2	3	4	...	n
Term, a_n	a_1	$a_1 r$	$a_1 r^2$	$a_1 r^3$	...	$a_1 r^{n-1}$

Notice that the common ratio between the terms is r. The table shows that to get the nth term, you multiply the first term by the common ratio r raised to the power $n - 1$. A geometric sequence can be defined by an exponential function in which n is the independent variable, a_n is the dependent variable, and r is the base. The domain is the counting numbers.

Key Concept nth term of a Geometric Sequence

The nth term a_n of a geometric sequence with first term a_1 and common ratio r is given by the following formula, where n is any positive integer and $a_1, r \neq 0$.

$$a_n = a_1 r^{n-1}$$

Example 3   Find the nth Term of a Geometric Sequence

a. Write an equation for the nth term of the sequence $-6, 12, -24, 48, \ldots$.

The first term of the sequence is -6. So, $a_1 = -6$. Now find the common ratio.

The common ratio is -2.

$$\frac{12}{-6} = -2 \qquad \frac{-24}{12} = -2 \qquad \frac{48}{-24} = -2$$

$a_n = a_1 r^{n-1}$ Formula for nth term

$a_n = -6(-2)^{n-1}$ $a_1 = -6$ and $r = 2$

b. Find the ninth term of this sequence.

$a_n = a_1 r^{n-1}$ Formula for nth term

$a_9 = -6(-2)^{9-1}$ For the nth term, $n = 9$.

$\quad = -6(-2)^8$ Simplify.

$\quad = -6(256)$ $(-2)^8 = 256$

$\quad = -1536$

> ### Watch Out!
> **Negative Common Ratio** If the common ratio is negative, as in Example 3, make sure to enclose the common ratio in parentheses. $(-2)^8 \neq -2^8$

> **Guided Practice**
>
> **3.** Write an equation for the nth term of the geometric sequence $96, 48, 24, 12, \ldots$. Then find the tenth term of the sequence.

Real-World Example 4   Graph a Geometric Sequence

BASKETBALL The NCAA women's basketball tournament begins with 64 teams. In each round, one half of the teams are left to compete, until only one team remains. Draw a graph to represent how many teams are left in each round.

Compared to the previous rounds, one half of the teams remain. So, $r = \frac{1}{2}$. Therefore, the geometric sequence that models this situation is 64, 32, 16, 8, 4, 2, 1. So in round two, 32 teams compete, in round three 16 teams compete, and so forth. Use this information to draw a graph.

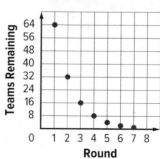

> **Guided Practice**
>
> **4. TENNIS** A tennis ball is dropped from a height of 12 feet. Each time the ball bounces back to 80% of the height from which it fell. Draw a graph to represent the height of the ball after each bounce.

> ### Real-World Link
> The first NCAA Division I women's basketball tournament was held in 1982. The Final Four ticket prices were $5 and $7 in 1982. They now cost as much as $800.
>
> **Source:** NCAA Sports

Check Your Understanding

◯ = Step-by-Step Solutions begin on page R11.

Example 1 Determine whether each sequence is *arithmetic*, *geometric*, or *neither*. **Explain.**

1. 200, 40, 8, ...   **2.** 2, 4, 16, ...   **3.** −6, −3, 0, 3, ...   **4.** 1, −1, 1, −1, ...

Example 2 Find the next three terms in each geometric sequence.

5. 10, 20, 40, 80, ...   **6.** 100, 50, 25, ...   **7.** $4, -1, \frac{1}{4}, \ldots$ **8.** −7, 21, −63, ...

Example 3 Write an equation for the *n*th term of each geometric sequence, and find the indicated term.

9. the fifth term of −6, −24, −96, ...

10. the seventh term of −1, 5, −25, ...

11. the tenth term of 72, 48, 32, ...

12. the ninth term of 112, 84, 63, ...

Example 4 **13. EXPERIMENT** In a physics class experiment, Diana drops a ball from a height of 16 feet. Each bounce has 70% the height of the previous bounce. Draw a graph to represent the height of the ball after each bounce.

Practice and Problem Solving

Extra Practice is on page R7

Example 1 Determine whether each sequence is *arithmetic*, *geometric*, or *neither*. **Explain.**

14. 4, 1, 2, ...   **15.** 10, 20, 30, 40, ...   **16.** 4, 20, 100, ...

17. 212, 106, 53, ...   **18.** −10, −8, −6, −4, ...   **19.** 5, −10, 20, 40, ...

Example 2 Find the next three terms in each geometric sequence.

20. 2, −10, 50, ...   **㉑** 36, 12, 4, ...   **22.** 4, 12, 36, ...

23. 400, 100, 25, ...   **24.** −6, −42, −294, ...   **25.** 1024, −128, 16, ...

Example 3 **26.** The first term of a geometric series is 1 and the common ratio is 9. What is the 8th term of the sequence?

27. The first term of a geometric series is 2 and the common ratio is 4. What is the 14th term of the sequence?

28. What is the 15th term of the geometric sequence −9, 27, −81, ...?

29. What is the 10th term of the geometric sequence 6, −24, 96, ...?

Example 4 **30. PENDULUM** The first swing of a pendulum is shown. On each swing after that, the arc length is 60% of the length of the previous swing. Draw a graph that represents the arc length after each swing.

24 ft

31. Find the eighth term of a geometric sequence for which $a_3 = 81$ and $r = 3$.

32. **MP REASONING** At an online mapping site, Mr. Mosley notices that when he clicks a spot on the map, the map zooms in on that spot. The magnification increases by 20% each time.

a. Write a formula for the *n*th term of the geometric sequence that represents the magnification of each zoom level for the map.

(*Hint:* The common ratio is not just 0.2.)

b. What is the fourth term of this sequence? What does it represent?

33. MULTI-STEP Sarina's parents have offered her two different allowance options for her 9-week summer vacation. She can get paid $30 a week, or she can get paid $1 the first week, $2 the second week, $4 the third week, and so on.

   a. Which option should Sarina choose?

   b. Explain your solution process.

34. SIERPINSKI'S TRIANGLE Consider the inscribed equilateral triangles at the right. The perimeter of each triangle is one half of the perimeter of the next larger triangle. What is the perimeter of the smallest triangle?

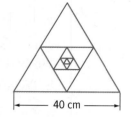

40 cm

35 If the second term of a geometric sequence is 3 and the third term is 1, find the first and fourth terms of the sequence.

36. If the third term of a geometric sequence is −12 and the fourth term is 24, find the first and fifth terms of the sequence.

37. EARTHQUAKES The Richter scale is used to measure the force of an earthquake. The table shows the increase in magnitude for the values on the Richter scale.

Richter Number (x)	Increase in Magnitude (y)	Rate of Change (slope)
1	1	–
2	10	9
3	100	
4	1000	
5	10,000	

   a. Copy and complete the table. Remember that the rate of change is the change in y divided by the change in x.

   b. Plot the ordered pairs (Richter number, increase in magnitude).

   c. Describe the graph that you made of the Richter scale data. Is the rate of change between any two points the same?

   d. Write an exponential equation that represents the Richter scale.

H.O.T. Problems   Use Higher-Order Thinking Skills

38. CHALLENGE Write a sequence that is both geometric and arithmetic. Explain your answer.

39. MP CRITIQUE ARGUMENTS Haro and Matthew are finding the ninth term of the geometric sequence −5, 10, −20, ... Is either of them correct? Explain your reasoning.

Haro	Matthew
$r = \dfrac{10}{-5}$ or −2	$r = \dfrac{10}{-5}$ or −2
$a_9 = -5\,(-2)^{9-1}$	$a_9 = -5 \cdot (-2)^{9-1}$
$= -5(512)$	$= -5 \cdot -256$
$= -2560$	$= 1280$

40. MP STRUCTURE Write a sequence of numbers that form a pattern but are neither arithmetic nor geometric. Explain the pattern.

41. ℮ WRITING IN MATH How are graphs of geometric sequences and exponential functions similar? different?

42. WRITING IN MATH Summarize how to find a specific term of a geometric sequence.

43. The formula for the nth term of a sequence is $a_n = 5(3)^{n-1}$. If the sequence were graphed on a coordinate plane, which point would be on the curve? **MP** 7

○ **A** (3, 225)

○ **B** (4, 400)

○ **C** (5, 405)

○ **D** (6, 243)

44. Find the formula for the sequence 5125, 1025, 205, 41 . . . **MP** 7

○ **A** $a_n = 5125\left(\frac{1}{5}\right)^n$

○ **B** $a_n = 5125\left(\frac{1}{5}\right)^{n-1}$

○ **C** $a_n = 5125(5)^n$

○ **D** $a_n = 41(5)^{n-1}$

45. Which formula represents the terms of this table? **MP** 7

n	1	2	3	4
a_n	3	27	243	2187

○ **A** $a_n = 3(9)^n$

○ **B** $a_n = 3(9)^{n-1}$

○ **C** $a_n = 9(3)^n$

○ **D** $a_n = 9(3)^{n-1}$

46. What is the 8th term of the sequence 64,768, 16,192, 4048, 1012. . .? **MP** 7

○ A 63.25

○ B 15.8125

○ C 3.953125

○ D 0.25

47. A coffee shop increases the price of their coffee by 15% each year. If the initial price of a cup of coffee is $1.32, which points are on the graph that represents the price of coffee over time? **MP** 4, 7

☐ **A** (1, 1.32)

☐ **B** (2, 1.65)

☐ **C** (3, 1.83)

☐ **D** (4, 2.01)

☐ **E** (5, 2.31)

☐ **F** (6, 2.84)

48. MULTI-STEP In 2015, the average attendance for a college's basketball games was 6500. Since then, the attendance has dropped by an average of 7% each year. **MP** 4, 7

a. Which formula can be used to find the nth term of a geometric sequence that represents the attendance over time?

○ **A** $a_n = 6500(1.07)^n$

○ **B** $a_n = 6500(1.07)^{n-1}$

○ **C** $a_n = 6500(0.93)^n$

○ **D** $a_n = 6500(0.93)^{n-1}$

b. What is the sixth term in the sequence?

c. What does the eighth term of the sequence represent?

d. Sketch a graph of the sequence.

49. The population of a town was 10,250 in 2016 and is expected to decline by 1.5% each year. What is the predicted population in 2030? **MP** 4, 7

Algebra Lab
Average Rate of Change of Exponential Functions

You know that the rate of change of a linear function is the same for any two points on the graph. The rate of change of an exponential function is not constant.

Mathematical Practices
MP 4 Model with Mathematics

Activity   Evaluating Investment Plans

John has $2000 to invest in one of two plans. Plan 1 offers to increase his principal by $75 each year, while Plan 2 offers to pay 3.6% interest compounded monthly. The dollar value of each investment after t years is given by $A_1 = 2000 + 75t$ and $A_2 = 2000(1.003)^{12t}$, respectively. Use the function values, the average rate of change, and the graphs of the equations to interpret and compare the plans.

Step 1 Copy and complete the table below by finding the missing values for A_1 and A_2.

t	0	1	2	3	4	5
A_1						
A_2						

Step 2 Find the average rate of change for each plan from $t = 0$ to 1, $t = 3$ to 4, and $t = 0$ to 5.

Plan 1: $\dfrac{2075 - 2000}{1 - 0}$ or 75 $\dfrac{2300 - 2225}{4 - 3}$ or 75 $\dfrac{2375 - 2000}{5 - 0}$ or 75

Plan 2: $\dfrac{2073.2 - 2000}{1 - 0}$ or 73.2 $\dfrac{2309.27 - 2227.74}{4 - 3}$ or about 82 $\dfrac{2393.79 - 2000}{5 - 0}$ or about 79

Step 3 Graph the ordered pairs for each function. Connect each set of points with a smooth curve.

Step 4 Use the graph and the rates of change to compare the plans. Both graphs have a rate of change for the first year of about $75 per year. From year 3 to 4, Plan 1 continues to increase at $75 per year, but Plan 2 grows at a rate of more than $81 per year. The average rate of change over the first five years for Plan 1 is $75 per year and for Plan 2 is over $78 per year. This indicates that as the number of years increases, the investment in Plan 2 grows at an increasingly faster pace. This is supported by the widening gap between their graphs.

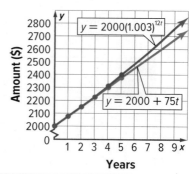

Exercises

The value of a company's piece of equipment decreases over time due to depreciation. The function $y = 16{,}000(0.985)^{2t}$ represents the value after t years.

1. What is the average rate of change over the first five years?

2. What is the average rate of change of the value from year 5 to year 10?

3. What conclusion about the value can we make based on these average rates of change?

4. **MP** **REGULARITY** Copy and complete the table for $y = x^4$.

x	−3	−2	−1	0	1	2	3
y							

Compare and interpret the average rate of change for $x = -3$ to 0 and for $x = 0$ to 3.

Recursive Formulas

:: Then	:: Now	:: Why?
● You wrote explicit formulas to represent arithmetic and geometric sequences.	**1** Use a recursive formula to list terms in a sequence. **2** Write recursive formulas for arithmetic and geometric sequences.	● Clients of a shuttle service get picked up from their homes and driven to premium outlet stores for shopping. The total cost of the service depends on the total number of customers. The costs for the first six customers are shown.

Number of Customers	Cost ($)
1	25
2	35
3	45
4	55
5	65
6	75

New Vocabulary

recursive formula

Mathematical Practices

3 Construct viable arguments and critique the reasoning of others.

1 Using Recursive Formulas An explicit formula allows you to find any term a_n of a sequence by using a formula written in terms of n. For example, $a_n = 2n$ can be used to find the fifth term of the sequence 2, 4, 6, 8, 10 or $a_5 = 2(5) = 10$.

A **recursive formula** allows you to find the nth term of a sequence by performing operations to one or more of the preceding terms. Since each term in the sequence above is 2 greater than the term that preceded it, we can add 2 to the fourth term to find that the fifth term is $8 + 2$ or 10. We can then write a recursive formula for a_n.

$$
\begin{aligned}
a_1 &= & &= 2 \\
a_2 &= & a_1 + 2 \text{ or } 2 + 2 &= 4 \\
a_3 &= & a_2 + 2 \text{ or } 4 + 2 &= 6 \\
a_4 &= & a_3 + 2 \text{ or } 6 + 2 &= 8 \\
&\vdots & \vdots & \\
a_n &= & a_{n-1} + 2 &
\end{aligned}
$$

A recursive formula for the sequence above is $a_1 = 2$, $a_n = a_{n-1} + 2$, for $n \geq 2$ where n is an integer. The term denoted a_{n-1} represents the term immediately before a_n. Notice that the first term a_1 is given, along with the domain for n.

Example 1   Use a Recursive Formula

Find the first five terms of each sequence.

a. $a_1 = 7$ and $a_n = a_{n-1} - 12$, if $n \geq 2$

Use $a_1 = 7$ and the recursive formula. A table can help organize the results.

n	$a_n = a_{n-1} - 12$	a_n
1	—	7
2	$a_2 = 7 - 12$	-5
3	$a_3 = -5 - 12$	-17
4	$a_4 = -17 - 12$	-29
5	$a_5 = -29 - 12$	-41

The first five terms are 7, -5, -17, -29, and -41.

PhotoAlto

b. $a_1 = 3$ and $a_n = 4a_{n-1}$, if $n \geq 2$

Use $a_1 = 3$ and the recursive formula to find the next four terms.

n	$a_n = 4a_{n-1}$	a_n
1	—	3
2	$a_2 = 4(3)$	12
3	$a_3 = 4(12)$	48
4	$a_4 = 4(48)$	192
5	$a_5 = 4(192)$	768

The first five terms are 3, 12, 48, 192, and 768.

▶ **Guided Practice**

1A. $a_1 = -2$ and $a_n = a_{n-1} + 4$, if $n \geq 2$

1B. $a_1 = -4$ and $a_n = \dfrac{3}{2} a_{n-1}$, if $n \geq 2$

2 Writing Recursive Formulas To write a recursive formula for an arithmetic or geometric sequence, complete the following steps.

Study Tip

Defining n For the nth term of a sequence, the value of n must be a positive integer. Although we must still state the domain of n, from this point forward, we will assume that n is an integer.

> ### 🔑 Key Concept Writing Recursive Formulas
>
> **Step 1** Determine whether the sequence is arithmetic or geometric by finding a common difference or a common ratio.
>
> **Step 2** Write a recursive formula.
>
> **Arithmetic Sequences** $a_n = a_{n-1} + d$, where d is the common difference
>
> **Geometric Sequences** $a_n = r \cdot a_{n-1}$, where r is the common ratio
>
> **Step 3** State the first term and domain for n.

Study Tip

🅜🅟 **Perseverance** The domain for n is decided by the given terms. Since the first term is already given, it makes sense that the first term to which the formula would apply is the 2nd term of the sequence, or when $n = 2$.

Example 2   Write Recursive Formulas

Write a recursive formula for the sequence 6, 24, 96, 384... .

Step 1 First subtract each term from the term that follows it.

$24 - 6 = 18$ $96 - 24 = 72$ $384 - 96 = 288$

There is no common difference. Check for a common ratio by dividing each term by the term that precedes it.

$\dfrac{24}{6} = 4$ $\dfrac{96}{24} = 4$ $\dfrac{384}{96} = 4$

There is a common ratio of 4. The sequence is geometric.

Step 2 Use the formula for a geometric sequence.

$a_n = r \cdot a_{n-1}$ Recursive formula for geometric sequence

$a_n = 4a_{n-1}$ $r = 4$

Step 3 The first term a_1 is 6, and $n \geq 2$.

A recursive formula for the sequence is $a_1 = 6$, $a_n = 4a_{n-1}$, $n \geq 2$.

▶ **Guided Practice**

2. Write a recursive formula for the sequence 4, 10, 25, 62.5, ...

A sequence can be represented by both an explicit formula and a recursive formula.

Example 3   Write Recursive and Explicit Formulas

COST Refer to the beginning of the lesson. Let n be the number of customers.

a. Write a recursive formula for the sequence.

> **Steps 1 and 2** First subtract each term from the term that follows it.
>
> $$35 - 25 = 10 \qquad 45 - 35 = 10 \qquad 55 - 45 = 10$$
>
> There is a common difference of 10. The sequence is arithmetic.

> **Step 3** Use the formula for an arithmetic sequence.
>
> $a_n = a_{n-1} + d$ Recursive formula for arithmetic sequence
> $a_n = a_{n-1} + 10$ $d = 10$

> **Step 4** The first term a_1 is 25, and $n \geq 2$.

A recursive formula for the sequence is $a_1 = 25$, $a_n = a_{n-1} + 10$, $n \geq 2$.

b. Write an explicit formula for the sequence.

> **Step 1** The common difference is 10.

> **Step 2** Use the formula for the nth term of an arithmetic sequence.
>
> $a_n = a_1 + (n-1)d$ Formula for the nth term
> $ = 25 + (n-1)10$ $a_1 = 25$ and $d = 10$
> $ = 25 + 10n - 10$ Distributive Property
> $ = 10n + 15$ Simplify.

An explicit formula for the sequence is $a_n = 10n + 15$.

> **Guided Practice**
>
> **3. SAVINGS** The money that Ronald has in his savings account earns interest each year. He does not make any withdrawals or additional deposits. The account balance at the beginning of each year is $10,000, $10,300, $10,609, $10,927.27, and so on. Write a recursive formula and an explicit formula for the sequence.

If several successive terms of a sequence are needed, a recursive formula may be useful, whereas if just the nth term of a sequence is needed, an explicit formula may be useful. Thus, it is sometimes beneficial to translate between the two forms.

Example 4   Translate between Recursive and Explicit Formulas

a. Write a recursive formula for $a_n = 6n + 3$.

$a_n = 6n + 3$ is an explicit formula for an arithmetic sequence with $d = 6$ and $a_1 = 6(1) + 3$ or 9. Therefore, a recursive formula for a_n is $a_1 = 9$, $a_n = a_{n-1} + 6$, $n \geq 2$.

b. Write an explicit formula for $a_1 = 120$, $a_n = 0.8a_{n-1}$, $n \geq 2$.

$a_n = 0.8a_{n-1}$ is a recursive formula for a geometric sequence with $a_1 = 120$ and $r = 0.8$. Therefore, an explicit formula for a_n is $a_n = 120(0.8)^{n-1}$.

> **Guided Practice**
>
> **4A.** Write a recursive formula for $a_n = 4(3)^{n-1}$.
>
> **4B.** Write an explicit formula for $a_1 = -16$, $a_n = a_{n-1} - 7$, $n \geq 2$.

Check Your Understanding   ◯ = Step-by-Step Solutions begin on page R11.

Go Online! for a
Self-Check Quiz

Example 1   Find the first five terms of each sequence.

    1. $a_1 = 16, a_n = a_{n-1} - 3, n \geq 2$ **2.** $a_1 = -5, a_n = 4a_{n-1} + 10, n \geq 2$

Example 2   Write a recursive formula for each sequence.

    3. 1, 6, 11, 16, ...   **4.** 4, 12, 36, 108, ...

Example 3   **5. BALL** A ball is dropped from an initial height of 10 feet. The maximum heights the ball reaches on the first three bounces are shown.

    a. Write a recursive formula for the sequence.

    b. Write an explicit formula for the sequence.

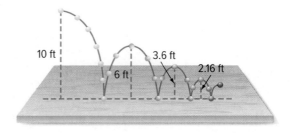

Example 4   For each recursive formula, write an explicit formula. For each explicit formula, write a recursive formula.

    6. $a_1 = 4, a_n = a_{n-1} + 16, n \geq 2$ **7** $a_n = 5n + 8$

    8. $a_n = 15(2)^{n-1}$ **9.** $a_1 = 22, a_n = 4a_{n-1}, n \geq 2$

Practice and Problem Solving

Extra Practice is on page R7.

Example 1   Find the first five terms of each sequence.

    10. $a_1 = 23, a_n = a_{n-1} + 7, n \geq 2$ **11.** $a_1 = 48, a_n = -0.5a_{n-1} + 8, n \geq 2$

    12. $a_1 = 8, a_n = 2.5a_{n-1}, n \geq 2$ **13.** $a_1 = 12, a_n = 3a_{n-1} - 21, n \geq 2$

    14. $a_1 = 13, a_n = -2a_{n-1} - 3, n \geq 2$ **15.** $a_1 = \frac{1}{2}, a_n = a_{n-1} + \frac{3}{2}, n \geq 2$

Example 2   Write a recursive formula for each sequence.

    16. 12, −1, −14, −27, ...   **17.** 27, 41, 55, 69, ...

    18. 2, 11, 20, 29, ...   **19.** 100, 80, 64, 51.2, ...

    20. 40, −60, 90, −135, ...   **21.** 81, 27, 9, 3, ...

Example 3   **22.** Ⓜ **MODELING** A landscaper is building a brick patio. Part of the patio includes a pattern constructed from triangles. The first four rows of the pattern are shown.

    a. Write a recursive formula for the sequence.

    b. Write an explicit formula for the sequence.

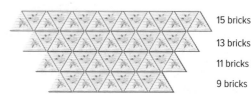

15 bricks

13 bricks

11 bricks

9 bricks

Example 4   For each recursive formula, write an explicit formula. For each explicit formula, write a recursive formula.

    23. $a_n = 3(4)^{n-1}$ **24.** $a_1 = -2, a_n = a_{n-1} - 12, n \geq 2$

    25. $a_1 = 38, a_n = \frac{1}{2}a_{n-1}, n \geq 2$ **26.** $a_n = -7n + 52$

27 **PHOTO SHARING** Barbara shares a photo with five of her friends. Each of her friends shared the photo with five more friends, and so on.

    a. Find the first five terms of the sequence representing the number of people who receive the photo in the nth round.

    b. Write a recursive formula for the sequence.

    c. If Barbara represents a_1, find a_8.

28. GEOMETRY Consider the pattern below. The number of blue boxes increases according to a specific pattern.

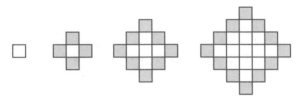

    a. Write a recursive formula for the sequence of the number of blue boxes in each figure.

    b. If the first box represents a_1, find the number of blue boxes in a_8.

29. TREE The heights of a certain type of tree over the past four years are shown.

10 ft   11 ft   12.1 ft   13.31 ft

    a. Write a recursive formula for the height of the tree.

    b. If the pattern continues, how tall will the tree be in two more years? Round your answer to the nearest tenth of a foot.

30. MULTIPLE REPRESENTATIONS The Fibonacci sequence is neither arithmetic nor geometric and can be defined by a recursive formula. The first terms are 1, 1, 2, 3, 5, 8, ...

    a. Logical Determine the relationship between the terms of the sequence. What are the next five terms in the sequence?

    b. Algebraic Write a formula for the nth term if $a_1 = 1$, $a_2 = 1$, and $n \geq 3$.

    c. Algebraic Find the 15th term.

    d. Analytical Explain why the Fibonacci sequence is not an arithmetic sequence.

H.O.T. Problems   Use **H**igher-**O**rder **T**hinking Skills

31. ERROR ANALYSIS Patrick and Lynda are working on a math problem that involves the sequence 2, −2, 2, −2, 2, ... Patrick thinks that the sequence can be written as a recursive formula. Lynda believes that the sequence can be written as an explicit formula. Is either of them correct? Explain.

32. CHALLENGE Find a_1 for the sequence in which $a_4 = 1104$ and $a_n = 4a_{n-1} + 16$.

33. MP CONSTRUCT ARGUMENTS Determine whether the following statement is *true* or *false*. Justify your reasoning.

There is only one recursive formula for every sequence.

34. CHALLENGE Find a recursive formula for 4, 9, 19, 39, 79, ...

35. WRITING IN MATH Explain the difference between an explicit formula and a recursive formula.

36. MULTI-STEP The table shows several terms of a sequence.  1, 2, 7

n	1	2	3	4
a_n	−12	25	−49	99

a. Which type of sequence is shown?

   ○ **A** recursive

   ○ **B** explicit

   ○ **C** arithmetic

   ○ **D** geometric

   ○ **E** none of the above

b. What is the next term in the sequence?

c. Which formula describes the sequence?

   ○ **A** $a_1 = -12$, $a_n = -12n - 1$, $n \geq 2$

   ○ **B** $a_1 = -12$, $a_n = -2a_{n-1}$, $n \geq 2$

   ○ **C** $a_1 = -12$, $a_n = -2a_{n-1}$, $n \geq 2$

   ○ **D** $a_1 = -12$, $a_n = a_{n-1} + 37$, $n \geq 2$

d. Can an explicit formula be written as a linear equation for the sequence? Explain.

37. The number of students at Superior Middle School over a three-year period is shown in the table.

Year	1	2	3
Number of Students	700	770	847

Use a recursive formula to find the number of students that will be at the school in the next year. MP 2

○ **A** 70

○ **B** 630

○ **C** 932

○ **D** 1632

38. Which of the following are formulas for the sequence 8, 16, 24, 32, ...? MP 2

   ☐ **A** $a_1 = 8$, $a_n = 2a_{n-1}$, $n \geq 2$

   ☐ **B** $a_1 = 8$, $a_n = 2a_{n-1} + 8$, $n \geq 2$

   ☐ **C** $a_n = 8n$

   ☐ **D** $a_n = 8n + 8$

   ☐ **E** $a_n = 8^{3n}$

39. The table shows several terms of a sequence.

n	1	2	3	4
a_n	−2	4	−8	16

a. Which of the following is a recursive formula for the sequence? MP 2

   ○ **A** $a_1 = -2$, $a_n = -2a_{n-1}$

   ○ **B** $a_1 = -2$, $a_n = 2a_{n-1}$

   ○ **C** $a_1 = -2$, $a_n = -1(a_{n-1})^2$

   ○ **D** $a_1 = -2$, $a_n = -1(2^n)$

b. What is the tenth term of the given sequence from the table? MP 1, 2

40. Mrs. Rodriguez wrote the first four terms of a sequence in a table on the board.

n	1	2	3	4
a_n	−2	−9	−37	−149

What is the sixth term of the given sequence from the table? MP 1, 2

○ **A** −2389

○ **B** −2388

○ **C** −597

○ **D** −596

Study Guide

Key Concepts

Multiplication and Division Properties of Exponents (Lessons 7-1 and 7-2)

For any nonzero real numbers a and b and any integers m, n, and p, the following are true.

- Product of Powers: $a^m \cdot a^n = a^{m+n}$
- Power of a Power: $(a^m)^n = a^{m \cdot n}$
- Power of a Product: $(ab)^m = a^m b^m$
- Quotient of Powers: $\dfrac{a^m}{a^p} = a^{m-p}$
- Power of a Quotient: $\left(\dfrac{a}{b}\right)^m = \dfrac{a^m}{b^m}$
- Zero Exponent: $a^0 = 1$
- Negative Exponent: $a^{-n} = \dfrac{1}{a^n}$ and $\dfrac{1}{a^{-n}} = a^n$

Rational Exponents (Lesson 7-3)

For any positive real number b and any integers m and $n > 1$, the following are true:

$$b^{\frac{1}{2}} = \sqrt{b} \qquad b^{\frac{1}{n}} = \sqrt[n]{b} \qquad b^{\frac{m}{n}} = \left(\sqrt[n]{b}\right)^m \text{ or } \sqrt[n]{b^m}$$

Radical Expressions (Lesson 7-4)

For real numbers a, b, c, and d,

- $\sqrt{ab} = \sqrt{a} \cdot \sqrt{b}$, if $a \geq 0$ and $b \geq 0$.
- $\sqrt{\dfrac{a}{b}} = \dfrac{\sqrt{a}}{\sqrt{b}}$, if $a \geq 0$ and $b > 0$.
- $a\sqrt{b} \pm c\sqrt{b} = (a \pm c)\sqrt{b}$, if $b \geq 0$.
- $a\sqrt{b}(c\sqrt{d}) = (ac)\sqrt{bd}$, if $b \geq 0$ and $d \geq 0$.

Transformations of Exponential Functions (Lesson 7-6)

- An exponential function can be written as $f(x) = ab^{x-h} + k$. The parameters a, h, and k are parameters that dilate, reflect, or translate a parent function with base b.

Exponential Functions (Lesson 7-7)

- Exponential growth: $y = a(1 + r)^t$, where $r > 0$
- Exponential decay: $y = a(1 - r)^t$, where $0 < r < 1$

FOLDABLES Study Organizer

Use your Foldable to review the chapter. Working with a partner can be helpful. Ask for clarification of concepts as needed.

Key Vocabulary

asymptote (p. 430)	geometric sequence (p. 462)
common ratio (p. 462)	monomial (p. 395)
compound interest (p. 451)	negative exponent (p. 404)
conjugates (p. 421)	nth root (p. 411)
constant (p. 395)	order of magnitude (p. 405)
cube root (p. 411)	radical expression (p. 419)
exponential decay (p. 430)	rational exponent (p. 410)
exponential equation (p. 413)	rationalizing the denominator (p. 421)
exponential function (p. 430)	
exponential growth (p. 430)	recursive formula (p. 469)
	zero exponent (p. 403)

Vocabulary Check

Choose the word or term that best completes each sentence.

1. 2 is a(n) _____ of 8.

2. The rules for operations with exponents can be extended to apply to expressions with a(n) _____ such as $7^{\frac{2}{3}}$.

3. $f(x) = 3^x$ is an example of a(n) _____.

4. $a_1 = 4$ and $a_n = 3a_{n-1} - 12$, if $n \geq 2$, is a(n) _____ for the sequence 4, -8, -20, -32, ...

5. $2^{3x-1} = 16$ is an example of a(n) _____.

6. The equation for _____ is $y = C(1 - r)^t$.

Concept Check

7. How can you determine whether a set of data in a table display exponential behavior?

8. Explain how to find the common ratio of a geometric sequence.

Lesson-by-Lesson Review

7-1 Multiplication Properties of Exponents

Simplify each expression.

9. $x \cdot x^3 \cdot x^5$

10. $(2xy)(-3x^2y^5)$

11. $(-4ab^4)(-5a^5b^2)$

12. $(6x^3y^2)^2$

13. $\left[(2r^3t)^3\right]^2$

14. $(-2u^3)(5u)$

15. $(2x^2)^3(x^3)^3$

16. $\frac{1}{2}(2x^3)^3$

17. **GEOMETRY** Use the formula $V = \pi r^2 h$ to find the volume of the cylinder.

Example 1

Simplify $(5x^2y^3)(2x^4y)$.

$(5x^2y^3)(2x^4y)$

$= (5 \cdot 2)(x^2 \cdot x^4)(y^3 \cdot y)$ Commutative Property

$= 10x^6y^4$ Product of Powers

Example 2

Simplify $(3a^2b^4)^3$.

$(3a^2b^4)^3 = 3^3(a^2)^3(b^4)^3$ Power of a Product

$= 27a^6b^{12}$ Simplify.

7-2 Division Properties of Exponents

Simplify each expression. Assume that no denominator equals zero.

18. $\dfrac{(3x)^0}{2a}$

19. $\left(\dfrac{3xy^3}{2z}\right)^3$

20. $\dfrac{12y^{-4}}{3y^{-5}}$

21. $a^{-3}b^0c^6$

22. $\dfrac{-15x^7y^8z^4}{-45x^3y^5z^3}$

23. $\dfrac{(3x^{-1})^{-2}}{(3x^2)^{-2}}$

24. $\left(\dfrac{6xy^{11}z^9}{48x^6yz^{-7}}\right)^0$

25. $\left(\dfrac{12}{2}\right)\left(\dfrac{x}{y^5}\right)\left(\dfrac{y^4}{x^4}\right)$

26. **GEOMETRY** The area of a rectangle is $25x^2y^4$ square feet. The width of the rectangle is $5xy$ feet. What is the length of the rectangle?

5xy

Example 3

Simplify $\dfrac{2k^4m^3}{4k^2m}$. Assume that no denominator equals zero.

$\dfrac{2k^4m^3}{4k^2m} = \left(\dfrac{2}{4}\right)\left(\dfrac{k^4}{k^2}\right)\left(\dfrac{m^3}{m}\right)$ Group powers with the same base.

$= \left(\dfrac{1}{2}\right)k^{4-2}\,m^{3-1}$ Quotient of Powers

$= \dfrac{k^2m^2}{2}$ Simplify.

Example 4

Simplify $\dfrac{t^4uv^{-2}}{t^{-3}u^7}$. Assume that no denominator equals zero.

$\dfrac{t^4uv^{-2}}{t^{-3}u^7} = \left(\dfrac{t^4}{t^{-3}}\right)\left(\dfrac{u}{u^7}\right)(v^{-2})$ Group the powers with the same base.

$= (t^{4+3})(u^{1-7})(v^{-2})$ Quotient of Powers

$= t^7u^{-6}v^{-2}$ Simplify.

$= \dfrac{t^7}{u^6v^2}$ Simplify.

7-3 Rational Exponents

Simplify.

27. $\sqrt[3]{343}$

28. $\sqrt[6]{729}$

29. $625^{\frac{1}{4}}$

30. $\left(\frac{8}{27}\right)^{\frac{1}{3}}$

31. $256^{\frac{3}{4}}$

32. $32^{\frac{2}{5}}$

33. $343^{\frac{4}{3}}$

34. $\left(\frac{4}{49}\right)^{\frac{3}{2}}$

Solve each equation.

35. $6^x = 7776$

36. $4^{4x-1} = 32$

Example 5

Simplify $125^{\frac{2}{3}}$.

$$125^{\frac{2}{3}} = \left(\sqrt[3]{125}\right)^2 \qquad b^{\frac{m}{n}} = \left(\sqrt[n]{b}\right)^m$$
$$= \left(\sqrt[3]{5 \cdot 5 \cdot 5}\right)^2 \qquad 125 = 5^3$$
$$= 5^2 \text{ or } 25 \qquad \text{Simplify.}$$

Example 6

Solve $9^{x-1} = 729$.

$$9^{x-1} = 729 \qquad \text{Original equation}$$
$$9^{x-1} = 9^3 \qquad \text{Rewrite 729 as } 9^3.$$
$$x - 1 = 3 \qquad \text{Power Property of Equality}$$
$$x = 4 \qquad \text{Add 1 to each side.}$$

7-4 Radical Expressions

Simplify.

37. $\sqrt{36x^2y^7}$

38. $\sqrt{20ab^3}$

39. $\sqrt{3} \cdot \sqrt{6}$

40. $2\sqrt{3} \cdot 3\sqrt{12}$

41. $(4 - \sqrt{5})^2$

42. $(1 + \sqrt{2})^2$

43. $\sqrt{\frac{50}{a^2}}$

44. $\sqrt{\frac{2}{5}} \cdot \sqrt{\frac{3}{4}}$

45. $\frac{3}{2 - \sqrt{5}}$

46. $\frac{5}{\sqrt{7} + 6}$

47. $\sqrt{6} - \sqrt{54} + 3\sqrt{12} + 5\sqrt{3}$

48. $2\sqrt{6} - \sqrt{48}$

49. $\sqrt{2}(5 + 3\sqrt{3})$

50. $(2\sqrt{3} - \sqrt{5})(\sqrt{10} + 4\sqrt{6})$

51. $(6\sqrt{5} + 2)(4\sqrt{2} + \sqrt{3})$

52. **MOTION** The velocity of a dropped object when it hits the ground can be found using $v = \sqrt{2gd}$, where v is the velocity in feet per second, g is the acceleration due to gravity, and d is the distance in feet the object drops. Find the speed of a penny when it hits the ground, after being dropped from 984 feet. Use 32 feet per second squared for g.

Example 7

Simplify $\frac{2}{4 + \sqrt{3}}$.

$$\frac{2}{4 + \sqrt{3}} \qquad \text{Original expression}$$

$$= \frac{2}{4 + \sqrt{3}} \cdot \frac{4 - \sqrt{3}}{4 - \sqrt{3}} \qquad \text{Rationalize the denominator.}$$

$$= \frac{2(4) - 2\sqrt{3}}{4^2 - (\sqrt{3})^2} \qquad (a - b)(a + b) = a^2 - b^2$$

$$= \frac{8 - 2\sqrt{3}}{16 - 3} \qquad (\sqrt{3})^2 = 3$$

$$= \frac{8 - 2\sqrt{3}}{13} \qquad \text{Simplify.}$$

Example 8

Simplify $2\sqrt{6} - \sqrt{24}$.

$$2\sqrt{6} - \sqrt{24} = 2\sqrt{6} - \sqrt{4 \cdot 6} \qquad \text{Product Property}$$
$$= 2\sqrt{6} - 2\sqrt{6} \qquad \text{Simplify.}$$
$$= 0 \qquad \text{Simplify.}$$

Example 9

Simplify $\left(\sqrt{3} - \sqrt{2}\right)\left(\sqrt{3} + 2\sqrt{2}\right)$.

$$\left(\sqrt{3} - \sqrt{2}\right)\left(\sqrt{3} + 2\sqrt{2}\right)$$
$$= \left(\sqrt{3}\right)\left(\sqrt{3}\right) + \left(\sqrt{3}\right)\left(2\sqrt{2}\right) + \left(-\sqrt{2}\right)\left(\sqrt{3}\right) + \left(\sqrt{2}\right)\left(2\sqrt{2}\right)$$
$$= 3 + 2\sqrt{6} - \sqrt{6} + 4$$
$$= 7 + \sqrt{6}$$

7-5 Exponential Functions

Graph each function. Find the y-intercept, and state the domain, range, and the equation of the asymptote.

53. $y = 2^x$

54. $y = 3^x + 1$

55. $y = 4^x + 2$

56. $y = 2^x - 3$

57. BIOLOGY The population of bacteria in a petri dish increases according to the model $p = 550(2.7)^{0.008t}$, where t is the number of hours and $t = 0$ corresponds to 1:00 P.M. Use this model to estimate the number of bacteria in the dish at 5:00 P.M.

Example 10

Graph $y = 3^x + 6$. Find the y-intercept, and state the domain and range.

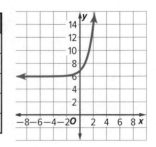

x	$3^x + 6$	y
-3	$3^{-3} + 6$	6.04
-2	$3^{-2} + 6$	6.11
-1	$3^{-1} + 6$	6.33
0	$3^0 + 6$	7
1	$3^1 + 6$	9

The y-intercept is $(0, 7)$. The domain is all real numbers, and the range is all real numbers greater than 6.

7-6 Transformations of Exponential Functions

Describe the transformation $g(x)$ as it relates to the parent function $f(x)$.

58. $f(x) = 2^x$; $g(x) = 5(2)^x$

59. $f(x) = 3^x$; $g(x) = -0.5(3)^x - 2$

60. Compare the key features of $f(x) = 2^x$, $g(x) = 2^x - 3$, and $p(x) = 2^x + 4$.

61. Write the function $g(x)$ given that $g(x)$ is a reflection and vertical dilation of $f(x) = 4^x$.

Example 11

Compare the graphs of $f(x) = 2^x$ and $g(x) = \frac{1}{3}(2^x)$.

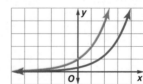

For both functions, the asymptote is $y = 0$. The function $f(x)$ has y-intercept 1 and $g(x)$ has y-intercept $\frac{1}{3}$. Both functions are increasing as $x \longrightarrow \infty$. Compared to $f(x)$, $g(x)$ is vertically compressed.

7-7 Writing Exponential Functions

62. Write an exponential function in the form $y = ab^x$ for the graph.

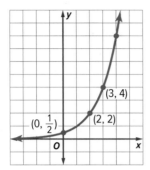

(3, 4)

(0, $\frac{1}{2}$) (2, 2)

Write an exponential function for a graph that passes through the given points.

63. (1, 12) and (2, 36)

64. (2, 20) and (4, 80)

65. Find the final value of $2500 invested at an annual interest rate of 2% compounded monthly for 10 years.

66. **COMPUTERS** Zita's computer is depreciating at a rate of 3% per year. She bought the computer for $1200.

   a. Write an equation to represent this situation.

   b. What will the computer's value be after 5 years?

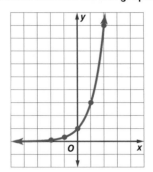
7-8 Transforming Exponential Expressions

67. First Bank's savings accounts have an interest rate of 0.15% compounded monthly. Main Street Bank has a 1.5% yearly rate. Compare the yearly rates for both banks.

68. What is the effective daily interest rate for a credit card that has a 24% yearly interest rate?

69. Mark deposits $2000 in a savings account with an interest rate of 1.75% compounded annually. Find the effective quarterly interest rate.

7-9 Geometric Sequences as Exponential Functions

Find the next three terms in each geometric sequence.

70. $-1, 1, -1, 1, \ldots$

71. $3, 9, 27, \ldots$

72. $256, 128, 64, \ldots$

Write the equation for the nth term of each geometric sequence.

73. $-1, 1, -1, 1, \ldots$

74. $3, 9, 27, \ldots$

75. $256, 128, 64, \ldots$

76. SPORTS A basketball is dropped from a height of 20 feet. It bounces to $\frac{1}{2}$ its height after each bounce. Draw a graph to represent the situation.

Example 15

Find the next three terms in the geometric sequence $2, 6, 18, \ldots$.

Step 1 Find the common ratio. Each number is 3 times the previous number, so $r = 3$.

Step 2 Multiply each term by the common ratio to find the next three terms.

$18 \times 3 = 54, 54 \times 3 = 162, 162 \times 3 = 486$

The next three terms are 54, 162, and 486.

Example 16

Write the equation for the nth term of the geometric sequence $-3, 12, -48, \ldots$.

The common ratio is -4. So $r = -4$.

$a_n = a_1 r^{n-1}$ Formula for the nth term

$a_n = -3(-4)^{n-1}$ $a_1 = -3$ and $r = -4$

7-10 Recursive Formulas

Find the first five terms of each sequence.

77. $a_1 = 11, a_n = a_{n-1} - 4, n \geq 2$

78. $a_1 = 3, a_n = 2a_{n-1} + 6, n \geq 2$

Write a recursive formula for each sequence.

79. $2, 7, 12, 17, \ldots$

80. $32, 16, 8, 4, \ldots$

81. $2, 5, 11, 23, \ldots$

Example 17

Write a recursive formula for $3, 1, -1, -3, \ldots$.

Step 1 First subtract each term from the term that follows it.

$1 - 3 = -2, -1 - 1 = -2, -3 - (-1) = -2$

There is a common difference of -2. The sequence is arithmetic.

Step 2 Use the formula for an arithmetic sequence.

$a_n = a_{n-1} + d$ Recursive formula

$a_n = a_{n-1} + (-2)$ $d = -2$

Step 3 The first term a_1 is 3, and $n \geq 2$.

A recursive formula is $a_1 = 3, a_n = a_{n-1} - 2, n \geq 2$.

Simplify each expression.

1. $(x^2)(7x^8)$

2. $(5a^7bc^2)(-6a^2bc^5)$

3. **MULTIPLE CHOICE** Express the volume of the solid as a monomial.

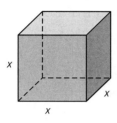

    A x^3 **C** $6x^3$

    B $6x$ **D** x^6

Simplify each expression. Assume that no denominator equals 0.

4. $\dfrac{x^6y^8}{x^2}$

5. $\left(\dfrac{2a^4b^3}{c^6}\right)^0$

6. $\dfrac{2xy^{-7}}{8x}$

Simplify.

7. $\sqrt[3]{1000}$ 8. $\sqrt[5]{3125}$

9. $1728^{\frac{1}{3}}$ 10. $\left(\dfrac{16}{81}\right)^{\frac{1}{2}}$

11. $27^{\frac{2}{3}}$ 12. $10{,}000^{\frac{3}{4}}$

13. $27^{\frac{5}{3}}$ 14. $\left(\dfrac{1}{121}\right)^{\frac{3}{2}}$

Solve each equation.

15. $12^x = 1728$

16. $7^{x-1} = 2401$

17. $9^{x-3} = 729$

Simplify.

18. $\sqrt{98}$ 19. $\sqrt{6} \cdot \sqrt{18}$

20. $\sqrt{50x^3y^2}$ 21. $\sqrt{\dfrac{10}{25}}$

Simplify.

22. $5\sqrt{3} - 15\sqrt{3} + 8\sqrt{5}$

23. $7\sqrt{2} + 4\sqrt{18} - 6\sqrt{50}$

24. $(2\sqrt{3} + 5\sqrt{2})(4\sqrt{3} - 6\sqrt{2})$

25. $(3\sqrt{8} - 5\sqrt{12})^2$

Graph each function. Find the y-intercept, and state the domain, range, and the equation of the asymptote.

26. $y = 2(5)^x$

27. $y = -3(11)^x$

28. $y = 3^x + 2$

29. Describe the transformations in $g(x) = -0.15(3)^{x+2}$ and $p(x) = (3)^{-x} - 1$ as each relates to the parent function $f(x) = 3^x$.

Find the next three terms in each geometric sequence.

30. $2, -6, 18, \ldots$

31. $1000, 500, 250, \ldots$

32. $32, 8, 2, \ldots$

33. **MULTIPLE CHOICE** Lynne invested $500 into an account with a 6.5% interest rate compounded monthly. How much will Lynne's investment be worth in 10 years?

    A $600.00

    B $938.57

    C $956.09

    D $957.02

34. **INVESTMENTS** Shelly's investment of $3000 has been losing value at a rate of 3% each year. What will her investment be worth in 6 years?

35. **RATES** Which offers a better rate for a savings account, 2% compounded yearly or 0.2% compounded monthly? Why?

Find the first five terms of each sequence.

36. $a_1 = 18,\ a_n = a_{n-1} - 4,\ n \geq 2$

37. $a_1 = -2,\ a_n = 4a_{n-1} + 5,\ n \geq 2$

Performance Task

Provide a clear solution to each part of the task. Be sure to show all of your work, include all relevant drawings, and justify your answers.

BACTERIAL GROWTH A scientist is studying a newly discovered bacterium thought to be responsible for a recent decline in a certain kind of tree.

Part A

The scientist creates a culture of the bacteria in a petri dish and monitors it over time. She notices that the number of bacterial cells triples every hour.

1. Determine what kind of expression would accurately model the bacteria's growth over time.

2. **Sense-Making** Write an equation that models the bacteria's growth over time.

3. **Regularity** Graph the equation.

4. Assuming the scientist started with one bacterial cell, determine the number of cells that will be in the petri dish after 6 hours.

Part B

The scientist determines that the tree population is decreasing by 20% every year due to this bacterium.

5. Write an equation that models the tree population over time.

6. Determine after about how many years will the tree population be about half of its starting population.

Part C

The scientist formulates an antibiotic to counteract the bacterium. She and her team apply it to a sample region of the affected trees and record the tree population over the course of four years. Results are gathered one year after her experiment begins and are recorded in the table below.

Year	1	2	3	4
Number of Trees	1,000	900	810	729

7. Assuming the pattern continues, write an equation that represents the geometric sequence in the table.

8. Determine whether or not the antibiotic is effective. Explain your answer.

Test-Taking Strategy

Example

Read the problem. Identify what you need to know. Then use the information in the problem to solve.

According to one Website, the value of a new smart TV depreciates by 30% each year. David bought a 65-inch smart TV for \$1149. What is the value of his TV after 6 years?

A \$135.18 **C** \$344.70

B \$804.30 **D** \$5546.00

Step 1 **What do you need to find?**
The value of David's TV 6 years after he purchased it.

Step 2 **Is there enough information to solve the problem?**
Yes, there is.

Step 3 **What information, if any, is not needed to solve the problem?**
The size of the TV is not needed to solve the problem.

Step 4 **Are there any obvious wrong answers?**
D is obviously wrong because the value is greater than the original cost.

Step 5 **What is the correct answer?**
To solve the problem, I need to write an equation. Because the value is depreciating, I will use the equation for exponential decay, $y = a(1 - r)^t$. In this problem, the initial amount a is \$1149, the rate r of depreciation is 30% or 0.30, the time t is 6 years, and y is the value of the TV. So, $y = 1149(1 - 0.30)^6$. I can enter the expression into a calculator, and the result is 135.178701. If I round to the nearest cent, the value of the TV after 6 years is \$135.18.

The correct answer is A.

Test-Taking Tip

Using a Scientific or Graphing Calculator
Knowing *when* to use a calculator to solve a problem is just as important as knowing *how* to use one.

Apply the Strategy

In 2017, the estimated population of the United States was 325,900,000, and the population was increasing at an annual rate of 0.73%. Estimate the population of the United States in 2030.

A 296,291,126

B 328,279,070

C 358,219,482

D 405,367,922

Read each question. Then fill in the correct answer on the answer document provided by your teacher or on a sheet of paper.

1. Solve $3^{2x} = 27^{\frac{4}{3}}$.

   ○ A $\frac{2}{3}$

   ○ B 2

   ○ C 4

   ○ D 8

2. To find the number of games the team needs to win in order to get to the playoffs, Robert solved the following inequality. Let x be the number of games. Which statement describes the solution?

   $$-3x < -15$$

   ○ A The team must win at least 5 games.

   ○ B The team must win more than 5 games.

   ○ C The team must win less than 5 games.

   ○ D The team has already won 5 more games than they need to in order to make the playoffs.

3. A rectangular prism has length $3t^7$, width t^2z^4, and height $9z^8$. What is the length of a side of a cube with the same volume?

   ○ A $3t^3z^4$

   ○ B $9t^3z^4$

   ○ C $12t^{14}z^{32}$

   ○ D $27t^9z^{12}$

4. Simplify $\sqrt{32x^4y}$.

   [ ]

> **Test-Taking Tip**
>
> **Question 1** Rewrite each power with the same base before solving the equation. Substitute the answer choice into the original equation to check your answer.

5. Select all true statements about the graphs of $y = 3^x$ and $y = \left(\frac{1}{3}\right)^x$

   ☐ A Both graphs have a domain of all real numbers.

   ☐ B Both graphs have a range of all real numbers.

   ☐ C Both graphs have a y-intercept of 1.

   ☐ D The graph of $y = 3^x$ increases as x increases.

   ☐ E The graph of $y = \left(\frac{1}{3}\right)^x$ shows exponential growth.

6. Simplfy the expression. Assume the denominator is not 0.

   $$\left(\frac{3b^6r^8}{24b^{11}r}\right)^4$$

   ○ A $\frac{1}{b^{20}}$

   ○ B $\frac{r^{32}}{4096b^{20}}$

   ○ C $\frac{r^{28}}{8b^{20}}$

   ○ D $\frac{r^{28}}{4096b^{20}}$

7. Juaquim's starting salary at a company is $42,000 per year. On average, employees receive a 6.5% increase in salary every year. Which equation shows what his salary will be in x years?

   ○ A $s = (1 + 0.065)^x$

   ○ B $s = 42{,}000(0.065)^x$

   ○ C $s = 42{,}000(1 + 0.065)^x$

   ○ D $s = 42{,}000(1 - 0.065)^x$

8. Which function represents the data in the table?

x	0	1	2	3
y	6	12	24	48

   ○ A $y = 6(2^x)$ ○ C $y = 2^x + 6$

   ○ B $y = 2(6^x)$ ○ D $y = 6^x$

9. For which expression is $3xy^3z^6$ the square root?

- **A** $3x^2y^6z^{12}$
- **B** $6xy^3z^6$
- **C** $9x^3y^5z^8$
- **D** $9x^2y^6z^{12}$

10. What is the 7th term of the sequence?

$$13, 52, 208, 832, \ldots$$

11. On the coordinate plane, Roger draws a line that is parallel to the given line and passes through the point $(-4, 1)$.

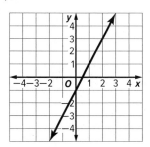

What is the equation of Roger's line?

- **A** $y = 2x + 9$
- **B** $y = 2x - 1$
- **C** $y = \frac{1}{2}x + 3$
- **D** $y = -\frac{1}{2}x - 1$

12. Find the value of the exponent.

$$\left(\frac{a^5bc^3}{b^4c^2}\right)^? = \frac{b^{27}}{a^{45}c^9}$$

13. Which equation represents the nth term of the arithmetic sequence?

$$-6, -1, 4, 9, \ldots$$

- **A** $a_n = -5n + 11$
- **B** $a_n = -5n - 1$
- **C** $a_n = 5n - 1$
- **D** $a_n = 5n - 11$

14. What is the recursive formula for the sequence represented in the table?

n	1	2	3	4
a_n	405	135	45	15

- **A** $a_1 = 405, a_n = 3a_{n-1}$
- **B** $a_1 = 405, a_n = \frac{1}{3}a_1$
- **C** $a_1 = 405, a_n = \frac{1}{3}a_{n-1}$
- **D** $a_1 = 405, a_n = 405 - \left(\frac{1}{3}\right)^{n-1}$

15. There were originally 460 spores in a petri dish. After each inspection, there were 10% fewer spores.

A Write a function for the number of spores $S(t)$ after t inspections.

B Approximately how many spores were in the petri dish after 4 inspections?

C **MP** What mathematical practice did you use to solve this problem?

Need Extra Help?

If you missed Question...	1	2	3	4	5	6	7	8	9	10	11	12	13	14	15
Go to Lesson...	7-3	5-2	7-1	7-4	7-5	7-2	7-7	7-7	7-1	7-9	4-3	7-2	3-6	7-10	7-9

CHAPTER 8
Polynomials

THEN

You applied the laws of exponents and explored exponential functions.

NOW

In this chapter, you will:

- Add, subtract, and multiply polynomials.
- Factor trinomials.
- Factor differences of squares.
- Factor perfect squares.

(MP) WHY

PROJECTILES Polynomial expressions, including quadratic expressions, can be used to model the heights of objects during a period of motion. These models are used to model many situations such as firing fireworks off of the ground or off of a structure like a bridge. The quadratic expression $-16t^2 + v_0 t + h_0$ is used model the height in feet of an object where v_0 is the initial velocity of the object in feet per second, h_0 is the initial height in feet, and t is time in seconds. These expressions can then be factored.

Use the mathematical practices to complete the activity.

1. Use Tools Use the Internet to find a bridge. Choose one to model and find its height.

2. Model with Mathematics Suppose a firework was fired off of the bridge you chose with an initial velocity of 30 feet per second. Write a quadratic expression that represents the situation.

3. Discuss Compare your expression with that of your classmates. What similarities and differences do you notice?

4. Use Structure Is there a common factor between the terms in your expression? If so, what is it?

Spaces Images/Blend Images

Go Online to Guide Your Learning

Explore & Explain

 Product Mat and Algebra Tiles

Use the Product Mat and the Algebra Tiles tool to model multiplying polynomials. This will enhance your understanding of the math concepts discussed in Lesson 8-3.

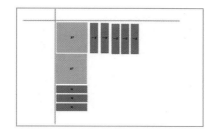

 The Geometer's Sketchpad

Use The Geometer's Sketchpad to find squares and products of sums and differences.

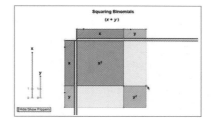

eBook

Interactive Student Guide

Before starting the chapter, answer the **Chapter Focus** preview questions. Check your answers as you complete each lesson. At the end of the chapter, try the **Performance Task**.

Organize

 **Foldables**

Make this Foldable to help you organize your notes about quadratic expressions. Fold 5 sheets of grid paper in half. Label the first page with the title of the chapter. Label the rest of the pages with lesson numbers.

Collaborate

 Chapter Project

In the **Greens Going Green** project, you will use what you have learned about factoring and multiplying polynomials to complete a project that addresses environmental literacy .

Focus

 LEARNSMART

Need help studying? Complete the **Expressions and Equations** domain in LearnSmart to review for the chapter test.

ALEKS

You can use the **Polynomials and Factoring** topic in ALEKS to find out what you know about linear functions and what you are ready to learn.*

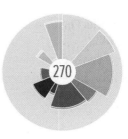

* Ask your teacher if this is part of your program.

Get Ready for the Chapter

Go Online! for Vocabulary Review Games and key vocabulary in 13 languages.

Connecting Concepts

Concept Check

Review the concepts used in this chapter by answering each question below.

1. State the Distributive Property.

2. How would you use the Distributive Property to simplify $6x(3+x)$?

3. How do you determine the area of a rectangle? Write that equation for the rectangle shown.

   $b + 3c$

   a

4. Use the Distributive Property to simplify the expression for the area of the rectangle shown.

5. Two numbers have a product of -12 and a sum of 4. What are the numbers?

6. Can $5a - 2 + 6a$ be simplified? If so, find the simplified expression.

7. Can $(-2y^3)(9y^4)$ be simplified? If so, find the simplified expression.

8. Write the exponent law for multiplying powers, for example $m^3(m^4)$.

Performance Task Preview

You will use the concepts and skills in this chapter to solve problems about installing hot tubs and swimming pools. Understanding polynomials will help you finish the Performance Task at the end of the chapter.

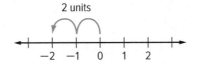

MP **In this Performance Task you will:**

- make sense of problems
- look for and express regularity in repeated reasoning

New Vocabulary

English		Español
polynomial	p. 491	polinomio
binomial	p. 491	binomio
trinomial	p. 491	trinomio
degree of a monomial	p. 491	grado de un monomio
degree of a polynomial	p. 491	grado de un polinomio
standard form of a polynomial	p. 492	forma estándar de polinomio
leading coefficient	p. 492	coeficiente lider
FOIL method	p. 507	método foil
quadratic expression	p. 507	expression cuadrática
factoring	p. 520	factorización
factoring by grouping	p. 521	factorización por agrupamiento
prime polynomial	p. 535	polinomio primo
difference of two squares	p. 539	diferencia de cuadrados
perfect square trinomial	p. 540	trinomio cuadrado perfecto

Review Vocabulary

absolute value valor absoluto the absolute value of any number n is the distance the number is from zero on a number line and is written $|n|$

2 units

$$\begin{array}{ccccccc} & & & & & & \\ -2 & -1 & 0 & 1 & 2 & & \end{array}$$

The absolute value of -2 is 2 because it is 2 units from 0.

perfect square cuadrado perfecto a number with a square root that is a rational number

Algebra Lab
Adding and Subtracting Polynomials

Algebra tiles can be used to model polynomials. A polynomial is a monomial or the sum of monomials. The diagram below shows the models.

Mathematical Practices
 5 Use appropriate tools strategically.

Polynomial Models

- Polynomials are modeled using three types of tiles.

- Each tile has an opposite.

Activity 1   Model Polynomials

Work cooperatively. Use algebra tiles to model each polynomial.

$5x$

To model this polynomial, you will need five green x-tiles.

$-2x^2 + x + 3$

To model this polynomial, you will need two red $-x^2$-tiles, one green x-tile, and three yellow 1-tiles.

Monomials such as $3x$ and $-2x$ are called *like terms* because they have the same variable to the same power.

Polynomial Models

- Like terms are represented by tiles that have the same shape and size.

like terms

- A *zero pair* may be formed by pairing one tile with its opposite. You can remove or add zero pairs without changing the polynomial.

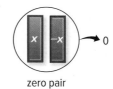
zero pair

Activity 2   Add Polynomials

Work cooperatively. Use algebra tiles to find $(2x^2 - 3x + 5) + (x^2 + 6x - 4)$.

Step 1

Model each polynomial.

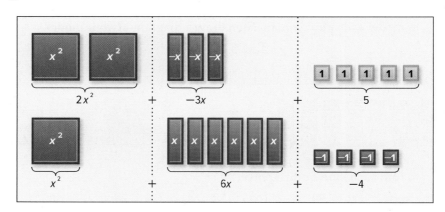

Adding and Subtracting Polynomials Continued

Step 2

Combine like terms and remove zero pairs.

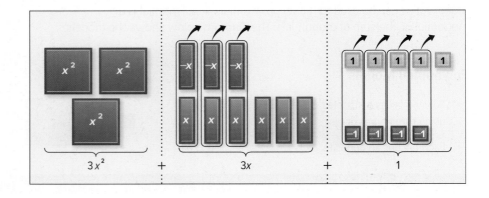

$3x^2$ + $3x$ + 1

Step 3 Write the polynomial. $(2x^2 - 3x + 5) + (x^2 + 6x - 4) = 3x^2 + 3x + 1$

Activity 3   Subtract Polynomials

Work cooperatively. Use algebra tiles to find $(4x + 5) - (-3x + 1)$**.**

Step 1 Model the polynomial $4x + 5$.

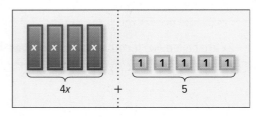

$4x$ + 5

Step 2 To subtract $-3x + 1$, remove three red $-x$-tiles and one yellow 1-tile. You can remove the 1-tile, but there are no $-x$-tiles. Add 3 zero pairs of x-tiles. Then remove the three red $-x$-tiles.

Step 3 Write the polynomial.
$(4x + 5) - (-3x + 1) = 7x + 4$

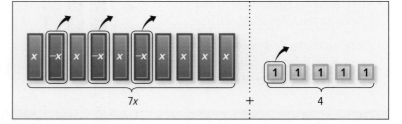

$7x$ + 4

Model and Analyze

Use algebra tiles to model each polynomial. Then draw a diagram of your model.

1. $-2x^2$
2. $5x - 4$
3. $x^2 - 4x$

Write an algebraic expression for each model.

4.

5.

Use algebra tiles to find each sum or difference.

6. $(x^2 + 5x - 2) + (3x^2 - 2x + 6)$ **7.** $(2x^2 + 8x + 1) - (x^2 - 4x - 2)$ **8.** $(-4x^2 + x) - (x^2 + 5x)$

Adding and Subtracting Polynomials

- You identified monomials and their characteristics.

- **1** Write polynomials in standard form.
- **2** Add and subtract polynomials.

- The sales data of digital audio players can be modeled by the equation

$U = -2.7t^2 + 49.4t + 128.7$, where U is the number of units shipped in millions and t is the number of years since 2005.

The expression $-2.7t^2 + 49.4t + 128.7$ is an example of a polynomial. Polynomials can be used to model situations.

New Vocabulary

polynomial
binomial
trinomial
degree of a monomial
degree of a polynomial
standard form of a polynomial
leading coefficient

Mathematical Practices

3 Construct viable arguments and critique the reasoning of others.

1 Polynomials in Standard Form A **polynomial** is a monomial or the sum of monomials, each called a *term* of the polynomial. Some polynomials have special names. A **binomial** is the sum of *two* monomials, and a **trinomial** is the sum of *three* monomials.

Monomial	Binomial	Trinomial
$5x$	$2x^2 + 7$	$x^3 - 10x + 1$

The **degree of a monomial** is the sum of the exponents of all its variables. A nonzero constant term has degree 0, and zero has no degree.

The **degree of a polynomial** is the greatest degree of any term in the polynomial. You can find the degree of a polynomial by finding the degree of each term. Polynomials are named based on their degree.

Degree	Name
0	constant
1	linear
2	quadratic
3	cubic
4	quartic
5	quintic
6 or more	6th degree, 7th degree, and so on

Example 1   Identify Polynomials

Determine whether each expression is a polynomial. If it is a polynomial, find the degree and determine whether it is a *monomial*, *binomial*, or *trinomial*.

Expression	Is it a polynomial?	Degree	Monomial, binomial, or trinomial?
a. $4y - 5$	Yes; $4y - 5$ is the sum of $4y$ and -5.	1	binomial
b. -6.5	Yes; -6.5 is a real number.	0	monomial
c. $7a^{-3} + 9b$	No; $7a^{-3} = \frac{7}{a^3}$, which is not a monomial.	—	———
d. $6x^3 + 4x + x + 3$	Yes; $6x^3 + 4x + x + 3 = 6x^3 + 5x + 3$, the sum of three monomials.	3	trinomial

▶ **Guided Practice**

1A. x

1B. $-3y^2 - 2y + 4y - 1$

1C. $5rx + 7tuv$

1D. $10x^{-4} - 8x^a$

The terms of a polynomial can be written in any order. However, polynomials in one variable are usually written in standard form. The **standard form of a polynomial** has the terms in order from greatest to least degree. In this form, the coefficient of the first term is called the **leading coefficient**.

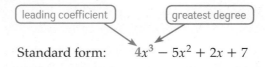

| leading coefficient | | greatest degree |

Standard form: $4x^3 - 5x^2 + 2x + 7$

Example 2   Standard Form of a Polynomial

Write each polynomial in standard form. Identify the leading coefficient.

a. $3x^2 + 4x^5 - 7x$

Find the degree of each term.

Degree: 2   5   1

Polynomial: $3x^2 + 4x^5 - 7x$

The greatest degree is 5. Therefore, the polynomial can be rewritten as $4x^5 + 3x^2 - 7x$, with a leading coefficient of 4.

b. $5y - 9 - 2y^4 - 6y^3$

Find the degree of each term.

Degree: 1   0   4   3

Polynomial: $5y - 9 - 2y^4 - 6y^3$

The greatest degree is 4. Therefore, the polynomial can be rewritten as $-2y^4 - 6y^3 + 5y - 9$, with a leading coefficient of -2.

▶ **Guided Practice**

2A. $8 - 2x^2 + 4x^4 - 3x$

2B. $y + 5y^3 - 2y^2 - 7y^6 + 10$

2 Add and Subtract Polynomials Adding polynomials involves adding like terms. You can group like terms by using a horizontal or vertical format.

Example 3   Add Polynomials

Find each sum.

a. $(2x^2 + 5x - 7) + (3 - 4x^2 + 6x)$

Horizontal Method

Group and combine like terms.

$(2x^2 + 5x - 7) + (3 - 4x^2 + 6x)$

$\qquad = [2x^2 + (-4x^2)] + [5x + 6x] + [-7 + 3]$ Group like terms.

$\qquad = -2x^2 + 11x - 4$ Combine like terms.

b. $(3y + y^3 - 5) + (4y^2 - 4y + 2y^3 + 8)$

Vertical Method

Align like terms in columns and combine.

$\qquad y^3 + 0y^2 + 3y - 5$ Insert a placeholder to help align the terms.

$\underline{(+)\ 2y^3 + 4y^2 - 4y + 8}$ Align and combine like terms.

$\qquad 3y^3 + 4y^2 - \ y + 3$

▶ **Guided Practice**

3A. $(5x^2 - 3x + 4) + (6x - 3x^2 - 3)$

3B. $(y^4 - 3y + 7) + (2y^3 + 2y - 2y^4 - 11)$

Study Tip

MP Sense-Making When finding the additive inverse of a polynomial, you are multiplying every term by −1.

You can subtract a polynomial by adding its additive inverse. To find the additive inverse of a polynomial, write the opposite of each term, as shown.

$$-(3x^2 + 2x - 6) = \underbrace{-3x^2 - 2x + 6}_{\text{Additive Inverse}}$$

Example 4   Subtract Polynomials

Find each difference.

a. $(3 - 2x + 2x^2) - (4x - 5 + 3x^2)$

Horizontal Method

Subtract $4x - 5 + 3x^2$ by adding its additive inverse.

$(3 - 2x + 2x^2) - (4x - 5 + 3x^2)$

$= (3 - 2x + 2x^2) + (-4x + 5 - 3x^2)$ — The additive inverse of $4x - 5 + 3x^2$ is $-4x + 5 - 3x^2$

$= [2x^2 + (-3x^2)] + [(-2x) + (-4x)] + [3 + 5]$ — Group like terms.

$= -x^2 - 6x + 8$ — Combine like terms.

b. $(7p + 4p^3 - 8) - (3p^2 + 2 - 9p)$

Vertical Method

Align like terms in columns and subtract by adding the additive inverse.

$$\begin{array}{r} 4p^3 + 0p^2 + 7p - 8 \\ (-) \qquad 3p^2 - 9p + 2 \\ \hline \end{array}$$

Add the opposite.

$$\begin{array}{r} 4p^3 + 0p^2 + 7p - 8 \\ (+) \quad -3p^2 + 9p - 2 \\ \hline 4p^3 - 3p^2 + 16p - 10 \end{array}$$

Guided Practice

4. $(4x^3 - 3x^2 + 6x - 4) - (-2x^3 + x^2 - 2)$

Adding or subtracting integers results in an integer, so the set of integers is closed under addition and subtraction. Similarly, adding or subtracting polynomials results in a polynomial, so the set of polynomials is closed under addition and subtraction.

Real-World Example 5   Add and Subtract Polynomials

ELECTRONICS The equations $P = 7m + 137$ and $C = 4m + 78$ represent the number of smartphones P and gaming consoles C sold in m months at an electronics store. Write an equation for the total monthly sales T of phones and gaming consoles. Then predict the number of phones and gaming consoles sold in 10 months.

To write an equation that represents the total sales T, add the equations that represent the number of smartphones P and gaming consoles C.

$T = 7m + 137 + 4m + 78$

$\quad = 11m + 215$

Substitute 10 for m to predict the number of phones and gaming consoles sold in 10 months.

$T = 11(10) + 215$

$\quad = 110 + 215$ or 325

Therefore, a total of 325 smartphones and gaming consoles will be sold in 10 months.

Guided Practice

5. Use the information above to write an equation that represents the difference in the monthly sales of smartphones and the monthly sales of gaming consoles. Use the equation to predict the difference in monthly sales in 24 months.

Real-World Link

Sales of smartphones have been steadily increasing by a rate of 4% each year.

Source: Business Wire

Ron Levine/Photodisc/Getty Images

Check Your Understanding ○ = Step-by-Step Solutions begin on page R11.

✓ *Go Online!* for a Self-Check Quiz

Example 1 Determine whether each expression is a polynomial. If it is a polynomial, find the degree and determine whether it is a *monomial, binomial,* or *trinomial.*

1. $7ab + 6b^2 - 2a^3$

2. $2y - 5 + 3y^2$

3. $3x^2$

4. $\dfrac{4m}{3p}$

5. $5m^2p^3 + 6$

6. $5q^{-4} + 6q$

Example 2 Write each polynomial in standard form. Identify the leading coefficient.

7. $2x^5 - 12 + 3x$

8. $-4d^4 + 1 - d^2$

9. $4z - 2z^2 - 5z^4$

10. $2a + 4a^3 - 5a^2 - 1$

Examples 3–4 Find each sum or difference.

11. $(6x^3 - 4) + (-2x^3 + 9)$

12. $(g^3 - 2g^2 + 5g + 6) - (g^2 + 2g)$

13. $(4 + 2a^2 - 2a) - (3a^2 - 8a + 7)$

14. $(8y - 4y^2) + (3y - 9y^2)$

15. $(-4z^3 - 2z + 8) - (4z^3 + 3z^2 - 5)$

16. $(-3d^2 - 8 + 2d) + (4d - 12 + d^2)$

17. $(y + 5) + (2y + 4y^2 - 2)$

18. $(3n^3 - 5n + n^2) - (-8n^2 + 3n^3)$

Example 5 **19.** **MP** **SENSE-MAKING** The total number of students T who traveled for spring break consists of two groups: students who flew to their destinations F and students who drove to their destination D. The number (in thousands) of students who flew and the total number of students who flew or drove can be modeled by the following equations, where n is the number of years since 2010.

$$T = 14n + 21 \qquad F = 8n + 7$$

a. Write an equation that models the number of students who drove to their destination for this time period.

b. Predict the number of students who will drive to their destination in 2027.

c. How many students will drive or fly to their destination in 2030?

Practice and Problem Solving

Extra Practice is on page R8.

Example 1 Determine whether each expression is a polynomial. If it is a polynomial, find the degree and determine whether it is a *monomial, binomial,* or *trinomial.*

20. $\dfrac{5y^3}{x^2} + 4x$

21. 21

22. $c^4 - 2c^2 + 1$

23. $d + 3d^c$

24. $a - a^2$

25. $5n^3 + nq^3$

Example 2 Write each polynomial in standard form. Identify the leading coefficient.

26. $5x^2 - 2 + 3x$

27. $8y + 7y^3$

28. $4 - 3c - 5c^2$

29. $-y^3 + 3y - 3y^2 + 2$

30. $11t + 2t^2 - 3 + t^5$

31. $2 + r - r^3$

32. $\dfrac{1}{2}x - 3x^4 + 7$

33. $-9b^2 + 10b - b^6$

Examples 3-4 **Find each sum or difference.**

34. $(2c^2 + 6c + 4) + (5c^2 - 7)$

35 $(2x + 3x^2) - (7 - 8x^2)$

36. $(3c^3 - c + 11) - (c^2 + 2c + 8)$

37. $(z^2 + z) + (z^2 - 11)$

38. $(2x - 2y + 1) - (3y + 4x)$

39. $(4a - 5b^2 + 3) + (6 - 2a + 3b^2)$

40. $(x^2y - 3x^2 + y) + (3y - 2x^2y)$

41. $(-8xy + 3x^2 - 5y) + (4x^2 - 2y + 6xy)$

42. $(5n - 2p^2 + 2np) - (4p^2 + 4n)$

43. $(4rxt - 8r^2x + x^2) - (6rx^2 + 5rxt - 2x^2)$

Example 5 **44. PETS** From 2006 through 2016, suppose the number of dogs D and the number of cats C (in hundreds) adopted from animal shelters in a region of the United States are modeled by the equations $D = 2n + 3$ and $C = n + 4$, where n is the number of years since 2006.

    a. Write a function that models the total number T of dogs and cats adopted in hundreds for this time period.

    b. If this trend continues, how many dogs and cats will be adopted in 2020?

Classify each polynomial according to its degree and number of terms.

45. $4x - 3x^2 + 5$

46. $11z^3$

47. $9 + y^4$

48. $3x^3 - 7$

49. $-2x^5 - x^2 + 5x - 8$

50. $10t - 4t^2 + 6t^3$

51. ENROLLMENT In a rapidly growing school system, the number (in hundreds) of total students is represented by N and the number of students in kindergarten through fifth grade is represented by P. The equations $N = 1.25t^2 - t + 7.5$ and $P = 0.7t^2 - 0.95t + 3.8$ model the number of students enrolled from 2006 to 2015, where t is the number of years since 2006.

    a. Write an equation modeling the number of students S in grades 6 through 12 enrolled for this time period.

    b. How many students were enrolled in grades 6 through 12 in the school system in 2013?

52. MP REASONING The perimeter of the triangle can be represented by the expression $3x^2 - 7x + 2$. Write a polynomial that represents the measure of the third side.

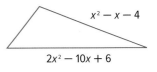

53. GEOMETRY Consider the rectangle.

    a. What does $(4x^2 + 2x - 1)(2x^2 - x + 3)$ represent?

    b. What does $2(4x^2 + 2x - 1) + 2(2x^2 - x + 3)$ represent?

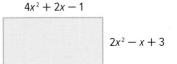

Find each sum or difference.

54. $(4x + 2y - 6z) + (5y - 2z + 7x) + (-9z - 2x - 3y)$

55. $(5a^2 - 4) + (a^2 - 2a + 12) + (4a^2 - 6a + 8)$

56. $(3c^2 - 7) + (4c + 7) - (c^2 + 5c - 8)$

57. $(3n^3 + 3n - 10) - (4n^2 - 5n) + (4n^3 - 3n^2 - 9n + 4)$

58. FOOTBALL A school district has two high schools, North and South. From 2010 through 2015, the total attendance T at games for both schools and at games for North High School N can be modeled by the following equations, where x is the number of years since 2010.

$$T = -0.69x^3 + 55.38x^2 + 643.31x + 10,538 \qquad N = -3.78x^3 + 58.96x^2 + 265.96x + 5257$$

Estimate how many people attended South High School football games in 2015.

59 **CAR RENTAL** The cost to rent a car for a day is $25 plus $0.35 for each mile driven.

    a. Write a polynomial that represents the cost of renting a car for m miles.

    b. If a car is driven 145 miles for one day, how much would it cost to rent?

    c. If a car is driven 105 miles each day for four days, how much would it cost to rent a car?

    d. If a car is driven 220 miles each day for seven days, how much would it cost to rent a car?

60. **MULTIPLE REPRESENTATIONS** In this problem, you will explore perimeter and area.

    a. **Geometric** Draw three rectangles that each have a perimeter of 400 feet.

    b. **Tabular** Record the width and length of each rectangle in a table like the one shown below. Find the area of each rectangle.

Rectangle	Length	Width	Area
1	100 ft		
2	50 ft		
3	75 ft		
4	x ft		

    c. **Graphical** On a coordinate system, graph the area of rectangle 4 in terms of the length, x. Use the graph to determine the largest area possible.

    d. **Analytical** Determine the length and width that produce the largest area.

H.O.T. Problems   Use **H**igher-**O**rder **T**hinking Skills

61. **ERROR ANALYSIS** Cheyenne and Sebastian are finding $(2x^2 - x) - (3x + 3x^2 - 2)$. Is either of them correct? Explain your reasoning.

Cheyenne	Sebastian
$(2x^2 - x) - (3x + 3x^2 - 2)$	$(2x^2 - x) - (3x + 3x^2 - 2)$
$= (2x^2 - x) + (-3x + 3x^2 - 2)$	$= (2x^2 - x) + (-3x - 3x^2 - 2)$
$= 5x^2 - 4x - 2$	$= -x^2 - 4x - 2$

62. **MP REASONING** Determine whether each of the following statements is *true* or *false*. Explain your reasoning.

    a. A binomial can have a degree of zero.

    b. The order in which polynomials are subtracted does not matter.

63. **CHALLENGE** Write a polynomial that represents the sum of an odd integer $2n + 1$ and the next two consecutive odd integers.

64. **WRITING IN MATH** Why would you add or subtract equations that represent real-world situations? Explain.

65. **WRITING IN MATH** Describe how to add and subtract polynomials using both the vertical and horizontal formats.

66. An equilateral triangle is shown below.

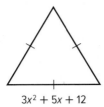

$3x^2 + 5x + 12$

What is its perimeter? **MP** 2

- **A** $24x + 36$
- **B** $6x^2 + 10x + 24$
- **C** $9x^2 + 15x + 36$
- **D** $9x^6 + 15x^3 + 36$

67. Which expression is equivalent to $3x^2 + 14 - (7x - 6) + 29 + (3x^2 + 5x) + 9x$?
MP 7

- **A** $3x^2 + 21x + 37$
- **B** $6x^2 - 2x + 37$
- **C** $6x^2 + 7x + 49$
- **D** $6x^2 + 21x + 49$

68. The perimeter of a triangle is $3x^2 + 6x - 9$. If two of the side lengths add to $2x^2 + 4x - 6$, what is the length of the third side? **MP** 1

69. The age of a child is $2x - 5$. The age of her mother is $x^2 + 4$. Find an expression for the sum of their ages 5 years from now. **MP** 7

70. MULTI-STEP Two rectangles are shown. **MP** 1, 3

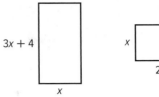

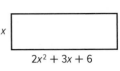

$3x + 4$

x

x

$2x^2 + 3x + 6$

a. What is an expression for the perimeter of the rectangle on the left? **MP** 2

- **A** $3x + 5$
- **B** $4x + 4$
- **C** $6x + 8$
- **D** $8x + 8$

b. Write an expression for the perimeter of the rectangle on the right.

c. Write an expression for the total perimeter of both rectangles.

d. The two rectangles are combined to form a longer rectangle with width x. Find the perimeter of the new rectangle. Show your work.

e. Is the total perimeter of both rectangles you found in **c** the same as the perimeter you found after combining the two rectangles in **d**? Why or why not?

f. If $x = 1.5$ cm, what is the perimeter of the new rectangle you found in part **d**?

71. A board game has square and equilateral triangle pieces. If the square and equilateral triangle pieces both have the same side length of $2x + 1$, find the sum of the perimeters of one square and one equilateral triangle. **MP** 2

72. A library holds $x^2 - 2x + 10$ books. Alice checks out $3x - 5$ of them. How many books are left in the library? **MP** 1

Multiplying a Polynomial by a Monomial

:: **Then**
- You multiplied monomials.

:: **Now**
1. Multiply a polynomial by a monomial.
2. Solve equations involving the products of monomials and polynomials.

:: **Why?**
- Charmaine Brooks is opening a fitness club. She tells the contractor that the length of the fitness room should be three times the width plus 8 feet.

  To cover the floor with mats for exercise classes, Ms. Brooks needs to know the area of the floor. So she multiplies the width times the length, $w(3w + 8)$.

 Mathematical Practices

3 Construct viable arguments and critique the reasoning of others.

8 Look for and express regularity in repeated reasoning.

1 Polynomial Multiplied by Monomial To find the product of a polynomial and a monomial, you can use the Distributive Property.

Example 1   Multiply a Polynomial by a Monomial

Find $-3x^2(7x^2 - x + 4)$.

Horizontal Method

$-3x^2(7x^2 - x + 4)$ Original expression

$= -3x^2(7x^2) - (-3x^2)(x) + (-3x^2)(4)$ Distributive Property

$= -21x^4 - (-3x^3) + (-12x^2)$ Multiply.

$= -21x^4 + 3x^3 - 12x^2$ Simplify.

Vertical Method

$$\begin{array}{r} 7x^2 - x + 4 \\ (\times) \qquad\quad -3x^2 \\ \hline -21x^4 + 3x^3 - 12x^2 \end{array}$$

Distributive Property
Multiply.

> **Guided Practice**

Find each product.

1A. $5a^2(-4a^2 + 2a - 7)$ **1B.** $-6d^3(3d^4 - 2d^3 - d + 9)$

We can use this same method more than once to simplify large expressions.

Example 2   Simplify Expressions

Simplify $2p(-4p^2 + 5p) - 5(2p^2 + 20)$.

$2p(-4p^2 + 5p) - 5(2p^2 + 20)$ Original expression

$= (2p)(-4p^2) + (2p)(5p) + (-5)(2p^2) + (-5)(20)$ Distributive Property

$= -8p^3 + 10p^2 - 10p^2 - 100$ Multiply.

$= -8p^3 + (10p^2 - 10p^2) - 100$ Commutative and Associative Properties

$= -8p^3 - 100$ Combine like terms.

Simplify each expression.

2A. $3(5x^2 + 2x - 4) - x(7x^2 + 2x - 3)$ **2B.** $15t(10y^3t^5 + 5y^2t) - 2y(yt^2 + 4y^2)$

We can use the Distributive Property to multiply monomials by polynomials and solve real-world problems.

Real-World Example 3   Write and Evaluate a Polynomial Expression

DANCE The theme for a school dance is "Solid Gold." For one decoration, Kana is covering a trapezoid-shaped piece of poster board with metallic gold paper to look like a bar of gold. If the height of the poster board is 18 inches, how much metallic paper will Kana need in square inches?

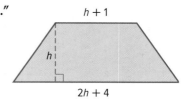

Understand   You want to find the area of the poster board.

Plan   Since the poster board is a trapezoid, Area $= \frac{1}{2} \cdot$ height $\cdot$ (base$_1$ + base$_2$).

Solve   Let $h =$ the height of the poster, $b_1 = h + 1$.
Let $b_1 = h + 1$, let $b_2 = 2h + 4$ and let $h =$ height of the trapezoid.

$$A = \frac{1}{2}h(b_1 + b_2) \quad\quad \text{Area of a trapezoid}$$

$$= \frac{1}{2}h[(h + 1) + (2h + 4)] \quad\quad b_1 = h + 1 \text{ and } b_2 = 2h + 4$$

$$= \frac{1}{2}h(3h + 5) \quad\quad \text{Add and simplify.}$$

$$= \frac{3}{2}h^2 + \frac{5}{2}h \qu\quad \text{Distributive Property}$$

$$= \frac{3}{2}(18)^2 + \frac{5}{2}(18) \quad\quad h = 18$$

$$= 531 \quad\quad \text{Simplify.}$$

Kana will need 531 square inches of metallic paper.

Check $b_1 = h + 1 = 18 + 1 = 19$ and $b_2 = 2h + 4 = 2(18) + 4 = 40$

$$A = \frac{1}{2}h(b_1 + b_2) = \frac{1}{2}(18)(19 + 40) = \frac{1}{2}(18)(59) = 531 \checkmark$$

Study Tip

MP Tools Choosing the right formula is an important step in solving problems. If you don't already have a formula sheet, you might want to make one of your own to use while studying.

Guided Practice

3. DANCE Kachima is making triangular bandanas for the dogs and cats in her pet club. The base of the bandana is the length of the collar with 4 inches added to each end to tie it on. The height is $\frac{1}{2}$ of the collar length.

A. If Kachima's dog has a collar length of 12 inches, how much fabric does she need in square inches?

B. If Kachima makes a bandana for her friend's cat with a 6-inch collar, how much fabric does Kachima need in square inches?

Real-World Link

In a recent year, the pet supply business hit an estimated $7.05 billion in sales. This business ranges from gourmet food to rhinestone tiaras, pearl collars, and cashmere coats.

Source: *Entrepreneur Magazine*

Big Cheese Photo/SuperStock

2 **Solve Equations with Polynomial Expressions** We can use the Distributive Property to solve equations that involve the products of monomials and polynomials.

Go Online!

When simplifying a line expression, it may be helpful to mark like terms with similar shapes or highlighting. Log into your **eStudent Edition** to use the pen or highlighting tools with your text.

Example 4   Equations with Polynomials on Both Sides

Solve $2a(5a - 2) + 3a(2a + 6) + 8 = a(4a + 1) + 2a(6a - 4) + 50$.

$2a(5a - 2) + 3a(2a + 6) + 8 = a(4a + 1) + 2a(6a - 4) + 50$	Original equation
$10a^2 - 4a + 6a^2 + 18a + 8 = 4a^2 + a + 12a^2 - 8a + 50$	Distributive Property
$16a^2 + 14a + 8 = 16a^2 - 7a + 50$	Combine like terms.
$14a + 8 = -7a + 50$	Subtract $16a^2$ from each side.
$21a + 8 = 50$	Add $7a$ to each side.
$21a = 42$	Subtract 8 from each side.
$a = 2$	Divide each side by 21.

CHECK

$$2a(5a - 2) + 3a(2a + 6) + 8 = a(4a + 1) + 2a(6a - 4) + 50$$
$$2(2)[5(2) - 2] + 3(2)[2(2) + 6] + 8 \stackrel{?}{=} 2[4(2) + 1] + 2(2)[6(2) - 4] + 50$$
$$4(8) + 6(10) + 8 \stackrel{?}{=} 2(9) + 4(8) + 50 \quad \text{Simplify.}$$
$$32 + 60 + 8 \stackrel{?}{=} 18 + 32 + 50 \quad \text{Multiply.}$$
$$100 = 100 \checkmark \quad \text{Add and subtract.}$$

▶ **Guided Practice**

Solve each equation.

4A. $2x(x + 4) + 7 = (x + 8) + 2x(x + 1) + 12$

4B. $d(d + 3) - d(d - 4) = 9d - 16$

Check Your Understanding   = Step-by-Step Solutions begin on page R11.

Go Online! for a Self-Check Quiz

Example 1   Find each product.

  1. $5w(-3w^2 + 2w - 4)$ **2.** $6g^2(3g^3 + 4g^2 + 10g - 1)$

  3. $4km^2(8km^2 + 2k^2m + 5k)$ **4.** $-3p^4r^3(2p^2r^4 - 6p^6r^3 - 5)$

  (5) $2ab(7a^4b^2 + a^5b - 2a)$ **6.** $c^2d^3(5cd^7 - 3c^3d^2 - 4d^3)$

Example 2   Simplify each expression.

  7. $t(4t^2 + 15t + 4) - 4(3t - 1)$ **8.** $x(3x^2 + 4) + 2(7x - 3)$

  9. $-2d(d^3c^2 - 4dc^2 + 2d^2c) + c^2(dc^2 - 3d^4)$

 10. $-5w^2(8w^2x - 11wx^2) + 6x(9wx^4 - 4w - 3x^2)$

Example 3   **11. TELEVISIONS** Marlene is buying a new LED television. The height of the screen of the television is one half the width plus 10 inches. The width is 40 inches. Find the height of the screen in inches.

Example 4   Solve each equation.

 12. $-6(11 - 2c) = 7(-2 - 2c)$ **13.** $t(2t + 3) + 20 = 2t(t - 3)$

 14. $-2(w + 1) + w = 7 - 4w$ **15.** $3(y - 2) + 2y = 4y + 14$

 16. $a(a + 3) + a(a - 6) + 35 = a(a - 5) + a(a + 7)$

 17. $n(n - 4) + n(n + 8) = n(n - 13) + n(n + 1) + 16$

Example 1   Find each product.

    18. $b(b^2 - 12b + 1)$ **19.** $f(f^2 + 2f + 25)$

    20. $-3m^3(2m^3 - 12m^2 + 2m + 25)$ **21.** $2j^2(5j^3 - 15j^2 + 2j + 2)$

    22. $2pr^2(2pr + 5p^2r - 15p)$ **23.** $4t^3u(2t^2u^2 - 10tu^4 + 2)$

Example 2   Simplify each expression.

    24. $-3(5x^2 + 2x + 9) + x(2x - 3)$ **25.** $a(-8a^2 + 2a + 4) + 3(6a^2 - 4)$

    26. $-4d(5d^2 - 12) + 7(d + 5)$ **27.** $-9g(-2g + g^2) + 3(g^2 + 4)$

    28. $2j(7j^2k^2 + jk^2 + 5k) - 9k(-2j^2k^2 + 2k^2 + 3j)$

    29. $4n(2n^3p^2 - 3np^2 + 5n) + 4p(6n^2p - 2np^2 + 3p)$

Example 3   **30. DAMS** A new dam being built has the shape of a trapezoid. The length of the base at the bottom of the dam is 2 times the height. The length of the base at the top of the dam is $\frac{1}{5}$ times the height minus 30 feet.

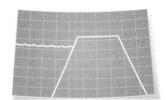

    a. Write an expression to find the area of the trapezoidal cross section of the dam.

    b. If the height of the dam is 180 feet, find the area of this cross section.

Example 4   Solve each equation.

    31 $7(t^2 + 5t - 9) + t = t(7t - 2) + 13$

    32. $w(4w + 6) + 2w = 2(2w^2 + 7w - 3)$

    33. $5(4z + 6) - 2(z - 4) = 7z(z + 4) - z(7z - 2) - 48$

    34. $9c(c - 11) + 10(5c - 3) = 3c(c + 5) + c(6c - 3) - 30$

    35. $2f(5f - 2) - 10(f^2 - 3f + 6) = -8f(f + 4) + 4(2f^2 - 7f)$

    36. $2k(-3k + 4) + 6(k^2 + 10) = k(4k + 8) - 2k(2k + 5)$

Simplify each expression.

    37. $\frac{2}{3}np^2(30p^2 + 9n^2p - 12)$ **38.** $\frac{3}{5}r^2t(10r^3 + 5rt^3 + 15t^2)$

    39. $-5q^2w^3(4q + 7w) + 4qw^2(7q^2w + 2q) - 3qw(3q^2w^2 + 9)$

    40. $-x^2z(2z^2 + 4xz^3) + xz^2(xz + 5x^3z) + x^2z^3(3x^2z + 4xz)$

    41. PARKING A parking garage charges $30 per month plus $1.50 per daytime hour and $1.00 per hour during nights and weekends. Suppose Trent parks in the garage for 47 hours in January and h of those are night and weekend hours.

    a. Find an expression for Trent's January bill.

    b. Find the cost if Trent had 12 hours of night and weekend hours.

    42. **MP MODELING** Che is building a dog house for his new puppy. The upper face of the dog house is a trapezoid. If the height of the trapezoid is 12 inches, find the area of the face of this piece of the dog house.

43 **TENNIS** The tennis club is building a new tennis court with a path around it.

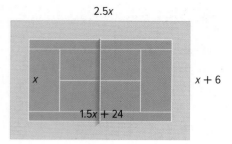

   a. Write an expression for the area of the tennis court.

   b. Write an expression for the area of the path.

   c. If $x = 36$ feet, what is the perimeter of the outside of the path?

44. MULTIPLE REPRESENTATIONS In this problem, you will investigate the degree of the product of a monomial and a polynomial.

   a. Tabular Write three monomials of different degrees and three polynomials of different degrees. Determine the degree of each monomial and polynomial. Multiply the monomials by the polynomials. Determine the degree of each product. Record your results in a table like the one shown below.

Monomial	Degree	Polynomial	Degree	Product of Monomial and Polynomial	Degree

   b. Verbal Make a conjecture about the degree of the product of a monomial and a polynomial. What is the degree of the product of a monomial of degree a and a polynomial of degree b?

H.O.T. Problems   Use Higher-Order Thinking Skills

45. ERROR ANALYSIS Pearl and Ted both worked on this problem. Is either of them correct? Explain your reasoning.

Pearl
$2x^2(3x^2 + 4x + 2)$
$6x^4 + 8x^2 + 4x^2$
$6x^4 + 12x^2$

Ted
$2x^2(3x^2 + 4x + 2)$
$6x^4 + 8x^3 + 4x^2$

46. **MP** **PERSEVERANCE** Find p such that $3x^p(4x^{2p + 3} + 2x^{3p - 2}) = 12x^{12} + 6x^{10}$.

47. CHALLENGE Simplify $4x^{-3}y^2(2x^5y^{-4} + 6x^{-7}y^6 - 4x^0y^{-2})$.

48. REASONING Is there a value for x that makes the statement $(x + 2)^2 = x^2 + 2^2$ true? If so, find a value for x. Explain your reasoning.

49. OPEN-ENDED Write a monomial and a polynomial using n as the variable. Find their product.

50. WRITING IN MATH Describe the steps to multiply a polynomial by a monomial.

51. Brian has 3 less than 5 times as many quarters as his sister has dimes. Brian used n for his sister's number of dimes and wrote the expression below to represent the total amount of money they have.

$$0.25(5n - 3) + 0.10n$$

Which expression is an equivalent form for this total? **MP** 2, 4

- ○ **A** $1.35n - 3$
- ○ **B** $1.35n - 0.75$
- ○ **C** $0.115n + 0.85$
- ○ **D** $0.135n^2 - 0.075n$

52. A student is simplifying an expression.

Step 1: $3x(x^2 + 5x + 12) - 4x(2 - x)$

Step 2: ?

Which expression could be an equivalent form written for Step 2? **MP** 2

- ○ **A** $3x(18x^3) + 4x(x)$
- ○ **B** $3x(60x^3) + 4x(-2x)$
- ○ **C** $3x^3 + 15x^2 + 36 - 8x - 4x^2$
- ○ **D** $3x^3 + 15x^2 + 36x - 8x + 4x^2$

53. **MULTI-STEP** Katie is making different colored pennants for her 30 classmates. Each pennant is the same size with a base of length x and a height of $2x^2 + 3x + 6$. **MP** 2, 4

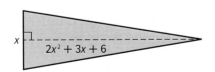

a. Which expression represents the number of square units of fabric Katie needs to make each pennant?

- ○ **A** $x^3 + 1.5x^2 + 3x$
- ○ **B** $2x^3 + 3x^2 + 6x$
- ○ **C** $3x^3 + 4.5x^2 + 9x$
- ○ **D** $4x^3 + 6x^2 + 12x$

b. Which expression represents the number of square units of fabric Katie needs to make all the pennants?

- ○ **A** $10x^3 + 15x^2 + 30x$
- ○ **B** $20x^3 + 30x^2 + 60x$
- ○ **C** $30x^3 + 45x^2 + 90x$
- ○ **D** $60x^3 + 90x^2 + 180x$

c. The city's sports team asked Katie to make 30 larger pennants to hang in the city's sports dome. If the pennants have a base of 2 meters what is the area of one of the larger pennants?

[   ]

d. What is the total area of the fabric that Katie will need to make the 30 pennants for the city? Show your work.

[   ]

e. If fabric costs $10/m^2$, what is the total cost of the fabric that Katie will need to make pennants for the city? Show your work.

[   ]

54. Which statements about the expressions are true? **MP** 2

I. $2.5x^2(2x^3 - 10x^2 + x)$

II. $6yz^3(9y^2 - 4z)$

III. $v^2w(w^3 + 3v^2 - 180v^5)$

- ☐ **A** The value of expression I when $x = 1$ is greater than 0.
- ☐ **B** Two of the expressions have a degree greater than 5.
- ☐ **C** When $v = -2$ and $w = 2$, the value of the expression III is negative.
- ☐ **D** Expression III has the greatest degree.
- ☐ **E** The simplified form of expression II is $54y^3z^3 - 24yz^4$.
- ☐ **F** Expression I has a degree of 5.

Multiplying Polynomials

You can use algebra tiles to find the product of two binomials.

Mathematical Practices

MP 3 Construct viable arguments and critique the reasoning of others.
8 Look for and express regularity in repeated reasoning.

Activity 1   Multiply Binomials

Work cooperatively. Use algebra tiles to find $(x + 3)(x + 4)$.

The rectangle will have a width of $x + 3$ and a length of $x + 4$. Use algebra tiles to mark off the dimensions on a product mat. Then complete the rectangle with algebra tiles.

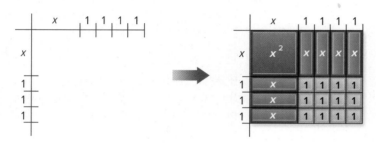

The rectangle consists of one blue x^2-tile, seven green x-tiles, and 12 yellow 1-tiles. The area of the rectangle is $x^2 + 7x + 12$. So, $(x + 3)(x + 4) = x^2 + 7x + 12$.

Activity 2   Multiply Binomials

Work cooperatively. Use algebra tiles to find $(x - 2)(x - 5)$.

Step 1 The rectangle will have a width of $x - 2$ and a length of $x - 5$. Use algebra tiles to mark off the dimensions on a product mat. Then begin to make the rectangle with algebra tiles.

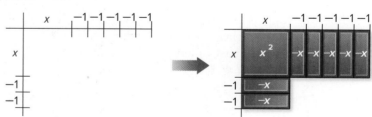

Step 2 Determine whether to use 10 yellow 1-tiles or 10 red −1-tiles to complete the rectangle. The area of each yellow tile is the product of −1 and −1. Fill in the space with 10 yellow 1-tiles to complete the rectangle.

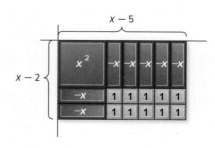

The rectangle consists of one blue x^2-tile, seven red $-x$-tiles, and 10 yellow 1-tiles. The area of the rectangle is $x^2 - 7x + 10$. So, $(x - 2)(x - 5) = x^2 - 7x + 10$.

(continued on the next page)

Activity 3 Multiply Binomials

Work cooperatively. Use algebra tiles to find $(x - 4)(2x + 3)$.

Step 1 The rectangle will have a width of $x - 4$ and a length of $2x + 3$. Use algebra tiles to mark off the dimensions on a product mat. Then begin to make the rectangle with algebra tiles.

Step 2 Determine what color x-tiles and what color 1-tiles to use to complete the rectangle. The area of each red x-tile is the product of x and -1. The area of each red -1-tile is represented by $1(-1)$ or -1.

Complete the rectangle with four red x-tiles and 12 red -1-tiles.

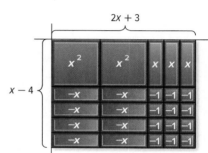

Step 3 Rearrange the tiles to simplify the polynomial you have formed. Notice that 3 zero pairs are formed by three positive and three negative x-tiles.

There are two blue x^2-tiles, five red $-x$-tiles, and 12 red -1-tiles left. In simplest form, $(x - 4)(2x + 3) = 2x^2 - 5x - 12$.

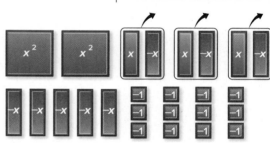

Model and Analyze

Work cooperatively. Use algebra tiles to find each product.

1. $(x + 1)(x + 4)$
2. $(x - 3)(x - 2)$
3. $(x + 5)(x - 1)$
4. $(x + 2)(2x + 3)$
5. $(x - 1)(2x - 1)$
6. $(x + 4)(2x - 5)$

Is each statement _true_ or _false_? Justify your answer with a drawing of algebra tiles.

7. $(x - 4)(x - 2) = x^2 - 6x + 8$
8. $(x + 3)(x + 5) = x^2 + 15$

9. **WRITING IN MATH** You can also use the Distributive Property to find the product of two binomials. The figure at the right shows the model for $(x + 4)(x + 5)$ separated into four parts. Write a sentence or two explaining how this model shows the use of the Distributive Property.

Multiplying Polynomials

⋮⋮Then	⋮⋮Now	⋮⋮Why?

⋮⋮Then

- You multiplied polynomials by monomials.

⋮⋮Now

1 Multiply binomials by using the FOIL method.

2 Multiply polynomials by using the Distributive Property.

⋮⋮Why?

Bodyboards, which are used to ride waves, are made of foam and are more rectangular than surfboards. A bodyboard's dimensions are determined by the height and skill level of the user.

The length of Ann's bodyboard should be Ann's height h minus 32 inches or $h - 32$. The board's width should be half of Ann's height plus 11 inches or $\frac{1}{2}h + 11$. To approximate the area of the bodyboard, you need to find $(h - 32)\left(\frac{1}{2}h + 11\right)$.

 New Vocabulary

FOIL method
quadratic expression

MP **Mathematical Practices**

6 Attend to precision.

8 Look for and express regularity in repeated reasoning.

1 **Multiply Binomials** To multiply two binomials such as $h - 32$ and $\frac{1}{2}h + 11$, the Distributive Property is used. Binomials can be multiplied horizontally or vertically.

Example 1   The Distributive Property

Find each product.

a. $(2x + 3)(x + 5)$

Vertical Method

Multiply by 5.

$$\begin{array}{r} 2x + 3 \\ (\times)\ x + 5 \\ \hline 10x + 15 \end{array}$$

$5(2x + 3) = 10x + 15$

Multiply by x.

$$\begin{array}{r} 2x + 3 \\ (\times)\ x + 5 \\ \hline 10x + 15 \\ 2x^2 + 3x \\ \hline \end{array}$$

$x(2x + 3) = 2x^2 + 3x$

Combine like terms.

$$\begin{array}{r} 2x + 3 \\ (\times)\ x + 5 \\ \hline 10x + 15 \\ 2x^2 + 3x \\ \hline \end{array}$$

$2x^2 + 13x + 15$

Horizontal Method

$$\begin{aligned} (2x + 3)(x + 5) &= 2x(x + 5) + 3(x + 5) && \text{Rewrite as the sum of two products.} \\ &= 2x^2 + 10x + 3x + 15 && \text{Distributive Property} \\ &= 2x^2 + 13x + 15 && \text{Combine like terms.} \end{aligned}$$

b. $(x - 2)(3x + 4)$

Vertical Method

Multiply by 4.

$$\begin{array}{r} x - 2 \\ (\times)\ 3x + 4 \\ \hline 4x - 8 \end{array}$$

$4(x - 2) = 4x - 8$

Multiply by $3x$.

$$\begin{array}{r} x - 2 \\ (\times)\ 3x + 4 \\ \hline 4x - 8 \\ 3x^2 - 6x \\ \hline \end{array}$$

$3x(x - 2) = 3x^2 - 6x$

Combine like terms.

$$\begin{array}{r} x - 2 \\ (\times)\ 3x + 4 \\ \hline 4x - 8 \\ 3x^2 - 6x \\ \hline \end{array}$$

$3x^2 - 2x - 8$

Horizontal Method

$$\begin{aligned} (x - 2)(3x + 4) &= x(3x + 4) - 2(3x + 4) && \text{Rewrite as the difference of two products.} \\ &= 3x^2 + 4x - 6x - 8 && \text{Distributive Property} \\ &= 3x^2 - 2x - 8 && \text{Combine like terms.} \end{aligned}$$

1A. $(3m + 4)(m + 5)$ **1B.** $(5y - 2)(y + 8)$

A shortcut version of the Distributive Property for multiplying binomials is called the **FOIL method**.

🔄 Key Concept FOIL Method

Words   To multiply two binomials, find the sum of the products of **F** the *First* terms, **O** the *Outer* terms, **I** the *Inner* terms, and **L** the *Last* terms.

Example

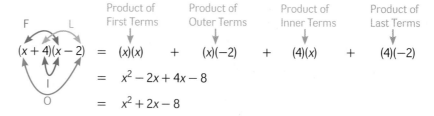

$$(x + 4)(x - 2) = (x)(x) + (x)(-2) + (4)(x) + (4)(-2)$$
$$= x^2 - 2x + 4x - 8$$
$$= x^2 + 2x - 8$$

Example 2   FOIL Method

Find each product.

a. $(2y - 7)(3y + 5)$

$$(2y - 7)(3y + 5) = (2y)(3y) + (2y)(5) + (-7)(3y) + (-7)(5) \quad \text{FOIL method}$$
$$= 6y^2 + 10y - 21y - 35 \quad \text{Multiply.}$$
$$= 6y^2 - 11y - 35 \quad \text{Combine like terms.}$$

b. $(4a - 5)(2a - 9)$

$$(4a - 5)(2a - 9)$$
$$= (4a)(2a) + (4a)(-9) + (-5)(2a) + (-5)(-9) \quad \text{FOIL method}$$
$$= 8a^2 - 36a - 10a + 45 \quad \text{Multiply.}$$
$$= 8a^2 - 46a + 45 \quad \text{Combine like terms.}$$

▶ **Guided Practice**

2A. $(x + 3)(x - 4)$ **2B.** $(4b - 5)(3b + 2)$

2C. $(2y - 5)(y - 6)$ **2D.** $(5a + 2)(3a - 4)$

Notice that when two linear expressions are multiplied, the result is a quadratic expression. A **quadratic expression** is an expression in one variable with a degree of 2. When three linear expressions are multiplied, the result has a degree of 3.

The FOIL method can be used to find an expression that represents the area of a rectangular object when the lengths of the sides are given as binomials.

Real-World Example 3   FOIL Method

SWIMMING POOL A contractor is building a deck around a rectangular swimming pool. The deck is x feet from every side of the pool. Write an expression for the total area of the pool and deck.

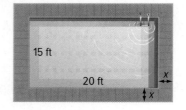

15 ft

20 ft

x

x

Understand We need to find an expression for the total area of the pool and deck.

Plan Find the product of the length and width of the pool with the deck.

Solve Since the deck is the same distance from every side of the pool, the length and width of the pool are $2x$ longer. So, the length can be represented by $2x + 20$ and the width can be represented by $2x + 15$.

Area = length · width	Area of a rectangle
$= (2x + 20)(2x + 15)$	Substitution
$= (2x)(2x) + (2x)(15) + (20)(2x) + (20)(15)$	FOIL Method
$= 4x^2 + 30x + 40x + 300$	Multiply.
$= 4x^2 + 70x + 300$	Combine like terms.

So, the total area of the deck and pool is $4x^2 + 70x + 300$.

Check Choose a value for x. Substitute this value into $(2x + 20)(2x + 15)$ and $4x^2 + 70x + 300$. The result should be the same for both expressions.

Guided Practice

3. If the pool is 25 feet long and 20 feet wide, find the area of the pool and deck.

2 Multiply Polynomials The Distributive Property can also be used to multiply any two polynomials.

Example 4   The Distributive Property

Find each product.

a. $(6x + 5)(2x^2 - 3x - 5)$

$(6x + 5)(2x^2 - 3x - 5)$

$= 6x(2x^2 - 3x - 5) + 5(2x^2 - 3x - 5)$	Distributive Property
$= 12x^3 - 18x^2 - 30x + 10x^2 - 15x - 25$	Multiply.
$= 12x^3 - 8x^2 - 45x - 25$	Combine like terms.

b. $(2y^2 + 3y - 1)(3y^2 - 5y + 2)$

$(2y^2 + 3y - 1)(3y^2 - 5y + 2)$

$= 2y^2(3y^2 - 5y + 2) + 3y(3y^2 - 5y + 2) - 1(3y^2 - 5y + 2)$	Distributive Property
$= 6y^4 - 10y^3 + 4y^2 + 9y^3 - 15y^2 + 6y - 3y^2 + 5y - 2$	Multiply.
$= 6y^4 - y^3 - 14y^2 + 11y - 2$	Combine like terms.

Study Tip

MP **Structure** If a polynomial with c terms and a polynomial with d terms are multiplied together, there will be $c \cdot d$ terms before simplifying. In Example 4a, there are 2 · 3 or 6 terms before simplifying.

Guided Practice

4A. $(3x - 5)(2x^2 + 7x - 8)$

4B. $(m^2 + 2m - 3)(4m^2 - 7m + 5)$

Examples 1–2 Find each product.

1. $(x + 5)(x + 2)$ **2.** $(y - 2)(y + 4)$ **3.** $(b - 7)(b + 3)$

4. $(4n + 3)(n + 9)$ **5.** $(8h - 1)(2h - 3)$ **6.** $(2a + 9)(5a - 6)$

Example 3   **7. FRAME** Hugo is designing a frame as shown at the right. The frame has a width of x inches all the way around. Write an expression that represents the total area of the picture and frame.

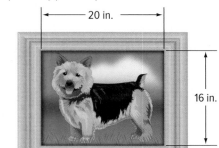

20 in.

16 in.

Example 4 Find each product.

8. $(2a - 9)(3a^2 + 4a - 4)$

9. $(4y^2 - 3)(4y^2 + 7y + 2)$

10. $(x^2 - 4x + 5)(5x^2 + 3x - 4)$

11. $(2n^2 + 3n - 6)(5n^2 - 2n - 8)$

Practice and Problem Solving

Extra Practice is on page R8.

Examples 1–2 Find each product.

12. $(3c - 5)(c + 3)$ **13.** $(g + 10)(2g - 5)$ **14.** $(6a + 5)(5a + 3)$

⑮ $(4x + 1)(6x + 3)$ **16.** $(5y - 4)(3y - 1)$ **17.** $(6d - 5)(4d - 7)$

18. $(3m + 5)(2m + 3)$ **19.** $(7n - 6)(7n - 6)$ **20.** $(12t - 5)(12t + 5)$

21. $(5r + 7)(5r - 7)$ **22.** $(8w + 4x)(5w - 6x)$ **23.** $(11z - 5y)(3z + 2y)$

Example 3   **24. GARDEN** A walkway surrounds a rectangular garden. The width of the garden is 8 feet, and the length is 6 feet. The width x of the walkway around the garden is the same on every side. Write an expression that represents the total area of the garden and walkway.

Example 4 Find each product.

25. $(2y - 11)(y^2 - 3y + 2)$ **26.** $(4a + 7)(9a^2 + 2a - 7)$

27. $(m^2 - 5m + 4)(m^2 + 7m - 3)$ **28.** $(x^2 + 5x - 1)(5x^2 - 6x + 1)$

29. $(3b^3 - 4b - 7)(2b^2 - b - 9)$ **30.** $(6z^2 - 5z - 2)(3z^3 - 2z - 4)$

Simplify.

31. $(m + 2)[(m^2 + 3m - 6) + (m^2 - 2m + 4)]$

32. $[(t^2 + 3t - 8) - (t^2 - 2t + 6)](t - 4)$

MP **STRUCTURE** Find an expression to represent the area of each shaded region.

33.

2x + 3

x + 1

3x + 2

34.

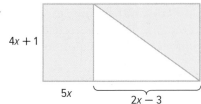

4x + 1

5x

2x − 3

35 **VOLLEYBALL** The dimensions of a sand volleyball court are represented by a width of $6y - 5$ feet and a length of $3y + 4$ feet.

    a. Write an expression that represents the area of the court.

    b. The length of a sand volleyball court is 31 feet. Find the area of the court.

36. GEOMETRY Write an expression for the area of a triangle with a base of $2x + 3$ and a height of $3x - 1$.

Find each product.

37. $(a - 2b)^2$ **38.** $(3c + 4d)^2$ **39.** $(x - 5y)^2$

40. $(2r - 3t)^3$ **41.** $(5g + 2h)^3$ **42.** $(4y + 3z)(4y - 3z)^2$

43. CONSTRUCTION A sandbox kit allows you to build a square sandbox or a rectangular sandbox as shown.

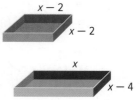

    a. What are the possible values of x? Explain.

    b. Which shape has the greater area?

    c. What is the difference in areas between the two?

44. MULTIPLE REPRESENTATIONS In this problem, you will investigate the square of a sum.

    a. Tabular Copy and complete the table for each sum.

Expression	(Expression)^2
$x + 5$	
$3y + 1$	
$z + q$	

    b. Verbal Make a conjecture about the terms of the square of a sum.

    c. Symbolic For a sum of the form $a + b$, write an expression for the square of the sum.

H.O.T. Problems   Use **H**igher-**O**rder **T**hinking Skills

45. **MP ARGUMENTS** Determine if the following statement is *sometimes*, *always*, or *never* true. Explain your reasoning.

    The FOIL method can be used to multiply a binomial and a trinomial.

46. **MP STRUCTURE** Find $(x^m + x^p)(x^{m-1} - x^{1-p} + x^p)$.

47. CHALLENGE Write a binomial and a trinomial involving a single variable. Then find their product.

48. **MP REGULARITY** Compare and contrast the procedure used to multiply a trinomial by a binomial using the vertical method with the procedure used to multiply a three-digit number by a two-digit number.

49. WRITING IN MATH Summarize the methods that can be used to multiply polynomials.

50. Maria is framing an 8 inch-by-10 inch photograph with matting as shown.

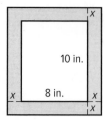

Which is an expression for the area of the photograph with the mat? **MP** 4

- ○ **A** $4x + 18$

- ○ **B** $4x^2 + 80$

- ○ **C** $4x^2 + 28x + 80$

- ○ **D** $4x^2 + 36x + 80$

51. MULTI-STEP Consider a square with side length x and another square with side length $x + 3$. **MP** 6

a. Write an expression for the area of each square.

b. The area of the larger square is 15 square units more than the area of the smaller square. Find the value of x.

c. Use the value of x you found in **b** to find the area of each square.

52. The area of the shaded region shown below is 214 units².

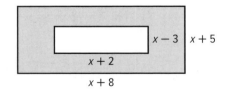

What is the value of x? **MP** 7

53. Which expression is equivalent to the square of the quantity $4x$ plus 8? **MP** 7

- ○ **A** $8x + 16$

- ○ **B** $16x^2 + 64$

- ○ **C** $16x^2 + 24x + 64$

- ○ **D** $16x^2 + 64x + 64$

54. What is the product of $(2x + 5)$ and $(4x - 7)$? **MP** 8

- ○ **A** $6x - 2$

- ○ **B** $8x^2 - 27x - 35$

- ○ **C** $8x^2 - 35$

- ○ **D** $8x^2 + 6x - 35$

55. The sum of two polynomials is $4x^2 + 3x - 1$. One of the polynomials is $2x + 1$. What is the product of the polynomials? **MP** 8

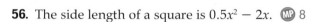

56. The side length of a square is $0.5x^2 - 2x$. **MP** 8

a. Find the perimeter of the square.

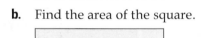

b. Find the area of the square.

57. Find each product. **MP** 8

a. $(2x - 3)(x^2 - 4x + 1)$

b. $(3x - 3)(2x^2 - 4x + 1)$

58. A rectangle has a length of $x^3 - 2x^2 + 5x$ and a width of $2x^2 - 9x - 5$. **MP** 8

a. Find the perimeter of the rectangle.

b. Find the area of the rectangle.

59. Bob is twice as old as his sister. The sum of their ages is $6x + 12$. Find the product of their ages. **MP** 1

Special Products

- You multiplied binomials by using the FOIL method.

1 Find squares of sums and differences.

2 Find the product of a sum and a difference.

- Colby wants to attach a dartboard to a square piece of corkboard. If the radius of the dartboard is $r + 12$, how large does the square corkboard need to be?

Colby knows that the diameter of the dartboard is $2(r + 12)$ or $2r + 24$. Each side of the square also measures $2r + 24$. To find how much corkboard is needed, Colby must find the area of the square: $A = (2r + 24)^2$.

 Mathematical Practices

8 Look for and express regularity in repeated reasoning.

1 **Squares of Sums and Differences** Some pairs of binomials, such as squares like $(2r + 24)^2$, have products that follow a specific pattern. Using the pattern can make multiplying easier. The square of a sum, $(a + b)^2$ or $(a + b)(a + b)$, is one of those products.

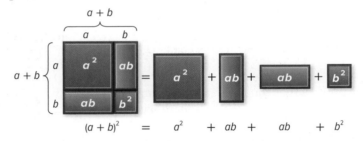

$$(a + b)^2 = a^2 + ab + ab + b^2$$

🔑 Key Concept   Square of a Sum

Words	The square of $a + b$ is the square of a plus twice the product of a and b plus the square of b.

Symbols $(a + b)^2 = (a + b)(a + b)$ **Example** $(x + 4)^2 = (x + 4)(x + 4)$

$$= a^2 + 2ab + b^2$$ $$= x^2 + 8x + 16$$

Example 1   Square of a Sum

Find $(3x + 5)^2$.

$(a + b)^2 = a^2 + 2ab + b^2$ Square of a sum

$(3x + 5)^2 = (3x)^2 + 2(3x)(5) + 5^2$ $a = 3x, b = 5$

$\qquad\qquad = 9x^2 + 30x + 25$ Simplify. Use FOIL to check your solution.

▶ **Guided Practice**

Find each product.

1A. $(8c + 3d)^2$ **1B.** $(3x + 4y)^2$

There is also a pattern for the *square of a difference*. Write $a - b$ as $a + (-b)$ and square it using the square of a sum pattern.

$$(a - b)^2 = [a + (-b)]^2$$
$$= a^2 + 2(a)(-b) + (-b)^2 \quad \text{Square of a sum}$$
$$= a^2 - 2ab + b^2 \quad \text{Simplify.}$$

Key Concept Square of a Difference

Words   The square of $a - b$ is the square of a minus twice the product of a and b plus the square of b.

Symbols $(a - b)^2 = (a - b)(a - b)$ **Example** $(x - 3)^2 = (x - 3)(x - 3)$
$$= a^2 - 2ab + b^2 \qquad\qquad\qquad = x^2 - 6x + 9$$

Watch Out!

MP **Regularity** Remember that $(x - 7)^2$ does not equal $x^2 - 7^2$, or $x^2 - 49$.

$(x - 7)^2$
$= (x - 7)(x - 7)$
$= x^2 - 14x + 49$

Example 2   **Square of a Difference**

Find $(2x - 5y)^2$.

$$(a - b)^2 = a^2 - 2ab + b^2 \qquad \text{Square of a difference}$$
$$(2x - 5y)^2 = (2x)^2 - 2(2x)(5y) + (5y)^2 \qquad a = 2x \text{ and } b = 5y$$
$$= 4x^2 - 20xy + 25y^2 \qquad \text{Simplify.}$$

Guided Practice

Find each product.

2A. $(6p - 1)^2$ **2B.** $(a - 2b)^2$

The product of the square of a sum or the square of a difference is called a *perfect square trinomial*. We can use these to find patterns to solve real-world problems.

Real-World Example 3   **Square of a Difference**

PHYSICAL SCIENCE Each edge of a cube of aluminum is 4 centimeters less than each edge of a cube of copper. Write an equation to model the surface area of the aluminum cube.

Let $c =$ the length of each edge of the cube of copper. So, each edge of the cube of aluminum is $c - 4$.

$$SA = 6s^2 \qquad \text{Formula for surface area of a cube}$$
$$SA = 6(c - 4)^2 \qquad \text{Replace } s \text{ with } c - 4.$$
$$SA = 6[c^2 - 2(4)(c) + 4^2] \qquad \text{Square of a difference}$$
$$SA = 6(c^2 - 8c + 16) \qquad \text{Simplify.}$$

Guided Practice

3. GARDENING Alano has a garden that is g feet long and g feet wide. He wants to add 3 feet to the length and the width.

A. Show how the new area of the garden can be modeled by the square of a binomial.

B. Find the square of this binomial.

2 Product of a Sum and a Difference

Now we will see what the result is when we multiply a sum and a difference, or $(a + b)(a - b)$. Recall that $a - b$ can be written as $a + (-b)$.

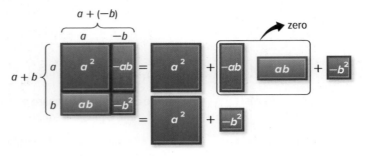

Notice that the middle terms are opposites and add to a zero pair. So $(a + b)(a - b) = a^2 - ab + ab - b^2 = a^2 - b^2$.

Study Tip

Patterns When using any of these patterns, a and b can be numbers, variables, or expressions with numbers and variables.

🔑 Key Concept Product of a Sum and a Difference

Words The product of $a + b$ and $a - b$ is the square of a minus the square of b.

Symbols $(a + b)(a - b) = (a - b)(a + b)$
$$= a^2 - b^2$$

Example 4 Product of a Sum and a Difference

Find $(2x^2 + 3)(2x^2 - 3)$.

$(a + b)(a - b) = a^2 - b^2$ Product of a sum and a difference

$(2x^2 + 3)(2x^2 - 3) = (2x^2)^2 - (3)^2$ $a = 2x^2$ and $b = 3$

$= 4x^4 - 9$ Simplify.

Guided Practice

Find each product.

4A. $(3n + 2)(3n - 2)$ **4B.** $(4c - 7d)(4c + 7d)$

Check Your Understanding   = Step-by-Step Solutions begin on page R11.

Go Online! for a Self-Check Quiz

Examples 1–2 Find each product.

1. $(x + 5)^2$ **2.** $(11 - a)^2$ **3** $(2x + 7y)^2$

4. $(3m - 4)(3m - 4)$ **5.** $(g - 4h)(g - 4h)$ **6.** $(3c + 6d)^2$

Example 3

7. GENETICS The color of a Labrador retriever's fur is genetic. Dark genes D are dominant over yellow genes y. A dog with genes DD or Dy will have dark fur. A dog with genes yy will have yellow fur. Pepper's genes for fur color are Dy, and Ramiro's are Dy.

	D	y
D	DD	Dy
y	Dy	yy

   a. Use the square of a sum to write an expression that models the possible fur colors of the dogs' puppies.

   b. What is the probability that a puppy will have dark fur?

Example 4   Find each product.

     8. $(a - 3)(a + 3)$ **9.** $(x + 5)(x - 5)$

     10. $(6y - 7)(6y + 7)$ **11.** $(9t + 6)(9t - 6)$

Practice and Problem Solving

Extra Practice is on page R8.

Examples 1–2   Find each product.

     12. $(a + 10)(a + 10)$ **13.** $(b - 6)(b - 6)$

     14. $(h + 7)^2$ **15.** $(x + 6)^2$

     16. $(8 - m)^2$ **17.** $(9 - 2y)^2$

     18. $(2b + 3)^2$ **19.** $(5t - 2)^2$

     20. $(8h - 4n)^2$

Example 3   **21. GENETICS** The ability to roll your tongue is inherited genetically if either parent has the dominant trait T.

	T	t
T	TT	Tt
t	Tt	tt

       a. Show how the combinations can be modeled by the square of a sum.

       b. Predict the percent of children that will have both dominant genes, one dominant gene, and both recessive genes.

Example 4   Find each product.

     22. $(u + 3)(u - 3)$ **23** $(b + 7)(b - 7)$ **24.** $(2 + x)(2 - x)$

     25. $(4 - x)(4 + x)$ **26.** $(2q + 5r)(2q - 5r)$ **27.** $(3a^2 + 7b)(3a^2 - 7b)$

     28. $(5y + 7)^2$ **29.** $(8 - 10a)^2$ **30.** $(10x - 2)(10x + 2)$

     31. $(3t + 12)(3t - 12)$ **32.** $(a + 4b)^2$ **33.** $(3q - 5r)^2$

     34. $(2c - 9d)^2$ **35.** $(g + 5h)^2$ **36.** $(6y - 13)(6y + 13)$

     37. $(3a^4 - b)(3a^4 + b)$ **38.** $(5x^2 - y^2)^2$ **39.** $(8a^2 - 9b^3)(8a^2 + 9b^3)$

     40. $\left(\frac{3}{4}k + 8\right)^2$ **41.** $\left(\frac{2}{5}y - 4\right)^2$ **42.** $(7z^2 + 5y^2)(7z^2 - 5y^2)$

     43. $(2m + 3)(2m - 3)(m + 4)$ **44.** $(r + 2)(r - 5)(r - 2)(r + 5)$

     45. **MP** **SENSE-MAKING** Write a polynomial that represents the area of the figure at the right.

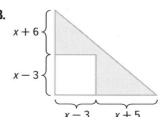

     46. **MULTI-STEP** A flying disk has a radius of 4.9 inches. Jolanda discovers that adding a hole with a radius of 3.75 inches to the center of the disk reduces the weight of the disk, so it travels farther. Jolanda wants to experiment with the size of the disk to increase flying distance. In the new disks she creates, the sizes of the disks and holes are increased or decreased by the same amount.

       a. Write an expression for the area of the top of the new flying disks.

       b. Explain your solution process.

GEOMETRY Find the area of each shaded region.

     47.

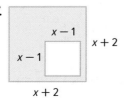

     48.

Find each product.

49. $(c + d)(c + d)(c + d)$ **50.** $(2a - b)^3$ **51.** $(f + g)(f - g)(f + g)$

52. $(k - m)(k + m)(k - m)$ **53.** $(n - p)^2(n + p)$ **54.** $(q + r)^2(q - r)$

55. **WRESTLING** A high school wrestling mat must be a square with 38-foot sides and contain two circles as shown. Suppose the inner circle has a radius of r feet, and the radius of the outer circle is 9 feet longer than the inner circle.

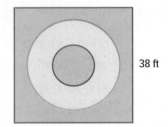

38 ft

    a. Write an expression for the area of the larger circle.

    b. Write an expression for the area of the portion of the square outside the larger circle.

56. **MULTIPLE REPRESENTATIONS** In this problem, you will investigate a pattern. Begin with a square piece of construction paper. Label each edge of the paper a. In any of the corners, draw a smaller square and label the edges b.

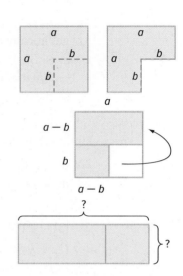

    a. **Numerical** Find the area of each of the squares.

    b. **Concrete** Cut the smaller square out of the corner. What is the area of the shape?

    c. **Analytical** Remove the smaller rectangle on the bottom. Turn it and slide it next to the top rectangle. What is the length of the new arrangement? What is the width? What is the area?

    d. **Analytical** What pattern does this verify?

H.O.T. Problems   Use **H**igher-**O**rder **T**hinking Skills

57. **WHICH ONE DOESN'T BELONG?** Which expression does not belong? Explain.

$(2c - d)(2c - d)$	$(2c + d)(2c - d)$	$(2c + d)(2c + d)$	$(c + d)(c + d)$

58. **CHALLENGE** Does a pattern exist for $(a + b)^3$?

    a. Investigate this question by finding the product $(a + b)(a + b)(a + b)$.

    b. Use the pattern you discovered in part **a** to find $(x + 2)^3$.

    c. Draw a diagram of a geometric model for $(a + b)^3$.

    d. What is the pattern for the cube of a difference, $(a - b)^3$?

59. **CHALLENGE** Find c such that $25x^2 - 90x + c$ is a perfect square trinomial.

60. **OPEN-ENDED** Write two binomials with a product that is a binomial. Then write two binomials with a product that is not a binomial.

61. **WRITING IN MATH** Describe how to square the sum of two quantities, square the difference of two quantities, and how to find the product of a sum of two quantities and a difference of two quantities.

62. Danielle is making a quilt with triangles and squares. There are 4 colored triangles inside each square. The base and height of each triangle is $x + 4$ inches as shown. Which polynomial represents the area of the colored triangles in square inches? (MP) 4

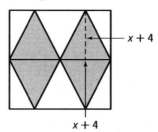

- **A** $\frac{1}{2}x + 2$
- **B** $\frac{1}{2}x^2 + 4x + 8$
- **C** $2x^2 + 16x + 32$
- **D** $4x^2 + 32x + 64$

63. MULTI-STEP Write an expression the for the area of each figure. (MP) 1

- **a.** a square with a side length of $2x + 4$ units

- **b.** a rectangle with a length of $5x + 3$ units and a width of $5x - 3$ units

- **c.** What is the difference in area between the square and rectangle?

64. How do the patterns used to find the square of a sum and the square of a difference differ? (MP) 3, 6

65. For a right triangle, the Pythagorean theorem says the sum of the squares of the lengths of the legs equals the square of the hypotenuse, or $a^2 + b^2 = c^2$.

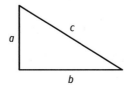

The legs of a right triangle have lengths of $2x + 3$ and $4x - 1$. Which polynomial is equivalent to c^2? (MP) 8

- **A** $6x + 2$
- **B** $4x^2 + 12x + 9$
- **C** $20x^2 + 4x + 10$
- **D** $36x^2 + 24x + 4$

66. The top of a square coffee table has an area of $64x^2 - 48x + 9$. Which is the perimeter of the tabletop? (MP) 7

- **A** $8x - 3$
- **B** $16x - 9$
- **C** $32x - 12$
- **D** $20x$

67. What is the expanded form of $\left(\frac{1}{5} + 4x\right)^2$? (MP) 8

- **A** $8x + \frac{2}{5}$
- **B** $16x^2 + \frac{8}{10}x + \frac{1}{10}$
- **C** $16x^2 + \frac{8}{5}x + \frac{1}{25}$
- **D** $16x^2 + \frac{8}{5}x + \frac{1}{10}$

68. Find each product. (MP) 1, 8

- **a.** $(2x - 3)^2$
- **b.** $(2x + 3)^2$
- **c.** $(2x + 3)(2x - 3)$

69. A square field has perimeter $4x^2 + 8$ yards. Find the area of this field. (MP) 7

70. The area of an isosceles right triangle is $8x^2 + 4x + 0.5$. Find the square of the length of the hypotenuse. (MP) 2

Determine whether each expression is a polynomial. If it is a polynomial, find the degree and determine whether it is a *monomial, binomial,* or *trinomial*. (Lesson 8-1)

1. $3y^2 - 2$

2. $4t^5 + 3t^2 + t$

3. $\dfrac{3x}{5y}$

4. ax^{-3}

5. $3b^2$

6. $2x^{-3} - 4x + 1$

7. POPULATION The table shows the population density for Nevada for various years. (Lesson 8-1)

Year	Years Since 1930	People/ Square Mile
1930	0	0.8
1960	30	2.6
1980	50	7.3
1990	60	10.9
2010	80	24.6

a. The population density d of Nevada from 1930 to 2010 can be modeled by $d = 0.005n^2 - 0.142n + 1$, where n represents the number of years since 1930. Classify the polynomial.

b. What is the degree of the polynomial?

c. Predict the population density of Nevada for 2020 and for 2030. Explain your method.

d. **MP** What mathematical practice did you use to solve this problem?

Find each sum or difference. (Lesson 8-1)

8. $(y^2 + 2y + 3) + (y^2 + 3y - 1)$

9. $(3n^3 - 2n + 7) - (n^2 - 2n + 8)$

10. $(5d + d^2) - (4 - 4d^2)$

11. $(x + 4) + (3x + 2x^2 - 7)$

12. $(3a - 3b + 2) - (4a + 5b)$

13. $(8x - y^2 + 3) + (9 - 3x + 2y^2)$

Find each product. (Lesson 8-2)

14. $6y(y^2 + 3y + 1)$

15. $3n(n^2 - 5n + 2)$

16. $d^2(-4 - 3d + 2d^2)$

17. $-2xy(3x^2 + 2xy - 4y^2)$

18. $ab^2(12a + 5b - ab)$

19. $x^2y^4(3xy^2 - x + 2y^2)$

20. MULTIPLE CHOICE Simplify $x(4x + 5) + 3(2x^2 - 4x + 1)$. (Lesson 8-2)

A $10x^2 + 17x + 3$ **C** $2x^2 - 7x + 3$

B $10x^2 - 7x + 3$ **D** $2x^2 + 17x + 3$

Find each product. (Lesson 8-3)

21. $(x + 2)(x + 5)$

22. $(3b - 2)(b - 4)$

23. $(n - 5)(n + 3)$

24. $(4c - 2)(c + 2)$

25. $(k - 1)(k - 3k^2)$

26. $(8d - 3)(2d^2 + d + 1)$

27. MANUFACTURING A company is designing a box for dry pasta in the shape of a rectangular prism. The length is 2 inches more than twice the width, and the height is 3 inches more than the length. Write an expression, in terms of the width, for the volume of the box. (Lesson 8-3)

Find each product. (Lesson 8-4)

28. $(x + 2)^2$

29. $(n - 11)^2$

30. $(4b - 2)^2$

31. $(6c + 3)^2$

32. $(5d - 3)(5d + 3)$

33. $(9k + 1)(9k - 1)$

34. DISC GOLF The discs approved for use in disc golf vary in size. (Lesson 8-4)

Smallest disc   **Largest disc**

x in.

$(x + 3.25)$ in.

a. Write two different expressions for the area of the largest disc.

b. If x is 10.5, what are the areas of the smallest and largest discs?

Algebra Lab
Factoring Using the Distributive Property

When two or more numbers are multiplied, these numbers are *factors* of the product. Sometimes you know the product of binomials and are asked to find the factors. This is called factoring. You can use algebra tiles and a product mat to factor binomials.

Mathematical Practices
MP **3** Construct viable arguments and critique the reasoning of others.

Work cooperatively to complete Activities 1 and 2.

Activity 1   Use Algebra Tiles to Factor $2x - 8$

Step 1   Model $2x - 8$.

Step 2   Arrange the tiles into a rectangle. The total area of the rectangle represents the product, and its length and width represent the factors.

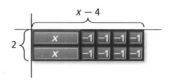

The rectangle has a width of 2 and a length of $x - 4$. Therefore, $2x - 8 = 2(x - 4)$.

Activity 2   Use Algebra Tiles to Factor $x^2 + 3x$

Step 1   Model $x^2 + 3x$.

Step 2   Arrange the tiles into a rectangle.

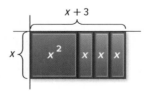

The rectangle has a width of x and a length of $x + 3$. Therefore, $x^2 + 3x = x(x + 3)$.

Model and Analyze
Work cooperatively. Use algebra tiles to factor each binomial.

1. $4x + 12$ **2.** $4x - 6$ **3.** $3x^2 + 4x$ **4.** $10 - 2x$

Determine whether each binomial can be factored. Justify your answer with a drawing.

5. $6x - 9$ **6.** $5x - 4$ **7.** $4x^2 + 7$ **8.** $x^2 + 3x$

9. **WRITING IN MATH** Write a paragraph that explains how you can use algebra tiles to determine whether a binomial can be factored. Include an example of one binomial that can be factored and one that cannot.

Using the Distributive Property

:: Then	:: Now	:: Why?
• Used the Distributive Property to evaluate expressions.	**1** Use the Distributive Property to factor polynomials. **2** Factor polynomials by grouping.	• The cost of rent for Mr. Cole's store is determined by the square footage of the space. The area of the store can be modeled by the expression $1.6w^2 + 6w$, where w is the width of the store in feet. We can use factoring to simplify the expression.

 New Vocabulary

factoring

factoring by grouping

Zero Product Property

 Mathematical Practices

2 Reason abstractly and quantitatively.

1 **Use the Distributive Property to Factor** You have used the Distributive Property to multiply a monomial by a polynomial. You can work backward to express a polynomial as the product of a monomial factor and a polynomial factor.

$$1.6w^2 + 6w = 1.6w(w) + 6(w)$$
$$= w(1.6w + 6)$$

So, $w(1.6w + 6)$ is the *factored form* of $1.6w^2 + 6w$. **Factoring** a polynomial involves finding the *completely* factored form.

Example 1   Use the Distributive Property

Use the Distributive Property to factor each polynomial.

a. $27y^2 + 18y$

Find the GCF of each term.

$27y^2 = ③ \cdot ③ \cdot 3 \cdot ⓨ \cdot y$ Factor each term.

$18y = 2 \cdot ③ \cdot ③ \cdot ⓨ$ Circle common factors.

$GCF = 3 \cdot 3 \cdot y$ or $9y$

Write each term as the product of the GCF and the remaining factors. Use the Distributive Property to *factor out* the GCF.

$27y^2 + 18y = 9y(3y) + 9y(2)$ Rewrite each term using the GCF.

$\qquad\qquad = 9y(3y + 2)$ Distributive Property

b. $-4a^2b - 8ab^2 + 2ab$

$-4a^2b = -1 \cdot ② \cdot 2 \cdot ⓐ \cdot a \cdot ⓑ$ Factor each term.

$-8ab^2 = -1 \cdot ② \cdot 2 \cdot 2 \cdot ⓐ \cdot ⓑ \cdot b$ Circle common factors.

$2ab = ② \cdot ⓐ \cdot ⓑ$

$GCF = 2 \cdot a \cdot b$ or $2ab$

$-4a^2b - 8ab^2 + 2ab = 2ab(-2a) - 2ab(4b) + 2ab(1)$ Rewrite each term using the GCF.

$\qquad\qquad\qquad\qquad = 2ab(-2a - 4b + 1)$ Distributive Property

▶ **Guided Practice**

1A. $15w - 3v$ **1B.** $7u^2t^2 + 21ut^2 - ut$

Real-World Example 2 Use Factoring

AGILITY Penny is a Fox Terrier who competes with her trainer in the agility course. Within the course, Penny must leap over a hurdle. The expression $20t - 16t^2$ models the height of the leap in inches after t seconds. Factor the expression. Then find the height of Penny's leap after 0.5 second.

Find the GCF of each term.

$20t = 2 \cdot 2 \cdot 5 \cdot t$ Factor each term.

$-16t^2 = -1 \cdot 2 \cdot 2 \cdot 2 \cdot 2 \cdot t \cdot t$ Circle common factors.

$GCF = 2 \cdot 2 \cdot t$ or $4t$

$20t - 16t^2 = 4t(5) + 4t(-4t)$ Rewrite each term using the GCF.

$\quad\quad\quad\quad = 4t(5 - 4t)$ Distributive Property

The height of Penny's leap after t seconds can be modeled by $4t(5 - 4t)$.

Substitute 0.5 for t to find the height after 0.5 second.
$4t(5 - 4t)$
$\quad = 4(0.5)[5 - 4(0.5)]$ Substitution, $t = 0.5$
$\quad = 2(5 - 2)$ Simplify.
$\quad = 2(3)$ Subtract.
$\quad = 6$ Multiply.

After 0.5 second, Penny will be 6 inches above the ground.

Real-World Link

Dog agility tests a person's skills as a trainer and handler. Competitors race through an obstacle course that includes hurdles, tunnels, a see-saw, and line poles.

Source: United States Dog Agility Association

▶ **Guided Practice**

2. **KANGAROOS** The height of a kangaroo's hop in inches after t seconds can be modeled by $24t - 16t^2$. Factor the expression. Then find the height of the kangaroo after 1 second.

2 Factoring by Grouping Using the Distributive Property to factor polynomials with four or more terms is called **factoring by grouping** because terms are put into groups and then factored. The Distributive Property is then applied to a common binomial factor.

Key Concept Factoring by Grouping

Words	A polynomial can be factored by grouping only if all of the following conditions exist. • There are four of more terms. • Terms have common factors that can be grouped together. • There are two common factors that are identical or additive inverses of each other.
Symbols	$ax + bx + ay + by = (ax + bx) + (ay + by)$ $\quad\quad\quad\quad\quad\quad\quad = x(a + b) + y(a + b)$ $\quad\quad\quad\quad\quad\quad\quad = (x + y)(a + b)$

Example 3 Factor by Grouping

Factor each polynomial.

a. $4qr + 8r + 3q + 6$

$$
\begin{aligned}
4qr + 8r + 3q + 6 & \qquad \text{Original expression} \\
= (4qr + 8r) + (3q + 6) & \qquad \text{Group terms with common factors.} \\
= 4r(q + 2) + 3(q + 2) & \qquad \text{Factor the GCF from each group.}
\end{aligned}
$$

Notice that $(q + 2)$ is common in both groups, so it becomes the GCF.

$$= (4r + 3) + (q + 2)$$

Study Tip

MP Reasoning To check your factored answers, multiply your factors out. You should get your original expression as a result.

Check You can check this result by multiplying the two factors. The product should be equal to the original expression.

$$(4r + 3)(q + 2) = 4qr + 8r + 3q + 6 \checkmark \text{ FOIL Method}$$

b. $2u^2v - 15 - 6u^2 + 5v$

$$
\begin{aligned}
2u^2v - 15 - 6u^2 + 5v & \qquad \text{Original expression} \\
= 2u^2v - 6u^2 + 5v - 15 & \qquad \text{Commutative Property} \\
= (2u^2v - 6u^2) + (5v - 15) & \qquad \text{Group terms with common factors.} \\
= 2u^2(v - 3) + 5(v + 3) & \qquad \text{Factor the GCF from each group.} \\
= (2u^2 + 5)(v - 3) & \qquad \text{Distributive Property}
\end{aligned}
$$

▶ **Guided Practice**

Factor each polynomial.

3A. $rn + 5n - r - 5$

3B. $3np + 15p - 4n - 20$

3C. $tw^3 - 2w^3 + 10t - 20$

3D. $4ab^2 + 21 + 12b^2 + 7a$

It can be helpful to recognize when binomials are additive inverses of each other. For example, $6 - a = -1(a - 6)$.

Example 4 Factor by Grouping with Additive Inverses

Factor each polynomial.

a. $2mk - 12m + 42 - 7k$

$$
\begin{aligned}
2mk - 12m + 42 - 7k & \\
= (2mk - 12m) + (42 - 7k) & \qquad \text{Group terms with common factors.} \\
= 2m(k - 6) + 7(6 - k) & \qquad \text{Factor the GCF from each group.} \\
= 2m(k - 6) + 7[(-1)(k - 6)] & \qquad 6 - k = -1(k - 6) \\
= 2m(k - 6) - 7(k - 6) & \qquad \text{Associative Property} \\
= (2m + 7)(k - 6) & \qquad \text{Distributive Property}
\end{aligned}
$$

b. $21b^4 - 3ab^4 + 4a - 28$

$21b^4 - 3ab^4 + 4a - 28$

$= (21b^4 - 3ab^4) + (4a - 28)$ Group terms with common factors.

$= 3b^4(7 - a) + 4(a - 7)$ Factor the GCF from each group.

$= 3b^4[(-1)(a - 7)] + 4(a - 7)$ $7 - k = -1(k - 7)$

$= -3b^4(a - 7) + 4(a - 7)$ Associative Property

$= (-3b^4 + 4)(a - 7)$ Distributive Property

▶ **Guided Practice**

Factor each polynomial.

4A. $c - 2cd + 8d - 4$ **4B.** $3p - 2p^2 - 18p + 27$

Check Your Understanding   ◯ = Step-by-Step Solutions begin on page R11.

Go Online! for a Self-Check Quiz

Example 1   **Use the Distributive Property to factor each polynomial.**

1. $21b - 15a$ **2.** $14c^2 + 2c$

3. $10g^2h^2 + 9gh^2 - g^2h$ **4.** $12jk^2 + 6j^2k + 2j^2k^2$

Examples 3–4   **Factor each polynomial.**

⑤ $np + 2n + 8p + 16$ **6.** $xy - 7x + 7y - 49$

7. $3bc - 2b - 10 + 15c$ **8.** $9fg - 45f - 7g + 35$

9. $3km - 21k + 2m - 14$ **10.** $4cd^2 + 3cd - 20d - 15$

11. $10p^3q - 5p^4 + 3pq - 6q^2$ **12.** $5ab - b - 6 + 30a$

Example 2   **13. SPIDERS** Jumping spiders can commonly be found in homes and barns throughout the United States. The height of a jumping spider's jump in feet after t seconds can be modeled by the expression $12t - 16t^2$.

     a. Write the factored form of the expression for the height of the spider after t seconds.

     b. What is the spider's height after 0.1 second?

     c. What is the spider's height after 0.5 second?

14. **REASONING** At a Fourth of July celebration, a rocket is launched straight up with an initial velocity of 128 feet per second. The height of the rocket in feet above sea level is modeled by the expression $128t - 16t^2$, where t is the time in seconds after the rocket is launched.

     a. What is the GCF of each term of the expression?

     b. Write the factored form of the expression for the height of the rocket after t seconds.

     c. Will the height of the rocket be greater after 4 seconds or after 6 seconds?

Example 1   Use the Distributive Property to factor each polynomial.

15. $16t - 40y$

16. $30v + 50x$

17. $2k^2 + 4k$

18. $5z^2 + 10z$

19. $4a^2b^2 + 2a^2b - 10ab^2$

20. $5c^2v - 15c^2v^2 + 5c^2v^3$

Examples 3–4 Factor each polynomial.

21. $fg - 5g + 4f - 20$

22. $a^2 - 4a - 24 + 6a$

23. $hj - 2h + 5j - 10$

24. $xy - 2x - 2 + y$

25. $45pq - 27q - 50p + 30$

26. $24ty - 18t + 4y - 3$

27. $3dt - 21d + 35 - 5t$

28. $8r^2 + 12r$

29. $21th - 3t - 35h + 5$

30. $vp + 12v + 8p + 96$

31. $5br - 25b + 2r - 10$

32. $2nu - 8u + 3n - 12$

33. $5gf^2 + g^2f + 15gf$

34. $rp - 9r + 9p - 81$

35. $27cd^2 - 18c^2d^2 + 3cd$

36. $18r^3t^2 + 12r^2t^2 - 6r^2t$

37. $48tu - 90t + 32u - 60$

38. $16gh + 24g - 2h - 3$

39. $20p^2 + 15p + 8pr + 6r$

40. $4bc - 10b + 5c - 2c^2$

41. $9mk^2 - 15k^2 + 10 - 6m$

42. $x^4y^2 - 7x^3y + 3xy - 21$

43. $48fg + 8g - 18f - 3$

44. $2a^3b - 8a^2 + 12b - 3ab^2$

Example 2   **45.** MP **SENSE-MAKING** Use the drawing at the right.

   a. Write an expression in factored form to represent the area of the blue section.

   b. Write an expression in factored form to represent the area of the region formed by the outer edge.

   c. Write an expression in factored form to represent the yellow region.

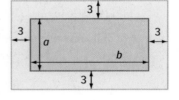

46. FIREWORKS 10-inch fireworks shell is fired from ground level. The height of the shell in feet can be modeled by the expression $264t - 16t^2$, where t is the time in seconds after launch.

   a. Write the expression that represents the height in factored form.

   b. What is the height of the shell 0 seconds and 16.5 seconds after launch?

   c. Based on your solution in part **b**, where was the shell 0 seconds and 16.5 seconds after launch?

   d. The shell reached its maximum height 8.25 seconds after launch. What was the maximum height of the shell?

47. VOLCANOS The 1980 eruption of Washington's Mt. St. Helens had an initial lateral blast with a velocity of 440 feet per second. The expression $440t - 16t^2$ models the height of the rock erupted from the volcano after t seconds.

   a. Write the expression that represents the height of the rock after t seconds in factored form.

   b. What is the height of the rock 10 seconds after it is erupted from the volcano? 20 seconds after it is erupted?

   c. The rock reached its maximum height 13.75 seconds after it is erupted. What was the maximum height of the rock?

48. RIDES Suppose the height in feet of a rider t seconds after being dropped on a drop tower thrill ride can be modeled by the expression $-16t^2 - 96t + 160$.

    a. Write an expression to represent the height in factored form.

    b. From what height is the rider initially dropped?

    c. At what height will the rider be after 3 seconds of falling? Is this possible? Explain.

49 ARCHERY The height in feet of an arrow can be modeled by the expression $72t - 16t^2$, where t is time in seconds after the arrow is released. Write the expression in factored form and find the height of the arrow after 3 seconds.

50. TENNIS A tennis player hits a tennis ball upward with an initial velocity of 80 feet per second. The height in feet of the tennis ball can be modeled by the expression $80t - 16t^2$, where t is time in seconds. Write the expression in factored form and find the height of the tennis ball 2 seconds and 4 seconds after being hit.

51. MULTIPLE REPRESENTATIONS In this problem, you will explore the *box method* of factoring. To factor $x^2 + x - 6$, write the first term in the top left-hand corner of the box, and then write the last term in the lower right-hand corner.

	?	?
?	x^2	?
?	?	-6

    a. Analytical Determine which two factors have a product of -6 and a sum of 1.

    b. Symbolic Write each factor in an empty square in the box. Include the positive or negative sign and variable.

    c. Analytical Find the factor for each row and column of the box. What are the factors of $x^2 + x - 6$?

    d. Verbal Describe how you would use the box method to factor $x^2 - 3x - 40$.

H.O.T. Problems   Use **H**igher-**O**rder **T**hinking Skills

52. ERROR ANALYSIS Hernando and Rachel are factoring $2mp - 6p + 27 - 9m$. Is either of them correct? Explain your reasoning.

Hernando
$2mp - 6p + 27 - 9m$
$= (2mp - 6p) + (27 - 9m)$
$= 2p(m - 3) + 9(3 - m)$
$= (2p + 9)(m - 3)(3 - m)$

Rachel
$2mp - 6p + 27 - 9m$
$= (2mp - 6p) + (27 - 9m)$
$= 2p(m - 3) + 9(3 - m)$
$= 2p(m - 3) + 9[(-1)(m - 3)]$
$= 2p(m - 3) - 9(m - 3)$
$= (2p - 9)(m - 3)$

53. CHALLENGE Given the expression $4yz^2 + 24z + 5yz + 30$, show two different ways to group the terms and factor the polynomial.

54. OPEN-ENDED Write a four-term polynomial that can be factored by grouping. Then factor the polynomial.

55. REASONING Given the expression $ab^2c^3 + a^7bc^2 + a^3c^6$, identify the GCF of the terms. Then factor the expression.

56. WRITING IN MATH Explain why you cannot use factoring by grouping on a polynomial with three terms.

57. The area of a rectangle is represented by $10x^3 + 15x^2 + 4x + 6$. Its dimensions are binomials in x with prime coefficients. What are the dimensions of the rectangle? **MP** 7

58. MULTI-STEP A rectangle has a length of $4x$ units and a width of $(6x + 7)$ units. A right triangle has legs of length x and $(8x - 4)$ units. **MP** 1, 2

  a. Express the areas of the rectangle and the right triangle, respectively.

  b. Find the factored form of the difference of the area of the rectangle and triangle.

59. Jill is designing her name using geometric shapes. She is using a vertical rectangle with length $2x$ for the letter "i," as shown.

$2x$

If the area of the rectangle is represented by $6x^2 - 14x$, which expression represents the width? **MP** 4

  ◯ **A** $3x - 7$

  ◯ **B** $3x - 14$

  ◯ **C** $2x$

  ◯ **D** $3x^2 - 7x$

60. Factor $6b^2c - 9bc + 22bd - 33d$. **MP** 7

  ◯ **A** $(3bc + 11d)(2b - 3)^2$

  ◯ **B** $(3bc + 11d)(2b - 3)$

  ◯ **C** $(2b - 3)$

  ◯ **D** $(3bc - 11d)(2b + 3)$

61. The height of a golf ball in feet after t seconds can be modeled by $104t - 16t^2$. Which is the GCF of the terms in the polynomial? **MP** 4

  ◯ **A** $2t$

  ◯ **B** $4t$

  ◯ **C** 8

  ◯ **D** $8t$

62. Which expression is equivalent to the following polynomial? **MP** 6

$$2x^3 - 10x^2 + 15 - 3x$$

  ◯ **A** $(2x^2 + 3)(x - 5)$

  ◯ **B** $(2x^2 + 3)(x - 5)(5 - x)$

  ◯ **C** $(2x^2 - 3)(x - 5)$

  ◯ **D** $(2x^2 - 3)(x - 5)^2$

63. Factor the following expressions completely: **MP** 1, 7

  a. $6y^3 + 21y^2 + 4y + 14$

  b. $12a^2 - 20ab + 9ay - 15by$

  c. $4w^3 + 3wz - 8w^2 - 6z$

  d. $15m^2n + 27mn^2$

  e. $8wy + 12xy + 10wz + 15xz$

64. The area of two rectangles can be represented by $x^3 + 4x^2 + 2x + 8$ and $x^3 - 3x^2 + 2x - 6$. If the widths of the rectangles are equal, find the width of the rectangles. **MP** 2, 7

65. The product of two binomials $2x^3 - 3x^2 - 10x + 15$. Find the sum of these two binomials. **MP** 7

Algebra Lab
Proving the Elimination Method

The elimination method for solving systems of equations requires adding one equation in a system with a multiple of the other equation. You can prove that this process does not affect the solution and is therefore is a valid method by using the steps outlined below.

Mathematical Practices

 3 Look for viable arguments and critique the reasoning of others.
7 Look for and make use of structure.

Claim: Given a system of linear equations in two variables, replacing one equation by the sum of that equation and a multiple of the other produces a system with the same solutions.

Step 1 Solve a General System of Equations

Solve the general system of linear equations by using substitution.

$$ax + by = c$$

$$dx + ey = f$$

Solve each equation for x to get $x = \dfrac{c - by}{a}$ and $x = \dfrac{f - ey}{d}$. Use substitution to set the right sides of the equations equal and find an expression for y. Solving $\dfrac{c - by}{a} = \dfrac{f + ey}{d}$ for y gives $y = \dfrac{af - dc}{-bd + ae}$.

Repeat this process to find x. Solving each equation in the system for y gives $y = \dfrac{c - ax}{b}$ and $y = \dfrac{f - dx}{e}$.

Substitute to find an expression for x. Solving $\dfrac{c - ax}{b} = \dfrac{f - dx}{e}$ for x gives $x = \dfrac{bf - ec}{-ae + bd}$.

The solution of the general system is $\left(\dfrac{bf - ec}{-ae + bd}, \dfrac{af - dc}{-bd + ae} \right)$.

Now that you know the solution of the general system, replace one of the equations with the sum of that equation and a multiple of the other to create a new system of equations. If the systems have the same solution, then the method of elimination is valid.

Step 2 Create a New System of Equations

Replace an equation in the general system of equations with the sum of that equation and a multiple of the other.

Replace $dx + ey = f$ with the sum of $dx + ey = f$ and a multiple of $ax + by = c$. To ensure that this will be true for any case, use a general real number g as the factor.

$dx + ey = f$	First equation
$(+)\ agx + bgy = gc$	Multiple of second equation
$dx + agx + ey + bgy = f + gc$	Add.
$(d + ag)x + (e + bg)y = f + gc$	Factor.

So, the new system is $\begin{cases} ax + by = c \\ (d + ag)x + (e + bg)y = f + gc. \end{cases}$

Step 3 Solve the New System of Equations

Solve the new system of linear equations by using substitution.

$$ax + by = c$$

$$(d + ag)x + (e + bg)y = f + gc$$

Solve each equation for x to get $x = \dfrac{c - by}{a}$ and $x = \dfrac{f + gc - ey - bgy}{d + ag}$. Substitute to find y.

$\dfrac{c - by}{a} = \dfrac{f + gc - ey - bgy}{d + ag}$	Substitution
$dc - bdy + acg - abgy = af + acg - aey - abgy$	Cross multiply and distribute.
$dc - bdy = af - aey$	Subtract *acg* and add *abgy* to each side.
$-bdy + aey = af - dc$	Isolate the *y*-terms.
$y(-bd + ae) = af - dc$	Factor out *y*.
$y = \dfrac{af - dc}{-bd + ae}$	Divide each side by $-bd + ae$.

Repeat this process to find x. Solving each equation in the system for y gives $y = \dfrac{c - ax}{b}$ and

$y = \dfrac{f + gc - dx - agx}{e + bg}$. Substitute to find an expression for x. Solving

$\dfrac{c - ax}{b} = \dfrac{f + gc - dx - agx}{e + bg}$ for x gives $x = \dfrac{bf - ec}{-ae + bd}$.

The solution of the new system is $\left(\dfrac{bf - ec}{-ae + bd}, \dfrac{af - dc}{-bd + ae} \right)$.

The solutions of the original system in Step 1 and the new system are the same. Therefore, the claim is true.

Exercises

Solve the system of equations by elimination. Verify the solution using a different method.

1. $2x - 3y = 4$
$\ x + 2y = 9$

2. $4x - 3y = 4$
$\ 3x + 2y = 37$

3. $2x + 5y = 26$
$\ -3x - 4y = -25$

4. $5x + y = 0$
$\ 5x + 2y = 30$

5. $2x - 3y = -7$
$\ 3x + y = -5$

6. $4x - 3y = -1$
$\ 3x - y = -2$

7. Consider the system $\begin{matrix} ax - 3y = 4 \\ 3x + 2y = 3a \end{matrix}$.

   a. Solve the system of equations by the process of elimination. Leave your solution in terms of a.

   b. Is your solution in part **a** the same as the solution you get using the substitution method?

You can use algebra tiles to factor trinomials. If a polynomial represents the area of a rectangle formed by algebra tiles, then the rectangle's length and width are *factors* of the area. If a rectangle cannot be formed to represent the trinomial, then the trinomial is not factorable.

Mathematical Practices
MP **3** Construct viable arguments and critique the reasoning of others.

Activity 1 Factor $x^2 + bx + c$

Work cooperatively. Use algebra tiles to factor $x^2 + 4x + 3$.

Step 1 Model $x^2 + 4x + 3$.

Step 2 Place the x^2-tile at the corner of the product mat. Arrange the 1-tiles into a rectangular array. Because 3 is prime, the three tiles can be arranged in a rectangle in one way, a 1-by-3 rectangle.

Step 3 Complete the rectangle with the x-tiles.

The rectangle has a width of $x + 1$ and a length of $x + 3$.

Therefore, $x^2 + 4x + 3 = (x + 1)(x + 3)$.

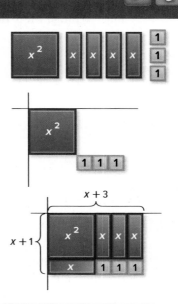

Activity 2 Factor $x^2 + bx + c$

Use algebra tiles to factor $x^2 + 8x + 12$.

Step 1 Model $x^2 + 8x + 12$.

Step 2 Place the x^2-tile at the corner of the product mat. Arrange the 1-tiles into a rectangular array. Since $12 = 3 \times 4$, try a 3-by-4 rectangle. Try to complete the rectangle. Notice that there is an extra x-tile.

Step 3 Arrange the 1-tiles into a 2-by-6 rectangular array. This time you can complete the rectangle with the x-tiles.

The rectangle has a width of $x + 2$ and a length of $x + 6$.

Therefore, $x^2 + 8x + 12 = (x + 2)(x + 6)$.

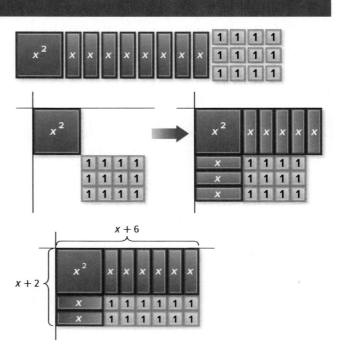

(continued on the next page)

Activity 3 Factor $x^2 - bx + c$

Work cooperatively. Use algebra tiles to factor $x^2 - 5x + 6$.

Step 1 Model $x^2 - 5x + 6$.

Step 2 Place the x^2-tile at the corner of the product mat. Arrange the 1-tiles into a 2-by-3 rectangular array as shown.

Step 3 Complete the rectangle with the x-tiles. The rectangle has a width of $x - 2$ and a length of $x - 3$.

Therefore, $x^2 - 5x + 6 = (x - 2)(x - 3)$.

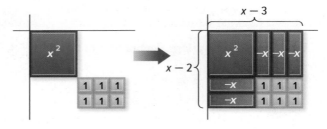

Activity 4 Factor $x^2 - bx - c$

Use algebra tiles to factor $x^2 - 4x - 5$.

Step 1 Model $x^2 - 4x - 5$.

Step 2 Place the x^2-tile at the corner of the product mat. Arrange the 1-tiles into a 1-by-5 rectangular array as shown.

Step 3 Place the x-tiles as shown. Recall that you can add zero pairs without changing the value of the polynomial. In this case, add a zero pair of x-tiles.

The rectangle has a width of $x + 1$ and a length of $x - 5$.

Therefore, $x^2 - 4x - 5 = (x + 1)(x - 5)$.

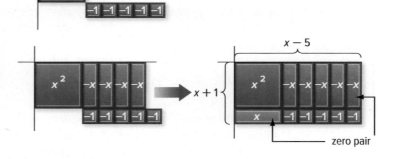

Model and Analyze

Work cooperatively. Use algebra tiles to factor each trinomial.

1. $x^2 + 3x + 2$ **2.** $x^2 + 6x + 8$ **3.** $x^2 + 3x - 4$ **4.** $x^2 - 7x + 12$

5. $x^2 + 7x + 10$ **6.** $x^2 - 2x + 1$ **7.** $x^2 + x - 12$ **8.** $x^2 - 8x + 15$

Tell whether each trinomial can be factored. Justify your answer with a drawing.

9. $x^2 + 3x + 6$ **10.** $x^2 - 5x - 6$ **11.** $x^2 - x - 4$ **12.** $x^2 - 4$

13. WRITING IN MATH How can you use algebra tiles to determine whether a trinomial can be factored?

Factoring Quadratic Trinomials

- You multiplied binomials by using the FOIL method.

1 Factor trinomials of the form $x^2 + bx + c$.

2 Factor trinomials of the form $ax^2 + bx + c$.

- Diana is having a rectangular hot tub installed near her pool. A 24-foot fence is to surround the hot tub. If the hot tub will cover an area of 36 square feet, what will be the dimensions of the hot tub?

To solve this problem, the landscape architect needs to find two numbers that have a product of 36 and a sum of 12, half the perimeter of the hot tub.

 New Vocabulary

prime polynomial

 Mathematical Practices

8 Look for and express regularity in repeated reasoning.

1 **Factor $x^2 + bx + c$** You have learned how to multiply two binomials using the FOIL method. Each of the binomials was a factor of the product. The pattern for multiplying two binomials can be used to factor certain types of trinomials.

$$\begin{aligned}(x + 3)(x + 4) &= x^2 + 4x + 3x + 3 \cdot 4 && \text{Use the FOIL method.} \\ &= x^2 + (4 + 3)x + 3 \cdot 4 && \text{Distributive Property} \\ &= x^2 + 7x + 12 && \text{Simplify.}\end{aligned}$$

Notice that the coefficient of the middle term, $7x$, is the sum of 3 and 4, and the last term, 12, is the product of 3 and 4.

Observe the following pattern in this multiplication.

$$\begin{aligned}(x + 3)(x + 4) &= x^2 + (4 + 3)x + (3 \cdot 4) \\ (x + m)(x + p) &= x^2 + (p + m)x + mp && \text{Let } 3 = m \text{ and } 4 = p. \\ &= x^2 + \underbrace{(m + p)}x + \underbrace{mp} && \text{Commutative } (+) \\ &\quad\;\; x^2 + \quad bx \quad + \quad c && b = m + p \text{ and } c = mp\end{aligned}$$

Notice that the coefficient of the middle term is the sum of m and p, and the last term is the product of m and p. This pattern can be used to factor trinomials of the form $x^2 + bx + c$.

⚿ Key Concept Factoring $x^2 + bx + c$

Words	To factor trinomials in the form $x^2 + bx + c$, find two integers, m and p, with a sum of b and a product of c. Then write $x^2 + bx + c$ as $(x + m)(x + p)$.
Symbols	$x^2 + bx + c = (x + m)(x + p)$ when $m + p = b$ and $mp = c$.
Example	$x^2 + 6x + 8 = (x + 2)(x + 4)$, because $2 + 4 = 6$ and $2 \cdot 4 = 8$.

When c is positive, its factors have the same signs. Both of the factors are positive or negative based upon the sign of b. If b is positive, the factors are positive. If b is negative, the factors are negative.

Example 1   *b* and *c* are Positive

Factor $x^2 + 9x + 20$.

In this trinomial, $b = 9$ and $c = 20$. Since c is positive and b is positive, you need to find two positive factors with a sum of 9 and a product of 20. Make an organized list of the factors of 20, and look for the pair of factors with a sum of 9.

Factors of 20	Sum of Factors
1, 20	21
2, 10	12
4, 5	9

The correct factors are 4 and 5.

$$x^2 + 9x + 20 = (x + m)(x + p) \qquad \text{Write the pattern.}$$
$$= (x + 4)(x + 5) \qquad m = 4 \text{ and } p = 5$$

CHECK You can check this result by multiplying the two factors. The product should be equal to the original expression.

$$(x + 4)(x + 5) = x^2 + 5x + 4x + 20 \qquad \text{FOIL Method}$$
$$= x^2 + 9x + 20 \checkmark \qquad \text{Simplify.}$$

▶ **Guided Practice**

Factor each polynomial.

1A. $d^2 + 11d + 24$ **1B.** $9 + 10t + t^2$

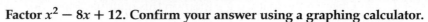

When factoring a trinomial in which b is negative and c is positive, use what you know about the product of binomials to narrow the list of possible factors.

Example 2   *b* is Negative and *c* is Positive

Factor $x^2 - 8x + 12$. Confirm your answer using a graphing calculator.

In this trinomial, $b = -8$ and $c = 12$. Since c is positive and b is negative, you need to find two negative factors with a sum of -8 and a product of 12.

Factors of 12	Sum of Factors
$-1, -12$	-13
$-2, -6$	-8
$-3, -4$	-7

The correct factors are -2 and -6.

$$x^2 - 8x + 12 = (x + m)(x + p) \qquad \text{Write the pattern.}$$
$$= (x - 2)(x - 6) \qquad m = -2 \text{ and } p = -6$$

CHECK Graph $y = x^2 - 8x + 12$ and $y = (x - 2)(x - 6)$ on the same screen. Since only one graph appears, the two graphs must coincide. Therefore, the trinomial has been factored correctly. ✓

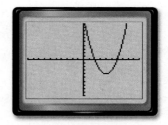

$[-10, 10]$ scl: 1 by $[-10, 10]$ scl: 1

▶ **Guided Practice**

Factor each polynomial.

2A. $21 - 22m + m^2$ **2B.** $w^2 - 11w + 28$

Problem-Solving Tip

Guess and Check When factoring a trinomial, make an educated guess, check for reasonableness, and then adjust the guess until the correct answer is found.

Study Tip

(MP) Regularity Once the correct factors are found, it is not necessary to test any other factors. In Example 2, -2 and -6 are the correct factors, so -3 and -4 do not need to be tested.

When c is negative, its factors have opposite signs. To determine which factor is positive and which is negative, look at the sign of b. The factor with the greater absolute value has the same sign as b.

Example 3 c is Negative

Factor each polynomial. Confirm your answers using a graphing calculator.

a. $x^2 + 2x - 15$

In this trinomial, $b = 2$ and $c = -15$. Since c is negative, the factors m and p have opposite signs. So either m or p is negative, but not both. Since b is positive, the factor with the greater absolute value is also positive.

List the factors of -15, where one factor of each pair is negative. Look for the pair of factors with a sum of 2.

Factors of -15	Sum of Factors
$-1, 15$	14
$-3, 5$	2

$$x^2 + 2x - 15 = (x + m)(x + p)$$ The correct factors are -3 and 5.

Write the pattern.

$$= (x - 3)(x + 5)$$ $m = -3$ and $p = 5$

CHECK $(x - 3)(x + 5) = x^2 + 5x - 3x - 15$ FOIL Method

$$= x^2 + 2x - 15 \checkmark$$ Simplify.

b. $x^2 - 7x - 18$

In this trinomial, $b = -7$ and $c = -18$. Either m or p is negative, but not both. Since b is negative, the factor with the greater absolute value is also negative.

List the factors of -18, where one factor of each pair is negative. Look for the pair of factors with a sum of -7.

Factors of -18	Sum of Factors
$1, -18$	-17
$2, -9$	-7
$3, -6$	-3

$$x^2 - 7x - 18 = (x + m)(x + p)$$ The correct factors are 2 and -9.

Write the pattern.

$$= (x + 2)(x - 9)$$ $m = 2$ and $p = -9$

CHECK Graph $y = x^2 - 7x - 18$ and $y = (x + 2)(x - 9)$ on the same screen.

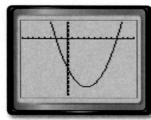

[-10, 15] scl: 1 by [-40, 20] scl: 1

The graphs coincide. Therefore, the trinomial has been factored correctly. $\checkmark$

▶ **Guided Practice**

3A. $y^2 + 13y - 48$ **3B.** $r^2 - 2r - 24$

2 Factor $ax^2 + bx + c$

You can apply the factoring methods to quadratic expressions where a is not 1. The factoring by grouping method is necessary when factoring these quadratic expressions.

Key Concept Factoring $ax^2 + bx + c$

Words To factor trinomials of the form $ax^2 + bx + c$, find two integers, m and p, with a sum of b and a product of ac. Then write $ax^2 + bx + c$ as $ax^2 + mx + px + c$, and factor by grouping.

Example $5x^2 - 13x + 6 = 5x^2 - 10x - 3x + 6$ $m = -10$ and $p = -3$
$$= 5x(x - 2) + (-3)(x - 2)$$
$$= (5x - 3)(x - 2)$$

Example 4   Factor $ax^2 + bx + c$

Factor each trinomial.

a. $7x^2 + 29x + 4$

In this trinomial, $a = 7$, $b = 29$, and $c = 4$. You need to find two numbers with a sum of 29 and a product of $7 \cdot 4$ or 28. Make a list of the factors of 28 and look for the pair of factors with the sum of 29.

Factors of 28	Sum of Factors
1, 28	29

The correct factors are 1 and 28.

$7x^2 + 29x + 4 = 7x^2 + mx + px + 4$ Write the pattern.
$\quad\quad\quad\quad\quad\quad = 7x^2 + 1x + 28x + 4$ $m = 1$ and $p = 28$
$\quad\quad\quad\quad\quad\quad = (7x^2 + 1x) + (28x + 4)$ Group terms with common factors.
$\quad\quad\quad\quad\quad\quad = x(7x + 1) + 4(7x + 1)$ Factor the GCF.
$\quad\quad\quad\quad\quad\quad = (x + 4)(7x + 1)$ $7x + 1$ is the common factor.

Study Tip

MP **Regularity** Always look for a GCF of the terms of a polynomial before you factor.

b. $5x^2 + 12x - 9$

In this trinomial, $a = 5$, $b = 12$, and $c = -9$. You need to find two numbers with a sum of 12 and a product of $5(-9)$ or -45. Make a list of factors of -45 and look for the pair of factors with the sum of 12.

Factors of -45	Sum of Factors
-1, 45	44
-3, 15	12

The correct factors are -3 and 15.

$5x^2 + 12x - 9 = 5x^2 + 15x - 3x - 9$ $m = 15$ and $p = -3$
$\quad\quad\quad\quad\quad\quad = (5x^2 + 15x) + (-3x - 9)$ Group terms with common factors.
$\quad\quad\quad\quad\quad\quad = 5x(x + 3) + (-3)(x + 3)$ Factor the GCF.
$\quad\quad\quad\quad\quad\quad = (5x - 3)(x + 3)$ Distributive Property

c. $3x^2 + 15x + 18$

The GCF of the terms $3x^2$, $15x$, and 18 is 3. Factor this first.

$3x^2 + 15x + 18 = 3(x^2 + 5x + 6)$ Distributive Property
$\quad\quad\quad\quad\quad\quad = 3(x + 3)(x + 2)$ Find two factors of 6 with a sum of 5.

▶ **Guided Practice**

4A. $5x^2 + 13x + 6$ **4B.** $6x^2 + 22x - 8$

Sometimes the coefficient of the x-term is negative.

Example 5   Factor $ax^2 - bx + c$

Factor $3x^2 - 17x + 20$.

In this trinomial, $a = 3$, $b = -17$, and $c = 20$. Since b is negative, $m + p$ will be negative. Since c is positive, mp will be positive.

To determine m and p, list the negative factors of ac or 60. The sum of m and p should be -17.

Factors of 60	Sum of Factors
$-2, -30$	-32
$-3, -20$	-23
$-4, -15$	-19
$-5, -12$	-17

The correct factors are -5 and -12.

$$
\begin{aligned}
3x^2 - 17x + 20 &= 3x^2 - 12x - 5x + 20 && m = -12 \text{ and } p = -5 \\
&= (3x^2 - 12x) + (-5x + 20) && \text{Group terms with common factors.} \\
&= 3x(x - 4) + (-5)(x - 4) && \text{Factor the GCF.} \\
&= (3x - 5)(x - 4) && \text{Distributive Property}
\end{aligned}
$$

▶ **Guided Practice**

5A. $2n^2 - n - 1$ **5B.** $10y^2 - 35y + 30$

A polynomial that cannot be written as a product of two polynomials with integral coefficients is called a **prime polynomial**.

Example 6   Determine Whether a Polynomial Is Prime

Factor $4x^2 - 3x + 5$, if possible. If the polynomial cannot be factored using integers, write *prime*.

Factors of 20	Sum of Factors
$-20, -1$	-21
$-4, -5$	-9
$-2, -10$	-12

In this trinomial, $a = 4$, $b = -3$, and $c = 5$. Since b is negative, $m + p$ is negative. Since c is positive, mp is positive. So, m and p are both negative. Next, list the factors of 20. Look for the pair with a sum of -3.

There are no factors with a sum of -3. So the quadratic expression cannot be factored using integers. Therefore, $4x^2 - 3x + 5$ is prime.

▶ **Guided Practice**

Factor each polynomial, if possible. If the polynomial cannot be factored using integers, write *prime*.

6A. $4r^2 - r + 7$ **6B.** $2x^2 + 3x - 5$

Examples 1–3 Factor each polynomial. Confirm your answers using a graphing calculator.

1. $x^2 + 14x + 24$

2. $y^2 - 7y - 30$

3. $n^2 + 4n - 21$

4. $m^2 - 15m + 50$

Examples 4–6 Factor each polynomial, if possible. If the polynomial cannot be factored using integers, write *prime*.

5. $4x^2 - 30x + 36$

6. $6x^2 - x - 14$

7. $3x^2 + 17x + 10$

8. $2x^2 + 22x + 56$

9. $5x^2 - 3x + 4$

10. $3x^2 - 11x - 20$

Practice and Problem Solving Extra Practice is on page R8.

Examples 1–3 Factor each polynomial. Confirm your answers using a graphing calculator.

11. $x^2 + 17x + 42$

12. $y^2 - 17y + 72$

13. $a^2 + 8a - 48$

14. $n^2 - 2n - 35$

15. $44 + 15h + h^2$

16. $40 - 22x + x^2$

17. $-24 - 10x + x^2$

18. $-42 - m + m^2$

Examples 4–6 Factor each polynomial, if possible. If the polynomial cannot be factored using integers, write *prime*.

19 $2x^2 + 19x + 24$

20. $5x^2 + 34x + 24$

21. $4x^2 + 22x + 10$

22. $4x^2 + 38x + 70$

23. $2x^2 - 3x - 9$

24. $4x^2 - 13x + 10$

25. $2x^2 + 3x + 6$

26. $5x^2 + 3x + 4$

27. $12x^2 + 69x + 45$

28. $4x^2 - 5x + 7$

29. $5x^2 + 23x + 24$

30. $3x^2 - 8x + 15$

MP **STRUCTURE** Factor each polynomial.

31. $q^2 + 11qr + 18r^2$

32. $x^2 - 14xy - 51y^2$

33. $x^2 - 6xy + 5y^2$

34. $a^2 + 10ab - 39b^2$

GEOMETRY Find an expression for the perimeter of a rectangle with the given area.

35. $A = x^2 + 24x - 81$

36. $A = x^2 + 13x - 90$

Factor each polynomial, if possible. If the polynomial cannot be factored using integers, write *prime*.

37. $-6x^2 - 23x - 20$

38. $-4x^2 - 15x - 14$

39. $-5x^2 + 18x + 8$

40. $-6x^2 + 31x - 35$

41. $-4x^2 + 5x - 12$

42. $-12x^2 + x + 20$

43. **MULTIPLE REPRESENTATIONS** In this problem, you will explore factoring a special type of polynomial.

    a. **Geometric** Draw a square and label the sides *a*. Within this square, draw a smaller square that shares a vertex with the first square. Label the sides *b*. What are the areas of the two squares?

b. Geometric Cut and remove the small square. What is the area of the remaining region?

c. Analytical Draw a diagonal line between the inside corner and outside corner of the figure, and cut along this line to make two congruent pieces. Then rearrange the two pieces to form a rectangle. What are the dimensions?

d. Analytical Write the area of the rectangle as the product of two binomials.

e. Verbal Complete this statement: $a^2 - b^2 = \ldots$ Why is this statement true?

H.O.T. Problems   Use Higher–Order Thinking Skills

44. ERROR ANALYSIS Jerome and Charles have factored $x^2 + 6x - 16$. Is either of them correct? Explain your reasoning.

Jerome
$x^2 + 6x - 16 = (x + 2)(x - 8)$

Charles
$x^2 + 6x - 16 = (x - 2)(x + 8)$

CONSTRUCT ARGUMENTS Find all values of k so that each polynomial can be factored using integers.

45. $x^2 + kx + 14$

46. $2x^2 + kx + 12$

47. $x^2 - 8x + k, k > 0$

48. $2x^2 - 5x + k, k > 0$

49. (MP) **REASONING** For any factorable trinomial, $x^2 + bx + c$, will the absolute value of b *sometimes*, *always*, or *never* be less than the absolute value of c? Explain.

50. REASONING A square has an area of $9x^2 + 30xy + 25y^2$ square inches. The dimensions are binomials with positive integer coefficients. What is the perimeter of the square? Explain.

51. CHALLENGE Factor $(4y - 5)^2 + 3(4y - 5) - 70$.

52. WRITING IN MATH Compare the method for factoring trinomials of the form $ax^2 + bx + c$ to the method for factoring trinomials of the form $x^2 + bx + c$. Discuss how to find the signs of the factors of c.

53. FLAGS The official flag of most countries is rectangular in shape, but the flag of Switzerland is a square. However, Swiss naval vessels fly a rectangular flag. The area of the square flag is $x^2 - 6x + 9$ square feet and the area of the rectangular naval flag is $x^2 - 2x - 3$. How many feet longer is the naval flag than the square flag?

54. Which expression shows the factored form of $x^2 + 10x - 56$? **MP** 8

- **A** $(x - 7)(x + 8)$
- **B** $(x - 14)(x + 4)$
- **C** $(x + 14)(x - 4)$
- **D** $(x + 7)(x - 8)$

55. Which expression is a factor of $x^2 - 3x - 108$? **MP** 8

- **A** $(x - 9)$
- **B** $(x - 12)$
- **C** $(x + 4)$
- **D** $(x - 27)$

56. The area of a rectangle is represented by the polynomial $6x^2 + 3x - 30$. Which expressions could represent the sides of the rectangle? **MP** 2

- **A** $2x - 5$ and $3x + 6$
- **B** $2x + 5$ and $3x - 6$
- **C** $2x + 5$ and $3x + 6$
- **D** $2x - 5$ and $3x - 6$

57. Which is a factor of the polynomial $5x^2 + 13xy - 6y^2$? **MP** 8

- **A** $(5x - 2y)$
- **B** $(5x + 2y)$
- **C** $(x - 3y)$
- **D** $(5x - 2)$

58. Factor each polynomial. **MP** 2

- **a.** $x^2 - 7x + 10$ ⬚
- **b.** $3x^2 - 6x + 3$ ⬚

59. MULTI-STEP A city has commissioned the building of a rectangular park. The area of the park will be $660x^2 + 524x + 85$ square feet. **MP** 1, 4

- **a.** To factor the expression for the area of the park, what must be the sum of m and p? the product?
  ⬚

- **b.** Which expressions could represent the dimensions of the park?
  - **A** $22x(30x + 17)$ and $5(30x + 17)$
  - **B** $22x - 5$ and $30x - 17$
  - **C** $30x + 17$ and $30x + 17$
  - **D** $22x + 5$ and $30x + 17$

- **c.** The city plans to build a pavilion in the park that will require an area of $27x^2 - 30x - 13$. Factor the expression to find the dimensions of the pavilion.
  ⬚

- **d.** If $x = 5$, what will be the dimensions of the park and the pavilion?
  ⬚

60. The perimeter of a rectangle is the same as the perimeter of a square. If the area of this rectangle is $x^2 + 4x - 21$ and the side lengths are integers, find: **MP** 7

- **a.** the perimeter of the rectangle ⬚
- **b.** the area of the square ⬚

61. Find a common factor between the polynomials $4x^2 + 5x + 9$ and $x^2 - 4x + 3$. **MP** 7
⬚

62. Factor each polynomial. **MP** 7

- **a.** $z^2 + 2yz - 3y^2$ ⬚
- **b.** $3z^2 + 2yz - y^2$ ⬚
- **c.** $27y^2 - 12z^2$ ⬚

Factoring Special Products

∷Then
- You factored trinomials into two binomials.

∷Now

1 Factor binomials that are the difference of squares.

2 Factor trinomials that are perfect squares.

∷Why?
- Computer graphics designers use a combination of art and mathematics skills to design images and videos. Factoring can help to determine the dimensions and shapes of the figures they design.

New Vocabulary
difference of two squares
perfect square trinomial

MP Mathematical Practices
1 Make sense of problems.
4 Model with mathematics.
7 Make use of structure.
8 Look for repeated reasoning.

1 **Factor Differences of Squares** You have previously learned about the product of the sum and difference of two quantities. This resulting product is referred to as the **difference of two squares**. So, the factored form of the difference of squares is called the product of the sum and difference of the two quantities.

> #### Key Concept Factoring Differences of Squares
>
> **Symbols** $a^2 - b^2 = (a + b)(a - b)$ or $(a - b)(a + b)$
>
> **Examples** $x^2 - 25 = (x + 5)(x - 5)$ or $(x - 5)(x + 5)$
>
> $t^2 - 64 = (t + 8)(t - 8)$ or $(t - 8)(t + 8)$

Example 1   Factor Differences of Squares

Determine whether each polynomial is a difference of squares. Write *yes* or *no*. If so, factor it.

a. $16h^2 - 9a^2$

Yes; Since $16h^2 = (4h)^2$ and $9a^2 = (3a)^2$, this is the difference of squares.

$16h^2 - 9a^2 = (4h)^2 - (3a)^2$ Write in the form of $a^2 - b^2$.

$\qquad = (4h + 3a)(4h - 3a)$ Factor the difference of squares.

b. $121c^4 - 25d^3$

No; While $121c^4 = (11c^2)^2$, $25d^3$ does not equal the square of any monomial.

> **Guided Practice**
>
> **1A.** $81 - c^2$ **1B.** $36n^2 - 27m^4$
>
> **1C.** $25y^2 + 1$ **1D.** $64g^2 - h^2$

To factor a polynomial completely, a technique may need to be applied more than once. This also applies to the difference of squares pattern.

2 **Factor Perfect Square Trinomials** You have learned the patterns for the products of the binomials $(a + b)^2$ and $(a - b)^2$. Recall that these are special products that follow specific patterns.

$$(a + b)^2 = (a + b)(a + b) \qquad\qquad (a - b)^2 = (a - b)(a - b)$$

$$= a^2 + ab + ab + b^2 \qquad\qquad = a^2 - ab - ab + b^2$$

$$= a^2 + 2ab + b^2 \qquad\qquad = a^2 - 2ab + b^2$$

These products are called **perfect square trinomials**, because they are the squares of binomials. The above patterns can help you factor perfect square trinomials.

For a trinomial to be factorable as a perfect square, the first and last terms must be perfect squares and the middle term must be two times the square roots of the first and last terms.

⬙ Key Concept Factoring Perfect Square Trinomials

Symbols $a^2 + 2ab + b^2 = (a + b)(a + b) = (a + b)^2$

$a^2 - 2ab + b^2 = (a - b)(a - b) = (a - b)^2$

Examples $x^2 + 8x + 16 = (x + 4)(x + 4)$ or $(x + 4)^2$

$x^2 - 6x + 9 = (x - 3)(x - 3)$ or $(x - 3)^2$

Study Tip

MP **Reasoning** If the constant term of the trinomial is negative, the trinomial is not a perfect square trinomial, so it is not necessary to check the other conditions.

Example 2 Recognize and Factor Perfect Square Trinomials

Determine whether each trinomial is a perfect square trinomial. Write *yes* or *no*. If so, factor it.

a. $4y^2 + 12y + 9$

① Is the first term a perfect square? Yes, $4y^2 = (2y)^2$

② Is the last term a perfect square? Yes, $9 = 3^2$

③ Is the middle term equal to $2(2y)(3)$? Yes, $12y = 2(2y)(3)$

Since all three conditions are satisfied, $4y^2 + 12y + 9$ is a perfect square trinomial.

$4y^2 + 12y + 9 = (2y)^2 + 2(2y)(3) + 3^2$ Write as $a^2 + 2ab + b^2$.

$= (2y + 3)^2$ Factor using the pattern.

b. $9x^2 - 6x + 4$

① Is the first term a perfect square? Yes, $9x^2 = (3x)^2$

② Is the last term a perfect square? Yes, $4 = 2^2$

③ Is the middle term equal to $-2(3x)(2)$? No, $-6x \neq -2(3x)(2)$

Since the middle term does not satisfy the required condition, $9x^2 - 6x + 4$ is not a perfect square trinomial.

▶ **Guided Practice**

2A. $9y^2 + 24y + 16$ **2B.** $2a^2 + 10a + 25$

A polynomial is completely factored when it is written as a product of prime polynomials. More than one method might be needed to factor a polynomial completely. When completely factoring a polynomial, the Concept Summary can help you decide where to start.

Remember, if the polynomial does not fit any pattern or cannot be factored, the polynomial is prime.

Go Online!

You will use the factoring methods in this Concept Summary frequently. Log into your **eStudent Edition** to bookmark this page.

Concept Summary Factoring Methods

Steps	Number of Terms	Examples
Step 1 Factor out the GCF.	any	$4x^3 + 2x^2 - 6x = 2x(2x^2 + x - 3)$
Step 2 Check for a difference of squares or a perfect square trinomial.	2 or 3	$9x^2 - 16 = (3x + 4)(3x - 4)$ $16x^2 + 24x + 9 = (4x + 3)^2$ $25x^2 - 10x + 1 = (5x - 1)^2$
Step 3 Apply the factoring patterns for $x^2 + bx + c$ or $ax^2 + bx + c$ (general trinomials), or factor by grouping.	3 or 4	$x^2 - 8x + 12 = (x - 2)(x - 6)$ $2x^2 + 13x + 6 = (2x + 1)(x + 6)$ $12y^2 + 17y + 6$ $= 12y^2 + 9y + 8y + 6$ $= (12y^2 + 9y) + (8y + 6)$ $= 3y(4y + 3) + 2(4y + 3)$ $= (4y + 3)(3y + 2)$

Real-World Example 3   **Find Dimensions**

DESIGN **An artist is designing a stained glass window. It will be made up of pieces of colored glass and 2 clear squares as shown. Each clear square has a side length of 8 inches.**

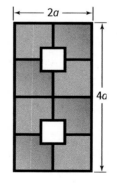

a. Write a polynomial that represents the area of the colored glass.

The total area of the window is $4a \cdot 2a$, or $8a^2$ square inches.

The area of the clear squares is $2 \cdot 8^2$, or 128 square inches.

The area of the colored glass is the total area of the window minus the area of the clear glass. So, the area of the colored glass is $8a^2 - 128$ square inches.

b. Factor the polynomial completely.

$8a^2 - 128 = 8(a^2 - 16)$ Factor out the GCF.

$= 8(a - 4)(a + 4)$ Factor the difference of squares.

▶ **Guided Practice**

3. Write the factored form of a polynomial that represents the area of the colored glass if the clear squares have a side length of 6 inches.

Example 4 Factor Completely

Factor each polynomial, if possible. If the polynomial cannot be factored, write *prime*.

a. $5x^2 - 80$

> **Step 1** The GCF of $5x^2$ and -80 is 5, so factor it out.
>
> **Step 2** Since there are two terms, check for a difference of squares.
>
> $$\begin{aligned} 5x^2 - 80 &= 5(x^2 - 16) &&\text{5 is the GCF of the terms.} \\ &= 5(x^2 - 4^2) &&x^2 = x \cdot x \text{ and } 16 = 4 \cdot 4 \\ &= 5(x - 4)(x + 4) &&\text{Factor the difference of squares.} \end{aligned}$$

b. $9x^2 - 6x - 35$

> **Step 1** The GCF of $9x^2$, $-6x$, and -35 is 1.
>
> **Step 2** Since 35 is not a perfect square, this is not a perfect square trinomial.
>
> **Step 3** Factor using the pattern $ax^2 + bx + c$. Are there two numbers with a product of $9(-35)$ or -315 and a sum of -6? Yes, the product of 15 and -21 is -315, and the sum is -6.
>
> $$\begin{aligned} 9x^2 - 6x - 35 &= 9x^2 + mx + px - 35 &&\text{Write the pattern.} \\ &= 9x^2 + 15x - 21x - 35 &&m = 15 \text{ and } p = -21 \\ &= (9x^2 + 15x) + (-21x - 35) &&\text{Group terms with common factors.} \\ &= 3x(3x + 5) - 7(3x + 5) &&\text{Factor out the GCF from each grouping.} \\ &= (3x + 5)(3x - 7) &&3x + 5 \text{ is the common factor.} \end{aligned}$$

c. $18x^4 + 24x^2 + 8$

> **Step 1** The GCF of $18x^4$, $24x^2$, and 8 is 2, so factor it out.
>
> **Step 2** Since there are three terms, check for a perfect square trinomial.
>
> $$\begin{aligned} 18x^4 + 24x^2 + 8 &= 2(9x^4 + 12x^2 + 4) &&\text{2 is the GCF of the terms.} \\ &= 2[(3x^2)^2 + 2(3x^2)(2) + 2^2] &&\text{Write as } a^2 + 2ab + b^2. \\ &= 2(3x^2 + 2)^2 &&\text{Factor the perfect square trinomial.} \end{aligned}$$

> **Guided Practice**
>
> **4A.** $2x^2 - 32$ **4B.** $12x^2 + 5x - 25$

Study Tip

MP Perseverance You can check your answer by:

- Using the FOIL method.
- Using the Distributive Property.
- Graphing the original expression and factored expression and comparing the graphs.

If the product of the factors does not match the original expression exactly, the answer is incorrect.

Example 5 Apply a Technique More Than Once

Factor $625 - x^4$.

$$\begin{aligned} 625 - x^4 &= (25)^2 - (x^2)^2 &&\text{Write } 625 - x^4 \text{ in } a^2 - b^2 \text{ form.} \\ &= (25 + x^2)(25 - x^2) &&\text{Factor the difference of squares.} \\ &= (25 + x^2)(5^2 - x^2) &&\text{Write } 25 - x^2 \text{ in } a^2 - b^2 \text{ form.} \\ &= (25 + x^2)(5 - x)(5 + x) &&\text{Factor the difference of squares.} \end{aligned}$$

> **Guided Practice**
>
> **5A.** $2y^4 - 50$ **5B.** $6x^4 - 96$
>
> **5C.** $2m^3 + m^2 - 50m - 25$ **5D.** $16y^4 - 1$

Watch Out!

Sum of Squares The sum of squares, $a^2 + b^2$, does not factor into $(a + b)(a + b)$. The sum of squares is a prime polynomial and cannot be factored.

Example 1 Determine whether each polynomial is a difference of squares. Write *yes* or *no*. If so, factor it.

   1. $q^2 - 121$ **2.** $4a^2 - 25$ **3.** $9n^2 + 1$

   4. $x^2 - 16y^3$ **5.** $16m^2 - k^4$ **6.** $r^2 - 9t^2$

Example 2 Determine whether each polynomial is a perfect square trinomial. Write *yes* or *no*. If so, factor it.

   7. $25x^2 + 60x + 36$ **8.** $6x^2 + 30x + 36$

   9. $y^4 + 2y^2 + 1$ **10.** $25x^2y^2 - 20xy + 4y^2$

Examples 4-5 Factor each polynomial, if possible. If the polynomial cannot be factored, write *prime*.

   11. $u^4 - 81$ **12.** $2d^4 - 32f^4$

   ⑬ $20r^4 - 45n^4$ **14.** $256n^4 - c^4$

   15. $2c^3 + 3c^2 - 2c - 3$ **16.** $3f^2 - 24f + 48$

   17. $3t^3 + 2t^2 - 48t - 32$ **18.** $w^3 - 3w^2 - 9w + 27$

Example 3 **19.** **MODELING** The drawing at the right is a square with a square cut out of it.

   a. Write an expression that represents the area of the shaded region.

   b. Find the dimensions of a rectangle with the same area as the shaded region in the drawing. Assume that the dimensions of the rectangle must be represented by binomials with integral coefficients.

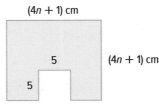
$(4n + 1)$ cm

5    $(4n + 1)$ cm

5

Practice and Problem Solving   Extra Practice begins on page R8.

Examples 1, 2, 4, 5 Factor each polynomial, if possible. If the polynomial cannot be factored, write *prime*.

   20. $16a^2 - 121b^2$ **21.** $12m^3 - 22m^2 - 70m$

   22. $8c^2 - 88c + 242$ **23.** $12x^2 - 84x + 147$

   24. $w^4 - 625$ **25.** $12p^3 - 3p$

   26. $8x^2 + 10x - 21$ **27.** $a^2 - 49$

   28. $4m^3 + 9m^2 - 36m - 81$ **29.** $3m^4 + 243$

   30. $3x^3 + x^2 - 75x - 25$ **31.** $12a^3 + 2a^2 - 192a - 32$

Example 3 **32.** **REASONING** The area of a square is represented by $9x^2 - 42x + 49$. Find the length of each side.

   33. **MODELING** Zelda is building a deck in her backyard. The area of the deck in square feet can be represented by the expression $576 - x^2$. Find expressions for the dimensions of the deck. Then, find the perimeter of the deck when $x = 8$ feet.

   34. **REASONING** The volume of a rectangular prism is represented by the expression $8y^3 + 40y^2 + 50y$. Find the possible dimensions of the prism if the dimensions are represented by polynomials with integer coefficients.

Factor each polynomial, if possible. If the polynomial cannot be factored, write *prime*.

35. $x^3 + 2x^2y - 4x - 8y$

36. $2a^2b^2 - 2a^2 - 2ab^3 + 2ab$

37. $2r^3 - r^2 - 72r + 36$

38. $3k^3 - 24k^2 + 48k$

39. $4c^4d - 10c^3d + 4c^2d^3 - 10cd^3$

40. $g^2 + 2g - 3h^2 + 4h$

41 $x^4 + 6x^3 - 36x^2 - 216x$

42. $15m^3 + 12m^2 - 375m - 300$

43. $y^8 - 256$

44. $k^3 - 5k^2 - 100k + 500$

45. MODELING For the student council elections, Franco is building a voting box shown. The volume of the box in cubic inches is given by the polynomial $h^3 - 12h^2 + 36h$.

   a. Factor the polynomial. What are the possible dimensions of the box in terms of h?

   b. Franco says that your value of h must be greater than 6 inches. Is he correct? Justify your reasoning.

46. **MP** **SENSE-MAKING** Determine whether each of the following statements are *true* or *false*. Given an example or counterexample to justify your answer.
   a. All binomials that have a perfect square in each of the two terms can be factored.
   b. All trinomials that have a perfect square in the first and last terms can be factored.

H.O.T. Problems   Use Higher-Order Thinking Skills

47. ERROR ANALYSIS Elizabeth and Lorenzo are factoring an expression. Is either of them correct? Explain your reasoning.

Elizabeth	Lorenzo
$16x^4 - 25y^2 =$	$16x^4 - 25y^2 =$
$(4x - 5y)(4x + 5y)$	$(4x^2 - 5y)(4x^2 + 5y)$

48. **MP** **PERSEVERENCE** Factor $x^{n+6} + x^{n+2} + x^n$ completely.

49. **MP** **STRUCTURE** Write a perfect square trinomial equation in which the coefficient of the middle term is negative and the last term is a fraction. Solve the equation.

50. **MP** **STRUCTURE** Factor and simplify $9 - (k + 3)^2$, a difference of squares.

51. WRITING IN MATH Describe why the difference of squares pattern has no middle term with a variable.

52. WRITING IN MATH Explain how to determine whether a trinomial is a perfect square trinomial.

53. WHICH ONE DOESN'T BELONG Identify the trinomial that does not belong. Explain.

$4x^2 - 36x + 81$	$25x^2 + 10x + 1$	$4x^2 + 10x + 4$	$9x^2 - 24x + 16$

54. MULTI-STEP A water tank has a volume represented by $2x^2y^4 - 32x^2$.

   a. Is the polynomial a difference of two squares, *yes* or *no*? **MP** 2, 7

   b. Which expression shows how to factor out the GCF?

     ○ **A** $2(x^2y^4 - 16x^2)$

     ○ **B** $2x^2(y^4 - 16)$

     ○ **C** $2x^2(y^4 - 32)$

     ○ **D** $2x(xy^4 - 32x)$

   c. Which, if any, pattern can be used to factor the resulting expression in part **b**?

   d. Write the complete factorization of the polynomial.

55. Carli is factoring the expression $6x^5 - 54x$. Which of the following statements is true? **MP** 7

   ○ **A** $x^2 - 3$ is not a difference of squares.

   ○ **B** $x^4 - 9$ is not a difference of squares.

   ○ **C** The complete factored form of the expression is $6x(x^4 - 9)$.

   ○ **D** The expression cannot be factored.

56. The volume of the prism shown below is represented by the polynomial $\ell^3 - 16\ell^2 + 60\ell$ where ℓ is the length of the prism. **MP** 2

   a. Write the polynomial in factored form.

   b. Use the factors to write an expression for the height and width in terms of the length.

57. The area of a square is represented by $16x^2 + 40x + 25$. Find the length of each side. **MP** 2

58. The area of a circle is $A = \pi(25x^2 - 40x + 16)$. Which represents the radius of the circle? **MP** 2

   ○ **A** $5x + 4$

   ○ **B** $(5x + 4)^2$

   ○ **C** $5x - 4$

   ○ **D** $(5x - 4)^2$

59. Which expression is equivalent to $x^{16} - y^{16}$? **MP** 1, 2, 7

   ○ **A** $(x^4 + y^4)(x^4 - y^4)$

   ○ **B** $(x^8 + y^8)(x^4 + y^4)(x^2 + y^2)(x + y)(x - y)$

   ○ **C** $(x^8 - y^8)(x^8 - y^8)$

   ○ **D** $(x^8 + y^8)(x^4 + y^4)(x^2 + y^2)(x + y)(x + y)$

60. Determine whether the following statement is *true* or *false*. Give an example or counterexample to justify your answer. **MP** 2, 3

For a trinomial to be a perfect square trinomial, the middle term must be positive.

61. The volume of a swimming pool is represented by the expression $12x^3 - 84x^2 + 147x$. Find the possible dimensions of the pool if the dimensions are represented by polynomials with integer coefficients. **MP** 2, 4

62. Which expression is equivalent to $3r^3 - 7r^2 - 3r + 7$? **MP** 7

   ○ **A** $(r - 1)(r - 1)(3r + 7)$

   ○ **B** $(r - 1)(r + 1)(3r + 7)$

   ○ **C** $(r - 1)(r + 1)(3r - 7)$

   ○ **D** $r(3r - 7)(r - 1)$

Study Guide and Review

abc **Go Online!** for Vocabulary Review Games and key vocabulary in 13 languages

Study Guide

Key Concepts

Operations with Polynomials (Lessons 8-1 through 8-4)

- To add or subtract polynomials, add or subtract like terms.
- To multiply polynomials, use the Distributive Property.
- Special products:
  $(a + b)^2 = a^2 + 2ab + b^2$
  $(a - b)^2 = a^2 - 2ab + b^2$
  $(a + b)(a - b) = a^2 - b^2$

Factoring Using the Distributive Property (Lesson 8-5)

- Using the Distributive Property to factor polynomials with four or more terms is called factoring by grouping.
  $ax + bx + ay + by = (ax + bx) + (ay + by)$
  $= x(a + b) + y(a + b)$
  $= (a + b)(x + y)$

Factoring Quadratic Trinomials (Lesson 8-6)

- To factor $x^2 + bx + c$, find two integers, m and p, with a sum of b and a product of c. Then write $x^2 + bx + c$ as $(x + m)(x + p)$.
- To factor $ax^2 + bx + c$, find two integers, m and p, with a sum of b and a product of ac. Then write as $ax^2 + mx + px + c$ and factor by grouping.

Factoring Special Products (Lesson 8-7)

- $a^2 - b^2 = (a - b)(a + b)$
- For a trinomial to be a perfect square, the first and last terms must be perfect squares, and the middle term must be twice the product of the square roots of the first and last terms.
- When factoring, begin by looking for a GCF that can be factored out.

FOLDABLES® Study Organizer

Use your Foldable to review the chapter. Working with a partner can be helpful. Ask for clarification of concepts as needed.

Key Vocabulary

binomial (p. 491)

degree of a monomial (p. 491)

degree of a polynomial (p. 491)

difference of two squares (p. 539)

factoring (p. 520)

factoring by grouping (p. 521)

FOIL method (p. 507)

leading coefficient (p. 492)

perfect square trinomial (p. 540)

polynomial (p. 491)

prime polynomial (p. 535)

quadratic expression (p. 507)

standard form of a polynomial (p. 492)

trinomial (p. 491)

Vocabulary Check

State whether each sentence is *true* or *false*. If *false*, replace the underlined phrase or expression to make a true sentence.

1. $x^2 + 5x + 6$ is an example of a prime polynomial.

2. $(x + 5)(x - 5)$ is the factorization of a difference of squares.

3. $4x^2 - 2x + 7$ is a polynomial of degree 2.

4. $(x + 5)(x - 2)$ is the factored form of $x^2 - 3x - 10$.

5. Expressions with four or more unlike terms can sometimes be factored by grouping.

6. A polynomial is in standard form when the terms are in order from least to greatest.

7. $x^2 - 12x + 36$ is an example of a perfect square trinomial.

8. The leading coefficient of $1 + 6a + 9a^2$ is 1.

9. $x^2 - 16$ is an example of a(n) perfect square trinomial.

10. The FOIL method is used to multiply two trinomials.

Concept Check

Classify each polynomial as *prime polynomial, difference of squares,* or *perfect square trinomial*.

11. $x^2 - 25$

12. $(16x^2 + 40x + 25)$

13. $x^2 + 25$

Lesson-by-Lesson Review

8-1 Adding and Subtracting Polynomials

Write each polynomial in standard form.

14. $x + 2 + 3x^2$

15. $1 - x^4$

16. $2 + 3x + x^2$

17. $3x^5 - 2 + 6x - 2x^2 + x^3$

Find each sum or difference.

18. $(x^3 + 2) + (-3x^3 - 5)$

19. $a^2 + 5a - 3 - (2a^2 - 4a + 3)$

20. $(4x - 3x^2 + 5) + (2x^2 - 5x + 1)$

21. **PICTURE FRAMES** Jean is framing a painting that is a rectangle. What is the perimeter of the frame?

$5x + 3$

$2x^2 - 3x + 1$

Example 1

Write $3 - x^2 + 4x$ in standard form.

Step 1 Find the degree of each term.

3:	degree 0
$-x^2$:	degree 2
$4x$:	degree 1

Step 2 Write the terms in descending order of degree.

$3 - x^2 + 4x = -x^2 + 4x + 3$

Example 2

Find $(8r^2 + 3r) - (10r^2 - 5)$.

$(8r^2 + 3r) - (10r^2 - 5)$

$= (8r^2 + 3r) + (-10r^2 + 5)$ Use the additive inverse.

$= (8r^2 - 10r^2) + 3r + 5$ Group like terms.

$= -2r^2 + 3r + 5$ Add like terms.

8-2 Multiplying a Polynomial by a Monomial

Solve each equation.

22. $x^2(x + 2) = x(x^2 + 2x + 1)$

23. $2x(x + 3) = 2(x^2 + 3)$

24. $2(4w + w^2) - 6 = 2w(w - 4) + 10$

25. **GEOMETRY** Find the area of the rectangle.

$3x$

$x^2 + x - 7$

Example 3

Solve $m(2m - 5) + m = 2m(m - 6) + 16$.

$m(2m - 5) + m = 2m(m - 6) + 16$

$2m^2 - 5m + m = 2m^2 - 12m + 16$

$2m^2 - 4m = 2m^2 - 12m + 16$

$-4m = -12m + 16$

$8m = 16$

$m = 2$

8-3 Multiplying Polynomials

Find each product.

26. $(x - 3)(x + 7)$

27. $(3a - 2)(6a + 5)$

28. $(3r - 7t)(2r + 5t)$

29. $(2x + 5)(5x + 2)$

30. **PARKING LOT** The parking lot shown is to be paved. What is the area to be paved?

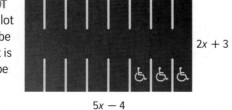

$2x + 3$

$5x - 4$

Example 4

Find $(6x - 5)(x + 4)$.

$(6x - 5)(x + 4)$

         F   O   I   L

$= (6x)(x) + (6x)(4) + (-5)(x) + (-5)(4)$

$= 6x^2 + 24x - 5x - 20$ Multiply.

$= 6x^2 + 19x - 20$ Combine like terms.

8-4 Special Products

Find each product.

31. $(x + 5)(x - 5)$ **32.** $(3x - 2)^2$

33. $(5x + 4)^2$ **34.** $(2x - 3)(2x + 3)$

35. $(2r + 5t)^2$ **36.** $(3m - 2)(3m + 2)$

37. GEOMETRY Write an expression to represent the area of the shaded region.

2x + 5

x + 2

x − 2

2x − 5

Example 5

Find $(x - 7)^2$.

$$(a - b)^2 = a^2 - 2ab + b^2 \qquad \text{Square of a Difference}$$
$$(x - 7)^2 = x^2 - 2(x)(7) + (-7)^2 \qquad a = x \text{ and } b = 7$$
$$= x^2 - 14x + 49 \qquad \text{Simplify.}$$

Example 6

Find $(5a - 4)(5a + 4)$.

$$(a + b)(a - b) = a^2 - b^2 \qquad \text{Product of a Sum and Difference}$$
$$(5a - 4)(5a + 4) = (5a)^2 - (4)^2 \qquad a = 5a \text{ and } b = 4$$
$$= 25a^2 - 16 \qquad \text{Simplify.}$$

8-5 Using the Distributive Property

Use the Distributive Property to factor each polynomial.

38. $12x + 24y$

39. $14x^2y - 21xy + 35xy^2$

40. $8xy - 16x^3y + 10y$

41. $a^2 - 4ac + ab - 4bc$

42. $2x^2 - 3xz - 2xy + 3yz$

43. $24am - 9an + 40bm - 15bn$

Factor each expression.

44. $24ab + 54a - 20b - 45$

45. $3r^3 - 12r^2 + 4p - pr$

46. $6c^2d + 30cd$

47. $18f^4g^5 - 3f^6g^2 + 9f^4g^3$

48. GEOMETRY The area of the rectangle shown is $x^3 - 2x^2 + 5x$ square units. What is the length?

x

Example 7

Factor $12y^2 + 9y + 8y + 6$.

$12y^2 + 9y + 8y + 6$

$$= (12y^2 + 9y) + (8y + 6) \qquad \text{Group terms with common factors.}$$
$$= 3y(4y + 3) + 2(4y + 3) \qquad \text{Factor the GCF from each group.}$$
$$= (4y + 3)(3y + 2) \qquad \text{Distributive Property}$$

Example 8

Factor $5a^2 - 10ab + 6b - 3a$.

$5a^2 - 10ab + 6b - 3a$

$$= (5a^2 - 10ab) + (6b - 3a) \qquad \text{Group terms with common factors.}$$
$$= 5a(a - 2b) + 3(2b - a) \qquad \text{Factor the GCF from each group.}$$
$$= 5a(a - 2b) + 3[(-1)(a - 2b)] \qquad 2b - a = -1(a - 2b)$$
$$= 5a(a - 2b) - 3(a - 2b) \qquad \text{Associative Property}$$
$$= (5a - 3)(a - 2b) \qquad \text{Distributive Property}$$

8-6 Factoring Quadratic Trinomials

Factor each polynomial.

49. $x^2 - 8x + 15$

50. $x^2 + 9x + 20$

51. $x^2 - 5x - 6$

52. $x^2 + 3x - 18$

53. $x^2 + 5x - 50$

54. $x^2 - 6x + 8$

55. $x^2 + 12x + 32$

56. $x^2 - 2x - 48$

57. $x^2 + 11x + 10$

58. ART An artist is working on a painting. If the area of the canvas is represented by $x^2 + 2x - 24$, what are the dimensions of the canvas?

Factor each trinomial, if possible. If the trinomial cannot be factored, write *prime*.

59. $12x^2 + 22x - 14$

60. $2y^2 - 9y + 3$

61. $3x^2 - 6x - 45$

62. $2a^2 + 13a - 24$

63. $20x^2 + x - 12$

64. $2x^2 - 3x - 20$

65. $3x^2 - 13x - 10$

66. $6x^2 - 7x - 5$

67. GEOMETRY The area of the rectangle shown is $6x^2 + 11x - 7$ square units. What is the width of the rectangle?

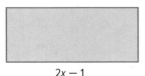

$2x - 1$

Example 9

Factor $x^2 + 10x + 21$.

$b = 10$ and $c = 21$, so $m + p$ is positive and mp is positive. Therefore, m and p must both be positive. List the positive factors of 21, and look for the pair of factors with a sum of 10.

Factors of 21	Sum of 10
1, 21	22
3, 7	10

The correct factors are 3 and 7.

$$x^2 + 10x + 21 = (x + m)(x + p) \qquad \text{Write the pattern.}$$
$$= (x + 3)(x + 7) \qquad m = 3 \text{ and } p = 7$$

Example 10

Factor $12a^2 + 17a + 6$.

$a = 12$, $b = 17$, and $c = 6$. Since b is positive, $m + p$ is positive. Since c is positive, mp is positive. So, m and p are both positive. List the factors of 12(6) or 72, where both factors are positive.

Factors of 72	Sum of 17
1, 72	73
2, 36	38
3, 24	27
4, 18	22
6, 12	18
8, 9	17

The correct factors are 8 and 9.

$$12a^2 + 17a + 6 = 12a^2 + ma + pa + 6$$
$$= 12a^2 + 8a + 9a + 6$$
$$= (12a^2 + 8a) + (9a + 6)$$
$$= 4a(3a + 2) + 3(3a + 2)$$
$$= (3a + 2)(4a + 3)$$

So, $12a^2 + 17a + 6 = (3a + 2)(4a + 3)$.

8-7 Factoring Special Products

Factor each polynomial.

68. $y^2 - 81$

69. $64 - 25x^2$

70. $16a^2 - 21b^2$

71. $3x^2 - 3$

72. $a^2 - 25$

73. $9x^2 - 25$

Factor each polynomial, if possible. If the polynomial cannot be factored, write *prime*.

74. $x^2 + 12x + 36$

75. $x^2 + 5x + 25$

76. $9y^2 - 12y + 4$

77. $4 - 28a + 49a^2$

78. $x^4 - 1$

79. $x^4 - 16x^2$

80. $9x^2 + 25$

81. $-3x^2 - 12x - 12$

Example 11

Factor $x^4 - 16$.

$x^4 - 16$	Original expression
$= (x^2)^2 - 4^2$	Difference of squares
$= (x^2 - 4)(x^2 + 4)$	Factor the difference of squares.
$= (x + 2)(x - 2)(x^2 + 4)$	Factor the second difference of squares.

Example 12

Factor $4x^2 + 16x + 16$.

$4x^2 + 16x + 16 = (2x)^2 + 2(2x)(4) + 4^2$	Write as $a^2 + 2ab + b^2$.
$= (2x + 4)^2$	Factor using the pattern.

CHAPTER 8
Practice Test

Go Online! for another Chapter Test

Find each sum or difference.

1. $(x + 5) + (x^2 - 3x + 7)$

2. $(7m - 8n^2 + 3n) - (-2n^2 + 4m - 3n)$

3. **MULTIPLE CHOICE** Antonia is carpeting two of the rooms in her house. The dimensions are shown. Which expression represents the total area to be carpeted?

  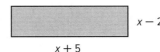

x $x - 2$

$x + 3$ $x + 5$

A $x^2 + 3x$ C $x^2 + 3x - 5$

B $2x^2 + 6x - 10$ D $8x + 12$

Find each product.

4. $a(a^2 + 2a - 10)$

5. $(2a - 5)(3a + 5)$

6. $(x - 3)(x^2 + 5x - 6)$

7. $(x + 3)^2$

8. $(2b - 5)(2b + 5)$

9. **FINANCIAL LITERACY** Suppose you invest $4000 in a 2-year certificate of deposit (CD).

    a. If the interest rate is 5% per year, the expression $4000(1 + 0.05)^2$ can be evaluated to find the total amount of money after two years. Explain the numbers in this expression.

    b. Find the amount at the end of two years.

    c. Suppose you invest $10,000 in a CD for 4 years at an annual rate of 6.25%. What is the total amount of money you will have after 4 years?

10. **MULTIPLE CHOICE** The area of the rectangle shown below is $2x^2 - x - 15$ square units. What is the width of the rectangle?

A $x - 5$

B $x + 3$

C $x - 3$

D $2x - 3$

$2x + 5$

Solve each equation.

11. $5(t^2 - 3t + 2) = t(5t - 2)$

12. $3x(x + 2) = 3(x^2 - 2)$

Factor each polynomial.

13. $5xy - 10x$

14. $7ab + 14ab^2 + 21a^2b$

15. $4x^2 + 8x + x + 2$

16. $10a^2 - 50a - a + 5$

Factor each polynomial.

17. $x^3 + x^2 - x - 1$

18. $x^2 - x - 2$

19. $a^2 - 2a$

20. **MULTIPLE CHOICE** Chantel is carpeting a room that has an area of $x^2 - 100$ square feet. If the width of the room is $x - 10$ feet, what is the length of the room?

A $x - 10$ ft

B $x + 10$ ft

C $x - 100$ ft

D 10 ft

Factor each trinomial.

21. $x^2 + 7x + 6$ 22. $x^2 - 3x - 28$

23. $10x^2 - x - 3$ 24. $15x^2 + 7x - 2$

25. $x^2 - 25$ 26. $4x^2 - 81$

27. $9x^2 - 12x + 4$ 28. $16x^2 + 40x + 25$

Factor each polynomial.

29. $x^3 + 4x^2 - x - 4$ 30. $x^4 - x^3 - 2x^2$

31. $a^2 - 2a + 1$ 32. $2x^2 - 13x + 20$

33. **MULTIPLE CHOICE** Which choice is a factor of $x^4 - 1$ when it is factored completely?

A $x^2 - 1$ C x

B $x - 1$ D 1

Performance Task

Provide a clear solution to each part of the task. Be sure to show all of your work, include all relevant drawings, and justify your answers.

DESIGN Henri owns a company that builds swimming pools. He is currently working on several projects.

Part A

One of Henri's customers is thinking about having a pool installed in their backyard. Henri marks off the maximum area the pool could occupy, which is in the shape of a square, with side length s, measured in feet. The client decides they want to maintain some of their existing yard space, so she asks Henri to subtract 2 feet from the width and 8 feet from the length.

1. **Model** Write expressions that represent the perimeter and area of the pool the client has requested.

Part B

Another of Henri's clients wants his pool to have a length that is 6 less than 4 times its width. He wants a hot tub installed directly beside the pool, such that it overflows and heats the pool. The hot tub will be a square with a side length that is 3 feet less than the width of the pool.

2. Write an expression that represents the area of the pool.

3. Write an expression that represents the area of the pool and hot tub together.

4. **Reasoning** If the area of the hot tub is 36 square feet, find the area of the pool.

Part C

Another client only wants a hot tub installed. The expression that represents the total area of the square hot tub is $x^2 + 8x + 16$.

5. Write an expression that represents the side length of the hot tub.

Part D

Another client is a professional swimmer and wants a lap pool installed. The lap pool will occupy an area of $9x^2 - 400$ square feet.

6. Find the dimensions of the lap pool if $x = 9$.

Part E

Construct an Argument The cost of Henri's services is directly proportional to the area of the installation. The client in Part B wants to save some money so he decides to make the size of his pool smaller. He decides to either reduce the length of the pool so that it is twice the width plus 8 feet or not build the hot tub.

7. If the width of the pool remains the same as in Part B, which option should the client choose? Justify your answer.

Test-Taking Strategy

Example

Read the problem. Identify what you need to know. Then use the information in the problem to solve.

The city wants to pave a walking path around the rectangular pond shown. If $x = 21$, what will be the length of the path in yards?

$6x^2 - 5x - 56 \text{ yd}^2$

A 106 yards

C 212 yards

B 208 yards

D 2485 yards

Test-Taking Tip

Solve Multi-Step Problems Some problems that you will encounter on standardized tests require you to solve multiple parts in order to come up with the final solution. To solve these problems, you'll need to read the problem carefully and organize your approach.

Step 1 **What do you need to find?**
The length of the walking path, which is the perimeter of the pond.

Step 2 **What are the steps needed to solve the problem?**
I need to factor the given expression to find the dimensions of the pond. Then, I need to write an expression for the perimeter and evaluate the expression for $x = 21$.

Step 3 **What is the correct answer?**
Factoring the expression for area gives $6x^2 - 5x - 56 = (2x - 7)(3x + 8)$. So, the pond is $2x - 7$ yards by $3x + 8$ yards. The perimeter is twice the width plus twice the length. So, $P = 2(2x - 7) + 2(3x + 8)$, or $10x + 2$ yards. Substituting 21 for x in the expression gives a perimeter of 212 yards.

Step 4 **How can you check that your answer is correct?**
I can substitute 21 for x in the expressions for the length and width to find that the pond is 35 yards by 71 yards. Using these dimensions, the perimeter should be $2(35) + 2(71)$, or 212 yards. The correct answer is C.

Apply the Strategy

Read the problem. Identify what you need to know. Then use the information in the problem to solve.

What is the area of the square?

A $x^2 + 16$

C $x^2 - 8x - 16$

B $4x - 16$

D $x^2 - 8x + 16$

$x - 4$

Answer the questions below.

a. What do you need to find?

b. What are the steps needed to solve the problem?

c. What is the correct answer?

d. How can you check that your answer is correct?

Read each question. Then fill in the correct answer on the answer document provided by your teacher or on a sheet of paper.

1. Two thirds of x plus 16 is as much as 24 times x minus the quantity x plus 10. Which equation models this relationship?

   ○ **A** $\frac{2}{3}x + 16 = 24x - x + 10$

   ○ **B** $\frac{2}{3}x + 16 = 24x - (x + 10)$

   ○ **C** $\frac{2}{3} + x + 16 = 24x - (x + 10)$

   ○ **D** $\frac{2}{3}x + 16 = 24x - 10$

2. A rectangle has a length of $2x^2 + 3x + 7$ and a width of $x^2 + 5$. What is the perimeter of the rectangle?

   ○ **A** $2(2x^2 + 3x + 12)$

   ○ **B** $3(x^2 + x + 4)$

   ○ **C** $6(x^2 + x + 4)$

   ○ **D** $5x^2 + 6x + 19$

3. A rectangular garden has a perimeter of 108 feet. The length of the garden is 8 feet more than the width.

   a. Write expressions that represent the perimeter and area of the garden in terms of the width, w.

   > [ ]

   b. What is the width of the garden?

   > [ ] feet

   c. **MP** What mathematical practice did you use to solve this problem?

4. Which expression is a factor of $8x^2 - 26x - 45$?

   ○ **A** $4x - 5$ ○ **C** $2x + 9$

   ○ **B** $(2x + 9)^2$ ○ **D** $2x - 9$

5. Solve: $3x(x - 6) + 4 = x(3x + 6) + 18$

   ○ **A** $-\frac{11}{12}$

   ○ **B** $-\frac{7}{12}$

   ○ **C** $-\frac{7}{6}$

   ○ **D** $\frac{7}{12}$

6. Jennifer plans a triangular garden to grow fruits and vegetables. She needs to know the area. The measurements of the garden are shown below.

   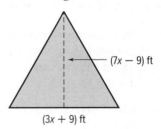

   $(7x - 9)$ ft

   $(3x + 9)$ ft

   Which expression should Jennifer use to find the area of the garden?

   ○ **A** $\frac{1}{2}(21x^2 + 90x - 81)$

   ○ **B** $\frac{1}{2}(21x^2 - 90x - 81)$

   ○ **C** $\frac{1}{2}(21x^2 + 36x - 81)$

   ○ **D** $\frac{1}{2}(21x^2 + 36x)$

7. What is the slope of the line $3y = 12x - 6$?

   ○ **A** -6

   ○ **B** -2

   ○ **C** 4

   ○ **D** 12

Test-Taking Tip

Question 7 Rewrite a linear equation in slope-intercept form to find the slope of the line. In this form, the coefficient of y is 1.

8. A prism has a volume of 243 cubic centimeters. If it has a height of 3 centimeters and its base is a square with sides of $(x + 3)$ centimeters, which equation represents the area of the base of the prism?

   ○ **A** $81 = x^2 + 6x + 9$

   ○ **B** $243 = x^2 + 6x + 9$

   ○ **C** $81 = x^2 + 6x + 6$

   ○ **D** $81 = x^2 + 3x + 9$

9. The area of a rectangular room is $x^2 + 2x - 224$. The length of the room is $x + 16$. Which expression represents the perimeter of the room?

   ○ **A** $(x + 16)(x - 14)$

   ○ **B** $2(x + 1)$

   ○ **C** $4(x + 1)$

   ○ **D** $4(x - 1)$

10. Some values for two linear equations are shown in the tables below.

x	y
−2	13
−1	14
0	15
3	18

x	y
−6	8
0	5
4	3
6	2

   What are the two equations that make up the system represented by these tables?

   ○ **A** $x - y = 15, x + y = 5$

   ○ **B** $2x + y = 13, x + 2y = 10$

   ○ **C** $x - y = -15, x + 2y = 10$

   ○ **D** $x - y = -15, x + y = 5$

11. Consider the polynomial expression below.

$$3x^2 + 18x + 27$$

   Select all of the factoring methods needed to completely factor the polynomial.

   ☐ **A** Factor a difference of squares.

   ☐ **B** Factor a perfect square trinomial.

   ☐ **C** Factor out the GCF.

   ☐ **D** Factor by grouping.

   ☐ **E** The polynomial cannot be factored.

12. During the coin toss, the referee at a football game dropped the coin to the ground from a height of 4 feet. The expression $-16t^2 + h_0$ can be used to represents the height of the coin in feet t seconds after it is tossed, where h_0 is the initial height. Which is the factored form of the expression that represents the height of the coin tossed by the referee after t seconds?

   ○ **A** $-16t^2 + 4$

   ○ **B** $-4(4t^2 - 1)$

   ○ **C** $-4t(4t - 1)$

   ○ **D** $-2(8t^2 - 2)$

13. Which expression is equivalent to $8x^4 - 128$?

   ○ **A** $8(x + 2)(x - 2)(x^2 + 4)$

   ○ **B** $8(x + 2)^3(x - 2)$

   ○ **C** $(x + 2)(x - 2)(x^2 + 4)$

   ○ **D** $(x - 2)^2(x^2 + 4)$

Need Extra Help?

If you missed Question...	1	2	3	4	5	6	7	8	9	10	11	12	13
Go to Lesson...	2-1	8-1	8-2	8-6	8-2	8-3	4-1	8-4	8-6	4-2	8-7	8-5	8-7

Quadratic Functions and Equations

THEN

You factored polynomial expressions, including quadratic expressions.

NOW

In this chapter, you will:

- Solve quadratic equations by factoring, graphing, completing the square, and using the Quadratic Formula.
- Analyze functions with successive differences and ratios.
- Identify and graph special functions.
- Solve systems of linear and quadratic equations.

(MP) WHY

FINANCE The value of a company's stock is extremely important to the company as well as to shareholders.

Use the mathematical practices to complete the activity.

1. Use Tools Use the Internet to find information on financial markets and functions that have been used to model them. Then, select a stock that increased and then decreased over a period of time and record its value each day during this interval.

2. Model with Mathematics Use graphing technology to plot the data you recorded. Let x be the time, in days, since the beginning of the interval, and y be the value, in dollars, of the stock. Describe the shape of the graph.

3. Discuss Compare your graphs to those of your classmates. What do you think the graph would look like if the value of the stock decreased and then increased over a period of time?

Getty Images

Go Online to Guide Your Learning

Explore & Explain

 Equation Chart Mat

Use the Equation Chart Mat and the Algebra Tiles tool to explore and enhance your understanding of solving quadratic equations by completing the square, as discussed in Lesson 9-5.

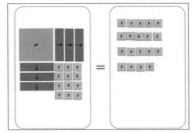

 The Geometer's Sketchpad

Use The Geometer's Sketchpad to apply dilations and reflections to quadratic functions (Lesson 9-2) and to solve quadratic equations by graphing (Lesson 9-3).

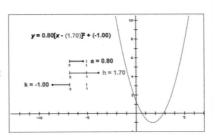

$y = 0.80[x - (1.70)]^2 + (-1.00)$

$a = 0.80$
$h = 1.70$
$k = -1.00$

eBook

Interactive Student Guide

Before starting the chapter, answer the **Chapter Focus** preview questions. Check your answers as you complete each lesson. At the end of the chapter, try the **Performance Task**.

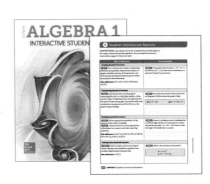

Organize

 Foldables

Make this Foldable to help you organize your notes about quadratic functions. Begin with a sheet of notebook paper folded lengthwise to the margin rule. Fold it 4 times, unfold, cut and label as shown.

maximum or minimum
vertex
axis of symmetry
y-intercept

FOLDABLES

Collaborate

 Chapter Project

In the **3... 2... 1... Blast Off!** project, you can use what you learn about quadratic equations to complete a project that addresses business literacy.

Focus

 LEARNSMART

Need help studying? Complete the **Quadratic Functions and Modeling** and **Expressions and Equations** domains in LearnSmart to review for the chapter test.

ALEKS

You can use the **Functions and Lines** and **Quadratic Functions and Equations** topics in ALEKS to find out what you know about linear functions and what you are ready to learn.*

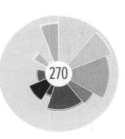

270

* Ask your teacher if this is part of your program.

Get Ready for the Chapter

Go Online! for Vocabulary Review Games and key vocabulary in 13 languages.

Connecting Concepts

Concept Check

Review the concepts used in this chapter by answering each question below.

1. Explain how a table of values could be used to graph a function.

2. How do you create a table of values from a new equation?

3. Explain how to draw the graph of a line from a table of values.

4. Explain how to draw the graph of a line from an equation without using a table of values.

5. Refer to the graph at the right. Create a table of values for this graph.

6. Write an equation of the line graphed at the right.

7. Explain how to factor a trinomial with a coefficient of 1.

8. Describe the difference between solving an absolute value equation and a linear equation.

Performance Task Preview

You will use the concepts and skills in this chapter to solve problems about the mechanics of baseball. Understanding quadratic equations will help you finish the Performance Task at the end of the chapter.

MP **In this Performance Task you will:**

- model with mathematics
- make sense of problems and persevere in solving them
- reason abstractly and quantitatively
- look for and make use of structure

New Vocabulary

English		Español
quadratic function	p. 559	función cuadrática
parabola	p. 559	parábola
axis of symmetry	p. 559	eje de simetría
vertex	p. 559	vértice
minimum	p. 559	mínimo
maximum	p. 559	máximo
vertex form	p. 575	forma de vértice
double root	p. 581	doble raíz
completing the square	p. 596	completar el cuadrado
Quadratic Formula	p. 606	Formula cuadrática
discriminant	p. 610	discriminante

Review Vocabulary

domain dominio all the possible values of the independent variable, x

factoring factorización to express a polynomial as the product of monomials and polynomials

leading coefficient coeficiente delantero the coefficient of the first term of a polynomial written in standard form

range rango all the possible values of the dependent variable, y

In the function represented by the table, the domain is $\{0, 2, 4, 6\}$, and the range is $\{3, 5, 7, 9\}$.

x	y
0	3
2	5
4	7
6	9

transformation transformación the movement of a graph on the coordinate plane

Graphing Quadratic Functions

:: **Then**

- You graphed linear and exponential functions.

:: **Now**

1. Analyze the characteristics of graphs of quadratic functions.

2. Graph quadratic functions.

:: **Why?**

- The Innovention Fountain in Epcot's Futureworld in Orlando, Florida, is an elaborate display of water, light, and music. The sprayers shoot water in shapes that can be modeled by quadratic equations. You can use graphs of these equations to show the path of the water.

 New Vocabulary

quadratic function
standard form
parabola
axis of symmetry
vertex
minimum
maximum

 Mathematical Practices

2 Reason abstractly and quantitatively.

1 Characteristics of Quadratic Functions **Quadratic functions** are nonlinear and can be written in the form $f(x) = ax^2 + bx + c$, where $a \neq 0$. This form is called the **standard form** of a quadratic function.

The shape of the graph of a quadratic function is called a **parabola**. Parabolas are symmetric about a central line called the **axis of symmetry**. The axis of symmetry intersects a parabola at only one point, called the **vertex**.

> 🔑 **Key Concept** Quadratic Functions
>
> | | |
> |---|---|
> | **Parent Function:** | $f(x) = x^2$ |
> | **Standard Form:** | $f(x) = ax^2 + bx + c$ |
> | **Type of Graph:** | parabola |
> | **Axis of Symmetry:** | $x = -\dfrac{b}{2a}$ |
> | **y-intercept:** | c |
>
>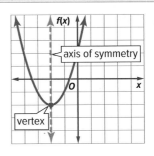

When $a > 0$, the graph of $y = ax^2 + bx + c$ opens upward. The lowest point on the graph is the **minimum**. When $a < 0$, the graph opens downward. The highest point is the **maximum**. The maximum or minimum is the vertex.

Example 1   **Graph a Parabola**

Use a table of values to graph $y = 3x^2 + 6x - 4$. State the domain and range.

x	y
1	5
0	−4
−1	−7
−2	−4
−3	5

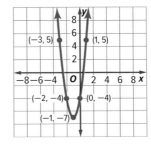

Graph the ordered pairs, and connect them to create a smooth curve. The parabola extends to infinity. The domain is all real numbers or $\{-\infty < x < \infty\}$. The range is $\{y \mid y \geq -7\}$, because −7 is the minimum.

▶ **Guided Practice**

1. Use a table of values to graph $y = x^2 + 3$. State the domain and range.

Recall that figures with symmetry are those in which each half of the figure matches exactly. A parabola is symmetric about the axis of symmetry. Every point on the parabola to the left of the axis of symmetry has a corresponding point on the other half. The function is increasing on one side of the axis of symmetry and decreasing on the other side.

$y = x^2 + 2x - 5$

zeros
zeros
$x = -1$ axis of symmetry
$(-1, -6)$ vertex

When identifying characteristics from a graph, it is often easiest to locate the vertex first. It is either the maximum or minimum point of the graph. Then locate the x-intercepts. They are the zeros of a quadratic function.

Example 2   Identify Characteristics from Graphs

Find the vertex, the equation of the axis of symmetry, the y-intercept, and the zeros of each graph.

a.

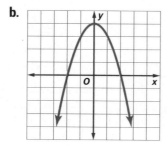

Step 1 Find the vertex. The parabola opens upward, so the vertex is located at the minimum point, $(-1, 0)$.

Step 2 Find the axis of symmetry. The axis of symmetry is the line that goes through the vertex and divides the parabola into congruent halves. It is located at $x = -1$.

Step 3 Find the y-intercept. The parabola crosses the y-axis at $(0, 1)$, so the y-intercept is 1.

Step 4 Find the zeros. The zeros are the points where the graph intersects the x-axis. This graph only intersects the x-axis once, at $(-1, 0)$.

b.

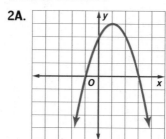

Step 1 Find the vertex. The parabola opens downward, so the vertex is located at its maximum point, $(0, 4)$.

Step 2 Find the axis of symmetry. The axis of symmetry is located at $x = 0$.

Step 3 Find the y-intercept. The parabola crosses the y-axis at $(0, 4)$, so the y-intercept is 4.

Step 4 Find the zeros. The zeros are the points where the graph intersects the x-axis, $(-2, 0)$ and $(2, 0)$.

 Guided Practice

2A.

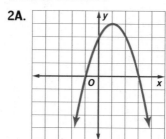

2B.

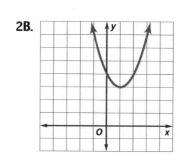

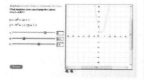

Go Online!

Investigate how changing the values of a, b, and c in the equation of a quadratic function affects the graph by using the **Graphing Tools** in ConnectED.

Example 3   **Identify Characteristics from Functions**

Find the vertex, the equation of the axis of symmetry, and the y-intercept of each function.

a. $y = 2x^2 + 4x - 3$

$$x = -\frac{b}{2a} \qquad \text{Formula for the equation of the axis of symmetry}$$

$$x = -\frac{4}{2 \cdot 2} \qquad a = 2 \text{ and } b = 4$$

$$x = -1 \qquad \text{Simplify.}$$

The equation for the axis of symmetry is $x = -1$.

To find the vertex, use the value you found for the axis of symmetry as the x-coordinate of the vertex. Find the y-coordinate using the original equation.

$$y = 2x^2 + 4x - 3 \qquad \text{Original equation}$$

$$= 2(-1)^2 + 4(-1) - 3 \qquad x = -1$$

$$= -5 \qquad \text{Simplify.}$$

The vertex is at $(-1, -5)$.

The y-intercept always occurs at $(0, c)$. So, the y-intercept is -3.

b. $y = -x^2 + 6x + 4$

$$x = -\frac{b}{2a} \qquad \text{Formula for the equation of the axis of symmetry}$$

$$x = -\frac{6}{2(-1)} \qquad a = -1 \text{ and } b = 6$$

$$x = 3 \qquad \text{Simplify.}$$

The equation of the axis of symmetry is $x = 3$.

$$y = -x^2 + 6x + 4 \qquad \text{Original equation}$$

$$= -(3)^2 + 6(3) + 4 \qquad x = 3$$

$$= 13 \qquad \text{Simplify.}$$

The vertex is at $(3, 13)$.

The y-intercept is 4.

▶ **Guided Practice**

3A. $y = -3x^2 + 6x - 5$ **3B.** $y = 2x^2 + 2x + 2$

Next you will learn how to identify whether the vertex is a maximum or a minimum.

Key Concept **Maximum and Minimum Values**

Words	The graph of $f(x) = ax^2 + bx + c$, where $a \neq 0$: • opens upward and has a minimum value when $a > 0$, and • opens downward and has a maximum value when $a < 0$. The range of a quadratic function is all real numbers greater than or equal to the minimum, or all real numbers less than or equal to the maximum.
Examples	a is positive. 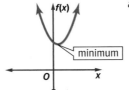 a is negative. 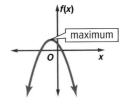

Example 4   Maximum and Minimum Values

Consider $f(x) = -2x^2 - 4x + 6$.

a. Determine whether the function has a *maximum* or *minimum* value.

For $f(x) = -2x^2 - 4x + 6$, $a = -2$, $b = -4$, and $c = 6$.

Because a is negative the graph opens down, so the function has a maximum value.

b. State the maximum or minimum value of the function.

The maximum value is the y-coordinate of the vertex.

The x-coordinate of the vertex is $\frac{-b}{2a}$ or $\frac{4}{2(-2)}$ or -1.

$$f(x) = -2x^2 - 4x + 6 \qquad \text{Original function}$$

$$f(-1) = -2(-1)^2 - 4(-1) + 6 \qquad x = -1$$

$$f(-1) = 8 \qquad \text{Simplify.}$$

The maximum value is 8.

c. State the domain and range of the function.

The domain is all real numbers. The range is all real numbers less than or equal to the maximum value, or $\{y \mid y \leq 8\}$.

▶ **Guided Practice**

Consider $g(x) = 2x^2 - 4x - 1$.

4A. Determine whether the function has a *maximum* or *minimum* value.

4B. State the maximum or minimum value.

4C. State the domain and range of the function.

Watch Out!
Minimum and Maximum Values Don't forget to find both coordinates of the vertex (x, y). The minimum or maximum value is the y-coordinate.

Review Vocabulary
Domain and Range The domain is the set of all of the possible values of the independent variable x. The range is the set of all of the possible values of the dependent variable y.

2 **Graph Quadratic Functions** You have learned how to find several important characteristics of quadratic functions.

🔑 Key Concept Graph Quadratic Functions

Step 1	Find the equation of the axis of symmetry.
Step 2	Find the vertex, and determine whether it is a maximum or minimum.
Step 3	Find the y-intercept.
Step 4	Use symmetry to find additional points on the graph, if necessary.
Step 5	Connect the points with a smooth curve.

Study Tip

MP Sense-Making When locating points that are on opposite sides of the axis of symmetry, not only are the points equidistant from the axis of symmetry, they are also equidistant from the vertex.

Example 5 Graph Quadratic Functions

Graph $f(x) = x^2 + 4x + 3$.

Step 1 Find the equation of the axis of symmetry.

$x = \dfrac{-b}{2a}$ Formula for the equation of the axis of symmetry

$x = \dfrac{-4}{2 \cdot 1}$ or -2 $a = 1$ and $b = 4$

Step 2 Find the vertex, and determine whether it is a maximum or minimum.

$f(x) = x^2 + 4x + 3$ Original equation

$\quad = (-2)^2 + 4(-2) + 3$ $x = -2$

$\quad = -1$ Simplify.

The vertex lies at $(-2, -1)$. Because a is positive the graph opens up, and the vertex is a minimum.

Step 3 Find the y-intercept.

$f(x) = x^2 + 4x + 3$ Original equation

$\quad = (0)^2 + 4(0) + 3$ $x = 0$

$\quad = 3$ The y-intercept is 3.

Step 4 The axis of symmetry divides the parabola into two equal parts. So if there is a point on one side, there is a corresponding point on the other side that is the same distance from the axis of symmetry and has the same y-value.

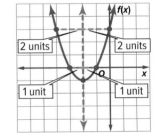

Step 5 Connect the points with a smooth curve.

▶ **Guided Practice** Graph each function.

5A. $f(x) = -2x^2 + 2x - 1$

5B. $f(x) = 3x^2 - 6x + 2$

There are general differences between linear, exponential, and quadratic functions.

	Linear Functions	Exponential Functions	Quadratic Functions
Equation	$y = mx + b$	$y = ab^x, a \neq 0, b > 0, b \neq 1$	$y = ax^2 + bx + c, a \neq 0$
Degree	1	x	2
Graph	line	curve	parabola
Increasing / Decreasing	$m > 0$: y is increasing on the entire domain. $m < 0$: y is decreasing on the entire domain.	$a > 0, b > 1$ or $a < 0$, $0 < b < 1$: y is increasing on the entire domain. $a > 0, 0 < b < 1$ or $a < 0$, $b > 1$: y is decreasing on the entire domain.	$a > 0$: y is decreasing to the left of the axis of symmetry and increasing on the right. $a < 0$: y is increasing to the left of the axis of symmetry and decreasing on the right.
End Behavior	$m > 0$: as x increases, y increases; as x decreases, y decreases. $m < 0$: as x increases, y decreases; as x decreases, y increases	$b > 1$: as x decreases, y approaches 0; $a > 0$, as x increases, y increases; $a < 0$, as x increases, y decreases. $0 < b < 1$: as x increases, y approaches 0; $a > 0$, as x decreases, y increases; $a < 0$, as x decreases, y decreases.	$a > 0$: as x increases, y increases; as x decreases, y increases. $a < 0$: as x increases, y decreases; as x decreases, y decreases

You have used what you know about quadratic functions, parabolas, and symmetry to create graphs. You can analyze these graphs to solve real-world problems.

Real-World Example 6 Use a Graph of a Quadratic Function

SCHOOL SPIRIT The cheerleaders at Lake High School launch T-shirts into the crowd every time the Lakers score a touchdown. The height of the T-shirt can be modeled by the function $h(x) = -16x^2 + 48x + 6$, where $h(x)$ represents the height in feet of the T-shirt after x seconds.

a. Graph the function.

$$x = -\frac{b}{2a} \qquad \text{Equation of the axis of symmetry}$$

$$x = -\frac{48}{2(-16)} \text{ or } \frac{3}{2} \qquad a = -16 \text{ and } b = 48$$

The equation of the axis of symmetry is $x = \frac{3}{2}$. Thus, the x-coordinate for the vertex is $\frac{3}{2}$.

$$y = -16x^2 + 48x + 6 \qquad \text{Original equation}$$

$$= -16\left(\frac{3}{2}\right)^2 + 48\left(\frac{3}{2}\right) + 6 \qquad x = \frac{3}{2}$$

$$= -16\left(\frac{9}{4}\right) + 48\left(\frac{3}{2}\right) + 6 \qquad \left(\frac{3}{2}\right)^2 = \frac{9}{4}$$

$$= -36 + 72 + 6 \text{ or } 42 \qquad \text{Simplify.}$$

The vertex is at $\left(\frac{3}{2}, 42\right)$.

Let's find another point. Choose an x-value of 0 and substitute. Our new point is at (0, 6). The point paired with it on the other side of the axis of symmetry is (3, 6).

Repeat this and choose an x-value of 1 to get (1, 38) and its corresponding point (2, 38). Connect these points and create a smooth curve.

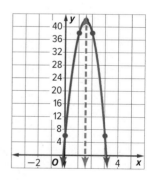

b. At what height was the T-shirt launched?

The T-shirt is launched when time equals 0, or at the y-intercept.

So, the T-shirt was launched 6 feet from the ground.

c. What is the maximum height of the T-shirt? When was the maximum height reached?

The maximum height of the T-shirt occurs at the vertex.

So the T-shirt reaches a maximum height of 42 feet. The time was $\frac{3}{2}$ or 1.5 seconds after launch.

Guided Practice

6. TRACK Emilio is competing in the javelin throw. The height of the javelin can be modeled by the equation $y = -16x^2 + 64x + 6$, where y represents the height in feet of the javelin after x seconds.

  A. Graph the path of the javelin.

  B. At what height is the javelin thrown?

  C. What is the maximum height of the javelin?

Check Your Understanding   ◯ = Step-by-Step Solutions begin on page R11.

Go Online! for a
Self-Check Quiz

Example 1   Use a table of values to graph each equation. State the domain and range.

1. $y = 2x^2 + 4x - 6$

2. $y = x^2 + 2x - 1$

3. $y = x^2 - 6x - 3$

4. $y = 3x^2 - 6x - 5$

Example 2   Find the vertex, the equation of the axis of symmetry, the y-intercept, and the zeros of each graph.

5.
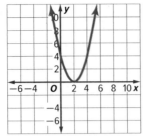

6.
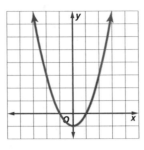

7.
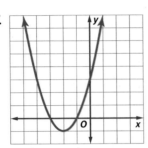

8.
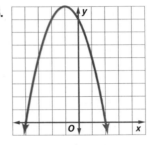

Example 3   Find the vertex, the equation of the axis of symmetry, and the y-intercept of the graph of each function.

9. $y = -3x^2 + 6x - 1$

10. $y = -x^2 + 2x + 1$

11. $y = x^2 - 4x + 5$

12. $y = 4x^2 - 8x + 9$

Example 4   Consider each function.

   a. Determine whether the function has *maximum* or *minimum* value.

   b. State the maximum or minimum value.

   c. What are the domain and range of the function?

13. $y = -x^2 + 4x - 3$

14. $y = -x^2 - 2x + 2$

15. $y = -3x^2 + 6x + 3$

16. $y = -2x^2 + 8x - 6$

Example 5   Graph each function.

17. $f(x) = -3x^2 + 6x + 3$

18. $f(x) = -2x^2 + 4x + 1$

19. $f(x) = 2x^2 - 8x - 4$

20. $f(x) = 3x^2 - 6x - 1$

Example 6   21. **MP REASONING** A juggler is tossing a ball into the air. The height of the ball in feet can be modeled by the equation $y = -16x^2 + 16x + 5$, where y represents the height of the ball at x seconds.

   a. Graph this equation.

   b. At what height is the ball thrown?

   c. What is the maximum height of the ball?

Example 1   Use a table of values to graph each equation. State the domain and range.

22. $y = x^2 + 4x + 6$ **23.** $y = 2x^2 + 4x + 7$ **24.** $y = 2x^2 - 8x - 5$

25. $y = 3x^2 + 12x + 5$ **26.** $y = 3x^2 - 6x - 2$ **27.** $y = x^2 - 2x - 1$

Example 2   Find the vertex, the equation of the axis of symmetry, the y-intercept, and the zeros of each graph.

28.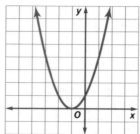

29.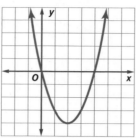

30.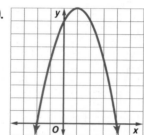

31.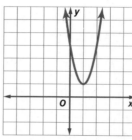

32.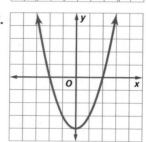

33.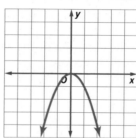

Example 3   Find the vertex, the equation of the axis of symmetry, and the y-intercept of each function.

34. $y = x^2 + 8x + 10$ **35** $y = 2x^2 + 12x + 10$ **36.** $y = -3x^2 - 6x + 7$

37. $y = -x^2 - 6x - 5$ **38.** $y = 5x^2 + 20x + 10$ **39.** $y = 7x^2 - 28x + 14$

40. $y = 2x^2 - 12x + 6$ **41.** $y = -3x^2 + 6x - 18$ **42.** $y = -x^2 + 10x - 13$

Example 4   Consider each function.

  a. Determine whether the function has a *maximum* or *minimum* value.

  b. State the maximum or minimum value.

  c. What are the domain and range of the function?

43. $y = -2x^2 - 8x + 1$ **44.** $y = x^2 + 4x - 5$ **45.** $y = 3x^2 + 18x - 21$

46. $y = -2x^2 - 16x + 18$ **47.** $y = -x^2 - 14x - 16$ **48.** $y = 4x^2 + 40x + 44$

49. $y = -x^2 - 6x - 5$ **50.** $y = 2x^2 + 4x + 6$ **51.** $y = -3x^2 - 12x - 9$

Example 5   Graph each function.

52. $y = -3x^2 + 6x - 4$ **53.** $y = -2x^2 - 4x - 3$ **54.** $y = -2x^2 - 8x + 2$

55. $y = x^2 + 6x - 6$ **56.** $y = x^2 - 2x + 2$ **57.** $y = 3x^2 - 12x + 5$

Example 6

58. BOATING Miranda has her boat docked on the west side of Casper Point. She is boating over to Casper Marina, which is located directly east of where her boat is docked. The equation $d = -16t^2 + 66t$ models the distance she travels north of her starting point, where d is the number of feet and t is the time traveled in minutes.

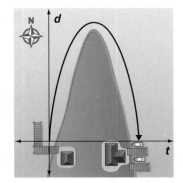

  a. Graph this equation.

  b. What is the maximum number of feet north that she traveled?

  c. How long did it take her to reach Casper Marina?

GRAPHING CALCULATOR Graph each equation. Use the **TRACE** feature to find the vertex on the graph. Round to the nearest thousandth if necessary.

59. $y = 4x^2 + 10x + 6$ **60.** $y = 8x^2 - 8x + 8$

61. $y = -5x^2 - 3x - 8$ **62.** $y = -7x^2 + 12x - 10$

63. GOLF The average amateur golfer can hit a ball with an initial upward velocity of 31.3 meters per second. The height can be modeled by the equation $h = -4.9t^2 + 31.3t$, where h is the height of the ball, in meters, after t seconds.

  a. Graph this equation. What do the portions of the graph where $h > 0$ represent in the context of the situation? What does the end behavior of the graph represent?

  b. At what height is the ball hit?

  c. What is the maximum height of the ball?

  d. How long did it take for the ball to hit the ground?

  e. State a reasonable range and domain for this situation.

64. MULTI-STEP The marching band is selling poinsettias to raise money for new uniforms. In last year's sale, the band charged $5 each, and they sold 150 poinsettias. They want to increase the price this year, and they expect to sell 10 fewer poinsettias for each $1 that they increase the price. The nursery will donate 50 poinsettias and charge them $1.50 for each additional poinsettia.

  a. If new band uniforms cost $925, will the band be able to raise enough money to buy them? If so, what is the lowest price they could charge for each poinsettia and still have enough money to buy the new uniforms?

  b. Describe your solution process.

65 FOOTBALL A football is kicked up from ground level at an initial upward velocity of 90 feet per second. The equation $h = -16t^2 + 90t$ gives the height h of the football after t seconds.

  a. What is the height of the ball after one second?

  b. When is the ball 126 feet high?

  c. When is the height of the ball 0 feet? What do these points represent in the context of the situation?

66. MP STRUCTURE Let $f(x) = x^2 - 9$.

  a. What is the domain of $f(x)$?

  b. What is the range of $f(x)$?

  c. For what values of x is $f(x)$ negative?

  d. When x is a real number, what are the domain and range of $f(x) = \sqrt{x^2 - 9}$?

67 **MULTIPLE REPRESENTATIONS** In this problem, you will investigate solving quadratic equations using tables.

a. Algebraic Determine the related function for each equation. Copy and complete the first two columns of the table below.

Equation	Related Function	Zeros	y-Values
$x^2 - x = 12$			
$x^2 + 8x = 9$			
$x^2 = 14x - 24$			
$x^2 + 16x = -28$			

b. Graphical Graph each related function with a graphing calculator.

c. Analytical The number of zeros is equal to the degree of the related function. Use the table feature on your calculator to determine the zeros of each related function. Record the zeros in the table above. Also record the values of the function one unit less than and one unit more than each zero.

d. Verbal Examine the function values for x-values just before and just after a zero. What happens to the sign of the function value before and after a zero?

H.O.T. Problems   Use **H**igher-**O**rder **T**hinking Skills

68. **MP** **SENSE-MAKING** Write a quadratic function for which the graph has an axis of symmetry of $x = -\dfrac{3}{8}$. Summarize your steps.

69. **ERROR ANALYSIS** Jade thinks that the parabolas represented by the graph and the description have the same axis of symmetry. Chase disagrees. Who is correct? Explain your reasoning.

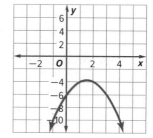

> a parabola that opens downward, passing
> through (0, 6) and having a vertex at (2, 2)

70. **CHALLENGE** Using the axis of symmetry, the y-intercept, and one x-intercept, write an equation for the graph shown.

71. **MP** **SENSE-MAKING** The graph of a quadratic function has a vertex at (2, 0). One point on the graph is (5, 9). Find another point on the graph. Explain how you found it.

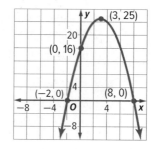

72. **MP** **REASONING** Describe a real-world situation that involves a quadratic equation. Explain what the vertex represents.

73. **MP** **CONSTRUCT ARGUMENTS** Provide a counterexample that is a specific case to show that the following statement is false. *The vertex of a parabola is always the minimum of the graph.*

74. **WRITING IN MATH** Use tables and graphs to compare and contrast an exponential function $f(x) = ab^x + c$, where $a \neq 0$, $b > 0$, and $b \neq 1$, a quadratic function $g(x) = ax^2 + c$, and a linear function $h(x) = ax + c$. Include intercepts, portions of the graph where the functions are increasing, decreasing, positive, or negative, relative maxima and minima, symmetries, and end behavior. Which function eventually increases at a faster rate than the others?

75. Which function has a range of $\{y \mid y \leq 4\}$? **MP** 1

- A $\quad y = x^2 - 8x + 20$
- B $\quad y = -x^2 + 8x - 12$
- C $\quad y = x^2 - 8x + 12$
- D $\quad y = -x^2 + 8x - 20$

76. Which statement best describes the function graphed below? **MP** 1

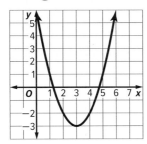

- A The equation of the axis of symmetry is $x = -3$.
- B The y-intercept is 1.
- C The maximum value is 6.
- D The range is $\{y \mid y \geq -3\}$.

77. What is the y-intercept of $f(x) = x^2 - 2x + 3$? **MP** 1, 7

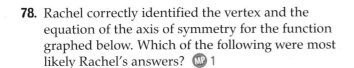

78. Rachel correctly identified the vertex and the equation of the axis of symmetry for the function graphed below. Which of the following were most likely Rachel's answers? **MP** 1

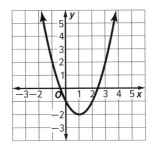

- A vertex: $(-2, 1)$; axis of symmetry: $x = 1$
- B vertex: $(-2, 1)$; axis of symmetry: $y = -2$
- C vertex: $(1, -2)$; axis of symmetry: $y = -2$
- D vertex: $(1, -2)$; axis of symmetry: $x = 1$

79. **MULTI-STEP** Use the function $y = x^2 + 4x + 3$ to answer the questions. **MP** 1, 7

a. What is the domain of the function?

- A $\{x \mid x \geq 1\}$
- B $\{x \mid x \leq -3\}$
- C $\{x \mid x \geq -1\}$
- D all real numbers

b. What is the range of the function?

- A $\{y \mid y \geq -1\}$
- B $\{y \mid y \leq -1\}$
- C $\{y \mid y \geq 1\}$
- D all real numbers

c. What is the vertex of the function?

- A $(-4, 3)$
- B $(-2, 1)$
- C $(-2, -1)$
- D $(0, 3)$

d. Which points lie on the parabola? Select all that apply.

- A $(-4, 0)$
- B $(-3, 0)$
- C $(-2, 1)$
- D $(-1, 0)$
- E $(0, 3)$
- F $(1, 8)$
- G $(2, 12)$

80. Ed is graphing the function $y = x^2 - 2x + c$. For what value of c will the range of the function be $\{y \mid y \geq -7\}$? **MP** 1, 7

Algebra Lab
Rate of Change of a Quadratic Function

Mathematical Practices

MP 4 Model with mathematics

A model rocket is launched from the ground with an upward velocity of 144 feet per second. The function $y = -16x^2 + 144x$ models the height y of the rocket in feet after x seconds. Using this function, we can investigate the rate of change of a quadratic function.

Activity

Step 1 Copy the table below.

x	0	0.5	1.0	1.5	...	9.0
y	0					
Rate of Change	—					

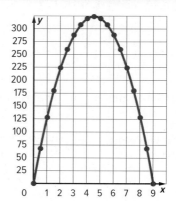

Step 2 Find the value of y for each value of x from 0 through 9.

Step 3 Graph the ordered pairs (x, y) on grid paper. Connect the points with a smooth curve. Notice that the function *increases* when $0 < x < 4.5$ and *decreases* when $4.5 < x < 9$.

Step 4 Recall that the *rate of change* is the change in y divided by the change in x. Find the rate of change for each half second interval of x and y.

Exercises

Use the quadratic function $y = x^2$.

1. Make a table, similar to the one in the Activity, for the function using $x = -4, -3, -2, -1, 0, 1, 2, 3,$ and 4. Find the values of y for each x-value.

2. Graph the ordered pairs on grid paper. Connect the points with a smooth curve. Describe where the function is increasing and where it is decreasing.

3. Find the rate of change for each column starting with $x = -3$. Compare the rates of change when the function is increasing and when it is decreasing.

4. **CHALLENGE** If an object is dropped from 100 feet in the air and air resistance is ignored, the object will fall at a rate that can be modeled by the function $f(x) = -16x^2 + 100$, where $f(x)$ represents the object's height in feet after x seconds. Make a table like that in Exercise 1, selecting appropriate values for x. Fill in the x-values, the y-values, and rates of change. Compare the rates of change. Describe any patterns that you see.

Transformations of Quadratic Functions

:: Then

- You graphed quadratic functions by using the vertex and axis of symmetry.

:: Now

1 Apply translations to quadratic functions.

2 Apply dilations and reflections to quadratic functions.

:: Why?

- The graphs of the parabolas shown at the right are the same size and shape, but notice that the vertex of the red parabola is higher on the *y*-axis than the vertex of the blue parabola. Shifting a parabola up and down is an example of a transformation.

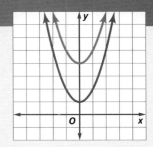

 New Vocabulary

vertex form

MP Mathematical Practices

1 Make sense of problems and persevere in solving them.

8 Look for and express regularity in repeated reasoning.

1 Translations The graph of $f(x) = x^2$ represents the parent graph of the quadratic functions. The parent graph can be translated up, down, left, right, or in two directions.

When a constant k is added to the quadratic function $f(x)$, the result is a vertical translation.

Key Concept Vertical Translations

The graph of $g(x) = x^2 + k$ is the graph of $f(x) = x^2$ translated vertically.

If $k > 0$, the graph of $f(x) = x^2$ is translated $|k|$ units **up**.

If $k < 0$, the graph of $f(x) = x^2$ is translated $|k|$ units **down**.

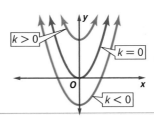

Example 1 Vertical Translations of Quadratic Functions

Describe the translation in each function as it relates to the graph of $f(x) = x^2$.

a. $h(x) = x^2 + 3$

$k = 3$ and $3 > 0$
$h(x)$ is a translation of the graph of $f(x) = x^2$ up 3 units.

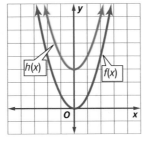

b. $g(x) = x^2 - 4$

$k = -4$ and $-4 < 0$
$g(x)$ is a translation of the graph of $f(x) = x^2$ down 4 units.

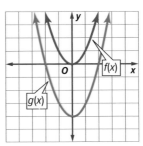

▶ **Guided Practice**

1A. $g(x) = x^2 - 7$ **1B.** $g(x) = 5 + x^2$ **1C.** $g(x) = -5 + x^2$ **1D.** $g(x) = x^2 + 1$

When a constant h is subtracted from the x-value before the function $f(x)$ is performed, the result is a horizontal translation.

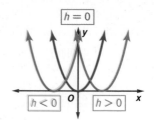

Key Concept Horizontal Translations

The graph of $g(x) = (x - h)^2$ is the graph of $f(x) = x^2$ translated horizontally.

If $h > 0$, the graph of $f(x) = x^2$ is translated h units to the **right**.

If $h < 0$, the graph of $f(x) = x^2$ is translated $|h|$ units to the **left**.

Example 2 Horizontal Translations of Quadratic Functions

Describe the translation in each function as it relates to the graph of $f(x) = x^2$.

a. $g(x) = (x - 2)^2$

$k = 0$, $h = 2$ and $2 > 0$
$g(x)$ is a translation of the graph of $f(x) = x^2$ to the right 2 units.

b. $g(x) = (x + 1)^2$

$k = 0$, $h = -1$ and $-1 < 0$
$g(x)$ is a translation of the graph of $f(x) = x^2$ to the left 1 unit.

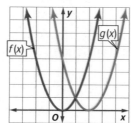

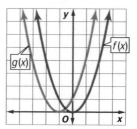

▶ **Guided Practice**

2A. $g(x) = (x - 3)^2$

2B. $g(x) = (x + 2)^2$

When constants h and k are present in the function, the parent function is translated in both the horizontal and vertical directions.

Go Online!

Investigate transformations of the graphs of quadratic functions by using the **Graphing Tools** in ConnectED.

Example 3 Multiple Translations of Quadratic Functions

Describe the translations in each function as it relates to the graph of $f(x) = x^2$.

a. $g(x) = (x - 3)^2 + 2$

$k = 2$, $h = 3$ and $3 > 0$
$g(x)$ is a translation of the graph of $f(x) = x^2$ to the right 3 units and up 2 units.

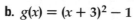
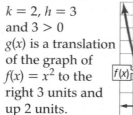

b. $g(x) = (x + 3)^2 - 1$

$k = -1$, $h = -3$ and $-3 < 0$
$g(x)$ is a translation of the graph of $f(x) = x^2$ to the left 3 units and down 1 unit.

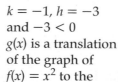
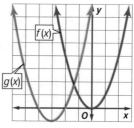

▶ **Guided Practice**

3A. $g(x) = (x + 2)^2 + 3$

3B. $g(x) = (x - 4)^2 - 4$

2 **Dilations and Reflections** When a quadratic function $f(x)$ is multiplied by a positive constant a, the result $a \cdot f(x)$ is a vertical dilation. The function is stretched or compressed vertically by a factor of $|a|$. When x is multiplied by a positive constant a before the quadratic function $f(x)$ is evaluated, the result $f(ax)$ is a horizontal dilation. The function is stretched or compressed vertically by a factor of $\frac{1}{|a|}$.

🎵 Key Concept Dilations of Quadratic Functions

Vertical Dilations

For $a \cdot f(x)$, if $|a| > 1$, the graph of $f(x)$ is stretched vertically.	For $a \cdot f(x)$, if $0 < |a| < 1$, the graph of $f(x)$ is compressed vertically.
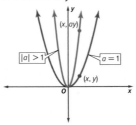	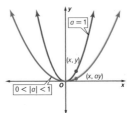
Every point on the graph of $f(x)$ is farther from the x-axis.	Every point on the graph of $f(x)$ is closer to the x-axis.

Horizontal Dilations

For $f(a \cdot x)$, if $|a| > 1$, the graph of $f(x)$ is compressed horizontally.	For $f(a \cdot x)$, if $0 < |a| < 1$, the graph of $f(x)$ is stretched horizontally.
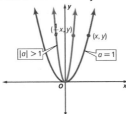	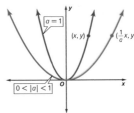
Every point on the graph of $f(x)$ is closer to the y-axis.	Every point on the graph of $f(x)$ is farther from the y-axis.

Study Tip

MP **Sense-Making** When the graph of a quadratic function is stretched vertically, the shape of the graph is narrower than that of the parent function. When it is compressed vertically, the graph is wider than the parent function.

Example 4 **Dilations of Quadratic Functions**

Describe the dilation in each function as it relates to the graph of $f(x) = x^2$.

a. $p(x) = \frac{1}{2}x^2$

$a = \frac{1}{2}$ and $0 < \frac{1}{2} < 1$
$p(x)$ is a dilation of the graph of $f(x) = x^2$ that is compressed vertically.

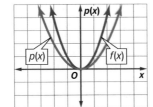

b. $g(x) = (3x)^2$

$a = 3$ and $3 > 1$
$g(x)$ is a dilation of the graph of $f(x) = x^2$ that is compressed horizontally.

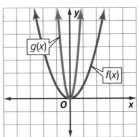

▶ **Guided Practice**

4A. $j(x) = 2x^2$

4B. $g(x) = \left(\frac{2}{5}x\right)^2$

When a quadratic function $f(x)$ is multiplied by -1 after the function has been evaluated, the result is a reflection across the x-axis. When $f(x)$ is multiplied by -1 before the function has been evaluated, the result is a reflection across the y-axis.

Key Concept Reflections

The graph of $-f(x)$ is the reflection of the graph of $f(x) = x^2$ across the x-axis.

The graph of $f(-x)$ is the reflection of the graph of $f(x) = x^2$ across the y-axis. Because $f(x)$ is symmetric about the y-axis, $f(-x)$ appears the same as $f(x)$.

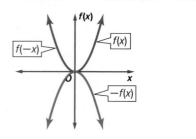

Example 5   Reflections and Dilations of Quadratic Functions

Describe the transformation in each function as it relates to the graph of $f(x) = x^2$.

a. $g(x) = -2x^2$

$a = -2$, $-2 < 0$, and $|-2| > 1$, so the graph is reflected across the x-axis and the graph is vertically stretched.

b. $p(x) = (-2x)^2$

$a = -2$, $-2 < 0$, so the graph is reflected across the y-axis and horizontally compresssed.

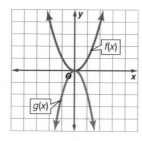

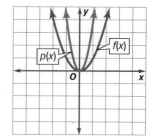

Guided Practice

5A. $r(x) = 2(-x)^2$

5B. $g(x) = -\frac{1}{5}x^2$

5C. $j(x) = (-2x)^2$

Real-World Example 6   Analyze a Function

BRIDGES The lower arch of the Sydney Harbor Bridge can be modeled by $g(x) = -0.0018(x - 251.5)^2 + 118$, where x is horizontal distance in meters and $g(x)$ is height in meters. Describe the transformations in $g(x)$ as it relates to the graph of the parent function.

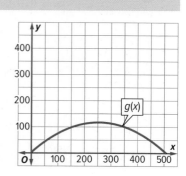

The values of a, h, and k are -0.0018, 251.5, and 118, respectively. Because a is negative and $0 < |a| < 1$, $g(x)$ is reflected across the x-axis and compressed vertically. The graph is translated right 251.5 units and up 118 units.

6. **FOOTBALL** In order to allow water to drain, football fields rise from each sideline to the center of the field. The cross section of a football field can be modeled by $g(x) = -0.000234(x - 80)^2 + 1.5$ where $g(x)$ is the height of the field and x is the distance from the sideline in feet. Describe how $g(x)$ is related to the graph of $f(x) = x^2$.

A quadratic function written in the form $f(x) = a(x - h)^2 + k$ is in **vertex form**.

♪ **Concept Summary** Transformations of Quadratic Functions

$$f(x) = a(x - h)^2 + k$$

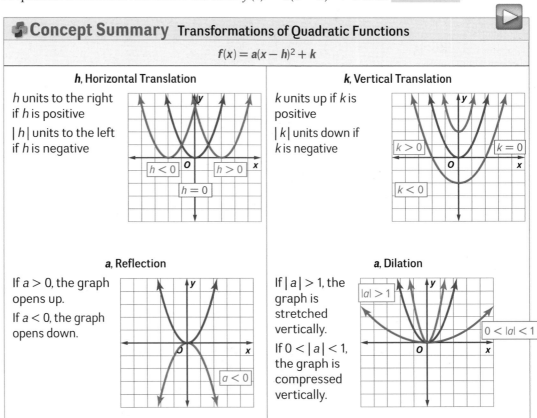

h, Horizontal Translation

h units to the right if *h* is positive

|*h*| units to the left if *h* is negative

k, Vertical Translation

k units up if *k* is positive

|*k*| units down if *k* is negative

a, Reflection

If $a > 0$, the graph opens up.

If $a < 0$, the graph opens down.

a, Dilation

If $|a| > 1$, the graph is stretched vertically.

If $0 < |a| < 1$, the graph is compressed vertically.

You can use a graphing calculator to verify descriptions of transformations.

Example 7   Compare Graphs

Compare the graphs to the parent function $f(x) = x^2$. Then, verify using technology.

$$g(x) = -2.1(x + 3)^2 \qquad\qquad j(x) = (-0.4x)^2 + 6$$

$g(x)$ is reflected across the x-axis, stretched vertically, and translated left 3 units. $j(x)$ is reflected across the y-axis, stretched horizontally and translated up 6 units.

To verify using a graphing calculator, enter each equation in the **Y=** list and graph. Use the features of the calculator to investigate the transformations. Note the change in the vertex, intercepts, and relation to the axes.

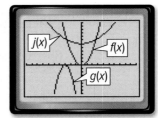

[−10, 10] scl: 1 by [−8, 12] scl: 1

7. Compare the graphs to the parent function $f(x) = x^2$. Then, verify using technology.

$$p(x) = -\frac{3}{4}(x - 2)^2 \qquad\qquad q(x) = (-2x)^2 - 5$$

To identify a quadratic equation, when given the vertex and a point, solve for a in the vertex form.

> **Study Tip**
>
> The vertex form of any quadratic function is $f(x) = a(x - h)^2 + k$, where the vertex is (h, k). The vertex being at $(2, 4)$ means that the graph was translated 2 units right and 4 units up.

Example 8   Identify Equations Given the Vertex and a Point

Write a quadratic function in vertex and standard form that contains (3, 6) and vertex (2, 4).

$$\begin{aligned}
f(x) &= a(x - h)^2 + k & \text{Vertex form of a quadratic function}\\
6 &= a(3 - 2)^2 + 4 & [x, f(x)] = [(3, 6), (h, k)] = (2, 4)\\
6 &= a + 4 & (3 - 2)^2 = 1\\
2 &= a & \text{Subtract 4 from each side.}
\end{aligned}$$

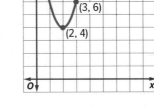

The vertex form of the function is $f(x) = 2(x - 2)^2 + 4$.

$$\begin{aligned}
f(x) &= 2(x - 2)^2 + 4 & \text{Vertex form}\\
f(x) &= 2(x^2 - 4x + 4) + 4 & \text{Expand } (x - 2)^2.\\
f(x) &= 2x^2 - 8x + 12 & \text{Simplify.}
\end{aligned}$$

The standard form of the function is $f(x) = 2x^2 - 8x + 12$.

Guided Practice

Write a quadratic function in vertex and standard form that contains vertex V and point P.

8A. $V(5, -1)$, $P(3, -5)$ **8B.** $V(0, 3)$, $P(4, 8)$ **8C.** $V(1, 2)$, $P(-2, 8)$

Go Online! for a
Self-Check Quiz

Check Your Understanding

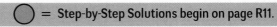 = Step-by-Step Solutions begin on page R11.

Examples 1–5   **Describe the transformations in each function as it relates to the graph of $f(x) = x^2$.**

1. $g(x) = x^2 - 11$ **2.** $h(x) = \frac{1}{2}(x - 2)^2$ **3.** $h(x) = -x^2 + 8$

4. $g(x) = (6x)^2$ **5.** $g(x) = -4(x + 3)^2$ **6.** $h(x) = -x^2 - 2$

Example 6–7   **7. BASKETBALL** The path of Reggie's basketball as he attempts to make a basket is modeled by $g(x) = -0.204(x - 6.2)^2 + 13.8$, where x is the horizontal distance from Reggie in feet and $g(x)$ is the height of the ball. Describe the transformations in $g(x)$ as it relates to the graph of the parent function. Then, verify using technology.

Example 8   **Write a quadratic function in vertex and standard form that contains vertex V and point P.**

8. $V(3, -2)$, $P(0, 4)$ **9.** $V(1, -1)$, $P(2, 3)$

Practice and Problem Solving

Extra Practice is on page R9.

Examples 1–5   **Describe the transformations in each function as it relates to the graph of $f(x) = x^2$.**

10. $g(x) = -10 + x^2$ **⑪** $h(x) = -7 - x^2$ **12.** $g(x) = 2(x - 3)^2 + 8$

13. $h(x) = 6 + \frac{2}{3}x^2$ **14.** $g(x) = -5 - \frac{4}{3}x^2$ **15.** $h(x) = 3 + \frac{5}{2}x^2$

16. $g(x) = 0.25x^2 - 1.1$ **17.** $h(x) = 1.35(x + 1)^2 + 2.6$

Example 6-7

18. DIVING The height of a diver in meters after x seconds is modeled by the function $g(x) = -7.2(x - 0.75)^2 + 10$.

  a. Find the values a, h, and k in $g(x)$.

  b. Describe the transformations in $g(x)$ as it relates to the graph of the parent function.

  c. Use technology to estimate the height of the diver after 1 second.

19. ROCKETS Candice and Paulo both launch a rocket during science club. The path of Candice's rocket is modeled by $c(x) = -16(x - 2.5)^2 + 105$ and Paulo's rocket is modeled by $p(x) = -16(x - 2.8)^2 + 126.5$, where $c(x)$ and $p(x)$ represent the height of each rocket after x seconds.

  a. Use technology to compare the graphs of $c(x)$ and $p(x)$ to the parent function.

  b. Whose rocket went higher?

  c. Whose rocket was in the air for a longer time?

Example 8

Write a quadratic function in vertex and standard forms that contains vertex V and point P.

20. $V(-2, -2)$, $P(-4, -10)$ **21.** $V(-1, -4)$, $P(2, 0)$

22. $V(7, -2)$, $P(4, 4)$ **23.** $V(-5, -9)$, $P(-2, -6)$

24. SQUIRRELS A squirrel 12 feet above the ground drops an acorn from a tree. The function $h = -16t^2 + 12$ models the height of the acorn above the ground in feet after t seconds. Graph the function, and compare it to its parent graph.

25 **ROCKS** A rock falls from a cliff 300 feet above the ground. At the same time, another rock falls from a cliff 700 feet above the ground.

  a. Write functions that model the height h of each rock after t seconds.

  b. If the rocks fall at the same time, how much sooner will the first rock reach the ground?

26. SPRINKLERS The path of water from a sprinkler can be modeled by quadratic functions. The following functions model paths for three different sprinklers.

Sprinkler A: $y = -0.35x^2 + 3.5$ Sprinkler B: $y = -0.21x^2 + 1.7$
Sprinkler C: $y = -0.08x^2 + 2.4$

  a. Which sprinkler will send water the farthest? Explain.

  b. Which sprinkler will send water the highest? Explain.

  c. Which sprinkler will produce the narrowest path? Explain.

27. GOLF The path of a drive can be modeled by a quadratic function where $g(x)$ is the vertical distance in yards of the ball from the ground and x is the horizontal distance in yards.

  a. How can you obtain $g(x)$ from the graph of $f(x) = x^2$.

  b. A second golfer hits a ball from the red tee, which is 30 yards closer to the hole. What function $h(x)$ can be used to describe the second golfer's shot?

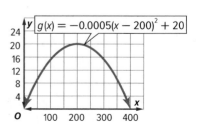

Describe the transformations to obtain the graph of $g(x)$ from the graph of $f(x)$.

28. $f(x) = x^2 + 3$
$g(x) = x^2 - 2$

29. $f(x) = x^2 - 4$
$g(x) = (x - 2)^2 + 7$

30. $f(x) = -6x^2$
$g(x) = -3x^2$

31. MULTIPLE REPRESENTATIONS In this problem you will analyze $f(bx)$ if $f(x) = x^2$.

   a. Graphical Graph $f(bx)$ for $b = -3, -1, 0.25, 0.5, 1, 2,$ and 3.

   b. Analytical Describe the transformation from $f(x)$ to $f(bx)$ when $b > 1$.

   c. Analytical Describe the transformation from $f(x)$ to $f(bx)$ when $0 < b < 1$.

   d. Analytical Describe the transformation from $f(x)$ to $f(bx)$ when $b < 0$.

   e. Graphical Graph $af(x)$ for $a = -2, -0.5, 0.5,$ and 3.

Tell which equation matches Graph 1 and Graph 2.

32.

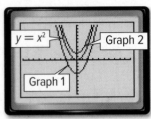

$[-10, 10]$ scl: 1 by $[-10, 10]$ scl: 1

$y = x^2 + 2$

$y = x^2 - 4$

33.

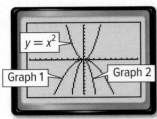

$[-10, 10]$ scl: 1 by $[-10, 10]$ scl: 1

$y = -\frac{1}{3}x^2$

$y = -2x^2$

Compare and contrast each pair of functions. Use a graphing calculator to confirm.

34. $y = x^2$, $y = x^2 + 3$

35. $y = \frac{1}{2}x^2$, $y = 3x^2$

36. $y = x^2$, $y = (x - 5)^2$

37. $y = 3x^2$, $y = -3x^2$

38. $y = x^2$, $y = -4x^2$

39. $y = x^2 - 1$, $y = x^2 + 2$

40. $y = \frac{1}{2}x^2 + 3$, $y = -2x^2$

41. $y = x^2 - 4$, $y = (x - 4)^2$

H.O.T. Problems   Use **Higher-Order Thinking Skills**

42. MP CONSTRUCT ARGUMENTS Are the following statements *sometimes*, *always*, or *never* true? Explain.

   a. The graph of $y = x^2 + k$ has its vertex at the origin.

   b. The graphs of $y = ax^2$ and its reflection over the x-axis are the same width.

   c. The graph of $y = x^2 + k$, where $k \geq 0$, and the graph of a quadratic with vertex at $(0, -3)$ have the same maximum or minimum point.

43. CHALLENGE Write a function of the form $y = ax^2 + k$ with a graph that passes through the points $(-2, 3)$ and $(4, 15)$.

44. MP CONSTRUCT ARGUMENTS Determine whether all quadratic functions that are reflected across the y-axis produce the same graph. Explain your answer.

45. OPEN-ENDED Write a quadratic function that opens downward and is wider than the parent graph.

46. WRITING IN MATH Describe how the values of a and k affect the graphical and tabular representations for the functions $y = ax^2$, $y = x^2 + k$, and $y = ax^2 + k$.

47. Luis graphed the parent quadratic function as shown. Then he graphed a second function that is a translation of the parent graph 2 units down and 3 units to the left. Which is an equation for the second graph? **MP** 1, 8 F.IF.7a, F.BF.3

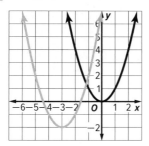

- ○ **A** $f(x) = x^2 - 6x + 7$
- ○ **B** $f(x) = x^2 - 6x + 11$
- ○ **C** $f(x) = x^2 + 6x + 7$
- ○ **D** $f(x) = x^2 + 6x + 11$

48. The graph of the function $f(x) = x^2$ is reflected across the x-axis and compressed vertically. Which of the following could be the equation for the graph? Selct all that apply. **MP** 1, 8 F.BF.3

- ☐ **A** $f(x) = \frac{3}{2}x^2$
- ☐ **B** $f(x) = -\frac{3}{2}x^2$
- ☐ **C** $f(x) = \frac{1}{3}x^2$
- ☐ **D** $f(x) = -\frac{1}{3}x^2$
- ☐ **E** $f(x) = -\frac{2}{3}x^2$
- ☐ **F** $f(x) = \frac{2}{3}x^2$

49. The graph of $f(x) = x^2$ is reflected across the x-axis and translated to the left 4 units. What is the value of h when the equation of the transformed graph is written in vertex form? **MP** 1, 8 F.BF.3

50. The graph of a quadratic function $g(x)$ is shown below.

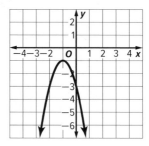

Which statement about the relationship between the graph of $g(x)$ and the graph of the parent function $f(x) = x^2$ is not true? **MP** 1, 8 F.IF.7a, F.BF.3

- ○ **A** $g(x)$ is a vertical stretch of the graph of $f(x) = x^2$.
- ○ **B** $g(x)$ is a reflection across the x-axis of the graph of $f(x) = x^2$.
- ○ **C** In vertex form, the equation of the function is $g(x) = -2(x - 1)^2 - 1$.
- ○ **D** In standard form, the equation of the function is $g(x) = -2x^2 - 4x - 3$.

51. **MULTI-STEP** The ideal weight of a kitten in pounds is modeled by the function $g(x) = 0.0009(x + 7.05)^2 - 0.071$, where x is the age of the kitten in weeks. **MP** 1, 7 F.BF.3

- **a.** Determine the value of a in $g(x)$.
- **b.** Determine the value of h in $g(x)$.
- **c.** Determine the value of k in $g(x)$.
- **d.** Select all transformations in $g(x)$ as it relates to the parent function.
  - ☐ **A** reflected across the x-axis
  - ☐ **B** compressed vertically
  - ☐ **C** stretched vertically
  - ☐ **D** translated right
  - ☐ **E** translated left
  - ☐ **F** translated up

Solving Quadratic Equations by Graphing

:: **Then**

- You solved quadratic equations by factoring.

:: **Now**

1 Solve quadratic equations by graphing.

2 Estimate solutions of quadratic equations by graphing.

:: **Why?**

- Dorton Arena at the state fairgrounds in Raleigh, North Carolina, has a shape created by two intersecting parabolas. The shape of one of the parabolas can be modeled by $y = -x^2 + 127x$, where x is the width of the parabola in feet, and y is the length of the parabola in feet. The x-intercepts of the graph of this function can be used to find the distance between the points where the parabola meets the ground.

 New Vocabulary

double root

 Mathematical Practices

3 Construct viable arguments and critique the reasoning of others.

5 Use appropriate tools strategically.

---

1 **Solve by Graphing** A quadratic equation can be written in the standard form $ax^2 + bx + c = 0$, where $a \neq 0$. To write a quadratic function as an equation, replace y or $f(x)$ with 0. Quadratic equations may have two, one, or no real solutions.

Key Concept   Solutions of Quadratic Equations

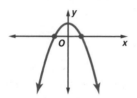

two unique real solutions

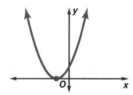

one unique real solution

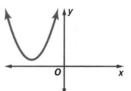

no real solutions

Recall that the solutions or roots of an equation can be identified by finding the x-intercepts of the related graph.

Quadratic Function

$$f(x) = x^2 - x - 6$$

$$f(-2) = (-2)^2 - (-2) - 6 \text{ or } 0$$
$$f(3) = 3^2 - 3 - 6 \text{ or } 0$$

-2 and 3 are zeros of the function.

Quadratic Equation

$$x^2 - x - 6 = 0$$

$$(-2)^2 - (-2) - 6 \text{ or } 0$$
$$3^2 - 3 - 6 \text{ or } 0$$

-2 and 3 are roots of the equation.

Graph of Function

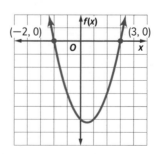

The x-intercepts are -2 and 3.

Example 1   Two Roots

Solve each equation by graphing.

a. $x^2 - 2x - 8 = 0$

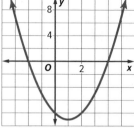

Step 1 Rewrite the equation in standard form.
This equation is written in standard form.

Step 2 Graph the related function $f(x) = x^2 - 2x - 8$.

> **Watch Out!**
>
> **MP** **Precision** Solutions found from the graph of an equation may appear to be exact. Check them in the original equation to be sure.

Step 3 Locate the x-intercepts of the graph. The x-intercepts of the graph appear to be at -2 and 4, so the solutions are -2 and 4.

CHECK Check each solution in the original equation.

$$x^2 - 2x - 8 = 0 \qquad \text{Original equation} \qquad x^2 - 2x - 8 = 0$$
$$(-2)^2 - 2(-2) - 8 \overset{?}{=} 0 \qquad x = -2 \text{ or } x = 4 \qquad (4)^2 - 2(4) - 8 \overset{?}{=} 0$$
$$0 = 0 \checkmark \qquad \text{Simplify.} \qquad 0 = 0 \checkmark$$

b. $-2x^2 + 8 = -6x$

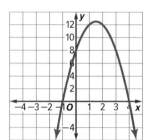

Step 1 Rewrite the equation in standard form.

$$-2x^2 + 8 = -6x \qquad \text{Original equation}$$
$$-2x^2 + 6x + 8 = 0 \qquad \text{Add } 6x \text{ to each side.}$$

Step 2 Graph the related function $f(x) = -2x^2 + 6x + 8$.

Step 3 Locate the x-intercepts of the graph. The x-intercepts of the graph appear to be at -1 and 4, so the solutions are -1 and 4.

> **Guided Practice**

Solve each equation by graphing.

1A. $-x^2 - 3x + 18 = 0$ **1B.** $x^2 - 4x + 3 = 0$

The solutions in Example 1 were two distinct numbers. Sometimes the two roots are the same number, called a **double root**.

Example 2   Double Root

Solve each equation by graphing.

a. $-x^2 + 4x - 4 = 0$

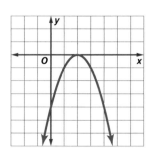

Step 1 Rewrite the equation in standard form.
This equation is written in standard form.

Step 2 Graph the related function $f(x) = -x^2 + 4x - 4$.

Step 3 Locate the x-intercepts of the graph. Notice that the vertex of the parabola is the only x-intercept. Therefore, there is only one solution, 2.

CHECK Check the solution in the original equation.

$$-x^2 + 4x - 4 = 0 \qquad \text{Original equation}$$
$$-(2)^2 + 4(2) - 4 = 0 \qquad x = 2$$
$$0 = 0 \checkmark \qquad \text{Simplify.}$$

b. $x^2 - 6x = -9$

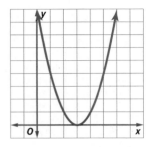

Step 1 Rewrite the equation in standard form.

$$x^2 - 6x = -9 \qquad \text{Original equation}$$
$$x^2 - 6x + 9 = 0 \qquad \text{Add 9 to each side.}$$

Step 2 Graph the related function $f(x) = x^2 - 6x + 9$.

Step 3 Locate the x-intercepts of the graph. Notice that the vertex of the parabola is the only x-intercept. Therefore, there is only one solution, 3.

▸ **Guided Practice**

Solve each equation by graphing.

2A. $x^2 + 25 = 10x$ **2B.** $x^2 = -8x - 16$

Sometimes the roots are not real numbers. Quadratic equations with solutions that are not real numbers lead us to extend the number system to allow for solutions of these equations. These numbers are called *complex numbers*. You will study complex numbers in Algebra 2.

Example 3   No Real Roots

Solve $2x^2 - 3x + 5 = 0$ by graphing.

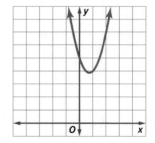

Step 1 Rewrite the equation in standard form.
This equation is written in standard form.

Step 2 Graph the related function
$f(x) = 2x^2 - 3x + 5$.

Step 3 Locate the x-intercepts of the graph. This graph has no x-intercepts. Therefore, this equation has no real number solutions. The solution set is ∅.

▸ **Guided Practice**

Solve each equation by graphing.

3A. $-x^2 - 3x = 5$ **3B.** $-2x^2 - 8 = 6x$

2 Estimate Solutions
The real roots found thus far have been integers. However, the roots of quadratic equations are usually not integers. In these cases, use estimation to approximate the roots of the equation.

Example 4 — Approximate Roots with a Table

Solve $x^2 + 6x + 6 = 0$ by graphing. If integral roots cannot be found, estimate the roots to the nearest tenth.

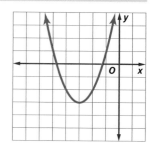

Graph the related function $f(x) = x^2 + 6x + 6$.

The x-intercepts are located between -5 and -4 and between -2 and -1.

Make a table using an increment of 0.1 for the x-values located between -5 and -4 and between -2 and -1.

Look for a change in the signs of the function values. The function value that is closest to zero is the best approximation for a zero of the function.

Study Tip

Location of Zeros Since quadratic functions are continuous, there must be a zero between two x-values for which the corresponding y-values have opposite signs.

x	−4.9	−4.8	−4.7	−4.6	−4.5	−4.4	−4.3	−4.2	−4.1
y	0.61	0.24	−0.11	−0.44	−0.75	−1.04	−1.31	−1.56	−1.79

x	−1.9	−1.8	−1.7	−1.6	−1.5	−1.4	−1.3	−1.2	−1.1
y	−1.79	−1.56	−1.31	−1.04	−0.75	−0.44	−0.11	0.24	0.61

For each table, the function value that is closest to zero when the sign changes is -0.11. Thus, the roots are approximately -4.7 and -1.3.

Guided Practice

4. Solve $2x^2 + 6x - 3 = 0$ by graphing. If integral roots cannot be found, estimate the roots to the nearest tenth.

Approximating the x-intercepts of graphs is helpful for real-world applications.

Real-World Example 5 — Approximate Roots with a Calculator

SOCCER A goalie kicks a soccer ball with an upward velocity of 65 feet per second, and her foot meets the ball 1 foot off the ground. The quadratic function $h = -16t^2 + 65t + 1$ represents the height of the ball h in feet after t seconds. Approximately how long is the ball in the air?

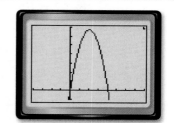

$[-4, 7]$ scl: 1 by $[-10, 70]$ scl: 10

You need to find the roots of the equation $-16t^2 + 65t + 1 = 0$. Use a graphing calculator to graph the related function $f(x) = -16t^2 + 65t + 1$.

The positive x-intercept of the graph is approximately 4. Therefore, the ball is in the air for approximately 4 seconds.

Guided Practice

5. If the goalie kicks the soccer ball with an upward velocity of 55 feet per second and his foot meets the ball 2 feet off the ground, approximately how long is the ball in the air?

Real-World Link

The game of soccer, called "football" outside of North America, began in 1863 in Britain when the Football Association was founded. Soccer is played on every continent of the world.

Source: Sports Know How

Check Your Understanding ◯ = Step-by-Step Solutions begin on page R11.

Go Online! for a
Self-Check Quiz

Examples 1–3 Solve each equation by graphing.

1. $x^2 + 3x - 10 = 0$

2. $2x^2 - 8x = 0$

3. $x^2 + 4x = -4$

4. $x^2 + 12 = -8x$

Solve each equation by examining the given graph of the related function.

5. $2x^2 + 14x - 16 = 0$

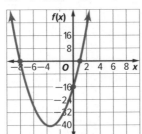

6. $\frac{1}{4}x^2 - 2x + 3 = 0$

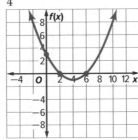

Example 4 Solve each equation by graphing. If integral roots cannot be found, estimate the roots to the nearest tenth.

7. $x^2 = 25$

8. $x^2 - 8x = -9$

Example 5 **9. SCIENCE FAIR** Ricky built a model rocket. Its flight can be modeled by the equation shown, where h is the height of the rocket in feet after t seconds. About how long was Ricky's rocket in the air?

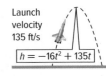

Launch velocity 135 ft/s

$h = -16t^2 + 135t$

Practice and Problem Solving Extra Practice is on page R9.

Examples 1–3 Solve each equation by graphing.

10. $x^2 + 7x + 14 = 0$

⑪ $x^2 + 2x - 24 = 0$

12. $x^2 - 16x + 64 = 0$

13. $x^2 - 5x + 12 = 0$

14. $x^2 + 14x = -49$

15. $x^2 = 2x - 1$

16. $x^2 - 10x = -16$

17. $-2x^2 - 8x = 13$

18. $2x^2 - 16x = -30$

19. $2x^2 = -24x - 72$

20. $-3x^2 + 2x = 15$

21. $x^2 = -2x + 80$

22. $3x^2 - 6 = 3x$

23. $4x^2 + 24x = -36$

24. $-2x^2 - 9 = 8x$

Example 4 Solve each equation by graphing. If integral roots cannot be found, estimate the roots to the nearest tenth.

25. $x^2 + 2x - 9 = 0$

26. $x^2 - 4x = 20$

27. $x^2 + 3x = 18$

Example 5 **28. SOFTBALL** The equation $h = -16t^2 + 47t + 3$ models the height h, in feet, of a ball that Sofia hits after t seconds.

   a. How long is the ball in the air?

   b. Find the y-intercept and describe its meaning in the context of the situation.

29. RIDES The Terror Tower launches riders straight up and returns straight down. The equation $h = -16t^2 + 122t$ models the height h, in feet, of the riders from their starting position after t seconds. How long is it until the riders return to the bottom?

Use factoring to determine how many times the graph of each function intersects the *x*-axis. Identify each zero.

30. $y = x^2 - 8x + 16$

31. $y = x^2 + 4x + 4$

32. $y = x^2 + 2x - 24$

33. $y = x^2 + 12x + 32$

34. NUMBER THEORY Use a quadratic equation to find two numbers that have a sum of 9 and a product of 20.

35. NUMBER THEORY Use a quadratic equation to find two numbers that have a sum of 1 and a product of -12.

36. Ⓜ️ **MODELING** The height of a golf ball in the air can be modeled by the equation $h = -16t^2 + 76t$, where *h* is the height in feet of the ball after *t* seconds.

   a. How long was the ball in the air?

   b. What is the ball's maximum height?

   c. When will the ball reach its maximum height?

37 SKIING Stefanie is in a freestyle aerial competition. The equation $h = -16t^2 + 30t + 10$ models Stefanie's height *h*, in feet, *t* seconds after leaving the ramp.

   a. How long is Stefanie in the air?

   b. When will Stefanie reach a height of 15 feet?

   c. To earn bonus points in the competition, you must reach a height of 20 feet. Will Stefanie earn bonus points?

38. MULTIPLE REPRESENTATIONS In this problem, you will explore the relationship between the factors of a quadratic equation and the zeros of the related graph.

   a. Graphical Graph $y = x^2 - 2x - 3$.

   b. Analytical Name the zeros of the function.

   c. Algebraic Factor the related equation $x^2 - 2x - 3 = 0$.

   d. Analytical Set each factor equal to zero and solve. What are the values of *x*?

   e. Analytical What conclusions can you draw about the factors of quadratic equations?

GRAPHING CALCULATOR Solve each equation by graphing.

39. $x^3 - 3x^2 - 6x + 8 = 0$

40. $x^3 - 8x^2 + 15x = 0$

H.O.T. Problems　　　Use **H**igher-**O**rder **T**hinking Skills

41. CHALLENGE Describe a real-world situation in which a thrown object travels in the air. Write an equation that models the height of the object with respect to time, and determine how long the object travels in the air.

42. Ⓜ️ **STRUCTURE** For the quadratic function $y = ax^2 + bx + c$, determine the number of *x*-intercepts of the function if:

   a. $a < 0$ and the vertex lies below the *x*-axis

   b. $a > 0$ and the vertex lies below the *x*-axis

   c. $a < 0$ and the vertex lies above the *x*-axis

   d. $a > 0$ and the vertex lies above the *x*-axis

43. CHALLENGE Find the roots of $x^2 = 2.25$ without using a calculator. Explain your strategy.

44. WRITING IN MATH Explain how the roots of a quadratic equation are related to the graph of a quadratic function.

45. Consider $x^2 - x = 6$. **MP** 1, 6

    a. Which is the related graph of the equation?

      ○ **A**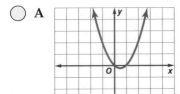

      ○ **B**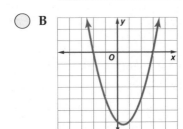

      ○ **C**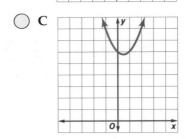

      ○ **D**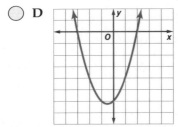

    b. Select all of the solutions of the equation.

      ☐ **A**   −3

      ☐ **B**   −2

      ☐ **C**   0

      ☐ **D**   1

      ☐ **E**   0

      ☐ **F**   ∅

46. Solve $x^2 - 12x = -36$ by graphing. **MP** 1

    [   ]

47. MULTI-STEP Use the function $y = x^2 - 2x + c$ to answer the questions. **MP** 1, 7

    a. If $c = -3$ and $y = 0$, how many solutions are there to the related equation?

    b. If $c = -2$ and $y = -3$, how many solutions are there to the related equation?

    c. For what values of c will there be no solution when $y = 0$? Check all that apply.

      ☐ **A** $c = -1$ ☐ **C** $c = 2$ ☐ **E** $c = 5$

      ☐ **B** $c = 0$ ☐ **D** $c = 1$

48. Sean solved a quadratic equation by graphing. The equation has no real solution. Which statement best describes Sean's graph? **MP** 6, 7

    ○ **A**   The graph has exactly two x-intercepts.

    ○ **B**   The graph has exactly one x-intercept.

    ○ **C**   The graph has exactly one x-intercept of 0.

    ○ **D**   The graph has no x-intercepts.

49. Which quadratic equations have solutions of −2 and 5? Select all that apply. **MP** 1

    ☐ **A** $x^2 - 3x - 10 = 0$

    ☐ **B** $x^2 + 3x = 10$

    ☐ **C** $2x^2 - 6x = 20$

    ☐ **D** $-x^2 + 10 = 3x$

    ☐ **E** $-x^2 + 3x + 10 = 0$

50. Neil is using the graph shown to solve $x^2 - 5x + 4 = 0$. What are the solutions of the quadratic equation? **MP** 1, 6

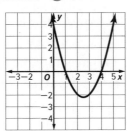

    ○ **A**   0

    ○ **B**   2.5

    ○ **C**   1, 4

    ○ **D**   ∅

Graphing Technology Lab
Quadratic Inequalities

Recall that the graph of a linear inequality consists of the boundary and the shaded half-plane. The solution set of the inequality lies in the shaded region of the graph. Graphing quadratic inequalities is similar to graphing linear inequalities.

Mathematical Practices

MP **5** Use appropriate tools strategically.

Activity 1 Shade Inside a Parabola

Work cooperatively. Graph $y \geq x^2 - 5x + 4$ in a standard viewing window.

First, clear all functions from the **Y=** list.

To graph $y \geq x^2 - 5x + 4$, enter the equation in the **Y=** list. Then use the left arrow to select =. Press ENTER until shading above the line is selected.

KEYSTROKES: ◄ ◄ ENTER ENTER ► ► X,T,θ,n x² − 5 X,T,θ,n + 4 ZOOM 6

All ordered pairs for which y is *greater than or equal* to $x^2 - 5x + 4$ lie *above or on* the line and are solutions.

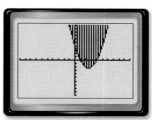

[−10, 10] scl: 1 by [−10, 10] scl: 1

A similar procedure will be used to graph an inequality in which the shading is outside of the parabola.

Activity 2 Shade Outside a Parabola

Work cooperatively. Graph $y - 4 \leq x^2 - 5x$ in a standard viewing window.

First, clear the graph that is displayed.

KEYSTROKES: Y CLEAR

Then rewrite $y - 4 \leq x^2 - 5x$ as $y \leq x^2 - 5x + 4$, and graph it.

KEYSTROKES: ◄ ◄ ENTER ENTER ENTER ► ► X,T,θ,n x² − 5 X,T,θ,n + 4 GRAPH

All ordered pairs for which y is *less than or equal* to $x^2 - 5x + 4$ lie *below or on* the line and are solutions.

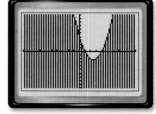

[−10, 10] scl: 1 by [−10, 10] scl: 1

Exercises

1. Compare and contrast the two graphs shown above.

2. Graph $y - 2x + 6 \geq 5x^2$ in the standard viewing window. Name three solutions of the inequality.

3. Graph $y - 6x \leq -x^2 - 3$ in the standard viewing window. Name three solutions of the inequality.

Solving Quadratic Equations by Factoring

:: **Then**

- You solved quadratic equations by graphing.

:: **Now**

1 Solve quadratic equations by using the Square Root Property.

2 Solve quadratic equations by factoring.

:: **Why?**

- Froghoppers are insects commonly found in Africa, Europe, and North America. They are only about 6 millimeters long, but they can jump up to 70 times their body height. A froghopper's jump can be modeled by the equation $h = 12t - 16t^2$, where t is the time in seconds and h is the height in feet. You can use factoring and the Zero Product Property to determine when the froghopper will complete its jump.

New Vocabulary
Square Root Property
Zero Product Property

Mathematical Practices
1 Make sense of problems and persevere in solving them.
6 Attend to precision.

Quadratic equations can be solved using a variety of methods. In addition to solving quadratic equations by graphing, you can solve quadratic equations algebraically.

1 **Square Root Property** A quadratic equation in the form $x^2 = n$ can be solved by using the **square root property**. The equation can be solved by applying the square root to each sides of the equation.

⚙ Key Concept Square Root Property

Words	To solve a quadratic equation in the form $x^2 = n$, take the square root of each side.
Symbols	For any number $n \geq 0$, if $x^2 = n$, then $x = \pm\sqrt{n}$.
Example	$x^2 = 25$
	$x = \pm\sqrt{25}$ or ± 5

In the equation $x^2 = n$, if n is not a perfect square, you need to approximate the square root. Use a calculator to find an approximation. If n is a perfect square, you will have an exact answer.

Example 1   Use the Square Root Property

Solve each equation. Check your solutions.

a. $(x - 3)^2 = 64$

$(x - 3)^2 = 64$	Original Equation
$x - 3 = \pm\sqrt{64}$	Square Root Property
$x - 3 = \pm 8$	$64 = 8(8)$ or $-8(-8)$
$x = 3 \pm 8$	Add 3 to each side.
$x = 3 + 8$ or $x = 3 - 8$	Separate into two equations.
$x = 11 \qquad x = -5$	Simplify.

The roots are 11 and -5.

CHECK Substitute -11 and 5 for x in the original equation.

$$(x - 3)^2 = 64 \qquad\qquad (x - 3)^2 = 64$$
$$(11 - 3)^2 \overset{?}{=} 64 \qquad\qquad [(-5) - 3]^2 \overset{?}{=} 64$$
$$(8)^2 \overset{?}{=} 64 \qquad\qquad (-8)^2 \overset{?}{=} 64$$
$$64 = 64 \checkmark \qquad\qquad 64 = 64 \checkmark$$

Study Tip

MP **Persevere** Equations involving square roots can often be solved mentally. For $x^2 = n$, think *The square of what number is* n? When n is a perfect square, x is rational. Otherwise, x is irrational.

b. $(y + 5)^2 + 7 = 28$

$(y + 5)^2 + 7 = 28$	Original Equation
$(y + 5)^2 = 21$	Subtract 7 from each side.
$y + 5 = \pm\sqrt{21}$	Square Root Property
$y = -5 \pm \sqrt{21}$	Subtract –5 from each side.
$y = -5 + \sqrt{21}$ or $y = -5 - \sqrt{21}$	Separate into two equations.

The roots are $-5 + \sqrt{21}$ and $-5 - \sqrt{21}$
Using a calculator, $-5 + \sqrt{21} \approx -0.42$ and $-5 - \sqrt{21} \approx -9.58$.

Guided Practice

Solve each equation. Round to the nearest hundredth if necessary.

1A. $(x + 6)^2 = 81$

1B. $(a - 8)^2 - 3 = 10$

1C. $4(m + 1)^2 = 36$

When solving real-world problems using the Square Root Property, it is important to determine whether both solutions make sense in the context of the situation.

Real-World Example 2   Solve an Equation by Using the Square Root Property

PHYSICAL SCIENCE During an experiment, a ball is dropped from a height of 205 feet. The formula $h = -16t^2 + h_0$ can be used to approximate the number of seconds t it takes for the ball to reach height h from an initial height of h_0 in feet. Find the time it takes the ball to reach the ground.

At ground level, $h = 0$ and the initial height is 205, so $h_0 = 205$.

$h = -16t^2 + h_0$	Original formula
$0 = -16t^2 + 205$	Replace h with 0 and h_0 with 205.
$-205 = -16t^2$	Subtract 205 from each side.
$12.8125 = t^2$	Divide each side by -16.
$\pm 3.6 \approx t$	Use the Square Root Property.

Since a negative number does not make sense in this situation, the solution is 3.6. It takes about 3.6 seconds for the ball to reach the ground.

Guided Practice

2. Find the time it takes a ball to reach the ground if it is dropped from a height that is half as high as the one described above.

2 Solve Quadratic Equations by Factoring

Some equations can be solved by factoring. A quadratic equation will have roots if its factored form has real numbers. Consider the following:

$$7(0) = 0 \qquad 0(4 - 3 - 1) = 0 \qquad -71(0) = 0 \qquad (3.59)0 = 0$$

Notice that in each case, at least one of the factors is 0. These examples demonstrate the **Zero Product Property**.

Key Concept Zero Product Property

Words   If the product of two factors is 0, then at least one of the factors must be 0.

Symbols   For any numbers a and b, if $ab = 0$, then $a = 0$, $b = 0$, or both a and b equal zero.

Example 3   Solve Equations by Factoring

Solve each equation. Check your solutions.

a. $(2d + 6)(3d - 15) = 0$

$(2d + 6)(3d - 15) = 0$		Original equation
$2d + 6 = 0 \quad$ or $\quad 3d - 15 = 0$		Zero Product Property
$2d = -6 \qquad\qquad 3d = 15$		Solve each equation.
$d = -3 \qquad\qquad d = 5$		Divide.

The roots are -3 and 5.

CHECK Substitute -3 and 5 for d in the original equation.

$$(2d + 6)(3d - 15) = 0 \qquad\qquad (2d + 6)(3d - 15) = 0$$
$$[2(-3) + 6][3(-3) - 15] \stackrel{?}{=} 0 \qquad [2(5) + 6][3(5) - 15] \stackrel{?}{=} 0$$
$$(-6 + 6)(-9 - 15) \stackrel{?}{=} 0 \qquad (10 + 6)(15 - 15) \stackrel{?}{=} 0$$
$$(0)(-24) \stackrel{?}{=} 0 \qquad\qquad 16(0) \stackrel{?}{=} 0$$
$$0 = 0 \checkmark \qquad\qquad\qquad 0 = 0 \checkmark$$

b. $c^2 = 3c$

$c^2 = 3c$	Original equation
$c^2 - 3c = 0$	Subtract $3c$ from each side to get 0 on one side of the equation.
$c(c - 3) = 0$	Factor by using the GCF to get the form $ab = 0$.
$c = 0 \quad$ or $\quad c - 3 = 0$	Zero Product Property
$c = 3$	Solve each equation.

The roots are 0 and 3.   Check by substituting 0 and 3 for c.

Watch Out!

Unknown Value It may be tempting to solve an equation by dividing each side by the variable. However, the variable has an unknown value, so you may be dividing by 0, which is undefined.

▶ Guided Practice

4A. $3n(n + 2) = 0$ **4B.** $8b^2 - 40b = 0$ **4C.** $x^2 = -10x$

Recall that there are different ways to factor an equation. After an equation is factored, it can be solved. First factor the equation. Then use the properties of equality to isolate the variable for which you are trying to solve the equation. Check by substituting in the original equation.

Key Concept Methods of Factoring

Method	Symbols
Factor Using the Distributive Property	$ax + bx + ay + by = x(a + b) + y(a + b) = (a + b)(x + y)$
Factor Quadratic Trinomials	$x^2 + bx + c = (x + m)(x + p)$
Factor Differences of Squares	$a^2 - b^2 = (a + b)(a - b)$
Factor Perfect Squares	$a^2 + 2ab + b^2 = (a + b)^2$ $a^2 - 2ab + b^2 = (a - b)^2$

Example 4   Solve Quadratic Equations by Factoring

Solve each equation. Check your solutions.

a. $y^2 + 5y - 24 = 0$

$\quad y^2 + 5y - 24 = 0$ Original equation

$\quad (y + 8)(y - 3) = 0$ Factor.

$\quad\quad y + 8 = 0 \quad \text{or} \quad y - 3 = 0$ Zero Product Property

$\quad\quad\quad y = -8 \quad \text{or} \quad\quad y = 3$ Solve each equation.

The roots are -8 and 3.

CHECK Substitute -8 and 3 for y in the original equation.

$$y^2 + 5y - 24 = 0 \quad\quad\quad\quad y^2 + 5y - 24 = 0$$

$$(-8)^2 + 5(-8) - 24 \stackrel{?}{=} 0 \quad\quad (3)^2 + 5(3) - 24 \stackrel{?}{=} 0$$

$$64 - 40 - 24 \stackrel{?}{=} 0 \quad\quad\quad 9 + 15 - 24 \stackrel{?}{=} 0$$

$$0 = 0 \checkmark \quad\quad\quad\quad\quad\quad\quad 0 = 0 \checkmark$$

b. $3x^2 - 12 = 9x$

$\quad\quad 3x^2 - 12 = 9x$ Original equation

$\quad\quad 3x^2 - 9x - 12 = 0$ Subtract $9x$ from each side.

$\quad\quad (3x - 12)(x + 1) = 0$ Factor.

$\quad\quad\quad 3x - 12 = 0 \text{ or } x + 1 = 0$ Zero Product Property

$\quad\quad\quad\quad x = 4 \quad\quad\quad x = -1$ Solve each equation.

The roots are 4 and -1.

▶ **Guided Practice**

Solve each equation. Check your solutions.

4A. $49 - x^2 = 0$

4B. $25y^2 - 60y + 81 = 45$

4C. $x^2 - 13x + 15 = -21$

Example 5   Write Quadratic Functions Given Their Graphs

Write a quadratic function for the given graph.

Step 1 Find the factors of the related expression.

The two zeros of the graph are -4 and 3, so $(x + 4)$ and $(x - 3)$ are factors of the related expression.

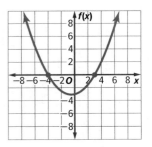

There are many quadratic functions of different shapes that have these factors, so we need to identify this specific shape. The function $f(x) = a(x + 4)(x - 3)$ represents the graph.

Step 2 Determine whether a is positive or negative.

The graph opens upward, so a must be positive.

Step 3 Determine the value of a.

Use another point on the graph to determine the value of a. The point $(4, 2)$ is on the graph.

$$f(x) = a(x + 4)(x - 3) \qquad \text{Quadratic function with roots of } -4 \text{ and } 3$$
$$2 = a(4 + 4)(4 - 3) \qquad [x, f(x)] = (4, 2)$$
$$2 = 8a \qquad \text{Simplify.}$$
$$\frac{1}{4} = a \qquad \text{Divide each side by 8.}$$

The function is $f(x) = \frac{1}{4}(x + 4)(x - 3)$ or $f(x) = \frac{1}{4}x^2 + \frac{1}{4}x - 3$.

▶ **Guided Practice**

Write a quadratic function that has a graph that contains the given points.

5A. $(-2, 0), (4, -12), (5, 0)$

5B. $(-5, 0), (-1, 0), (1, 6)$

5C. $(-4, 0), (0, 4), (4, 0)$

Check Your Understanding   ◯ = Step-by-Step Solutions begin on page R11.

Example 1   **Solve each equation. Check your solutions.**

① $x^2 = 88$ **2.** $4x^2 = 36$

3. $(x + 1)^2 = 16$ **4.** $(x - 3)^2 = 10$

Example 2   **5.** **REASONING** While painting his bedroom, Nick drops his paintbrush off his ladder from a height of 6 feet. Use the formula $h = -16t^2 + h_0$ to approximate the number of seconds to the nearest tenth it takes for the paintbrush to hit the floor.

6. SEWING Stefani is making a square quilt with a side length of $x + 2$ feet. If the quilt will have an area of 36 feet, what is the value of x?

Examples 3-4 Solve each equation. Check your solutions.

7. $3k(k + 10) = 0$

8. $(4m + 2)(3m + 9) = 0$

9. $20p^2 - 15p = 0$

10. $r^2 = 14r$

11. $a^2 - 10a + 9 = 0$

12. $b^2 + 7b - 30 = 0$

13. $5x^2 - 12x - 7 = 14$

14. $2y^2 = y + 1$

Example 5 **MP** **SENSE-MAKING** Write a quadratic function that has a graph that contains the given points.

15. $(-3, 0), (4, 0), (8, -11)$

16. $(-6, 0), (6, 24), (9, 0)$

Practice and Problem Solving

Extra Practice is found on page R9.

Example 1 Solve each equation. Check your solutions.

17. $9a^2 = 81$

18. $2b^2 = 66$

19. $c^2 - 10 = 90$

20. $d^2 + 16 = 60$

21. $m^2 = 72$

22. $n^2 = 169$

23. $(j + 5)^2 = 20$

24. $(p - 15)^2 = 121$

25. $(x - 2)^2 = 16$

26. $(y + 18)^2 = 77$

Examples 3-4 Solve each equation. Check your solutions.

27. $3b(9b - 27) = 0$

28. $2n(3n + 3) = 0$

29. $(8z + 4)(5z + 10) = 0$

30. $(7x + 3)(2x - 6) = 0$

31. $b^2 = -3b$

32. $a^2 = 4a$

33. $x^2 - 18x + 80 = 0$

34. $2y^2 - 26y + 80 = 0$

35. $z^2 - 5z - 66 = 0$

36. $3a^2 + 18a = 81$

37. $16b^2 + 24b + 20 = 15$

38. $8c^2 + 7c = 1$

39. $48x^2 + 68x + 24 = 0$

40. $14y^2 = -2y + 16$

Example 5 Write a quadratic function that has a graph that contains the given points.

41. $(-2, 0), (6, 0), (8, 15)$

42. $(-8, 0), (0, -16), (1, 0)$

Example 2 43. **SCREENS** The area A in square feet of a projected picture on a movie screen can be modeled by the equation $A = 0.25\ d^2$, where d represents the distance from a projector to a movie screen. At what distance will the projected picture have an area of 100 square feet?

44. **CHECKERBOARD** A standard checkerboard is square and made up of 64 equally sized smaller squares. The equation $576 = (x + 4)^2$ can be used to model the total area of a checkerboard. Find the length of the side of the checkerboard.

Solve each equation. Check your solutions.

45. $a^2 + 8a + 16 = 25$

46. $4b^2 = 80b - 400$

47. Write a quadratic function for the given graph.

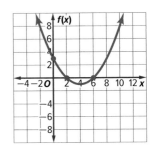

48. MULTI-STEP Ichiro wants to build an indoor swimming pool. Model A is 42 inches deep and holds 1750 cubic feet of water. The length of the pool is 5 feet more than the width. Ichiro has budgeted $200 dollars per month to heat the pool. His neighbor owns Model A, and she spends about $150 per month to heat her pool. Ichiro wants a larger pool with a depth of 42 inches. Changing the length and width by a combined 10 feet increases the heating cost by 25%. The depth does not affect the cost of heating.

    a. What size pool should Ichiro have built to fit his budget? (*Hint:* Find the percent of increase in cost.)

    b. Explain your solution process.

    c. What assumptions did you make?

49 **PROM** For prom court voting, Jen is building a ballot box that is h inches tall. The width of the box is 2 inches shorter than the height and the length is 8 inches longer than the height. If the volume is 96 cubic inches, what are the dimensions of the box?

H.O.T. Problems   Use **H**igher-**O**rder **T**hinking Skills

50. ERROR ANALYSIS Kerry says that the given graphs can be represented by the same quadratic function because they have the same zeros. Is she correct? Explain.

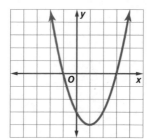 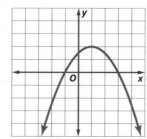

51. **MP** **SENSE-MAKING** Explain how you could you solve $x^2 + 8x + 16 = 0$ by using the Square Root Property.

52. **MP** **STRUCTURE** Given the equation $c = a^2 - ab$, for what values of a and b does $c = 0$?

53. **MP** **SENSE-MAKING** Given the equation $(ax + b)(ax - b) = 0$, solve for x. What do we know about the values of a and b?

54. WRITING IN MATH Explain how to solve a quadratic equation by using the Zero Product Property.

55. **MP** **SENSE-MAKING** The polynomial $2x^2 - 5x - 3$ has $(2x + 1)$ and $(x - 3)$ as its factors. What are the solutions to the equation $2x^2 - 5x - 3 = 0$?

Write an equation that has the given roots.

56. $-6, 4$ **57.** $0, 7$ **58.** $-3, 1, 6$

59. SHOTPUT An athlete throws a shot put with an initial upward velocity of 29 feet per second and from an initial height of 6 feet.

    a. Write an equation that models the height of the shot put in feet with respect to time in seconds.

    b. After how many seconds will the shot put hit the ground?

60. For what value of c does $4x^2 - 32x + c = 0$ have only one real solution?

61. Solve $x^4 - 18x^2 + 81 = 0$.

Preparing for Assessment

62. The rectangle shown has an area of 60 square centimeters. **MP** 5

x

$x - 7$

a. What is the length x of the rectangle?

- **A**   5 cm
- **B**   12 cm
- **C**   33.5 cm
- **D**   46 cm

b. Find the perimeter of the rectangle.

[   ]

63. A soccer ball is kicked into the air. The height of the soccer ball can be modeled by the equation $h = -16t^2 + 24t$, where h is the height of the ball at t seconds. What is the value of t when $h = 0$? Select all that apply. **MP** 1, 4

- **A**   0 s
- **B**   0.5 s
- **C**   1 s
- **D**   1.5 s
- **E**   2 s

64. Write a quadratic function that has a graph that contains the points $(-7, 0)$, $(4, 0)$, and $(5, 18)$. **MP** 6, 7

[   ]

65. Solve $2(x^2 + 8) = 16$. **MP** 6, 7

[   ]

66. Suppose a maple tree has a leaf that is 60 feet from the ground. The equation $h = -16t^2 + 60$ describes the height h, in feet, of the leaf t seconds after it falls from the tree. How many seconds will it take the leaf to fall to the ground? **MP** 4, 6

- **A** $-\dfrac{\sqrt{15}}{2}$ s
- **B** $\dfrac{15}{4}$ s
- **C** $\dfrac{\sqrt{15}}{4}$ s
- **D** $\dfrac{\sqrt{15}}{2}$ s

67. MULTI-STEP The area of the triangle shown, in square inches, is equivalent to the area of another triangle represented by the expression $2x^2 - 16x - 31$. **MP** 1, 6

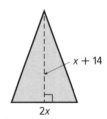

$x + 14$

$2x$

a. Write an equation relating the areas of the triangles.

[   ]

b. Find the height of the triangle shown.

[   ]

c. Are both solutions valid? Explain.

68. Which are solutions of the equation $(x - 2)^2 = 16$? Select all that apply. **MP** 7

- **A** -4
- **B** -2
- **C** 2
- **D** 4
- **E** 6

Then	Now	Why?
● You solved quadratic equations by using the Square Root Property and by factoring.	**1** Solve quadratic equations by completing the square. **2** Identify key features of quadratic functions by writing quadratic equations in vertex form.	● In competitions, skateboarders may launch themselves from a half pipe into the air to perform tricks. The equation $h = -16t^2 + 20t + 12$ can be used to model their height, in feet, after t seconds. To find how long a skateboarder is in the air if he is 25 feet above the half pipe, you can solve $25 = -16t^2 + 20t + 12$ by using a method called completing the square.

New Vocabulary

completing the square

Mathematical Practices

1 Make sense of problems and persevere in solving them.

2 Reason abstractly and quantitatively.

4 Model with mathematics.

7 Look for and make use of structure.

1 Complete the Square You have previously solved equations by taking the square root of each side. This method worked only because the expression on the left-hand side was a perfect square. In perfect square trinomials in which the leading coefficient is 1, there is a relationship between the **coefficient of the x-term** and the **constant term**.

$$(x + 5)^2 = x^2 + 2(5)(x) + 5^2$$
$$= x^2 + 10x + 25$$

Notice that $\left(\frac{10}{2}\right)^2 = 25$. To get the constant term, divide the coefficient of the x-term by 2 and square the result. Any quadratic expression in the form $x^2 + bx$ can be made into a perfect square by using a method called **completing the square**.

Key Concept Completing the Square

Words	To complete the square for any quadratic expression of the form $x^2 + bx$, follow the steps below.
	Step 1 Find one half of b, the coefficient of x.
	Step 2 Square the result in Step 1.
	Step 3 Add the result of Step 2 to $x^2 + bx$.
Symbols	$x^2 + bx + \left(\frac{b}{2}\right)^2 = \left(x + \frac{b}{2}\right)^2$

Example 1 Complete the Square

Find the value of c that makes $x^2 + 4x + c$ a perfect square trinomial.

Method 1 Use algebra tiles.

Arrange the tiles for $x^2 + 4x$ so that the two sides of the figure are congruent.

To make the figure a square, add 4 positive 1-tiles.

Method 2 Use the complete the square algorithm.

Step 1	Find $\frac{1}{2}$ of 4.	$\frac{4}{2} = 2$
Step 2	Square the result in Step 1.	$2^2 = 4$
Step 3	Add the result of Step 2 to $x^2 + 4x$.	$x^2 + 4x + 4$

Thus, $c = 4$. Notice that $x^2 + 4x + 4 = (x + 2)^2$.

▶ **Guided Practice**

  1. Find the value of c that makes $r^2 - 8r + c$ a perfect square trinomial.

You can complete the square to solve quadratic equations. First, you must isolate the x^2- and bx-terms.

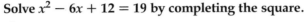

Example 2   Solve an Equation by Completing the Square

Solve $x^2 - 6x + 12 = 19$ by completing the square.

$x^2 - 6x + 12 = 19$	Original equation
$x^2 - 6x = 7$	Subtract 12 from each side.
$x^2 - 6x + 9 = 7 + 9$	Since $\left(\frac{-6}{2}\right)^2 = 9$, add 9 to each side.
$(x - 3)^2 = 16$	Factor $x^2 - 6x + 9$.
$x - 3 = \pm 4$	Take the square root of each side.
$x = 3 \pm 4$	Add 3 to each side.

$x = 3 + 4$ or $x = 3 - 4$	Separate the solutions.
$= 7$ $= -1$	The solutions are 7 and -1.

▶ **Guided Practice**

  2. Solve $x^2 - 12x + 3 = 8$ by completing the square.

To solve a quadratic equation in which the leading coefficient is not 1, divide each term by the coefficient. Then isolate the x^2- and x-terms and complete the square.

Example 3   Equation with a ≠ 1

Solve $-2x^2 + 8x - 18 = 0$ by completing the square.

$-2x^2 + 8x - 18 = 0$	Original equation
$\dfrac{-2x^2 + 8x - 18}{-2} = \dfrac{0}{-2}$	Divide each side by -2.
$x^2 - 4x + 9 = 0$	Simplify.
$x^2 - 4x = -9$	Subtract 9 from each side.
$x^2 - 4x + 4 = -9 + 4$	Since $\left(\frac{-4}{2}\right)^2 = 4$, add 4 to each side.
$(x - 2)^2 = -5$	Factor $x^2 - 4x + 4$.

No real number has a negative square. So, this equation has no real solutions.

▶ **Guided Practice**

  3. Solve $3x^2 - 9x - 3 = 21$ by completing the square.

JERSEYS The senior class at Bay High School buys jerseys to wear to the football games. The cost of the jerseys can be modeled by the equation $C = 0.1x^2 + 2.4x + 25$, where C is the amount it costs to buy x jerseys. How many jerseys can they purchase for \$430?

The seniors have \$430, so set the equation equal to 430 and complete the square.

$0.1x^2 + 2.4x + 25 = 430$	Original equation
$\dfrac{0.1x^2 + 2.4x + 25}{0.1} = \dfrac{430}{0.1}$	Divide each side by 0.1.
$x^2 + 24x + 250 = 4300$	Simplify.
$x^2 + 24x + 250 - 250 = 4300 - 250$	Subtract 250 from each side.
$x^2 + 24x = 4050$	Simplify.
$x^2 + 24x + 144 = 4050 + 144$	Since $\left(\frac{24}{2}\right)^2 = 144$, add 144 to each side.
$x^2 + 24x + 144 = 4194$	Simplify.
$(x + 12)^2 = 4194$	Factor $x^2 + 24x + 144$.
$x + 12 = \pm\sqrt{4194}$	Take the square root of each side.
$x = -12 \pm\sqrt{4194}$	Subtract 12 from each side.

Use a calculator to approximate each value of x.

$x = -12 + \sqrt{4194}$	or $x = -12 - \sqrt{4194}$	Separate the solutions.
≈ 52.8	≈ -76.8	Evaluate.

Since you cannot buy a negative number of jerseys, the negative solution is not reasonable. The seniors can afford to buy 52 jerseys.

▶ **Guided Practice**

4. If the senior class were able to raise \$620, how many jerseys could they buy?

Real-World Link

The annual "Battle for Paul Bunyan's Axe" is one of the oldest Division 1-A college football rivalry games. It takes place between the University of Minnesota and the University of Wisconsin. The rivalry began in 1890 and the trophy was introduced in 1948.

Source: Bleacher Report

2 Vertex Form In Lesson 9-1, you graphed quadratic functions written in standard form. A quadratic function can also be written in vertex form, $y = a(x - h)^2 + k$, where $a \neq 0$. In this form, you can identify key features of the graph of the function.

🔑 Key Concept Vertex Form

For a quadratic function in vertex form, $y = a(x - h)^2 + k$, the following are true.

- The vertex of the graph is the point (h, k).

- The graph opens up and has a minimum value of k when $a > 0$.

- The graph opens down and has a maximum value of k when $a < 0$.

- The axis of symmetry is the line $x = h$.

- The zeros are the x-intercepts of the graph.

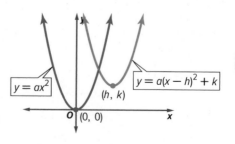

Example 5   Write Functions in Vertex Form

Write $y = -2x^2 + 20x - 42$ in vertex form.

$$y = -2x^2 + 20x - 42 \qquad \text{Original Equation}$$

$$y + 42 = -2x^2 + 20x \qquad \text{Add 42 to each side.}$$

$$y + 42 = -2(x^2 - 10x) \qquad \text{Factor out } -2.$$

$$y + 42 - 50 = -2(x^2 - 10x + 25) \qquad \text{Since } -2\left(\frac{10}{2}\right)^2 = -50, \text{ add } -50 \text{ to each side.}$$

$$y - 8 = -2(x - 5)^2 \qquad \text{Factor } x^2 - 10x + 25.$$

$$y = -2(x - 5)^2 + 8 \qquad \text{Add 8 to each side.}$$

The vertex form of the function is $y = -2(x - 5)^2 + 8$.

▶ **Guided Practice**

5. $y = x^2 + 2x + 4$

Check Your Understanding

⬤ = Step-by-Step Solutions begin on page R11.

Example 1   Find the value of c that makes each trinomial a perfect square.

1 $x^2 - 18x + c$ **2.** $x^2 + 22x + c$

3. $x^2 + 9x + c$ **4.** $x^2 - 7x + c$

Examples 2–3   Solve each equation by completing the square. Round to the nearest tenth if necessary.

5. $x^2 + 4x = 6$ **6.** $x^2 - 8x = -9$

7. $4x^2 + 9x - 1 = 0$ **8.** $-2x^2 + 10x + 22 = 4$

Example 4   **9.** **MODELING** Collin is building a deck on the back of his family's house. He has enough lumber for the deck to be 144 square feet. The length should be 10 feet more than its width. What should the dimensions of the deck be?

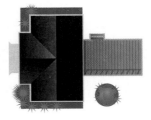

Example 5   Write each quadratic function in vertex form.

10. $y = x^2 - 12x + 16$

11. $y = x^2 + 18x + 36$

12. $y = 3x^2 + 12x - 39$

Practice and Problem Solving

Extra Practice is on page R9.

Example 1   Find the value of c that makes each trinomial a perfect square.

13. $x^2 + 26x + c$ **14.** $x^2 - 24x + c$ **15.** $x^2 - 19x + c$

16. $x^2 + 17x + c$ **17.** $x^2 + 5x + c$ **18.** $x^2 - 13x + c$

Examples 2-3 Solve each equation by completing the square. Round to the nearest tenth if necessary.

19. $x^2 + 6x - 16 = 0$

20. $x^2 - 2x - 14 = 0$

21. $x^2 - 8x - 1 = 8$

22. $x^2 + 3x + 21 = 22$

23. $x^2 - 11x + 3 = 5$

24. $5x^2 - 10x = 23$

25. $2x^2 - 2x + 7 = 5$

26. $3x^2 + 12x + 81 = 15$

Example 4

27. **(MP) MODELING** The price p in dollars for a particular stock can be modeled by the quadratic equation $p = 3.5t - 0.05t^2$, where t represents the number of days after the stock is purchased. When is the stock worth $60?

MODELING Find the value of x for each figure. Round to the nearest tenth if necessary.

28. $A = 45$ in^2

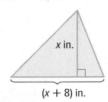

x in.

$(x + 8)$ in.

29. $A = 110$ ft^2

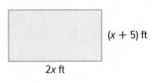

$(x + 5)$ ft

$2x$ ft

30. **(MP) MODELING** Find the area of the triangle below.

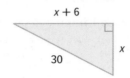

$x + 6$

x

30

Example 5 Write each quadratic function in vertex form.

31. $y = x^2 - 8x - 10$

32. $y = -x^2 + 2x - 3$

Solve each equation by completing the square. Round to the nearest tenth if necessary.

33. $0.2x^2 - 0.2x - 0.4 = 0$

34. $0.5x^2 = 2x - 0.3$

35. $2x^2 - \dfrac{11}{5}x = -\dfrac{3}{10}$

36. $\dfrac{2}{3}x^2 - \dfrac{4}{3}x = \dfrac{5}{6}$

37. MODELING The height of an object t seconds after it is dropped is given by the equation $h = -\dfrac{1}{2}gt^2 + h_0$, where h_0 is the initial height and g is the acceleration due to gravity. The acceleration due to gravity near the surface of Mars is 3.73 m/s^2, while on Earth it is 9.8 m/s^2. Suppose an object is dropped from an initial height of 120 meters above the surface of each planet.

a. On which planet would the object reach the ground first?

b. How long would it take the object to reach the ground on each planet? Round each answer to the nearest tenth.

c. Do the times that it takes the object to reach the ground seem reasonable? Explain your reasoning.

Write each quadratic function in vertex form.

38. $y = -2x^2 + 16x - 29$

39. $y = 3x^2 + 24x + 45$

40. MODELING Before she begins painting a picture, Donna stretches her canvas over a wood frame. The frame has a length of 60 inches and a width of 4 inches. She has enough canvas to cover 480 square inches. Donna decides to increase the dimensions of the frame. If the increase in the length is 10 times the increase in the width, what will the dimensions of the frame be?

41. MULTIPLE REPRESENTATIONS In this problem, you will investigate a property of quadratic equations.

   a. Tabular Copy the table shown and complete the second column.

   b. Algebraic Set each trinomial equal to zero, and solve the equation by completing the square. Complete the last column of the table with the number of real roots of each equation.

   c. Verbal Compare the number of real roots of each equation to the result in the $b^2 - 4ac$ column. Is there a relationship between these values? If so, describe it.

   d. Analytical Predict how many real solutions $2x^2 - 9x + 15 = 0$ will have. Verify your prediction by solving the equation.

Trinomial	$b^2 - 4ac$	Number of Roots
$x^2 - 8x + 16$	0	1
$2x^2 - 11x + 3$		
$3x^2 + 6x + 9$		
$x^2 - 2x + 7$		
$x^2 + 10x + 25$		
$x^2 + 3x - 12$		

H.O.T. Problems   Use **H**igher-**O**rder **T**hinking Skills

42. **MP** **ARGUMENTS** Given $y = ax^2 + bx + c$ with $a \neq 0$, derive the equation for the axis of symmetry by completing the square and rewriting the equation in the form $y = a(x - h)^2 + k$.

43 **CHALLENGE** Determine the number of solutions $x^2 + bx = c$ has if $c < -\left(\dfrac{b}{2}\right)^2$. Explain.

44. WHICH ONE DOESN'T BELONG? Identify the expression that does not belong with the other three. Explain your reasoning.

$n^2 - n + \dfrac{1}{4}$	$n^2 + n + \dfrac{1}{4}$	$n^2 - \dfrac{2}{3}n + \dfrac{1}{9}$	$n^2 + \dfrac{1}{3}n + \dfrac{1}{9}$

45. **MP** **REASONING** Write an equation in vertex form for the parabola shown below. Justify your reasoning.

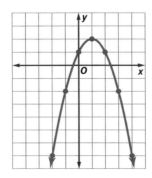

46. MULTI-STEP Consider $3x^2 - 24x = 51$. ⓂⓅ 1,7

a. What is the first step to solve the equation by completing the square?

b. What value should be added to each side to complete the square.

◯ **A** -16

◯ **B** -4

◯ **C** 16

◯ **D** 144

c. Which of the following are solutions to the equation? Select all solutions.

☐ **A** 3

☐ **B** $\dfrac{17}{3}$

☐ **C** $4 + \sqrt{33}$

☐ **D** $-4 + \sqrt{33}$

☐ **E** $4 - \sqrt{33}$

☐ **F** $-4 - \sqrt{33}$

d. Which is a key feature of the graph of the equation?

◯ **A** The axis of symmetry is $x = 4$.

◯ **B** The vertex is $(-4, -33)$.

◯ **C** The graph opens down.

◯ **D** The minimum value is 33.

47. Which of the following is least likely to be a step in solving the equation $x^2 + 10x + 10 = 66$ by completing the square? ⓂⓅ 3,7

◯ **A** Subtract 10 from both sides of the equation.

◯ **B** Subtract 25 from both sides of the equation.

◯ **C** Add $\left(\dfrac{10}{2}\right)^2$ to both sides of the equation.

◯ **D** Factor $x^2 + 10x + 25$.

48. Which equation has no real solutions? ⓂⓅ 1,7

◯ **A** $x^2 - 2x = 2$

◯ **B** $x^2 - 2x = 3$

◯ **C** $x^2 - 3x = -2$

◯ **D** $x^2 + 3x = -3$

49. Which equation has solutions of -1 and 3? ⓂⓅ 1,7

◯ **A** $x^2 - x - 7 = -4$

◯ **B** $x^2 - 2x + 7 = -4$

◯ **C** $x^2 - 2x - 7 = -4$

◯ **D** $2x^2 - 4x - 14 = -4$

50. What value should be added to $x^2 - 20$ in order to complete the square? ⓂⓅ 2

51. Solve $-3x + 30x - 72 = 0$ by completing the square. ⓂⓅ 2

52. Determine whether the following statement is *true* or *false*. Justify your answer. ⓂⓅ 3

To complete the square for $x^2 + bx$, add $\dfrac{b}{2}$ to the expression.

53. Chris solved the equation $x^2 - 6x - 2 = 6$ by completing the square. Which best describes the solution or solutions? ⓂⓅ 6

◯ **A** Rounded to the nearest tenth, the solutions are -4.1 and 4.1.

◯ **B** The solution is 3.

◯ **C** Rounded to the nearest tenth, the solutions are -1.1 and 7.1.

◯ **D** This equation has no real solutions.

Algebra Lab
Finding the Maximum or Minimum Value

In Lesson 9-5, we learned about the vertex form of the equation of a quadratic function. You will now learn how to write equations in vertex form and use them to identify key characteristics of the graphs of quadratic functions.

Mathematical Practices
 8 Look for and express regularity in repeated reasoning

Activity 1   Find a Minimum

Work cooperatively. Write $y = x^2 + 4x - 10$ in vertex form. Identify the axis of symmetry, extrema, and zeros. Then graph the function.

Step 1 Complete the square to write the function in vertex form.

$y = x^2 + 4x - 10$	Original function
$y + 10 = x^2 + 4x$	Add 10 to each side.
$y + 10 + 4 = x^2 + 4x + 4$	Since $\left(\frac{4}{2}\right)^2 = 4$, add 4 to each side.
$y + 14 = (x + 2)^2$	Factor $x^2 + 4x + 4$.
$y = (x + 2)^2 - 14$	Subtract 14 from each side to write in vertex form.

Step 2 Identify the axis of symmetry and extrema based on the equation in vertex form. The vertex is at (h, k) or $(-2, -14)$. Since there is no negative sign before the x^2-term, the parabola opens up and has a minimum at $(-2, -14)$. The equation of the axis of symmetry is $x = -2$.

Step 3 Solve for x to find the zeros.

$(x + 2)^2 - 14 = 0$	Vertex form, $y = 0$
$(x + 2)^2 = 14$	Add 14 to each side.
$x + 2 = \pm\sqrt{14}$	Take square root of each side.
$x \approx -5.74$ or 1.74	Subtract 2 from each side.

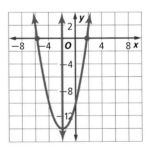

The zeros are approximately -5.74 and 1.74.

Step 4 Use the key features to graph the function.

There may be a negative coefficient before the quadratic term. When this is the case, the parabola will open down and have a maximum.

Activity 2   Find a Maximum

Work cooperatively. Write $y = -x^2 + 6x - 5$ in vertex form. Identify the axis of symmetry, extrema, and zeros. Then graph the function.

Step 1 Complete the square to write the equation of the function in vertex form.

$y = -x^2 + 6x - 5$	Original function
$y + 5 = -x^2 + 6x$	Add 5 to each side.
$y + 5 = -(x^2 - 6x)$	Factor out -1.
$y + 5 - 9 = -(x^2 - 6x + 9)$	Since $\left(\frac{6}{2}\right)^2 = 9$, add -9 to each side.
$y - 4 = -(x - 3)^2$	Factor $x^2 - 6x + 9$.
$y = -(x - 3)^2 + 4$	Add 4 to each side to write in vertex form.

Algebra Lab
Finding the Maximum or Minimum Value *Continued*

Step 2 Identify the axis of symmetry and extrema based on the equation in vertex form. The vertex is at (h, k) or $(3, 4)$. Since there is a negative sign before the x^2-term, the parabola opens down and has a maximum at $(3, 4)$. The equation of the axis of symmetry is $x = 3$.

Step 3 Solve for x to find the zeros.

$$0 = -(x - 3)^2 + 4 \qquad \text{Vertex form, } y = 0$$
$$(x - 3)^2 = 4 \qquad \text{Add } (x - 3)^2 \text{ to each side.}$$
$$x - 3 = \pm 2 \qquad \text{Take the square root of each side.}$$
$$x = 5 \text{ or } 1 \qquad \text{Add 3 to each.}$$

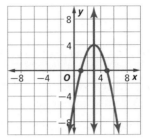

Step 4 Use the key features to graph the function.

Analyze the Results
Work cooperatively.

1. Why do you need to complete the square to write the equation of a quadratic function in vertex form?

Write each function in vertex form. Identify the axis of symmetry, extrema, and zeros. Then graph the function.

2. $y = x^2 + 6x$

3. $y = x^2 - 8x + 6$

4. $y = x^2 + 2x - 12$

5. $y = x^2 + 6x + 8$

6. $y = x^2 - 4x + 3$

7. $y = x^2 - 2.4x - 2.2$

8. $y = -4x^2 + 16x - 11$

9. $y = 3x^2 - 12x + 5$

10. $y = -x^2 + 6x - 5$

Activity 3 — Use Extrema in the Real World

DIVING Alexis jumps from a diving platform upward and outward before diving into the pool. The function $h = -9.8t^2 + 4.9t + 10$, where h is the height of the diver in meters above the pool after t seconds approximates Alexis's dive. Graph the function, then find the maximum height that she reaches and the equation of the axis of symmetry.

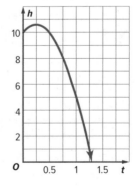

Step 1 Graph the function.

Step 2 Complete the square to write the eqution of the function in vertex form.
$$h = -9.8t^2 + 4.9t + 10$$
$$h = -9.8(t - 0.25)^2 + 10.6125$$

Step 3 The vertex is at $(0.25, 10.6125)$, so the maximum height is 10.6125 meters. The equation of the axis of symmetry is $x = 0.25$.

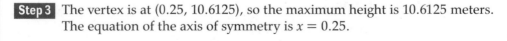
Exercises

11. **SOFTBALL** Jenna throws a ball in the air. The function $h = -16t^2 + 40t + 5$, where h is the height in feet and t represents the time in seconds, approximates Jenna's throw. Graph the function, then find the maximum height of the ball and the equation of the axis of symmetry. When does the ball hit the ground?

Use a table of values to graph each equation. State the domain and range. (Lesson 9-1)

1. $y = x^2 + 3x + 1$

2. $y = 2x^2 - 4x + 3$

3. $y = -x^2 - 3x - 3$

4. $y = -3x^2 - x + 1$

Consider $y = x^2 - 5x + 4$. (Lesson 9-1)

5. Write the equation of the axis of symmetry.

6. Find the coordinates of the vertex. Is it a maximum or minimum point?

7. Graph the function.

8. **SOCCER** A soccer ball is kicked from ground level with an initial upward velocity of 90 feet per second. The equation $h = -16t^2 + 90t$ gives the height h of the ball after t seconds. (Lesson 9-1)

a. What is the height of the ball after one second?

b. How many seconds will it take for the ball to reach its maximum height?

c. When is the height of the ball 0 feet? What do these points represent in this situation?

d. 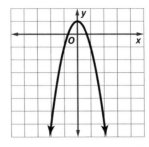 What mathematical practice did you use to solve this problem?

Describe how the graph of each function is related to the graph of $f(x) = x^2$. (Lesson 9-2)

9. $g(x) = x^2 + 3$

10. $h(x) = 2x^2$

11. $g(x) = x^2 - 6$

12. $h(x) = \frac{1}{5}x^2$

13. $g(x) = -x^2 + 1$

14. $h(x) = -\frac{5}{8}x^2$

15. **CONSTRUCTION** Christopher is repairing the roof on a shed. He accidentally dropped a box of nails from a height of 14 feet. This is represented by the equation $h = -16t^2 + 14$, where h is the height in feet and t is the time in seconds. Describe how the graph is related to $h = t^2$. (Lesson 9-2)

16. **MULTIPLE CHOICE** Which is an equation for the function shown in the graph? (Lesson 9-2)

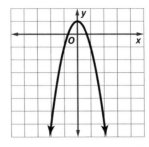

A $y = -2x^2$

B $y = 2x^2 + 1$

C $y = x^2 - 1$

D $y = -2x^2 + 1$

Solve each equation by graphing. If integral roots cannot be found, estimate the roots to the nearest tenth. (Lesson 9-3)

17. $x^2 + 5x + 6 = 0$

18. $x^2 + 8 = -6x$

19. $-x^2 + 3x - 1 = 0$

20. $x^2 = 12$

21. **PARTIES** Della's parents are throwing a Sweet 16 party for her. At 10:00, a ball will slide 25 feet down a pole and light up. A function that models the drop is $h = -t^2 + 5t + 25$, where h is height in feet of the ball after t seconds. How many seconds will it take for the ball to reach the bottom of the pole? (Lesson 9-3)

25 ft

Solve each equation by factoring. (Lesson 9-4)

22. $x^2 + 14x + 49 = 0$

23. $x^2 - 2x = 48$

24. $2x^2 + 5x - 12 = 0$

Solve each equation by completing the square. Round to the nearest tenth. (Lesson 9-5)

25. $x^2 + 4x + 2 = 0$

26. $x^2 - 2x - 10 = 0$

27. $2x^2 + 4x - 5 = 7$

Solving Quadratic Equations by Using the Quadratic Formula

:: **Then** ::**Now** :: **Why?**

- You solved quadratic equations by completing the square.

1 Solve quadratic equations by using the Quadratic Formula.

2 Use the discriminant to determine the number of solutions of a quadratic equation.

- For adult women, the normal systolic blood pressure P in millimeters of mercury (mm Hg) can be modeled by $P = 0.01a^2 + 0.05a + 107$, where a is age in years. This equation can be used to approximate the age of a woman with a certain systolic blood pressure. However, it would be difficult to solve by factoring, graphing, or completing the square.

 New Vocabulary
Quadratic Formula

 Mathematical Practices
6 Attend to precision.

1 **Quadratic Formula** By completing the square for the quadratic equation $ax^2 + bx + c = 0$, $a \neq 0$, you can derive the Quadratic Formula. The **Quadratic Formula** can be used to find the solutions of any quadratic equation.

$$ax^2 + bx + c = 0$$
  Standard quadratic equation

$$x^2 + \frac{b}{a}x + \frac{c}{a} = 0$$
  Divide each side by a.

$$x^2 + \frac{b}{a}x = -\frac{c}{a}$$
  Subtract $\frac{c}{a}$ from each side.

$$x^2 + \frac{b}{a}x + \left(\frac{b}{2a}\right)^2 = -\frac{c}{a} + \left(\frac{b}{2a}\right)^2$$
  Complete the square by adding $\left(\frac{b}{2a}\right)^2$ to each side.

$$x^2 + \frac{b}{a}x + \left(\frac{b}{2a}\right)^2 = -\frac{c}{a} + \frac{b^2}{4a^2}$$
  Simplify the right side.

$$\left(x + \frac{b}{2a}\right)^2 = -\frac{c}{a} + \frac{b^2}{4a^2}$$
  Factor the left side.

$$\left(x + \frac{b}{2a}\right)^2 = \frac{b^2 - 4ac}{4a^2}$$
  Find a common denominator and simplify the right side.

$$x + \frac{b}{2a} = \pm \frac{\sqrt{b^2 - 4ac}}{2a}$$
  Square Root Property

$$x = -\frac{b}{2a} \pm \frac{\sqrt{b^2 - 4ac}}{2a}$$
  Subtract $\frac{b}{2a}$ from each side.

$$x = \frac{-b \pm \sqrt{b^2 - 4ac}}{2a}$$
  Simplify.

The equation $x = \dfrac{-b \pm \sqrt{b^2 - 4ac}}{2a}$ is the Quadratic Formula.

Key Concept The Quadratic Formula

The solutions of a quadratic equation $ax^2 + bx + c = 0$, where $a \neq 0$, are given by the Quadratic Formula.

$$x = \frac{-b \pm \sqrt{b^2 - 4ac}}{2a}$$

Example 1   Use the Quadratic Formula

Solve each equation by using the Quadratic Formula.

a. $x^2 - 12x = -20$

Step 1 Rewrite the equation in standard form.

$$x^2 - 12x = -20 \qquad \text{Original equation}$$

$$x^2 - 12x + 20 = 0 \qquad \text{Add 20 to each side.}$$

Step 2 Apply the Quadratic Formula.

$$x = \frac{-b \pm \sqrt{b^2 - 4ac}}{2a} \qquad \text{Quadratic Formula}$$

$$= \frac{-(-12) \pm \sqrt{(-12)^2 - 4(1)(20)}}{2(1)} \qquad a = 1, b = -12, \text{ and } c = 20$$

$$= \frac{12 \pm \sqrt{144 - 80}}{2} \qquad \text{Multiply.}$$

$$= \frac{12 \pm \sqrt{64}}{2} \text{ or } \frac{12 \pm 8}{2} \qquad \text{Subtract and take the square root.}$$

$$x = \frac{12 - 8}{2} \text{ or } x = \frac{12 + 8}{2} \qquad \text{Separate the solutions.}$$

$$= 2 \qquad\qquad = 10 \qquad \text{The solutions are 2 and 10.}$$

b. $x^2 - 6x = 7$

Step 1 Rewrite the equation in standard form.

$$x^2 - 6x = 7 \qquad \text{Original equation}$$

$$x^2 - 6x - 7 = 0 \qquad \text{Subtract 7 from each side.}$$

Step 2 Apply the Quadratic Formula.

$$x = \frac{-b \pm \sqrt{b^2 - 4ac}}{2a} \qquad \text{Quadratic Formula}$$

$$= \frac{-(-6) \pm \sqrt{(-6)^2 - 4(1)(7)}}{2(1)} \qquad a = 1, b = -6, \text{ and } c = -7$$

$$= \frac{6 \pm \sqrt{36 - 28}}{2} \qquad \text{Multiply.}$$

$$= \frac{6 \pm \sqrt{64}}{2} \text{ or } \frac{6 \pm 8}{2} \qquad \text{Subtract and take the square root.}$$

$$x = \frac{6 + 8}{2} \text{ or } x = \frac{6 - 8}{2} \qquad \text{Separate the solutions.}$$

$$x = 7 \text{ or } x = -1 \qquad \text{The solutions are } -1 \text{ and 7.}$$

▶ **Guided Practice**

1A. $2x^2 + 9x = 18$ **1B.** $-3x^2 + 5x = -2$

1C. $x^2 + 14x + 40 = 0$ **1D.** $3x^2 + 2x - 1 = 0$

The solutions of quadratic equations are not always integers.

Example 2   Use the Quadratic Formula

Solve each equation by using the Quadratic Formula. Round to the nearest tenth if necessary.

a. $3x^2 + 5x - 12 = 0$

For this equation, $a = 3$, $b = 5$, and $c = -12$.

$$x = \frac{-b \pm \sqrt{b^2 - 4ac}}{2a}$$ Quadratic Formula

$$= \frac{-(5) \pm \sqrt{(5)^2 - 4(3)(-12)}}{2(3)}$$ $a = 3$, $b = 5$, and $c = -12$

$$= \frac{-5 \pm \sqrt{25 + 144}}{6}$$ Multiply.

$$= \frac{-5 \pm \sqrt{169}}{6} \text{ or } \frac{-5 \pm 13}{6}$$ Add and simplify.

$$x = \frac{-5 - 13}{6} \text{ or } x = \frac{-5 + 13}{6}$$ Separate the solutions.

$$= -3 \qquad\qquad = \frac{4}{3}$$ Simplify.

The solutions are -3 and $\frac{4}{3}$.

b. $10x^2 - 5x = 25$

Step 1   Rewrite the equation in standard form.

$$10x^2 - 5x = 25$$ Original equation

$$10x^2 - 5x - 25 = 0$$ Subtract 25 from each side.

Step 2   Apply the Quadratic Formula.

$$x = \frac{-b \pm \sqrt{b^2 - 4ac}}{2a}$$ Quadratic Formula

$$= \frac{-(-5) \pm \sqrt{(-5)^2 - 4(10)(-25)}}{2(10)}$$ $a = 10$, $b = -5$, and $c = -25$

$$= \frac{5 \pm \sqrt{25 + 1000}}{20}$$ Multiply.

$$= \frac{5 \pm \sqrt{1025}}{20}$$ Add.

$$= \frac{5 - \sqrt{1025}}{20} \text{ or } \frac{5 + \sqrt{1025}}{20}$$ Separate the solutions.

$$\approx -1.4 \qquad\qquad \approx 1.9$$ Simplify.

The solutions are about -1.4 and 1.9.

> **Go Online!**
>
> You will want to reference the Quadratic Formula often. Log into your eStudent Edition to bookmark this lesson.

> **Study Tip**
>
> **MP** Precision In Example 2, the number $\sqrt{1025}$ is irrational, so the calculator can only give you an approximation of its value.
> So, the exact answer in Example 2 is $\frac{5 \pm \sqrt{1025}}{20}$. The numbers -1.4 and 1.9 are approximations.

▸ **Guided Practice**

2A. $4x^2 - 24x + 35 = 0$ **2B.** $3x^2 - 2x - 9 = 0$

You can solve quadratic equations by using one of many equivalent methods. No one way is always best.

> ## Watch Out!
>
> **Solutions** Remember that the methods of factoring, completing the square, or using the quadractic formula to solve quadratic equations are equivalent. Each method should produce the same solution.

Example 3   Solve Quadratic Equations Using Different Methods

Solve $x^2 - 4x = 12$.

Method 1   Graphing

Rewrite the equation in standard form.

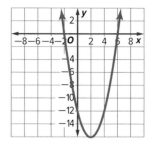

$$x^2 - 4x = 12 \qquad \text{Original equation}$$

$$x^2 - 4x - 12 = 0 \qquad \text{Subtract 12 from each side.}$$

Graph the related function $f(x) = x^2 - 4x - 12$.
Locate the x-intercepts of the graph.

The solutions are -2 and 6.

Method 2   Factoring

$$\begin{aligned} x^2 - 4x &= 12 & &\text{Original equation} \\ x^2 - 4x - 12 &= 0 & &\text{Subtract 12 from each side.} \\ (x - 6)(x + 2) &= 0 & &\text{Factor.} \\ x - 6 = 0 \text{ or } x + 2 &= 0 & &\text{Zero Product Property} \\ x = 6 \qquad x &= -2 & &\text{Solve for } x. \end{aligned}$$

Method 3   Completing the Square

The equation is in the correct form to complete the square, since the leading coefficient is 1 and the x^2 and x terms are isolated.

$$\begin{aligned} x^2 - 4x &= 12 & &\text{Original equation} \\ x^2 - 4x + 4 &= 12 + 4 & &\text{Since } \left(\tfrac{-4}{2}\right)^2 = 4 \text{, add 4 to each side.} \\ (x - 2)^2 &= 16 & &\text{Factor } x^2 - 4x + 4. \\ x - 2 &= \pm 4 & &\text{Take the square root of each side.} \\ x &= 2 \pm 4 & &\text{Add 2 to each side.} \\ x = 2 + 4 \text{ or } x &= 2 - 4 & &\text{Separate the solutions.} \\ = 6 \qquad &= -2 & &\text{Simplify.} \end{aligned}$$

Method 4   Quadratic Formula

From Method 1, the standard form of the equation is $x^2 - 4x - 12 = 0$.

$$\begin{aligned} x &= \frac{-b \pm \sqrt{b^2 - 4ac}}{2a} & &\text{Quadratic Formula} \\ &= \frac{-(-4) \pm \sqrt{(-4)^2 - 4(1)(-12)}}{2(1)} & &a = 1, b = -4, \text{ and } c = -12 \\ &= \frac{4 \pm \sqrt{16 + 48}}{2} & &\text{Multiply.} \\ &= \frac{4 \pm \sqrt{64}}{2} \text{ or } \frac{4 \pm 8}{2} & &\text{Add and simplify.} \\ x &= \frac{4 - 8}{2} \text{ or } x = \frac{4 + 8}{2} & &\text{Separate the solutions.} \\ &= -2 \qquad\qquad = 6 & &\text{Simplify.} \end{aligned}$$

> ## Guided Practice
>
> **Solve each equation.**
>
> **3A.** $2x^2 - 17x + 8 = 0$ **3B.** $4x^2 - 4x - 11 = 0$

Concept Summary Solving Quadratic Equations

Method	When to Use
Factoring	Use when the constant term is 0 or if the factors are easily determined. Not all equations are factorable.
Graphing	Use when an approximate solution is sufficient.
Using Square Roots	Use when an equation can be written in the form $x^2 = n$. Can only be used if the equation has no x-term.
Completing the Square	Can be used for any equation $ax^2 + bx + c = 0$, but is simplest to apply when b is even and $a = 1$.
Quadratic Formula	Can be used for any equation $ax^2 + bx + c = 0$.

2 The Discriminant In the Quadratic Formula, the expression under the radical sign, $b^2 - 4ac$, is called the **discriminant**. The discriminant can be used to determine the number of real solutions of a quadratic equation.

Key Concept Using the Discriminant

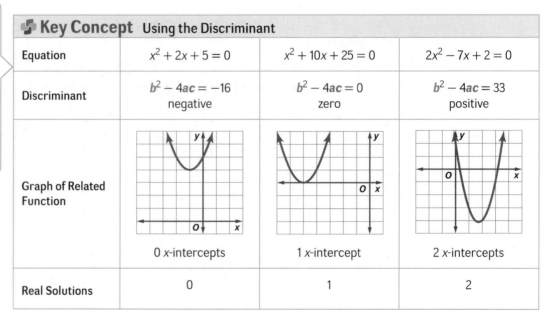

Equation	$x^2 + 2x + 5 = 0$	$x^2 + 10x + 25 = 0$	$2x^2 - 7x + 2 = 0$
Discriminant	$b^2 - 4ac = -16$ negative	$b^2 - 4ac = 0$ zero	$b^2 - 4ac = 33$ positive
Graph of Related Function	0 x-intercepts	1 x-intercept	2 x-intercepts
Real Solutions	0	1	2

Example 4 Use the Discriminant

State the value of the discriminant of $4x^2 + 5x = -3$. Then determine the number of real solutions of the equation.

Step 1 Rewrite in standard form. $4x^2 - 5x = -3 \longrightarrow 4x^2 - 5x + 3 = 0$

Step 2 Find the discriminant.

$$b^2 - 4ac = (-5)^2 - 4(4)(3) \qquad a = 4, b = -5, \text{ and } c = 3$$

$$= -23 \qquad \text{Simplify.}$$

Since the discriminant is negative, the equation has no real solutions.

Guided Practice

4A. $2x^2 + 11x + 15 = 0$ **4B.** $9x^2 - 30x + 25 = 0$

✓ *Go Online!* for a
Self-Check Quiz

Examples 1–2 Solve each equation by using the Quadratic Formula. Round to the nearest tenth.

1. $x^2 - 2x - 15 = 0$ **2.** $x^2 - 10x + 16 = 0$ **3.** $x^2 - 8x = -10$

4. $x^2 + 3x = 12$ **5.** $10x^2 - 31x + 15 = 0$ **6.** $5x^2 + 5 = -13x$

Example 3 Solve each equation. State which method you used.

7. $2x^2 + 11x - 6 = 0$ **8.** $2x^2 - 3x - 6 = 0$

9. $9x^2 = 25$ **10.** $x^2 - 9x = -19$

Example 4 State the value of the discriminant for each equation. Then determine the number of real solutions of the equation.

11. $x^2 - 9x + 21 = 0$ **12.** $2x^2 - 11x + 10 = 0$

13. $9x^2 + 24x = -16$ **14.** $3x^2 - x = 8$

15. **JAGUARUNDI** A jaguarundi springs from a fence post to swat at a low flying bird. Her height h in feet can be modeled by the equation $h = -16t^2 + 22.3t + 2$, where t is time in seconds. Use the discriminant to determine if the jaguarundi will reach the bird if the bird is flying at a height of 10 feet. Explain.

Practice and Problem Solving

Extra Practice is on page R9.

Examples 1–2 Solve each equation by using the Quadratic Formula. Round to the nearest tenth if necessary.

16. $4x^2 + 5x - 6 = 0$ **⟨17⟩** $x^2 + 16 = 0$ **18.** $6x^2 - 12x + 1 = 0$

19. $5x^2 - 8x = 6$ **20.** $2x^2 - 5x = -7$ **21.** $5x^2 + 21x = -18$

22. $81x^2 = 9$ **23.** $8x^2 + 12x = 8$ **24.** $4x^2 = -16x - 16$

25. $10x^2 = -7x + 6$ **26.** $-3x^2 = 8x - 12$ **27.** $2x^2 = 12x - 18$

28. **SQUIRRELS** A flying squirrel drops 60 feet from a tree before leveling off. A function that approximates this drop is $h = -16t^2 + 60$, where h is the distance it drops in feet and t is the time in seconds. About how many seconds does it take for the squirrel to drop 60 feet?

Example 3 Solve each equation. State which method you used.

29. $2x^2 - 8x = 12$ **30.** $3x^2 - 24x = -36$ **31.** $x^2 - 3x = 10$

32. $4x^2 + 100 = 0$ **33.** $x^2 = -7x - 5$ **34.** $12 - 12x = -3x^2$

Example 4 State the value of the discriminant for each equation. Then determine the number of real solutions of the equation.

35. $0.2x^2 - 1.5x + 2.9 = 0$ **36.** $2x^2 - 5x + 20 = 0$ **37.** $x^2 - \frac{4}{5}x = 3$

38. $0.5x^2 - 2x = -2$ **39.** $2.25x^2 - 3x = -1$ **40.** $2x^2 = \frac{5}{2}x + \frac{3}{2}$

41. **(MP) MODELING** The percent of customers who have a store's loyalty card h can be estimated by $h = -0.2n^2 + 7.2n + 1.5$, where n is the number of years since 2000.

a. Use the Quadratic Formula to determine when 20% of the store's customers will have a loyalty card.

b. Is a quadratic equation a good model for this information? Explain.

42. TRAFFIC The equation $d = 0.05v^2 + 1.1v$ models the distance d in feet it takes a car traveling at a speed of v miles per hour to come to a complete stop. If Hannah's car stopped after 250 feet on a highway with a speed limit of 65 miles per hour, was she speeding? Explain your reasoning.

Without graphing, determine the number of x-intercepts of the graph of the related function for each equation.

43. $4.25x + 3 = -3x^2$ **44.** $x^2 + \dfrac{2}{25} = \dfrac{3}{5}x$ **45.** $0.25x^2 + x = -1$

Solve each equation by using the Quadratic Formula. Round to the nearest tenth

46. $-2x^2 - 7x = -1.5$ **47.** $2.3x^2 - 1.4x = 6.8$ **48.** $x^2 - 2x = 5$

49 POSTER Bartolo is making a poster for the dance. He wants to cover three fourths of the area with text.

a. Write an equation for the area of the section with text.

b. Solve the equation by using the Quadratic Formula.

c. What should be the margins of the poster?

50. MULTIPLE REPRESENTATIONS In this problem, you will investigate writing a quadratic equation with given roots. If p is a root of $0 = ax^2 + bx + c$, then $(x - p)$ is a factor of $ax^2 + bx + c$.

a. **Tabular** Copy and complete the first two columns of the table.

b. **Algebraic** Multiply the factors to write each equation with integral coefficients. Use the equations to complete the last column of the table. Write each equation.

c. **Analytical** How could you write an equation with three roots? Test your conjecture by writing an equation with roots 1, 2, and 3. Is the equation quadratic? Explain.

Roots	Factors	Equation
$2, 5$	$(x - 2), (x - 5)$	$(x - 2)(x - 5) = 0$ $x^2 - 7x + 10 = 0$
$1, 9$		
$-1, 3$		
$0, 6$		
$\dfrac{1}{2}, 7$		
$-\dfrac{2}{3}, 4$		

H.O.T. Problems   Use Higher-Order Thinking Skills

51. CHALLENGE Find all values of k such that $2x^2 - 3x + 5k = 0$ has two solutions.

52. MP SENSE-MAKING Use factoring techniques to determine the number of real zeros of $f(x) = x^2 - 8x + 16$. Compare this method to using the discriminant.

CHALLENGE Determine whether there are *two*, *one*, or *no* real solutions of each equation.

53. The graph of the related quadratic function does not have an x-intercept.

54. The graph of the related quadratic function touches but does not cross the x-axis.

55. The graph of the related quadratic function intersects the x-axis twice.

56. Both a and b are greater than 0 and c is less than 0 in a quadratic equation.

57. **WRITING IN MATH** Why can the discriminant be used to confirm the number of real solutions of a quadratic equation?

58. WRITING IN MATH Describe the advantages and disadvantages of each method of solving quadratic equations. Why are the methods equivalent? Which method do you prefer, and why?

59. MULTI-STEP Gabrielle is using the Quadratic Formula to solve the equation $8x^2 - 3x = 9$. Her work so far is shown below.
MP 3, 6

$$\frac{-b \pm \sqrt{b^2 - 4ac}}{2a}$$

$$\frac{-(-3) \pm \sqrt{(-3)^2 - 4(8)(9)}}{2(8)}$$

$$\frac{3 \pm \sqrt{-9 - 288}}{16}$$

a. What errors, if any, did Gabrielle make when using the Quadratic Formula to solve this equation? Select all that apply.

- [ ] **A** There are no errors.
- [ ] **B** She made a sign error when substituting for a.
- [ ] **C** She made a sign error when substituting for b.
- [ ] **D** She made a sign error when substituting for c.
- [ ] **E** She mixed up the values of a, b, and c when substituting.
- [ ] **F** She squared the value of b incorrectly.

b. Which are the correct solutions of $8x^2 - 3x = 9$? Select all that apply.

- [ ] **A** 0.9
- [ ] **B** 1.3
- [ ] **C** 0
- [ ] **D** −1.3
- [ ] **E** −0.9
- [ ] **F** no solution

c. What are two strategies Gabrielle could have used to reduce errors in this calculation?

60. Which equation has no real solutions? **MP** 1

- ○ **A** $2x^2 - 4x = 5$
- ○ **B** $2x^2 + 4x = 5$
- ○ **C** $-2x^2 + 4x = 5$
- ○ **D** $-2x^2 + 4x = -5$

61. Examine the quadratic equation $3x^2 - 4x + 2 = 0$.
MP 6

a. What do you know about the discriminant of $3x^2 - 4x + 2$?

- ○ **A** It is a negative integer.
- ○ **B** It is a positive integer.
- ○ **C** It is an irrational number.
- ○ **D** It is a rational number.

b. What is one solution of $3x^2 - 4x + 2$?

- ○ **A** about 0.2
- ○ **B** about 0.3
- ○ **C** about 1.7
- ○ **D** There is no solution.

62. Which equation has solutions, rounded to the nearest tenth, of −2.1 and 2.4? **MP** 6

- ○ **A** $3x^2 - x - 15 = 0$
- ○ **B** $2x^2 - x - 15 = 0$
- ○ **C** $3x^2 - 4x + 2 = 0$
- ○ **D** $2x^2 - 4x + 2 = 0$

63. For what value of c does the quadratic equation $-2x^2 + 12x - c = 0$ have exactly one real solution? **MP** 6, 7

- ○ **A** $c = 3$
- ○ **B** $c = 18$
- ○ **C** $c = -18$
- ○ **D** $c = 72$

64. A piece of scaffolding falls to the ground from a height of 50 feet. The situation is modeled by the function $h = -16t^2 + 50$, where h is the height in feet and t is the time in seconds. About how many seconds is the best estimate for the time it will take for the scaffolding to hit the ground? **MP** 4, 6

Solving Systems of Linear and Quadratic Equations

::Then

- You graphed and solved linear and quadratic functions.

::**Now**

1 Solve systems of linear and quadratic equations by graphing.

2 Solve systems of linear and quadratic equations by using algebraic methods.

::**Why?**

The height of a kicked football can be modeled by a quadratic equation. The height of a player running up a hill to catch the ball during a practice drill can be modeled by a linear equation. The intersection of the graphs of the equations tells the time and height when the player should catch the ball.

 Mathematical Practices

2 Reason abstractly and quantitatively.

4 Model with mathematics.

1 Graph a System of Linear and Quadratic Equations Like solving systems of linear equations, you can solve systems of linear and quadratic equations by graphing the equations on the same coordinate plane. A system with one linear equation and one quadratic equation can have two solutions, zero solutions, or one solution. The number of solutions depends upon the number of intersections of the two graphs.

Key Concept   Number of Solutions

Number of Solutions	Description	Graph
one solution	two equations with graphs that intersect at one point	
two solutions	two equations with graphs that intersect at two points	
no solution	two equations with graphs that do not intersect	

You can solve a system of linear and quadratic equations by using the following steps. The points of intersection of the graphs are the solutions to the system.

Key Concept   Solving a Linear and Quadratic System   by Graphing

Step 1   Graph the equations on the same coordinate plane.

Step 2   Find the coordinates of each point of intersection.

Step 3   Check the solutions. Substitute the x- and y-values of each solution in the original equation.

Example 1   Solve a System by Graphing

Solve the system of equations by graphing.
$$y = x^2 + 4x - 1$$

$$y = 2x + 2$$

Step 1   Graph $y = x^2 + 4x - 1$ and $y = 2x + 2$ on the same coordinate plane.

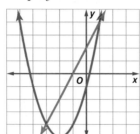

Study Tip

MP **Sense-Making**   The order in which you graph the linear and quadratic equations does not matter.

Step 2   Find the coordinates of each point of intersection.
The graphs intersect at $(-3, -4)$ and $(1, 4)$.

Step 3   To check your solutions, substitute each solution into both of the original equations to see if the values make both equations true.

$y = x^2 + 4x - 1$	Original equation	$y = 2x + 2$
$-4 \stackrel{?}{=} (-3)^2 + 4(-3) - 1$	$x = -3, y = -4$	$-4 \stackrel{?}{=} 2(-3) + 2$
$-4 \stackrel{?}{=} 9 - 12 - 1$	Simplify.	$-4 \stackrel{?}{=} -6 + 2$
$-4 = -4 \checkmark$	Simplify.	$-4 = -4 \checkmark$

$y = x^2 + 4x - 1$	Original equation	$y = 2x + 2$
$4 \stackrel{?}{=} 1^2 + 4(1) - 1$	$x = 1, y = 4$	$4 \stackrel{?}{=} 2(1) + 2$
$4 \stackrel{?}{=} 1 + 4 - 1$	Simplify.	$4 \stackrel{?}{=} 2 + 2$
$4 = 4 \checkmark$	Simplify.	$4 = 4 \checkmark$

▶ **Guided Practice**

    1. $y = x^2 - x - 5$
        $y = x + 3$

2 Solve a Linear and Quadratic System Algebraically

Linear and quadratic systems can also be solved algebraically. Just as you can use subsitution to solve systems of linear equations, you can use this method to solve systems of linear and quadratic equations.

> ### Key Concept Solving a Linear and Quadratic System Algebraically
>
> **Step 1** Solve each equation for y.
>
> **Step 2** Substitute one expression for y in the other equation.
>
> **Step 3** Solve for x.
>
> **Step 4** Substitute the x-value(s) in either of the original equations.
>
> **Step 5** Solve for y.
>
> **Step 6** Graph the equations to check the solution(s).

Example 2 Solve a System Algebraically

Solve the system of equations algebraically.

$$y = x^2 - 2x - 3$$

$$x + y = 3$$

Step 1 Solve each equation for y.

The first equation is already solved for y. Solve the second equation for y.

$x + y = 3$	Second equation
$-x + x + y = -x + 3$	Subtract x from to each side.
$y = -x + 3$	Simplify.

Step 2 Substitute the quadratic expression for y in the linear equation.

$x^2 - 2x - 3 = -x + 3$	Substitute $x^2 - 2x - 3$ for y

Step 3 Solve for x.

$x^2 - 2x - 3 = -x + 3$	Equation from Step 2
$x^2 - x - 3 = 3$	Add x to each side.
$x^2 - x - 6 = 0$	Subtract 3 from each side.
$(x - 3)(x + 2) = 0$	Factor.
$x - 3 = 0 \text{ or } x + 2 = 0$	Zero Product Property
$x = 3 \text{ or } x = -2$	Solve for x.

Step 4 Substitute the x-value(s) in either of the original equations.

$x + y = 3$	$x + y = 3$	Original equation
$3 + y = 3$	$-2 + y = 3$	$x = 3, x = -2.$

Step 5 Solve for y.

$3 + y = 3$	$-2 + y = 3$
$3 + y - 3 = 3 - 3$	$-2 + y + 2 = 3 + 2$
$y = 0$	$y = 5$

The solutions of the system are $(3, 0)$ and $(-2, 5)$.

Study Tip

MP **Structure** After you have found the values of x, you can substitute them in either equation to find the values of y. Most of the time, it will be easier to use the linear function instead of the quadratic function.

Study Tip

MP **Sense-Making** After you solve algebraically, you can check by graphing. You can also check by substituting the x- and y-values into both equations in the system to make sure the values result in true equations.

Step 6 Graph the equations to check the solutions.

The graphs of the equations intersect at (3, 0) and (−2, 5), so (3, 0) and (−2, 5) are the solutions of this system.

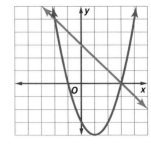

Guided Practice

2. $y = x^2 - 5x - 12$
$2y - 16 = 6x$

You can use systems of linear and quadratic equations to model revenue and look for trends in sales data.

Real-World Example 3   Apply a System of Linear and Quadratic Equations

SALES Camping equipment sales at a sporting goods store can be modeled by the function $y = -x^2 + 12x + 25$ and gift card sales can be modeled by the function $y = 5x + 7$, where x represents the number of months past January and y represents the revenue in thousands of dollars. Solve a system of equations algebraically to find the month in which the revenue from camping equipment sales is equal to the revenue from gift card sales.

Substitute one expression for y in the other equation. Then solve for x.

$-x^2 + 12x + 25 = 5x + 7$	Substitute.
$12x + 25 = x^2 + 5x + 7$	Add x^2 to each side.
$25 = x^2 - 7x + 7$	Subtract $12x$ from each side.
$0 = x^2 - 7x - 18$	Subtract 25 from each side.
$0 = (x - 9)(x + 2)$	Factor.
$x - 9 = 0 \quad x + 2 = 0$	Zero Product Property
$x = 9 \quad x = -2$	Solve for x.

A solution of the system occurs when $x = 9$, so the revenue from camping equipment sales and from gift card sales are equal 9 months past January, or in October.

Because the functions only model data for months past January, $x = -2$ is not a viable solution as it represents a month prior to January.

Graph the equations to check the solutions.

The graphs intersect at $x = 9$, so the solution is correct.

Guided Practice

3. **AMUSEMENT PARKS** Revenue from single-day ticket sales at a local amusement park can be modeled by the function $y = -x^2 + 11x + 42$ and revenue from season pass sales can be modeled by the function $y = -5x + 81$, where x represents the number of months past January and y represents the total revenue in tens of thousands of dollars. Solve a system of equations algebraically to find the first month when the revenue from single-day ticket sales is equal to the revenue from season pass sales.

When you solve a linear and quadratic system algebraically, you may arrive at a an equation in one variable that has no real solutions. This means the original system has no real solutions.

Example 4   Solve a System of Linear and Quadratic Equations Algebraically

Solve the system of equations algebraically.

$y = x + 1$

$y = x^2 + 2$

The equations are already solved for y.

Substitute one expression for y in the other equation. Then solve for x.

$x^2 + 2 = x + 1$	Substitute.
$x^2 - x + 2 = 1$	Subtract x from each side.
$x^2 - x + 1 = 0$	Subtract 1 from each side.

Find the discriminant, $b^2 - 4ac$, of the quadratic equation $x^2 - x + 1 = 0$.

$b^2 - 4ac = (-1)^2 - 4(1)(1)$	$a = 1, b = -1,$ and $c = 1$
$= 1 - 4$	Simplify.
$= -3$	Subtract.

Study Tip

MP Structure If the discriminant of a quadratic equation is negative, then the quadratic equation has no real solutions.

The discriminant is negative, so $x^2 - x + 1 = 0$ has no real solutions. Therefore, the system does not have a solution.

Graph the equations to check your solution.

The graphs do not intersect, so the system has no solution.

▶ **Guided Practice**

4. $y = -x^2 + x - 1$

$-3 + y = x$

○ = Step-by-Step Solutions begin on page R11.

Go Online! for a
Self-Check Quiz

Check Your Understanding

Example 1   **Solve each system of equations by graphing.**

1. $y = x^2 - 2x - 3$

$y = x - 3$

2. $y = x^2 - 4$

$y = -3$

3. $y = -x^2 + 2x + 1$

$y = -x + 1$

4. $y = -2x^2 + 5$

$y = 2x + 1$

Examples 2–4   **5. TICKET PRICES** The revenue for a school play can be modeled by $y = -x^2 + 8x$, where x is the ticket price in dollars. The cost to produce the school play can be modeled by $y = -20 + 9x$. Determine the ticket price that will allow the school play to break even. (*Hint*: Breaking even means that the revenue equals the cost.)

Solve each system of equations algebraically.

6. $y = -x + 3$

$y = x^2 + 1$

7 $y = x^2 - 5x + 5$

$y = 5x - 20$

8. $-x^2 - x + 19 = y$

$x = y + 80$

9. $y = 3x^2 + 21x - 5$

$-10x + y = -1$

Example 1 Solve each system of equations by graphing.

10. $y = x^2 - 2x - 2$

$y = 4x - 11$

11. $y = x - 5$

$y = x^2 - 6x + 5$

12. $y = x + 1$

$y = x^2 + 2x + 4$

13. $y = 2x$

$y = -x^2 + 2x + 4$

Examples 2–4 **14. VIDEO GAMES** Jamar is using a coordinate plane to program a video game in which a fish jumps out of the water to catch a fly. The path of the fish is modeled by $y = -x^2 + 8x - 10$ and the path of the fly is modeled by $y = -x + 8$, where y is the height in centimeters above the water and x is the distance in centimeters from the left edge of the screen.

a. How far is the fish from left edge of the screen when it catches the fly? Is there more than one possible distance? Explain.

b. What is the height of the fish when it catches the fly? Is there more than one possible height? Explain.

Solve each system algebraically.

15. $y = 6x^2$

$3x + y = 3$

16. $y = x^2 + 1$

$8x + y = 10$

17 $y = x^2$

$y - 9 = 0$

18. $y = x^2 - 8x + 19$

$x - 0.5y = 3$

19. $y = x^2 + 11$

$y = -12x$

20. $y = 5x - 20$

$y = x^2 - 5x + 5$

21. CITY PLANNING A city planner is using the first quadrant of a coordinate plane to lay out new streets in a suburban development. Franklin Road is modeled $y = -5x^2 + 32x + 2$ and Jefferson Street is modeled by $y = 28x + 1$.

a. At what point do Franklin Road and Jefferson Street intersect?

b. The city planner changes the equation for Jefferson Street to $y = 28x + 4$. How does this affect the intersection of the streets? Explain.

22. PARACHUTING A parachutist jumps off a tall building. She falls freely for a few seconds before releasing her parachute. Her height during the free fall can be modeled by $y = -4.9x^2 + x + 360$, where y is the height in meters and x is the number of seconds after jumping. Her height after she releases the parachute can be modeled by $y = -4x + 142$.

a. How long after jumping did she release her parachute?

b. What was her height when she released her parachute?

23. **MP** **STRUCTURE** The graph shows a quadratic function and a linear function $y = k$.

a. If the linear function were changed to $y = k + 4$, how many solutions would the system have?

b. If the linear function were changed to $y = k - 1$, how many solutions would the system have?

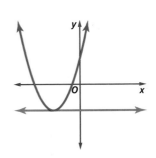

24. **MP SENSE-MAKING** Solve $x^2 + 3x - 5 = 2x + 1$ by using a system of linear and quadratic equations. Describe your solution method.

Each graph shows the graph of a linear equation that is part of a system of equations. Write a quadratic equation that could be part of the system so that the system has the given number of solutions.

25. one solution

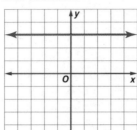

26. two solutions

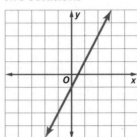

27. no solution

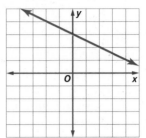

H.O.T. Problems   Use **H**igher-**O**rder **T**hinking Skills

28. **CHALLENGE** How many solutions does the system of equations have? Explain.

$$y = -2x^2 + 4x - 1$$

$$-5x + y = 5$$

29. **MP TOOLS** Use a graphing calculator to solve the system involving two quadratic equations.

$$y = x^2 + 3x - 5$$

$$y = -x^2$$

30. **MP CRITIQUE ARGUMENTS** Sarah says that if a system of linear and quadratic equations involves a linear equation of the form $y = c$, where c is a constant, and there is only one solution, then the graph of the linear equation intersects that of the quadratic equation at the vertex. Is she correct? Explain your answer.

31. **ERROR ANALYSIS** Tomas solved the system of equations and found no solution. What error did he make? Explain.

$$y = 0$$

$$y - x^2 = 0$$

32. **WRITING IN MATH** Explain why a linear and quadratic system of equations can have no more than two solutions.

33. **MP TOOLS** Use a graphing calculator to determine the solutions of each system of equations.

    a. $y = x^2 - 2x + 1$

       $y = 2x - 3$

    b. $y = x^2 - 2x + 5$

       $y = 2x$

    c. $y = x^2 - 2x + 1$

       $y = 2x + 5$

34. Which ordered pair is a solution of the following system? **MP** 7

$y = x^2 + 7x + 10$

$y = x + 1$

- ○ **A** $(-3, -2)$
- ○ **C** $(3, 4)$
- ○ **B** $(3, 40)$
- ○ **D** $(-4, -3)$

35. Which of the following are solutions of the given system? Select all that apply. **MP** 7

$y = x^2 + 7x + 12$

$y = 2x + 8$

- ☐ **A** $(-4, 0)$
- ☐ **B** $(-1, 6)$
- ☐ **C** $(1, 10)$
- ☐ **D** $(4, 16)$
- ☐ **E** $(-1, -4)$
- ☐ **F** $(12, 8)$

36. In the system shown below, k is a constant. If the system has exactly one solution, what is the value of k? **MP** 2

$y = x^2 - 3x + 7$

$y = x + k$

- ○ **A** -4
- ○ **C** 3
- ○ **B** 2
- ○ **D** 7

37. The quadratic equation $y = x^2 + 1$ is part of a linear and quadratic system of equations. If the system has two solutions, select all of the linear equations that could complete the system. **MP** 1, 3

- ☐ **A** $y = 1$
- ☐ **B** $y = 3$
- ☐ **C** $y = x$
- ☐ **D** $y = x + 2$
- ☐ **E** $y = x - 2$
- ☐ **F** $y = -x + 3$

38. The graph below shows the system $y = x - 1$ and $y = -x^2 - x$. Based on the graph, which of the following is a solution of the system, assuming coordinates are rounded to the nearest tenth? **MP** 1, 6

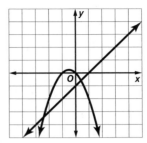

- ○ **A** $(0.4, -2.4)$
- ○ **B** $(-1.0, -3.4)$
- ○ **C** $(-0.5, 0.25)$
- ○ **D** $(-2.4, -3.4)$

39. **MULTI-STEP** An animator is using a coordinate plane to create a cartoon about outer space. The path of a comet is represented by $y = x^2 - 3x + 4$. The path of a rocket is represented by $y - x = 1$. **MP** 4

- **a.** Graph the equations in the system on the same coordinate plane.

- **b.** Is it possible that the rocket and the comet will collide? If so, what are the coordinates at which this might happen? If not, why not?

- **c.** The animator wants to change the path of the rocket so that there is exactly one point at which the rocket and the comet could collide. How can the animator change the equation for the path of the rocket?

40. Which system has $(1, -1)$ as a solution? **MP** 7

- ○ **A** $y = x^2 - 3x + 1$
       $y = x + 2$
- ○ **B** $y = x^2 + 3x - 1$
       $y = x - 2$
- ○ **C** $y = x^2 - 2$
       $y = x - 2$
- ○ **D** $y = x^2 - 2$
       $y = -x + 2$

Analyzing Functions with Successive Differences

∷Then

● You graphed linear, quadratic, and exponential functions.

∷Now

1 Identify linear, quadratic, and exponential functions from given data.

2 Write equations that model data.

∷Why?

● Every year, the golf team sells candy to raise money for charity. By knowing what type of function models the sales of the candy, they can determine the best price for the candy.

 Mathematical Practices

7 Look for and make use of structure.

1 **Identify Functions** You can use linear functions, quadratic functions, and exponential functions to model data. The general forms of the equations and a graph of each function type are listed below.

⚡ Key Concept Linear and Nonlinear Functions

Linear Function	Quadratic Function	Exponential Function
$y = mx + b$	$y = ax^2 + bx + c$	$y = ab^x$, when $b > 0$

Example 1 Choose a Model Using Graphs

Graph each set of ordered pairs. Determine whether the ordered pairs represent a *linear* **function, a** *quadratic* **function, or an** *exponential* **function.**

a. $\{(-2, 5), (-1, 2), (0, 1), (1, 2), (2, 5)\}$

The ordered pairs appear to represent a quadratic function.

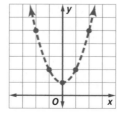

b. $\left\{\left(-2, \frac{1}{4}\right), \left(-1, \frac{1}{2}\right), (0, 1), (1, 2), (2, 4)\right\}$

The ordered pairs appear to represent an exponential function.

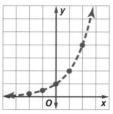

▶ **Guided Practice**

1A. $(-2, -3), (-1, -1), (0, 1), (1, 3)$

1B. $(-1, 0.25), (0, 1), (1, 4), (2, 16)$

Another way to determine which model best describes data is to use patterns. The differences of successive y-values are called *first differences*. The differences of successive first differences are called *second differences*.

- If the differences of successive y-values are all equal, the data represent a linear function.

- If the second differences are all equal, but the first differences are not equal, the data represent a quadratic function.

- If the ratios of successive y-values are all equal and $r \neq 1$, the data represent an exponential function.

Watch Out!

x-Values Before you check for successive differences or ratios, make sure the *x*-values are increasing by the same amount.

Go Online!

Review with your graphing calculator. Ask your teacher to assign the Graphing Calculator Easy File™ 5-Minute Check to you in ConnectED.

Example 2   Choose a Model Using Differences or Ratios

Look for a pattern in each table of values to determine which kind of model best describes the data.

a.

x	−2	−1	0	1	2
y	−8	−3	2	7	12

First differences:

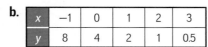

Since the first differences are all equal, the table of values represents a linear function.

b.

x	−1	0	1	2	3
y	8	4	2	1	0.5

First differences:

The first differences are not all equal. So, the table of values does not represent a linear function. Find the second differences and compare.

First differences: −4   −2   −1   −0.5

Second differences: 2   1   0.5

The second differences are not all equal. So, the table of values does not represent a quadratic function. Find the ratios of the y-values and compare.

Ratios: 8   4   2   1   0.5

$\frac{4}{8} = \frac{1}{2}$ $\frac{2}{4} = \frac{1}{2}$ $\frac{1}{2}$ $\frac{0.5}{1} = \frac{1}{2}$

The ratios of successive y-values are equal. Therefore, the table of values can be modeled by an exponential function.

Guided Practice

2A.

x	−3	−2	−1	0	1
y	−3	−7	−9	−9	−7

2B.

x	−2	−1	0	1	2
y	−18	−13	−8	−3	2

2 Write Equations Once you find the model that best describes the data, you can write an equation for the function. For a quadratic function in this lesson, the equation will have the form $y = ax^2$.

Example 3   Write an Equation

Determine which kind of model best describes the data. Then write an equation for the function that models the data.

x	-4	-3	-2	-1	0
y	32	18	8	2	0

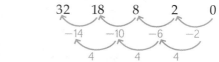

Step 1 Determine which model fits the data.

First differences are not equal.

Second differences are equal.

A quadratic function models the data.

> **Watch Out!**
>
> **Finding a** In Example 3, the point $(0, 0)$ cannot be used to find the value of a. You will have to divide each side by 0, giving you an undefined value for a.

Step 2 Write an equation for the function that models the data.

The equation has the form $y = ax^2$. Find the value of a by choosing one of the ordered pairs from the table of values. Let's use $(-1, 2)$.

$y = ax^2$ Equation for quadratic function

$2 = a(-1)^2$ $x = -1$ and $y = 2$

$2 = a$ An equation that models the data is $y = 2x^2$.

▶ **Guided Practice**

3A.

x	-2	-1	0	1	2
y	11	7	3	-1	-5

3B.

x	-3	-2	-1	0	1
y	0.375	0.75	1.5	3	6

Real-World Example 4   Write an Equation for a Real-World Situation

BOOK CLUB The table shows the number of book club members for four consecutive years. Determine which model best represents the data. Then write a function that models the data.

Understand Find a model for the data, and then write a function.

Time (years)	0	1	2	3	4
Members	5	10	20	40	80

Plan Find a pattern using successive differences or ratios. Then use the general form of the equation to write a function.

Solve The constant ratio is 2. This is the value of the base. An exponential function of the form $y = ab^x$ models the data.

$y = ab^x$ Equation for exponential function

$5 = a(2)^0$ $x = 0, y = 5$, and $b = 2$

$5 = a$ The equation that models the data is $y = 5(2)^x$.

Check You used $(0, 5)$ to write the function. Verify that every other ordered pair satisfies the equation.

> **Real-World Link**
>
> A poll by the National Education Association found that 87% of all teens polled found reading relaxing, 85% viewed reading as rewarding, and 79% found reading exciting.
>
> **Source:** American Demographics

▶ **Guided Practice**

4. ADVERTISING The table shows the cost of placing an ad in a newspaper. Determine a model that best represents the data and write a function that models the data.

No. of Lines	5	6	7	8
Total Cost ($)	14.50	16.60	18.70	20.80

Check Your Understanding ◯ = Step-by-Step Solutions begin on page R11.

✓ **Go Online!** for a Self-Check Quiz

Example 1 Graph each set of ordered pairs. Determine whether the ordered pairs represent a *linear* function, a *quadratic* function, or an *exponential* function.

1. $(-2, 8), (-1, 5), (0, 2), (1, -1)$

2. $(-3, 7), (-2, 3), (-1, 1), (0, 1), (1, 3)$

3. $(-3, 8), (-2, 4), (-1, 2), (0, 1), (1, 0.5)$

4. $(0, 2), (1, 2.5), (2, 3), (3, 3.5)$

Example 2 Look for a pattern in each table of values to determine which kind of model best describes the data.

5.

x	0	1	2	3	4
y	5	8	17	32	53

6.

x	−3	−2	−1	0
y	−6.75	−7.5	−8.25	−9

7.

x	−1	0	1	2	3
y	3	6	12	24	48

8.

x	3	4	5	6	7
y	−1.5	0	2.5	6	10.5

Example 3 Determine which kind of model best describes the data. Then write an equation for the function that models the data.

9.

x	−1	0	1	2	3
y	1	3	9	27	81

10.

x	−5	−4	−3	−2	−1
y	125	80	45	20	5

11.

x	−3	−2	−1	0	1
y	1	1.5	2	2.5	3

12.

x	−1	0	1	2
y	−1.25	−1	−0.75	−0.5

Example 4 **13. PLANTS** The table shows the height of a plant for four consecutive weeks. Determine which kind of function best models the height. Then write a function that models the data.

Week	0	1	2	3	4
Height (in.)	3	3.5	4	4.5	5

Practice and Problem Solving

Extra Practice is on page R9.

Example 1 Graph each set of ordered pairs. Determine whether the ordered pairs represent a *linear* function, a *quadratic* function, or an *exponential* function.

14. $(-1, 1), (0, -2), (1, -3), (2, -2), (3, 1)$

15. $(1, 2.75), (2, 2.5), (3, 2.25), (4, 2)$

16. $(-3, 0.25), (-2, 0.5), (-1, 1), (0, 2)$

17. $(-3, -11), (-2, -5), (-1, -3), (0, -5)$

18. $(-2, 6), (-1, 1), (0, -4), (1, -9)$

19. $(-1, 8), (0, 2), (1, 0.5), (2, 0.125)$

Examples 2–3 Look for a pattern in each table of values to determine which kind of model best describes the data. Then write an equation for the function that models the data.

20.

x	−3	−2	−1	0
y	−8.8	−8.6	−8.4	−8.2

21

x	−2	−1	0	1	2
y	10	2.5	0	2.5	10

22.

x	−1	0	1	2	3
y	0.75	3	12	48	192

23.

x	−2	−1	0	1	2
y	0.008	0.04	0.2	1	5

24.

x	0	1	2	3	4
y	0	4.2	16.8	37.8	67.2

25.

x	−3	−2	−1	0	1
y	14.75	9.75	4.75	−0.25	−5.25

Example 4

26. **WEBSITES** A company tracked the number of visitors to its website over four days. Determine which kind of model best represents the number of visitors to the website with respect to time. Then write a function that models the data.

Day	0	1	2	3	4
Visitors (in thousands)	0	0.9	3.6	8.1	14.4

27. **FROZEN YOGURT** The cost of a build-your-own cup of frozen yogurt depends on the weight of the contents. The table shows the cost for up to 6 ounces.

Ounces	1	2	3	4	5	6
Cost ($)	0.49	0.98	1.47	1.96	2.45	2.94

   a. Graph the data and determine which kind of function best models the data.

   b. Write an equation for the function that models the data.

   c. Use your equation to determine how much 10 ounces of yogurt would cost.

28. **DEPRECIATION** The value of a car depreciates over time. The table shows the value of a car over a period of time.

Year	0	1	2	3	4
Value ($)	18,500	15,910	13,682.60	11,767.04	10,119.65

   a. Determine which kind of function best models the data.

   b. Write an equation for the function that models the data.

   c. Use your equation to determine how much the car is worth after 7 years.

29. **BACTERIA** A scientist estimates that a bacteria culture with an initial population of 12 will triple every hour.

   a. Make a table to show the bacteria population for the first 4 hours.

   b. Which kind of model best represents the data?

   c. Write a function that models the data.

   d. How many bacteria will there be after 8 hours?

30. **PRINTING** A printing company charges the fees shown to print flyers. Write a function that models the total cost of the flyers, and determine how much 30 flyers would cost.

Quick 2 U Printing
Set Up Fee $25
15¢ each flyer

H.O.T. Problems   Use **H**igher-**O**rder **T**hinking Skills

31. **MP REASONING** Write a function that has constant second differences, first differences that are not constant, a y-intercept of -5, and contains the point $(2, 3)$.

32. **MP CONSTRUCT ARGUMENTS** What type of function will have constant third differences but not constant second differences? Explain.

33. **MP STRUCTURE** Write a linear function that has a constant first difference of 4.

34. **PROOF** Write a paragraph proof to show that linear functions grow by equal differences over equal intervals, and exponential functions grow by equal factors over equal intervals. (*Hint:* Let $y = ax$ represent a linear function and let $y = a^x$ represent an exponential function.)

35. **WRITING IN MATH** How can you determine whether a given set of data should be modeled by a *linear* function, a *quadratic* function, or an *exponential* function?

36a. MULTI-STEP Which function best models the data in the table of values? **MP** 7, 8

x	0	1	2	3	4
y	3	0	−1	0	3

○ **A** $y = x - 3$

○ **B** $y = -\left(\frac{1}{2}\right)^x$

○ **C** $y = x^2 + 4x + 3$

○ **D** $y = x^2 - 4x + 3$

b. Copy and complete the table below so that the data is best modeled by a linear equation. Write this linear equation.

x	0	1	2	3	4
y	3	0	−3		

c. Copy and complete the table below so that the data is best modeled by an exponential equation. Write this exponential equation.

x	0	1	2	3	4
y	1	4	16		

37. Which tables of values are best modeled by an exponential equation? Select all that apply. **MP** 8

☐ **A**

x	−2	−1	0	1	2
y	−4	−1	0	1	4

☐ **B**

x	−2	−1	0	1	2
y	0.5	1.5	4.5	13.5	40.5

☐ **C**

x	−2	−1	0	1	2
y	1	0	1	4	9

☐ **D**

x	−2	−1	0	1	2
y	2	4	8	16	32

☐ **E**

x	−2	−1	0	1	2
y	20	10	5	2.5	1.25

☐ **F**

x	−2	−1	0	1	2
y	−1	0	−1	−4	−9

38. Jake graphed the sets of ordered pairs in the table below.

x	−2	−1	0	1	2
y	10	6	2	−2	−6

Which equation best models the data in the table? **MP** 8

○ **A** $y = -4x + 2$

○ **B** $y = -4x - 2$

○ **C** $y = -2x^2 + 2$

○ **D** $y = -2x^2 - 2$

39. For what value of c does the equation $y = x^2 + 6x + c$ model the data in the table? **MP** 7

x	−4	−3	−2	−1	0
y	3	2	3	6	11

40. What type of model best fits the data? **MP** 3, 7

x	−1	0	1	2	3
y	5	6	9	13	17

○ **A** linear

○ **B** quadratic

○ **C** exponential

○ **D** none of the above

41. What type of model best fits the data? **MP** 3, 7

x	−2	−1	0	1	2
y	4	3	4	7	12

○ **A** linear

○ **B** quadratic

○ **C** exponential

○ **D** none of the above

If there is a constant increase or decrease in data values, there is a linear trend. If the values are increasing or decreasing more and more rapidly, there may be a quadratic or exponential trend.

Mathematical Practices
MP **5** Use tools strategically

With a graphing calculator, you can find the appropriate regression equation and an R^2 value. R^2 is the **coefficient of determination.** The closer R^2 is to 1, the better the model fits the data.

Activity 1

CHARTER AIRLINE The table shows the average monthly number of flights made each year by a charter airline that was founded in 2010. Work cooperatively to predict the number of flights in 2030 and when the airline will meet its goal of 200 flights per month.

Year	2010	2011	2012	2013	2014	2015	2016	2017
Flights	17	20	24	28	33	38	44	50

Step 1 Make a scatter plot.

Enter the number of years since 2010 in **L1** and the number of flights in **L2**. Graph the scatter plot. It appears that the data may have either a quadratic or exponential trend.

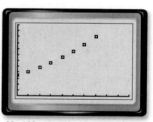

[0, 10] scl: 1 by [0, 60] scl: 5

Step 2 Find the regression equation.

Check both trends by examining their regression equations. To acquire the exponential or quadratic equation select **ExpReg** or **QuadReg** on the **STAT** menu. To choose, fit both and use the one with the R^2 value closest to 1. In this case the quadratic is a better fit.

Step 3 Graph the quadratic regression equation.

Copy the equation to the **Y=** list and graph.

Step 4 Predict using the equation.

Use the graph to predict the number of flights in 2030. There will be approximately 177 flights per month if this trend continues.

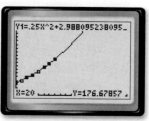

Step 5 Find a solution using the model.

Use the graph to determine when the airline will reach its goal of 200 flights a month. Graph the equation $y = 200$ and find the intersection. If this trend continues the airline will reach their goal in 2032.

[0, 25] scl: 1 by [0, 200] scl: 5

(continued on the next page)

FROGS The table shows the frog population in a small pond. Work cooperatively to predict what the frog population will be in year 11 and when the pond will reach its capacity of 500 frogs.

Year	0	1	2	3	4	5	6	7	8
Population	80	90	102	115	130	147	166	188	212

Step 1 Make a scatter plot. Enter the years 1–8 in **L1** and the population in **L2**. Graph the scatter plot. The data appear to have either a quadratic or exponential trend.

Step 2 Find the regression equation. Check both trends by examining their regression equations. To acquire the exponential or quadratic equation select **ExpReg** or **QuadReg** on the **STAT** menu. To choose, fit both and use the one with the R^2 value closest to 1.

Step 3 Graph the exponential regression equation. Copy the equation to the **Y=** list and graph.

Step 4 Predict using the equation.

Use the graph to predict the number of frogs in year 11. There will be approximately 306 frogs in the pond if this trend continues.

Step 5 Find a solution using the model. The pond cannot maintain more than 500 frogs. Use the graph to determine when the pond will reach capacity. Graph the equation $y = 500$ and find the intersection. If this trend continues the pond will reach capacity in year 15.

Exercises

Plot each set of data points. Determine whether to use a *linear*, *quadratic* or *exponential* regression equation. State the coefficient of determination.

1.

x	y
1	30
2	40
3	50
4	55
5	50
6	40

2.

x	y
0.0	12.1
0.1	9.6
0.2	6.3
0.3	5.5
0.4	4.8
0.5	1.9

3.

x	y
0	1.1
2	3.3
4	2.9
6	5.6
8	11.9
10	19.8

4.

x	y
1	1.67
5	2.59
9	4.37
13	6.12
17	5.48
21	3.12

5. BAKING Alyssa baked a cake and is waiting for it to cool so she can ice it. The table shows the temperature of the cake every 5 minutes after Alyssa took it out of the oven.

a. Make a scatter plot of the data.

b. Which regression equation has an R^2 value closest to 1? Is this the equation that best fits the context of the problem? Explain your reasoning.

c. Find an appropriate regression equation, and state the coefficient of determination. What is the domain and range?

d. Alyssa will ice the cake when it reaches room temperature (70°F). Use the regression equation to predict when she can ice her cake.

Time (min)	Temperature (°F)
0	350
5	244
10	178
15	137
20	112
25	96
30	89

Combining Functions

- You wrote and graphed linear, exponential, and quadratic functions.

1 Combine functions by using addition and subtraction.

2 Combine functions by using multiplication.

- On February 15, 2013, a meteor exploded over Chelyabinsk, Russia, sending a 1400-pound meteorite to Earth's surface. You can write a function that models the temperature of the meteorite by combining an exponential decay function and a constant function that represents the local air temperature.

 Mathematical Practices

4 Model with mathematics.

7 Look for and make use of structure.

1 **Add and Subtract Functions** You can perform operations, such as addition and subtraction, with functions just as you perform operations with real numbers.

> ### Key Concept Adding and Subtracting Functions
>
> **Words** Given two functions, $f(x)$ and $g(x)$, you can form new functions, $(f + g)(x)$, by adding the two functions, and $(f - g)(x)$, by subtracting the two functions.
>
> **Symbols** $(f + g)(x) = f(x) + g(x)$
>
> $(f - g)(x) = f(x) - g(x)$

Example 1   Add and Subtract Functions

Given that $f(x) = x^2 + 3x - 7$ and $g(x) = x - 5$, find each function.

a. $(f + g)(x)$

$$(f + g)(x) = f(x) + g(x) \qquad \text{Addition of functions}$$
$$= (x^2 + 3x - 7) + (x - 5) \qquad \text{Substitution}$$
$$= x^2 + 3x + x - 7 - 5 \qquad \text{Combine like terms.}$$
$$= x^2 + 4x - 12 \qquad \text{Simplify.}$$

b. $(f - g)(x)$

$$(f - g)(x) = f(x) - g(x) \qquad \text{Subtraction of functions}$$
$$= (x^2 + 3x - 7) - (x - 5) \qquad \text{Substitution}$$
$$= x^2 + 3x - 7 - x + 5 \qquad \text{Distributive Property}$$
$$= x^2 + 3x - x - 7 + 5 \qquad \text{Combine like terms.}$$
$$= x^2 + 2x - 2 \qquad \text{Simplify.}$$

▶ **Guided Practice**

Given that $f(x) = 4^x + 1$ and $g(x) = 5x^2 + x - 1$, find each function.

1A. $(f + g)(x)$

1B. $(f - g)(x)$

Real-World Example 2 Build a New Function by Adding

FINANCIAL LITERACY
Hugo graduates
from college with
$28,500 in student
loan debt. He
decides to defer
his payments while
he is in graduate
school. During that
time, he still accrues
interest on his
student loans at an

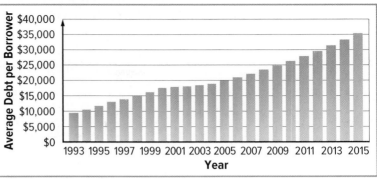

Source: Wall Street Journal

annual rate of 5.65%. While he is in graduate school, Hugo's parents also lend him
$400 per month for rent. His parents decide not to charge him interest on this loan.

a. Write an exponential function $f(t)$ to represent the amount of money Hugo owes
on his student loans t years after interest begins to accrue.

$f(t) = a(1 + r)^t$ Equation for exponential growth

$\quad = 28,500(1 + 0.0565)^t$ Substitution with $a = 28,500$ and $r = 5.65\%$ or 0.0565

$\quad = 28,500(1.0565)^t$ Simplify.

b. Write a function $g(t)$ to represent the amount of money Hugo owes his parents
after t years.

Since t is the time in years, first find the amount of money Hugo borrows from his
parents each year.

$12(400) = 4800$

Hugo borrows $4800 from his parents each year.

So, $g(t) = 4800t$ represents the total amount Hugo owes his parents after t years.

c. Find $C(t) = f(t) + g(t)$ and explain what this function represents.

$C(t) = f(t) + g(t)$ Addition of functions

$\quad = 28,500(1.0565)^t + 4800t$ Substitution

$C(t)$ represents the total amount of money Hugo has to repay after t years.

d. If Hugo spends 3 years in graduate school, find the total amount of money he will
have to repay.

Because Hugo spends 3 years in graduate school, $t = 3$.

$C(t) = 28,500(1.0565)^t + 4800t$

$C(3) = 28,500(1.0565)^3 + 4800(3)$

$\quad = 33,608.83 + 14,400$

$\quad = \$48,008.83$

Guided Practice

BANKING Malia deposits $350 in a new savings account and $2100 in a new checking
account. The savings account pays 2% interest, compounded annually. Malia's weekly
paycheck of $148 is deposited directly into her checking account and earns no interest.

2A. Write functions $S(t)$ and $C(t)$ to represent the amount of money Malia has in her
savings account and in her checking account, respectively, after t years. Assume
she makes no withdrawals from or additional deposits into the accounts.

2B. Find $M(t) = S(t) + C(t)$ and explain what this function represents. Then find $M(2)$
and explain what this represents.

2 Multiply Functions

Just as you can add and subtract functions, you can also multiply them. When you multiply functions that have more than one term, use the Distributive Property to multiply each term in the first function by each term in the second function.

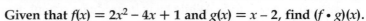

Key Concept Multiplying Functions

Words	Given two functions, $f(x)$ and $g(x)$, you can form a new function, $(f \cdot g)(x)$, by multiplying the two functions.
Symbols	$(f \cdot g)(x) = f(x) \cdot g(x)$

Example 3   Multiply Functions

Given that $f(x) = 2x^2 - 4x + 1$ and $g(x) = x - 2$, find $(f \cdot g)(x)$.

$$
\begin{aligned}
(f \cdot g)(x) &= f(x) \cdot g(x) && \text{Multiplication of functions} \\
&= (2x^2 - 4x + 1)(x - 2) && \text{Substitution} \\
&= (2x^2)(x) + (2x^2)(-2) + (-4x)(x) + \\
&\quad\; (-4x)(-2) + (1)(x) + (1)(-2) && \text{Distributive Property} \\
&= 2x^3 - 4x^2 - 4x^2 + 8x + x - 2 && \text{Simplify.} \\
&= 2x^3 - 8x^2 + 9x - 2 && \text{Simplify.}
\end{aligned}
$$

Study Tip

MP **Structure** Be sure to distribute all terms of one function to all terms of the other function. When a function has more than two terms, you may find it helpful to draw arrows from one term to the others.

Guided Practice

Given that $f(x) = 3x - 2$, $g(x) = 4x^2 + 5x - 1$, and $h(x) = 8^x + x$ find each function.

3A. $(f \cdot g)(x)$

3B. $(f \cdot h)(x)$

3C. $(h \cdot g)(x)$

Some real-world situations are best modeled by the product of functions.

Real-World Example 4   Build a New Function by Multiplying

BUSINESS A sandwich shop charges $7 for a large sub and sells an average of 360 large subs per day. The shop's owner predicts that they will sell 20 fewer subs per day for every $0.25 increase in the price.

a. Let x represent the number of $0.25 price increases. Write a function $P(x)$ to represent the price of a large sub.

The price of a large sub is $7 plus $0.25 times the number of price increases.
$P(x) = 7 + 0.25x$

b. Write a function $T(x)$ to represent the number of large subs sold per day.

The number of large subs sold is 360 minus 20 times the number of price increases.
$T(x) = 360 - 20x$

c. Write a function $R(x)$ that can be used to find the revenue from sales of large subs.

The revenue from sales of large subs is equal to the price times the number of large subs sold.

$$
\begin{aligned}
R(x) &= P(x) \cdot T(x) && \text{Multiplication of functions} \\
&= (7 + 0.25x)(360 - 20x) && \text{Substitution} \\
&= -5x^2 - 50x + 2520 && \text{Multiply}
\end{aligned}
$$

Real World Link

The submarine sandwich, or sub, may have gotten its name because its shape is similar to that of a submarine ship. However, this style of sandwich goes by many different names, including hoagie, hero, and grinder.

Gustaf Brundin/E+/Getty Images

d. **If the owner decides to charge \$8.50 for a large sub, find the revenue from sales of large subs.**

Since x represents the number of \$0.25 price increases, and the price increased from \$7.00 to \$8.50, $x = 6$. Substitute 6 for x in $R(x)$.

$$R(6) = -5(6)^2 - 50(6) + 2520$$

$$= -180 - 300 + 2520$$

$$= \$2040$$

▶ **Guided Practice**

THEATERS **A theater currently sells tickets for \$18 and they sell an average of 250 tickets per show. The box office manager estimates that they can sell 25 more tickets for every \$1.50 decrease in the price.**

4A. Let x represent the number of \$1.50 price decreases. Write a function $P(x)$ to represent the price of a ticket, a function $T(x)$ to represent the number of tickets sold, and a function $R(x)$ that can be used to find the revenue from ticket sales.

4B. If the manager decides to sell tickets at \$13.50, find the revenue from ticket sales.

Check Your Understanding   ◯ = Step-by-Step Solutions begin on page R11.

Go Online! for a Self-Check Quiz

Example 1
F.BF.1b

Given that $f(x) = x^2 - 5x - 9$, $g(x) = 4x + 1$, and $h(x) = 3x$, find each function.

1. $(f + g)(x)$　　　**2.** $(f - g)(x)$　　　**3.** $(f + h)(x)$

4. $(g - f)(x)$　　　**5.** $(g - h)(x)$　　　**6.** $(g + h)(x)$

Example 2
F.BF.1b

7. **COINS** A coin collector buys a rare nickel and a rare quarter. She pays \$20 for the nickel and \$42 for the quarter. The value of the nickel increases by 4.5% per year. The value of the quarter increases by \$2 per year.

a. Write a function $f(t)$ to represent the value of the nickel after t years.

b. Write a function $g(t)$ to represent the value of the quarter after t years.

c. Find $V(t) = f(t) + g(t)$ and explain what this function represents.

d. How much will the coins we worth after 5 years?

Example 3
F.BF.1b

Given that $f(x) = 3x + 10$, $g(x) = x^2 - 6x - 2$, and $h(x) = 2x - 5$, find each function.

8. $(g \cdot h)(x)$　　　**9** $(f \cdot g)(x)$　　　**10.** $(f \cdot h)(x)$

Example 4
F.BF.1b

11. **FARMERS' MARKETS** A vendor at a farmers' market sells jars of strawberry jam for \$4 per jar. At this price, he sells an average of 160 jars of jam per day. The vendor predicts that he will sell 10 fewer jars of jam for each \$0.15 increase in the price.

a. Let x represent the number of \$0.15 price increases. Write a function $P(x)$ to represent the price of a jar of strawberry jam.

b. Write a function $T(x)$ to represent the number of jars of jam sold per day.

c. Write a function $R(x)$ that can be used to find the revenue from sales of strawberry jam.

d. If the vendor decides to charge \$4.60 for a jar of jam, find the revenue from sales of strawberry jam.

Example 1   **Given that $f(x) = -x^2 + 9$, $g(x) = 2^x + 2$, and $h(x) = 3x^2 - 6x - 9$, find each function.**

12. $(f + g)(x)$ **13.** $(f + h)(x)$ **14.** $(f - g)(x)$

15. $(g - f)(x)$ **16.** $(g + h)(x)$ **17.** $(f - h)(x)$

Example 4   **18. MUSEUMS** An art museum has an admission price of $15. An average of 620 people visit the museum each day. The museum's director predicts that each $0.50 decrease in the price of admission will result in 40 additional visitors per day.

     a. Let x represent the number of $0.50 price decreases. Write a function $A(x)$ to represent the museum's admission price.

     b. Write a function $V(x)$ to represent the number of visitors per day.

     c. Write a function $R(x)$ that can be used to find the museum's daily revenue.

     d. If the director decides to change the museum's admission price to $13, what daily revenue can the museum expect?

Example 3   **Given that $f(x) = -x^2 + 2x + 1$, $g(x) = 5x - 8$, $h(x) = -3x - 3$, and $k(x) = 5x^2$, find each function.**

19. $(f \cdot g)(x)$ **20.** $(f \cdot h)(x)$ **21.** $(h \cdot k)(x)$

22. $(g \cdot h)(x)$ **23.** $(g \cdot k)(x)$ **24.** $(f \cdot k)(x)$

Example 2   **25 BLOGS** Aurelio writes a sports blog and a photography blog. The sports blog has 385 subscribers and Aurelio expects the number of subscribers to increase at a rate of 30 subscribers per month. The photography blog has 590 subscribers and Aurelio expects the number of subscribers to increase by 15% per year.

     a. Write a function $f(t)$ to represent the number of subscribers to the sports blog after t years.

     b. Write a function $g(t)$ to represent the number of subscribers to the photography blog after t years.

     c. Find $(f + g)(t)$ and explain what this function represents.

     d. Find the total number of subscribers Aurelio expects to have after 6 years.

Use the table of values to find each of the following.

26. $(f + g)(-1)$

27. $(f \cdot g)(2)$

28. $(f - g)(0)$

29. $(f \cdot g)(0)$

30. $(g - f)(1)$

x	f(x)	g(x)
−2	2	3
−1	0	−5
0	−4	1
1	8	5
2	−7	−3

31. MANUFACTURING A company makes boxes in the shape of a rectangular prism with an open top. They start with a rectangular piece of cardboard that is 10 inches long and 8 inches wide. Then they cut identical squares from each corner, as shown, and fold up the sides.

     a. Let x represent the side length of the squares that are cut from the corners of the cardboard. Write a function $f(x)$ to represent the length of the resulting box.

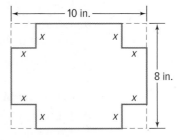

b. Write a function $g(x)$ to represent the width of the resulting box.

c. Find $(f \cdot g)(x)$ and explain what this function represents.

d. What is the domain of $(f \cdot g)(x)$? Explain.

32. **MP** **PERSEVERANCE** Suppose $f(x) = x^2 - 2x + 14$ and $(f + g)(x) = 4x^2 - 3x - 5$. Find $g(x)$.

Use the graph of $f(x)$ and $g(x)$ to find each of the following.

33. $(f + g)(1)$ **34.** $(f - g)(1)$ **35.** $(f + g)(-2)$

36. $(f \cdot g)(0)$ **37.** $(f \cdot g)(-1)$ **38.** $(f \cdot g)(-3)$

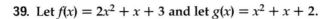

39. Let $f(x) = 2x^2 + x + 3$ and let $g(x) = x^2 + x + 2$.

a. Find $(f - g)(x)$ and $(g - f)(x)$. Explain how these two functions are related.

b. Graph $(f - g)(x)$ and $(g - f)(x)$.

c. How are the graphs of $(f - g)(x)$ and $(g - f)(x)$ related?

d. Do you think the relationship you noticed between the graphs of $(f - g)(x)$ and $(g - f)(x)$ will hold true given any functions $f(x)$ and $g(x)$? Explain.

40. **TABLETS** A telecommunications company offers a plan in which you purchase a tablet for $149 and then pay a monthly fee for online access. The function $C(x) = 29x + 149$ models the total cost after x months. Explain how you can think of $C(x)$ as the sum of two functions. What do each of these functions represent?

Given that $f(x) = 3^x$ and $g(x) = 2(3^x)$, find each function.

41 $(f + g)(x)$ **42.** $(f - g)(x)$ **43.** $(f \cdot g)(x)$

Describe each statement as *sometimes*, *always*, or *never* true.

44. If $f(x)$ and $g(x)$ are both linear functions, then the domain of $(f + g)(x)$ is all real numbers.

45. If $f(x)$ and $g(x)$ are both quadratic functions, then $(f - g)(x)$ is a linear function.

46. If $f(0) = 1$ and $g(0) = 1$, then the graph of $(f + g)(x)$ passes through the origin.

47. The function $(f \cdot g)(x)$ is a quadratic function.

H.O.T. Problems   Use **H**igher-**O**rder **T**hinking Skills

48. **OPEN-ENDED** Write two quadratic functions, $f(x)$ and $g(x)$, so that $(f + g)(x) = x^2 - 2x + 1$.

49. **MP** **CRITIQUE ARGUMENTS** Jeremy said that if $f(x)$ and $g(x)$ are both linear functions, then $(f \cdot g)(x)$ cannot be a linear function. Is he correct? Explain your answer.

50. **MP** **SENSE-MAKING** If $f(x)$ is a linear function and $g(x)$ is a quadratic function, what type of function is $(f + g)(x)$? Explain.

51. **ERROR ANALYSIS** Mikayla was asked to find $(f - g)(x)$ given that $f(x) = 3x^2 + 7x - 1$ and $g(x) = 2x^2 - 3x - 4$. She wrote $(f - g)(x) = x^2 + 4x - 5$.

a. What error did Mikayla make?

b. Find the correct function $(f - g)(x)$.

52. **WRITING IN MATH** Describe an example of a real-world situation in which you might write two functions, $f(x)$ and $g(x)$, and then subtract the functions to find $(f - g)(x)$. Explain what $f(x)$, $g(x)$, and $(f - g)(x)$ would represent in the real-world situation.

53. Given that $f(x) = 2x^2 - 4x$ and $g(x) = 6x - 5$, which of the following is $(f \cdot g)(x)$? **MP** 7

   ○ **A** $(f \cdot g)(x) = 2x^2 + 2x - 5$

   ○ **B** $(f \cdot g)(x) = 12x^3 - 34x^2 + 20x$

   ○ **C** $(f \cdot g)(x) = 12x^3 - 24x^2 + 20x$

   ○ **D** $(f \cdot g)(x) = 2x^2 - 10x + 5$

54. The graphs of $f(x)$ and $g(x)$ are shown. What is $(f - g)(-2)$? **MP** 1

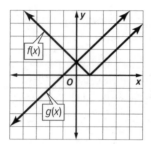

   ○ **A** -4

   ○ **B** -2

   ○ **C** 3

   ○ **D** 4

55. Given that $f(x) = 3x + 4$, $g(x) = -x^2 + 2x - 1$, and $h(x) = 10$, which of the following are quadratic functions? **MP** 7

   ☐ **A** $(f + g)(x)$

   ☐ **B** $(f - g)(x)$

   ☐ **C** $(f \cdot g)(x)$

   ☐ **D** $(f + h)(x)$

   ☐ **E** $(g \cdot h)(x)$

   ☐ **F** $(g - h)(x)$

56. What is the value of $(f + g)(3)$ if $f(x) = 2^x - 1$ and $g(x) = 3^x$? **MP** 7

   [   ]

57. **MULTI-STEP** In 2010, the population of Oakville was 82,400 and increasing at a rate of 2.5% per year. In 2010, the population of Elmwood was 75,600 and decreasing by an average of 300 residents each month. **MP** 4

   a.   Write a function $f(x)$ to represent the population of Oakville x years after 2010.

   b.   Write a function $g(x)$ to represent the population of Elmwood x years after 2010.

   c.   Find $(f - g)(x)$ and explain what this function represents.

   d.   Find $(f - g)(6)$ and explain what it represents.

58. Based on the table of values, what is $(f \cdot g)(2)$? **MP** 1, 7

x	$f(x)$	$g(x)$
-2	2	-8
-1	-3	3
0	2	-5
1	0	2
2	4	-3

   ○ **A** -16

   ○ **B** -12

   ○ **C** 1

   ○ **D** 7

59. For which pair of functions does $(f - g)(x) = x$? **MP** 7

   ○ **A** $f(x) = 2x^2 + 2x - 5$; $g(x) = -2x^2 - x + 5$

   ○ **B** $f(x) = -4x^2 - x + 1$; $g(x) = -4x^2 + x + 1$

   ○ **C** $f(x) = -x^2 + 3x - 2$; $g(x) = -x^2 + 2x - 2$

   ○ **D** $f(x) = x^2 + 10x + 4$; $g(x) = -x^2 + 9x + 4$

Go Online! for Vocabulary Review Games and key vocabulary in 13 languages

Study Guide

Key Concepts

Graphing Quadratic Functions (Lesson 9-1)

- A quadratic function can be described by an equation of the form $y = ax^2 + bx + c$, where $a \neq 0$.
- The axis of symmetry for the graph of $y = ax^2 + bx + c$, where $a \neq 0$, is $x = -\dfrac{b}{2a}$.

Transformations of Quadratic Functions (Lesson 9-2)

- $f(x) = x^2 + k$ translates the graph up or down.
- $f(x) = ax^2$ compresses or expands the graph vertically.

Solving Quadratic Equations (Lessons 9-3 through 9-6)

- Quadratic equations can be solved by graphing. The solutions are the x-intercepts or zeros of the related quadratic function.
- Some quadratic equations of the form $ax^2 + bx + c = 0$ can be solved by factoring and then using the Zero Product Property.

Solving Systems of Linear and Quadratic Equations (Lesson 9-7)

- To solve a system graphically, determine the point(s) of intersection of the graphs.

Analyzing Functions with Successive Differences (Lesson 9-8)

- If the differences of successive y-values are all equal, the data represent a linear function.
- If the second differences are all equal, but the first differences are not, the data represent a quadratic function.

Combining Functions (Lesson 9-9)

- To add or subtract functions, combine like terms.
- To multiply functions, apply the distributive property and combine like terms.

FOLDABLES® Study Organizer

Use your Foldable to review the chapter. Working with a partner can be helpful. Ask for clarification of concepts as needed.

Key Vocabulary

axis of symmetry (p. 559)	Quadratic Formula (p. 606)
completing the square (p. 596)	quadratic function (p. 559)
discriminant (p. 610)	standard form (p. 559)
double root (p. 581)	vertex (p. 559)
maximum (p. 559)	vertex form (p. 575)
minimum (p. 559)	
parabola (p. 559)	

Vocabulary Check

State whether each sentence is *true* or *false*. If *false*, replace the underlined term to make a true sentence.

1. The <u>axis of symmetry</u> of a quadratic function can be found by using the equation $x = -\dfrac{b}{2a}$.

2. The <u>vertex</u> is the maximum or minimum point of a parabola.

3. The graph of a quadratic function is a(n) <u>straight line</u>.

4. The graph of a quadratic function has a(n) <u>maximum</u> if the coefficient of the x^2-term is positive.

5. A quadratic equation with a graph that has two x-intercepts has <u>one</u> real root.

6. The expression $b^2 - 4ac$ is called the <u>discriminant</u>.

7. The solutions of a quadratic equation are called <u>roots</u>.

8. The graph of the parent function is <u>translated down</u> to form the graph of $f(x) = x^2 + 5$.

Concept Check

9. Explain the relationship between the solutions of a quadratic equation and the graph of the related quadratic function.

10. Explain how functions can be combined to form new functions.

Lesson-by-Lesson Review

9-1 Graphing Quadratic Functions

Consider each equation.

a. Determine whether the function has a *maximum* or *minimum* value.

b. State the maximum or minimum value.

c. What are the domain and range of the function?

11. $y = x^2 - 4x + 4$

12. $y = -x^2 + 3x$

13. $y = x^2 - 2x - 3$

14. $y = -x^2 + 2$.

15. **ROCKET** A toy rocket is launched with an upward velocity of 32 feet per second. The equation $h = -16t^2 + 32t$ gives the height of the ball t seconds after it is launched.

a. Determine whether the function has a *maximum* or *minimum* value.

b. State the maximum or minimum value.

c. State a reasonable domain and range for this situation.

Example 1

Consider $f(x) = x^2 + 6x + 5$.

a. Determine whether the function has a *maximum* or *minimum* value.

For $f(x) = x^2 + 6x + 5$, $a = 1$, $b = 6$, and $c = 5$.

Because a is positive, the graph opens up, so the function has a minimum value.

b. State the *maximum* or *minimum* value of the function.

The minimum value is the y-coordinate of the vertex. The x-coordinate of the vertex is $\frac{-b}{2a}$ or $\frac{-6}{2(1)}$ or -3.

$f(x) = x^2 + 6x + 5$ Original function

$f(-3) = (-3)^2 + 6(-3) + 5$ $x = -3$

$f(-3) = -4$ Simplify.

The minimum value is -4.

c. State the domain and range of the function.

The domain is all real numbers. The range is all real numbers greater than or equal to the minimum value, or $\{y \mid y \geq -4\}$.

9-2 Transformations of Quadratic Functions

Describe the transformations in each function as it relates to the graph of $f(x) = x^2$.

16. $f(x) = x^2 + 8$

17. $f(x) = x^2 - 3$

18. $f(x) = 2(x - 1)^2$

19. $f(x) = 4x^2 - 18$

20. $f(x) = \frac{1}{3}x^2$

21. $f(x) = \frac{1}{4}x^2$

Example 2

Describe the transformations in $f(x) = x^2 - 2$ as it relates to the graph of $f(x) = x^2$.

The graph of $f(x) = x^2 + k$ represents a translation up or down of the parent graph.

Since $k = -2$, the translation is down.

So, the graph is translated down 2 units from the parent function.

22. Write an equation for the function shown in the graph.

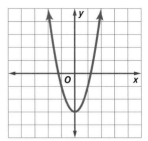

23. PHYSICS A ball is dropped off a cliff that is 100 feet high. The function $h = -16t^2 + 100$ models the height h of the ball after t seconds. Compare the graph of this function to the graph of $h = t^2$.

Example 3

Write an equation for the function shown in the graph.

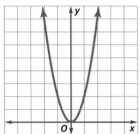

Since the graph opens upward, the leading coefficient must be positive. The parabola has not been translated up or down, so $c = 0$. Since the graph is stretched vertically, it must be of the form of $f(x) = ax^2$ where $a > 1$. The equation for the function is $y = 2x^2$.

9-3 Solving Quadratic Equations by Graphing

Solve each equation by graphing. If integral roots cannot be found, estimate the roots to the nearest tenth.

24. $x^2 - 3x - 4 = 0$

25. $-x^2 + 6x - 9 = 0$

26. $x^2 - x - 12 = 0$

27. $x^2 + 4x - 3 = 0$

28. $x^2 - 10x = -21$

29. $6x^2 - 13x = 15$

30. NUMBER THEORY Find two numbers that have a sum of 2 and a product of -15.

Example 4

Solve $x^2 - x - 6 = 0$ by graphing.

Graph the related function $f(x) = x^2 - x - 6$.

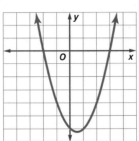

The x-intercepts of the graph appear to be at -2 and 3, so the solutions are -2 and 3.

9-4 Solving Quadratic Equations by Factoring

Solve each equation. Check your solutions.

31. $x^2 + 6x - 55 = 0$

32. $(g + 8)^2 = 49$

33. $y^2 - 4y = 32$

34. $3k^2 - 8k = 3$

35. $2n^2 + 4n = 16$

36. $4w^2 + 9 = 12w$

Example 5

Solve $x^2 - 2x = 120$ by factoring.

Write the equation in standard form and factor.

$x^2 - 2x = 120$	Original equation
$x^2 - 2x - 120 = 0$	Subtract 120 from each side.
$(x - 12)(x + 10) = 0$	Factor.
$x - 12 = 0$ or $x + 10 = 0$	Zero Product Property
$x = 12 \qquad x = -10$	Solve each equation

The solutions are 12 and -10.

9-5 Solving Quadratic Equations by Completing the Square

Solve each equation by completing the square. Round to the nearest tenth if necessary.

37. $x^2 + 6x + 9 = 16$

38. $-a^2 - 10a + 25 = 25$

39. $y^2 - 8y + 16 = 36$

40. $y^2 - 6y + 2 = 0$

41. $n^2 - 7n = 5$

42. $-3x^2 + 4 = 0$

43. $a^2 - 4a + 9 = 0$

44. $2a^2 - 4a + 1 = 0$

45. **NUMBER THEORY** Find two numbers that have a sum of -2 and a product of -48.

Example 6

Solve $x^2 - 16x + 32 = 0$ by completing the square. Round to the nearest tenth if necessary.

Isolate the x^2- and x-terms. Then complete the square and solve.

$x^2 - 16x + 32 = 0$	Original equation
$x^2 - 16x = -32$	Isolate the x^2- and x-terms.
$x^2 - 16x + 64 = -32 + 64$	Complete the square.
$(x - 8)^2 = 32$	Factor.
$x - 8 = \pm\sqrt{32}$	Take the square root.
$x = 8 \pm\sqrt{32}$	Add 8 to each side.
$x = 8 \pm 4\sqrt{2}$	Simplify.

The solutions are about 2.3 and 13.7.

9-6 Solving Quadratic Equations by Using the Quadratic Formula

Solve each equation by using the Quadratic Formula. Round to the nearest tenth if necessary.

46. $x^2 - 8x = 20$

47. $21x^2 + 5x - 7 = 0$

48. $d^2 - 5d + 6 = 0$

49. $2f^2 + 7f - 15 = 0$

50. $2h^2 + 8h + 3 = 3$

51. $4x^2 + 4x = 15$

52. **GEOMETRY** The area of a square can be quadrupled by increasing the side length and width by 4 inches. What is the side length?

State the discriminant for each equation. Then determine the number of real solutions of the equation.

53. $a^2 - 4a + 5 = 0$

54. $-6x^2 + 2x + 3 = 0$

Example 7

Solve $x^2 + 10x + 9 = 0$ by using the Quadratic Formula.

$x = \dfrac{-b \pm \sqrt{b^2 - 4ac}}{2a}$	Quadratic Formula
$= \dfrac{-10 \pm \sqrt{10^2 - 4(1)(9)}}{2(1)}$	$a = 1, b = 10, c = 9$
$= \dfrac{-10 \pm \sqrt{64}}{2}$	Simplify.
$x = \dfrac{-10 + 8}{2}$ or $x = \dfrac{-10 - 8}{2}$	Separate the solutions.
$= -1 \qquad\qquad = -9$	Simplify.

9-7 Solving Systems of Linear and Quadratic Equations

Solve each system of equations.

55. $y = x^2 - 4x + 4$
$y = 2x - 1$

56. $y = x^2 + 5x - 3$
$y = 15 - 2x$

57. $y = 2x^2 - 4x + 1$
$y - 5x = -3$

58. $y = 4x^2 + 4x - 3$
$x + y = 3$

59. $y = x^2 - 4x + 9$
$y = 2x$

60. $y = x^2 + 9$
$y = 10x$

61. $y = x^2 + 9$
$y = x$

Example 8

Solve the system of equations $\begin{cases} y = x^2 + 2x \\ y = x + 2 \end{cases}$.

The equations are solved for y. Substitute and solve for x.

$x^2 + 2x = x + 2$ Substitute.
$x^2 + x = 2$ Subtract x from each side.
$x^2 + x - 2 = 0$ Subtract 2 from each side.
$(x + 2)(x - 1) = 0$ Factor.
$x + 2 = 0$ or $x - 1 = 0$ Zero Product Property
$x = -2$ or $x = 1$ Solve for x.

Substitute to find the values of y.

$y = x + 2$ $y = x + 2$
$y = (-2) + 2$ $y = 1 + 2$
$y = 0$ $y = 3$

The solutions are $(-2, 0)$ and $(1, 3)$.

9-8 Analyzing Functions with Successive Differences

Look for a pattern in each table of values to determine which kind of model best describes the data. Then write an equation for the function that models the data.

62.

x	0	1	2	3	4
y	0	3	12	27	48

63.

x	0	1	2	3	4
y	1	2	4	8	16

64.

x	0	1	2	3	4
y	0	-1	-4	-9	-16

65.

x	0	1	2	3	4
y	8	5	2	-1	-4

Example 9

Determine the model that best describes the data. Then write an equation for the function that models the data.

x	0	1	2	3	4
y	3	4	5	6	7

Step 1 First differences: 3 4 5 6 7
                              1 1 1 1

A linear function models the data.

Step 2 The slope is 1 and the y-intercept is 3, so the equation is $y = x + 3$.

9-9 Combining Functions

Consider the following functions.

$f(x) = 4x^2 - 8x + 2$

$g(x) = -5x^2 + x + 1$

$h(x) = 3 - x^2$

Determine each of the following.

66. $(f + g)(x)$

67. $(g + h)(x)$

68. $(f - h)(x)$

69. $(h - g)(x)$

70. $(f \cdot h)(x)$

71. $(g \cdot h)(x)$

Example 10

Let $f(x) = -2x^2 - x + 5$ and $g(x) = x^2 - 1$. Determine $(f \cdot g)(x)$ and $(f - g)(x)$.

$(f \cdot g)(x)$	Original expression
$= (-2x^2 - x + 5)(x^2 - 1)$	Substitute.
$= -2x^4 + 2x^2 - x^3 + x + 5x^2 - 5$	Distribute.
$= -2x^4 - x^3 + 7x^2 + x - 5$	Combine like terms.
$(f - g)(x)$	Original expression
$= (-2x^2 - x + 5) - (x^2 - 1)$	Substitute.
$= -2x^2 - x + 5 - x^2 + 1$	Distribute.
$= -3x^2 - x + 6$	Combine like terms.

Go Online! for another Chapter Test

Use a table of values to graph the following functions. State the domain and range.

1. $y = x^2 + 2x + 5$

2. $y = 2x^2 - 3x + 1$

Consider $y = x^2 - 7x + 6$.

3. Determine whether the function has a *maximum* or *minimum* value.

4. State the maximum or minimum value.

5. What are the domain and range?

Describe how the graph of each function is related to the graph of $f(x) = x^2$.

6. $g(x) = x^2 - 5$

7. $g(x) = -3x^2$

8. $h(x) = \frac{1}{2}x^2 + 4$

9. MULTIPLE CHOICE Which is an equation for the function shown in the graph?

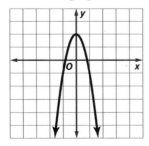

A $y = -3x^2$

B $y = 3x^2 + 1$

C $y = x^2 + 2$

D $y = -3x^2 + 2$

Solve each equation by graphing. If integral roots cannot be found, estimate the roots to the nearest tenth.

10. $x^2 + 7x + 10 = 0$

11. $x^2 - 5 = -3x$

Solve each equation by factoring.

12. $x^2 + 18x + 81 = 0$

13. $x^2 - 18x = 40$

14. $5x^2 - 31x + 6 = 0$

Solve each equation by completing the square. Round to the nearest tenth if necessary.

15. $x^2 + 2x + 5 = 0$

16. $x^2 + 5x - 8 = 12$

17. $2x^2 - 36 = -6x$

Solve each equation by using the Quadratic Formula. Round to the nearest tenth if necessary.

18. $x^2 - x - 30 = 0$

19. $x^2 - 10x = -15$

20. $2x^2 + x - 15 = 0$

21. BASEBALL Elias hits a baseball into the air. The equation $h = -16t^2 + 60t + 3$ models the height h in feet of the ball after t seconds. How long is the ball in the air?

Solve each system of equations algebraically.

22. $y = x^2 - 7x + 3$
$y = -2x - 1$

23. $y = 3x^2 - 8x - 1$
$3x + y = 1$

24. $y = 1 - x^2$
$y = 4x + 5$

25. Graph $\{(-2, 4), (-1, 1), (0, 0), (1, 1), (2, 4)\}$. Determine whether the ordered pairs represent a *linear function*, a *quadratic function*, or an *exponential function*.

26. CAR CLUB The table shows the number of car club members for four consecutive years after it began.

Time (years)	0	1	2	3	4
Members	10	20	40	80	160

a. Determine which model best represents the data.

b. Write a function that models the data.

c. Predict the number of car club members after 6 years.

Consider the functions $f(x) = 2x^2 - 4x + 1$ **and** $g(x) = -6x^2 + 9x - 8$.

27. Determine $(f + g)(x)$.

28. Determine $(f - g)(x)$.

29. Determine $(f \cdot g)(x)$.

Performance Task

Provide a clear solution to each part of the task. Be sure to show all of your work, include all relevant drawings, and justify your answers.

BASEBALL A professional baseball player hits a baseball. Its height above the field can be modeled by the equation $y = -0.002x^2 + 0.78x + 4$, where y is height of the baseball, in feet, and x is the horizontal distance the baseball is from home plate, in feet.

Part A

1. What is the height of the baseball when it is 20 feet from home plate?

2. At what distance(s) is the baseball from home plate when its height is 15 feet? Round to the nearest tenth.

Part B

We can use the quadratic equation to find how far the baseball will travel before it hits the ground. If the baseball is hit towards left center field, where the fence is 370 feet from home plate, we can determine whether the baseball will travel far enough to be a home run.

3. **Modeling** Write an equation that could be used to determine if the baseball will travel far enough to be home run.

4. **Reasoning** Solve the equation and round to the nearest tenth, if necessary. Based on the solution, does the baseball travel far enough to be home run? Explain.

Part C

The graph of the related quadratic function reveals characteristics of the baseball's trajectory.

5. What is the maximum height of the baseball?

6. How far away from home plate does the baseball reach its maximum height?

7. How far above the ground was the baseball at the moment it was hit by the player?

Part D

Another player hits a foul ball. Its height y can be modeled by $y = -0.02x^2 + x + 4$ where x is the distance, in feet, from home plate. A fan runs up the stands to catch the foul ball. His height above the field y, in feet, is approximately modeled by $y = 0.7x - 21$, where x is his distance from home plate, in feet.

8. At what distance and height will the fan catch the baseball?

Part E

9. **Construct an Argument** Is it more reasonable to use a linear or quadratic model for the trajectory of a baseball after it is hit by the batter?

Test-Taking Strategy

Read the problem. Identify what you need to know. Then use the information in the problem to solve.

Find the exact roots of the quadratic equation $-2x^2 + 6x + 5 = 0$.

A $\dfrac{3 \pm \sqrt{17}}{4}$ **C** $\dfrac{3 \pm \sqrt{19}}{2}$

B $\dfrac{4 \pm \sqrt{17}}{3}$ **D** $\dfrac{3 \pm \sqrt{19}}{4}$

Step 1 **What quantities are given in the problem?**
A quadratic equation in the form $ax^2 + bx + c = 0$, where $a = -2$, $b = 6$, and $c = 5$.

Step 2 **What quantities do you need to find?**
The roots, which are the values of x that make the equation true.

Step 3 **Is there a formula that relates these quantities? If so, write it.**
Yes, the Quadratic Formula can be used. $x = \dfrac{-b \pm \sqrt{b^2 - 4ac}}{2a}$

Step 4 **Substitute the known quantities to solve for the unknown quantity.**

$$x = \dfrac{-6 \pm \sqrt{(6)^2 - 4(-2)(5)}}{2(-2)} = \dfrac{-6 \pm \sqrt{76}}{-4}$$
$$= \dfrac{3 \pm \sqrt{19}}{2}$$

The correct answer is C.

> **Test-Taking Tip**
> Using a Formula A *formula* is an equation that shows a relationship among certain quantities. The two most important things to remember when using a formula are to substitute the correct values for the appropriate variables, and be careful to simplify or solve correctly. If you do these two things, the formula will guide you to the correct answer.

Apply the Strategy

Read each problem. Identify what you need to know. Then use the information in the problem to solve.

1. Find the exact roots of the quadratic equation $x^2 + 5x - 12 = 0$.

   A $\dfrac{-5 \pm \sqrt{73}}{2}$ **C** $\dfrac{-3 \pm \sqrt{73}}{4}$

   B $\dfrac{4 \pm \sqrt{61}}{3}$ **D** $\dfrac{-1 \pm \sqrt{61}}{2}$

Answer the questions below.

1. What quantities are given in the problem?
2. What quantities do you need to find?
3. Is there a formula that relates these quantities? If so, write it.
4. Substitute the known quantities to solve for the unknown quantity. What is the correct answer?

Read each question. Then fill in the correct answer on the answer document provided by your teacher or on a sheet of paper.

1. Celia graphed the set of ordered pairs in the table below.

x	2	3	4	5	6
y	3	0	−1	0	3

Which equation best models the data in the table?

○ **A** $y = 2x − 1$

○ **B** $y = −2x + 7$

○ **C** $y = x^2 + 8x + 15$

○ **D** $y = x^2 − 8x + 15$

2. The graph of $f(x) = x^2$ is reflected across the x-axis and translated to the right 5 units. What is the value of b when the equation of the transformed graph is written in standard form $f(x) = ax^2 + bx + c$?

[ ]

3. **MULTI-STEP** Maria starts with $200 in her savings account and deposits $25 each month. Stefan starts with $275 in his savings account and deposits $20 each month.

a. Write an algebraic equation to represent the amount, a, in in dollars, that Maria and Stefan each have in their respective savings accounts after t months.

Maria: [ ]

Stefan: [ ]

b. In what month do Maria and Stefan have the same amount in their savings accounts?

[ ]

c. 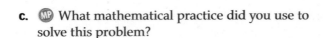 What mathematical practice did you use to solve this problem?

4. Which statements describe the function graphed below? Select all that apply.

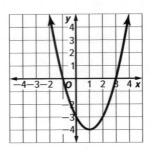

☐ **A** The range is $\{y \mid y \geq −4\}$.

☐ **B** The y-intercept is 3.

☐ **C** The maximum value is −4.

☐ **D** The equation of the axis of symmetry is $x = −4$.

☐ **E** The x-intercepts are −1 and 3.

5. Which equation has no real solutions?

○ **A** $x^2 − 4x = −4$

○ **B** $x^2 − 4x = −8$

○ **C** $x^2 + 4x = −2$

○ **D** $x^2 + 4x = 2$

6. Let $f(x) = 3x^2 − 5x + 1$ and $g(x) = x^2 − x − 6$. Write an expression for $(g − f)(x)$ in standard form.

[ ]

7. Which inequality has the solution set graphed below?

○ **A** $−\frac{1}{3}x < 1$ ○ **C** $−\frac{1}{3}x < −1$

○ **B** $−\frac{1}{3}x > 1$ ○ **D** $−\frac{1}{3}x > −1$

8. Karl solved a quadratic equation by graphing the related function as shown below.

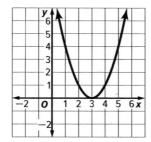

Which equation did he solve?

- **A** $x^2 + 6x = 9$
- **B** $x^2 + 6x = -9$
- **C** $x^2 - 6x = 9$
- **D** $x^2 - 6x = -9$

9. What are the solutions of the quadratic equation $5x^2 - 2x = 8$? Round to the nearest tenth if necessary.

- **A** no solution
- **C** 1.1 or −1.5
- **B** −1.1
- **D** −1.1 or 1.5

10. For what domain values of the function $y = -2x + 1$ is the range of the function $\{y \mid y < 5\}$?

- **A** $\{x \mid x < 2\}$
- **C** $\{x \mid x < -2\}$
- **B** $\{x \mid x > 2\}$
- **D** $\{x \mid x > -2\}$

Test-Taking Tip

Question 8 First, use the graph to find the zero of the function. Then, use what you know about the relationship between the zeros of a quadratic function and their corresponding linear factors to identify the correct equation.

---

11. If $b = 12$, what value of c makes $x^2 + bx + c$ a perfect square trinomial?

12. Which expression has a value of 25?

- **A** $5^{\frac{2}{3}}$
- **B** $5^{\frac{3}{2}}$
- **C** $125^{\frac{2}{3}}$
- **D** $125^{\frac{3}{2}}$

13. The graph of the quadratic function $h(x)$ is shown below.

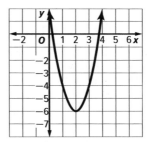

Which of the following statements best describes the function?

- **A** $h(x)$ is a vertical compression of the graph of $f(x) = x^2$.
- **B** $h(x)$ is a reflection across the x-axis of the graph of $f(x) = x^2$.
- **C** In vertex form, the equation of the function is $f(x) = 2(x - 2)^2 - 6$.
- **D** In standard form, the equation of the function is $f(x) = 2x^2 - 8x + 14$.

Need Extra Help?

If you missed Question...	1	2	3	4	5	6	7	8	9	10	11	12	13
Go to Lesson...	9-8	9-2	6-2	9-1	9-5	9-9	5-2	9-3	9-6	3-1	9-5	7-3	9-2

THEN

You calculated simple probability.

NOW

In this chapter, you will:

- Determine which measure of center best describes a set of data.
- Represent data using dot plots, histograms, bar graphs, and box plots and analyze their shapes.
- Summarize data in two-way frequency tables.
- Describe the effects linear transformations have on measures of center and spread.

(MP) WHY

FOOD SERVICE The food service industry includes everything from food trucks to five-star restaurants. Regardless of the venue, customer satisfaction is important. A restaurant owner can use surveys and analysis to improve the customer experience.

1. Use Tools A food truck owner surveys a random sampling of customers who rate their experiences from 1 (poor) to 10 (excellent).

Food Truck Customer Survey				
8	9	9	10	7
9	4	8	9	10

2. Apply Reasoning What can you infer from this data set? What is the average rating?

3. Create a Graph Create a histogram using the data from the survey.

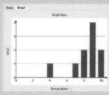

4. Discuss What rating did the food truck receive the most? What does this tell you about customer satisfaction?

Go Online to Guide Your Learning

Explore & Explain	Organize

 Normal Distribution

The Normal Distribution mat in eToolkit is a useful tool for enhancing your understanding of the distribution of data discussed in Lesson 10-4.

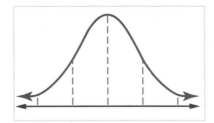

 Foldables

Get organized! Create this Foldable to help you organize your Chapter 10 notes about statistics.

 Box-and-Whisker Plots

Use two of the Box-and-Whisker Plot tools from the eToolkit to compare sets of data, as explored in Lesson 10-5.

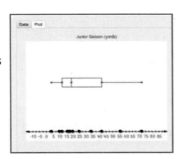

Collaborate

Chapter Project

In the **I Have the Best Idea for a Show** project, you will use what you have learned about distributions of data to complete a project that addresses business literacy.

eBook

Interactive Student Guide

Before starting the chapter, answer the **Chapter Focus** preview questions. Check your answers as you complete each lesson. At the end of the chapter, try the **Performance Task**.

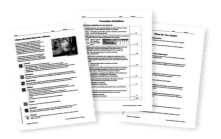

Focus

 LEARNSMART

Need help studying? Complete the **Descriptive Statistics** domain in LearnSmart to review for the chapter test.

 ALEKS

You can use the **Data Analysis and Probability** topic in ALEKS to find out what you know about statistics and what you are ready to learn.*

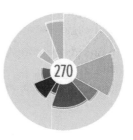

*Ask your teacher if this is part of your program.

Get Ready for the Chapter

abc *Go Online!* for Vocabulary Review Games and key vocabulary in 13 languages.

Connecting Concepts

Concept Check

Review the concepts used in this chapter by answering the questions below.

1. Describe how to determine the probability of selecting a specific color cube from a bag with 1 green cube, 6 red cubes, and 4 yellow cubes.

2. Define probability using the terms "event" and "outcomes."

3. What is the equation $a(b + c) = ab + ac$ an example of?

4. What is an absolute value?

5. If all the values in a data set are increased by the same amount, what happens to the range?

6. Describe how you would determine the volume of a cylinder.

7. Given the linear equation $y = 3x + 1$, illustrated here, how were the data points determined so that the line could be graphed?

$y = 3x + 1$

8. What do you call the sets of data points you create to graph a linear equation, as was done for the equation and graph in question 7?

Performance Task Preview

You can use the concepts and skills in this chapter to solve problems in a real-world setting. Understanding statistics will help you finish the Performance Task at the end of the chapter.

MP **In this Performance Task you will:**

- construct viable arguments
- attend to precision
- use appropriate tools strategically
- reason abstractly and quantitatively
- look for and make use of structure

New Vocabulary

English		Español
measures of central tendency	p. 651	medidas de tendencia central
mean	p. 651	media
median	p. 651	mediana
mode	p. 651	moda
frequency table	p. 659	tabla de frecuencias
bar graph	p. 659	gráfico de barra
cumulative frequency	p. 659	frecuencia acumulativa
histogram	p. 659	histograma
range	p. 665	rango
quartile	p. 665	cuartile
measures of position	p. 665	medidas de la posición
lower quartile	p. 665	cuartil inferior
upper quartile	p. 665	cuartil superior
interquartile range	p. 666	amplitud intercuartílica
outlier	p. 666	valores atípico
standard deviation	p. 667	desviación tipica
variance	p. 667	varianza
distribution	p. 673	distribución
linear transformation	p. 680	transformación lineal
relative frequency	p. 690	frecuencia relativa

Review Vocabulary

bivariate data datos bivariate **data with two variables**

box-and-whisker plot diagrama de caja patillas a diagram that divides a set of data into four parts using the median and quartiles; a box is drawn around the quartile values, and whiskers extend from each quartile to the extreme data points

Measures of Center

©Hill Street Studios/Blend Images LLC

:: Then

- You collected data and were introduced to the concepts of mean, median, and mode.

:: Now

1. Represent sets of data by using measures of center.

2. Represent sets of data by using percentiles.

:: Why?

- Each year, many students are given standardized achievement tests. Most of these test scores are reported as percentiles, and they may also include graphics with sliding scales to clarify scores that may otherwise be difficult to interpret.

 New Vocabulary

variable
quantitative data
qualitative data
measures of center
measures of central tendency
mean
median
mode
percentiles

MP Mathematical Practices

3 Construct viable arguments and critique the reasoning of others.

4 Model with mathematics.

7 Look for and make use of structure.

1 Measures of Center A **variable** is any characteristic, number, or quantity that can be counted or measured. A variable is an item of data. Data that can be measured are called **quantitative data**. Data that can be organized into different categories are called categorical or **qualitative data**. Quantitative data in one variable are often summarized using a single number to represent what is average or typical. Measures of what is average are also called **measures of center** or **central tendency**. The most common measures of center are mean, median, and mode.

Key Concept Measures of Center

- The **mean** is the sum of the values in a data set divided by the total number of values in the set.

- The **median** is the middle value or the mean of the two middle values in a set of data when the data are arranged in numerical order.

- The **mode** is the value or values that appear most often in a set of data. A set of data can have no mode, one mode, or more than one mode.

Real-World Example 1 Measures of Center

BASEBALL The table shows the number of hits Julius made for his baseball team. Find the mean, median, and mode.

Team Played	Hits
Badgers	3
Hornets	6
Bulldogs	4
Vikings	0
Rangers	3
Panthers	7

Mean: To find the mean, find the sum of all the hits and divide by the number of games in which he made these hits.

Julius's team played 6 other teams.

$$\text{Mean} = \frac{3+6+4+0+3+7}{6} = \frac{23}{6} \approx 3.83 \text{ or about 4 hits}$$

Median: To find the median, order the numbers from least to greatest and find the middle value or values.

0, 3, 3, 5, 6, 7

$$\text{median} = \frac{3+5}{2}, \text{ or 4 hits}$$

Because there is an even number of values, find the mean of the middle two.

Mode: From the arrangement of data values, you can see that the value that occurs most often is 3, so the mode of the data set is 3 hits.

Julius's mean and median number of hits for these baseball games was 4, and his mode was 3 hits.

Day	Tips ($)
Monday	47
Tuesday	52
Wednesday	68
Friday	90
Sunday	73

▶ **Guided Practice**

1. TIPS Gloria is a server at a popular restaurant. The table shows the total tips she received each day she worked. Find the mean, median, and mode.

In Example 1, the mean and median are close together, so they both represent the average of Julius's number of hits well. Notice that the median is greater than the mean. This indicates that the games with fewer hits than the median are more spread out than the games with more hits than the median. The mode is greater than most of the games.

Example 2   Use Measures of Center to Analyze Data

SALARY Compare and contrast the measures of center of the employee salaries for the two stores. Based on the statistics, which store pays its employees better?

Hourly Salaries ($)	
Big Win Games	Game Place
10.80, 11, 11.50, 10, 10.90, 13.90, 10.80, 11.20	9.50, 9.50, 10.40, 10.40, 9.50, 10.80, 20.50, 9.50

Big Win Games
Mean:

$$\frac{10.8 + 11 + 11.5 + 10 + 10.9 + 13.9 + 10.8 + 11.2}{8}$$

$\approx \$11.26$

Median: $\frac{10.9 + 11}{2}$ or 10.95

Mode: 10.80

Game Place
Mean:

$$\frac{9.5 + 9.5 + 10.4 + 10.4 + 9.5 + 10.8 + 20.5 + 9.5}{8}$$

$= \$9.95$

Median: $\frac{9.5 + 10.4}{2}$ or 9.95

Mode: 9.50

The mean salary for both stores is equal, but the median and mode for Big Win Games is greater. At Game Place, $20.50 is an outlier causing the mean to increase. So, the mean is not an accurate representation of the salary of most employees at that store.

Because the median and mode are greater, Big Win Games pays their employees better.

▶ **Guided Practice**

2. TOURNAMENT Mariana and Rachael are in a fishing tournament. Compare and contrast the measures of center of the fish they caught. Based on the statistics, who should win the tournament?

Fish Length (in.)	
Mariana	Rachael
12.5, 10.25, 10, 11.75, 12, 10.25, 10	10.5, 11.5, 8, 10.5, 8.5, 18.5, 9.25

You may find that certain measures of center do not give you the information you need to fully analyze a situation. In that case, you will need to determine which additional statistics would be useful to have.

Example 3   Determine Best Measures of Center

Analyze each situation. Which measure of center would best describe the data? Explain.

a. Researching the employee salary at a specific company

The median would be best. Because a few employees may have a significantly greater salary, the mean may not accurately describe the salary of most employees.

b. The attendance of a football game at a certain high school

The mean attendance for each game in a season would be best. While the attendance changes with each game, it is unlikely that there is an outlier.

Guided Practice

Which measure of center may be the best to use to describe the data?

3A. A professional basketball player negotiating his salary.

3B. Planning a food budget by analyzing the monthly costs for the previous year

---

2 Percentiles A **percentile** is a measure that is often used to report test data, such as standardized test scores. Percentiles measure rank from the bottom and tell us what percent of the scores were below a given score. The lowest score is the 1st percentile and the highest score is the 99th percentile. There is no 0 or 100th percentile rank.

> ### Key Concept   Finding Percentiles
>
> **To find the percentile rank of an element of a data set, use these steps.**
>
> **Step 1** Order the data values from greatest to least.
>
> **Step 2** Find the number of data values less than the chosen element. Divide that number by the number of values in the set.
>
> **Step 3** Multiply the value from Step 2 by 100.

Study Tip

Percent vs. Percentile
Percent and *percentile* have different meanings. For example, a score at the 40th percentile means 40% of the scores are either the same as the score of the 40th percentile or less than that rank. It does not mean a person scored 40% of the possible points.

Example 4   Find the Percentile Rank of a Data Value

A talent show was held for 20 finalists in the Teen Idol contest. Each performer received a score from 0 through 30, with 30 being the highest. Find Victor's percentile rank.

Step 1 Order the scores from greatest to least.

29	28	27	26	25	22	21	20	18	17
16	15	14	12	11	10	9	6	5	4

Steps 2 and 3 Find Victor's percentile rank.

Victor had a score of 28. There are 18 scores below his score. To find his percentile rank, use the following formula.

$$\frac{\text{number of scores below }28}{\text{total number of scores}} \cdot 100 = \frac{18}{20} \cdot 100 = 90$$

So, Victor scored at the 90th percentile in the contest.

Name	Score	Name	Score
Arnold	17	Ishi	27
Benito	9	James	20
Brooke	25	Kat	16
Carmen	21	Malik	10
Daniel	14	Natalie	26
Delia	29	Pearl	4
Fernando	15	Twyla	6
Heather	12	Victor	28
Horatio	5	Warren	22
Ingrid	11	Yolanda	18

Guided Practice

4. Find Fernando's percentile rank.

Check Your Understanding   ◯ = Step-by-Step Solutions begin on page R11.

Go Online! for a Self-Check Quiz

Example 1

1. The table shows the number of students working at the concession stand each hour. Find the mean, median, and mode of the number of students. Which measure represents the data the best? Explain.

Hour	1	2	3	4	5
Number of Students	3	8	6	4	2

Example 2

2. **READING** Lee and Romina have different English teachers. Both teachers have 5 books that are required reading for their class. The page counts for the books are shown in the table. Compare and contrast the measures of center for the books in each class. Based on the statistics, which class is reading longer books?

Book Page Counts	
Lee's Class	Romina's Class
103, 114, 708, 98, 122	237, 178, 225, 206, 232

Example 3

Analyze each situation. Which measure of center best describes the data? Explain.

3. The miles per gallon ratings for cars from all major car manufacturers

4. The property value of homes in a suburban town

Example 4

(5) **OLYMPICS** The table shows the total medal counts for some countries at the 2014 Sochi Olympics.

a. Find Canada's percentile rank.

b. Are there any countries at the 50th percentile mark? If so, which ones?

6. **OLYMPICS** Use the table for Exercise 5.

a. Find the percentile rank for the United States.

b. Which countries are below the 50th percentile in medal counts?

c. Which countries are above the 70th percentile?

Country	Total	Country	Total
Belarus	6	Switzerland	11
France	15	Netherlands	24
Austria	17	United States	28
Norway	26	Germany	19
Russia	33	Canada	25

Source: Business Insider

Practice and Problem Solving

Extra Practice is on page R10.

Example 1

7. **COMPUTER TABLETS** The table shows the prices of comparable computer tablets at ten different locations.

a. Find the mean, median, and mode for the data set.

b. You want to buy one of the tablets. Which is the better measure of center to consider? Justify your reasoning.

Tablet Prices ($)
289.95, 259.95, 310, 349.95, 459.95, 399.95, 259.95, 300, 389, 405

Examples 2, 3

8. **POPULATIONS** The table shows the populations of the six largest cities in North Carolina and Colorado, according to a recent census. Compare and contrast the measures of center for the two states. Based on the statistics, which state has more populated cities?

North Carolina		Colorado	
City	Population (thousands)	City	Population (thousands)
Charlotte	731	Aurora	325
Durham	228	Colorado Springs	416
Fayetteville	201	Denver	600
Greensboro	270	Fort Collins	144
Raleigh	404	Lakewood	143
Winston-Salem	230	Thornton	119

Example 4

9. The table below shows the results of a recent Algebra test in a class of 18 students.

Name	Abby	Brian	Cassie	Dan	Emma	Fritz	Gabby	Huang	Imogen
Score	84	96	80	72	80	80	84	92	84
Name	Jenny	Kelly	Larry	Marcus	Nathan	Owen	Patti	Rachel	Yolanda
Score	100	84	80	88	80	76	80	84	92

Find Marcus's percentile rank.

10. Give a real-world example of data for which the given measure of center best represents the data.

   a. mean

   b. median

   c. mode

11. Darla is training for a 5K race. Her practice times (m:s) are 20:45, 21:30, 21:15, 22:32, and 21:40.

   a. Find her mean practice time in minutes and seconds. (*Hint:* Change the practice times to seconds first.)

   b. Darla ran the actual race 30 seconds faster than her median practice time. What was her race time?

12. **HEIGHT** The heights of the students in class are given in the table.

   a. In what percentile is the tallest person in the class?

   b. In what percentile is the third shortest person in the class?

   c. Is there a person in the 50th percentile? If so, who?

   d. Is there someone in the 100th percentile? If so, who?

Student	Height (inches)
Charlene	61
Anthony	64
Bart	65
Renata	59
Pedro	63
Georgio	58
Jin	62
Essie	63
Caroline	60
Antoine	59

13. The table shows the scores from six judges at a gymnastics meet. The final score is calculated by dropping the highest and lowest scores and then finding the mean of the 4 remaining scores.

   a. Find the final score for each gymnast. Who won the meet?

   b. What is the mean score for each gymnast if all six judges are considered? Who would win?

   c. Compare your result for parts **a** and **b**. What can you conclude about the scoring for the meet?

	Judge 1	Judge 2	Judge 3	Judge 4	Judge 5	Judge 6
John	9.6	9.6	9.8	9.4	9.6	9.8
Raul	9.5	9.3	9.8	9.4	9.2	9.7
Laird	9.6	9.8	9.4	9.2	9.7	9.4
Boris	9.4	9.2	9.6	9.8	9.5	9.6
Han	9.7	9.3	9.7	9.4	9.3	9.7

14. A value that divides a set of data into ten parts of equal size is called a *decile*. The first decile contains data up to but not including the 10th percentile. The second decile contains data from the 10th percentile up to but not including the 20th decile, and so on.

   a. Use the table for Example 4. Which contestants' scores fall into the sixth decile?

   b. In which decile are Heather and Daniel?

H.O.T. Problems   Use Higher-Order Thinking Skills

15. **MP ARGUMENTS** A student in another class says that a set of data cannot have the same mean and mode. Do you agree or disagree? Explain. Include an example or counterexample.

16. **WRITING IN MATH** How can you describe a data value based on its position in the data set?

17. **MP REASONING** A set of data can only have one mean and one median. How many values can a set of data have for the mode? Explain your reasoning.

18. **CHALLENGE** Write an example of a data set with an equal number of scores at the 25th percentile, 50th percentile, and 75th percentile.

19. **WRITING IN MATH** Compare and contrast percentile rank and percent score.

20. **OPEN ENDED** Write an example of a data set whose mode is greater than the mean and median.

21. **MODELING** Create a data set of test scores with the median lower than the mean. Describe what would happen to the mean if an unusually high test score were added to the data set.

22. MULTI-STEP The top ten state populations, in millions, according to the 2010 Census, are shown in the table below. Ⓜ 1, 4, 6, 7

State	Population (Millions)
California	37.3
Florida	18.8
Georgia	9.7
Illinois	12.8
Michigan	9.9
New York	19.4
North Carolina	9.5
Ohio	11.5
Pennsylvania	12.7
Texas	25.1

a. What is the mean population for these states?

b. Which is the median population for these states?
- **A** 9.9 million
- **B** 12.7 million
- **C** 12.75 million
- **D** 12.8 million

c. Find the percentile rank of North Carolina among these ten states.

d. Find the percentile rank of Georgia.

e. Is there any state at the 50th percentile? If so, list it.

f. Which states are below the 20th percentile? Select all that apply.
- **A** Ohio
- **B** Georgia
- **C** Illinois
- **D** North Carolina

23. Amit scored 98, 80, 92, 79, 84, 88, and 80 for one semester. Find the mean, median, and mode for his test scores. Ⓜ 4, 6
- **A** mean = 86.83, median = 84, mode = 80
- **B** mean = 85.86, median = 84, mode = 80
- **C** mean = 85.86, median = 84, no mode
- **D** mean = 85.86, median = 86, mode = 80

24. You are researching the number of free throws the basketball team makes each game over the course of the regular season. Which measure of center may be used best to describe the data you collect? Ⓜ 1, 4, 7
- **A** mean   **B** median   **C** mode

25. Write an example of a data set consisting of five numbers in which the median is equal to the mean. Ⓜ 2, 6

26. Determine whether the following statement is *true* or *false*. Justify your answer. Ⓜ 3

The percentile rank of the team with the highest score is at the 100th percentile.

27. Which measure of center may best describe the number of participants in the school activities? Explain. Ⓜ 1, 3, 6, 4

Club	Number of Members
Student Council	48
Dance Committee	15
Student Government	60
Chemistry Club	7
Principal's Assistants	2
Band	120
Library Assistants	3
Culinary Club	18
Honors Society	90
Community Service	99

Representing Data

Corbis Premium RF/Alamy

:: Then

- You used measures of center to compare data sets.

:: Now

1. Represent data by using dot plots.

2. Determine whether a discrete or continuous graphical representation is appropriate, and then create the bar graph or histogram.

:: Why?

- The final scores for golfers in a tournament can be divided into equal intervals. A histogram is a visual way to see the frequency of each interval.

 New Vocabulary
dot plot
frequency table
bar graph
cumulative frequency
histogram

MP Mathematical Practices
1 Make sense of problems and persevere in solving them.
4 Model with mathematics.

1 Representing Data with Dot Plots A **dot plot** is a diagram that shows the frequency of data on a number line. Dot plots are also called *line plots*. When data are represented as a dot plot, the gaps and clusters of the data become more apparent.

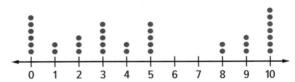

🔑 Key Concept Making a Dot Plot

Step 1 Write the data in order from least to greatest.

Step 2 Draw a number line that starts at the least data point and ends at the greatest data point. Choose an appropriate scale.

Step 3 Plot the dots on the number line. Stack the points when there is more than one data point with the same number.

Step 4 If appropriate, include a label for the number line and title for the dot plot.

Example 1 Make a Dot Plot

Represent the data as a dot plot.

11, 12, 14, 15, 12, 13, 15, 13, 9, 15, 12, 13, 15, 15, 11

Step 1 Write the data points in order from least to greatest.

9, 11, 11, 12, 12, 12, 13, 13, 13, 14, 15, 15, 15, 15, 15

Step 2 Make a number line that starts at the least data point and ends at the greatest data point. Choose an appropriate scale.

The data are whole numbers ranging from 9 to 15. So, make a number line starting at 9 with intervals of 1.

Step 3 Plot dots on the number line. Stack the dots when there is more than one of the same value.

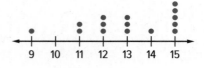

Step 4 If appropriate, include a label for the number line and title for the dot plot.

Since no information is given regarding what these data represent, no title is needed for this dot plot.

Guided Practice

1A. 42, 40, 40, 45, 50, 42, 50, 46, 50, 40, 45, 40, 43, 45

1B. 100, 101, 106, 105, 100, 102, 102, 101, 101, 100, 100, 108, 100, 101

Real-World Example 2   Make a Dot Plot with a Scaled Number Line

INTERNET USAGE The table shows Internet users of Middle Eastern countries as a percentage of their total population. Represent the data as a dot plot.

Step 1 Write the data points in order from least to greatest.

22.6, 28.1, 33.0, 57.2, 64.6, 65.9, 74.7, 78.6, 78.7, 80.4, 86.1, 91.9, 93.2, 96.4

Step 2 Make a number line. Choose an appropriate scale.

The data range from 22.6 to 96.4. The data represent a broad range with specific values, so it is unlikely that any data point is represented more than once. To represent the data in a meaningful way, divide the range into equal intervals.

Step 3 Plot the dots on the number line. Stack the points when there is more than one data point in the same interval.

Country	Internet Users (% of Population)
Bahrain	96.4
Iran	57.2
Iraq	33.0
Israel	74.7
Jordan	86.1
Kuwait	78.7
Lebanon	80.4
Oman	78.6
Palestine	64.6
Qatar	91.9
Saudi Arabia	65.9
Syria	28.1
United Arab Emirates	93.2
Yemen	22.6

Step 4 Include a label for the number line and title for the dot plot.

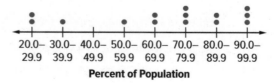

Internet Usage by Middle Eastern Countries

Percent of Population

Guided Practice

2A. The table shows salaries for employees at Julio's company. Represent the data as a dot plot.

38,150	40,500	42,750	43,685	57,890	37,550
41,235	78,990	66,000	44,435	45,775	39,800

2B. The table shows the number of miles people in Manda's fitness class drive from home to the gym. Represent the data as a dot plot.

11	21	14	9	15	16	25
26	5	22	13	22	15	8
22	19	16	10	4	19	17

2 Representing Data with Bar Graphs or Histograms

A **frequency table** uses tally marks to record and display frequencies of events. A **bar graph** compares categories of data with bars representing the frequencies. Bar graphs are used when the data are discrete, which means that the data belong in specific categories and there are no "in between" values. To indicate this, there is space between the bars.

Study Tip

MP **Modeling** A bar graph must have an appropriate scale with equal intervals on the *y*-axis that does not misrepresent the data.

Example 3 Make a Bar Graph

Make a bar graph to display the data gathered from a survey of students about their favorite sport.

Sport	Tally	Frequency
basketball	IIII IIII IIII	15
football	IIII IIII IIII IIII IIII	25
soccer	IIII IIII IIII III	18
baseball	IIII IIII IIII IIII I	21

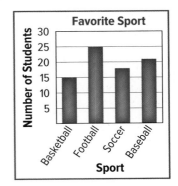

Step 1 Draw a horizontal axis and a vertical axis. Label the axes as shown. Add a title.

Step 2 Draw a bar to represent each sport. The vertical scale is the number of students who chose each sport. The horizontal scale identifies the sport.

Guided Practice

3. Make a bar graph to display the data regarding members of the orchestra.

Instrument	brass	percussion	strings	woodwinds
Frequency	16	3	31	17

The **cumulative frequency** for each event is the sum of its frequency and the frequencies of all preceding events. A **histogram** is a type of bar graph used to display numerical data that have been organized into equal intervals. Each interval is represented by an interval called a *bin*. A histogram represents continuous data, so the bins have no spaces between them

Example 4 Make a Histogram and a Cumulative Frequency Histogram

Make histograms of the frequency and the cumulative frequency.

Age at Inauguration	40–44	45–49	50–54	55–59	60–64	65–69
U.S. Presidents	2	7	13	12	7	3

Find the cumulative frequency for each interval.

Age	< 45	< 50	< 55	< 60	< 65	< 70
Presidents	2	2 + 7 = 9	9 + 13 = 22	22 + 12 = 34	34 + 7 = 41	41 + 3 = 44

Make each histogram like a bar graph but with no space between the bars.

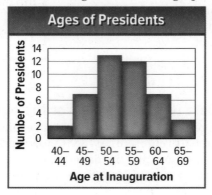

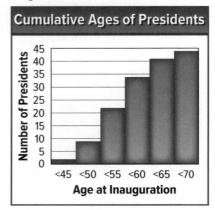

▶ **Guided Practice**

4. Make histograms of the frequency and the cumulative frequency.

Gas Price per Gallon	$1.80–1.84	$1.85–1.89	$1.90–1.94	$1.95–1.99	$2.00–2.04	$2.05–2.09	$2.10–2.14
Frequency	2	7	6	3	3	2	1

Given a set of data, you will need to decide the best way to graphically display it. Use a bar graph for discrete data and a histogram for continuous data.

Example 5   Determine an Appropriate Graph

Determine whether the data are *discrete* or *continuous*. Then, make a graph of the data to show the total medals in each sport.

a. **OLYMPICS** The table shows the total number of Olympic medals won by U.S. athletes from the first Summer Olympics in 1896 through 2012.

Event	Gold	Silver	Bronze
boxing	49	23	39
diving	48	41	43
swimming	230	164	126
track & field	319	247	193
wrestling	52	43	34

Step 1 Determine whether the data should be represented as a bar graph or histogram.

These data represent discrete, categorical data, so use a bar graph.

Step 2 Determine appropriate categories, and tally the data.

Each sport will represent a category of the bar graph.

Complete the table to find the total number of medals in each sport.

Event	Total
boxing	111
diving	132
swimming	520
track & field	759
wrestling	129

Steps 3 and 4 Draw a bar to represent each category. Label the axes, and include a title for the graph.

Label the *x*-axis as *Event* and the *y*-axis as *Medals*. Then title the graph.

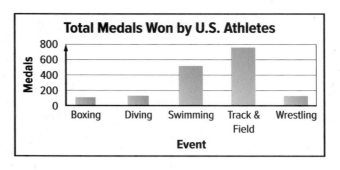

Total Medals Won by U.S. Athletes

Real-World Link

The first New York City Marathon was held in 1970. The official wheelchair division was introduced in 2000.

b. MARATHON The results of the top finishers of the 2015 New York City Marathon, wheelchair division, are given in the table.

Step 1 Determine whether the data should be represented as a bar graph or histogram.

Because racers can finish with any time, the data are continuous and you can use a histogram.

Step 2 Determine appropriate intervals and tally the data.

Since the data are spread over several minutes, group the data by the minute. Then, tally each interval.

Use a table to determine the frequency of each minute.

Steps 3 and 4 Draw a bar to represent each interval.

Label the axes, and include a title for the graph.

Place	Time (h:m:s)	Place	Time (h:m:s)
1	1:30:54	6	1:35:37
2	1:30:55	7	1:35:38
3	1:34:05	8	1:36:45
4	1:35:19	9	1:36:59
5	1:35:21	10	1:38:39

Time (h:m:s)	Frequency
1:30:00 – 1:30:59	2
1:31:00 – 1:31:59	0
1:32:00 – 1:32:59	0
1:33:00 – 1:33:59	0
1:34:00 – 1:34:59	1
1:35:00 – 1:35:59	4
1:36:00 – 1:36:49	2
1:37:00 – 1:37:59	0
1:38:00 – 1:38:59	1

Study Tip

MP Modeling The appropriate graph to represent example B can either be a dot plot or a bar graph since each type shows which interval has the most runners.

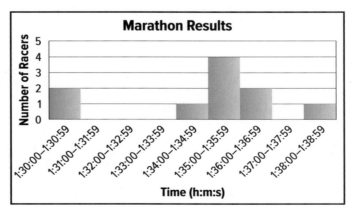

Guided Practice

5. The table shows the heights of players on the Cedar Ridge High School basketball team. Determine whether the data are *discrete* or *continuous*. Then make an appropriate graph.

67	70	69	73	75	72	68
70	70	71	68	73	69	71

Check Your Understanding ◯ = Step-by-Step Solutions begin on page R11.

Go Online! for a
Self-Check Quiz

Example 1

1. Represent the data as a dot plot.

   100, 80, 95, 90, 100, 95, 70, 95, 90, 90, 95, 85, 100, 100, 95

Example 2

2. **SOLAR POWER** Kellen tracked the energy output, in kilowatt hours, of the solar panels on his house for two weeks in June. Represent the data as a dot plot.

Energy Output in June						
16.8	14.4	15.2	16.6	14.0	16.9	12.8
13.8	12.3	15.9	16.4	15.6	14.2	10.2

Example 3

3. **SURVEYS** Alana surveyed several students to find how many hours of sleep they typically get each night. The results are shown in the table. Make a bar graph of the data.

Hours of Sleep					
Alana	8	Kwam	7.5	Tomas	7.75
Nick	8.25	Kate	7.25	Sharla	8.5

Example 4

4. **PLAYS** The frequency table at the right shows the ages of people attending a high school play.

   a. Make a histogram to display the data.
   b. Make a cumulative frequency histogram showing the number of people attending who were less than 20-, 40-, 60-, or 80-years of age.

Age	Tally	Frequency
0–19	IIII IIII IIII IIII IIII IIII IIII IIII IIII II	47
20–39	IIII IIII IIII IIII IIII IIII IIII IIII III	43
40–59	IIII IIII IIII IIII IIII IIII I	31
60–79	IIII III	8

Example 5

5. **PARKS** The table shows the areas of national parks in Alaska. Determine whether the data are *discrete* or *continuous*. Then graph the data.

Area (km²) of National Parks in Alaska			
13,050.5	2711.3	33,682.6	7084.9
14,870.3	10,601.7	19,185.8	30,448.1

Practice and Problem Solving

Extra Practice is found on page R10.

Example 1

6. Represent the data as a dot plot. 8, 6, 0, 2, 7, 1, 8, 1, 4, 8, 0, 1, 2, 8, 4, 7, 1, 5, 9, 1

Example 2

7. **MOUNTAINS** The table gives the elevation of the highest mountain peaks in the United States in feet. Represent the data as a dot plot.

Elevation (ft)				
20,308	17,402	16,391	15,325	14,829
18,009	16,421	16,237	14,951	14,573

Example 3

8. **VIDEO GAMES** The table shows the number of active video game players by country. Make a bar graph of the data.

Example 4

9. **PHOTO SHARING** The table shows the users of a photo-sharing app by age group. Use a histogram to graph the data.

Country	Players (millions)	Country	Players (millions)
Australia	9.5	Poland	11.8
Brazil	40.2	Spain	17
France	25.3	Turkey	21.8
Germany	38.5	United Kingdom	33.6
Italy	18.6	United States	157

Age	18–24	25–34	35–44	45–54	55–64	65+
Users (%)	45	26	13	10	6	1

Example 5

10. The table shows ages of students who are taking a college calculus course. Determine whether the data is *discrete or continuous*, and then make an appropriate graph for the data.

18	18	20	19	19	20	18	20	20
19	19	18	18	20	18	20	21	18

11. The table shows all the grades received on a biology test. Determine whether the data is *discrete or continuous*, and then make an appropriate graph for the data.

96.5	94.3	75	75.7	82	89	82.6	96.8	91	94.5	78	68.7
68	76	82.6	89.7	82.3	75.1	96.3	91.5	89.8	75.7	91.7	94.9

12. Shelby wrote down the day of the month on which her classmates were born. Use the data to make a graph.

2	4	7	11	1	18	12	3	9	28
4	17	10	2	15	30	20	25	26	8
6	19	23	28	16	31	24	12	6	31

13. SURVEY Harper took a survey of the families in her neighborhood to find out how many mobile devices they have. Determine whether her data are *discrete* or *continuous*, and then make an appropriate graph for the data.

Family	Mobile Devices	Family	Mobile Devices
Anderson	6	Patel	4
Clark	3	Perez	2
Davis	3	Roberts	4
Garcia	3	Turner	1
Li	4	Ward	2

14. Freddy is trying out for the wrestling team at his school in 20 days. In order to make the team, he has to gain a total of 9 pounds. Freddy records his daily weight gain on a journal. Determine whether the date is *discrete or continuous*, and then make an appropriate graph for the data.

+.35	+.55	+.9	+.25	+.35	+.4	+1	+.75	+.25	+.55
+.25	+.45	+.75	+.35	+.75	+.55	+.9	+.35	+.55	+1

15. A survey of students' favorite movie genres was conducted at school. Make a bar graph to display the data.

Comedy	Drama	Crime	Thriller	Action
58	23	47	38	32

H.O.T. Problems   Use Higher-Order Thinking Skills

16. MP SENSE MAKING The dot plot shows the number of hours students study per week.

a. How many students studied 5 hours per week?

b. What is the total number of students in this class?

c. How many of the students study 4 or more hours per week?

17. MP SENSE MAKING Henry recorded the daily temperatures in his town and displayed the data in a dot plot.

a. What is the most frequent temperature?

b. For how many days did Henry record the temperature?

c. How many days reached a temperature of 53 °F or higher?

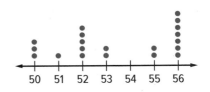

18. WRITING IN MATH Describe one difference and one similarity between a histogram and a bar graph.

19. The table shows the results of a survey about time spent on the Internet per month. Which statements are supported by the data? Select all that apply. **MP** 1, 3, 4

Time (h)	Frequency	Cumulative Frequency
0–4	7	7
5–9	4	11
10–14	11	22
15–19	17	39
20–24	15	54
25–29	6	60

☐ **A** The interval of 0–4 hours per month has the least frequency.

☐ **B** Eleven people answered that they spend less than 10 hours per month on the Internet.

☐ **C** Sixty people answered the survey.

☐ **D** More than half of the people who answered the survey spend less than 20 hours per month on the Internet.

20. Mr. Berkley asked his students how many books they read over summer vacation. Their responses are shown in the table. **MP** 1, 3, 4

Number of Books				
4	4	3	8	0
1	9	8	4	3
5	6	7	8	0
1	2	3	9	10
3	3	4	7	6

a. Make a display of the data.

b. Explain why you chose the type of display you made.

21. The dot plot shows how many miles members of the track team ran one week. List the number of miles from most frequent to least frequent. **MP** 1, 4

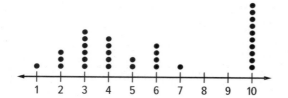

22. Explain whether a bar graph or histogram would better display data representing an organization's financial contributions to four different charities. **MP** 3

23. The table shows how Dezi spent her allowance of $40. Explain whether these data should be displayed as a bar graph or histogram. **MP** 3, 4

Allowance	
How Spent	Amount ($)
savings	15
downloaded music	8
snacks	5
T-shirt	12

24. The table shows the speeds (mph) of 20 of the fastest land animals. Which intervals are reasonable for a histogram of this data? **MP** 2, 4

42	40	40	35	50
32	50	36	50	40
45	70	43	45	32
40	35	61	48	35

Source: *The World Almanac*

○ **A** 30–40, 40–50, 50–60, 60–70

○ **B** 35–39, 40–49, 50–59, 60–70

○ **C** 30–39, 40–49, 50–59, 60–69, 70–79

○ **D** 35–44, 45–54, 55–64, 65–74

25. **MULTI-STEP** The histogram shows the height of basketball players on a high school team. **MP** 1, 4

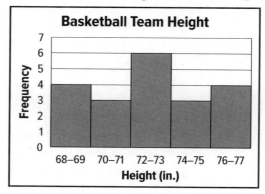

a. How many people have heights less than 72 in.?

b. In what interval is the most frequent height?

c. What percentage of the players have heights greater than 73 in.?

Measures of Spread

You analyzed data collection techniques.

1 Calculate measures of spread.

2 Analyze data sets using statistics.

At the start of every class period for one week, each of Mr. Day's algebra students randomly draws 9 pennies from a jar of 1000 pennies and note the dates on the coins. What measures of central tendency can each student find for the set of coins that they have?

How else can the data be compared?

New Vocabulary

measures of spread or variation

range

quartiles

measure of position

lower quartile

upper quartile

five-number summary

interquartile range

outlier

standard deviation

variance

Mathematical Practices

2 Reason abstractly and quantitatively.

6 Attend to precision.

1 **Variation** Two very different data sets can have the same mean, so statisticians also use **measures of spread** or **variation** to describe how widely the data values vary. One such measure that you studied in earlier courses is the **range**, which is the difference between the greatest and least values in a set of data.

Values in a data set can be described based on the position of a value relative to other values in a set. **Quartiles** are common **measures of position** that divide a data set arranged in ascending order into four groups, each containing about one fourth or 25% of the data. The median marks the second quartile Q_2 and separates the data into upper and lower halves. The first or **lower quartile** Q_1 is the median of the lower half, while the third or **upper quartile** Q_3 is the median of the upper half.

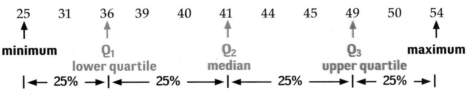

The three quartiles, along with the minimum and maximum values, are called a **five-number summary** of a data set. Note that when calculating quartiles, if the number of values in a set of data are odd, the median is not included in either half of the data when calculating Q_1 or Q_3.

Real-World Example 1   Five-Number Summary

FUNDRAISER **The number of boxes of donuts Aang sold for a fundraiser each day for the last 11 days were 22, 16, 35, 26, 14, 17, 28, 29, 21, 17, and 20. Find the minimum, lower quartile, median, upper quartile, and maximum of the data set. Then interpret this five-number summary.**

Order the data from least to greatest. Use the list to determine the quartiles.

14, 16, **17**, 17, 20, **21**, 22, 26, **28**, 29, 35

Min. Q_1 Q_2 Q_3 **Max.**

The minimum is 14, the lower quartile is 17, the median is 21, the upper quartile is 28, and the maximum is 35. Over the last 11 days, Aang sold a minimum of 14 boxes and a maximum of 35 boxes. He sold fewer than 17 boxes 25% of the time, fewer than 21 boxes 50% of the time, and fewer than 28 boxes 75% of the time.

1. The manager of a convenience store measured the amount of soda dispensed into 36-ounce cups of cola and gathered this data: 36.1, 35.8, 35.2, 36.5, 36.0, 36.2, 35.7, 35.8, 35.9, 36.4, 35.6. Find the minimum, lower quartile, median, upper quartile, and maximum. Then interpret this five-number summary.

The difference between the upper and lower quartiles is called the **interquartile range**. The interquartile range, or IQR, contains about 50% of the values.

$$14,\ 16,\ 17,\ 17,\ 20,\ 21,\ 22,\ 26,\ 28,\ 29,\ 35$$

$$Q_1 \qquad\qquad\qquad\qquad\qquad Q_3$$

$$|\!\leftarrow IQR = Q_3 - Q_1 \text{ or } 11 \rightarrow\!|$$

Before deciding on which measure of center best describes a data set, check for outliers. An **outlier** is an extremely high or extremely low value when compared with the rest of the values in the set. To check for outliers, look for data values that are beyond the upper or lower quartiles by more than 1.5 times the interquartile range.

Real-World Example 2   **Effect of Outliers**

TEST SCORES Students taking a test received the following scores: 88, 79, 94, 90, 45, 71, 82, and 88.

a. Identify any outliers in the data.

First determine the median and upper and lower quartiles of the data.

$$45,\qquad 71,\qquad 79,\qquad 82,\qquad 88,\qquad 88,\qquad 90,\qquad 94$$

$$Q_1 = \frac{71 + 79}{2} \text{ or } 75 \qquad Q_2 = \frac{82 + 88}{2} \text{ or } 85 \qquad Q_3 = \frac{88 + 90}{2} \text{ or } 89$$

Find the interquartile range.

$$IQR = Q_3 - Q_1 = 89 - 75 \text{ or } 14$$

Use the interquartile range to find the values beyond which any outliers would lie.

$Q_1 - 1.5(IQR)$	and	$Q_3 + 1.5(IQR)$	Values beyond which outliers lie
$75 - 1.5(14)$		$89 + 1.5(14)$	$Q_1 = 75, Q_3 = 89,$ and $IQR = 14$
54		110	Simplify.

There are no scores greater than 110, but there is one score less than 54. The score of 45 can be considered an outlier for this data set.

> **Study Tip**
>
> **Interquartile Range**
> When the interquartile range is a small value, the data in the set are close together. A large interquartile range means that the data are spread out.

b. Find the mean and median of the data set with and without the outlier. Describe what happens.

Data Set	Mean	Median
with outlier	$\dfrac{88 + 79 + 94 + 90 + 45 + 71 + 82 + 88}{8}$ or about 79.6	85
without outlier	$\dfrac{88 + 79 + 94 + 90 + 71 + 82 + 88}{7}$ or about 84.6	88

Removal of the outlier causes the mean and median to increase, but notice that the mean is affected more by the removal of the outlier than the median.

2. A student surveyed friends to find the amount of time in minutes that they spend on social networking Websites each day. The results were: 25, 35, 45, 30, 65, 50, 25, 100, 45, 35, 5, 105, 110, 190, 40, 30, 80. Find the mean and median and identify any outliers. If there is an outlier, find the mean and median without the outlier. State which measure is affected more by the removal of the outlier.

2 Statistical Analysis

In a set of data, the **standard deviation** shows how the data deviate from the mean. A low standard deviation indicates that the data tend to be very close to the mean, while a high standard deviation indicates that the data are spread out over a larger range of values.

The standard deviation is represented by the lowercase Greek letter sigma, σ. The **variance** σ^2 of the data is the square of the standard deviation.

🗐 Key Concept   Standard Deviation

Step 1 Find the mean, $\bar{x}$.

Step 2 Find the square of the difference between each data value x_n and the mean, $(\bar{x} - x_n)^2$.

Step 3 Find the sum of all of the values in Step 2.

Step 4 Divide the sum by the number of values in the set of data n. This value is the variance.

Step 5 Take the square root of the variance.

Formula $\sigma = \sqrt{\dfrac{(\bar{x} - x_1)^2 + (\bar{x} - x_2)^2 + \ldots + (\bar{x} - x_n)^2}{n}}$

Real-World Example 3   Variance and Standard Deviation

ELECTRONICS **Ed surveys his classmates to find out how many electronic devices each person has in their home. Find and interpret the standard deviation of the data set.**

$$\{9, 10, 11, 6, 9, 11, 9, 8, 11, 8, 7, 9, 11, 11, 5\}$$

Step 1 Find the mean.

$$\bar{x} = \frac{9 + 10 + 11 + 6 + 9 + 11 + 9 + 8 + 11 + 8 + 7 + 9 + 11 + 11 + 5}{15} \text{ or } 9$$

Step 2 Find the square of the differences, $(\bar{x} - x_n)^2$.

$(9-9)^2 = 0$	$(9-10)^2 = 1$	$(9-11)^2 = 4$	$(9-6)^2 = 9$	$(9-9)^2 = 0$
$(9-11)^2 = 4$	$(9-9)^2 = 0$	$(9-8)^2 = 1$	$(9-11)^2 = 4$	$(9-8)^2 = 1$
$(9-7)^2 = 4$	$(9-9)^2 = 0$	$(9-11)^2 = 4$	$(9-11)^2 = 4$	$(9-5)^2 = 16$

Step 3 Find the sum.

$$0 + 1 + 4 + 9 + 0 + 4 + 0 + 1 + 4 + 1 + 4 + 0 + 4 + 4 + 16 = 52$$

Step 4 Find the variance.

$$\sigma^2 = \frac{(\bar{x} - x_1)^2 + (\bar{x} - x_2)^2 + \ldots + (\bar{x} - x_n)^2}{n} \qquad \text{Formula for variance}$$

$$= \frac{52}{15} \text{ or about } 3.47 \qquad \text{The sum is 52 and } n = 15.$$

Step 5 Find the standard deviation.

$$\sigma = \sqrt{\sigma^2} \qquad \text{Square root of the variance}$$

$$\approx \sqrt{3.47} \text{ or about } 1.86$$

A standard deviation of 1.86 is small compared to the mean of 9. This suggests that most of the data values are relatively close to the mean.

▶ **Guided Practice**

3. NUTRITION Caleb tracked his Calorie intake for a week. Find and interpret the standard deviation of his Calorie intake.

1950, 2000, 2100, 2000, 1900, 2100, 2000

The mean and standard deviation can be used to compare data.

Real-World Example 4	Compare Two Sets of Data

GOLF Miguel plays golf at Redstone and Memorial Park golf courses. Compare the means and standard deviations of each set of Miguel's scores.

Redstone				
81	78	79	82	80
80	79	83	81	80

Memorial Park				
84	79	86	78	77
88	85	79	87	86

Use a graphing calculator to find the mean and standard deviation. Clear all lists. Then press [STAT] [ENTER], and enter each data value into **L1**. To view the statistics, press [STAT] [▶] 1 [ENTER].

Redstone

Memorial Park

Miguel's mean score at Redstone is 80.3 with a standard deviation of about 1.4. His mean score at Memorial Park is 82.9 with a standard deviation of about 4.0. Therefore, he tends to score lower at Redstone. The greater standard deviation at Memorial Park indicates that there is greater variability to his scores at that course.

▶ **Guided Practice**

4. SWIMMING Anna is considering two different lineups for her 4 × 100 relay team. Below are the times in minutes recorded for each lineup. Compare the means and standard deviations of each set of data.

Lineup A				
4.25	4.31	4.19	4.40	4.23
4.18	4.71	4.56	4.32	4.39

Lineup B				
4.47	4.68	4.25	4.41	4.49
4.18	4.27	4.69	4.32	4.44

Check Your Understanding   ◯ = Step—by—Step Solutions begin on page R11.

Example 1   **1. SMARTPHONES** A student surveyed the prices of smartphones at local stores and collected this data: $199.99, $99.99, $249.99, $399.99, $439.99, $349.99. Find the five-number summary.

Study Tip

Symbols The standard deviation of a sample *s* and the standard deviation of a population σ are calculated in different ways. In this text, you will calculate the standard deviation of a population.

Example 2

2. PRICING A department store manager recorded these prices for eight similar pairs of pants with different brands: $69.99, $41.99, $29.99, $24.99, $29.99, $33.99, $46.99, and $36.99. Find the mean and median of the data set, and identify any outliers. If the set has an outlier, find the mean and median without the outlier, and state which measure is affected more by the removal of this value.

Example 3

3 PART-TIME JOBS Ms. Johnson asks all of the members of the girls' tennis team to find the number of hours each week they work at part-time jobs: {10, 12, 0, 6, 9, 15, 12, 10, 11, 20}. Find and interpret the standard deviation of the data set.

Example 4

4. MODELING Mr. Jones recorded the number of pull-ups done by his students. Compare the means and standard deviations of each group.
Boys: {5, 16, 3, 8, 4, 12, 2, 15, 0, 1, 9, 3} Girls: {2, 4, 0, 3, 5, 4, 6, 1, 3, 8, 3, 4}

Practice and Problem Solving

Extra Practice is on page R10.

Example 1

Find the minimum, lower quartile, median, upper quartile, and maximum values for each data set.

5. ART MUSUEM The city art museum's annual gala event had the following attendance for the last nine years: 68, 99, 73, 65, 67, 62, 80, 81, 83.

6. AMUSEMENT RIDE The following ages of riders on a roller coaster were recorded for the last hour of the day: 45, 17, 16, 22, 25, 19, 20, 21, 32, 37, 19, 21, 24, 20, 18, 22, 23, 19.

Example 2

Find the mean and median of the data set, and then identify any outliers. If the set has an outlier, find the mean and median without the outlier, and state which measure is affected more by the removal of this value.

7. MOVIES A math teacher asked a group of his students to count the number of movies they owned.

Number of Movies					
26	39	5	82	12	14
0	3	15	19	41	6

8. SWIMMING The owner of a public swimming pool tracked the daily attendance.

Daily Attendance					
86	45	91	104	95	88
127	85	79	102	98	103

Example 3

9. MODELING Samantha earns $8.50 per hour for babysitting. She takes a survey of her friends to see what they charge per hour. The results are {$8.00, $8.50, $9.00, $7.50, $15.00, $8.25, $8.75}. Find and interpret the standard deviation of the data.

10. ARCHERY Carla participates in competitive archery. Each competition allows a maximum of 90 points. Carla's results for the last 8 competitions are {76, 78, 81, 75, 80, 80, 76, 77}. Find and interpret the standard deviation of the data.

Example 4

11. BASKETBALL The coach of the Wildcats basketball team is comparing the number of fouls called against his team with the number called against their rivals, the Trojans. He records the number of fouls called against each team for each game of the season. Compare the means and standard deviations of each set of data.

Wildcats			
15	12	13	9
11	12	14	12
8	16	9	9
11	13	12	14

Trojans			
9	10	14	13
7	8	10	10
9	7	11	9
12	11	13	8

12. BATTING AVERAGES The batting averages for the last 10 seasons for a baseball team are 0.267, 0.305, 0.304, 0.201, 0.284, 0.302, 0.311, 0.289, 0.300, and 0.292.

a. From reading the data, do you notice an obvious outlier? Explain.

b. Find the range and the five-number summary for the set of data and interpret the summary.

c. Identify any outlier(s) and find the mean and median without the outlier(s). State which measure was more affected by the outlier.

d. What is the lowest batting average that would result in no outlier?

13 **MOVIE RATINGS** Two movies were rated by the same group of students. Ratings were from 1 to 10, with 10 being the best.

a. Compare the means and standard deviations of each set of data.

b. Provide an argument for why movie A would be preferred, then an argument for movie B.

Movie A			
7	8	7	6
8	6	7	8
6	8	8	6
7	7	8	8

Movie B			
9	5	10	6
3	10	9	4
8	3	9	9
2	8	10	3

14. **RUNNING** The results of a 5-kilometer race are published in a local paper. Over a hundred people participated, but only the times of the top 15 finishers are listed.

15th Annual 5K Road Race					
Place	Time (min:s)	Place	Time (min:s)	Place	Time (min:s)
1	17:51	6	19:03	11	19:50
2	18:01	7	19:06	12	20:07
3	18:17	8	19:27	13	20:11
4	18:22	9	19:49	14	20:13
5	18:26	10	19:49	15	20:13

a. Find the mean and standard deviation of the top 15 running times. (*Hint*: Convert each time to seconds.)

b. Identify the sample and population.

c. Analyze the sample. Classify the data as *quantitative* or *qualitative*. Can a statistical analysis of the sample be applied to the population? Explain your reasoning.

H.O.T. Problems   Use Higher-Order Thinking Skills

15. **ERROR ANALYSIS** Jennifer and Megan are determining one way to decrease the size of the standard deviation of a set of data. Is either of them correct? Explain.

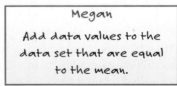

Jennifer
Remove the outliers from the data set.

Megan
Add data values to the data set that are equal to the mean.

16. **MP** **CONSTRUCT ARGUMENTS** Determine whether the statement *Two random samples taken from the same population will have the same mean and standard deviation* is *sometimes*, *always*, or *never* true. Explain.

17. **CHALLENGE** Describe a situation in which identifying an outlier and revising the mean to take the outlier into account results in a better descriptor for the data.

18. **CHALLENGE** Write a set of data with a standard deviation that is equal to the variance.

WRITING IN MATH Compare and contrast each of the following.

19. range and interquartile range

20. mean and standard deviation

21. Identify the outlier in the following data set.
(MP) 2, 6

{11, 13, 7, 9, 14, 18, 10, 13, 9}

- **A** 7
- **B** 9
- **C** 18
- **D** none

22. The worldwide grosses for the top 10 highest grossing films are shown below. (MP) 1, 4, 6

Rank	Gross (million $)	Rank	Gross (million $)
1	2788	6	1516
2	2187	7	1405
3	2060	8	1342
4	1670	9	1277
5	1520	10	1215

Top Grossing Films

a. Find the minimum, lower quartile, median, upper quartile, and maximum of the data set.

b. Interpret the five-number summary.

23. Which is the standard deviation of the following data set? (MP) 1

{0, 8, 14, 1, 12, 6, 6, 10}

- **A** 4.6
- **B** 7
- **C** 7.1
- **D** 21.4

24. Brandon records the temperature in Celsius at various times during a chemistry lab. His data is shown below. Find the mean and standard deviation.
(MP) 1, 4

{−16.0, −18.3, −15.8, −16.6, −18.2, −17.4, −16.5}

25. Find the upper and lower quartiles for the following data. (MP) 1

{37, 62, 10, 13, 54, 44, 47, 28, 30}

- **A** $Q_1 = 13$; $Q_3 = 54$
- **B** $Q_1 = 20.5$; $Q_3 = 50.5$
- **C** $Q_1 = 28$; $Q_3 = 47$
- **D** $Q_1 = 36$; $Q_3 = 37.5$

26. The members of a science club in New Orleans recorded the monthly rainfall in inches for one year. Their findings were:

5.87, 5.47, 5.24, 5.02, 4.62, 6.83, 6.2, 6.15, 5.55, 3.05, 5.09, 5.07

Select all the statements that are true.
(MP) 6

- **A** The data has an outlier, 3.05 inches.
- **B** A standard deviation of about 0.9114 is small compared to the mean of 5.3467. This suggests that most of the data values are relatively close to the mean.
- **C** The interquartile range for this data is 0.961 inches.
- **D** The median for the data is 5.355 inches.
- **E** The measure of central tendency affected most by the outlier is the median.

27. MULTI-STEP The weekly pay for 10 employees at a small sandwich shop is $54, $278, $70, $159, $482, $49, $205, $70, $386, and $63. (MP) 6

a. Find the range and the interquartile range of the data.

b. Find the mean, median, and mode of the data.

c. Identify any outliers.

d. What is the standard deviation, and what does this indicate about the data?

Find the mean, median, and mode for each set of data. (Lesson 10-1)

1. {10, 11, 18, 24, 30}

2. {4, 8, 9, 9, 10, 14, 16}

3. **TEMPERATURE** Taryn records the daily high temperature for two cities for one week. Find the mean, median, and mode of each data set. Which city should Taryn visit for warmer weather?
(Lesson 10-1)

Day	Sunnydale Temperature (°F)	Sun Valley Temperature (°F)
Sunday	95	80
Monday	88	86
Tuesday	86	91
Wednesday	90	102
Thursday	90	103
Friday	93	91
Saturday	89	85

4. **COMPETITION** The scores for a dance competition are shown in the table. (Lesson 10-1)

Name	Score	Name	Score
Adam	87	Emilio	79
Bella	91	Fran	82
Camila	90	Greg	94
Devonte	88	Holly	81

a. Who is the 50th percentile?

b. Which contestant(s) are below the 25th percentile?

c. Find Bella's percentile rank.

5. **CARNIVAL** The frequency table shows the ages of people attending a carnival. (Lesson 10-2)

Age	Frequency
0–19	66
20–39	49
40–59	54
60–79	16

a. Make a histogram to display the data.

b. Make a cumulative frequency histogram showing the number of people attending who were less than 20, 40, 60, or 80 years of age.

6. **MUSIC** The table shows the results of a survey in which students were asked to choose which of the four instruments they would like to learn. Make a bar graph of the data. (Lesson 10-2)

Favorite Instrument	
Instrument	Number of Students
drums	8
guitar	12
piano	5
trumpet	7

Find the range, median, lower quartile, and upper quartile for each set of data. (Lesson 10-3)

7. {16, 19, 21, 24, 25, 31, 35}

8. {77, 75, 72, 70, 79, 77, 70, 76}

9. **PLAY AREA** Ian listed the ages of the children playing at the play area at the mall. (Lesson 10-3)

$$\{2, 3, 2, 2, 4, 2, 3, 2, 8, 3, 4, 2\}$$

a. Find and interpret the standard deviation of the data set.

b. **MP** Which mathematical practice did you use?

10. **GIFTS** Several friends are chipping in to buy a gift for their teacher. Indigo is keeping track of how much each friend spends. Find the standard deviation. (Lesson 10-3)

$$\{\$10, \$5, \$3, \$6, \$7, \$8\}$$

A $2.22   C $4.92

B $3.00   D $6.50

11. Describe the center and spread of the data using the mean and standard deviation. (Lesson 10-3)

9, 11, 2, 6, 8, 10, 6, 3, 10, 11, 9, 8, 3,
8, 5, 11, 14, 6, 8, 6, 11, 5, 9, 10, 8

Distributions of Data

∷Then	∷Now	∷Why?
• You calculated measures of central tendency and variation.	**1** Describe the shape of a distribution. **2** Use the shapes of distributions to select appropriate statistics.	• Over many years, the number of bluebirds in the United States declined dramatically. In order to increase the bluebird population, volunteers, in some states, supply bluebird nestboxes and monitor bluebird trails. The population of bluebirds in each state can be displayed in a distribution of data.

 New Vocabulary

distribution
negatively skewed
   distribution
symmetric distribution
positively skewed
   distribution

 Mathematical Practices

5 Use appropriate tools strategically.

1 **Describing Distributions** A **distribution** of data shows the observed or theoretical frequency of each possible data value. Recall that a histogram is a type of bar graph used to display data that have been organized into equal intervals. A histogram is useful when viewing the overall distribution of the data within a set over its range. You can see the shape of the distribution by drawing a curve over the histogram.

Key Concept Symmetric and Skewed Distributions

Negatively Skewed Distribution	Symmetric Distribution	Positively Skewed Distribution
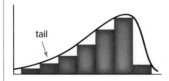		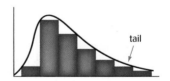
The majority of the data are on the right.	The data are evenly distributed.	The majority of the data are on the left.

Example 1 Distribution Using a Histogram

Use a graphing calculator to construct a histogram for the data, and use it to describe the shape of the distribution.

> 25, 22, 31, 25, 26, 35, 18, 39, 22, 32, 34, 26, 42, 23, 40, 36, 18, 30
> 26, 30, 37, 23, 19, 33, 24, 29, 39, 21, 43, 25, 34, 24, 26, 30, 21, 22

First, press $\boxed{\text{STAT}}$ $\boxed{\text{ENTER}}$ and enter each data value.
Then, press $\boxed{\text{2nd}}$ [STAT PLOT] $\boxed{\text{ENTER}}$ $\boxed{\text{ENTER}}$ and choose
⬛. Press $\boxed{\text{ZOOM}}$ [ZoomStat] to adjust the window.

The graph is high on the left and has a tail on the right. Therefore, the distribution is positively skewed.

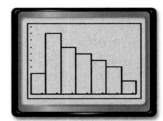

[17, 45] scl: 4 by [0, 10] scl: 1

▶ **Guided Practice**

1. Use a graphing calculator to construct a histogram for the data, and use it to describe the shape of the distribution.

> 8, 11, 15, 25, 21, 26, 20, 12, 32, 20, 31, 14, 19, 27, 22, 21, 14, 8
> 6, 23, 18, 16, 28, 25, 16, 20, 29, 24, 17, 35, 20, 27, 10, 16, 22, 12

A box-and-whisker plot can also be used to identify the shape of a distribution. Box-and-whisker plots are sometimes called box plots. Recall from Lesson 10-3 that a box-and-whisker plot displays the spread of a data set by dividing it into four quartiles. The data from Example 1 are displayed below.

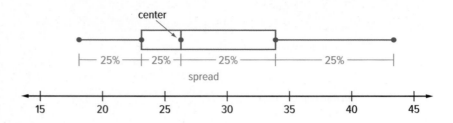

Notice that the left whisker is shorter than the right whisker, and that the line representing the median is closer to the left whisker. This represents a peak on the left and a tail to the right.

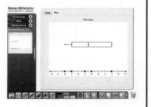
Key Concept Symmetric and Skewed Box-and-Whisker Plots

Negatively Skewed	Symmetric	Positively Skewed
The left whisker is longer than the right. The median is closer to the shorter whisker.	The whiskers are the same length. The median is in the center of the data.	The right whisker is longer than the left. The median is closer to the shorter whisker.

Study Tip

Outliers In Example 2, notice that the outlier does not affect the shape of the distribution.

Example 2   Distribution Using a Box-and-Whisker Plot

Use a graphing calculator to construct a box-and-whisker plot for the data, and use it to determine the shape of the distribution.

9, 17, 15, 10, 16, 2, 17, 19, 10, 18, 14, 8, 20, 20, 3, 21, 12, 11
5, 26, 15, 28, 12, 5, 27, 26, 15, 53, 12, 7, 22, 11, 8, 16, 22, 15

Enter the data as **L1**. Press [2nd] [STAT PLOT] [ENTER] [ENTER] and choose ⊡⊶. Adjust the window to the dimensions shown.

The lengths of the whiskers are approximately equal, and the median is in the middle of the data. This indicates that the data are equally distributed to the left and right of the median. Thus, the distribution is symmetric.

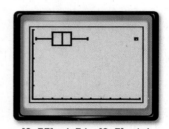

[0, 55] scl: 5 by [0, 5] scl: 1

▶ **Guided Practice**

2. Use a graphing calculator to construct a box-and-whisker plot for the data, and use it to describe the shape of the distribution.

40, 50, 35, 48, 43, 31, 52, 42, 54, 38, 50, 46, 49, 43, 40, 50, 32, 53
51, 43, 47, 41, 49, 50, 34, 54, 51, 44, 54, 39, 47, 35, 51, 44, 48, 37

2 **Analyzing Distributions** You have learned that data can be described using statistics. The mean and median describe the center. The standard deviation and quartiles describe the spread. You can use the shape of the distribution to choose the most appropriate statistics that describe the center and spread of a set of data.

When a distribution is symmetric, the mean accurately reflects the center of the data. However, when a distribution is skewed, this statistic is not as reliable.

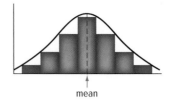

In Lesson 10-3, you discovered that outliers can have a strong effect on the mean of a data set, while the median is less affected. So, when a distribution is skewed, the mean lies away from the majority of the data toward the tail. The median is less affected and stays near the majority of the data.

Negatively Skewed Distribution

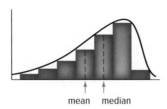

Positively Skewed Distribution

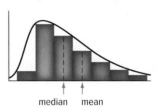

When choosing appropriate statistics to represent a set of data, first determine the shape of the distribution.

- If the distribution is relatively symmetric, the mean and standard deviation can be used.
- If the distribution is skewed or has outliers, use the five-number summary.

Example 3   **Choose Appropriate Statistics**

Describe the center and spread of the data using either the mean and standard deviation or the five-number summary. Justify your choice by constructing a histogram for the data.

21, 28, 16, 30, 25, 34, 21, 47, 18, 36, 24, 28, 30, 15, 33, 24, 32, 22
27, 38, 23, 29, 15, 27, 33, 19, 34, 29, 23, 26, 19, 30, 25, 13, 20, 25

Use a graphing calculator to create a histogram. The graph is high in the middle and low on the left and right. Therefore, the distribution is symmetric.

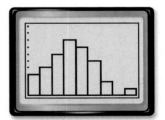

[12, 48] scl: 4 by [0, 10] scl: 1

The distribution is relatively symmetric, so use the mean and standard deviation to describe the center and spread. Press STAT ▶ ENTER ENTER .

The mean $\overline{x}$ is about 26.1 with standard deviation σ of about 7.1.

> **Guided Practice**

3. Describe the center and spread of the data using either the mean and standard deviation or the five-number summary. Justify your choice by creating a histogram for the data.

> 19, 2, 25, 14, 24, 20, 27, 30, 14, 25, 19, 32, 21, 31, 25, 16, 24, 22
> 29, 6, 26, 32, 17, 26, 24, 26, 32, 10, 28, 19, 26, 24, 11, 23, 19, 8

A box-and-whisker plot is helpful when viewing a skewed distribution since it is constructed using the five-number summary.

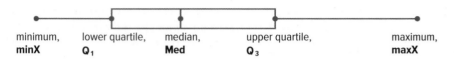

minimum, **minX**   lower quartile, **Q₁**   median, **Med**   upper quartile, **Q₃**   maximum, **maxX**

Real-World Example 4 Choose Appropriate Statistics

COMMUNITY SERVICE The number of community service hours each of Ms. Tucci's students completed is shown. Describe the center and spread of the data using either the mean and standard deviation or the five-number summary. Justify your choice by constructing a box-and-whisker plot for the data.

Community Service Hours												
6	13	8	7	19	12	2	19	11	22	7	33	13
3	8	10	5	25	16	6	14	7	20	10	30	

Use a graphing calculator to create a box-and-whisker plot. The right whisker is longer than the left and the median is closer to the left whisker. Therefore, the distribution is positively skewed.

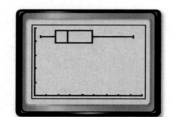

[0, 36] scl: 4 by [0, 5] scl: 1

The distribution is positively skewed, so use the five-number summary. The range is 33 − 2 or 31. The median number of hours completed is 11, and half of the students completed between 7 and 19 hours.

> **Guided Practice**

4. FUNDRAISER The money raised per student in Mr. Bulanda's fifth period class is shown. Describe the center and spread of the data using either the mean and standard deviation or the five-number summary. Justify your choice by creating a box-and-whisker plot for the data.

Money Raised per Student (dollars)									
41	27	52	18	42	32	16	95	27	65
36	45	5	34	50	15	62	38	57	20
38	21	33	58	25	42	31	8	40	28

Check Your Understanding   ◯ = Step-by-Step Solutions begin on page R11.

Go Online! for a
Self-Check Quiz

Examples 1–2 **Use a graphing calculator to construct a histogram and a box-and-whisker plot for the data. Then describe the shape of the distribution.**

1. 80, 84, 68, 64, 57, 88, 61, 72, 76, 80, 83, 77, 78, 82, 65, 70, 83, 78
   73, 79, 70, 62, 69, 66, 79, 80, 86, 82, 73, 75, 71, 81, 74, 83, 77, 73

2. 30, 24, 35, 84, 60, 42, 29, 16, 68, 47, 22, 74, 34, 21, 48, 91, 66, 51
   33, 29, 18, 31, 54, 75, 23, 45, 25, 32, 57, 40, 23, 32, 47, 67, 62, 23

Example 3 **Describe the center and spread of the data using either the mean and standard deviation or the five-number summary. Justify your choice by constructing a histogram for the data.**

3. 58, 66, 52, 75, 60, 56, 78, 63, 59, 54, 60, 67, 72, 80, 68, 88, 55, 60
   59, 61, 82, 70, 67, 60, 58, 86, 74, 61, 92, 76, 58, 62, 66, 74, 69, 64

Example 4 **4. PRESENTATIONS** The length of the students' presentations in Ms. Monroe's second period class are shown. Describe the center and spread of the data using either the mean and standard deviation or the five-number summary. Justify your choice by constructing a box-and-whisker plot for the data.

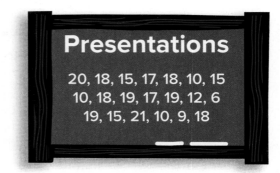

Presentations

20, 18, 15, 17, 18, 10, 15
10, 18, 19, 17, 19, 12, 6
19, 15, 21, 10, 9, 18

Practice and Problem Solving   Extra Practice is on page R10.

Examples 1–2 **Use a graphing calculator to construct a histogram and a box-and-whisker plot for the data. Then describe the shape of the distribution.**

5. 55, 65, 70, 73, 25, 36, 33, 47, 52, 54, 55, 60, 45, 39, 48, 55, 46, 38
   50, 54, 63, 31, 49, 54, 68, 35, 27, 45, 53, 62, 47, 41, 50, 76, 67, 49

6. 42, 48, 51, 39, 47, 50, 48, 51, 54, 46, 49, 36, 50, 55, 51, 43, 46, 37
   50, 52, 43, 40, 33, 51, 45, 53, 44, 40, 52, 54, 48, 51, 47, 43, 50, 46

Example 3 **Describe the center and spread of the data using either the mean and standard deviation or the five-number summary. Justify your choice by constructing a histogram for the data.**

⑦ 32, 44, 50, 49, 21, 12, 27, 41, 48, 30, 50, 23, 37, 16, 49, 53, 33, 25
   35, 40, 48, 39, 50, 24, 15, 29, 37, 50, 36, 43, 49, 44, 46, 27, 42, 47

8. 82, 86, 74, 90, 70, 81, 89, 88, 75, 72, 69, 91, 96, 82, 80, 78, 74, 94
   85, 77, 80, 67, 76, 84, 80, 83, 88, 92, 87, 79, 84, 96, 85, 73, 82, 83

Example 4 **9. WEATHER** The daily low temperatures for New Carlisle over a 30-day period are shown. Describe the center and spread of the data using either the mean and standard deviation or the five-number summary. Justify your choice by constructing a box-and-whisker plot for the data.

Temperature (°F)													
48	50	55	53	57	53	44	61	57	49	51	58	46	54	57
50	55	47	57	48	58	53	49	56	59	52	48	55	53	51

10. TRACK While training for the 100-meter dash, Sarah pulled a muscle. After being cleared for practice, she continued to train. Sarah's median time was about 12.34 seconds, but her average time dropped to about 12.53 seconds. Sarah's 100-meter dash times are shown.

100-meter dash (seconds)				
12.20	12.35	13.60	12.24	12.72
12.18	12.06	12.41	12.28	13.06
12.87	12.04	12.38	12.20	13.12
12.30	13.27	12.93	12.16	12.02
12.50	12.14	11.97	12.24	13.09
12.46	12.33	13.57	11.96	13.34

   a. Use a graphing calculator to create a box-and-whisker plot. Describe the center and spread of the data.

   b. Sarah's slowest time prior to her injury was 12.50 seconds. Use a graphing calculator to create a box-and-whisker plot that *does not* include the times that she ran after her injury. Then describe the center and spread of the new data set.

   c. What effect does removing the times recorded after Sarah pulled a muscle have on the shape of the distribution and on how you should describe the center and spread?

11. MENU The prices for entrees at a restaurant are shown.

Entree Prices ($)				
9.00	11.25	16.50	9.50	13.00
18.50	7.75	11.50	13.75	9.75
8.00	16.50	12.50	10.25	17.75
13.00	10.75	16.75	8.50	11.50

   a. Use a graphing calculator to create a box-and-whisker plot. Describe the center and spread of the data.

   b. The owner of the restaurant decides to eliminate all entrees that cost more than $15. Use a graphing calculator to create a box-and-whisker plot that reflects this change. Then describe the center and spread of the new data set.

H.O.T. Problems   Use **Higher-Order Thinking Skills**

MP REASONING Identify the box-and-whisker plot that corresponds to each of the following histograms.

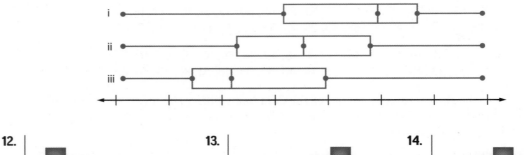

15. MP CONSTRUCT ARGUMENTS Research and write a definition for a *bimodal distribution*. How can the measures of center and spread of a bimodal distribution be described?

16. OPEN-ENDED Give an example of a set of real-world data with a distribution that is symmetric and one with a distribution that is not symmetric.

17. WRITING IN MATH Explain why the mean and standard deviation are used to describe the center and spread of a symmetrical distribution and the five-number summary is used to describe the center and spread of a skewed distribution.

18. The histogram at the right represents a data set.

What is the shape of the data? **MP** 1

○ **A** negatively skewed

○ **B** relatively symmetric

○ **C** positively skewed

○ **D** There is not enough information to determine the shape.

19. The following data represents the weights, in pounds, of a particular species of fish caught at random in a pond of a fish farm.

21, 19, 28, 23, 20, 18, 18, 24, 22, 27, 24, 26, 22, 26, 25, 28

What are the most appropriate statistics to use to describe the distribution of the weights of the fish, and why? **MP** 1, 2, 4

○ **A** mean and standard deviation because the distribution is relatively symmetric

○ **B** mean and standard deviation because the distribution is skewed

○ **C** five-number summary because the distribution is relatively symmetric

○ **D** five-number summary because the distribution is skewed

20. The box-and-whisker plot below represents a data set.

Which of the following statements are true? **MP** 1

☐ **A** The distribution is negatively skewed.

☐ **B** The distribution is relatively symmetric.

☐ **C** The distribution is positively skewed.

☐ **D** The mean and median are approximately the same.

☐ **E** The median is a more appropriate measure of center than the mean.

21. **MULTI-STEP** The following data represents the salaries of the employees of a company, in dollars. **MP** 1, 4

Employee Salaries ($)				
31,000	48,000	37,000	43,000	52,000
48,000	36,000	56,000	59,000	58,000
34,000	256,000	63,000	45,000	37,000

a. What is the median of the data?

b. What is the mean of the data?

c. Explain why the median is a more appropriate statistic than the mean to represent the data.

22. The data represent the of students in a class, in centimeters.

What are the most appropriate statistics to use to describe the distribution of the heights of the students, and why? **MP** 1, 2, 4

Heights (cm)				
151	158	209	153	197
150	178	155	209	217
148	164	158	162	158
195	168	178	186	190

○ **A** mean and standard deviation because the distribution is relatively symmetric

○ **B** mean and standard deviation because the distribution is skewed

○ **C** five-number summary because the distribution is relatively symmetric

○ **D** five-number summary because the distribution is skewed

23. The histogram at the right represents a data set.

What is the shape of the data? **MP** 1

○ **A** negatively skewed

○ **B** relatively symmetric

○ **C** positively skewed

○ **D** There is not enough information to determine the shape.

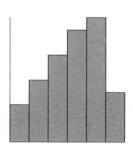

Comparing Sets of Data

- You calculated measures of central tendency and variation.

1 Determine the effect that transformations of data have on measures of central tendency and variation.

2 Compare data using measures of central tendency and variation.

- Tom gets paid hourly to do landscaping work. Because he is such a good employee, Tom is planning to ask his boss for a bonus. Tom's initial pay for a month is shown. He is trying to decide whether he should ask for an extra $5 per day or a 10% increase in his daily wages.

Tom's Pay ($)		
44	52	50
40	48	46
44	52	54
58	42	52
54	50	52
42	52	46
56	48	44
50	42	

New Vocabulary

linear transformation

Mathematical Practices

2 Reason abstractly and quantitatively.

1 **Transformations of Data** To see the effect that an extra $5 per day would have on Tom's pay, we can find the new daily pay values and compare the measures of center and variation for the two sets of data. The new data can be found by performing a *linear transformation*. A **linear transformation** is an operation performed on a data set that can be written as a linear function. Tom's daily pay after the $5 bonus can be found using $y = 5 + x$, where x represents his original daily pay and y represents his daily pay after the bonus.

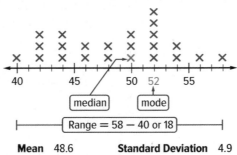

Tom's Earnings Before Extra $5

median   mode

Range = 58 − 40 or 18

Mean 48.6   **Standard Deviation** 4.9

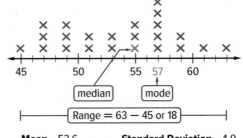

Tom's Earnings With Extra $5

median   mode

Range = 63 − 45 or 18

Mean 53.6   **Standard Deviation** 4.9

Notice that each value was translated 5 units to the right. Thus, the mean, median, and mode increased by 5. Since the new minimum and maximum values also increased by 5, the range remained the same. The standard deviation is unchanged because the amount by which each value deviates from the mean stayed the same.

These results occur when any positive or negative number is added to every value in a set of data.

Key Concept Transformations Using Addition

If a real number *k* is added to every value in a set of data, then:

- the mean, median, and mode of the new data set can be found by adding *k* to the mean, median, and mode of the original data set, and

- the range and standard deviation will not change.

Example 1   Transformation Using Addition

Find the mean, median, mode, range, and standard deviation of the data set obtained after adding 7 to each value.

13, 5, 8, 12, 7, 4, 5, 8, 14, 11, 13, 8

Technology Tip

1-Var Stats To quickly calculate the mean $\bar{x}$, median **Med**, standard deviation σ, and range of a data set, enter the data as **L1** in a graphing calculator, and then press

[STAT] [▶] [ENTER] [ENTER]. Subtract **minX** from **maxX** to find the range.

Method 1 Find the mean, median, mode, range, and standard deviation of the original data set.

| Mean | 9 | Mode | 8 | Standard Deviation | 3.3 |
| Median | 8 | Range | 10 | | |

Add 7 to the mean, median, and mode. The range and standard deviation are unchanged.

| Mean | 16 | Mode | 15 | Standard Deviation | 3.3 |
| Median | 15 | Range | 10 | | |

Method 2 Add 7 to each data value.

20, 12, 15, 19, 14, 11, 12, 15, 21, 18, 20, 15

Find the mean, median, mode, range, and standard deviation of the new data set.

| Mean | 16 | Mode | 15 | Standard Deviation | 3.3 |
| Median | 15 | Range | 10 | | |

▶ **Guided Practice**

1. Find the mean, median, mode, range, and standard deviation of the data set obtained after adding −4 to each value.

27, 41, 15, 36, 26, 40, 53, 38, 37, 24, 45, 26

To see the effect that a daily increase of 10% has on the data set, we can multiply each value by 1.10 and recalculate the measures of center and variation.

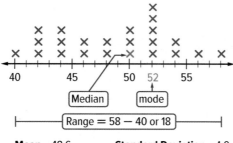

Tom's Earnings Before Extra 10%

Median   52   mode

Range = 58 − 40 or 18

Mean 48.6   **Standard Deviation** 4.9

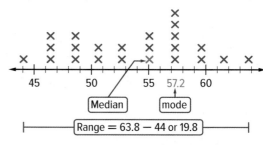

Tom's Earnings With Extra 10%

Median   57.2   mode

Range = 63.8 − 44 or 19.8

Mean 53.5   **Standard Deviation** 5.4

Notice that each value did not increase by the same amount, but did increase by a factor of 1.10. Thus, the mean, median, and mode increased by a factor of 1.10. Since each value was increased by a constant percent and not by a constant amount, the range and standard deviation both changed, also increasing by a factor of 1.10.

> 🔑 **Key Concept** Transformations Using Multiplication
>
> If every value in a set of data is multiplied by a constant k, $k > 0$, then the mean, median, mode, range, and standard deviation of the new data set can be found by multiplying each original statistic by k.

Go Online!

In **Personal Tutor** videos for this lesson, teachers describe how to compare sets of data. Watch with a partner, then try describing how to solve a problem for them. Have them ask questions to help your understanding.

Since the medians for both bonuses are equal and the means are approximately equal, Tom should ask for the bonus that he thinks he has the best chance of receiving.

Example 2   Transformation Using Multiplication

Find the mean, median, mode, range, and standard deviation of the data set obtained after multiplying each value by 3.

21, 12, 15, 18, 16, 10, 12, 19, 17, 18, 12, 22

Find the mean, median, mode, range, and standard deviation of the original data set.

Mean 16	Mode 12	Standard Deviation 3.7
Median 16.5	Range 12	

Multiply the mean, median, mode, range, and standard deviation by 3.

Mean 48	Mode 36	Standard Deviation 11.1
Median 49.5	Range 36	

▶ **Guided Practice**

2. Find the mean, median, mode, range, and standard deviation of the data set obtained after multiplying each value by 0.8.

$$63, 47, 54, 60, 55, 46, 51, 60, 58, 50, 56, 60$$

2 **Comparing Distributions**   Recall that when choosing appropriate statistics to represent data, you should first analyze the shape of the distribution. The same is true when comparing distributions.

- Use the mean and standard deviation to compare two symmetric distributions.

- Use the five-number summaries to compare two skewed distributions or a symmetric distribution and a skewed distribution.

Real-World Example 3   Compare Data Using Histograms

QUIZ SCORES Robert and Elaine's quiz scores for the first semester of Algebra 1 are shown below.

Robert's Quiz Scores
85, 95, 70, 87, 78, 82, 84, 84, 85, 99, 88, 74, 75, 89, 79, 80, 92, 91, 96, 81

Elaine's Quiz Scores
89, 76, 87, 86, 92, 77, 78, 83, 83, 82, 81, 82, 84, 85, 85, 86, 89, 93, 77, 85

a. **Use a graphing calculator to construct a histogram for each set of data. Then describe the shape of each distribution.**

Enter Robert's quiz scores as **L1** and Elaine's quiz scores as **L2**.

Technology Tip

Histograms   To create a histogram for a set of data in **L2**, press 2nd [STAT PLOT] ENTER ENTER , choose ⣿⣿⣿, and enter **L2** for **Xlist**.

Robert's Quiz Scores

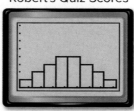

[69, 101] scl: 4 by [0, 8] scl: 1

Elaine's Quiz Scores

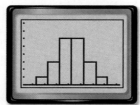

[69, 101] scl: 4 by [0, 8] scl: 1

Both distributions are high in the middle and low on the left and right. Therefore, both distributions are symmetric.

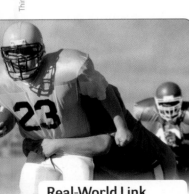

Thinkstock/Stockbyte/Getty Images

b. Compare the data sets using either the means and standard deviations or the five-number summaries. Justify your choice.

Both distributions are symmetric, so use the means and standard deviations to describe the centers and spreads.

Robert's Quiz Scores

Elaine's Quiz Scores

The means for the students' quiz scores are approximately equal, but Robert's quiz scores have a much higher standard deviation than Elaine's quiz scores. This means that Elaine's quiz scores are generally closer to her mean than Robert's quiz scores are to his mean.

Guided Practice

COMMUTE The students in two of Mr. Martin's classes found the average number of minutes that they each spent traveling to school each day.

3A. Use a graphing calculator to construct a histogram for each set of data. Then describe the shape of each distribution.

3B. Compare the data sets using either the means and standard deviations or the five-number summaries. Justify your choice.

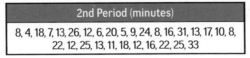

2nd Period (minutes)
8, 4, 18, 7, 13, 26, 12, 6, 20, 5, 9, 24, 8, 16, 31, 13, 17, 10, 8, 22, 12, 25, 13, 11, 18, 12, 16, 22, 25, 33

7th Period (minutes)
21, 4, 20, 13, 22, 6, 10, 23, 13, 25, 14, 16, 19, 21, 19, 8, 20, 18, 9, 14, 21, 17, 19, 22, 4, 19, 21, 26

Box-and-whisker plots are useful for comparisons of data because they can be displayed on the same screen.

Real-World Example 4   Compare Data Using Box-and-Whisker Plots

FOOTBALL Kurt's total rushing yards per game for his junior and senior seasons are shown.

Junior Season (yards)					
16	20	72	4	25	18
34	10	42	17	56	12

Senior Season (yards)					
77	54	109	60	156	72
39	83	73	101	46	80

a. Use a graphing calculator to construct a box-and-whisker plot for each set of data. Then describe the shape of each distribution.

Enter Kurt's rushing yards from his junior season as **L1** and his rushing yards from his senior season as **L2**. Graph both box-and-whisker plots on the same screen by graphing **L1** as **Plot1** and **L2** as **Plot2**.

For Kurt's junior season, the right whisker is longer than the left, and the median is closer to the left whisker. The distribution is positively skewed.

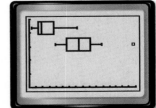

[0, 160] scl: 10 by [0, 5] scl: 1

For Kurt's senior season, the lengths of the whiskers are approximately equal, and the median is in the middle of the data. The distribution is symmetric.

b. Compare the data sets using either the means and standard deviations or the five-number summaries. Justify your choice.

One distribution is symmetric and the other is skewed, so use the five-number summaries to compare the data.

The upper quartile for Kurt's junior season was 38, while the minimum for his senior season was 39. This means that Kurt rushed for more yards in every game during his senior season than 75% of the games during his junior season.

The maximum for Kurt's junior season was 72, while his median for his senior season was 75. This means that in half of his games during his senior year, he rushed for more yards than in any game during his junior season. Overall, we can conclude that Kurt rushed for many more yards during his senior season than during his junior season.

Study Tip

Box-and-Whisker Plots
Recall that a box-and-whisker plot, also called a box plot, displays the spread of a data set by dividing it into four quartiles. Each quartile accounts for 25% of the data.

▶ **Guided Practice**

BASKETBALL The points Vanessa scored per game during her junior and senior seasons are shown.

4A. Use a graphing calculator to construct a histogram for each set of data. Then describe the shape of each distribution.

4B. Compare the data sets using either the means and standard deviations or the five-number summaries. Justify your choice.

Junior Season (points)
10, 12, 6, 10, 13, 8, 12, 3, 21, 14, 7, 0, 15, 6, 16, 8, 17, 3, 17, 2

Senior Season (points)
10, 32, 3, 22, 20, 30, 26, 24, 5, 22, 28, 32, 26, 21, 6, 20, 24, 18, 12, 25

Check Your Understanding   ◯ = Step-by-Step Solutions begin on page R11.

Go Online! for a
Self-Check Quiz

Example 1   **Find the mean, median, mode, range, and standard deviation of each data set that is obtained after adding the given constant to each value.**

1. 10, 13, 9, 8, 15, 8, 13, 12, 7, 8, 11, 12; + (−7)   **2.** 38, 36, 37, 42, 31, 44, 37, 45, 29, 42, 30, 42; + 23

Example 2   **Find the mean, median, mode, range, and standard deviation of each data set that is obtained after multiplying each value by the given constant.**

③ 6, 10, 3, 7, 4, 9, 3, 8, 5, 11, 2, 1; × 3   **4.** 42, 39, 45, 44, 37, 42, 38, 37, 41, 49, 42, 36; × 0.5

Example 3   **5. TRACK** Mark and Kyle's long jump distances are shown.

Kyle's Distances (ft)
17.2, 18.28, 18.56, 17.28, 17.36, 18.08, 17.43, 17.71, 17.46, 18.26, 17.51, 17.58, 17.41, 18.21, 17.34, 17.63, 17.55, 17.26, 17.18, 17.78, 17.51, 17.83, 17.92, 18.04, 17.91

Mark's Distances (ft)
18.88, 19.24, 17.63, 18.69, 17.74, 19.18, 17.92, 18.96, 18.19, 18.21, 18.46, 17.47, 18.49, 17.86, 18.93, 18.73, 18.34, 18.67, 18.56, 18.79, 18.47, 18.84, 18.87, 17.94, 18.7

a. Use a graphing calculator to construct a histogram for each set of data. Then describe the shape of each distribution.

b. Compare the data sets using either the means and standard deviations or the five-number summaries. Justify your choice.

Example 4

6. **TIPS** Miguel and Stephanie are servers at a restaurant. The tips that they earned to the nearest dollar over the past 15 workdays are shown.

Miguel's Tips ($)
14, 68, 52, 21, 63, 32, 43, 35, 70, 37, 42, 16, 47, 38, 48

Stephanie's Tips ($)
34, 52, 43, 39, 41, 50, 46, 36, 37, 47, 39, 49, 44, 36, 50

a. Use a graphing calculator to construct a box-and-whisker plot for each set of data. Then describe the shape of each distribution.

b. Compare the data sets using either the means and standard deviations or the five-number summaries. Justify your choice.

Practice and Problem Solving

Extra Practice is on page R10.

Example 1

Find the mean, median, mode, range, and standard deviation of each data set that is obtained after adding the given constant to each value.

7. 52, 53, 49, 61, 57, 52, 48, 60, 50, 47; $+ 8$ 8. 101, 99, 97, 88, 92, 100, 97, 89, 94, 90; $+ (-13)$

9. 27, 21, 34, 42, 20, 19, 18, 26, 25, 33; $+ (-4)$ 10. 72, 56, 71, 63, 68, 59, 77, 74, 76, 66; $+ 16$

Example 2

Find the mean, median, mode, range, and standard deviation of each data set that is obtained after multiplying each value by the given constant.

11. 11, 7, 3, 13, 16, 8, 3, 11, 17, 3; $\times 4$ 12. 64, 42, 58, 40, 61, 67, 58, 52, 51, 49; $\times 0.2$

13. 33, 37, 38, 29, 35, 37, 27, 40, 28, 31; $\times 0.8$ 14. 1, 5, 4, 2, 1, 3, 6, 2, 5, 1; $\times 6.5$

Example 3

15. **BOOKS** The page counts for the books that the students chose are shown.

1st Period
388, 439, 206, 438, 413, 253, 311, 427, 258, 511, 283, 578, 291, 358, 297, 303, 325, 506, 331, 482, 343, 372, 456, 267, 484, 227

6th Period
357, 294, 506, 392, 296, 467, 308, 319, 485, 333, 352, 405, 359, 451, 378, 490, 379, 401, 409, 421, 341, 438, 297, 440, 500, 312, 502

a. Use a graphing calculator to construct a histogram for each set of data. Then describe the shape of each distribution.

b. Compare the data sets using either the means and standard deviations or the five-number summaries. Justify your choice.

16. **MINIATURE GOLF** A sample of scores for the Blue and Red courses at Top Golf are shown.

Blue Course
46, 25, 62, 45, 30, 43, 40, 46, 33, 53, 35, 38, 39, 40, 52, 42, 44, 48, 50, 35, 32, 55, 28, 58

Red Course
53, 49, 26, 61, 40, 50, 42, 35, 45, 48, 31, 48, 33, 50, 35, 55, 38, 50, 42, 53, 44, 54, 48, 58

a. Use a graphing calculator to construct a histogram for each set of data. Then describe the shape of each distribution.

b. Compare the data sets using either the means and standard deviations or the five-number summaries. Justify your choice.

Example 4

17. **BRAINTEASERS** The amount of time (in minutes) that it took Leon and Cassie to complete puzzles is shown.

Leon's Times
4.5, 1.8, 3.2, 5.1, 2.0, 2.6, 4.8, 2.4, 2.2, 2.8, 1.8, 2.2, 3.9, 2.3, 3.3, 2.4

Cassie's Times
2.3, 5.8, 4.8, 3.3, 5.2, 4.6, 3.6, 5.7, 3.8, 4.2, 5.0, 4.3, 5.5, 4.9, 2.4, 5.2

a. Use a graphing calculator to construct a box-and-whisker plot for each set of data. Then describe the shape of each distribution.

b. Compare the data sets using either the means and standard deviations or the five-number summaries. Justify your choice.

18. **DANCE** The total amount of money that a sample of students spent to attend the homecoming dance is shown.

Boys (dollars)
114, 98, 131, 83, 91, 64, 94, 77, 96, 105, 72, 108, 87, 112, 58, 126

Girls (dollars)
124, 74, 105, 133, 85, 162, 90, 109, 94, 102, 98, 171, 138, 89, 154, 76

a. Use a graphing calculator to construct a box-and-whisker plot for each set of data. Then describe the shape of each distribution.

b. Compare the data sets using either the means and standard deviations or the five-number summaries. Justify your choice.

19. **LANDSCAPING** Refer to the beginning of the lesson. Rhonda, another employee that works with Tom, earned the following over the past month.

Rhonda's Pay ($)		
45	55	53
47	53	54
44	56	59
63	47	53
60	57	62
44	50	45
60	53	49
62	47	

a. Find the mean, median, mode, range, and standard deviation of Rhonda's earnings.

b. A $5 bonus had been added to each of Rhonda's daily earnings. Find the mean, median, mode, range, and standard deviation of Rhonda's earnings before the $5 bonus.

20. **SHOPPING** The items Lorenzo purchased are shown.

a. Find the mean, median, mode, range, and standard deviation of the prices.

b. A 7% sales tax was added to the price of each item. Find the mean, median, mode, range, and standard deviation of the items without the sales tax.

Baseball hat	$14.98
Jeans	$24.61
T-shirt	$12.84
T-shirt	$16.05
Backpack	$42.80
Folders	$2.14
Sweatshirt	$19.26

21. **MODELING** A salesperson has 15 SUVs priced between $33,000 and $37,000 and 5 luxury cars priced between $44,000 and $48,000 on a car lot. The average price for all of the vehicles is $39,250. The sales manager decides to reduce the prices of all 15 SUVs on the car lot by $2000 per vehicle. What is the new average price for the SUVs and luxury cars?

H.O.T. Problems   Use **H**igher-**O**rder **T**hinking Skills

22. **REASONING** If every value in a set of data is multiplied by a constant k, $k < 0$, then how is the mean, median, mode, range, and standard deviation of the new data set found?

23. **WRITING IN MATH** Compare and contrast the benefits of displaying data using histograms and box-and-whisker plots.

24. **REGULARITY** If k is added to every value in a set of data, and then each resulting value is multiplied by a constant m, $m > 0$, how can the mean, median, mode, range, and standard deviation of the new data set be found? Explain your reasoning.

25. **WRITING IN MATH** Explain why the mean and standard deviation are used to compare the center and spread of two symmetrical distributions and the five-number summary is used to compare the center and spread of two skewed distributions or a symmetric distribution and a skewed distribution.

26. Ms. Stein gave a quiz in which the total possible score was 50 points. The quiz scores are shown below.

32, 18, 12, 33, 20, 17, 23, 42, 28, 27, 12, 9, 21, 16, 32, 37

To increase scores, she decided to add 5 points to each quiz score, creating a new data set. **MP** 1, 4

a. What is the median of the original set of quiz scores?

b. What is the median of the new data set?

c. To the nearest tenth, what is the mean of the original set of quiz scores?

d. To the nearest tenth, what is the mean of the new data set?

e. What is the range of the original set of quiz scores?

f. What is the range of the new data set?

27. MULTI-STEP The histograms below represent the test scores in Ms. Gilmore's class and Mr. Vaught's class. **MP** 1, 2, 4

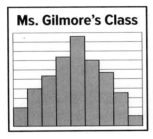 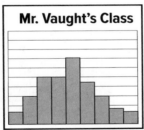

a. Which describes the shape of each distribution?

○ **A** Both distributions are relatively symmetric.

○ **B** Both distributions are skewed.

○ **C** The distribution for Ms. Gilmore's class is skewed. The distribution for Mr. Vaught's class is relatively symmetric.

○ **D** The distribution for Ms. Gilmore's class is relatively symmetric. The distribution for Mr. Vaught's class is skewed.

b. Which are the more appropriate statistics for comparing the two sets of data—the means and standard deviations, or the five-number summaries? Justify your choice.

28. MULTI-STEP The box-and-whisker plots below represent the heights of the baseball players on a high school team and the heights of the basketball players. **MP** 1, 2, 4

Baseball Players   **Basketball Players**

a. Which describes the shape of each distribution?

○ **A** Both distributions are relatively symmetric.

○ **B** Both distributions are skewed.

○ **C** The distribution of the heights of the baseball players is relatively symmetric. The distribution of the heights of the basketball players is skewed.

○ **D** The distribution of the baseball players is skewed. The distribution of the heights of the basketball players is relatively symmetric.

b. Which are the more appropriate statistics for comparing the two sets of data—the means and standard deviations, or the five-number summaries? Justify your choice.

29. MULTI-STEP The histograms below represent two data sets. **MP** 1, 2, 4

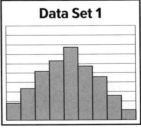 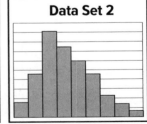

a. Which describes the shape of each distribution?

○ **A** Both distributions are relatively symmetric.

○ **B** Both distributions are skewed.

○ **C** Data set 1 is relatively symmetric. Data set 2 is skewed.

○ **D** Data set 1 is skewed. Data set 2 is relatively symmetric.

b. Which are the more appropriate statistics for comparing the two sets of data—the means and standard deviations, or the five-number summaries? Justify your choice.

Summarizing Categorical Data

Then	Now	Why?

 You measured central tendency and spread to describe data sets.

1 Summarize data in two-way frequency tables.

2 Summarize data in two-way relative frequency tables.

To make plans for a community park, the city may survey residents about how they use the park and what features they would like.

Age	Skate-boarding	Walking/ Running	Total
12–25	678	82	760
26–99	64	435	499
Total	742	517	1259

A two-way frequency table can help city officials quantify the residents' preferences and behavior patterns.

 New Vocabulary

two-way frequency table
relative frequency
two-way relative frequency table
marginal frequency
joint frequency
conditional relative frequency
association

MP **Mathematical Practices**

4 Model with mathematics.

7 Look for and make use of structure.

8 Look for and express regularity in repeated reasoning.

1 Two-Way Frequency Tables A **two-way frequency table** or *contingency* table is used to show the frequencies of data from a survey or experiment classified according to two categories, with rows indicating one category and columns indicating the other. The subcategories are the column and row headers of the table that represent the two different types of categories. Using a two-way frequency table allows people to improve their community planning, supply ordering, and other functions where they need to know the preferences of a variety of people.

Real-World Example 1 Interpret a Two-Way Frequency Table

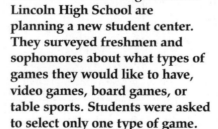

STUDENT CENTER Organizers at Lincoln High School are planning a new student center. They surveyed freshmen and sophomores about what types of games they would like to have, video games, board games, or table sports. Students were asked to select only one type of game. The results are shown in the table.

Student Center Preferences			
	Freshmen	Sophomores	Total
Video Games	20	170	190
Board Games	55	35	90
Table Sports	30	190	220
Total	105	395	500

a. Of the students who chose table sports, how many of them are freshmen?

Find the value that is in both the *Freshmen* column and *Table Sports* row. There are 30 students who chose table sports and are freshmen.

b. How many students chose video games?

Find the value that is in both the *Total* column and *Video Games* row. There are 190 students who chose video games.

c. How many students are freshmen?

Find the value that is in both the *Freshmen* column and *Total* row. There are 105 students who are freshmen.

d. How many students were surveyed in total?

Find the value that is in both the *Total* row and column. There are a total of 500 students who were surveyed.

e. One teacher suggested that the budget for the student center should be split evenly between the freshmen and sophomores. Another teacher disagrees. What argument can be made?

The table shows that only 105 out of 500, or about 20% of students, are freshmen. Therefore, it might make more sense for 80% of the money to be spent on the preferences of the sophomores.

> **Guided Practice**

CRAFTS The table shows craft options chosen by freshmen and sophomores at a camp event.

Craft	Freshmen	Sophomores	Total
Sand Art	20	17	37
Key Chain	15	18	33
Picture Frame	12	6	18
Total	47	41	88

1A. Which craft was most popular overall?

1B. Which craft was most popular with the sophomores?

1C. How many freshmen attended the camp event?

1D. How many attendees did crafts overall?

You can use a two-way frequency table and information that is known about some parts of a situation to deduce the rest of the data for the table.

Real-World Example 2   Complete a Two-Way Frequency Table

ACTIVITIES A small town conducted a survey to help decide on which community initiatives they should spend annual funding. The town has 2400 residents who responded to the survey. According to the survey, 45% of respondents are cyclists, of whom 80% enjoy spending time at the lake. Of the residents who are not cyclists, 75% enjoy spending time at the lake. Make a two-way frequency table to display the data.

How many residents reported that they enjoy spending time at the lake?

Step 1 Make a table and label the two sets of categories. Then, complete the table with the values from the given information.

Based on the only information given, only the total can be added to the table. The remaining values of the table must be calculated.

Survey Results			
Resident	Enjoy Lake	Do Not Enjoy Lake	Total
Cyclists			
Non-Cyclists			
Total			2400

Step 2 Find the values for the remaining cells of the table. 45% of the 2400 respondents are cyclists.

$(0.45)2400 = 1080$ There are 1080 total cyclists.

> **Study Tip**
>
> **MP Appropriate Tools**
> You can use spreadsheet software to make a two-way frequency table. Use formulas to calculate missing values.

Of the cyclists, 80% enjoy spending time at the lake.

(0.80)1080 = 864   There are **864** cyclists who enjoy the lake.

To find the number of cyclists who do not enjoy the lake, find the difference of the total number of cyclists and the number of cyclists who enjoy the lake.

1080 − 864 = 216   There are **216** cyclists who do not enjoy the lake.

To find the total number of non-cyclists, find the difference of the total respondents and the cyclists.

2400 − 1080 = 1320   There are **1320** non-cyclists.

Of the residents who are non-cyclists, 75% enjoy spending time at the lake.

(0.75)**1320** = 990   There are **990** non-cyclists who enjoy the lake.

To find the number of non-cyclists who do not enjoy the lake, find the difference of the total number of non-cyclists and the number of non-cyclists who enjoy the lake.

1320 − **990** = 330   There are **330** non-cyclists who do not enjoy the lake.

Find the totals for each column by adding cyclists and non-cyclists.

Enjoy Lake: **864** + **990** = 1854

Do Not Enjoy Lake: **216** + **330** = 546

Survey Results			
Resident	Enjoy Lake	Do Not Enjoy Lake	Total
Cyclists	**864**	**216**	1080
Non-Cyclists	**990**	**330**	1320
Total	1854	546	2400

Check that the totals of each column have a sum equal to the total number of respondents.

Study Tip

MP **Structure** When completing a frequency table, the *Total* row and the *Total* column must add up to the same value.

▶ **Guided Practice**

2. **BIRDS** A bird-watching club has counted all of the birds of a two certain species that they found in a park one afternoon. There are two color variations: plain wings and red-banded wings. They can tell the males and females apart by other color markings. The club found 18 males and a total of 42 birds. Of the females, 75% had plain wings, and 50% of the males had plain wings. Complete the two-way frequency table.

Gender	Red-Banded Wings	Plain Wings	Total
Males			18
Females			
Total			42

2 **Two-Way Relative Frequency Tables** A **relative frequency** is the ratio of the number of observations in a category to the total number of observations. To create a **two-way relative frequency table**, divide each of the values by the total number of observations and replace them with their corresponding decimals or percents.

Relative frequency tables express the categories of information as percentages. Then the data can be compared, to help decide if there is an association between two events.

Some parts of a relative frequency table have specific names. The subtotals of each subcategory, are called **marginal frequencies**.

Data which shows percentages for joint situations where two categories intersect, are called **joint frequencies**.

Real-World Example 3   Two-Way Relative Frequency Table

GYM A manager of a gym conducted a survey to determine whether the gym members prefer rock climbing, weight lifting, or neither. She also asked members whether they regularly bring guests to the gym. The frequency table below shows the surveyed responses of the surveyed members.

Activity	Bring Guests	Do Not Bring Guests	Total
Weight Lifting	86	2	88
Cooking Class	24	40	64
Neither	16	32	48
Total	126	74	200

a. Make a relative frequency table by converting the data in the table to percentages.

Divide each value in the table by the total number of members and multiply by 100% to find each frequency. For example:

Percentage of total members who bring guests: $\frac{126}{200} \cdot 100\% = 63\%$

Percentage of total members who neither prefer rock climbing nor weight lifting: $\frac{48}{200} \cdot 100\% = 24\%$

Activity	Bring Guests	Do Not Bring Guests	Total
Rock Climbing	43%	1%	44%
Weight Lifting	12%	20%	32%
Neither	8%	16%	24%
Total	63%	37%	100%

b. Find the joint frequency of rock climbing and bringing guests.

Find the value of both the *Rock Climbing* row and the *Bring Guests* column, which is 43%.

> **Reading Math**
>
> **MP Finding Patterns**
> In a relative frequency table, the sum of *Total* row and column should be equal 100%. You can use this fact to check that your math is correct.

▶ Guided Practice

PETS The table shows data about pet owners having dogs and cats.

Pet Owners	Owns a Dog	Does Not Own a Dog	Total
Owns a Cat	6	11	
Does Not Own a Cat	9	14	
Total			

3A. Fill in the missing numbers for the frequency table. Then convert the frequency table to a relative frequency table by changing the numbers to percentages of the whole.

3B. What is the joint frequency data for someone owning neither a dog nor a cat?

You can refine the comparisons between data points even more by examining the conditional relative frequency. A **conditional relative frequency** is the ratio of the joint frequency to the marginal frequency. Because each two-way frequency table has two categories, each two-way relative frequency table can provide two different conditional relative frequency tables.

Conditional frequencies can be used to compare data in a meaningful way and to eliminate bias that might appear because of skewed data sets. Bias can result when one group has a larger population than another.

Real-World Example 4 Conditional Relative Frequency

DANCE The dance squad is planning to buy new uniforms. The coach asked members of the varsity and junior varsity squads which color of uniform they would prefer. The results are recorded in a relative frequency table.

Color	Junior Varsity	Varsity	Total
Black	22%	16%	38%
Blue	11%	19%	30%
Red	28%	4%	32%
Total	61%	39%	100%

a. What percentage of members who prefer blue uniforms are on the varsity squad?

$\frac{19}{30} \cdot 100\% \approx 63.3\%$

b. What percentage of the junior varsity squad prefers red uniforms?

$\frac{28}{61} \cdot 100\% \approx 45.9\%$

c. What percentage of members who prefer black uniforms are on the junior varsity squad?

$\frac{22}{38} \cdot 100\% \approx 57.9\%$

d. What percentage of the varsity squad prefers blue uniforms?

$\frac{19}{39} \cdot 100\% \approx 48.7\%$

▶ **Guided Practice**

Use the relative frequency table in Example 4 to answer the following question.

4A. What percentage of members who prefer red uniforms are on the varsity squad?

4B. What percentage of members who prefer blue uniforms are on the junior varsity squad?

4C. What percentage of the varsity squad prefers black uniforms?

4D. What percentage of the junior varsity squad prefers blue uniforms?

Study Tip

MP Precision When rounding percentages in your calculations, check that sums of the percentages are close to the expected totals.

If the conditional frequency of one group is very different from that of another group, then there is an **association** between the two categories of data.

If the conditional frequencies of two groups are similar, then there may not be an association between the two data categories, and the combinations are simply formed by random chance.

Study Tip

MP Structure Note that there are different ways to compare data when looking for associations. You can calculate each joint data point as a percentage of the marginal data point to the right of the row, or at the bottom of the column.

Real-World Example 5   Associations Between Data in Relative Frequency Tables

LIBRARY The library staff is conducting a survey of local students in efforts to attract more patrons to the library. They asked students which programs they would be interested in attending and how far away from the library they live. The survey data is shown in the relative frequency table.

Distance (mi)	Weekend Book Club	After School Tutoring	Total
< 1	7%	36%	43%
1–3	30%	1%	31%
> 3	16%	10%	26%
Total	53%	47%	100%

a. What percentage of students who are interested in book club live between 1 and 3 miles from the library?

$\frac{30}{53} \cdot 100\% \approx 57\%$

b. What percentage of students who are interested in tutoring live less than 1 mile from the library?

$\frac{36}{47} \cdot 100\% \approx 77\%$

c. Is there an association between interest in book club or tutoring and the distance they live from the library?

Yes, there seems to be a stronger association between living very close to the library and after school tutoring than there is between weekend book club and living farther from the library.

▶ **Guided Practice**

5A. GARDENING Ronald noticed that several homes in his neighborhood have flowerbeds and gardens. He also noticed that some neighbors seemed to participate more in community activities. He collected the following information. Is there an association between having flowerbeds and participating in community events?

Home Landscape	Participate in at Least Half of Community Events	Participate in Less than Half of Community Events	Total
Flowerbeds/Gardens	30	10	40
Plain Lawn	20	80	100
Total	50	90	140

 **Go Online!** for a Self-Check Quiz

Check Your Understanding   = Step-by-Step Solutions begin on page R11.

Example 1   **EVENTS** Medieval reenactors at an event were dressed either as Normans or Saxons. Some were dressed as peasants, and others were dressed as nobles. Use the table for Exercises 1–8.

Role	Normans	Saxons	Total
Peasants	42	36	78
Nobles	29	14	43
Total	71	50	121

1 How many Norman nobles were there?

2. How many Saxons were there in total?

3. How many peasants were there in total?

Example 3

4. How many reenactors were at the event?

5. Make a two-way relative frequency table. Round to the nearest percent.

6. Which numbers are joint frequencies?

7 Which numbers are marginal frequencies?

8. What is the joint frequency of Norman nobles?

Example 2 **9.** At a dog show, there were 80 dogs competing. Of which the competitors, 60% were retrievers, while the rest were small dogs. Thirty percent of the dogs were male retrievers and 25% were female lap dogs. Make a two-way frequency table.

Example 4 **SKIING** At an international ski race, Giorgio and Susie recorded the nationality and skiwear for 24 skiers. Twelve skiers were Norwegian, 8 were Swiss, and the rest were American. Seventy-five percent of the Americans were wearing spring skiwear. Five of the Norwegians were wearing full ski suits. In all, 15 people were wearing spring skiwear.

10. Make a relative frequency table. Round to the nearest percent.

11. What percent of the skiers wearing full ski suits were American?

12. What percent of the skiers wearing spring skiwear were Swiss?

Example 5 **13.** Explain whether there is an association between nationality and skiwear at this event.

14. **REASONING** How reliable is the association? Explain your answer.

Practice and Problem Solving

Extra Practice is found on page R10.

Example 1 **BEACH** Jordan made a survey of the activities of people at the beach and recorded the data in a two-way frequency table. Use the table for Exercises 15–20.

Activities	Swimming Suits	Shorts and T-shirts	Dresses	Total
Swimming	18	0	0	18
Sitting or Lying on the Sand	21	3	3	27
Walking	5	9	5	19
Playing Sports	2	8	1	11
Total	46	20	9	75

15. How many people were swimming in total?

16. How many people were walking on the beach while wearing a dress?

17. How many people were walking on the beach with clothes other than their swimming suits?

18. How many people were swimming in shorts and T-shirts?

19. How many people were playing sports?

20. How many people were at the beach?

Example 2 **MEALS** Mary manages a banquet center, and she is compiling data on a meal options that were served. Of 24 vegetarian meals, 18 people left some food on the plate. Of the 18 chicken meals, 9 had leftovers. Thirty people ordered the beef option and 21 had nothing left of their plates.

21. Make a two-way frequency table.

22. Describe two ways to find the total number of diners.

23. Make a two-way relative frequency table of the data. Round to the nearest percent.

24. What percentage is the joint data point for ate everything and beef?

25. Did more people overall eat everything or have leftovers?

26. What percent of people who had the vegetarian meal ate everything from their plate?

27. Explain whether there an association between the type of meal a diner chose and whether they ate all of their food.

28. (MP) **REASONING** Discuss some possible reasons that the vegetarian meals had more leftovers than the meat meals. What actions might the manager consider?

29. (MP) **STRUCTURE** Elizabeth and her extended family went to the zoo. Complete the two-way frequency table about the frozen yogurt they bought.

Flavor	Waffle Cone	Regular Cone	Total
Strawberry	2		
Vanilla	1	1	
Chocolate	3		8
Total			14

Example 4

30. What is the conditional relative frequency of chocolate in waffle cones compared to total chocolate cones?

31. What is the marginal frequency for total strawberry cones bought?

32. What is the relative frequency for regular cones?

33. What is the frequency for waffle cones?

34. What is the joint frequency for vanilla regular cones?

Example 5

35. (MP) **STRUCTURE** A manager takes count of the level of preparedness of 24 employees on their first day on a construction job. Four have their paperwork ready but forgot to bring safety gear. Six have forgotten both safety gear and their paperwork.

a. Complete the two-way frequency table.

Safety Gear	Paperwork Ready	Paperwork Not Ready	Total
Have Safety Gear	8	6	
Do Not Have Safety Gear			
Total			24

b. What percentage of new employees in total had their paperwork ready?

c. What percentage of new employees were completely prepared?

d. What percentage of new employees had all of their safety gear ready?

e. Complete the two-way relative frequency table.

Safety Gear	Paperwork Ready	Paperwork Not Ready	Total
Have Safety Gear		25%	
Do Not Have Safety Gear			
Total	50%		100%

f. What is the conditional relative frequency for having paperwork but missing safety glasses, compared to the total people who had their paperwork ready?

g. What is the percentage of people who did not have their paperwork ready?

36. **MP** **REASONING** A manager of an Indian restaurant decided to keep track of customer orders for a week to determine which dishes are ordered. Complete the two-way frequency table.

Curry Flavor	With Lentils and Rice	With Plain Rice	With Naan Bread	Total
Green Curry	285			845
Red Curry	156	180		506
Yellow Curry	346		427	1284
Total			853	

   a. How many dishes were served overall?

   b. What is the conditional relative frequency of green curry with naan bread compared to green curry dishes served overall?

   c. What is the relative frequency of red curry with lentils and rice?

   d. What is the marginal frequency of dishes served with white rice?

   e. What is the marginal frequency for dishes served with lentils and rice?

   f. What is the conditional relative frequency of yellow curry with plain rice compared to plain rice dishes served overall?

37. **MP** **REGULARITY** Ms. Ramos is a preschool teacher, and she collected data about the allergies of students in her class. This would allow her to remember what foods are safe in her classroom. Copy and complete the two-way frequency table, and copy and complete the two-way relative frequency table.

Two-Way Frequency Table

Allergy	Eat Gluten	Should Not Eat Gluten	Total
Peanuts	4		7
Soy	2	1	
Milk	1		3
Total			13

Two-Way Relative Frequency Table

Allergy	Eat Gluten	Should Not Eat Gluten	Total
Peanuts	31%		
Soy	15%		
Milk			
Total			100%

H.O.T. Problems Use Higher-Order Thinking Skills

38. **MP** **CONSTRUCT ARGUMENTS** Discuss the difference between two-way frequency tables and two-way relative frequency tables. Which type of frequency table makes comparison easier?

39. **OPEN-ENDED** Discuss how conditional frequencies help to compare data points better than simple relative frequencies. Give an example.

40. **WRITING IN MATH** Describe the processes that you use to fill in two-way frequency tables, starting from minimal data.

41. PETS A pet store has three new litters of guinea pigs. The animal caretaker records the weight in grams of each new guinea pig and their colors and markings. **MP** 4, 7, 8

Weight (g)	Brown	Black and White	Total
< 80	4	5	9
> 80	2	1	3
Total	6	6	12

a. How many brown guinea pigs are greater than 80 g?

- ◯ **A** 2
- ◯ **B** 3
- ◯ **C** 4
- ◯ **D** 5

b. What is the relative frequency for black and white guinea pigs that weigh less than 80 g?

- ◯ **A** 8%
- ◯ **B** 33%
- ◯ **C** 42%
- ◯ **D** 50%

c. What percent of the black and white guinea pigs weigh greater than 80 g?

- ◯ **A** 8%
- ◯ **B** 17%
- ◯ **C** 42%
- ◯ **D** 83%

42. ACTITIVIES James surveyed 150 classmates about their involvement in band and sports. He found that 48% are in band and 25% of those in band also play sports. Eighteen percent of the surveyed students are neither in band nor play sports. Make a two-way frequency table to display the data. **MP** 1, 2, 4, 7

43. MULTI-STEP After heavy storm, Maria evaluates the vegetables in her garden. Damaged vegetables must be harvested right away, while undamaged vegetables can stay in the garden to further mature. **MP** 4, 7, 8

Vegetables	Damaged	Undamaged	Total
Tomatoes	46	12	
Peas	200	300	
Squash	6	9	
Total			

a. Which of the following statements are true?

- ☐ **A** The total number of peas is 600.
- ☐ **B** The marginal frequency for squash is 15.
- ☐ **C** The joint frequency of damaged peas is 500.
- ☐ **D** The marginal frequency for vegetables that are undamaged is 321.
- ☐ **E** The total number of vegetables is 573.
- ☐ **F** The joint frequency of undamaged tomatoes is 58.
- ☐ **G** The marginal frequency of vegetables that are damaged is 252.

b. What percentage of the tomatoes are damaged?

c. What percentage of the vegetables are damaged?

d. Calculate the conditional relative frequencies of each damaged vegetable. Is there an association between the type of vegetable and its likelihood to have been damaged in the storm?

e. What math operation explains how to find the marginal frequency data that would go in the last column, third cell down?

- ◯ **A** add the two cells above it
- ◯ **B** add the two cells below it
- ◯ **C** add the two cells to the left of it
- ◯ **D** subtract the cell in the middle from the cell at the left

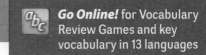

Go Online! for Vocabulary Review Games and key vocabulary in 13 languages

Study Guide

Key Concepts

Measures of Center (Lesson 10-1)

- The mean, median, and mode are measures of center, which summarize data with a single number.
- A percentile gives the percent of the data values in a set that are below a certain data value.

Representing Data (Lesson 10-2)

- Bar graphs, histograms, dot plots, and box plots can be used to represent data.

Measures of Spread (Lesson 10-3)

- A low standard deviation indicates that the data tend to be very close to the mean, while a high standard deviation indicates that the data are spread out over a larger range.

Distributions of Data and Comparing Sets of Data (Lessons 10-4 and 10-5)

- In a negatively skewed distribution, the majority of the data are on the right. In a positively skewed distribution, the majority of the data are on the left. In a symmetric distribution, the data are evenly distributed.
- Two symmetric distributions can be compared using mean and standard deviation. If one of the distributions is skewed, compare the five-number summary.

Summarizing Categorical Data (Lesson 10-6)

- A two-way frequency table is used to show the frequencies of data classified according to two categories, with rows indicating one category and columns indicating the other.
- Relative frequencies are found by dividing each value in a two-way table by the total number. Conditional relative frequencies are found by dividing each joint frequency by the marginal frequency.

FOLDABLES® Study Organizer

Use your Foldable to review the chapter. Working with a partner can be helpful. Ask for clarification of concepts as needed.

Key Vocabulary

bar graph (p. 659)	measures of center (p. 651)
cumulative frequency (p. 659)	median (p. 651)
distribution (p. 673)	mode (p. 651)
dot plot (p. 657)	percentile (p. 653)
frequency table (p. 659)	qualitative data (p. 651)
histogram (p. 659)	quantitative data (p. 651)
joint frequencies (p. 691)	relative frequency (p. 690)
linear transformation (p. 680)	skewed distribution (p. 673)
marginal frequencies (p. 691)	standard deviation (p. 667)
mean (p. 651)	symmetric distribution (p. 673)
	two-way frequency table (p. 688)

Vocabulary Check

Choose the term that best completes each sentence.

1. The (mean, median, mode) of a data set is the average data value.

2. A (symmetric, skewed) distribution is one in which there are more data values on one side than the other.

3. Data that can be measured are (qualitative, quantitative) data.

4. The totals of each subcategory in a two-way table are the (joint, marginal) frequencies.

Concept Check

5. Explain the difference between a bar graph and a histogram.

6. Explain how to determine the shape of a data distribution.

7. Explain how to find the mean, median, and mode.

8. Create examples for when the mean, median and mode are the best measures of central tendency.

Lesson-by-Lesson Review

10-1   Measures of Center

Find the mean, median, and mode of each data set.

9. {6, 20, 55, 20, 7, 18}

10. {10, 81, 66, 15}

11. **TESTS** Aliyah's first four science test scores are 91, 72, 97, and 82. Find the mean, median, and mode of the test scores. Which statistic best describes Aliyah's science tests?

12. **BOWLING** The scores for a bowling team are shown in the table.

Team Scores			
Bowler	Score	Bowler	Score
Lamar	112	Paige	251
Mike	136	Quinn	74
Nicole	177	Roman	68
Opal	65	Sergio	103

    a. Which bowlers are above the 50th percentile?

    b. Which bowler is in the 38th percentile?

Example 1

Find the mean, median, and mode of the set {34, 5, 1, 22, 18, 29, 1}.

To find the mean, find the sum and divide by the number of values.

$$\frac{34 + 5 + 1 + 22 + 18 + 29 + 1}{7} = \frac{110}{7} \approx 15.7$$

To find the median, order the numbers from least to greatest and find the middle value.

1, 1, 5, (18,) 22, 29, 34

To find the mode, find the number that occurs most often. The value 1 occurs most often.

The mean is about 15.7, the median is 18, and the mode is 1.

10-2   Representing Data

13. **MOVIES** The table shows the results of a survey in which students were asked to choose which of four genres of movies they like best. Make a bar graph of the data.

Genre	Frequency
Action	25
Romance	16
Comedy	30
Documentary	2

14. **CARS** The table shows the number of each type of vehicle that passes an intersection during an hour. Make a bar graph of the data.

Type	Frequency
Sedan	54
Truck	21
Bus	4
SUV	16

Example 2

Make a bar graph to display the data collected regarding students' favorite color.

Step 1: Decide whether the data is continuous or discrete. Bar graphs are used for discrete data and histograms are used for continuous data.

Step 2: Draw a horizontal axis and a vertical axis. Label the axes as shown. Add a title.

Step 3: Draw a bar to represent each category.

Color	Frequency
Red	12
Yellow	8
Green	15
Blue	9

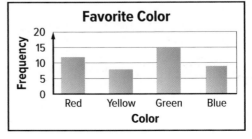

(continued on the next page)

Make a histogram of the frequency and the cumulative frequency.

15. BOOKS The frequency table shows the number of pages in the books Amelia borrows from a library.

Pages	Books
0–99	1
100–199	3
200–299	2
300–399	5
400–499	1

16. DINING The frequency table shows the ages of people at a restaurant.

Age	Frequency
0–19	4
20–39	12
40–59	18
60-79	3

Example 3

Make a histogram of the frequency and the cumulative frequency.

Height (ft)	0–0.9	1–1.9	2–2.9
Plants	5	12	7

Find the cumulative frequency for each interval.

Height (ft)	<1	<2	<3
Plants	5	17	24

Make each histogram like a bar graph but without spaces between the bars.

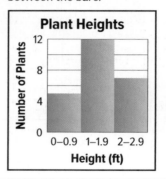

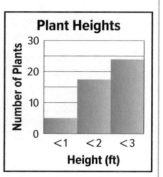

10-3 Measures of Spread

17. SHOVELING Ben mows lawns to earn money. The number of lawns he cuts for five weeks is {2, 4, 3, 5, 3}. Find and interpret the mean and standard deviation.

18. CANDY BARS Luci is keeping track of the number of candy bars each member of the drill team sold. The results are {20, 25, 30, 50, 40, 60, 20, 10, 42}. Find and interpret the mean and standard deviation.

19. FOOD A fast food company polls a random sample of its day and night customers to find how many times a month they eat out. Compare the means and standard deviations of each data set.

Day Customers	Night Customers
10, 3, 12, 15, 7, 8, 4, 12, 9, 14, 12, 9	15, 12, 13, 9, 11, 12, 14, 12, 8, 16, 9, 9

Example 4

GIFTS Joshua is collecting money from his family for a Mother's Day gift. He keeps track of how much each person has donated: {10, 5, 20, 15, 10}. Find and interpret the mean absolute deviation.

Step 1 Find the mean: $\bar{x} = \dfrac{10 + 5 + 20 + 15 + 10}{5}$ or 12

Step 2 Find the squares of the differences.

$(12 - 10)^2 = 4$ $(12 - 5)^2 = 49$
$(12 - 20)^2 = 64$ $(12 - 15)^2 = 9$
$(12 - 10)^2 = 4$

Step 3 Find the sum.
$4 + 49 + 64 + 9 + 4 = 130$

Step 4 Find the variance.
$\sigma^2 = \dfrac{130}{5}$ or 26

Step 5 Find the standard deviation.
$\sigma = \sqrt{\sigma^2} = \sqrt{26}$ or about 5.10

The standard deviation is fairly great compared to the mean. This suggests that many of the data values are not close to the mean.

10-4 Distributions of Data

Use a graphing calculator to construct a histogram for the data. Then describe the shape of the distribution.

20. 55, 62, 32, 56, 31, 59, 19, 61, 8, 48, 41, 69, 32, 63, 48, 60, 43, 66, 71, 70, 49, 56, 21, 67

21. 4, 19, 62, 28, 26, 59, 33, 39, 36, 72, 46, 48, 49, 44, 72, 76, 55, 53, 55, 62, 66, 69, 71, 74

22. MILK A grocery store manager tracked the amount of milk in gallons sold each day. Describe the center and spread of the data using either the mean and standard deviation or the five-number summary. Justify your choice by constructing a box-and-whisker plot for the data.

Gallons of Milk Sold per Day					
383	296	354	288	195	372
421	367	411	355	296	321
403	357	432	229	180	266

Example 5

DRIVING TESTS Several driving test results are shown. Describe the center and spread of the data using either the mean and standard deviation or the five-number summary. Justify your choice by constructing a box-and-whisker plot for the data.

Driving Test Scores					
80	95	100	95	95	100
100	90	75	60	90	80

Use a graphing calculator to create a box-and-whisker plot.

The left whisker is longer than the right and the median is closer to the right whisker. Therefore, the distribution is negatively skewed.

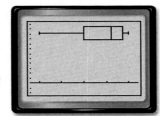

[56, 104] scl: 10 by [−2, 12] scl: 1

Use the five-number summary. The range is 40. The median score is 92.5, and half of the drivers scored between 80 and 97.5.

10-5 Comparing Sets of Data

Find the mean, median, mode, range, and standard deviation of each data set that is obtained after adding the given constant to each value.

23. 27, 21, 34, 42, 20, 19, 18, 26, 25, 33; +(−4)

24. 72, 56, 71, 63, 68, 59, 77, 74, 76, 66; +16

25. SCHOOL Principal Andrews tracked the number of disciplinary actions given by Ms. Miller and Ms. Anderson to their students each week.

Ms. Miller
9, 16, 12, 11, 12, 9, 10, 14, 13, 10, 9, 10, 11, 9, 12, 10, 11, 12

Ms. Anderson
7, 1, 0, 4, 2, 1, 6, 2, 2, 1, 4, 3, 0, 7, 0, 2, 5, 0

a. Use a graphing calculator to construct a histogram for each set of data. Then describe the shape of each distribution.

b. Compare the data sets using either the means and standard deviations or the five-number summaries. Justify your choice.

Example 6

Find the mean, median, mode, range, and standard deviation of the data set obtained after adding 6 to each value.

12, 15, 11, 12, 14, 16, 15, 12, 10, 13

Find the mean, median, mode, range, and standard deviation of the original data set.

Mean 13   Mode 12   Standard Deviation 1.8

Median 12.5   Range 6

Add 6 to the mean, median, and mode. The range and standard deviation are unchanged.

Mean 19   Mode 18   Standard Deviation 1.8

Median 18.5   Range 6

10-6   Summarizing Categorical Data

26. TRANSPORTATION A school conducts a survey to determine how students get to school. Of the 92 seniors, 34 responded that they drive to school and 9 responded that they ride the school bus. Of the 83 juniors, 12 responded that they drive to school and 42 responded that they walk to school. Create a two-way frequency table of the data.

27. CLUBS Hiroshi sent out a survey asking whether anyone would be interested in starting an environmental club. Of the 28 boys that responded, 8 said yes. Of the 34 girls that responded, 15 said no.

   a. Create a two-way frequency table.

   b. Convert the two-way frequency table into a relative frequency table.

   c. Create two conditional relative frequency tables: one for the interest in the club and one for gender.

Example 7

ELECTIONS Of the eighty-four 18–24 year olds in a survey, 62 responded that they would vote in the next election. Of the seventy-seven 25–34 year olds in the survey, 54 responded that they would not vote in the next election. Create a two-way frequency table of the data.

Step 1 Find the values for every combination of subcategories.

Step 2 Place every combination in the corresponding cell.

Step 3 Find the totals of each subcategory.

Step 4 Find and record the sum of the set of marginal frequencies. These two sums should be equal.

Age Group	Yes	No	Totals
18-24	62	22	84
25-34	23	54	77
Totals	85	76	161

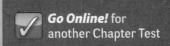
Find the mean, median, and mode for each set of data.

1. {99, 88, 88, 92, 100}

2. {30, 22, 38, 41, 33, 41, 30, 24}

3. TESTS Kevin's scores on the first four science tests are 88, 92, 82, and 94. What score must he earn on the fifth test so that the mean will be 90?

4. FOOD The table shows the results of a survey in which students were asked to choose their favorite food. Make a bar graph of the data.

Favorite Foods	
Food	Number of Students
pizza	15
chicken nuggets	10
cheesy potatoes	8
ice cream	5

5. SALES Nate is keeping track of how much people spent at the school store in one day. Find and interpret the mean and standard deviation for the data: 1, 1, 2, 3, 4, 5, 12.

6. MULTIPLE CHOICE Use a graphing calculator to construct a histogram for the data, and use it to describe the shape of the distribution.

$$16, 18, 14, 31, 19, 18, 10, 29,$$
$$12, 12, 28, 19, 17, 26, 15, 20$$

A positively skewed

B negatively skewed

C symmetric

D none of the above

7. Use a graphing calculator to construct a histogram for the data, and use it to describe the shape of the distribution.

$$19, 36, 26, 36, 40, 31, 30, 33, 23, 38, 23, 46$$

Find the mean, median, mode, range, and standard deviation of each data set that is obtained after multiplying each value by the given constant.

8. 9, 17, 31, 21, 17, 25, 13, 9, 12, 9; × 3

9. 16, 14, 23, 41, 38, 29, 18, 13, 16; × 0.25

Find the mean, median, mode, range, and standard deviation of each data set that is obtained after adding the given constant to each value.

10. 6, 9, 0, 15, 9, 14, 11, 13, 9, 5, 8, 6; + (−3)

11. 19, 22, 10, 17, 26, 24, 12, 22, 18, 17; + 8

12. MULTIPLE CHOICE Which pair of box-and-whisker plots depicts two positively skewed sets of data in which 75% of one set of data is greater than 75% of the other set of data?

A
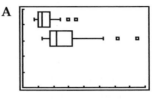

B
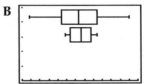

C
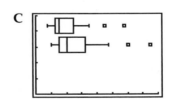

D
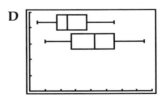

13. Two school districts send out a survey to residents asking for their opinions on a potential merger. Of the 225 residents in District 1 who responded to the survey, 178 said they are in favor of the merger. Of the 306 residents in District 2, 145 said they are against the merger.

a. Create a two-way frequency table.

b. Convert the two-way frequency table into a relative frequency table.

c. Create a conditional relative frequency table for the districts and for the responses.

Performance Task

Provide a clear solution to each part of the task. Be sure to show all of your work. Include all relevant drawings and justify your answers.

DIVING TEAM Kate Rodriguez is the coach of a diving team. She has been asked to choose one of the members of the team to compete at a state diving meet. Coach Rodriquez is considering two of the team's divers. To help her make the decision, she collects the divers' scores from their last 20 diving meets. The scores are shown in the table.

Part A

1. Make a histogram for each data set.
2. For each data set, describe the shape of the distribution and explain what the plots tell you about each diver.
3. Compare the data sets using either the means and standard deviations or the five-number summaries.

Diver	Scores
Amani	120.0, 135.2, 102.3, 120.6, 95.2, 99.3, 125.5, 136.1, 115.3, 124.6, 136.0, 75.6, 102.3, 126.0, 104.7, 90.3, 130.6, 126.9, 92.4, 80.3
Isabelle	97.5, 115.3, 104.6, 103.9, 108.3, 97.5, 117.3, 92.6, 125.3, 106.5, 108.3, 103.3, 121.1, 124.0, 91.3, 96.6, 116.3, 111.5, 103.3, 109.3

Part B

4. A third member of the team, Tamiko, tells Coach Rodriguez that she should also be considered for the state diving meet. She tells the coach that over her last 10 diving meets her median score is 110 and her interquartile range is about the same as Isabelle's interquartile range. Give a set of 10 scores that could represent Tamiko's data and justify your answer.

Part C

5. Coach Rodriguez wants to choose the diver with the best scores, but she also wants to choose a diver who is consistent. Before she makes her decision, she discovers that Isabelle's scores were incorrectly reported, and each score should be 10 points greater than what is shown.

   Taking this new information into consideration, who should Coach Rodriguez send to the competition?

Part D

6. During the competition, a survey was conducted to find out more about the crowd watching and what they might purchase. Of the 300 people surveyed, 114 of the people were greater than 45 years old. And of the 92 that said they would not purchase a snack while at the competition, 45 of them were 45 or younger.

   Construct a two-way frequency table to show the results of the survey and describe the categories and subcategories.

Test-Taking Strategy

Example

Read the problem. Identify what you need to know. Then use the information in the problem to solve. Show your work.

Determine which data point, if any, is an outlier in the following data set.

{0.85, 0.18, 1.04, 0.26, 2.09, 0.98, 0.40, 0.51, 0.79}

A 0.18   **C** 2.09

B 1.04   **D** none

Step 1 **What do you need to find?**
Any outliers in the data set

Step 2 **How can you solve the problem?**
I will start by arranging the data in ascending order and finding the interquartile range IQR. Then, I can use the IQR to find the values beyond which any outliers would lie.

Step 3 **What is the correct answer?**

$$0.18 \quad 0.26 \quad 0.40 \quad 0.51 \quad 0.79 \quad 0.85 \quad 0.98 \quad 1.04 \quad 2.09$$
$$\uparrow \qquad\qquad \uparrow \qquad\qquad \uparrow$$
$$Q_1 = 0.33 \qquad Q_2 = 0.79 \qquad Q_3 = 1.01$$

$IQR = Q_3 - Q_1 = 0.68$

$Q_1 - 1.5(IQR)$ and $Q_3 + 1.5(IQR)$

$0.33 - 1.5(0.68) = -0.69$ $1.01 + 1.5(0.68) = 2.03$

Because 2.09 is greater than 2.03, the answer is C.

> ## Test-Taking Tip
> **Strategy for Organizing Data** Sometimes you may be given a set of data that you need to analyze in order to solve problems. When you are given a problem statement containing data, consider:
> - making a list of the data
> - using a table to organize the data
> - using a data display (such as a bar graph, Venn diagram, circle graph, line graph, or box-and-whisker plot) to organize the data

Apply the Strategy

Read the problem. Identify what you need to know. Then use the information in the problem to solve. Show your work.

Determine which data point, if any, is an outlier in the following data set.
{22, 1.2, 12.5, 11.5, 12, 11, 15.5, 14, 10, 6.5, 23}

A 1.2

B 12

C 23

D none

Answer the questions below.

a. What do you need to find?

b. How can you solve the problem?

c. What is the correct answer?

Read each question. Then fill in the correct answer on the answer document provided by your teacher or on a sheet of paper.

1. Solve the following equation.
   $2x - 3(4 - x) = 5x + 3$

   ○ **A** -2

   ○ **B** 1

   ○ **C** 9

   ○ **D** no solution

2. Charlene graphed the linear system below.

   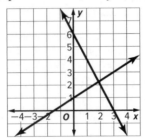

   Which is the best estimate of the solution?

   ○ **A** $(-1.5, 0)$

   ○ **B** $(0, 1)$

   ○ **C** $(2, 2)$

   ○ **D** $(3, 0)$

3. There were 24 cars that were either red or black in the parking lot. There were 7 times as many black cars as red cars.

   a. How many black cars were in the parking lot?

   [   ] black cars

   b. (MP) What mathematical practice did you use to solve this problem?

> **Test-Taking Tip**
>
> **Question 1** If solving an equation results in a true statement, such as $4 = 4$, then the solution is all real numbers. If a false statement results, such as $4 = 5$, then there is no solution.

4. How does the graph of $g(x) = f(x - 5)$ compare to the graph of the linear parent function $f(x) = x$?

   ○ **A** $f(x)$ is translated left 5 units.

   ○ **B** $f(x)$ is translated right 5 units.

   ○ **C** $f(x)$ is translated up 5 units.

   ○ **D** $f(x)$ is reflected across the x-axis.

5. The table shows the total cost for two different trips to a movie theater.

Adult Tickets	Child Tickets	Total Cost
2	3	$48
4	1	$56

   Let a represent the cost of an adult ticket and c represent the cost of a child ticket. Which system of equations best represents this situation?

   ○ **A** $2a + 4c = 48$
   $3a + c = 56$

   ○ **B** $2a + 3c = 48$
   $4a + c = 56$

   ○ **C** $2a + 4c = 56$
   $4a + c = 48$

   ○ **D** $2a + 3c = 56$
   $4a + c = 48$

6. Each value in a data set is increased by 4. Select the statistics of the new data set that can be found by adding 4 to the statistic of the original data set.

   ☐ **A** mean

   ☐ **B** median

   ☐ **C** mode

   ☐ **D** range

   ☐ **E** standard deviation

7. Let $f(x) = -3x - 2$. For what value of x does $f(x) = 19$?

○ **A** $x = -59$

○ **B** $x = -7$

○ **C** $x = 7$

○ **D** $x = 55$

8. Franco graphed the function $g(x)$ below.

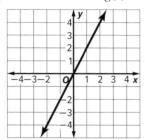

Select all of the descriptions of $g(x)$ as a transformation of the following parent function $f(x) = x$.

☐ **A** $g(x) = 2f(x)$

☐ **B** $g(x) = f(2x)$

☐ **C** $g(x) = f(x - 2)$

☐ **D** $g(x) = f(x + 2)$

☐ **E** $g(x) = f(x) - 2$

☐ **F** $g(x) = f(x) + 2$

9. What is the 5th term of the following arithmetic sequence?

$$a_1 = 5, a_{n+1} = 1.5 + a_n \boxed{}$$

10. Select all of the equations that have a graph symmetric to the y-axis.

☐ **A** $y = -3x + 9$

☐ **B** $y = 3x^2 + 9$

☐ **C** $y = -3x^2 + 9$

☐ **D** $y = 3^x + 9$

☐ **E** $y = (-3)^x + 9$

11. If $4x^2 - 256$ is factored as $4(x + k)(x - k)$, where k is a positive integer, what is the value of k that makes the expression true?

$\boxed{}$

12. Nina graphed $y = x^2$ on a coordinate plane. Brandi graphed $y = 4x^2$ on a coordinate plane. How is Brandi's graph different from Nina's graph?

○ **A** Brandi's graph is vertically compressed compared to Nina's graph.

○ **B** Brandi's graph is vertically stretched compared to Nina's graph.

○ **C** Brandi's graph is translated right 4 units.

○ **D** Brandi's graph is translated left 4 units.

13. Which of the following expressions is equivalent to $\sqrt{20} + \sqrt{5}$?

○ **A** 5

○ **B** $3\sqrt{5}$

○ **C** $5\sqrt{5}$

○ **D** $4\sqrt{10}$

Need Extra Help?

If you missed Question...	1	2	3	4	5	6	7	8	9	10	11	12	13
Go to Lesson...	2-3	6-1	2-1	3-1	6-5	10-5	4-1	3-5	3-6	9-1	8-7	9-2	7-4

Student Handbook

This **Student Handbook** can help you answer these questions.

What if I Need More Practice?

Extra Practice   R1

The **Extra Practice** section provides additional problems for each lesson so you have ample opportunity to practice new skills.

What if I Need to Check a Homework Answer?

Selected Answers and Solutions   R11

The answers to odd-numbered problems are included in **Selected Answers and Solutions**.

What if I Forget a Vocabulary Word?

Glossary/Glosario   R82

The **English-Spanish Glossary** provides definitions and page numbers of important or difficult words used throughout the textbook.

What if I Need to Find Something Quickly?

Index   R100

The **Index** alphabetically lists the subjects covered throughout the entire textbook and the pages on which each subject can be found.

What if I Forget a Formula?

Formulas and Measures,   **Inside Back Cover**
Symbols and Properties

Inside the back cover of your math book is a list of **Formulas and Symbols** that are used in the book.

Extra Practice

Write an algebraic expression for each verbal expression. (Lesson 1-1)

1. 6 times a number m

2. a number t less twelve

Evaluate each expression if $m = 2$, $t = 6$, and $z = 5$. (Lesson 1-2)

3. $2(t - z) + \frac{14}{m}$

4. $(m + 2z)^2 + 12tz$

5. **SPORTS** Adam mows lawns at an average rate of 40 minutes per lawn. Write and evaluate an expression to find the number of hours Adam spent mowing last weekend. (Lesson 1-2)

Lawns Per Day	
Friday	3
Saturday	11
Sunday	4

Evaluate each expression. Name the property used in each step. (Lesson 1-3)

6. $14\left(5 - \frac{1}{5} \cdot 25\right) + 2 \div (4 \cdot 1)$

7. $3(14 + 8 + 6) - 1 \cdot 18$

8. **PETS** Rosa takes two dogs and one cat to the veterinarian to be boarded during her vacation. (Lesson 1-3)

Pet	Board	Bath
dog	$25/day	$12
cat	$15/day	$8

   a. Use the table to find the total cost of boarding both dogs and the cat for 5 days.

   b. If Rosa has the veterinarian give all of the pets a bath while she is on vacation, what is the new total cost?

Use the Distributive Property to rewrite each expression. Then evaluate. (Lesson 1-4)

9. $14(102)$

10. $5\frac{1}{6}(30)$

11. **ARTS** Logan sells handmade wooden products. He charges $25 per bowl, $14 per picture frame, and $30 per jewelry box. On Friday, he sells 3 bowls, 6 picture frames, and 2 jewelry boxes. On Saturday, he sells 6 bowls, 14 picture frames, and 3 jewelry boxes. Write and evaluate an expression for his total sales. (Lesson 1-4)

12. The dimensions of a crate are 2.4 feet, 3.33 feet, and 4.15 feet. If the product of the dimensions is 33.166, give the volume of the crate rounded to the appropriate place. (Lesson 1-5)

Identify the independent and dependent variables for each relation. (Lesson 1-6)

13. Increasing the amount of fertilizer put on a plant increases the rate at which it grows.

14. Pam babysits to save money. The more kids she babysits, the more money she makes.

15. **INCOME** Jonathan draws the graph at the right to describe his income throughout his career. Describe what is happening in the graph. (Lesson 1-6)

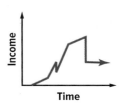

Determine whether the relation is a function. Explain. (Lesson 1-7)

16. $\{(2, 2), (-5, 2), (6, 6), (9, 4), (4, 9)\}$

17. $\{(0, 2), (1, 7), (0, -6), (4, 8), (-3, -1)\}$

18. Identify the function graphed as *linear* or *nonlinear*. Then estimate and interpret the intercepts of the graph, any symmetry, where the function is positive, negative, increasing, and decreasing, the x-coordinate of any relative extrema, and the end behavior of the graph. (Lesson 1-8)

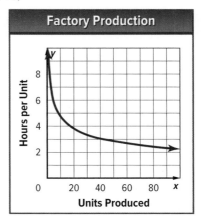

Extra Practice

Translate each sentence into a formula. (Lesson 2-1)

1. The number of seconds is 60 times the number of minutes.

2. A gallon contains 8 pints.

3. **THEATER** A theater has 1400 seats. Write and use an equation to find the number of rows if each row has 35 seats. (Lesson 2-1)

Solve each equation. Check your solution.
(Lesson 2-2)

4. $-18t + 26 = -19$ 5. $\frac{1}{2}b + \frac{3}{4} = \frac{1}{6}$

Write an equation and solve each problem.
(Lesson 2-3)

6. Find three consecutive even integers with a sum of 78.

7. Eight more than half a number is negative two.

Solve each equation. Check your solution.
(Lesson 2-3)

8. $-3w + 16 = 14.5$ 9. $\frac{4}{7} - \frac{k}{3} = \frac{1}{2}$

10. **ZOO** The Martin and Smith families went to the zoo. What is the cost of an adult ticket if they spent $75 total? (Lesson 2-3)

Family	Adults
Martin	2
Smith	3

Solve each equation. Check your solution.
(Lesson 2-4)

11. $16x - 8 = 21 - x$ 12. $\frac{x}{2} + \frac{1}{5} = 3x$

13. **GEOMETRY** Find the value of x so that the figures have the same perimeter. (Lesson 2-4)

Evaluate each expression if $m = 6$, $n = 15$, **and** $p = \frac{1}{2}$. (Lesson 2-5)

14. $|m - 12| - 3p$ 15. $18p + |n - m|$

Write an equation involving absolute value for each graph. (Lesson 2-5)

16.

17.

Solve each proportion. If necessary, round to the nearest hundredth. (Lesson 2-6)

18. $\frac{1.7}{n} = \frac{16}{30}$ 19. $\frac{418}{83} = \frac{b}{7}$ 20. $\frac{30}{y} = \frac{75}{135}$

21. **ART** Neil is enlarging a photo to hang on the wall. To keep the pictures proportional, what length is the unknown side? (Lesson 2-6)

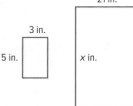

22. **CAMPING** A box with 4 days of food in it weighs 38.6 pounds. How much will a box with 7 days of food in it weigh? (Lesson 2-6)

23. **THEATER** At Carlson High School, 4 out of 15 students are in theater. If there are 1425 students at the school, how many are not in theater? (Lesson 2-6)

Solve each equation or formula for the variable indicated. (Lesson 2-7)

24. $\frac{2b - a}{c} = -\frac{1}{2}d$, for a

25. $y = w(h - 5)$, for h

26. $3x - 4z = 7 - xy$, for x

27. **WEIGHT** A watermelon weighs 6.3 pounds. One pound is approximately 0.454 kilogram. How many kilograms does the watermelon weigh? (Lesson 2-7)

28. **PAINT** Jeff painted 700 square feet in 75 minutes. There are 9 square feet in a square yard. What is his painting speed in square yards per hour? Round to the nearest tenth. (Lesson 2-7)

Graph each equation by making a table. (Lesson 3-1)

1. $2y - x = 5$

2. $x + y = 6$

Solve each equation. (Lesson 3-2)

3. $-5x + 14 = 7x - 28$

4. $-\frac{1}{2}x + 8 = 6x - 12 - \frac{13}{2}x$

5. **NURSERY** The function $b = 100 - 2.5f$ represents the remaining balance of store credit Louie has at Blooms Nursery. Find the zero and explain what it means in this situation. (Lesson 3-2)

6. Rewrite the equation $6x + 4y = 12$ in slope-intercept form. (Lesson 3-4)

7. State the slope and the y-intercept of the equation $2y - 4x = 14$. (Lesson 3-4)

8. What is the equation in slope-intercept form for the line shown? (Lesson 3-4)

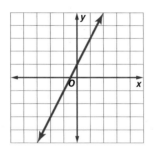

9. Graph $y = 4$. Then, state the slope and y-intercept. (Lesson 3-4)

10. Write a function to represent the graph $f(x) = x$ stretched vertically by a factor of 4 and translated up 3 units. (Lesson 3-5)

11. Describe the transformations in the graph of $g(x) = (-0.2x)$ as it relates to the parent function $f(x) = x$. (Lesson 3-5)

12. Describe the transformations in the graph of $g(x) = -3x - 1$ as it relates to the parent function $f(x) = x$. (Lesson 3-5)

Determine whether each sequence is an arithmetic sequence. Write *yes* or *no*. Explain. (Lesson 3-6)

13. $-2, 2, -4, 4, -6, 6 \ldots$

14. $-6, -3, 0, 3, 6 \ldots$

Write an equation for the nth term of each arithmetic sequence. Then graph the first five terms of the sequence. (Lesson 3-6)

15. $3, 3.5, 4, 4.5\ldots$

16. $1, -1.5, -4, -6.5\ldots$

Determine whether each function is linear. Write *yes* or *no*. Explain. (Lesson 3-3)

17.

x	−2	0	2	4	6
y	0	6	12	18	24

18.

x	7	4	1	−2	−5
y	14	2	12	4	10

19. **WEATHER** Refer to the graph. (Lesson 3-3)

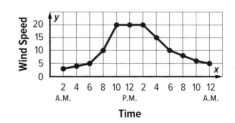

a. Find the rate of change in wind speed between 6 A.M. and 8 A.M.

b. Is there a greater change in wind speed during the day? If so, when does it occur?

c. The meteorologist says a storm came through at some point during the day. When do you think this may have happened? Explain your reasoning.

Graph each function. State the domain and range.

20. $f(x) = 2[\![x]\!]$ (Lesson 3-7)

21. $f(x) = |x + 5|$ (Lesson 3-8)

Extra Practice

Write an equation in slope-intercept form for the line that passes through the given point and has the given slope. (Lesson 4-1)

1. (4, 20), slope 6

2. (9, 6), slope $\frac{1}{3}$

Write an equation in slope-intercept form for the line that passes through each pair of points. (Lesson 4-1)

3. (0, −1), (1, 1)

5. (−3, 1), (2, 7)

4. (−4, −4), (2, 2)

6. (3, 0), (5, 4)

Write an equation in point-slope form and an equation in standard form for the line that passes through each pair of points. (Lesson 4-2)

7. (−2, 5), (−4, 13)

8. (3, 7), (9, 19)

Determine whether the graphs of the following equations are *parallel* or *perpendicular*. Explain. (Lesson 4-3)

9. $y = -2x + 6$, $2y = x - 3$

10. $y = 3x - 2$, $-3y = x + 6$

11. MAPS The director of street repairs wants to first replace curbs on streets that are parallel to each other. Which two streets will get new curbs first? (Lesson 4-3)

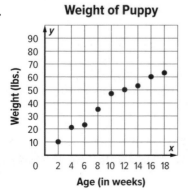

Determine whether each graph shows a *positive*, *negative*, or *no* correlation. If there is a positive or negative correlation, describe its meaning in the situation. (Lesson 4-4)

12. **Weight of Puppy**

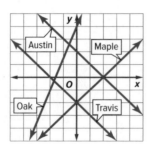

13. **Carnival Fundraiser**

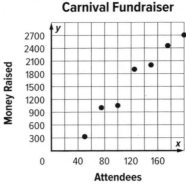

Write an equation in slope-intercept form for the line that passes through the given point and is parallel to the graph of the equation. (Lesson 4-3)

14. (−1, 6), $y = \frac{1}{4}x - 4$

15. (5, 7), $y = -x + 5$

16. Determine whether the situation indicates *correlation, causation,* or *both.* A landscaping company reports that there is a correlation between weeks where there is a severe thunderstorm and an increase in business. (Lesson 4-5)

17. Determine whether the situation indicates *correlation, causation,* or *both.* Giuseppe's Pizzeria noticed a positive correlation between pizza sales and days later in the week. (Lesson 4-5)

Write an equation of the regression line for the data in each table. Then find the correlation coefficient. (Lesson 4-6)

18. The table shows the numbers of plants in each garden and turnips produced.

Plants	2	3	4	5	6	7
Turnips	12	19	24	32	34	45

19. The table shows the amount of shrimp caught each day and the retail price of the shrimp.

Pounds (1000s)	48	52	60	65	73
Cost/lb ($)	3.50	3.42	3.35	3.15	3.08

Find the inverse of each function. (Lesson 4-7)

20. $f(x) = -3x + 8$

21. $f(x) = \frac{1}{2}x + 7$

Solve each inequality. Then graph the solution set on a number line. (Lesson 5-1)

1. $6t \leq 3$

2. $14 > k + 2$

3. $16 \leq 4n$

4. $6c < 5c + 3$

Define a variable, write an inequality, and solve each problem. Check your solution. (Lesson 5-1)

5. The sum of six times a number and three is less than the product of seven and a number.

6. Three times a number is greater than or equal to the sum of twice a number and 12.

Solve each inequality. Graph the solution on a number line. (Lesson 5-2)

7. $\frac{c}{8} > \frac{1}{4}$

8. $-3 \leq 4m$

Define a variable, write an inequality, and solve each problem. Then interpret your solution. (Lesson 5-2)

9. READING Thomas has a 432-page book to read in 12 days. At least how many pages must he read per day to finish the book on time?

10. DOGS Laura has a maximum of 91 minutes to walk 7 dogs. How much time can she spend walking each dog?

Solve each inequality. Graph the solution on a number line. (Lesson 5-3)

11. $1.2x + 6 < 4.6x - 3$

12. $-4\left(3g + \frac{1}{2}\right) \leq -6(3 + 2g)$

13. PIZZA Sam orders 3 large pizzas. Each pizza costs $12, and each topping costs $0.50. Sam has $38 to spend. Write and solve an inequality to find the greatest number of toppings Sam can afford. (Lesson 5-3)

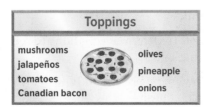

Toppings	
mushrooms	olives
jalapeños	pineapple
tomatoes	
Canadian bacon	onions

Solve each compound inequality. Then graph the solution set. (Lesson 5-4)

14. $6w + 3 < 9$ or $\frac{1}{2}w \geq 2$

15. $16 \geq 3x - 5$ and $3x - 2 > 2(x - 1)$

16. COUPON Victor has a coupon that is valid only for juice sold in containers between 16 and 32 ounces. (Lesson 5-4)

  a. Write a compound inequality that describes acceptable juice container sizes.

  b. Graph the inequality.

Solve each inequality. Then graph the solution set. (Lesson 5-5)

17. $\left|\frac{1}{2}z + 6\right| \leq 4$

18. $|3p + 3| > 1$

19. SHOP Kristi is shopping online for gifts for her friends. (Lesson 5-5)

  a. If prices ranged from $1.50 above and below the average CD price, find the range of prices.

Average Prices	
CD	$15.50
Book	$19
Shirt	$32

  b. Prices for the book varied $2.25 from the average. Write the range of average book prices.

  c. Graph the solution set for shirt prices if they varied $6 below to $4 above the average.

Graph each inequality. (Lesson 5-6)

20. $3(x + y) \geq 6$

21. $\frac{1}{2}y < 2(-1 - x)$

Use a graph to solve each inequality. (Lesson 5-6)

22. $5x + 3 > -2$

23. $y - 8 \leq -3$

24. DOG WASH It costs Pups and Suds $975 a week to operate their business. (Lesson 5-6)

Prices
Small/Med Dog $20
Large Dog $30

  a. The prices that they charge are shown. Write an inequality to describe how many of each type of dog they need to service to make a profit.

  b. How many dogs must they wash to make a profit each week?

Use the graph below to determine whether each system is *consistent* or *inconsistent* and if it is *independent* or *dependent*. (Lesson 6-1)

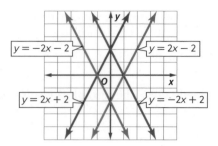

$y = -2x - 2$ $y = 2x - 2$

$y = 2x + 2$ $y = -2x + 2$

1. $y = 2x + 2$
$y = -2x - 2$

2. $y = -2x + 2$
$y = -2x - 2$

3. DANCES Mario and Tanesha are inflating balloons for the school dance. Mario has 12 balloons inflated and is inflating additional balloons at a rate of 3 balloons per minute. Tanesha has 16 balloons inflated and is inflating additional balloons at a rate of 2 balloons per minute. (Lesson 6-1)

   a. Write a system of equations to represent the situation.

   b. Graph each equation.

   c. How long will it take Mario to have more balloons filled than Tanesha?

Use substitution to solve each system of equations. (Lesson 6-2)

4. $x = -y + 3$
$3y + 2x = 10$

5. $-x + 2y = 6$
$4y - 2x = 11$

6. $y - 7x = 2$
$2x + 3 = 5y$

7. $-2y = x + 3$
$\frac{3}{2} = -\frac{1}{2}x - y$

8. FRUIT Sarah and Toni each bought fruit for a fundraiser. If Toni spent $4.30 and Sarah spent $2.80, how much does each type of fruit cost? (Lesson 6-3)

Girl	Apples	Oranges
Toni	6	5
Sarah	6	2

Use elimination to solve each system of equations. (Lesson 6-3)

9. $2m + 3n = 16$
$-3m - 3n = -4$

10. $-5k + 4j = 8$
$-5k - 6j = -12$

11. The difference of three times a number and a second number is two. The sum of the two numbers is fourteen. What are the two numbers? (Lesson 6-3)

Use elimination to solve each system of equations. (Lesson 6-4)

12. $1.6x + 2.2y = 5.4$
$-3.2x + 4y = -2.4$

13. $2x + 5y = -8$
$4x - 2y = 0$

14. CARNIVALS Scott and Isaac went to the school carnival. Use the table shown to determine how many tickets a ride and game each cost. (Lesson 6-4)

Rider	Rides	Games	Tickets
Scott	5	4	19
Isaac	7	2	23

15. SAVINGS Caleb made $105 mowing lawns and walking dogs, charging the rates shown. If he mowed half as many lawns as dogs walked, how many lawns did he mow and how many dogs did he walk? (Lesson 6-5)

$7.50 per dog   **$20 per lawn**

Determine the best method to solve each system of equations. Then solve the system. (Lesson 6-5)

16. $4x + 2y = 12$
$-y - 4x = 2$

17. $y + 3x = 11$
$-2x + 3y = 11$

18. BABYSITTING Kelsey and Emma babysit after school to earn extra money. Kelsey made $52 by charging $10 per hour and $4 per child. Emma made $67.50 by charging $15 per hour and $2.50 per child. (Lesson 6-5)

   a. Write a system of equations to represent the situation.

   b. How many hours and how many children did each babysit?

Solve each system of inequalities by graphing. (Lesson 6-6)

19. $0.5x - y \geq 3$
$x + y < 3$

20. $y < 2x - 1$
$y > 4(1 + 0.5x)$

Simplify each expression. (Lesson 7-1)

1. $(2xy^2)^3(2x^2y^3z)$ **2.** $(2ab^2c^3)(3ad^2)^2$

GEOMETRY **Express the area of each triangle as a monomial.** (Lesson 7-1)

3.

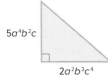

$5a^4b^2c$

$2a^2b^3c^4$

4.

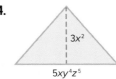

$3x^2$

$5xy^4z^5$

Simplify each expression. Assume that no denominator equals zero. (Lesson 7-2)

5. $\dfrac{3m^5n^3p^4}{5m^6n^4p^2q^5}$ **6.** $\left(\dfrac{x^2yz^3}{2xy^3z^3}\right)^3$

7. $\dfrac{2a^{-1}b^{-2}c^{-3}}{7a^2b^3c^4d^{-5}}$ **8.** $\left(\dfrac{4x^{-2}y^4z^5}{5x^5y^{-3}z^{-2}}\right)^{-1}$

Write each expression in radical form, or write each radical in exponential form. (Lesson 7-3)

9. $13^{\frac{1}{3}}$ **10.** $(7k)^{\frac{1}{2}}$

11. $\sqrt{17a}$ **12.** $3\sqrt{2xyz^2}$

Simplify. (Lesson 7-3)

13. $\sqrt[4]{\dfrac{81}{625}}$ **14.** $\sqrt[5]{0.00001}$

15. $4096^{\frac{1}{3}}$ **16.** $\left(\dfrac{125}{343}\right)^{\frac{4}{3}}$

Solve each equation. (Lesson 7-3)

17. $81^x = \dfrac{1}{3}$ **18.** $3^{4x} = 3^{x+1}$

Simplify each expression. (Lesson 7-4)

19. $\dfrac{6}{\sqrt{3}+2}$

20. $\left(3\sqrt{6}+2\sqrt{4}\right)\left(5\sqrt{2}-4\sqrt{3}\right)$

Graph each equation. Find the *y*-intercept, and state the domain and range. (Lesson 7-5)

21. $f(x) = -3^x - 1$ **22.** $f(x) = \left(\dfrac{1}{2}\right)^x + 3$

23. Determine whether the set of data shown below displays exponential behavior. Write *yes* or *no*. Explain why or why not. (Lesson 7-5)

x	-1	0	1	2	3	4
y	3	1	$\frac{1}{3}$	$\frac{1}{9}$	$\frac{1}{27}$	$\frac{1}{81}$

24. Write a function $g(x)$ to represent the graph of $f(x) = 9^x$ translated up 2 units and left 4 units. (Lesson 7-6)

25. Describe the transformations in the graph of $g(x) = 1.5^{-x} - 4$ as it relates to the graph of $f(x) = 1.5^x$. (Lesson 7-6)

26. **POPULATION** A neighborhood had 4518 residents in 2006. The number of residents has been declining by 3.5% each year. How many residents were there in 2012? (Lesson 7-7)

27. **MONEY** Sarah put $3000 in an investment that gets 6.2% compounded quarterly for 8 years. What will her investment be worth at the end of the 8 years? (Lesson 7-7)

28. **SOCCER** The Westside Soccer League has 186 players. They expect a 7.5% increase in players for at least the next 4 years. How many players will they have at that point? (Lesson 7-7)

29. Bank A offers a savings account with 4.5% interest compounded annually. Bank B offers a savings account with a monthly compounded interest rate of 0.39%. Which is the better plan? Explain. (Lesson 7-8)

Determine whether each sequence is *arithmetic*, *geometric*, or *neither*. Explain. (Lesson 7-9)

30. $-\dfrac{1}{2}, -\dfrac{1}{4}, 0, \dfrac{1}{4}, \dfrac{1}{2} \ldots$ **31.** $100, 90, 85, 75, 60\ldots$

Find the next three terms in each geometric sequence. (Lesson 7-9)

32. $48, -96, 192, -384, 768\ldots$

33. $150, 75, 37.5, 18.75, 9.375\ldots$

Find the first five terms of each sequence. (Lesson 7-10)

34. $a_1 = 5, a_n = 3.5a_{n-1} + 1, n \geq 2$

35. $a_1 = 12, a_n = -\dfrac{1}{2}a_{n-1} + \dfrac{5}{2}, n \geq 2$

Write a recursive formula for each sequence. (Lesson 7-10)

36. $7, 16, 43, 124, \ldots$ **37.** $729, 243, 81, 27, \ldots$

Extra Practice

Find each sum or difference. (Lesson 8-1)

1. $(7g^3 + 2g^2 - 12) - (-2g^3 - 4g)$

2. $(-3h^2 + 3h - 6) + (5h^2 - 3h - 10)$

Simplify each expression. (Lesson 8-2)

3. $-\frac{1}{2}n^3p^2(5np^3 - 3n^2p^2 + 8n)$

4. $6j^2(-3j + 3k^2) - 2k^2(2j + 10j^2)$

Solve each equation. (Lesson 8-2)

5. $-4(b + 3) + b(b - 3) = -b(6 - b) + 2(b - 3)$

6. $3(a - 3) + a(a - 1) + 12 = a(a - 2) + 3(a - 2) + 4$

Find each product. (Lesson 8-3)

7. $(-3t - 16)(5t + 2)$ **8.** $\left(4p + \frac{1}{2}\right)\left(\frac{1}{2}p + 4\right)$

9. SIDEWALKS Reynoldsville is repairing sidewalks. If the sidewalk is the same width around a city block, write an expression for the area of the block and the sidewalk. (Lesson 8-3)

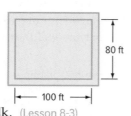
80 ft
|← 100 ft →|

Find each product. (Lesson 8-4)

10. $\left(\frac{1}{2}m + 3\right)^2$ **11.** $(2n - 6)(2n + 6)$

12. $(5a - 4)^2$ **13.** $(x - 2y)(x + 2y)$

Use the Distributive Property to factor each polynomial. (Lesson 8-5)

14. $4m^3n^2 + 16m^2n^3 - 8m^3n^4$

15. $12j^4k^4 + 36j^3k^2 - 3j^2k^5$

Factor each polynomial. (Lesson 8-5)

16. $x^2 - 4x + 3xy - 12y$

17. $4a - 10ab + 6b - 15b^2$

18. HEIGHT The height in feet of a ball bounced off the ground after t seconds is modeled by the expression $-16t^2 + 28t$. (Lesson 8-5)

a. Write the factored form of the expression for the height of the ball.

b. What is the height of the ball after 1.5 seconds?

Factor each polynomial. (Lesson 8-6)

19. $t^2 + 2t - 15$ **20.** $d^2 - 3d - 28$

21. $m^2 + 5m - 14$ **22.** $x^2 - 4x - 45$

23. GEOMETRY Find an expression for the perimeter of a rectangle with the area $6x^2 + x - 15$. (Lesson 8-6)

Factor each polynomial, if possible. If the polynomial cannot be factored using integers, write *prime*. (Lesson 8-6)

24. $6x^2 + 21x - 90$ **25.** $3x^2 - 11x - 42$

26. $6x^2 - 13x - 5$ **27.** $5y^2 - 3y + 11$

Factor each polynomial. (Lesson 8-7)

28. $\frac{1}{2}t^2 - 16^2$ **29.** $25d^3 - 49d$

30. $196t^2u^3 - 144u^3$ **31.** $169a^4b^6 - 121c^8$

32. $4g^2 - 1296h^2$ **33.** $18a^3 + 27a^2 - 50a - 75$

34. The area of a square is represented by $16x^2 - 40x + 25$. Find the length of each side. (Lesson 8-7)

35. The volume of a rectangular prism is represented by $6x^3 + x^2 - 2x$. Find the possible dimensions of the prism if the dimensions are represented by polynomials with integer coefficients. (Lesson 8-7)

36. FRUIT An apple fell 25 feet from a tree. The expression $25 - 16t^2$ models the height in feet of the apple after t seconds. (Lesson 8-7)

a. Write the factored form of the expression for the height of the apple.

b. What is the height of the apple after 0.5 second?

Determine whether each trinomial is a perfect square trinomial. Write *yes* or *no*. If so, factor it. (Lesson 8-7)

37. $64x^2 - 32x + 4$ **38.** $4a^2 - 12a + 16$

39. $12y^2 - 36y + 27$ **40.** $75b^3 - 60ab^2 + 12a^2b$

Find the vertex, the equation of the axis of symmetry, and the *y*-intercept of the graph of each equation. (Lesson 9-1)

1. $y = 4x^2 + 8x - 5$

2. $y = -2x^2 + 8x + 5$

3. $y = x^2 - 8x + 9$

4. $y = 4x^2 + 16x - 6$

5. KICKBALL A kickball is kicked into the air. The equation $h = -16t^2 + 60t$ gives the height h of the ball in feet after t seconds. (Lesson 9-1)

  a. What is the height of the ball after one second?

  b. When will the ball reach its maximum height?

  c. When will the ball hit the ground?

Describe the transformations in each function as it relates to the graph of $f(x) = x^2$. (Lesson 9-2)

6. $g(x) = -x^2 - 4$

7. $h(x) = 7x^2 + 2$

Match each equation to its graph. (Lesson 9-2)

A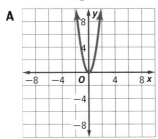

B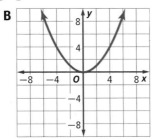

C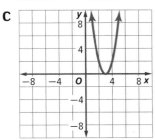

D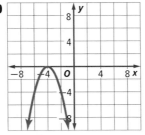

8. $y = 3x^2$

9. $y = \frac{1}{4}x^2$

10. $y = -(x + 4)^2$

11. $y = 2(x - 3)^2$

Solve each equation by graphing. (Lesson 9-3)

12. $-2x^2 - 2x + 4 = 0$

13. $x^2 = -2x + 3$

14. MARBLES Jason shot a marble straight up using a slingshot. The equation $h = -16t^2 + 42t + 5.5$ models the height h, in feet, of the marble after t seconds. After how long will the marble hit the ground? (Lesson 9-3)

15. Solve each equation. Check your solutions. (Lesson 9-4)

  a. $x^2 = 16$

  b. $x^2 = 144$

  c. $x^2 = 9$

16. Solve $(x - 2)^2 = 16$. (Lesson 9-4)

17. Solve $(2x + 8)(4x - 20) = 0$. (Lesson 9-4)

18. Solve $x^2 + 5x + 6 = 0$. (Lesson 9-4)

19. BIOLOGY The number of cells in a Petri dish can be modeled by the quadratic equation $n = 6t^2 - 4.5t + 74$, where t is the number of hours the cells have been in the dish. When will there be 200 cells in the Petri dish? (Lesson 9-5)

Solve each equation by completing the square. Round to the nearest tenth if necessary. (Lesson 9-5)

20. $x^2 + 4x - 8 = 5$

21. $3x^2 + 5x = 18$

22. Find the value of x in the figure if the area is 36 square inches. (Lesson 9-5)

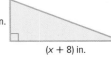
$(x + 2)$ in. $(x + 8)$ in.

Solve each equation by using the Quadratic Formula. Round to the nearest tenth if necessary. (Lesson 9-6)

23. $3x^2 + 10x = 15$

24. $\frac{1}{2}x^2 - 8x + 6 = 0$

State the value of the discriminant. Then determine the number of real solutions of the equation. (Lesson 9-6)

25. $4x^2 - 12x = -9$

26. $3x^2 + 8 = 9x$

27. Solve the system of equations algebraically. (Lesson 9-7)
$$y = x^2 + 2x + 1$$
$$y - x = 1$$

Look for a pattern in each table of values to determine which kind of model best describes the data. (Lesson 9-8)

28.

x	2	3	4	5	6
y	$\frac{9}{4}$	$\frac{27}{8}$	$\frac{81}{16}$	$\frac{243}{32}$	$\frac{729}{64}$

29.

x	−2	−1	0	1	2
y	−13	−6.25	0	5.75	11

30. Given that $f(x) = x^2 + 3x - 4$ and $g(x) = x - 4$, find each function. (Lesson 9-9)

  a. $(f + g)(x)$

  b. $(f - g)(x)$

31. Given that $f(x) = 2x^2 - x + 3$ and $g(x) = x + 1$, find $(f \bullet g)(x)$. (Lesson 9-9)

1.

66	66	60	59	69	71	62	63	64
67	64	65	66	67	68	62	69	70

The table represents the height, in inches, of the girls on the basketball team. (Lesson 10-1)

a. What is the mean height?

b. What is the median height?

c. What is the mode?

2. Which measure of center would best describe the data, *mean, median,* or *mode*? (Lesson 10-1)

a. Jamal scores mostly As on his math tests.

b. A teacher wants to explore how many minutes students study per day, with answers ranging from 15 minutes to 200 minutes and the most common answer being 100 minutes.

3. Kendra scored a 95% on her math test. Out of the 25 students in the class, 22 students scored less than Kendra. In what percentile did she score? (Lesson 10-1)

4. The table represents favorite ice cream flavors in Ms. Isbel's fourth grade class.

Chocolate	7
Vanilla	9
Strawberry	5
Cookies and Cream	6

Represent the data in a bar graph. (Lesson 10-2)

5. Which type of graph would best represent the scenario? (Lesson 10-2)

a. the height ranges of NBA basketball players

b. the types of music to which the students in high school listen

6. GAS The price of gasoline was recorded at 7 gas stations on the same day. They were: $2.63, $2.59, $2.70, $2.58, $2.83, $2.65, $2.71. Find the minimum, lower quartile, median, upper quartile, and maximum for the data set. (Lesson 10-3)

7. BOWLING Tina's results for 8 bowling games are shown below.
{110, 123, 147, 119, 153, 142, 113, 143}
Find and interpret the standard deviation of the data. (Lesson 10-3)

For Exercises 8 and 9, use these data.
{12, 18, 21, 18, 19, 18, 16, 23, 20, 15, 17, 18}

8. Describe the center and spread of the data using either the mean and standard deviation or the five-number summary. Justify your choice by constructing a histogram. (Lesson 10-4)

9. Find the mean, median, mode, range, and standard deviation of the data after multiplying each value by 3. (Lesson 10-5)

10. Use the table to complete the following exercises. (Lesson 10-6)

Favorite Subject	7th Grade	8th Grade	Total
Math	92	47	139
Science	79	51	130
English	68	25	93
Band	13	59	72
Total	252	182	434

a. How many seventh graders chose math as their favorite subject?

b. How many students are in the eighth grade?

c. How many more students prefer math than prefer science?

11. The PTA president is organizing a celebration and asked members whether they would prefer a brunch, lunch, or dinner celebration and whether they would bring an appetizer or dessert. Use the table to complete the following exercises. (Lesson 10-6)

Preference	Appetizer	Dessert	Total
Brunch	8	4	12
Lunch	7	2	9
Dinner	6	11	17
Total	21	17	38

a. Make a relative frequency table of the data.

b. What is the joint relative frequency of bringing an appetizer and preferring dinner?

Selected Answers and Solutions

Lesson 0-1

1. estimate; about 700 mi **3.** estimate; about 7 times **5.** exact; $98.75

Lesson 0-2

1. integers, rationals **3.** irrationals **5.** irrationals **7.** rationals **9.** rationals **11.** irrationals **13.** $-\frac{6}{5}, -\frac{3}{5}, \frac{3}{4},$ and $\frac{7}{5}$

15. $2\frac{1}{4}, 2.\overline{3}, \sqrt{7},$ and $\sqrt{8}$

17. $-3\frac{3}{4}, -3.5, -\sqrt{10},$ and $-\frac{15}{5}$

19. $\frac{5}{9}$ **21.** $\frac{13}{99}$ **23.** -5 **25.** ± 6 **27.** ± 1.2 **29.** $\frac{4}{7}$ **31.** $\frac{5}{18}$ **33.** 16 **35.** 26

Lesson 0-3

1. 5 **3.** -27 **5.** -22 **7.** -32 **9.** 22 **11.** 5 **13.** 8 **15.** -9 **17.** -115 **19.** $17°$ **21.** $150 **23.** $125

Lesson 0-4

1. $<$ **3.** $<$ **5.** $=$ **7.** $3.06, 3\frac{1}{6}, 3\frac{3}{4}, 3.8$ **9.** $-0.5, -\frac{1}{9}, \frac{1}{10}, 0.11$ **11.** $\frac{3}{5}$ **13.** $\frac{1}{16}$ **15.** 1 **17.** $2\frac{2}{3}$ **19.** $\frac{1}{9}$ **21.** $\frac{1}{6}$ **23.** $\frac{17}{30}$ **25.** $\frac{1}{4}$ **27.** -36.9 **29.** -19.33 **31.** 153.8 **33.** 93.3 **35.** $-\frac{5}{6}$ **37.** $\frac{9}{20}$ **39.** $\frac{2}{3}$ **41.** $\frac{3}{10}$

Lesson 0-5

1. 0.85 **3.** -7.05 **5.** 60 **7.** -4.8 **9.** -1.52 **11.** $\frac{6}{35}$ **13.** $\frac{2}{33}$ **15.** $\frac{21}{4}$ or $5\frac{1}{4}$ **17.** $-\frac{1}{2}$ **19.** $-\frac{1}{8}$ **21.** $\frac{10}{11}$ **23.** $\frac{5}{2}$ or $2\frac{1}{2}$ **25.** $\frac{7}{6}$ or $1\frac{1}{6}$ **27.** $-\frac{23}{14}$ or $-1\frac{9}{14}$ **29.** $-\frac{3}{16}$ **31.** 2 **33.** 3 **35.** $-\frac{3}{10}$ **37.** $\frac{9}{2}$ or $4\frac{1}{2}$ **39.** $\frac{11}{20}$ **41.** $\frac{5}{18}$ **43.** 3 slices **45.** 34 uniforms **47.** 6 ribbons

Lesson 0-6

1. $\frac{1}{20}$ **3.** $\frac{11}{100}$ **5.** $\frac{39}{50}$ **7.** $\frac{3}{500}$ **9.** 14 **11.** 40% **13.** 160 **15.** 9.5 **17.** 48 **19.** 0.25% **21.** 24.5 **23.** 150% **25.** 90% **27.** 5% **29a.** 20 g **29b.** 2350 mg **29c.** 44% **31.** 6 animals

Lesson 0-7

1. 20 m **3.** 90 in. **5.** 32 in. **7.** 29 ft **9.** 25.0 in. **11.** 31.4 in. **13.** 23.2 m **15.** 848.2 in. **17.** 13.4 cm **19.** 10.3 ft

Lesson 0-8

1. 6 cm^2 **3.** 120 m^2 **5.** 81 ft^2 **7.** 9 ft^2 **9.** 14.1 in^2 **11.** 12.6 ft^2 **13.** 50.3 cm^2 **15.** 201.1 in^2 **17.** 7 ft **19.** 20.5 $units^2$ **21.** 22.1 cm^2 **23.** 4.0 cm^2

Lesson 0-9

1. 30 cm^3 **3.** 48 yd^3 **5.** 1404 ft^3 **7.** 20 m^3 **9.** 27 m^3 **11.** 2070 in^3 **13.** 1 ft **15.** 4 cm **17.** 2770.9 in^3 **19a.** 128 ft^3 **19b.** 80 ft^3 **19c.** 5 ft 4 in.

Lesson 0-10

1. 68 in^2 **3.** 220 mm^2 **5.** 37 ft^2 **7.** 48 m^2 **9.** 216 in^2 **11.** 480.7 in^2 **13.** 24 m^2 **15.** 77 ft^2 **17.** 40.8 in^2

Lesson 0-11

1. $\frac{4}{15}$ **3.** $\frac{1}{2}$ **5.** $\frac{5}{6}$ **7.** $\frac{1}{2}$ **9.** $\frac{2}{3}$ **11.** 20 **13.** 12 codes **15.** $\frac{11}{24}$ **17.** 1:5 **19.** 13:11 **21.** 16 orders

Get Ready

1. 3 out of 10 is the simplest form of the fraction 54 out of 180.   **3.** You add up the three sides.   **5.** Because the figure is a rectangle, you can substitute 6 for l and 4 for w into the formula $P = 2l + 2w$ to find the perimeter. This will be the distance around the rectangle, so it is also equal to the amount of fencing you will need.
7. You would first do the calculation $72 \div 3 = 24$ so that you can measure the correct lengths before cutting the board.

Lesson 1-1

1. Sample answer: the product of 2 and m
3. Sample answer: a squared minus 18 times b
5. $6 - t$ **7.** $1 - \frac{r}{7}$ **9.** $n^3 + 5$ **11.** Sample answer: four times a number q **13.** Sample answer: 15 plus r **15.** Sample answer: 3 times x squared
17 Sample answer: 6 more than the product of 2 and a
19. $7 + x$ **21.** $5n$ **23.** $\frac{f}{10}$ **25.** $3n + 16$ **27.** $k^2 - 11$
29. $\pi r^2 h$ **31.** Sample answer: twenty-five plus six times a number squared
33. Sample answer: three times a number raised to the fifth power divided by two
35 **a.** Words: $\frac{3}{4}$ of the number of dreams
Expression: $\frac{3}{4} \cdot d$
The expression is $\frac{3}{4}d$.
b. $\frac{3}{4}(28) = 21$ dreams

37a.

10^2	$\cdot$	10^1	$=$	$10 \cdot 10 \cdot 10$	$=$	10^3
10^2	$\cdot$	10^2	$=$	$10 \cdot 10 \cdot 10 \cdot 10$	$=$	10^4
10^2	$\cdot$	10^3	$=$	$10 \cdot 10 \cdot 10 \cdot 10 \cdot 10$	$=$	10^5
10^2	$\cdot$	10^4	$=$	$10 \cdot 10 \cdot 10 \cdot 10 \cdot 10 \cdot 10$	$=$	10^6

37b. $10^2 \cdot 10^x = 10^{(2 + x)}$
37c. The exponent of the product of two powers is the sum of the exponents of the powers with the same base.   **39.** Sample answer: x is the number of minutes it takes to walk between my house and school. $2x + 15$ represents the amount of time in minutes I spend walking each day since I walk to and from school and I take my dog on a 15-minute walk.   **41.** 6   **43.** B   **45a.** C   **45b.** $32 + 2w$ **47.** A
49. B, C, E

Lesson 1-2

1. 81   **3.** 243
5 $5 \cdot 5 - 1 \cdot 3 = 25 - 3$
$= 22$
7. 28   **9.** 12   **11.** 20   **13.** $20 + 3 \times 4.95$; $34.85
15. 49   **17.** 64   **19.** 14   **21.** 36   **23.** 14
25. 142   **27.** 36   **29.** 3   **31.** 1

33 $(2t + 3g) \div 4 = (2(11) + 3(2)) \div 4$
$= (22 + 6) \div 4$
$= (28) \div 4$
$= 7$
35. 149   **37.** $3344 - 148 = 3196$ **39.** 16   **41.** 729
43. 177   **45.** 324   **47.** 29   **49.** 4080   **51.** $\frac{97}{31}$ **53.** 0
55. $28(15) + 12(20) + 30(15) + 15(20)$; $1410.00
57 **a.**

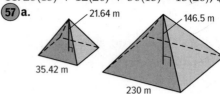

21.64 m
35.42 m
146.5 m
230 m

b. Words: one third times 230 squared times 146.5 minus one third times 35.42 squared times 21.64
c. Expression: $\frac{1}{3}(230)^2(146.5) - \frac{1}{3}(35.42)^2(21.64)$
$\approx 2583283.33 - 9049.68$
≈ 2574233.656 m^3
59. Curtis; Tara subtracted $10 - 9$ before multiplying 4 by 10.   **61.** Sample answer: $5 + 4 - 3 - 2 - 1$
63. Sample answer: Area of a trapezoid: $\frac{1}{2}h(b_1 + b_2)$; according to the order of operations, you have to add the lengths of the bases together first and then multiply by the height and by $\frac{1}{2}$.   **65.** C   **67.** C   **69.** A   **71a.** 2
71b. Sample answer: $6 + 12 \div (3 - 2) \times 4 = 54$
71c. Sample answer: $(6 + 12) \div 3 - 2 \times 4 = -2$

Lesson 1-3

1. $(1 \div 5)5 \cdot 14$
$= \frac{1}{5} \cdot 5 \cdot 14$ Substitution
$= (1) \cdot 14$ Multiplicative Inverse
$= 14$ Multiplicative Identity
3. $5(14 - 5) + 6(3 + 7) = 5(9) + 6(10)$ Substitution
$= 45 + 60$ Substitution
$= 105$ Substitution
5. $23 + 42 + 37$
$= 23 + 37 + 42$ Commutative (+)
$= (23 + 37) + 42$ Associative (+)
$= 60 + 42$ Substitution
$= 102$ Substitution
7. $3 \cdot 7 \cdot 10 \cdot 2$
$= 3 \cdot 2 \cdot 7 \cdot 10$ Commutative (×)
$= (3 \cdot 2) \cdot (7 \cdot 10)$ Associative (×)
$= 6 \cdot 70$ Substitution
$= 420$ Substitution
9 $3(22 - 3 \cdot 7) = 3(22 - 21)$ Substitution
$= 3(1)$ Substitution
$= 3$ Multiplicative Identity
11. $\frac{3}{4}[4 \div (7 - 4)]$
$= \frac{3}{4}[4 \div 3]$ Substitution
$= \frac{3}{4} \times \frac{4}{3}$ Substitution
$= 1$ Multiplicative Inverse

13. $2(3 \cdot 2 - 5) + 3 \cdot \frac{1}{3}$

$\begin{aligned}
&= 2(6 - 5) + 3 \cdot \frac{1}{3} &&\text{Substitution} \\
&= 2(1) + 3 \cdot \frac{1}{3} &&\text{Substitution} \\
&= 2 + 3 \cdot \frac{1}{3} &&\text{Multiplicative Identity} \\
&= 2 + 1 &&\text{Multiplicative Inverse} \\
&= 3 &&\text{Substitution}
\end{aligned}$

15. $2 \cdot \frac{22}{7} \cdot 14^2 + 2 \cdot \frac{22}{7} \cdot 14 \cdot 7$

$\begin{aligned}
&= 2 \cdot \frac{22}{7} \cdot 196 + 2 \cdot \frac{22}{7} \cdot 14 \cdot 7 &&\text{Substitution} \\
&= \frac{44}{7} \cdot 196 + \frac{44}{7} \cdot 14 \cdot 7 &&\text{Substitution} \\
&= 1232 + 616 &&\text{Substitution} \\
&= 1848 &&\text{Substitution}
\end{aligned}$

The surface area is 1848 in^2.

17. $25 + 14 + 15 + 36$

$\begin{aligned}
&= 25 + 15 + 14 + 36 &&\text{Commutative(+)} \\
&= (25 + 15) + (14 + 36) &&\text{Associative(+)} \\
&= 40 + 50 &&\text{Substitution} \\
&= 90 &&\text{Substitution}
\end{aligned}$

19. $\begin{aligned}
3\frac{2}{3} + 4 + 5\frac{1}{3} &= 3\frac{2}{3} + 5\frac{1}{3} + 4 &&\text{Commutative (+)} \\
&= \left(3\frac{2}{3} + 5\frac{1}{3}\right) + 4 &&\text{Associative (+)} \\
&= 9 + 4 &&\text{Substitution} \\
&= 13 &&\text{Substitution}
\end{aligned}$

21. $4.3 + 2.4 + 3.6 + 9.7$

$\begin{aligned}
&= 4.3 + 9.7 + 2.4 + 3.6 &&\text{Commutative (+)} \\
&= (4.3 + 9.7) + (2.4 + 3.6) &&\text{Associative (+)} \\
&= 14 + 6 &&\text{Substitution} \\
&= 20 &&\text{Substitution}
\end{aligned}$

23. $\begin{aligned}
12 \cdot 2 \cdot 6 \cdot 5 &= 12 \cdot 6 \cdot 2 \cdot 5 &&\text{Commutative (×)} \\
&= (12 \cdot 6) \cdot (2 \cdot 5) &&\text{Associative (×)} \\
&= 72 \cdot 10 &&\text{Substitution} \\
&= 720 &&\text{Substitution}
\end{aligned}$

25. $\begin{aligned}
0.2 \cdot 4.6 \cdot 5 &= (0.2 \cdot 4.6) \cdot 5 &&\text{Associative (×)} \\
&= 0.92 \cdot 5 &&\text{Substitution} \\
&= 4.6 &&\text{Substitution}
\end{aligned}$

27. $\begin{aligned}
1\frac{5}{6} \cdot 24 \cdot 3\frac{1}{11} &= 1\frac{5}{6}\left(24 \cdot 3\frac{1}{11}\right) &&\text{Associative (×)} \\
&= 1\frac{5}{6}\left(\frac{24}{1} \cdot \frac{34}{11}\right) \\
&= 1\frac{5}{6} \cdot \frac{816}{11} &&\text{Substitution} \\
&= \frac{8976}{66} &&\text{Substitution} \\
&= 136 &&\text{Substitution}
\end{aligned}$

29a. Sample answer: $2(17.25) + 3(15.50) + 2(5) + 5(18.99)$; $2(17.25 + 5) + 3(15.5) + 5(18.99)$
29b. $185.95

31. $\begin{aligned}
4(-1) + 9(4) - 2(6) &= -4 + 36 - 12 \\
&= 32 - 12 \\
&= 20
\end{aligned}$

33. -18 **35.** 192 **37.** Additive Identity; $35 + 0 = 35$
39. 0; Additive Identity **41.** 7; Reflexive Property
43. 3; Multiplicative Identity **45.** 2; Commutative Property **47.** 3; Multiplicative Inverse

49. $152 **51a.** 45.5 ft **51b.** Sample answer: An octagon has 8 sides, so the length of each side is $64 \div 8$ or 8 feet. The ledges are 18 inches, or 1.5 feet shorter than the sides, so each ledge is $8 - 1.5$ or 6.5 feet long. The side of the gazebo that contains the opening will not need a ledge, so the total length of the ledges is 6.5×7 or 45.5 feet. This is the minimum amount of wood that George needs. **51c.** Sample answer: Only seven sides will need shelves because one side of the gazebo needs to be the entrance.

53 a.

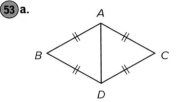

b. $\overline{AD} \cong \overline{AD}$ by the Reflexive Property. The Transitive Property shows that if $\overline{AB} \cong \overline{AC}$ and $\overline{AC} \cong \overline{DC}$, then $\overline{AB} \cong \overline{DC}$ and if $\overline{AB} \cong \overline{BD}$ and $\overline{AB} \cong \overline{AC}$, then $\overline{BD} \cong \overline{AC}$.

c. Because the sides are all congruent, each side has a length x. So, $P = x + x + x + x$.

55. Sample answer: You cannot divide by 0.
57. Sometimes; when a number is subtracted from itself then it holds but otherwise it does not.
59. $(2j)k = 2(jk)$; The other three equations illustrate the Commutative Property of Addition or Multiplication. This equation represents the Associative Property of Multiplication.
61. D **63.** A **65.** A **67a.** C **67b.** 28

Lesson 1-4

1. $55(12 + 15)$; $1485 **3.** $\left(6 + \frac{1}{9}\right)9$; 55
5. $2g^2(6) + 9g(6) - 3(6)$; $12g^2 + 54g - 18$

7. simplified

9. $\begin{aligned}
4(2x + 6) & \\
&= 4(2x) + 4(6) &&\text{Distributive Property} \\
&= 8x + 24 &&\text{Multiply.}
\end{aligned}$

11. $\begin{aligned}
4(5 + 3 + 4) &= 4(8 + 4) \\
&= 4(12) \\
&= 48 \text{ activities}
\end{aligned}$

13. $6(4) + 6(5)$; 54 **15.** $6(6) - 6(1)$; 30 **17.** $14(8) - 14(5)$; 42 **19.** $4(7) - 4(2)$; 20 **21.** $7(500 - 3)$; 3479
23. $36\left(3 + \frac{1}{4}\right)$; 117 **25.** $2(x) + 2(4)$; $2x + 8$
27. $2(-5) - 3m^2(-5)$; $-10 + 15m^2$ **29.** $18r$
31. $2m + 7$ **33.** $34 - 68n$ **35.** $13m + 5p$
37. $4fg + 17g$
39. $\begin{aligned}
7(a^2 + b) - 4(a^2 + b) & \\
&= 7a^2 + 7b - 4a^2 - 4b &&\text{Substitution} \\
&= 7a^2 - 4a^2 + 7b - 4b &&\text{Commutative (+)} \\
&= (7 - 4)a^2 + (7 - 4)b &&\text{Distributive Prop.} \\
&= 3a^2 + 3b &&\text{Substitution}
\end{aligned}$

41 A hexagon has six sides so an expression for the perimeter is $6(3x + 5)$.
$$6(3x + 5) = 6(3x) + 6(5)$$
$$= 18x + 30 \text{ units}$$

43. $14m + 11g$ **45.** $12k^3 + 12k$ **47.** $19x + 8$
49. $9 - 54b$ **51.** $12c - 6cd^2 + 6d$ **53.** $7y^3 + y^4$
55a. $2(x + 3)$

55b.

Area	Factored form
$2x + 6$	$2(x + 3)$
$3x + 3$	$3(x + 1)$
$3x - 12$	$3(x - 4)$
$5x + 10$	$5(x + 2)$

55c. Divide each term of the expression by the same number. Then write the expression as a product.
57. It should be considered a property of both. Both operations are used in $a(b + c) = ab + ac$.
59. Sample answer: You can use the Distributive Property to calculate quickly by expressing any number as a sum or difference of two more convenient numbers. Answers should include the following: Both methods result in the correct answer. In one method you multiply then add, and in the other you add then multiply.   **61.** C   **63.** B   **65.** B   **67a.** $3(2x + 4)$, $6x + 12$
67b. $78

Lesson 1-5

1. 2, 6, 9, 3   **3.** She should convert to a fraction and give each person a quarter of a pie.   **5.** 4
7 Because the radius is measured to the tenths, the volume would be rounded to tenths place, 137.3.
9. 1 ft; This is a large surface. Using the inch measure or smaller would most likely be more precise than necessary, so use the largest measure available.
11a. Mr. Menendez: 1.117; Ms. O'Toole: 1.272; Ms. Randall: 1.142; Mr. Fraser: 1.077   **11b.** Ms. O'Toole; she had a significantly greater score than the rest of the sales team.   **13.** 35 gallons   **15.** 240 thousand miles   **17a.** Sample answer: They can test at least 25 bags to find the actual weight and then base their report on the average weight of the bags.   **17b.** Sample answer: The test should be accurate to within $\frac{1}{8}$ pound, or 2 ounces. Fifty pounds is 800 ounces, so being off by 2 ounces is very accurate.   **17c.** Sample answer: If the bags they tested did not meet their standard of accuracy, they can report the results to the manufacturing company or to the government.
19 **a.** $Tax = 922.50 + 0.15(20,000 - 9225) = 922.50 + 0.15(10,775) = 922.50 + 1616.25 = 2538.75$; $2538.75   **b.** Sample answer: For the category married filing jointly, if the taxable income is between $18,450 and $74,900, the metric is $Tax = 1845 + 0.15(x - 18,450)$, where x is the taxable income.
21a. 199; yes   **21b.** Sample answer: Universities might also consider class rank, other test scores, and participation in other extracurricular or community activities.   **23.** Very

accurate; Sample answer: because an Olympic-sized pool has standard dimensions, the exact amount of water it contains can be calculated.   **25.** Not very accurate; Sample answer: While some male gorillas may weigh 152 kg, giving a specific value is not appropriate for all male gorillas.
27a. Sample answer: Test a sample of at least 20 cars, all the same model, for the city miles per gallon and the highway miles per gallon. Calculate their fuel economy ratings and take the average of the ratings for all the cars.   **27b.** 32 MPG   **29a.** B, C   **29b.** Sample: The ratio 0.653 will always be too high to be useful because the monthly salary minus the monthly debt is a low number.   **29c.** The ratios will all be 0. Becuase $0 < 0.3$, the loan should be made.
31a. centimeter   **31b.** millimeter   **33.** B   **35a.** 14.8731 g   **35b.** 28.2 g   **35c.** between 56.4 and 59.5 g; Find the range between the most accurate and the least accurate measurement. $14.8731 \times 4 = 59.4924$, and $28.2 \times 2 = 56.4$. Round each to the least accurate measurement.

Lesson 1-6

1.

x	y
4	3
-2	2
5	-6

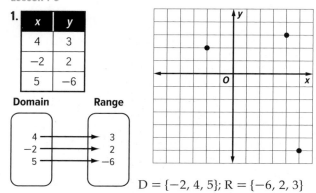

$D = \{-2, 4, 5\}$; $R = \{-6, 2, 3\}$
3. I: the temperature of the compound; D: the pressure of the compound   **5.** I: number of concert tickets, D: cost of tickets
7. The track team starts by running or walking, and then stops for a short period of time, then continues at the same pace. Finally, they run or walk at a slower pace.
9.

x	y
0	0
-3	2
6	4
-1	1

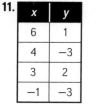

$D = \{0, -3, 6, -1\}$; $R = \{0, 2, 4, 1\}$
11.

x	y
6	1
4	-3
3	2
-1	-3

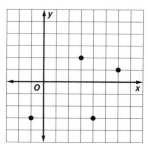

Domain **Range**

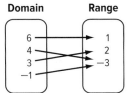

D = {6, 4, 3, −1};
R = {1, −3, 2}

b. The independent variable is the body weight b. The dependent variable is the water weight w.

c. The domain is the set of b values.
D = {100, 105, 110, 115, 120, 125, 130}. The range is the set of all w values. R = {66.7, 70, 73.3, 76.7, 80, 83.3, 86.7}

13.

x	y
6	7
3	−2
8	8
−6	2
2	−6

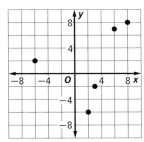

D = {−6, 2, 3, 6, 8};
R = {−6, −2, 2, 7, 8}

Domain **Range**

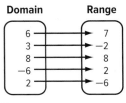

Water Weight Per Body Weight

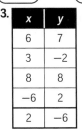

d. Graph the following ordered pairs: (66.7, 100), (70, 105), (73.3, 110), (76.7, 115), (80, 120), (83.3, 125), (86.7, 130).

Body Weight Per Water Weight

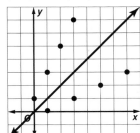

This graph shows what a person's body weight would be based on their water weight.

15 The number of students who attend is the independent variable because it does not depend on the amount of food there will be. The amount of food is the dependent variable because it depends on the number of students who attend.

17. The bungee jumper starts at the maximum height, and then jumps. After the initial jump, the jumper bounces up and down until coming to a rest.

19 Use the graph to determine what is happening to the value of the baseball card. The values are continually increasing.

21. (1, 5); The dog walker earns $5 for walking 1 dog.

23. I: number of dogs walked; D: amount earned

25. (5, 6); In the year 2015, sales were about $6 million. **27.** {(1, 2.50), (2, 4.50), (5, 10.50), (8, 16.50)}; D = {1, 2, 5, 8}; R = {2.50, 4.50, 10.50, 16.50}

29. {(4, −1), (8, 9), (−2, −6), (7, −3)}

31. {(4, −2), (−1, 3), (−2,−1), (1, 4)}

33. Sample answer: **35.** Sample answer:

   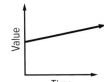

41. Reversing the coordinates gives (1, 0), (3, 1), (5, 2), and (7, 3).

Each point in the original relation is the same distance from the line as the corresponding points of the reverse relation. The graphs are symmetric about the line $y = x$.

43. D   **45.** C   **47a.** I: temperature, D: beach visitors
47b. B

Lesson 1-7

1. Yes; for each input there is exactly one output.
3. No; the domain value 2 is paired with 2 and −4.   **5.** No; when $x = 0$, $y = 1$ and $y = 6$.
7. Yes; its graph passes the vertical line test.
9a. {(0, 49,771), (1, 50,044), (2, 50,132), (3, 50,268)}

37 a.

b	$w=2\left(\frac{b}{3}\right)$	w	b	$w=2\left(\frac{b}{3}\right)$	w
100	$w=2\left(\frac{100}{3}\right)$	66.7	120	$w=2\left(\frac{120}{3}\right)$	80
105	$w=2\left(\frac{105}{3}\right)$	70	125	$w=2\left(\frac{125}{3}\right)$	83.3
110	$w=2\left(\frac{110}{3}\right)$	73.3	130	$w=2\left(\frac{130}{3}\right)$	86.7
115	$w=2\left(\frac{115}{3}\right)$	76.7			

9b.

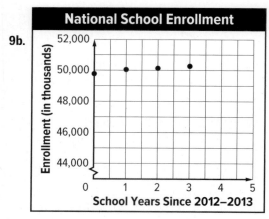

National School Enrollment

9c. The domain is the school year and the range is the enrollment; yes.

11. $f(-3) = 6(-3) + 7$
$= -18 + 7$
$= -11$

13. $6r - 5$ **15.** $a^2 + 5$ **17.** $6q + 13$ **19.** $b^2 - 4$

21. No; the domain value 4 is paired with both 5 and 6.
23. Yes; for each input there is exactly one output.
25. Yes; the graph passes the vertical line test.

27. yes **29.** yes **31.** yes **33.** -1 **35.** 14
37. -4 **39.** $-8y - 3$ **41.** $-2c + 7$ **43.** $-10d - 15$

45. a. Create a table using the rule given.

z	$f(z) = -\frac{5}{8}z + 87$	$f(z)$
0	$-\frac{5}{8}(0) + 87$	87
10	$-\frac{5}{8}(10) + 87$	80.75
20	$-\frac{5}{8}(20) + 87$	74.5
30	$-\frac{5}{8}(30) + 87$	68.25
40	$-\frac{5}{8}(40) + 87$	62

Plot the ordered pairs on a coordinate plane.

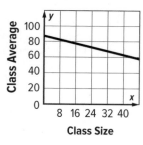

Class Size

When the class size is 0, the class average is 87. For each student added to the class, the class average decreases by 0.625 points.

b. $72 = -\frac{5}{8}z + 87$

$-15 = -\frac{5}{8}z$

$24 = z$

c. The domain can be any whole number; the range must be between 0 and 100.

47. The graph represents a function because each x-value is paired with only one y-value.

49. Sample answer: $\{(-2, 3), (0, 3), (2, 5)\}$

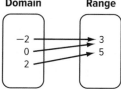

Domain   Range

51. $f(g + 3.5) = -4.3g - 17.05$

53. Sample answer: $f(x) = 3x + 2$

55. Sample answer: You can determine whether each element of the domain is paired with exactly one element of the range. For example, if given a graph, you could use the vertical line test; if a vertical line intersects the graph more than once, then the relation that the graph represents is not a function.

57. B **59.** D **61a.** 8; 5; 8; $a^2 - 2a + 5$ **61b.** Yes
61c. continuous

Lesson 1-8

1. Nonlinear; the y-intercept is 0, so there is no change in the stock value at the opening bell. The x-intercepts are 0, about 3.2, and about 4.5, so there is no change in the stock value after 0 hours, after about 3.2 hours, and after about 4.5 hours, respectively, after the opening bell. The graph has no line symmetry. The stock went up in value for the first 3.2 hours, then dropped below the starting value from about 3.2 hours until 4.5 hours, and finally went up again after 4.5 hours. The stock value starts the day increasing in value for the first 2 hours, then it goes down in value from 2 hours until 4 hours, and after 4 hours it goes up in value for the remainder of the day. The stock had a relative high value after 2 hours and then a relative low value after 4 hours. As the day goes on, the stock increases in value.

3. Linear; the y-intercept is about 45, so the temperature was about 45°F when the measurement started. The x-intercept is about 5.5, so after about 5.5 hours, the temperature was 0°F. The graph has no line symmetry. The temperature is above zero for the first 5.5 hours, and then below zero after 5.5 hours. The temperature is going down for the entire time. There are no extrema. As the time increases, the temperature will continue to drop forever, which is not very likely.

5. Because the graph is a curve and not a line, the graph is nonlinear. The graph intersects the y-axis at about (0, 20), so the y-intercept of the graph is about 20. This means that the purchase price of the vehicle was about $20,000. The graph approaches but never intersects the x-axis, so the graph has no x-intercept. This means that the value of the vehicle will never reach 0. There is no line over which the graph can be folded so that both halves match exactly.

Therefore the graph has no line symmetry. The graph lies entirely above the x-axis, therefore the function is positive for all values of x. This means that the value of the vehicle is always positive. When viewed from left to right, the graph goes down for all values of x, so the function is decreasing for all values of x, which means that the value of the vehicle is always decreasing. There are no relatively high or low points on the graph, so the function has no extrema. As you move right, the graph goes down, so the end behavior of the graph is that as x increases, y decreases. This means that as the number of years since the car was purchased increases, the value of the vehicle decreases.

7. Nonlinear; the y-intercept is about 5, so the company has a profit of about $5000 without spending any money on advertising. The x-intercept is about 21, so the company will make a profit of $0 if they spend $21,000 on advertising. Spending between $0 to $10,000 on advertising will produce the same profits as spending between $10,000 to $20,000. The company will make a profit if they spend between $0 and $21,000. If they spend more than $21,000 on advertising, they will lose money. The profits will increase until the company spends about $10,000, and then the profits will decrease for any amount greater than $10,000. Spending about $10,000 will produce the greatest profit. As more money is spent on advertising, the profits will decrease so that the company is losing money.   **9.** Nonlinear; the y-intercept is 0, which means that at the start, there was no medicine in the bloodstream. There appears to be no x-intercept, which means that the medicine does not ever fully leave the bloodstream for the time shown. The function is positive for all values of x, which means that after the medicine is taken, there is always some amount in the bloodstream. The function is increasing between about x = 0 and x = 8 and decreasing for x > 8, with a maximum value of about 2.5 at about x = 8. This means that the concentration of medicine increased over the first 8 hours to a maximum concentration of about 2.5 mg/mL, and then decreased. As x increases, the value of y decreases towards 0, which means that the concentration of medicine in the bloodstream becomes less and less, until there is practically none left.

11. Sample answer: Both the rural and urban populations can be modeled by linear functions. As the number of years since 2005 increases, the rural population of China decreases while the urban population increases.

13. Sample answer: The function has a y-intercept of 0 and an x-intercept of 0, indicating that the plant started with no height as a seed in the ground. The function is increasing over its domain, so that plant was always getting taller. The function has no relative extrema.

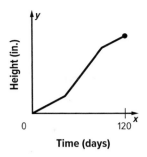

Time (days)

15. Sample answer: The function has a y-intercept of 27, indicating that the initial balance of the loan was $27,000. The x-intercept of 4 indicates that the loan was paid off after 4 years. The function is decreasing over its entire domain, indicating that the amount owed on the loan was always decreasing. The function has no relative extrema.

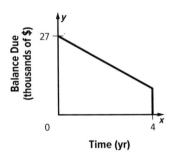

Time (yr)

(17) Plot the x-intercepts at (−2, 0) and (2, 0) and the y-intercept at (0, −4). Because the graph is nonlinear and decreasing for x < 0, draw a smooth curve starting somewhere to the left and above (−2, 0) that moves down through (−2, 0) to (0, −4). Because the graph has a relative minimum at x = 0 and is increasing for x > 0, turn at the point (0, −4) and draw a smooth curve moving up as you move right, through (2, 0) and continuing to the upper right portion of the graph. Sample graph:

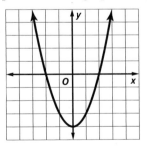

19. Sample graph:

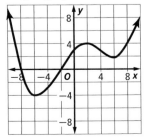

21. As x increases or decreases, y approaches 0.
23. The graph has a relative maximum at about $x = 2$ and a relative minimum at about $x = 4.5$. This means that the weekly gasoline price spiked around week 2 at a high of about \$3.50/gal and dipped around week 5 to a low of about \$1.50/gal.

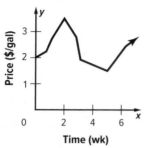

Average Weekly Gasoline Price

25. B   **27.** B   **29.** B

Chapter 1   Study Guide and Review

1. true   **3.** false; not in simplest form   **5.** Sample answer: After substituting the value(s) for the variable(s), you evaluate the resulting expression using the Order of Operations.
7. The additive identity states that the sum of a number and 0 is equal to the number. The additive inverse states that the sum of a number and its opposite is 0.   **9.** the product of 3 and x squared   **11.** $x + 9$ **13.** $4x - 5$ **15.** 216   **17.** $4.75 + 4.25g$ **19.** 18   **21.** 2   **23.** 3   **25.** 5
27. $4.75(3) + 5.25(2)$; \$24.75
29. $[5 \div (8 - 6)]$
$= [5 \div 2]\frac{2}{5}$ Substitution
$= \frac{5}{2} \cdot \frac{2}{5}$ Substitution
$= 1$ Multiplicative Inverse
31. $2 \cdot \frac{1}{2} + 4(4 \cdot 2 - 7)$
$= 2 \cdot \frac{1}{2} + 4(8 - 7)$ Substitution
$= 2 \cdot \frac{1}{2} + 4(1)$ Substitution
$= 1 + 4(1)$ Multiplicative Inverse
$= 1 + 4$ Multiplicative Identity
$= 5$ Substitution
33. $7\frac{2}{5} + 5 + 2\frac{3}{5}$
$= 7\frac{2}{5} + 2\frac{3}{5} + 5$ Commutative (+)
$= \left(7\frac{2}{5} + 2\frac{3}{5}\right) + 5$ Associative (+)
$= 10 + 5$ Substitution
$= 15$ Substitution
35. $5.3 + 2.8 + 3.7 + 6.2$
$= 5.3 + 3.7 + 2.8 + 6.2$ Commutative (+)
$= (5.3 + 3.7) + (2.8 + 6.2)$ Associative (+)
$= 9 + 9$ Substitution
$= 18$ Substitution
37. $(2)6 + (3)6$; 30   **39.** $8(6) - 8(2)$; 32
41. $-2(5) - (-2)(3)$; -4 **43.** $3(x) + 3(2)$; $3x + 6$

45. $6(d) - 6(3)$; $6d - 18$ **47.** $(9y)(-3) - (6)(-3)$; $-27y + 18$ **49.** $4(3 + 5 + 4)$; 48   **51.** hundreds of gallons   **53.** \$2.33
55. 139 cm to 145 cm

57.

x	y
1	3
2	4
3	5
4	6

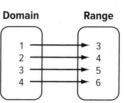

Domain   Range

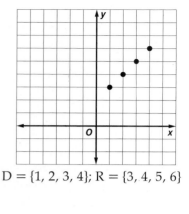

$D = \{1, 2, 3, 4\}$; $R = \{3, 4, 5, 6\}$

59.

x	y
−2	4
−1	3
0	2
−1	2

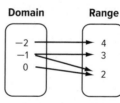

Domain   Range

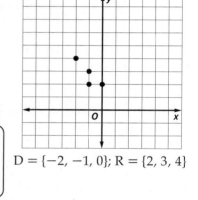

$D = \{-2, -1, 0\}$; $R = \{2, 3, 4\}$

61. $\{(-2, -2), (0, -3), (2, -2), (2, 0), (4, -1)\}$

63.

Planted	Growing
50	35
100	70
150	105
200	140

$D = \{50, 100, 150, 200\}$
$R = \{35, 70, 105, 140\}$

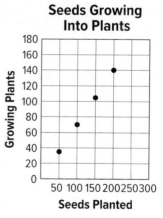

Seeds Growing Into Plants

65. function   **67.** -2 **69.** 4   **71.** $2m + 8$ **73.** Omar can carry the ball from 0 to 6 times, so $D = \{0 \le x \le 6\}$. The corresponding range is from 45 to 99 yards, so $R = \{45 \le y \le 99\}$.   **75.** Both runners begin and end at the high school. They both increase their distance from the high school to 3.2 miles. For Dante, this distance is reached in 18 minutes and for John, this distance takes about 20 minutes. John reaches a farther distance from the high school. His maximum distance is about 4.25 miles after 30 minutes. The distance from the high school begins to decrease at 30 minutes for John and at 32 minutes for Dante.

Linear Equations

Get Ready

1. $2b - 11$ **3.** $(9 - 4)$ **5.** 21%   **7.** Find the cross products.

Lesson 2-1

1. $15 - 3r = 6$

3 Words: A number squared plus 12 is the same as the quotient of p and 4.
Equation: $n^2 + 12 = p \div 4$
The equation is $n^2 + 12 = p \div 4$.

5. $8 + 3k = 5k - 3$ **7.** $\dfrac{25}{t} + 6 = 2t + 1$ **9.** $2000 + 40w = 3000$; 25   **11.** $P = 5s$ **13.** $4\pi r^2 = S$

15. Sample answer: The product of seven and m minus q is equal to 23.   **17.** Sample answer: Three times the sum of g and eight is the same as 4 times h minus 10.   **19.** Sample answer: A team of gymnasts competed in a regional meet. Each member of the team won 3 medals. There were a total of 45 medals won by the team. How many team members were there?
21. $f - 5g = 25 - f$ **23.** $4(14 + c) = a^2$ **25.** $3 \cdot 10 = 12f$; $2\frac{1}{2}$ flats

27 Words: C is five ninths times the difference of F and 32.
Equation: $C = \dfrac{5}{9} \cdot (F - 32)$
The equation is $C = \dfrac{5}{9}(F - 32)$.

29. $I = prt$ **31.** Sample answer: Four times m is equal to fifty-two.   **33.** Sample answer: Fifteen less than the square of r equals the sum of t and nineteen.

35. Sample answer: One third minus four fifths of z is four thirds of y cubed.   **37.** Sample answer: Ashley has a credit card that charges 12% interest on the principal balance. If Ashley's payment was $224, what was the principal balance on the credit card?   **39.** Sample answer: Fred was teaching his friends a new card game. Each player gets 5 cards, and 7 cards are placed in the center of the table. Because there are 52 cards in a deck, find how many players can play the game.
41. C   **43.** D

45 Words: the number of tent stakes + packets of drink mix + bottles of water = 17
$d = 3t$
$w = t + 2$
$$t + d + w = 17$$
$$t + 3t + (t + 2) = 17$$
$$5t + 2 = 17$$
$$5t + 2 - 2 = 17 - 2$$
$$5t = 15$$
$$t = 3$$
She brought 3 tent stakes.

47. Sample answer: My favorite television show has 30 new episodes each year. So far eight have aired. How many new episodes are left?
49. $\ell = \dfrac{P - 2w}{2}$ **51.** D   **53.** C   **55.** D

Lesson 2-2

1. 28   **3.** $\dfrac{5}{6}$ **5.** 9   **7.** -4.1 **9.** $-3\frac{1}{4}$ **11.** 16

13. $\dfrac{10}{9}$ or $1\frac{1}{9}$ **15.** $-\dfrac{4}{7}$ **17.** $22.75   **19.** 116   **21.** 22

23. -11

25
$$-16 - (-t) = -45$$
$$-16 + 16 - (-t) = -45 + 16$$
$$t = -29$$
Check: $-16 + (-29) = -45$
$$-45 = -45 \text{ Yes}$$

27. -32 **29.** -7 **31.** $1\frac{1}{8}$ **33.** $1\frac{2}{7}$ **35.** -708 **37.** 33

39. -2 **41.** $-1\frac{1}{9}$ **43.** $125 = t + 20$; 105 minutes

45. -77 **47.** $\dfrac{16}{3}$ **49.** -10 **51.** $-\dfrac{10}{7}$ or $-1\frac{3}{7}$ **53.** 18

55. 225   **57.** $\dfrac{2}{3} = -8n$; $n = -\dfrac{1}{12}$ **59.** $\dfrac{4}{5} = \dfrac{10}{16}n$; $n = \dfrac{32}{25}$

61 Words: Four and four fifths times a number is one and one fifth.
Equation: $4\dfrac{4}{5} \cdot n = 1\dfrac{1}{5}$
The equation is $4\dfrac{4}{5}n = 1\dfrac{1}{5}$.
Solve:
$$4\frac{4}{5}n = 1\frac{1}{5}$$
$$\frac{24}{5}n = \frac{6}{5}$$
$$5\left(\frac{24}{5}n\right) = 5\left(\frac{6}{5}\right)$$
$$24n = 6$$
$$n = \frac{6}{24}$$
$$n = \frac{1}{4}$$

63. $555 = 139 + p$; 416   **65.** $180 = t + 154$; 26 s

67. $1.6 - m = 0.8$; $0.8 million

69 Words: 45 million fewer than 57 million is the number who have blogs.
Equation: $57 - 45$
Solve: $57 - 45 = 12$ million

71a. $450 + m = 2500$; $2050   **71b.** $450 + 325 + m = 2500$; $1725   **71c.** $15t = 2500$; 167   **73.** Sample answer: $12 + n = 25$; subtract 12 from each side or add -12 to each side.   **75a.** Sometimes; $0 + 0 = 0$ but $2 + 2 \neq 2$.
75b. Always; this is the Addition Identity Property.
77a. $x = \dfrac{12}{a}$ **77b.** $x = 15 - a$ **77c.** $x = a - 5$
77d. $x = 10a$ **79a.** A   **79b.** C   **81.** A   **83a.** $120 = \dfrac{3}{8}t$
83b. D   **83c.** $120 = 0.4t$ **83d.** C   **83e.** Write 0.4 as $\dfrac{4}{10}$ and multiply each side by the reciprocal.   **83f.** 20

Lesson 2-3

1
$$3m + 4 = -11$$
$$3m + 4 - 4 = -11 - 4$$
$$3m = -15$$
$$m = -5$$
Check: $\quad 3m + 4 = -11$
$$3(-5) + 4 = -11$$
$$-15 + 4 = -11$$
$$-11 = -11 \text{ Yes}$$

3. -55 **5.** 61 **7.** $14 - 2n = -32$; 23
9. $n + (n + 2) + (n + 4) = 75$; $23, 25, 27$ **11.** -5
13. -5 **15.** 70 **17.** 27 **19.** 16 **21.** -61

23 **a.** Equation: $10d + 29.99 = 60$
$$10d + 29.99 - 29.99 = 60 - 29.99$$
$$10d = 30.01$$
$$d \approx 3$$
b. So, he can use $5 + 3$ or 8 GB.

25. $17 = 6x - 13$; 5 **27.** $n + (n + 2) + (n + 4) = 141$; 45, $47, 49$ **29.** $n + (n + 1) + (n + 2) + (n + 3) = -142$; $-37, -36, -35, -34$ **31.** $-7\frac{3}{5}$ **33.** -72

35. 108 **37.** $\frac{4}{5}$ **39.** $\frac{33}{14}$ **41.** $7\frac{1}{4}$ yr or 7 yr 3 mo

43
$$3.7q + 26.2 = 111.67$$
$$3.7q + 26.2 - 26.2 = 111.67 - 26.2$$
$$3.7q = 85.47$$
$$q = 23.1$$

45. 31.6 **47.** -3.5 **49.** 5 **51a.** Sample answer: If they plan to spend less than $80.13 at Riverfront, then the standard tickets are the best deal. **51b.** Sample answer: Assume that each voucher is worth a $5 meal, and the McAuleys would also have those 5 meals regardless of what tickets they purchase. The only other discount offered is the 15% and 20% discounts at Riverfront.

	Ticket Price	Meal Vouchers	Riverfront Cost	Total Cost
Standard Ticket	$3(25.99) =$ $77.97	$5(5) = \$25$	r	$102.97 + r$
Season Pass	$89.99	$5(5) = \$25$	$0.85r$	$114.99 + 0.85r$
Greatest Value	$119.99	$0	$0.80r$	$119.99 + 0.80r$

The standard tickets will be cheaper unless they plan to spend a lot of money at Riverfront. Furthermore, the *Greatest Value* pass offers a larger discount than the season pass. So, if they spend enough at Riverfront, the *Greatest Value* pass will eventually be cheaper. Create three equations relating the total costs of every option. Solving $102.97 + r = 114.99 + 0.85$ for r, we get $r = 80.13$.
Solving $102.97 + r = 119.99 + 0.80$ for r, we get $r = 85.10$.
Solving $114.99 + 0.85r = 119.99 + 0.80r$, we get $r = 100$.
Therefore, if they plan on spending more than $100 at Riverfront, they should purchase the *Greatest Value* pass. If they plan to spend between $80.13 and $100, then they should purchase the season pass. Otherwise, they should get standard tickets. **51c.** Sample answer: The free refills were ignored. They plan to eat 5 meals regardless of the vouchers. **53.** Sample answer: A pair of designer jeans costs $60. This is $40 more than twice the cost of a T-shirt. How much is the T-shirt? The T-shirt costs $10. **55a.** No; for there to be a solution there must be a number for which $a + 4 = a + 5$.

55b. Yes; for $b = 0$, $\frac{1 + b}{1 - b} = \frac{1 + 0}{1 - 0}$ or 1.

55c. No; $c - 5 = 5 - c$ when $c = 5$. However, $\frac{c - 5}{5 - c}$ is undefined for $c = 5$ because the fraction represents division by 0. **57.** Sample answer: In order to solve $4k + 20 = 236$, you would first subtract 20 from each side and then divide each side by 4. **59a.** $n, n + 1, n + 2, n + 3, n + 4$ **59b.** $n + (n + 1) + (n + 2) + (n + 3) + (n + 4) = -10$; $n = -4$ **59c.** The question asks for the greatest of the five integers, which is $n + 4$. Substitute -4 for n. $n + 4 = (-4) + 4 = 0$. **61.** C **63.** B

Lesson 2-4

1. 4 **3.** -7 **5.** no solution **7.** all numbers
9. A **11.** 4

13
$$6 + 3t = 8t - 14$$
$$6 + 3t - 3t = 8t - 3t - 14$$
$$6 = 5t - 14$$
$$6 + 14 = 5t - 14 + 14$$
$$20 = 5t$$
$$4 = t$$
Check: $6 + 3t = 8t - 14$
$$6 + 3(4) = 8(4) - 14$$
$$6 + 12 = 32 - 14$$
$$18 = 18 \text{ Yes}$$

15. $2\frac{2}{5}$ **17.** 6 **19.** -5 **21.** 1 **23.** $-4, -2$
25. no solution **27.** all numbers **29.** -25 **31.** 15
33. 3 **35.** -2 **37.** $-2, 0$
39 Equation: $5525 + 2.50x = 5.99x$
Solve: $5525 + 2.50x = 5.99x$
$$5525 + 2.50x - 2.50x = 5.99x - 2.50x$$
$$5525 = 3.49x$$
$$1583.09 \approx x$$
To determine the number of license plates the company needs to sell to make a profit each month, round up. So, they must sell 1584 license plates each month to make a profit.

41a. Sample answer: $y = 2x + 4$

x	−2	−1	0	1	2
y	0	2	4	6	8

$y = -x - 2$

x	−2	−1	0	1	2
y	0	−1	−2	−3	−4

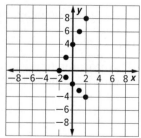

41b. -2 **41c.** Sample answer: The solution in part **b** is the x-coordinate for the point of intersection on the graph. **43.** Sample answer: $2x + 1 = \frac{3}{2}x - 2$; First I chose $\frac{3}{2}$ as the fractional coefficient. Then I chose 2 for the coefficient for the variable on the other side of the equation. After substituting -6 in for x on each side, 1 must be added to the left and 2 must be subtracted from the right to balance the equation. **45a.** Incorrect; the 2 must be distributed over both g and 5; 6. **45b.** correct

45c. Incorrect; to eliminate $-6z$ on the right side of the equal sign, $6z$ must be added to each side of the equation; 1.   **47.** Sample answer: If the equation has variables on each side of the equation, you must first add or subtract one of the terms from each side of the equation so that the variable is left on only one side of the equation. Then, solving the equation uses the same steps.   **49.** B   **51.** D   **53.** B   **55a.** $3n - 5 = \frac{1}{2}n + 9$
55b. 5.6

Lesson 2-5

1. 15   **3.** -4

5. $\{4, -2\}$;
$$-5\;-4\;-3\;-2\;-1\;\;0\;\;1\;\;2\;\;3\;\;4\;\;5$$

7. $\{-6, -2\}$
$$-7\;-6\;-5\;-4\;-3\;-2\;-1\;\;0\;\;1\;\;2\;\;3$$

9. $\varnothing$
$$-6\;-5\;-4\;-3\;-2\;-1\;\;0\;\;1\;\;2\;\;3\;\;4\;\;5\;\;6$$

11 Find the point that is the same distance from -2 and 4. This is the midpoint between -2 and 4, which is 1. The distance from -2 to 1 is 3 units. The distance from 4 to 1 is 3 units. So, an equation is $|x - 1| = 3$.

13 $|2x + z| + 2y = |2(2.1) + (-4.2)| + 2(3)$
$= |4.2 + (-4.2)| + 6$
$= |0| + 6$
$= 0 + 6$
$= 6$

15. -7.4 **17.** 8.4   **19.** -9.6 **21.** 0.4

23. $\{-11, -9\}$;
$$-13\;-12\;-11\;-10\;-9\;-8\;-7$$

25. $\{7, -3\}$;
$$-4\;-3\;-2\;-1\;\;0\;\;1\;\;2\;\;3\;\;4\;\;5\;\;6\;\;7\;\;8$$

27. $\varnothing$
$$-6\;-5\;-4\;-3\;-2\;-1\;\;0\;\;1\;\;2\;\;3\;\;4\;\;5\;\;6$$

29. $\{0, 6\}$
$$-3\;-2\;-1\;\;0\;\;1\;\;2\;\;3\;\;4\;\;5\;\;6\;\;7$$

31. 11% to 19%   **33.** $|x| = 4$ **35.** $|x - 1| = 4$

37. $\{-24, 16\}$
$$-24\;-20\;-16\;-12\;-8\;-4\;\;0\;\;4\;\;8\;\;12\;\;16\;\;20$$

39. $\left\{3, -\frac{9}{5}\right\}$
$$-5\;-4\;-3\;-2\;-1\;\;0\;\;1\;\;2\;\;3\;\;4\;\;5$$

41. no solution
$$-6\;-5\;-4\;-3\;-2\;-1\;\;0\;\;1\;\;2\;\;3\;\;4\;\;5\;\;6$$

43a. $|x - 52| = 2$; $\{50, 54\}$ **43b.** $|x - 53| = 1$; $\{52, 54\}$
43c. 203 and 214 seconds   **45a.** 47 to 53 mph
45b. Sample answer: The speedometer was calibrated more accurately than the speedometer in part **a**.
47. $|x| = 1\frac{1}{2}$ **49.** $\left|x - \frac{1}{4}\right| = \frac{1}{4}$ **51.** $\left|x + \frac{1}{3}\right| = 1$

53 **a.** Let h be the number of people who can clearly hear voices.
Equation: $|h - 20,000| = 1000$

b. $|h - 20,000| = 1000$
$h - 20,000 = 1000$ or $h - 20,000 = -1000$
$h - 20,000 + 20,000 = 1000 + 20,000$ or
$h - 20,000 + 20,000 = -1000 + 20,000$
$h = 21,000$ or $h = 19,000$
c. To find the range, find $21,000 - 19,000 = 2000$.
55 **a.** Each school could answer every question correctly, earning 50 points. They could also answer every question incorrectly, earning -50 points.
b. Let $m = $ the score on the math section. The maximum distance between m and the initial score of 160 is 50, so $|m - 160| = 50$. The maximum score is 210 and the minimum score is 110.
c. Every tenth value between -50 and 50 is possible. Let $c = $ correct, $n = $ incorrect, and $u = $ unanswered. Here are some combinations of answers. $50 = 5c$,
$40 = 4c + 1u, 30 = 3c + 2u, 20 = 2c + 3u$,
$10 = 2c + 1n + 2u, 0 = 2c + 2n + u$,
$-10 = 1n + 4u, -20 = 2n + 3u$,
$-30 = 3n + 2u, -40 = 4n + u, -50 = 5u$
57. Sometimes; when $x = -1$, the value is 0.
59. Sometimes; when c is a negative value the inequality is true.   **61.** An absolute value represents a distance from zero on a number line. A distance can never be a negative number.   **63.** Wesley; the absolute value of a number cannot be a negative number.   **65a.** C
65b. 5 and 9 inches   **67.** D   **69a.** $|x - 18500| = 1200$
69b. The first step is to write two equations with the absolute value sign. One equation equals 1200, and the second equation equals -1200. The last step for both equations is to add 18,500 to each side.   **69c.** 19,700 people   **69d.** 17,300 people   **71a.** $|x + 2| = 8$
71b. -10 **71c.** 6   **73.** $x = 1$ and $y = 3$

Lesson 2-6

1. no
3 $\frac{1.4}{2.1}$ is written in simplified form. $\frac{2.8}{4.4} = \frac{1.4}{2.2}$. Because the fractions are not equal, the ratios are not equivalent.
5. 5   **7.** about 253.3 min or 4 hours 13.3 min
9. yes   **11.** no   **13.** yes   **15.** 40   **17.** 29.25   **19.** 9.8
21. 1.32   **23.** 0.84
25 $\frac{t}{0.3} = \frac{1.7}{0.9}$
$0.9t = 1.7(0.3)$
$0.9t = 0.51$
$t \approx 0.57$
27. 6   **29.** 11   **31.** 156 mi   **33.** about \$262.59   **35.** 18
37. 0.8   **39.** 11   **41.** 130 students

43 **a.** Write each ratio.
for 2005, $\dfrac{\text{indoor theaters}}{\text{total theaters}} = \dfrac{37,040}{37,688}$
for 2006, $\dfrac{\text{indoor theaters}}{\text{total theaters}} = \dfrac{37,765}{38,415}$
for 2007, $\dfrac{\text{indoor theaters}}{\text{total theaters}} = \dfrac{38,159}{38,794}$

for 2008, $\dfrac{\text{indoor theaters}}{\text{total theaters}} = \dfrac{38,201}{38,834}$

for 2009, $\dfrac{\text{indoor theaters}}{\text{total theaters}} = \dfrac{38,605}{39,233}$

for 2010, $\dfrac{\text{indoor theaters}}{\text{total theaters}} = \dfrac{38,902}{39,520}$

for 2011, $\dfrac{\text{indoor theaters}}{\text{total theaters}} = \dfrac{38,974}{39,580}$

b. None of the ratios form a proportion.

45a.

45b.

ABCD		MNPQ		FGHJ	
Side length	2	Side length	4	Side length	1
Perimeter	8	Perimeter	16	Perimeter	4

45c. If the length of a side is increased by a factor, the perimeter is also increased by that factor. If the length of the sides are decreased by a factor, the perimeter is also decreased by the same factor.

47. Ratios and rates each compare two numbers by using division. However, rates compare two measurements that involve different units of measure.

49. If the tank is about $\dfrac{9}{16}$ full, he has about $\dfrac{9}{16} \times 10$ or $5\dfrac{5}{8}$ gallons of gas left. At 32 miles per gallon, he will be able to travel $32 \times 5\dfrac{5}{8}$ or 180 miles. Because Fort Worth is 200 miles away, he will run out of gas about 20 miles before reaching the city if he does not stop to get gas. **51.** C

53a. $\dfrac{14}{19} = \dfrac{x}{380}$ **53b.** B **55a.** $\dfrac{3}{5} = \dfrac{x}{x + 450}$ **55b.** The first step is to cross multiply and apply the distributive property. The next step is to get all x-terms on one side of the equation. The last step is to divide by the number in front of x. **55c.** 675 girls **55d.** 1125 people
57a. $\dfrac{4}{3} = \dfrac{x + 3}{x}$ **57b.** 9 **59.** 4

Lesson 2-7

1.
$$5a + c = -8a$$
$$5a - 5a + c = -8a - 5a$$
$$c = -13a$$
$$-\dfrac{c}{13} = a$$

3. $k = -7n - m$ **5a.** $h = \dfrac{V}{\pi r^2}$ **5b.** 8 in.
7. about 0.43875 ft

9.
$$x = b - cd$$
$$x - b = -cd$$
$$\dfrac{x - b}{-d} = c$$

11. $m = \dfrac{-n + p}{10}$ **13.** $v = \dfrac{9}{5}(z - w)$

15. $f = \dfrac{6g - 10}{d}$

17a. $v_f = at + v_i$ **17b.** 10 ft/s **19.** 49.8 L

21. $t = \dfrac{w - 11v}{31}$ **23.** $c = \dfrac{-13 + f}{10 - d}$ **25.** 1.0 mm/s

27. 3.9 km/s **29.** $t - 7 = r + 6$; $t = r + 13$

31. $\dfrac{9}{10}g = 7 + \dfrac{2}{3}k$; $k = \dfrac{3}{2}\left[\dfrac{9}{10}g - 7\right]$

33. $S = 2w(\ell + h) + 2\ell h$
$214 = 2(6)(7 + h) + 2(7)h$
$214 = 12(7 + h) + 14h$
$214 = 84 + 12h + 14h$
$130 = 26h$
$5 = h$
So, 5 inches.

35. about 364 in^3 **37.** Sandrea; she performed each step correctly; Fernando omitted the negative sign from $-5b$. **39a.** $x = \dfrac{y - 1}{yn - 1}$ **39b.** $y = -\dfrac{1}{3}x$ **41.** B
43. D **45a.** $P = \dfrac{A}{1 + r}$ **45b.** The step to solve the equation for P is to divide each side of the equation by $1 + r$. **45c.** $r = \dfrac{A - P}{P}$ **45d.** The first step is to apply the Distributive Property. The next step is to subtract P from both sides and then divide each side of the equation by P. **47.** $C = \dfrac{5(F - 32)}{9}$ **49.** $\dfrac{3y - 2}{2} = x$

Chapter 2 Study Guide and Review

1. true **3.** false, ratio **5.** false, variable **7.** An equation that involves several variables is called either a formula or a literal equation. **9.** $5x + 3 = 15$ **11.** $\dfrac{1}{2}m^3 = 4m - 9$

13. h squared minus five times h plus six is equal to zero. **15.** width: 8 ft, length: 19 ft **17.** -5 **19.** 2.1
21. 6 **23.** 14 **25.** 6 **27.** -11 **29.** 17 **31.** 2
33. 38.1 **35.** 19, 21, 23 **37.** 3 **39.** -2 **41.** 2
43. -8 **45.** 21 **47.** 28 **49.** -144
51. $\{-5, 17\}$

53. $\{-27, 63\}$

55. yes **57.** 20 **59.** 12 **61.** $y = \dfrac{9 - 3x}{2}$ **63.** $m = \dfrac{15 - 9n}{-5}$
65. $y = \dfrac{5}{2}(m - n)$ **67.** $h = \dfrac{2A}{a + b}$

Get Ready

1. Sample answer: The origin is the location of 0 on both axes. You can begin counting up or down from 0.
3. any point $(x, 5)$ **5.** Sample answer: add 5 to each side
7. Sample answer: Find the value of the expression by substituting values for all the variables and performing all the operations in the expression.

Lesson 3-1

1. yes; $x - y = -5$ **3.** yes; $y = 1$ **5.** 25, −4; The x-intercept 25 means that after 25 minutes, the temperature is 0°F. The y-intercept −4 means that at time 0, the temperature is −4°F.

7.

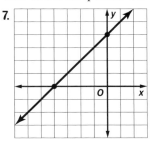

9.

x	$y = 2 - \dfrac{x}{2}$	y	(x, y)
−4	$y = 2 - \dfrac{(-4)}{2}$	4	(−4, 4)
−2	$y = 2 - \dfrac{(-2)}{2}$	3	(−2, 3)
0	$y = 2 - \dfrac{0}{2}$	2	(0, 2)
2	$y = 2 - \dfrac{2}{2}$	1	(2, 1)
4	$y = 2 - \dfrac{4}{2}$	0	(4, 0)

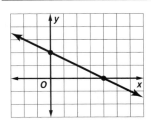

D = all real numbers;
R = all real numbers

11.

x	$y = 3$	y	(x, y)
−2	$y = 3$	3	(−2, 3)
−1	$y = 3$	3	(−1, 3)
0	$y = 3$	3	(0, 3)
1	$y = 3$	3	(1, 3)
2	$y = 3$	3	(2, 3)

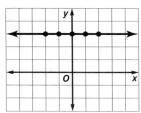

D = all real numbers;
R = {3}

13 $5x + y^2 = 25$
Because the y term is squared, this equation cannot be written in the form $Ax + By = C$. It is not a linear equation.

15. no **17.** yes; $4x + y = 0$ **19.** 3, 4
21. 6, 20; The x-intercept represents the number of seconds that it takes the eagle to land. The y-intercept represents the initial height of the eagle.

23.

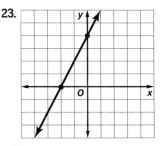

25.

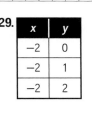

27.

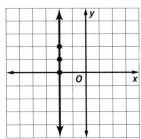

29.

x	y
−2	0
−2	1
−2	2

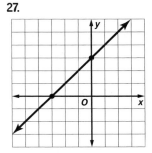

D = {−2};
R = all real numbers

31.

x	y
−1	8
0	0
1	−8

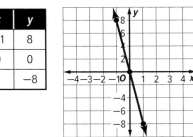

D = all real numbers;
R = all real numbers

33.

x	y
0	8
1	7
2	6

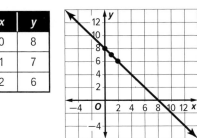

D = all real numbers;
R = all real numbers

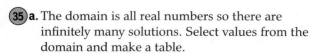

35 **a.** The domain is all real numbers so there are infinitely many solutions. Select values from the domain and make a table.

v	p = 0.15v	p	(v, p)
0	p = 0.15(0)	0	(0, 0)
2	p = 0.15(2)	0.3	(2, 0.3)
4	p = 0.15(4)	0.6	(4, 0.6)
6	p = 0.15(6)	0.9	(6, 0.9)
8	p = 0.15(8)	1.2	(8, 1.2)
10	p = 0.15(10)	1.5	(10, 1.5)

b. Create ordered pairs and graph them.

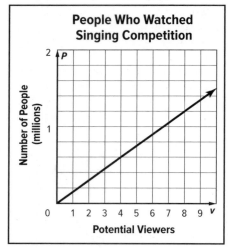

People Who Watched Singing Competition

c. Using the graph, when there are 14 million potential viewers, there will about 2.1 million people who watch.

d. A negative number does not make sense because you cannot have a negative number of viewers.

37. yes; $3x - 4y = 60$ **39.** yes; $3a = 2$

41. yes; $9m - 8n = -60$

43.
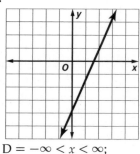
$D = -\infty < x < \infty$;
$R = -\infty < y < \infty$

45.

$D = -\infty < x < \infty$;
$R = -\infty < y < \infty$

47.
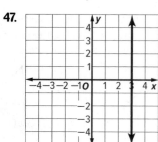
$D = 3$;
$R = -\infty < y < \infty$

49. No; Sample answer: The rental car would cost \$227. Mrs. Johnson has only \$210 to spend.

51 $5x + 3y = 15$

To find the x-intercept, let $y = 0$.

$$5x + 3y = 15$$
$$5x + 3(0) = 15$$
$$5x = 15$$
$$x = 3$$

The x-intercept is 3. This means the graph intersects the x-axis at (3, 0).
To find the y-intercept, let $x = 0$.

$$5x + 3y = 15$$
$$5(0) + 3y = 15$$
$$3y = 15$$
$$y = 5$$

The y-intercept is 5. This means the graph crosses the y-axis at (0, 5).

53. $2\frac{1}{2}$; $-1\frac{2}{3}$ **55.** 12; -3

57a. t-intercept: -9.19; s-intercept: 246,402

High School Graduates who scored 3+ on AP Exams

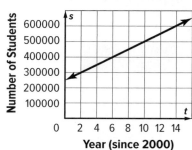

57b. 836,244

59. Sample answer: Yes; we used the formula $P = 4s$, which is linear.

Sample answer: No; we used the formula $A = s^2$, which is not linear.

Perimeter of a Square	
Side Length	Perimeter
1	4
2	8
3	12
4	16

Area of a Square	
Side Length	Area
1	1
2	4
3	9
4	16

Volume of a Cube	
Side Length	Volume
1	1
2	8
3	27
4	64

Sample answer: No; we used the formula $V = s^3$, which is not linear.

61. Sample answer: $y = 8$; horizontal line   **63.** Sample answer: $x - y = 0$; line through (0, 0)

65. B   **67.** A

1. 3   **3.** $\frac{1}{2}$ **5.** no zero   **7.** no zero

9. Tyrone must deliver 40 newspapers for the papers in his bag to weigh 0 pounds.   **11.** -3

13. no zero   **15.** $-\frac{10}{7}$ or $-1\frac{3}{7}$

17.
$$f(x) = 7$$
$$0 \neq 7$$
There are no zeros.

19. no zero   **21.** no zero   **23.** about 50; She can download a total of 50 songs before the gift card is completely used.   **25.** -8 **27.** $\frac{10}{3}$ or $3\frac{1}{3}$

29. $-\frac{34}{13}$ or $-2\frac{8}{13}$ **31.** $\frac{17}{25}$ **33.** $\frac{15}{8}$ or $1\frac{7}{8}$ **35.** 3

37. 4:00 P.M.

39. -3 **41.** -2

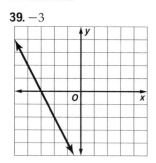

43. $\frac{9}{8}$ or $1\frac{1}{8}$

45. a. Sample answers given:

Number of Songs Downloaded	Total Cost ($)	Total Cost / Number Songs Downloaded
2	4	$\frac{4}{2} = 2$
4	8	$\frac{8}{4} = 2$
6	12	$\frac{12}{6} = 2$
8	16	$\frac{16}{8} = 2$
10	20	$\frac{20}{10} = 2$

b. As the number of songs downloaded increases by 2, the cost increases by 4.

c. The value of the total cost divided by the number of songs downloaded represents the cost per song. It costs $2 per song.

47. 3   **49.** Sample answer: $3 + 4x = 0$; $y = 3 + 4x$ or $f(x) = 3 + 4x$ **51.** A   **53.** B   **55.** 12.5   **57.** -1

1. $\frac{4}{3}$ **3a.** 1.035; There was an average increase in ticket price of $1.035 per year.   **3b.** Sample answer: 1998–2000; A steeper segment means a greater rate of change.   **3c.** Sample answer: 1998–2000; Ticket prices show a sharp increase.

5. No; the rate of change is not constant.   **7.** -1

9. $\frac{7}{9}$ **11.** 0   **13.** -8

15.
$$\text{rate of change} = \frac{\text{change in } y}{\text{change in } x}$$
$$= \frac{9 - 15}{2 - 1}$$
$$= \frac{-6}{1}$$
$$= -6$$

17. $\frac{1}{2}$ **19a.** Sample answer: $p = -1344t + 21{,}804$

19b. The car value depreciates by $1344 each year.
19c. $12,396   **21.** No; the rate of change is not constant.   **23.** Yes; the rate of change is constant.

25.
$$m = \frac{y_2 - y_1}{x_2 - x_1}$$
$$= \frac{1 - (-2)}{1 - 8}$$
$$= \frac{1 + 2}{1 - 8}$$
$$= -\frac{3}{7}$$

27. $\frac{1}{6}$ **29.** $\frac{4}{3}$ **31.** $-\frac{14}{15}$ **33.** 66   **35.** $\frac{1}{2}$ **37.** 3

39. $\frac{4}{5}$ **41.** Sample answer: about -1

43. $\frac{15}{4}$ **45.** $-\frac{2}{3}$

47. a. Plot the ordered pairs on a coordinate plane. Connect the points with a line.

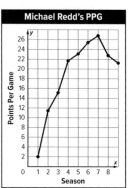

b. The graph is steepest between seasons 1 and 2, so his PPG increased the fastest from season 1 to season 2.

c. The rate of change was greater in the first four seasons and then slowed down between seasons 4 and 7. The rate of change was negative between seasons 7 and 9 when he started scoring fewer PPG each season.

49. The rate of change is $2\frac{1}{4}$ inches of growth per week.   **51.** Sample answer: Slope can be used to describe a rate of change. Rate of change is a ratio that describes how much one quantity changes with respect to a change in another quantity. The slope of a line is also a ratio and it is the ratio of the change in the y-coordinates to the change in the x-coordinates.

53. B **55.** $-\frac{3}{2}$ **57.** The rate of change is 3, so the perimeter is 3 times the side length. It must be a triangle.

Lesson 3-4

 $y = mx + b$
$y = 2x + 4$

To graph, plot the y-intercept $(0, 4)$. Then use the slope of 2 to move up 2 and right 1 from the y-intercept to find the next point. Connect the points with a straight line.

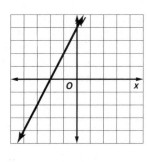

3. $y = \frac{3}{4}x - 1$ **5.** $4; 2$

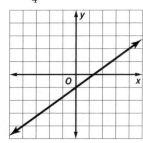 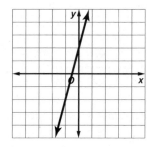

7. $\frac{3}{7}; 3$ **9.** $0; -1$

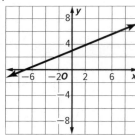 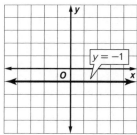

11. $y = \frac{2}{3}x + 2$ **13.** not possible **15a.** $S = 10w + 75$

15b. 10; 75

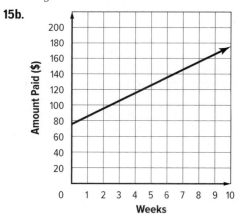

15c. $155

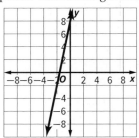 **17.** $y = mx + b$
$y = 5x + 8$

To graph, plot the y-intercept $(0, 8)$. Then use the slope of 5 to move up 5 and right 1 from the y-intercept to find the next point. Connect the points with a straight line.

19. $y = -4x + 6$ **21.** $y = 3x - 4$

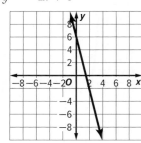 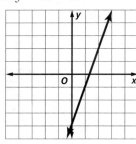

23. $3; 6$ **25.** $2; -4$

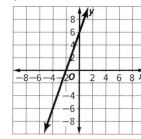 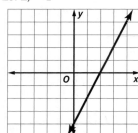

27. $-\frac{5}{2}; 4$ **29.** $0; 7$

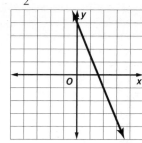 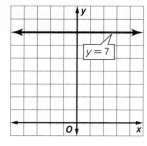

31. $0; 3$

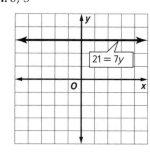

33. $y = -\frac{3}{5}x + 4$
35. $y = \frac{1}{2}x - 3$

37. a. Words: the population is 1267 plus 123 per year.
Equation: $P = 123t + 1267$

b. Graph the equation by plotting the y-intercept of (0, 1267). Then use the slope of 123 to move 123 up and 1 right.

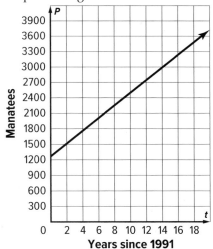

c. $P = 123t + 1267$
$P = 123(15) + 1267$
$P = 3112$

39. $y = \frac{2}{3}x - 5$ **41.** $y = -\frac{3}{7}x + 2$ **43.** $y = 5$

45.
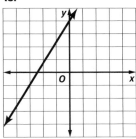

47.
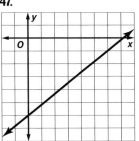

49.
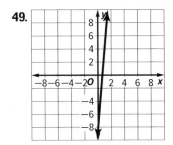

51a. $T = 157c + 218$
51b. $5242
53. $y = 0.5x + 7.5$
55. $y = -1.5x - 0.25$
57. $y = 3x$
59a. $C = 45m + 145$
59b. the cost per month to maintain the membership

59c. the start-up fee **59d.** $1225

61a. $A = 4200t + 3305$ **61b.** about 23 weeks

63. No; because a vertical line has no slope, it cannot be written in slope-intercept form.

65. Sample answer: Assume that the coefficient of y is not 0. We would first have to rewrite the equation in slope-intercept form. The rate of change is also the slope, so the coefficient for the x-variable is the rate of

change. **67.** A **69a.** $y = 4x - 12$ **69b.** $y = 0.25x + 1.5$
71a. 0 **71b.** $y = -3$ **73.** D

Lesson 3-5

1. translation 4 units left **3.** stretched vertically
5. reflected across the x-axis and stretched vertically
7. stretched vertically and translated down 9 units
9 The graph is reflected but not dilated. It is also translated 2 units up. So, the graph can be represented by $g(x) = -x + 2$. **11a.** $n(p) = 0.85(6.99p)$
11b. $34.95, $29.71 **13.** translation 3 units right
15. compressed horizontally **17.** translation 6 units left
19. translation 3 units up **21.** compressed horizontally
23. stretched horizontally **25.** compressed vertically
27. reflected across the x-axis and stretched vertically
29. reflected across the x-axis, stretched vertically, and translated down 5 units **31.** reflected across the y-axis and stretched horizontally **33** Because $a < 0$ and $|a| > 0$ in $g(x)$, the graph of $f(x)$ is reflected across the x-axis and stretched vertically. **35.** reflected across the x-axis, stretched vertically, and translated down 6 units
37. stretched horizontally and translated right 2 units
39. $g(x) = -x - 1$ **41.** $g(x) = x - 3$ **43.** $g(x) = -x - 3$
45. up **47.** right **49.** x-axis **51a.** $f(x) = 70x + 25$
51b. $h(x) = 70x + 30$ **51c.** translation
51d. $p(x) = 0.9(70x + 30)$ **51e.** vertical compression
53. Find a function that is equivalent to $g(x)$ and has a constant subtracted from x. $g(x) = mx + b + k = m\left(x + \frac{k}{m}\right) + b$. Subtracting $-\frac{k}{m}$ from x produces the same function as adding k to $f(x)$. Therefore, translating $f(x)$ up k units is the same as translating it $\frac{k}{m}$ units to the left. **55.** Sample answer: $y = 2x$; $y = -2x$ accompanied by a graph of both functions **57.** C **59.** C, E, F
61. $g(x) = -x + 3$ **63.** D

Lesson 3-6

1. No; there is no common difference. **3.** 0, -3, -6

5. $a_n = 17 - 2n$

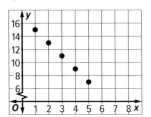

7. $a(n) = 55n + 525$

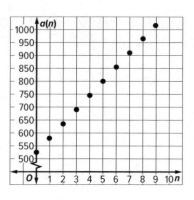

9. No; there is no common difference.

11. Yes; the common difference is 2.6.　**13.** 30, 36, 42

15 Step 1: Find the common difference by subtracting successive terms.

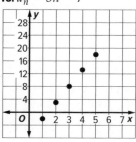

$$-\frac{1}{2}, \quad 0, \quad \frac{1}{2}, \quad 1, \, ...$$
The common difference is $\frac{1}{2}$.

Step 2: Add $\frac{1}{2}$ to the last term of the sequence to get the next term.

$$1 \quad 1\frac{1}{2} \quad 2 \quad 2\frac{1}{2}$$
The next three terms are $1\frac{1}{2}$, 2, $2\frac{1}{2}$.

17. $3\frac{7}{12}, 4\frac{1}{3}, 5\frac{1}{12}$

19. $a_n = 5n - 7$

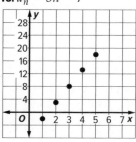

21. $a_n = 0.25n - 1$

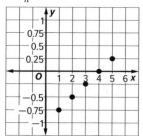

23a. $a(n) = 8n$

23b.

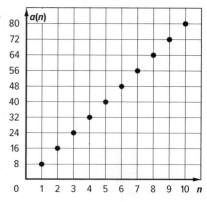

$D = \{1, 2, 3, 4, ...\}$;
$R = \{8, 16, 24, 32, ...\}$

25 The ordered pairs are (10, 7.50), (15, 8.75), (20, 10), (25, 11.25). So, the rate of change is

$$\frac{8.75 - 7.50}{15 - 10} = \frac{1.25}{5} = 0.25.$$

The cost is \$0.25 per word plus a flat fee.
$f(n) = 0.25n + b$
$7.50 = 0.25(10) + b$
$7.50 = 2.50 + b$
$5 = b$
So, the equation is $f(n) = 0.25n + 5$.

27. 77　**29.** 25,646　**31a.** $a_n = 2.5 + 0.5n$
31b. week 15　**31c.** Sample answer: No; eventually the number of miles run per day will become unrealistic.
33. -1
35a. Yes; there is a common difference; x;
$5x + 1, 6x + 1, 7x + 1$.
35b. No; unless $x = 0$, there is no common difference.　**37a.** 12　**37b.** 15　**37c.** 18　**39.** A　**41.** C
43a. $9 + 3n$　**43b.** 3
43c.

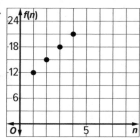

$D = \{1, 2, 3, 4, ...\}$;
$R = \{12, 15, 18, 21, ...\}$

43d. 12

1.

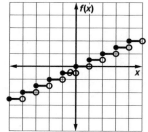

D = all real numbers;
R = all integer multiples of 0.5

3.

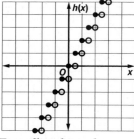

D = all real numbers;
R = all integers

5.

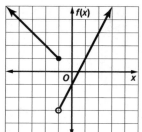

D = all real numbers;
$R = f(x) > -3$

7 First piece: The slope of 2 and the y-intercept of 0 show that elevator started at ground level and rose at a rate of 2 meters per second for 15 seconds.

shows that the elevator remained at a height of 30 meters for the time period between 15 and 30 seconds. Third piece: The slope of −2 and the x-intercept of 45 show that the elevator descended at a rate of 2 meters per second and reached ground level 45 seconds after timing began.

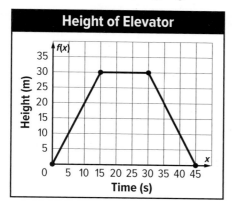

Height of Elevator

9.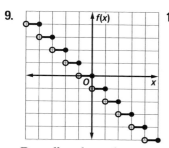

D = all real numbers;
R = all integers

11.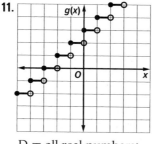

D = all real numbers;
R = all integers

13.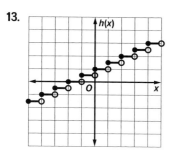

D = all real numbers;
R = all integer multiples of 0.5

15.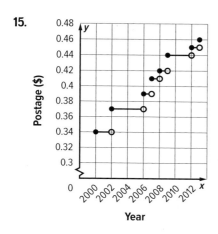

17.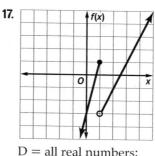

D = all real numbers;
R = all real numbers

19.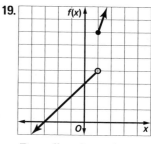

D = all real numbers;
R = f(x) < 4 or f(x) ≥ 7

21.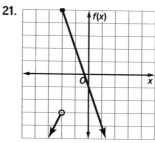

D = all real numbers;
R = f(x) ≤ 5

(23) First piece: The slope of –1000 and the y-intercept of 30,000 show that the plane starts at 30,000 feet and descends at a rate of 1000 feet per minute for 5 minutes. Second piece: The constant function value of 25,000 shows that the plane remains at an altitude of 25,000 feet after 5 minutes.

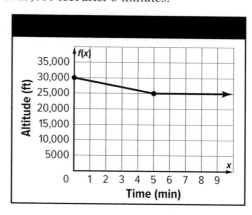

25.
D = all real numbers;
R = all integers

27.
D = all real numbers;
R = y > −4

29.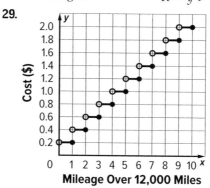

31 There are no instructors needed for 0 students, 1 instructor needed for 1 to 5 students, 2 instructors for 6 to 10 students, 3 instructors for 11 to 15 students, and so on.

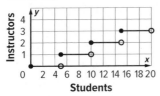

33.

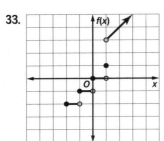

35. $f(x) = \left[\!\left[\dfrac{x}{3.79}\right]\!\right]$; 2 smoothies

37. No; the pieces of the graph overlap vertically, so the graph fails the vertical line test.

39. Sample answer:

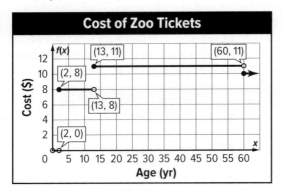

41a. true   **41b.** true   **41c.** false   **41d.** true   **41e.** false

43a.

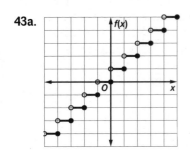

43b. D = all real numbers; R = all integers

43c. Sample answer: The graph consists of a series of horizontal steps. Each step has a length of 1 unit and includes the right endpoint but not the left endpoint. Each step is a translation 1 unit up and 1 unit right of the previous step. **43d.** The graph is a series of line segments. **45.** A

1.

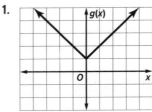

D = all real numbers;
R = g(x) ≥ 1; translation 1 unit up

3.

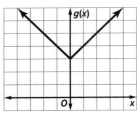

D = all real numbers;
R = g(x) ≥ 3; translation 3 units up

5.

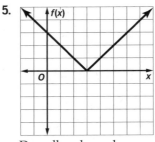

D = all real numbers;
R = f(x) ≥ 0; translated right 3 units

7.

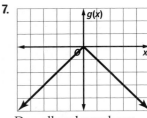

D = all real numbers;
R = g(x) ≤ 0; reflection across the x-axis

9.

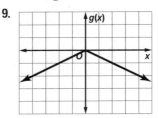

D = all real numbers;
R = g(x) ≤ 0; reflection across the x-axis and vertical compression by a factor of $\frac{1}{2}$

11.

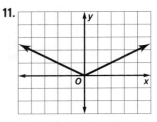

D = all real numbers;
R = g(x) ≥ 0; horizontal stretch by a factor of 2

13. 20 feet

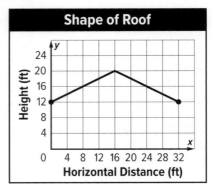

15. 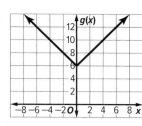 D = all real numbers; R = g(x) ≥ 6; translation 6 units up

17 x can be any number, but g(x) cannot be less than 0. The vertex of g(x) is 4 units left of the origin.

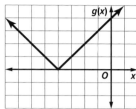 D = all real numbers; R = g(x) ≥ 0; translation 4 units left

19. 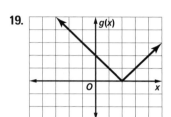 D = all real numbers; R = g(x) ≥ 0; translation 2 units right

21. 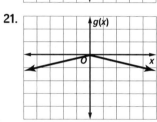 D = all real numbers; R = g(x) ≤ 0; reflection across the x-axis and vertical compression by a factor of $\frac{1}{4}$

23. 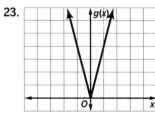 D = all real numbers; R = g(x) ≥ 0; horizontal compression by a factor of $\frac{1}{4}$

25. 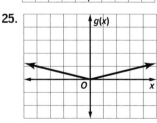 D = all real numbers; R = g(x) ≥ 0; horizontal stretch by a factor of 4

27 x can be any real number, but y cannot be less than 4; D = ; R = y ≥ 4

29. 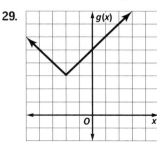 (−2, 3); translated left 2 units and up 3 units

31. 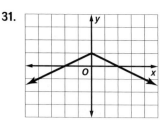 (0, 1); reflection across the x-axis, vertical compression by a factor of $\frac{1}{2}$ and translation 1 unit up

33. $f(x) = -4|x|$ or $f(x) = -|4x|$; explanation: The graph is a reflection across the x-axis and a vertical stretch by a factor of 4 of the graph of the parent function, so the value of the parameter a in the equation is −4.

35 **a.**

x	−5	−4	−3	−2	−1	0	1	2	3	4	5
f(x)	7	5	3	1	1	3	5	7	9	11	13

b. **c.**
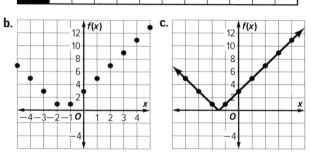

37. absolute value   **39.** piecewise   **41a.** (h, k)   **41b.** x = k
41c. a and −a   **43.** f(x) = {x − 3 if x ≥ 3; −x + 3 if x < 0}
45. They are additive inverses.   **47** If an absolute value function of the form g(x) = a|x − h| + k is reflected across the x-axis, a = −1. So, g(x) = −|2x|.
49. g(x) = −|3x| − 1   **51.** 100 square units
53. 16 square units

55.
$$f(x) = \begin{cases} \dfrac{1}{2}x - 3 \text{ if } x > 6 \\ -\dfrac{1}{2}x + 3 \text{ if } x \le 6 \end{cases}$$

57. Yes; Sample justification: If the vertex of an absolute value function is below the x-axis and the graph opens upward, then the function has two x-intercepts. Also, if the vertex of an absolute value function is above the x-axis and the graph opens downward, then the function will have two x-intercepts. For example, the graph of f(x) = |x| − 1 intersects the x-axis at (−1, 0) and at (1, 0), so it has two x-intercepts.   **59.** Sample answer: Write the function in the form f(x) = a|x − h| + k. If a > 0, the graph opens upward and has a minimum point. If a < 0, the graph opens downward and has a maximum point.   **61.** A, C   **63.** C   **65a.** false
65b. true   **65c.** true   **65d.** false   **65e.** true

1. true   **3.** true   **5.** true   **7.** false; domain   **9.** Find the difference between successive terms in the sequence.

11. −8, 6

13.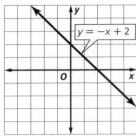

D = {all real numbers};
R = {all real numbers}

15.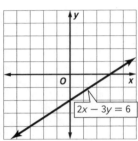

D = {all real numbers};
R = {all real numbers}

17a.

t	0	1	2	3	4	5
d	0	1.6	3.2	4.8	6.4	8

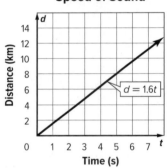

17b. about 11 km   **19.** 6   **21.** −$\frac{1}{2}$

23. −7

25. 9

27. 3   **29.** −$\frac{1}{2}$

31. −0.03; an average decrease in cost of $0.03 per year

33. $y = -2x - 9$

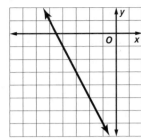

35. $y = -\frac{5}{8}x - 2$

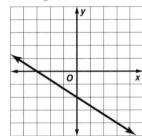

37. −3; 5

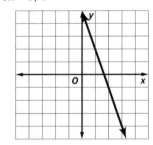

39. −$\frac{3}{4}$; 2

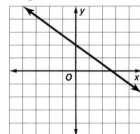

41. translation 7 units down   **43.** stretched vertically and translated right 2 units   **45.** translated up 50 units
47. 26, 31, 36   **49.** 64, 55, 46   **51.** 2, 6, 10
53. $a_n = 5n + 1$ **55.** $a_n = 3n + 9$ **57.** $a_n = 4820n$; 15 s

59.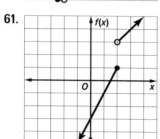
$D = \{-\infty < x < \infty\}$;
R = all integers

61.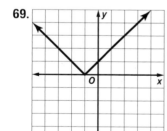
$D = \{-\infty < x < \infty\}$;
$R = \{f(x) \leq -1 \text{ or } f(x) > 3\}$

63. D = all real numbers, $R = \{f(x) \geq 6\}$

65. D = all real numbers, $R = \{10, 15\}$

67.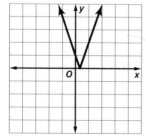
$D = \{-\infty < x < \infty\}$;
$R = \{f(x) \geq 0\}$

69.
$D = \{-\infty < x < \infty\}$;
$R = \{f(x) \geq 0\}$

71. $D = \{-\infty < x < \infty\}$; $R = \{f(x) \geq 0\}$

73. translated down 4 units
75. reflected across the x-axis
77. vertical compression by a factor of $\frac{1}{2}$

Get Ready

1. Begin at point A. Move vertically to the x-axis to find the x-coordinate, -4. Return to point A and move horizontally to the y-axis to find the y-coordinate, 2. Use the coordinates to write the ordered pair as $(-4, 2)$.
3. Subtraction within parentheses **5.** Subtract $15y$ from each side. Then divide each side by 5. **7.** 3 **9.** Horizontal lines have a slope of 0 and the slope of a vertical line is undefined.

Lesson 4-1

1. $y = 3x - 12$ **3.** $y = -x + 6$ **5.** $y = -3x + 9$
7. $y = 5x + 8$ **9a.** $C = 45p + 75$ **9b.** $750

11 $y = mx + b$
$4 = -1(-1) + b$
$4 = 1 + b$
$3 = b$
So, the equation is $y = -x + 3$.

13. $y = 8x - 55$ **15.** $y = 2x + 2$ **17.** $y = -x + 3$

19. $y = 7x - 16$ **21.** $x = -2$

23a. $y = 100{,}000x + 700{,}000$ **23b.** 2,100,000 **25.** $y = \frac{1}{2}x$

27. $y = -\frac{3}{4}x + 8\frac{1}{2}$ **29.** $y = -2$ **31a.** $G = 6.4t + 49.7$

31b. 222,500 **33a.** $2.75 **33b.** $35.40

35 First, find the slope.
$m = \dfrac{y_2 - y_1}{x_2 - x_1}$
$\quad = \dfrac{5 - (-3)}{2 - 5}$
$\quad = \dfrac{5 + 3}{2 - 5}$
$\quad = \dfrac{8}{-3}$
$\quad = -\dfrac{8}{3}$ or $-2\dfrac{2}{3}$

Next, use the slope-intercept formula.
$y = mx + b$
$5 = -\dfrac{8}{3}(2) + b$
$5 = -\dfrac{16}{3} + b$
$\dfrac{31}{3} = b$
$10\dfrac{1}{3} = b$
So, the equation is $y = -2\dfrac{2}{3}x + 10\dfrac{1}{3}$.

37. $y = -x - \dfrac{7}{12}$

39. Yes; substituting 6 and -2 for x and y, respectively, results in an equation that is true.

41. B; x represents the number of raffle tickets sold, y represents the total amount of money in the treasury. **43a.** 605.2 **43b.** 2038; In that year, the waste would be 0 tons. After that, the waste would be a negative amount, which is impossible.

45 a. $C = 76t + b$
$398 = 76(5) + b$
$398 = 380 + b$
$18 = b$
So, the equation is $C = 76t + 18$.

b.

Number of tickets	3	4	6	7
Cost ($)	246	322	474	550

c. Graph the equation by graphing the y-intercept $(0, 18)$ and use the slope of 76 to find the next point.

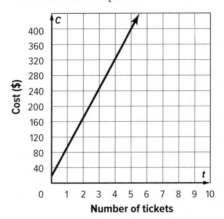

Eight tickets would be $626.

47. Jacinta; Tess switched the x- and y-coordinates on the point that she entered in step 3.

49a. $y = -\dfrac{A}{B}x + \dfrac{C}{B}$ **49b.** slope $= -\dfrac{A}{B}$

49c. y-intercept $= \dfrac{C}{B}$ **49d.** no, $B \neq 0$

51. Sample answer: If the problem is about something that could suddenly change, such as weather or prices, the graph could suddenly spike up. You need a constant rate of change to produce a linear graph. **53.** B **55.** D
57a. $y = 0.34x + 37.3$ **57b.** A, C, E

Lesson 4-2

1a. $40x + 5y = 200$

1b.

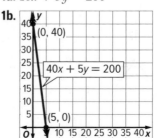

1c. 8 walks

(3) $y - y_1 = m(x - x_1)$
$y - 5 = -6[x - (-2)]$
$y - 5 = -6(x + 2)$

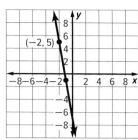

5. $y + 2 = \frac{7}{8}(x - 3)$, $y = \frac{7}{8}x - \frac{37}{8}$, $7x - 8y = 37$

7. $y + 2 = \frac{5}{3}(x + 6)$, $y = \frac{5}{3}x + 8$, $5x - 37 = -24$

9. $y - 7 = -\frac{3}{4}(x + 5)$, $y = -\frac{3}{4}x + \frac{13}{4}$, $3x + 4y = 13$

(11) a. Use $G(-3, 7)$ and $H(4, 1)$ to find the slope.
Slope $= m = \frac{y_2 - y_1}{x_2 - x_1} = \frac{1 - 7}{4 - (-3)} = -\frac{6}{7}$
Substitute the coordinates of G and the value of m
into the equation for point-slope form to get
$y - 7 = -\frac{6}{7}(x + 3)$.

b. Convert the equation in part **a** to standard form
by distributing $-\frac{6}{7}$ to get $y - 7 = -\frac{6}{7}x - \frac{18}{7}$.
Multiply both sides of the equation by 7 to get
$7y - 49 = -6x - 18$. Add $6x$ to each side and then
add 49 to each side. The equation in standard form is
$6x + 7y = 31$.

c. Convert the equation in part **a** to slope-intercept
form by distributing to get $y - 7 = -\frac{6}{7}x - \frac{18}{7}$. Then
add 7 to each side to get $y = -\frac{6}{7}x - \frac{67}{7}$.

d. Use $F(-3, 1)$ and $H(4, 1)$ to find the slope.
$m = \frac{1 - 1}{4 - (-3)} = \frac{0}{7}$ or 0
Substitute the coordinates of F and the value of m into
the equation for point-slope form to get
$y - 1 = 0(x + 3)$.

13. $y + 3 = -1(x + 6)$

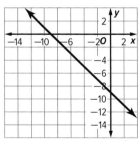

15. $x = -2$

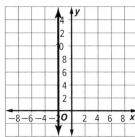

17. $y + 9 = -\frac{7}{5}(x + 2)$

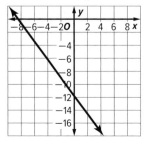

19. $y - 9 = -6(x + 9)$, $y = -6x - 45$, $6x + y = -45$

21. $y + 7 = \frac{9}{10}(x + 3)$, $y = \frac{9}{10}x - \frac{43}{10}$, $9x - 10y = 43$

23. $y + 3 = -\frac{1}{3}(x - 2)$, $y = -\frac{1}{3}x - \frac{7}{3}$, $x + 3y = -7$

25. $y - 6 = -2(x - 7)$, $y = -2x + 20$, $2x + y = 20$

27. $y + 5 = -6(x + 7)$, $y = -6x - 47$, $6x + y = -47$

29. $y + 2 = \frac{1}{6}(x - 4)$, $y = \frac{1}{6}x - \frac{8}{3}$, $x - 6y = 16$

31a. Let $x =$ the number of hours you rent the paddle
boat, and $y =$ the number of hours you rent the pontoon.
$100x + 150y = 1200$

31b.

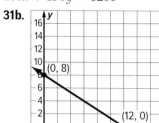

31c. 4 hours

33. $x + y = 6$ **35.** $5x + 4y = 20$

37. $y - 3 = 4(x - 1)$; $4x - y = 1$ **39.** $y - 7 = -\frac{4}{3}(x + 3)$;
$4x + 3y = 9$

(41) a. Since the barometric pressure is a function of the
altitude, the altitude is the independent variable x
and the pressure is the dependent variable, y. So the
points are $(1.8, 598)$ and $(2.1, 577)$.
Find the slope.
Slope $= m = \frac{y_2 - y_1}{x_2 - x_1} = \frac{577 - 598}{2.1 - 1.8} = -70$
Substitute the coordinates of one point and the value
of m into the point-slope form, $y - y_1 \, m(x - x_1)$ to get
$y - 598 = -70(x - 1.8)$. Convert this to slope-
intercept form, or $y = -70(x - 1.8) + 598$,
which simplifies to $y = -70x + 126 + 598$
or $y = -70x + 724$. Rewrite in function notation,
$f(x) = -70x + 724$.

b. Substitute 657 for y and solve for x, or $657 =$
$-70x + 724$; $70x = 67$. Solve for x, or $x = 0.96$ km.

43. Juana; $f(x)$ and $g(x)$ both have the same slope.
However, the x- and y-intercepts are different.

45. Sample answer: $y - g = \frac{j - g}{h - f}(x - f)$

47. Henry is correct. If you solve $Ax + By = C$, you get
$By = -Ax + C$, $y = \frac{Ax + C}{B}$, $y = -\frac{A}{B}x + \frac{C}{B}$. Since the
slope is the coefficient of x when the equation is written
in slope-intercept form, the slope is $-\frac{A}{B}$. So, Henry is
correct.   **49.** A   **51.** point-slope: $y - 8 = -4x + 6$;
slope-intercept: $y = -4x - 16$; standard: $4x + y = -16$

53. B   **55a.** $0.05x + 0.10y = 2.25$

55b.

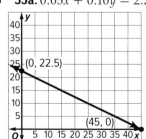

55c. 10 dimes

Selected Answers and Solutions

1. $y = \frac{1}{2}x + 2\frac{1}{2}$ **3.** Slope of $\overline{AC} = \frac{1-7}{-2-5}$ or $\frac{6}{7}$; slope of $\overline{BD} = \frac{-3-4}{3-(-3)}$ or $-\frac{7}{6}$; the paths are perpendicular.

5 Graph each line on a coordinate plane.
$y = -2x$ and $4y = 2x + 4$ are perpendicular to $2y = x$; $2y = x$ and $4y = 2x + 4$ are parallel.

7. $y = 2x + 7$ **9.** $x = 2$ **11.** $y = x - 5$

13. $y = -5x + 2$ **15.** $y = -\frac{3}{4}x + 1\frac{1}{2}$ **17.** Yes; the line containing $\overline{AD}$ and the line containing $\overline{BC}$ have the same slope, $\frac{1}{3}$. Therefore one pair of sides is parallel. The slope of $\overline{AB}$ is undefined and the slope of $\overline{CD}$ is $-\frac{5}{3}$.

19. Yes; the slopes are -6 and $\frac{1}{6}$. **21.** $2x - 8y = -24$ and $x - 4y = 4$ are perpendicular to $4x + y = -2$; $2x - 8y = -24$ and $x - 4y = 4$ are parallel.

23 The slope of the given line is -2. So, the slope of a line perpendicular is $\frac{1}{2}$.
$y = mx + b$
$-2 = \frac{1}{2}(-3) + b$
$-2 = -\frac{3}{2} + b$
$-\frac{1}{2} = b$
The equation is $y = \frac{1}{2}x - \frac{1}{2}$.

25. $y = -3x - 7$ **27.** $y = -\frac{1}{5}x + 8\frac{3}{5}$ **29.** $y = 2x + 16$

31. $y = -\frac{1}{5}x - \frac{3}{25}$ **33.** neither **35.** perpendicular

37. neither **39.** $y = 7x$

41 Find the slopes.
$$m = \frac{-1-6}{4-2} \qquad m = \frac{12-10}{14-7}$$
$$= \frac{-7}{2} \qquad\qquad = \frac{2}{7}$$
$$= -\frac{7}{2}$$
Because the slopes are opposite reciprocals, the pole is perpendicular to the line.

43a.

(coordinate grid with points A, B, C, O marked)

43b. Sample answer: (2, 2); $\overline{AB}$ and $\overline{CD}$ both have slope $\frac{1}{3}$, and $\overline{AC}$ and $\overline{BD}$ both have slope 3.

43c. Two; sample answer: Move C to $(-2, 0)$ and move D to $(4, 2)$. Moving C changes the slope of $\overline{AC}$ to -3. This is the opposite reciprocal of the slope $\overline{AB}$,

$\frac{1}{3}$. Moving D also changes the slope of $\overline{BD}$ so $\overline{BD}$ is perpendicular to $\overline{AB}$ and $\overline{CD}$ and it is parallel to $\overline{AC}$.

45. Sample answer: Parallel lines: similarities: The domain and range are all real numbers, the functions are both either increasing or decreasing on the entire domain, the end behavior is the same; differences: x- and y-intercepts are different. Perpendicular lines: similarities: The domain and range are all real numbers; differences: One function is increasing and the other is decreasing on the entire domain, as x decreases, y increases for one function and decreases for the other and as x increases, y increases for one function and decreases for the other. **47.** Carmen is correct; she correctly determined the slope of the perpendicular line. **49.** D **51.** D **53.** C

1. Positive; the longer you practice free throws, the more free throws you will make.

3a.

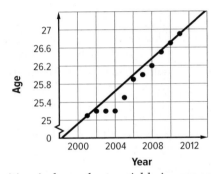

Median Age of Females When First Married

Positive; independent variable is year and dependent variable is median age of females when they were first married. **3b.** See above graph.
3c. Sample answer: Using (2001, 25.1) and (2011, 26.9) and rounding, $y = 0.18x - 335.08$
3d. Sample answer: 29.6 **3e.** Yes, according to the equation, the median age would be 35, which is likely.

5 As the height increases, the percentage decreases. The graph shows a slight negative correlation. This correlation means that the taller a player is, the lower their percentage of 3-point shots made is.

7 There is no pattern to the graph, so there is no correlation between the speed of a vehicle and the miles per gallon.

9a. $y = 224.25x + 68{,}837.5$ **9b.** 74,444 **9c.** No; the average attendance will fluctuate with other variables such as how good the team is that year.

11 a. The independent variable is the duration of the eruptions and the dependent variable is the interval of the eruptions. This is because the duration of the eruptions is not affected by the interval.

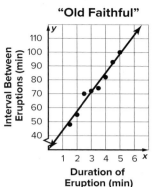

"Old Faithful"

Duration of Eruption (min)

Interval Between Eruptions (min)

The interval increases as the duration increases, so there is a positive correlation between the independent and dependent variables.

b. Use (2, 55) and (4, 82)

$$m = \frac{y_2 - y_1}{x_2 - x_1}$$

$$= \frac{82 - 55}{4 - 2}$$

$$= \frac{27}{2}$$

$$= 13.5$$

So, the slope is 13.5.

$y = 13.5x + b$	$y = 13.5x + 28$
$55 = 13.5(2) + b$	$y = 13.5(7.5) + 28$
$55 = 27 + b$	$y = 101.25 + 28$
$28 = b$	$y = 129.25$ min

c. Sample answer: The duration of an eruption is not dependent on the previous interval. Only the interval can be predicted by the length of the eruption.

13. Sample answer: The salary of an individual and the years of experience that they have; this would be a positive correlation because the more experience an individual has, the higher the salary would probably be.
15. Neither; line *g* has the same number of points above the line and below the line. Line *f* is close to 2 of the points; but for the rest of the data, there are 3 points above and 3 points below the line.   **17.** Sample answer: You can visualize a line to determine whether the data has a positive or negative correlation. The graph below shows the ages and heights of people. To predict a person's age given his or her height, write a linear equation for the line of fit. Then substitute the person's height and solve for the corresponding age. You can use the pattern in the scatter plot to make decisions.

Height

Age

19. B   **21.** B

Lesson 4-5

1a.

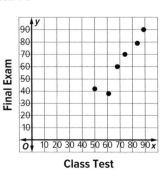

Final Exam

Class Test

1b. Positive; generally, the higher the class test score, the higher the final exam score.
1c. Sample answer: It is correlation only because the higher test scores do not directly cause the higher exam scores.
3. correlation, but not causation

 a.

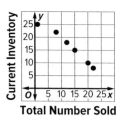

Current Inventory

Total Number Sold

5b. The correlation is negative because as sales increase, inventory decreases.

5c. Yes; as Mateo sells T-shirts, they are removed from his inventory.

7a.

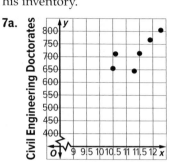

Civil Engineering Doctorates

Mozzarella Consumed (pounds)

7b. Positive; as the per capita amount of mozzarella consumed increases, the number of civil engineering doctorates awarded also increases.

7c. No; consumption of mozzarella does not cause anyone to obtain a doctorate degree in civil engineering.
9. causation

11 correlation, but not causation; The number of fast food places in a town does not necessarily affect the number of upscale restaurants.

13. yes; less

15. No; kids who like video games might also like construction sets.   **17.** Yes; Sample answer: There may be a third factor that is causing the correlation between the variables or the variables may be completely unrelated and still behave similarly.
19. Sample answer: A situation that illustrates causation is one that shows cause and effect, such as number of pages read and number of books completed.   **21.** B
23. A, C

25a.

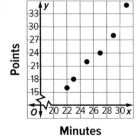

Points

Minutes

25b. There is a positive correlation between the variables because as the number of minutes increases, the number of points increases.

25c. There is no causation between the variables since playing more minutes may provide more opportunity to score, but does not cause the player to score more points. **27.** C

Lesson 4-6

1a. $y = 1.18x + 11$; The correlation coefficient is about 0.7181 which means that the equation models the data fairly well.
1b. The residuals appear to be randomly scattered, so the regression line fits the data reasonably well.

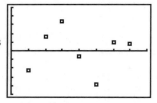

[0, 8] scl: 1.5 by [−5, 5] scl: 1.5

3a. $y = -271.88x + 554.48$
3b. $112.49

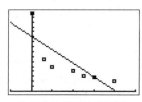

[−0.5, 2.5] scl: 1 by [0, 785] scl: 10

(5) Step 1: Enter the data by pressing ⬚ STAT ⬚ and selecting the Edit option. Let the year 2009 be represented by 0. Enter the years since 2009 into List 1 **(L1)**. These will represent the x-values. Enter the number of auditions into List 2 **(L2)**. These will represent the y-values.
Step 2: Perform the regression by pressing ⬚ STAT ⬚ and selecting the CALC option. Scroll down to LinReg(ax + b) and press ⬚ ENTER ⬚.
The equation is $y = 3.54x + 19.68$.
The correlation coefficient is about 0.9007, which means that the equation models the data very well.

(7) **a.** Enter the data using 0 for 1975. Use med-med to find $y = 601.44x + 1236.13$.
b. $2203 - 1975 = 28$. Substitute 28 into the equation in **a** to get 18,076. There were about 18,076 entrants in 2003.

9a. $y = 0.095x - 94.58$
9b.

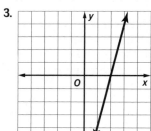

9c. about 48 tubs; about 380 tubs

11a. $y = 0.0326x + 1.598$

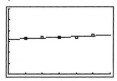

[0, 6] scl: 1 by [0, 3] scl: 0.5

11b. The regression line is a good fit as the residuals appear to be randomly scattered about the y-axis.

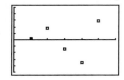

[0, 6] scl: 1 by [−0.05, 0.05] scl: 0.01

11c. Sample answer: about 2.28 million people; the prediction is reasonable because the equation models the data accurately.

13a. $y = 84,345.0x + 5,003,868.3$ **13b.** about 7,365,528.3

15.

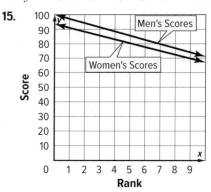

Sample answer: Men: $y = -2.9x + 100.0$; women: $y = -2.67x + 93.66$; men's scores have a steeper slope.

19. A   **21.** C   **23a.** $y = 250.4x + 1002.6$ **23b.** $4758.6 million   **23c.** 0.9971   **23d.** No; when the correlation coefficient is close to 1, the equation of the regression line models the data well.

Lesson 4-7

1. $\{(-15, 4), (-18, -8), (-16.5, -2), (-15.25, 3)\}$

3.

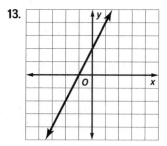

5. $f^{-1}(x) = -\frac{1}{2}x + \frac{7}{2}$
7a. $C^{-1}(x) = \frac{1}{70}x - \frac{60}{7}$
7b. x is Dwayne's total cost, and $C^{-1}(x)$ is the number of games Dwayne attended.
7c. 5

(9) Exchange the coordinates of the ordered pairs.
$(-4, -49) \rightarrow (-49, -4)$ $(8, 35) \rightarrow (35, 8)$
$(-1, -28) \rightarrow (-28, -1)$ $(4, 7) \rightarrow (7, 4)$
The inverse is $\{(-49, -4), (35, 8), (-28, -1), (7, 4)\}$.

11. $\{(7.4, -3), (4, -1), (0.6, 1), (-2.8, 3), (-6.2, 5)\}$

13.

15
$$f(x) = 17 - \frac{1}{3}x$$
$$y = 17 - \frac{1}{3}x$$
$$x = 17 - \frac{1}{3}y$$
$$x - 17 = -\frac{1}{3}y$$
$$-3(x - 17) = y$$
$$-3x + 51 = y$$

The inverse of $f(x)$ is $f^{-1}(x) = -3x + 51$.

17. $f^{-1}(x) = -\frac{1}{6}x + 2$ **19.** $f^{-1}(x) = -\frac{3}{4}x - 12$

21a. $C^{-1}(x) = \frac{1}{45}x - \frac{5}{9}$ **21b.** x is the total amount collected from the Fosters, and $C^{-1}(x)$ is the number of times Chuck mowed the Fosters' lawn.

21c. 22   **23.** $f^{-1}(x) = 15 - 5x$ **25.** $f^{-1}(x) = \frac{3}{2}x + 12$

27. $f^{-1}(x) = 3x - 3$ **29.** B   **31.** A

33 If the graph of $f(x)$ contains the points $(-3, 6)$ and $(6, 12)$, then the graph of $f^{-1}(x)$ contains the points $(6, -3)$ and $(12, 6)$. Find the slope of the line that passes through these points.
$$m = \frac{y_2 - y_1}{x_2 - x_1}$$
$$= \frac{6 - (-3)}{12 - 6}$$
$$= \frac{9}{6} \text{ or } \frac{3}{2}$$
Choose $(12, 6)$ and find the y-intercept of the line.
$$y = mx + b$$
$$6 = \frac{3}{2}(12) + b$$
$$6 = 18 + b$$
$$-12 = b$$
The line that passes through $(6, -3)$ and $(12, 6)$ is $y = \frac{3}{2}x - 12$. An equation for $f^{-1}(x)$ is $f^{-1}(x) = \frac{3}{2}x - 12$.

35. $f^{-1}(x) = \frac{1}{4}x + \frac{3}{4}$ **37a.** $A(x) = 8(x - 3)$ or $A(x) = 8x - 24$

37b.

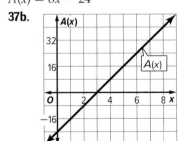

Sample answer: The domain represents possible values of x. The range represents the area of the rectangle and must be positive. This means that the domain of $A(x)$ is all real numbers greater than 3, and the range of $A(x)$ is all positive real numbers.

37c. $A^{-1}(x) = \frac{1}{8}x + 3$; x is the area of the rectangle and $A^{-1}(x)$ is the value of x in the expression for the length of the side of the rectangle $x - 3$.

37d.

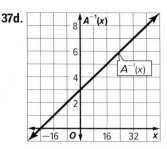

Sample answer: The domain represents the area of the rectangle and must be positive. The range represents possible values for x in the expression $x - 3$. This means that the domain of $A^{-1}(x)$ is all positive real numbers, and the range of $A^{-1}(x)$ is all real numbers greater than 3.

37e. Sample answer: The domain of $A(x)$ is the range of $A^{-1}(x)$, and the range of $A(x)$ is the domain of $A^{-1}(x)$.

39. $a = 2$; $b = 14$

41. Sometimes; sample answer: $f(x)$ and $g(x)$ do not need to be inverse functions for $f(a) = b$ and $g(b) = a$. For example, if $f(x) = 2x + 10$, then $f(2) = 14$ and if $g(x) = x - 12$, then $g(14) = 2$, but $f(x)$ and $g(x)$ are not inverse functions. However, if $f(x)$ and $g(x)$ are inverse functions, then $f(a) = b$ and $g(b) = a$.

43. Sample answer: A situation may require substituting values for the dependent variable into a function. By finding the inverse of the function, the dependent variable becomes the independent variable. This makes the substitution an easier process.

45. C   **47.** 10   **49.** A   **51.** D

Chapter 4   Study Guide and Review

1. true   **3.** true   **5.** true   **7.** No, two graphed lines can be neither parallel nor perpendicular.   **9.** They are used to esimate values that are not in the data.

11. $y = -4x + 2$ **13.** $y = -\frac{1}{3}x - \frac{1}{3}$ **15.** $y = 2x + 11$

17. $y = -\frac{2}{5}x - \frac{24}{5}$ **19.** $y - 3 = 5(x - 6)$ **21.** $y - 2 = 0$

23. $3x + y = 4$ **25.** $4x + 5y = 37$ **27.** $y = -2x + 18$

29. $y = \frac{1}{2}x + 5$ **31.** $y = 3x + 3$ **33.** $y = -\frac{1}{2}x - \frac{9}{2}$

35. $y = \frac{1}{2}x + \frac{5}{2}$ **37.** $y = 2x - 6$ **39.** $y = 210x + 230$

41. Positive: as the temperature increases, the amount of time Aiko spends outside increases.   **43.** $y = 5.36x + 11$; 65

45. {(3.5, 7), (8, 6.2), (2.7, −4), (1.4, −12)}   **47.** {(2.7, −4), (3.8, −1), (4.1, 0), (7.2, 3)}   **49.** $f^{-1}(x) = \frac{11}{5}x - 22$ **51.** $f^{-1}(x) = -\frac{1}{4}x - 3$ **53.** $f^{-1}(x) = -\frac{3}{2}x + \frac{3}{8}$

55. $f^{-1}(x) = \frac{x}{3}$ **57.** $f^{-1}(x) = \frac{x+7}{4}$ **59.** $f^{-1}(x) = x$

61. $f^{-1}(x) = \frac{35x + 5}{14}$ **63.** $f^{-1}(x) = -\frac{x}{3}$

65. $f^{-1}(x) = \frac{-8x - 1}{3}$ **67.** $f^{-1}(x) = x + 1$

Get Ready

1. $d = 100m + 325$ **3.** Sample answer: The goal in solving an inequality is to isolate the variable on one side of the inequality using properties and operations, just like when solving an equation. **5.** 24.6 mi/gal

Lesson 5-1

1. $\{x \mid x > 10\}$

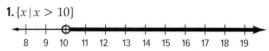

3. $\{g \mid g < -4\}$

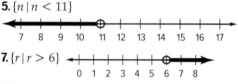

5. $\{n \mid n < 11\}$

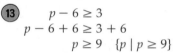

7. $\{r \mid r > 6\}$

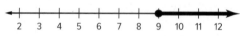

9. Sample answer: Let n = the number, $2n + 4 \geq n + 10$; $\{n \mid n \geq 6\}$. **11.** no more than 92 ft

13.
$$p - 6 \geq 3$$
$$p - 6 + 6 \geq 3 + 6$$
$$p \geq 9 \quad \{p \mid p \geq 9\}$$
Place a closed circle on 9 and an arrow to the right.

15. $\{t \mid t > -5\}$

17. $\{r \mid r < -5\}$

19. $\{q \mid q \leq 7\}$

21. $\{h \mid h < 30\}$

23. $\{c \mid c < -27\}$

25. $\{z \mid z \leq 4\}$

27. $\{y \mid y \leq -6\}$

29. $\{a \mid a > -9\}$

31. Sample answer: Let n = the number, $2n + 5 \leq n - 3$; $\{n \mid n \leq -8\}$. **33.** Sample answer: Let n = the number, $6n - 8 < 5n + 21$; $\{n \mid n < 29\}$.

35. Sample answer: Let n = the number of juniors who are Indiana Hoosiers fans who will not be attending a game; $n > 210 - 50$; $\{n \mid n > 160\}$; at least 160 juniors like the Indiana Hoosiers basketball team, but will not be attending a game this season.

37. Sample answer: Let t = the original water temperature; $t + 4 < 81$; $\{t \mid t < 77\}$; the water temperature was originally less than 77°.

39. Let m represent the amount left on the card.
$$m + 32 + 26 \leq 75$$
$$m + 58 \leq 75$$
$$m + 58 - 58 \leq 75 - 58$$
$$m \leq 17$$
She will have no more than $17 left on the card.

41. $\{c \mid c \geq 3.7\}$

43. $\left\{k \mid k > -\dfrac{5}{12}\right\}$

45a. 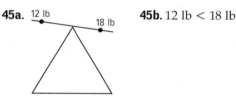 **45b.** 12 lb < 18 lb

45c.

	12	<	18
$2x$	24	<	36
$3x$	36	<	54
$4x$	48	<	72
$\frac{1}{2}x$	6	<	9
$\frac{1}{3}x$	4	<	6
$\frac{1}{4}x$	3	<	$4\frac{1}{2}$

45d. If a true inequality is multiplied by a positive number, the resulting inequality is also true. If a true inequality is divided by a positive number, the resulting inequality is also true.

47. 10 **49.** 3 **51.** 26 **53.** $c < a < d < b$ **55.** Solving linear inequalities is similar to solving linear equations. You must isolate the variable on one side of the inequality. To graph, if the problem is a less than or a greater than inequality, an open circle is used. Otherwise a dot is used. If the variable is on the left-hand side of the inequality, and the inequality sign is less than (or less than or equal to), the graph extends to the left; otherwise it extends to the right. **57.** C **59.** B **61.** A

1. Let t = the number of T-shirts sold; $12t > 5500$; $t > 458.33$; the band sold at least 459 T-shirts.

3. $\{r \mid r \geq 8\}$

number line from 0 to 12, closed circle at 8, shaded right

5. $\{h \mid h < -10\}$

number line from −18 to 0, open circle at −10, shaded left

7. $\{v \mid v > -12\}$

number line from −14 to 4, open circle at −12, shaded right

9. $\{z \mid z \geq -8\}$

number line from −9 to 2, closed circle at −8, shaded right

11. Let p = the number of pay periods for which Rodrigo will need to save; $25p \geq 560$; $p \geq 22.4$; Rodrigo will need to save for 23 weeks.

13.
$$\frac{1}{2}a < 20$$
$$2\left(\frac{1}{2}a\right) < 2(20)$$
$$a < 40 \quad \{a \mid a < 40\}$$

number line from 30 to 44, open circle at 40, shaded left

Check by substituting values less than 40.

15. $\{d \mid d \geq 68\}$

number line from 60 to 76, closed circle at 68, shaded right

17. $\{f \mid f < 432\}$

number line from 416 to 434, open circle at 432, shaded left

19. $\{j \mid j \leq -16\}$

number line from −20 to 2, closed circle at −16, shaded left

21. $\{p \mid p \leq 16\}$

number line from 0 to 18, closed circle at 16, shaded left

23. $\{y \mid y > -16\}$

number line from −18 to 0, open circle at −16, shaded right

25. $\{v \mid v < 12\}$

number line from 0 to 18, open circle at 12, shaded left

27. $\left\{b \mid b \leq -\frac{3}{4}\right\}$

number line from −8 to 8, closed circle at −3/4, shaded left

29. $\left\{f \mid f < -\frac{5}{7}\right\}$

number line from −5 to 5, open circle at about −1, shaded left

31. no more than 4

33. no more than 32 people
35. b   **37.** d

39.
$$\frac{2}{3}x < 42$$
$$3\left(\frac{2}{3}x\right) < 3(42)$$
$$2x < 126$$
$$x < 63 \qquad \text{fewer than 63 employees}$$

41a.

pyramid with height h cm and square base b cm by b cm

41b. $h = \dfrac{216}{b^2}$

41c.

b	1	3	6	9	12
h	216	24	6	$\frac{8}{3}$	$\frac{3}{2}$

41d. $b < h$ when $0 < b < 6$; $b > h$ when $h < 6$.

43a. $x > -\dfrac{5}{a}$ **43b.** $x \geq 8a$ **43c.** $x \leq -\dfrac{6}{a}$

45. Sometimes; the statement is true when $a > 0$ and $b < 0$.   **47.** Sample answer: The same processes are used when solving linear inequalities and equations that involve addition, subtraction, multiplication, or division by a positive number. However, when a linear inequality is multiplied or divided by a negative number, the inequality symbol must change directions so that the inequality remains true.   **49.** A   **51.** 150   **53.** A   **55.** B

Lesson 5-3

1. $15n + 225 \leq 3000$; $n \leq 185$; at most 185 lb per person

3.
$$6h - 10 \geq 32$$
$$6h - 10 + 10 \geq 32 + 10$$
$$6h \geq 42$$
$$h \geq 7$$
$$\{h \mid h \geq 7\}$$

number line from 0 to 18, closed circle at 7, shaded right

5. $\{x \mid x < -12\}$

number line from −20 to 4, open circle at −12, shaded left

7. Sample answer: Let n = the number; $4n - 6 > 8 + 2n$; $\{n \mid n > 7\}$.

Selected Answers and Solutions

9. $\{v \mid v \geq 0\}$

11. ∅

13
$$21 > 15 + 2a$$
$$21 - 15 > 2a$$
$$6 > 2a$$
$$3 > a \qquad \{a \mid a < 3\}$$

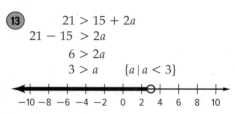

15. $\{w \mid w > 56\}$

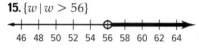

17. $\{w \mid w < -3\}$

19. $\left\{p \mid p > -\dfrac{24}{5}\right\}$

21. $\{h \mid h < -15\}$

23. Sample answer: Let $n =$ the number; $\dfrac{2}{3}n + 6 \geq 22$; $\{n \mid n \geq 24\}$. **25.** Sample answer: Let $n =$ the number; $8n - 27 \leq -n + 18$; $\{n \mid n \leq 5\}$. **27.** Sample answer: Let $n =$ the number; $3(n + 7) > 5n - 13$; $\{n \mid n < 17\}$

29. $\left\{n \mid n > -\dfrac{1}{3}\right\}$

31. ∅

33. $\{t \mid t \geq -1\}$

35. Sample answer: Let $s =$ the amount of sales made, $35{,}000 + 0.08s > 65{,}000$; $\{s \mid s > 375{,}000\}$; the sales must be more than \$375,000.

37.

$6(m - 3) > 5(2m + 4)$	Original inequality
$6m - 18 > 10m + 20$	Distributive Property
$6m - 18 - 6m > 10m + 20 - 6m$	Subtract $6m$ from each side.
$-18 > 4m + 20$	Simplify.
$-18 - 20 > 4m + 20 - 20$	Subtract 20 from each side.
$-38 > 4m$	Simplify.

$\dfrac{-38}{4} > \dfrac{4m}{4}$	Divide each side by 4.
$-9.5 > m$	Simplify.
$\{m \mid m < -9.5\}$	

39a. $5t + 565 \geq 1500$; $t \geq 187$

39b.

41 **a.** Words: temperature can be greater than 104
$$t > 104$$

b. $F > 104$
$$\dfrac{9}{5}C + 32 > 104$$
$$\dfrac{9}{5}C > 72$$
$$5\left(\dfrac{9}{5}C > 72\right)$$
$$9C > 360$$
$$C > 40$$

43. 1, 3, 5, 7; 3, 5, 7, 9; 5, 7, 9, 11; 7, 9, 11, 13

45. $\left\{x \mid x \geq \dfrac{1}{2}\right\}$ **47.** $\{m \mid m \geq 18\}$ **49.** $\{x \mid x \leq 8\}$

51. $\{x \mid x > -6\}$ **53.** $\{x \mid x \geq 1.5\}$ **55.** Add $3p$ and 2 to each side. The inequality becomes $9 \geq 3p$. Then divide each side by 3 to get $3 \geq p$.

57a. $\left\{x \mid x \geq -\dfrac{9}{2a}\right\}$ **57b.** $\left\{x \mid x > \dfrac{2}{1 + a}\right\}$

57c. $\{x \mid x < 6a\}$ **59.** Sample answer: The solution set for an inequality that results in a false statement is the empty set, as in $12 < -15$. The solution set for an inequality in which any value of x results in a true statement is all real numbers, as in $12 \leq 12$.
61. 100.5 **63.** C **65.** C

Lesson 5-4

1. $\{p \mid 12 \leq p \leq 16\}$

3. $\{a \mid a > 5\}$

5. $10 \text{ lb} \leq x \leq 22 \text{ lb}$

7

$n + 2 \leq -5$	$n + 6 \geq -6$
$n + 2 - 2 \leq -5 - 2 \quad$ and	$n + 6 - 6 \geq -6 - 6$
$n \leq -7$	$n \geq -12$

The solution set is $\{n \mid -12 \leq n \leq -7\}$.

9. $\{t \mid t \geq 1 \text{ or } t < -1\}$

11. $\{c \mid -1 \leq c < 2\}$

13. $\{m \mid m \text{ is a real number.}\}$

15. $\{y \mid y < -3\}$

17. Sample answer: Let $x =$ the smaller of two consecutive odd numbers, then $8 \leq 2x + 2 \leq 24$; $3 \leq x \leq 11$; 3, 5; 5, 7; 7, 9; 9, 11; 11, 13

19 The graph shows $x > -3$ and $x \leq 2$, so the inequality is $-3 < x \leq 2$.

21. $x < -4$ or $x > -3$ **23.** $x \leq -3$ or $x > 0$

25. $\left\{a \mid -3 < a \leq \dfrac{1}{2}\right\}$

27. $\{n \mid n < -3 \text{ or } n > -3\}$

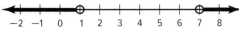

29. Sample answer: Let $n =$ the number; $5 \leq n - 8 \leq 14$; $\{n \mid 13 \leq n \leq 22\}$.   **31.** $-5n > 35$ or $-5n < 10$; $\{n \mid n < -7 \text{ or } n > -2\}$ **33.** $t < 75$ or $t > 90$
35. $t \geq 23$ and $t \leq 33$; $23 \leq t \leq 33$

37 a. Category 3 has wind speeds between 111 and 129: $111 \leq x \leq 129$.
Category 4 has wind speeds between 130 and 156: $130 \leq x \leq 156$.

b. The intersection is all values they have in common, which is none. The intersection is the empty set, $\varnothing$.

39. Neither; Chloe did not add 5 to 3, and Jonas did not add 5 to 7.   **41.** Sample answer: $x \leq 2$ or $x \geq 4$ **43.** Sample answer: The speed at which a roller coaster runs while staying on the track could represent a compound inequality that is an intersection.   **45.** C   **47.** C   **49.** B

Lesson 5-5

1. $\{a \mid 2 < a < 8\}$

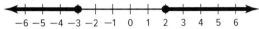

3. $\varnothing$

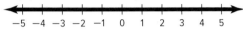

5. $\{n \mid n \leq -8 \text{ or } n \geq -2\}$

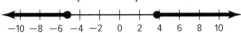

7. $\{m \mid 70.10 \leq m \leq 71.60\}$

9 $|r + 1| \leq 2$

Case 1:		Case 2:
$r + 1$ is positive.	and	$r + 1$ is negative.
$r + 1 \leq 2$	and	$r + 1 \geq -2$
$r \leq 1$	and	$r \geq -3$

The solution set is $\{r \mid -3 \leq r \leq 1\}$.

11. $\{h \mid -3 < h < 5\}$

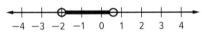

13. $\varnothing$

15. $\{k \mid k < 1 \text{ or } k > 7\}$

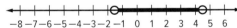

17. $\{p \mid p \leq -3 \text{ or } p \geq 2\}$

19. $\{c \mid c \text{ is a real number.}\}$

21. $\{n \mid n \leq -5\dfrac{1}{4} \text{ or } n \geq 3\dfrac{3}{4}\}$

23. $\{h \mid -5\dfrac{2}{3} < h < 5\}$

25. $\varnothing$

27. $\{r \mid -2 < r < \dfrac{2}{3}\}$

29. $\{h \mid -1.5 < h < 4.5\}$

31a. $\{t \mid t < 32 \text{ or } t > 212\}$

31b. $|t - 122| > 90$ **33.** $|x + 1| \leq 4$

35. $|x - 5.5| > 4.5$

37 $|g - 52| \leq 5$

Case 1:		Case 2:
$g - 52$ is positive.	and	$g - 52$ is negative.
$g - 52 \leq 5$	and	$g - 52 \geq -5$
$g \leq 57$	and	$g \geq 47$

The solution is $\{g \mid 47 \leq g \leq 57\}$.

39. $|t - 38| \leq 1.5$ **41.** $|c - 55| \leq 3$

43. No; Sample answer: Lucita forgot to change the direction of the inequality sign for the negative case of the absolute value.   **45.** Sample answer: If $t = 0$, then the absolute value is equal to 0, not greater than 0.

47. Sample answer: When an absolute value is on the left and the inequality symbol is $<$ or $\leq$, the compound sentence uses *and*, and if the inequality symbol is $>$ or $\geq$, the compound sentence used *or*. To solve, if $|x| < n$, then set up and solve the inequalities $x < n$ and $x > -n$, and if $|x| > n$, then set up and solve the inequalities $x > n$ or $x < -n$.   **49.** B   **51.** D   **53.** A   **55.** 3

Lesson 5-6

1.

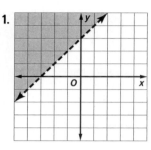

3.

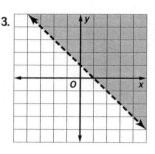

5.

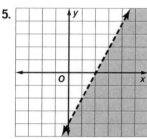

7. $y < x - 4$ **9a.** $115x + 685y > 2300$

9b. Sample answer: 1 skimboard and 4 longboards

11.

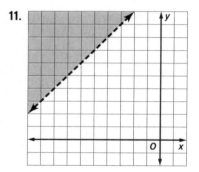

13.

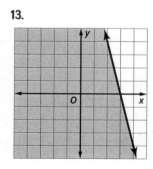

15.

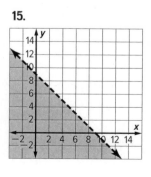

17.

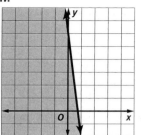

19. **a.** They need to make more than \$2000. Let x represent the number of hot dogs sold and let y represent the number of sodas sold.
$1x + 1.25y \geq 2000$

Step 1: First, solve for y in terms of x.

$$x + 1.25y \geq 2000$$
$$1.25y \geq -x + 2000$$
$$y \geq 0.80x + 2000$$

Then, graph $y = 0.80x + 2000$. Because the inequality involves $\geq$, graph the boundary with a solid line.

Step 2: Select a test point in either half-plane. A simple choice is $(0, 0)$.

$x + 1.25y \geq 2000$ Original inequality
$0 + 1.25(0) \stackrel{?}{\geq} 2000$ Substitution
$0 \not\geq 2000$ false

Step 3: Becausee this statement is not true, the half-plane containing the origin is not the solution. Shade the other half-plane.

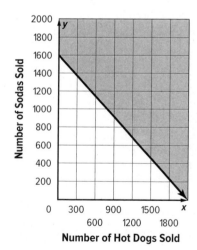

b. Sample answer: (400, 1600), (200, 1500), (300, 1400), (400, 1300), (1000, 1000)
Sample points should be in the shaded region of the graph in part **a**.

21. $y > \frac{1}{2}x + 1$

23. $x < -\frac{2}{3}$

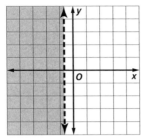

25. $x \le -\frac{2}{3}$

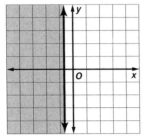

27. $x > \frac{3}{7}$

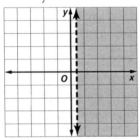

29a. Sample answer: 6 gallons of satin, 5 gallons of flat; First, I set up the inequality $40g + 20f \le 350$ to represent the amount Sybrina can spend. I also found that she must buy at least 10 gallons of paint to cover each of the rooms with two coats, so $g + f \ge 10$. Then, I solved both inequalities for f and graphed them. Next, I chose a test point to see that I should shade the graph below the first line and above the second line. If she paints her bedroom and her daughter's bedroom in satin, she will need 7 gallons of satin and 4 gallons of flat. This point is not in the shaded region, so it is not a solution. If Sybrina paints her bedroom and one of her son's bedrooms in satin, she will need 6 gallons of satin and 5 gallons of flat. This point is in the shaded region, so it is a solution. Sybrina can by 6 gallons of satin and 5 gallons of flat without exceeding her $350 budget.

29b. Sample answer: I assumed that Sybrina wanted to use satin in her bedroom. I also assumed that she could only buy paint in whole gallons.

31 $x < -4$
    Step 1: Graph $x = 4$ with a dotted line since it involves $<$.
    Step 2: Choose a test point in either half-plane. A simple choice is $(0, 0)$.
    $0 < -4$

Step 3: Because this statement is not true, the half-plane containing the origin is not the solution. Shade the other half-plane.

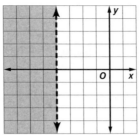 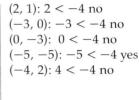

$(2, 1)$: $2 < -4$ no
$(-3, 0)$: $-3 < -4$ no
$(0, -3)$: $0 < -4$ no
$(-5, -5)$: $-5 < -4$ yes
$(-4, 2)$: $4 < -4$ no

33. Sample answer: $y \ge 4x - 2$

35. Sample answer: $y \le 3x - 2$

37a. $y \le 3x - 4$; $y \ge -x + 4$

37b.

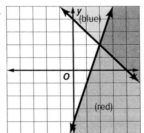

37c. The overlapping region represents the solutions that make both A and B true.

39. Sample answer: $y \le -x + 1$

41. Sample answer: The inequality $y > 10x + 45$ represents the cost of a monthly smartphone data plan with a flat rate of $45 for the first 2 GB of data used, plus $10 per each additional GB of data used. Both the domain and range are nonnegative real numbers because the GB used and the total cost cannot be negative.

43. D   **45a.** B   **45b.** -2 **47.** B

Chapter 5   Study Guide and Review

1. false; intersection   **3.** false; line   **5.** true
7. true   **9.** In math, the word *or* is inclusive, meaning the inequality is true if at least one inequality of the union is true.

11. $\{x \mid x \le -5\}$

$-7\ -6\ -5\ -4\ -3\ -2\ -1\ \ 0\ \ 1\ \ 2\ \ 3$

13. $\{a \mid a > -7\}$

$-9\ -8\ -7\ -6\ -5\ -4\ -3\ -2\ -1\ \ 0\ \ 1$

15. $\{y \mid y \le 7\}$

$0\ \ 1\ \ 2\ \ 3\ \ 4\ \ 5\ \ 6\ \ 7\ \ 8\ \ 9\ \ 10\ \ 11$

17. $\{x \mid x > 18\}$

$-10\ \ \ \ 0\ \ \ \ 10\ \ \ \ 20\ \ \ \ 30$

19. $\{p \mid p < 8\}$

$-12\ \ -6\ \ \ \ 0\ \ \ \ 6\ \ \ \ 12$

21. $\{m \mid m < -50\}$

$-120\ -80\ -40\ \ \ \ 0\ \ \ \ 40$

23. 24

25. $\{b \mid b > 6\}$

![number line with open circle at 6, shaded right, marks at -12, -6, 0, 6, 12]

27. $\{t \mid t > 6\}$

![number line with open circle at 6, shaded right, marks at -12, -6, 0, 6, 12]

29. at least 80 more

31. $\{t \mid 1 < t < 7\}$

![number line with open circles at 1 and 7, shaded between, marks -1 to 9]

33. $7 \leq x \leq 16$

35. $\{p \mid p < -9 \text{ or } p > 5\}$

![number line with open circles at -9 and 5, shaded outside, marks -10 to 10]

37. $\{f \mid f \leq 7 \text{ or } f \geq 11\}$

![number line with closed circles at 7 and 11, shaded outside, marks -2 to 18]

39. $\left\{b \mid -\dfrac{17}{2} < b < \dfrac{19}{2}\right\}$

![number line with open circles, shaded between, marks -10 to 10]

41. $\left\{y \mid -4 < y < \dfrac{5}{2}\right\}$

![number line with open circles at -4 and 5/2, shaded between, marks -5 to 5]

43. $\{k \mid k \geq -3 \text{ or } k \leq -11\}$

![number line with closed circles at -11 and -3, shaded outside, marks -14 to 8]

45.

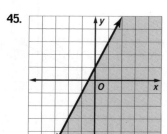

47.

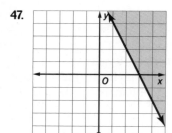

49.

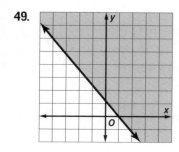

51. $(0, 4)$

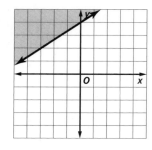

CHAPTER 6
Systems of Linear Equations and Inequalities

Get Ready

1. Yes, these expressions are the same.

3. Rewrite the expression as $\cdot \frac{7}{11}$.

5. Subtract $4y$ from each side to isolate the term with x on one side of the equation. **7.** $y = \frac{x-8}{2}$

Lesson 6-1

1. consistent and independent **3.** inconsistent

5. consistent and independent

7. 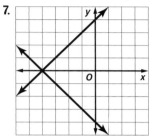 1 solution, $(-4, 0)$

9a. Alberto: $y = 20x + 35$; Ashanti: $y = 10x + 85$

9b.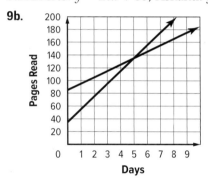

9c. (5, 135); Alberto will have read more after 5 days. **11.** consistent and independent

13 Because these two graphs intersect at one point, there is exactly one solution. Therefore, the system is consistent and independent.

15. consistent and independent

17. 1 solution; $(-1, -2)$ 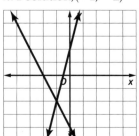 **19.** infinitely many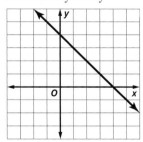

21. 1 solution; $(5, -1)$ 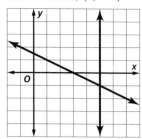 **23.** no solution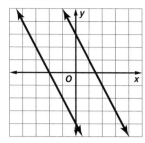

25a. Akira: $y = 30x + 22$; Jen: $y = 20x + 53$

25b.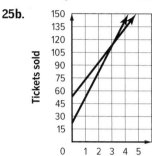

25c. (3.1, 115); After about 3 days Akira will have sold more tickets.

27 Graph the two equations on the same coordinate plane. The graphs appear to intersect at $(-4, -2)$. Check by substituting into the equations.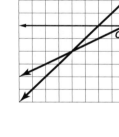

$y = \frac{1}{2}x$

$-2 = \frac{1}{2}(-4)$

$-2 = -2$ Yes

$y = x + 2$

$-2 = -4 + 2$

$-2 = -2$ Yes

So, the solution is $(-4, -2)$.

29. 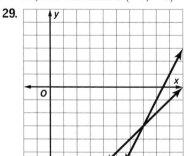 1 solution, $(7, -3)$

31. 1 solution, (5, 3)

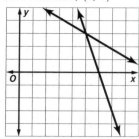

33. infinitely many

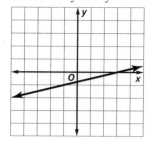

35. no solution

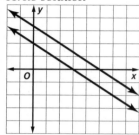

37. no solution

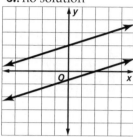

39a. Sample answers: At this rate, website A will have no visitors by 2069. Both websites will have the same number of visitors sometime during 2021. Website B will have 1 billion visitors by 2039.

39b. Sample answer: First, I found the average change in visitors per year for both websites: $\frac{476 - 512}{4}$ or $-9, \frac{251 - 131}{4}$ or 30. Next, I wrote equations to represent the situation: $y = -9x + 512$ and $y = 30x + 131$. Then, I graphed the equations and saw that they intersected between 9 and 10. So the number of visitors to the websites is the same sometime during the ninth year after 2012, or during 2021. I also saw from the graph when website A would reach 0 and website B would reach 1 billion.

39c. I assumed that these trends happened at a constant rate throughout the years, instead of the number of visitors going up and down over the years.

41.

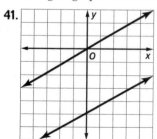

no solution

43. $y = 3x - 3$, $y = 3x + 4$; no solution

45. $y = -x + 2$, $y = 2x - 1$; (1, 1)

47 First solve each equation for y.

$2x + 3y = 5$	First equation
$3y = -2x + 5$	Subtract 2x.
$y = -\frac{2}{3}x + \frac{5}{3}$	Divide by 3.

$3x + 4y = 6$	Second equation
$4y = -3x + 6$	Subtract 3x.
$y = -\frac{3}{4}x + \frac{3}{2}$	Divide by 4.
$4x + 5y = 7$	Third equation
$5y = -4x + 7$	Subtract 4x.
$y = -\frac{4}{5}x + \frac{7}{5}$	Divide by 5.

Now graph all three new equations.

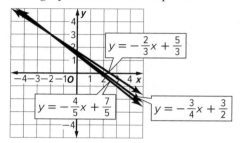

The graphs intersect at $(-2, 3)$, so this is the solution.

49. Always; if the equations are linear and have more than one common solution, they must be consistent and dependent, which means that they have an infinite number of solutions in common. **51.** Sample answers: $y = 5x + 3$; $y = -5x - 3$; $2y = 10x - 6$ **53.** A **55.** B
57a. $y = 7200 + 100x$; $y = 8850 - 50x$ **57b.** 2028
57c. 8300 people **57d.** Sample answer: Create a system of equations, graph the equations, and find the point of intersection. **57e.** Sample answer: I assumed that the populations continue to change at the same rate.
59. C

Lesson 6-2

1. (5, 10) **3.** (2, 0) **5.** infinitely many
7a. $x = m\angle X$, $y = m\angle Y$; $x + y = 180$, $x = 24 + y$
7b. $x = 102°$, $y = 78°$

9 Step 1: One equation is already solved for y.
$y = 4x + 5$
$2x + y = 17$
Step 2: Substitute $4x + 5$ for y in the second equation.
$2x + y = 17$
$2x + 4x + 5 = 17$
$6x + 5 = 17$
$6x = 12$
$x = 2$
Step 3: Substitute 2 for x in either equation to find y.
$y = 4x + 5$
$y = 4(2) + 5$
$y = 8 + 5$
$y = 13$
The solution is (2, 13).

11. $(-3, -11)$ **13.** $(-1, 0)$ **15.** infinitely many **17.** (2, 3)
19. no solution **21.** (2, 0) **23a.** Let d = demand for nurses; s = supply of nurses; t = number years; $d = 40{,}521t + 2{,}000{,}000$; $s = 5600t + 1{,}890{,}000$
23b. during 1996

25. a. Men: 8:21:00 = 8(60) + 21 = 501 and 8:18:37 = 8(60) + 18 = 498, then round up because the number of seconds is greater than 30. So, 8:18:37 rounds to 499. Women: 9:26:16 = 9(60) + 26 = 566, then because the number of seconds is less than 30, 9:26:16 rounds to 566 and 9:15:54 = 9(60) + 15 = 555, then round up because the number of seconds is greater than 30. So, 9:15:54 rounds to 556.

b. The y-intercept is (0, 501). Find the rate of change.

$$m = \frac{501 - 499}{0 - 12}$$
$$= \frac{2}{-12}$$
$$= -\frac{1}{6}$$

So, the equation is $y = -\frac{1}{6}x + 501$.

The y-intercept is (0, 566). Find the rate of change.

$$m = \frac{566 - 556}{0 - 12}$$
$$= \frac{10}{-12}$$
$$= -\frac{5}{6}$$

So, the equation is $y = -\frac{5}{6}x + 566$.

c. 2097; the graphs intersect around (97, 485)

27. Neither; Guillermo substituted incorrectly for b. Cara solved correctly for b but misinterpreted the pounds of apples bought. **29.** Sample answer: The solutions found by each of these methods should be the same. However, it may be necessary to estimate using a graph. So, when a precise solution is needed, you should use substitution. **31.** An equation containing a variable with a coefficient of 1 can easily be solved for the variable. That expression can then be substituted into the second equation for the variable. **33.** C
35. D **37a.** $x + y = 3(x - y)$; $x = y + 5$ **37b.** 5, 10
37c. The systems can be solved by substituting either $y + 5$ for x or $x - 5$ for y. **39.** C

Lesson 6-3

1. (2, 3)

3 Step 1: The like terms are already aligned.
$7f + 3g = -6$
$7f - 2g = -31$
Step 2: Subtract the equations.

$\begin{array}{r} 7f + 3g = -6 \\ (-)\ 7f - 2g = -31 \\ \hline 5g = 25 \\ g = 5 \end{array}$

Step 3: Substitute 5 for g in either equation to find f.
$7f + 3(5) = -6$
$7f + 15 = -6$
$7f = -21$
$f = -3$
The solution is (−3, 5).

5. 6, 18 **7.** (−3, 4) **9.** (−3, 1) **11.** (4, −2)
13. (8, −7) **15.** (4, 7) **17.** (4, 1.5) **19.** 5, 17

21
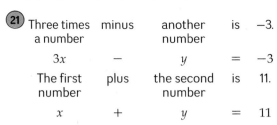

Three times a number	minus	another number	is	−3.
$3x$	−	y	=	−3

The first number	plus	the second number	is	11.
x	+	y	=	11

Steps 1 and 2: Write the equations vertically and add.

$\begin{array}{r} 3x - y = -3 \\ x + y = 11 \\ \hline 4x = 8 \\ x = 2 \end{array}$

Step 3: Substitute 2 for x in either equation to find y.
$x + y = 11$
$2 + y = 11$
$y = 9$
The numbers are 2 and 9.

23. adult, $17.95; children, $13.95 **25.** (2, −1)

27. $\left(-\frac{5}{6}, 3\right)$ **29.** $\left(2\frac{7}{9}, 13\frac{1}{3}\right)$

31a. $x + y = 66$; $x = 30 + y$

31b. (48, 18) **31c.** There are 48 teams that are not from the United States and 18 teams that are from the United States.

31d.
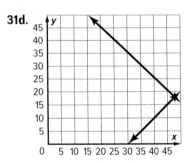

33 a. Sample answer: If you choose 4 pennies and 5 paper clips, the score will be 4(3) + 5 or 17.

b. The total number of objects is 9.
$p + c = 9$
Pennies are worth 3 points each and paper clips are worth 1 point each for a total of 15 points.
$3p + c = 15$
Solve:

$\begin{array}{r} p + c = 9 \\ (-)\ 3p + c = 15 \\ \hline -2p = -6 \\ p = 3 \end{array}$

Substitute 3 for p in either equation to find c.
$p + c = 9$
$3 + c = 9$
$c = 6$
So, $p = 3$ and $c = 6$.

p	c = 9 − p	3p + c
0	9	3(0) + 9 = 9
1	8	3(1) + 8 = 11
2	7	3(2) + 7 = 13
3	6	3(3) + 6 = 15
4	5	3(4) + 5 = 17
5	4	3(5) + 4 = 19

d. Yes; because the pennies are 3 points each, 3 of them makes 9 points. Add the 6 points from 6 paper clips and you get 15 points.

35. The result of the statement is false, so there is no solution. **37.** Sample answer: $-x + y = 5$; I used the solution to create another equation with the coefficient of the x-term being the opposite of its corresponding coefficient.

39. Sample answer: It would be most beneficial when one variable has either the same coefficient or opposite coefficients in each of the equations.

41. D **43.** E **45.** B

Lesson 6-4

1. (3, 2)

3 Eliminate y:

$(4x + 2y = -14)(-3)$ $-12x - 6y = 42$
$(5x + 3y = -17)(2)$ $\underline{10x + 6y = -34}$
 $-2x = 8$
 $x = -4$

Now, substitute -4 for x in either equation to find the value of y.

$4x + 2y = -14$
$4(-4) + 2y = -14$
$-16 + 2y = -14$
$2y = 2$
$y = 1$

The solution is $(-4, 1)$.

5. 6 mph **7.** $(-1, 3)$ **9.** $(-3, 4)$ **11.** $(-2, 3)$
13. (3, 5) **15.** $(1, -5)$ **17.** (0, 1)

19

Four times a number	minus	five times another number	equals	21.
$4x$	$-$	$5y$	$=$	21

Three times	the sum of the two numbers	is	36
3	$(x + y)$	$=$	36

$(4x - 5y = 21)(3)$ $12x - 15y = 63$
$(3(x + y) = 36)(5)$ $\underline{15x + 15y = 180}$
 $27x = 243$
 $x = 9$

Now substitute 9 for x in either equation to find y.

$4x - 5y = 21$
$4(9) - 5y = 21$

$36 - 5y = 21$
$-5y = -15$
$y = 3$ The two numbers are 9 and 3.

21. (2.5, 3.25) **23.** $\left(3, \dfrac{1}{2}\right)$

25a. Michelle should bake 7 tubes of cookies in 56 minutes and Julie should bake 3 tubes of cookies in 36 minutes. **25b.** Sample answer: Let $m =$ the number of tubes Michelle bakes. Let $j =$ the number of tubes Julie bakes. The number of cookies Michelle bakes plus the number of cookies Julie bakes should equal 264. Each batch Michelle makes produces 29 cookies. Each batch Julie makes produces 24 cookies. Therefore, $29m + 24j = 264$ relates the number of cookies baked to the number of tubes each girl uses. We also know that it takes Michelle's tubes 8 minutes to bake, and Julie's tubes take 12 minutes to bake. The girls have a little over 90 minutes to bake their cookies. So, $8m + 12j = 90$. Solve the system.

$29m + 24j = 264$ First equation
$\underline{(-)16m + 24j = 180}$ Multiply second equation by 2 and subtract.
$13m = 84$ j is eliminated.
$ m \approx 6.46$ Divide each side by 13.

It does not make sense for Michelle to bake part of a tube of cookie dough, so $m = 7$. Substituting 7 into the first equation, we determine that Julie bakes ≈ 2.54 or 3 tubes of cookies. **25c.** Sample answer: I assumed that Michelle and Julie cannot bake their cookies simultaneously. I assumed that one of the cookies that Michelle makes per tube is slightly smaller than the others. (You could have assumed that she ate the leftover cookie dough.) I assumed that the girls were able to bake an entire tube of cookie dough in one batch. I assumed that the amount of time needed to transfer baked cookies from the oven and unbaked cookies to the oven is negligible.

27 **a.** Let x be the cost of each trip to the batting cage and let y be the cost of each miniature golf game. For the first group, the equation is $16x + 3y = 44.81$. For the second group, the equation is $22x + 5y = 67.73$.
b. Solve.

$(16x + 3y = 44.81)(5)$ $80x + 15y = 224.05$
$(22x + 5y = 67.73)(-3)$ $\underline{-66x - 15y = -203.19}$
 $14x = 20.86$
 $x = 1.49$

Now, substitute 1.49 for x in either equation to find y.

$16x + 3y = 44.81$
$16(1.49) + 3y = 44.81$
$23.84 + 3y = 44.81$
$3y = 20.97$
$y = 6.99$

A trip to the batting cage costs $1.49 and a game of miniature golf costs $6.99.

29. One of the equations will be a multiple of the other. **31.** Sample answer: $2x + 3y = 6, 4x + 9y = 5$

33. Sample answer: It is more helpful to use substitution when one of the variables has a coefficient of 1 or if a coefficient can be reduced to 1 without turning other coefficients into fractions. Otherwise, elimination is more helpful because it will avoid the use of fractions when solving the system.   **35.** A   **37a.** x = number of cartons of cookies; y = number of cartons of candy bars; $9x + 12y = 1845$ and $8x + 15y = 2127.50$

37b. cookies: $\dfrac{\$55}{\text{carton}}$, candy: $\dfrac{\$112.50}{\text{carton}}$

37c. $5.50, $2.25

Lesson 6-5

1. elim ($\times$); $(2, -5)$ **3.** elim (+); $\left(-\dfrac{1}{3}, 1\right)$
5a. $y = -0.6x + 46.6$, $y = x - 11$
5b. The profit from pizza sales is equal to $46.60 minus the profit from sub sales. The number of pizzas sold is equal to the number of subs sold minus 11.
5c. $(36, 25)$; The debate team sold 36 subs and 25 pizzas.

7. subst.; $(2, -2)$ **9.** elim (−); $\left(1, -\dfrac{1}{2}\right)$

11 $-5x + 4y = 7$
$-5x - 3y = -14$

Because there are no coefficients of 1, elimination is the best method.

$(-5x + 4y = 7)(-1)$ $5x - 4y = -7$
$-5x - 3y = -14$ $\underline{-5x - 3y = -14}$
                                         $-7y = -21$
                                            $y = 3$

Now substitute 3 for y in either equation to find x.
$-5x + 4y = 7$
$-5x + 4(3) = 7$
$-5x + 12 = 7$
$-5x = -5$
$x = 1$
The solution is $(1, 3)$.
13. $g + b = 40$ and $g = 3b - 4$; 29 girls, 11 boys
15. 880 books; If they sell this number, then their income and expenses both equal $35,200.
17. $y = -2x + 3$, $y = x - 3$; $(2, -1)$

19 **a.** Let x be the cost per pound of the aluminum cans and y be the cost per pound of the newspapers.
For Mara, the equation is $9x + 26y = 3.77$.
For Ling, the equation is $9x + 114y = 4.65$.
b. Elimination is the best method for solving these equations.

$(9x + 26y = 3.77)(-1)$ $-9x - 26y = -3.77$
$9x + 114y = 4.65$ $\underline{9x + 114y = 4.65}$
                                        $88y = 0.88$
                                        $y = 0.01$

Now substitute 0.01 for y in either equation to find x.
$9x + 26y = 3.77$
$9x + 26(0.01) = 3.77$
$9x + 0.26 = 3.77$
$9x = 3.51$
$x = 0.39$

The aluminum cans are $0.39 per pound.
This solution is reasonable.
21a. $1.15   **21b.** $9.15   **23.** Sample answer: $x + y = 12$ and $3x + 2y = 29$, where x represents the cost of a student ticket for the basketball game and y represents the cost of an adult ticket; substitution could be used to solve the system; $(5, 7)$ means the cost of a student ticket is $5 and the cost of an adult ticket is $7.

25. Graphing: $(2, 5)$

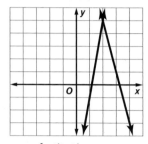

elimination by addition:	substitution:
$4x + y = 13$	$y = -4x + 13$
$6x - y = 7$	$6x - (-4x + 13) = 7$
$10x = 20$	$6x + 4x - 13 = 7$
$x = 2$	$10x = 20$
$4(2) + y = 13$	$x = 2$
$y = 5$	$4(2) + y = 13$
	$y = 5$

27. The third system; this system is the only one that is not a system of linear equations.   **29.** 12   **31.** C
33. C, D, G   **35.** 12, 17

Lesson 6-6

1. 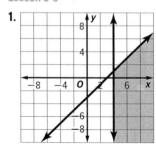   **3.**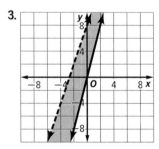

5. 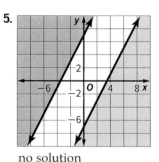   **7.**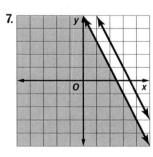

no solution

9a. Sample answer: Let h = the height of the driver in inches and w = the weight of the driver in pounds; $h < 79$ and $w < 295$

9a.

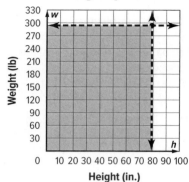

Driving Requirements

9b. Sample answer: 72 in. and 220 lb

9c. Yes, the point falls in the overlapping region.

11 Graph both inequalities on the same coordinate plane.

$y \geq 0$ has a solid line.

$y \leq x - 5$ has a solid line.

The solution is the intersection of the shading.

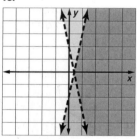

13.

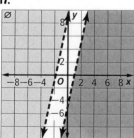

15.

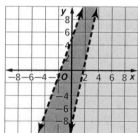

17.

no solution

19.

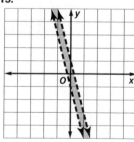

21.

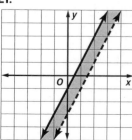

23.

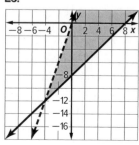

25a. Sample answer: Let f = square footage and let p = price; $1000 \leq f \leq 17{,}000$ and $10{,}000 \leq p \leq 150{,}000$

Ice Rink Resurfacers

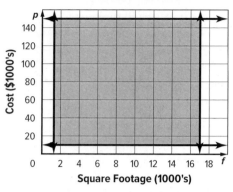

25b. Sample answer: an ice resurfacer for a rink of 5000 ft^2 and a price of \$20,000

25c. Yes; the point satisfies each inequality.

27.

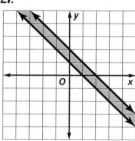

29.

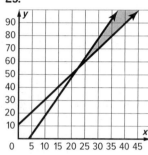

31.

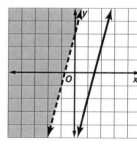

33.

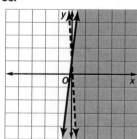

35.

37 **a.** Let x be the number of hours she works for a photographer and y be the number of hours she works coaching.

$x + y \leq 20$

$15x + 10y \geq 90$

b. Graph both inequalities on the same grid.
$x + y \le 20$ and $15x + 10y \ge 90$ have solid lines.
The solution is the intersection of the shading.

Earnings

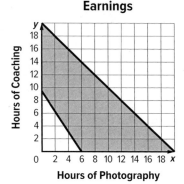

c. Two ordered pairs that are in the shaded area are (6, 10) and (8, 10). This means she could work for the photographer for 6 hours and coach for 10 or work for the photographer for 8 hours and coach for 10.

d. (2, 2) is not a solution because it does not fall in the shaded region. She would not earn enough money.

39. Sometimes; sample answer: $y > 3$, $y < -3$ will have no solution, but $y > -3$, $y < 3$ will have solutions.
41. Sample answer: $3x - y < -4$
43. Sample answer: The yellow region represents the beats per minute below the target heart rate. The blue region represents the beats per minute above the target heart rate. The green region represents the beats per minute within the target heart rate. Shading in different colors clearly shows the overlapping solution set of the system of inequalities.   **45.** A   **47.** C

Chapter 6   Study Guide and Review

1. true   **3.** false; dependent   **5.** true   **7.** false; system of inequalities   **9.** No, multiplication and division may also be used depending on the coefficients in the system of equations.

11. one; (3, 2)

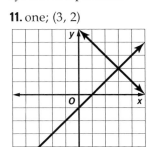

13. one; (0, 2)

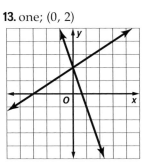

15.

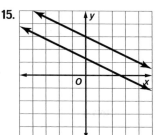

no solution

17. Sample answer: Let x be one number and y the other number; $x + y = 14$; $x - y = 4$; 9 and 5

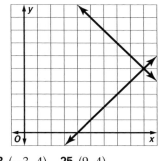

19. (2, −10)   **21.** (2, −6)   **23.** (−3, 4)   **25.** (9, 4)
27. (4, −2)   **29.** $\left(\frac{1}{2}, 6\right)$ **31.** (−3, 5)   **33.** Sample answer: Let f be the first type of card and let c be the second type of card; $f + c = 24$, $f + 3c = 50$; 11 \$1 cards and 13 \$3 cards.   **35.** (5, 7)   **37.** (2, 5)   **39.** (6, −1)
41. (1, −2)   **43.** Subs; (2, −6)   **45.** Subs; (24, −4)
47. Elim (−); (−2, 1)   **49.** Elim (×); (2, 5)   **51.** Sample answer: Let d represent the dimes and let q represent the quarters; $d + q = 25$, $0.10d + 0.25q = 4$; 15 dimes, 10 quarters

53.

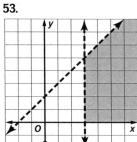

55.

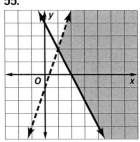

57.   **Jobs**

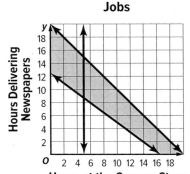

Exponents and Exponential Functions

Get Ready

1. Using exponents simplifies expressions.   **3.** The area would be in square inches.   **5.** The units would be feet cubed.   **7.** Yes, these expressions are different.

Lesson 7-1

1. Yes; constants are monomials.   **3.** No; there is a variable in the denominator.   **5.** Yes; this is a product of a number and variables.   **7.** k^4

9 $2q^2(9q^4) = (2 \cdot 9)(q^2 \cdot q^4)$
$$= 18q^{2+4}$$
$$= 18q^6$$

11. 3^8 or 6561   **13.** $16a^8b^{18}c^2$ **15.** $81p^{20}t^{24}$
17. $800x^8y^{12}z^4$ **19.** $-18g^7h^3j^{10}$ **21.** Yes; constants are monomials.   **23.** No; there is addition and more than one term.   **25.** Yes; this can be written as the product of a number and a variable.

27 $(q^2)(2q^4) = 2(q^2 \cdot q^4)$
$$= 2q^{2+4}$$
$$= 2q^6$$

29. $9w^8x^{12}$ **31.** $7b^{14}c^8d^6$ **33.** $j^{20}k^{28}$ **35.** 2^8 or 256
37. $4096r^{12}t^6$ **39.** $20c^5d^5$ **41.** $16a^{21}$ **43.** $512g^{27}h^{18}$
45. $294p^{27}r^{19}$ **47.** $30a^5b^7c^6$ **49.** $0.25x^6$ **51.** $-\frac{27}{64}c^3$
53. $-9x^3y^9$ **55.** $2{,}985{,}984r^{28}w^{32}$ **57a.** $0.12c$
57b. \$280   **59.** $15x^7$

61 a. $V = \pi r^2h$
$$= \pi(2p^3)^2(4p^3)$$
$$= \pi(22)(p^3)^2(4p^3)$$
$$= \pi(4)(p^6)(4p^3)$$
$$= \pi(4 \cdot 4)(p^6 \cdot p^3)$$
$$= \pi(16)(p^{6+3})$$
$$= 16\pi p^9$$

b.

Radius	Height	Volume
$4p$	p^7	$16\pi p^9$
$4p^2$	p^5	$16\pi p^9$
$2p^3$	$4p^3$	$16\pi p^9$
$2p^4$	$4p$	$16\pi p^9$
$2p$	$4p^7$	$16\pi p^9$

c. If the height of the container is doubled, the volume of the container is doubled. So, the volume is $32\pi p^9$.

63a.

Power	3^4	3^3	3^3	3^1	3^0	3^{-1}	3^{-3}	3^{-3}	3^{-4}
Value	81	27	9	3	1	$\frac{1}{3}$	$\frac{1}{9}$	$\frac{1}{27}$	$\frac{1}{81}$

63b. 1 and $\frac{1}{5}$ **63c.** $\frac{1}{a^n}$ **63d.** Any nonzero number raised to the zero power is 1.

65a.

Equation	Related Expression	Power of x	Linear or Nonlinear
$y = x$	x	1	linear
$y = x^2$	x^2	2	nonlinear
$y = x^3$	x^3	3	nonlinear

65b.

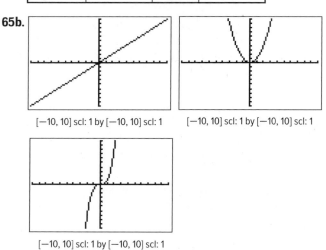

[−10, 10] scl: 1 by [−10, 10] scl: 1   [−10, 10] scl: 1 by [−10, 10] scl: 1

[−10, 10] scl: 1 by [−10, 10] scl: 1

65c. See chart for 65a.   **65d.** If the power of x is 1, the equation or its related expression is linear. Otherwise, it is nonlinear.   **67.** Sample answer: The area of a circle or $A = \pi r^2$, where r is the radius, can be used to find the area of any circle. The area of a rectangle or $A = w \times \ell$, where w is the width and ℓ is the length, can be used to find the area of any rectangle.   **69.** D   **71.** C   **73a.** B
73b. $30g^6h^3$

Lesson 7-2

1. t^3u^3

3 $\dfrac{m^6r^5p^3}{m^5r^2p^3} = \left(\dfrac{m^6}{m^5}\right)\left(\dfrac{r^5}{r^2}\right)\left(\dfrac{p^3}{p^3}\right)$
$$= m^{6-5}r^{5-2}p^{3-3}$$
$$= m^1r^3p^0$$
$$= mr^3$$

5. ghm **7.** xyz **9.** $\dfrac{4a^6b^{10}}{9}$ **11.** $\dfrac{32c^{15}d^{25}}{3125g^{10}}$ **13.** 1

15. $\dfrac{g^2h^4}{f^3}$ **17.** $\dfrac{a^5c^{13}}{3b^9}$ **19.** m^2p **21.** $\dfrac{r^4p^2}{4m^3t^4}$ **23.** $\dfrac{9x^2y^8}{25z^4}$

25. $\dfrac{p^6t^{21}}{1000}$ **27.** a^2b^7c

29 $\left(\dfrac{2r^3t^6}{5u^9}\right)^4 = \dfrac{2^4(r^3)^4(t^6)^4}{5^4(u^9)^4}$
$$= \dfrac{16r^{12}t^{24}}{625u^{36}}$$

31. 1   **33.** $\dfrac{p^4r^2}{t^3}$ **35.** $\dfrac{-f}{4}$ **37.** k^2mp^2 **39.** $\dfrac{3t^7}{u^6v^2}$
41. $\dfrac{r^3}{t^2x^{10}}$ **43.** 10^6; 10^9; about 10^3 or 1000 times as many users as servers   **45.** $-\dfrac{w^9}{3}$

47. $1600k^{13}$ **49.** $\dfrac{5q}{r^6 t^3}$ **51.** $\dfrac{4g^{12}}{h^4}$ **53.** $\dfrac{4x^8 y^4}{z^6}$

55. $\dfrac{16z^2}{y^8}$ **57.** 1000

59 a. The probability is $\dfrac{1}{6}$ multiplied d times, or $\left(\dfrac{1}{6}\right)^d$.

b. $\left(\dfrac{1}{6}\right)^d = \left(6^{-1}\right)^d$

$\qquad\qquad = 6^{-d}$

61. Sometimes; sample answer: The equation is true when $x = 1$, $y = 2$, and $z = 3$, but it is false when $x = 2$, $y = 2$, and $z = 3$.

63. $\dfrac{1}{x^n} = \dfrac{x^0}{x^n} = x^{0-n} = x^{-n}$

65. The Quotient of Powers Property is used when dividing two powers with the same base. The exponents are subtracted. The Power of a Quotient Property is used to find the power of a quotient. You find the power of the numerator and the power of the denominator.   **67.** C   **69.** D   **71.** C

Lesson 7-3

1. $\sqrt{12}$ **3.** $33^{\frac{1}{2}}$ **5.** 8 **7.** 7 **9.** 49

11 $216^{\frac{4}{3}} = \left(\sqrt[3]{216}\right)^4 = \left(\sqrt[3]{6 \cdot 6 \cdot 6}\right)^4 = 6^4$ or 1296

13. 4 **15.** 5.5 **17.** $\sqrt{15}$ **19.** $4\sqrt{k}$ **21.** $26^{\frac{1}{2}}$

23. $2(ab)^{\frac{1}{2}}$ **25.** 2 **27.** 6 **29.** 0.1 **31.** 11 **33.** 15

35. $\dfrac{1}{3}$ **37.** 4 **39.** 243 **41.** 625 **43.** $\dfrac{27}{1000}$ **45.** 5

47. $\dfrac{1}{2}$ **49.** $\dfrac{3}{2}$ **51.** 8 **53.** 8

55

$4^{3x} = 512$	Original equation
$(2^2)^{3x} = 2^9$	Write the expressions with a common base, 2.
$2^{6x} = 2^9$	Power of a Power Property
$6x = 9$	Power Property of Equality
$x = \dfrac{3}{2}$	Divide each side by 6.

57. 4 ft **59.** $\sqrt[3]{17}$ **61.** $7\sqrt[3]{b}$ **63.** $29^{\frac{1}{3}}$ **65.** $2a^{\frac{1}{3}}$

67. 0.3 **69.** a **71.** 16 **73.** $\dfrac{1}{3}$ **75.** $\dfrac{1}{27}$ **77.** $\dfrac{1}{\sqrt{k}}$

79. 12 **81.** -5 **83.** $-\dfrac{3}{2}$ **85a.** 440 Hz

85b. A below middle C, the 37th note

87

$r = 0.62V^{\frac{1}{3}}$	Original equation
$3.65 = 0.62V^{\frac{1}{3}}$	$r = \dfrac{7.3}{2}$ or 3.65
$\dfrac{3.65}{0.62} \approx V^{\frac{1}{3}}$	Divide each side by 0.62.
$\left(\dfrac{3.65}{0.62}\right)^{3 \cdot \frac{1}{3}} \approx V^{\frac{1}{3}}$	$\dfrac{3.65}{0.62} = \left(\dfrac{3.65}{0.62}\right)^{3 \cdot \frac{1}{3}}$
$\left(\dfrac{3.65}{0.62}\right)^3 \approx V$	Power Property of Equality
$204.0 \approx V$	Simplify.

$r = 0.62V^{\frac{1}{3}}$	Original equation
$3.8 = 0.62V^{\frac{1}{3}}$	$r = \dfrac{7.6}{2}$ or 3.8
$\dfrac{3.8}{0.62} \approx V^{\frac{1}{3}}$	Divide each side by 0.62.
$\left(\dfrac{3.8}{0.62}\right)^{3 \cdot \frac{1}{3}} \approx V^{\frac{1}{3}}$	$\dfrac{3.8}{0.62} = \left(\dfrac{3.8}{0.62}\right)^{3 \cdot \frac{1}{3}}$
$\left(\dfrac{3.8}{0.62}\right)^3 \approx V$	Power Property of Equality
$230.2 \approx V$	Simplify.

So the volume of a size 3 ball is 204.0 to 230.2 in³.

$r = 0.62V^{\frac{1}{3}}$	Original equation
$4.0 = 0.62V^{\frac{1}{3}}$	$r = \dfrac{8.0}{2}$ or 4.0
$\dfrac{4.0}{0.62} \approx V^{\frac{1}{3}}$	Divide each side by 0.62.
$\left(\dfrac{4.0}{0.62}\right)^{3 \cdot \frac{1}{3}} \approx V^{\frac{1}{3}}$	$\dfrac{4.0}{0.62} = \left(\dfrac{4.0}{0.62}\right)^{3 \cdot \frac{1}{3}}$
$\left(\dfrac{4.0}{0.62}\right)^3 \approx V$	Power Property of Equality
$268.5 \approx V$	Simplify.

$r = 0.62V^{\frac{1}{3}}$	Original equation
$4.15 = 0.62V^{\frac{1}{3}}$	$r = \dfrac{8.3}{2}$ or 4.15
$\dfrac{4.15}{0.62} \approx V^{\frac{1}{3}}$	Divide each side by 0.62.
$\left(\dfrac{4.15}{0.62}\right)^{3 \cdot \frac{1}{3}} \approx V^{\frac{1}{3}}$	$\dfrac{4.15}{0.62} = \left(\dfrac{4.15}{0.62}\right)^{3 \cdot \frac{1}{3}}$
$\left(\dfrac{4.15}{0.62}\right)^3 \approx V$	Power Property of Equality
$299.9 \approx V$	Simplify.

So the volume of a size 4 ball is 268.5 to 299.9 in³.

$r = 0.62V^{\frac{1}{3}}$	Original equation
$4.3 = 0.62V^{\frac{1}{3}}$	$r = \dfrac{8.6}{2}$ or 4.3
$\dfrac{4.3}{0.62} \approx V^{\frac{1}{3}}$	Divide each side by 0.62.
$\left(\dfrac{4.3}{0.62}\right)^{3 \cdot \frac{1}{3}} \approx V^{\frac{1}{3}}$	$\dfrac{4.3}{0.62} = \left(\dfrac{4.3}{0.62}\right)^{3 \cdot \frac{1}{3}}$
$\left(\dfrac{4.3}{0.62}\right)^3 \approx V$	Power Property of Equality
$333.6 \approx V$	Simplify.

$r = 0.62V^{\frac{1}{3}}$	Original equation
$4.5 = 0.62V^{\frac{1}{3}}$	$r = \dfrac{9.0}{2}$ or 4.5
$\dfrac{4.5}{0.62} \approx V^{\frac{1}{3}}$	Divide each side by 0.62.
$\left(\dfrac{4.5}{0.62}\right)^{3 \cdot \frac{1}{3}} \approx V^{\frac{1}{3}}$	$\dfrac{4.5}{0.62} = \left(\dfrac{4.5}{0.62}\right)^{3 \cdot \frac{1}{3}}$
$\left(\dfrac{4.5}{0.62}\right)^3 \approx V$	Power Property of Equality
$382.4 \approx V$	Simplify.

So the volume of a size 5 ball is 333.6 to 382.4 in³.

89. Sample answer: $2^{\frac{1}{2}}$ and $4^{\frac{1}{4}}$ **91.** $-1, 0, 1$
93. Sample answer: 2 is the principal fourth root of 16 because 2 is positive and $2^4 = 16$.   **95.** C   **97a.** 17.95 s
97b. $8.88 \dfrac{\text{m}}{\text{s}^2}$ **99.** B   **101.** A

Lesson 7-4

1. $2\sqrt{6}$ **3.** 10 **5.** $3\sqrt{6}$ **7.** $2x^2y^3\sqrt{15y}$

9. $3b^2|c|\sqrt{11ab}$ **11.** $\dfrac{9-3\sqrt{5}}{4}$ **13.** $\dfrac{2+2\sqrt{10}}{-9}$

15. $\dfrac{24+4\sqrt{7}}{29}$

17 $3\sqrt{5}+6\sqrt{5}=(3+6)\sqrt{5}$
$$=9\sqrt{5}$$

19. $-5\sqrt{7}$ **21.** $8\sqrt{5}$ **23.** $5\sqrt{2}+2\sqrt{3}$ **25.** $72\sqrt{3}$
27. $\sqrt{21}+3\sqrt{6}$ **29.** $14.5+3\sqrt{15}$ units²
31. $2\sqrt{14}$ **33.** 80 **35.** $45q^2\sqrt{q}$ **37.** $5r\sqrt{3qr}$
39. $4|g|h^2\sqrt{66}$ **41.** $4c^3d^2\sqrt{2}$

43 $\sqrt{\dfrac{32}{t^4}}=\dfrac{\sqrt{32}}{\sqrt{t^4}}$
$$=\dfrac{\sqrt{16\cdot2}}{t^2}$$
$$=\dfrac{\sqrt{16}\cdot\sqrt{2}}{t^2}$$
$$=\dfrac{4\sqrt{2}}{t^2}$$

45. $\dfrac{35-7\sqrt{3}}{22}$ **47.** $\dfrac{6\sqrt{3}+9\sqrt{2}}{2}$ **49.** $\dfrac{5\sqrt{6}-5\sqrt{3}}{3}$

51 **a.** $v=\sqrt{64h}$
$$=\sqrt{8\cdot8h}$$
$$=\sqrt{8^2}\sqrt{h}$$
$$=8\sqrt{h}$$

  b. $v=8\sqrt{h}$
$$=8\sqrt{134}$$
$$\approx92.6\text{ ft/s}$$

53. $11\sqrt{5}$ **55.** $5\sqrt{10}$ **57.** $3\sqrt{5}+6-\sqrt{30}-2\sqrt{6}$
59. $5\sqrt{5}+5\sqrt{2}$
61. $10\sqrt{7}+2\sqrt{5}$ units; 12 units² **63a.** $v=\dfrac{\sqrt{6k}}{3}$
63b. 73.5 joules **65.** $4-2\sqrt{3}$
67. $\dfrac{-4\sqrt{5}}{5}$ **69.** $\sqrt{2}$ **71.** $14-6\sqrt{5}$

73 **a.** $v_0=\sqrt{v^2-64h}$
$$=\sqrt{(120)^2-64(225)}$$
$$=\sqrt{0}$$
$$=0\text{ ft/s}$$

  b. Sample answer: In the formula, we are taking the square root of the difference, not the square root of each term.

75. Cross multiply and then divide. Rationalize the denominator to find that $x=\dfrac{5\sqrt{3}+9}{2}$.

77. Sample answer: $1+\sqrt{2}$ and $1-\sqrt{2}$; $(1+\sqrt{2})\cdot(1-\sqrt{2})=1-2=-1$

79. Irrational; irrational; no rational number could be added to or multiplied by an irrational number so that the result is rational. **81.** Sample answer: You can use the FOIL method. You multiply the first terms within the parentheses. Then you multiply the outer terms within the parentheses. Then you would multiply the inner terms within the parentheses. Then you would

R56

---

multiply the last terms within the parentheses. Combine any like terms and simplify any radicals. For example:
$$(\sqrt{2}+\sqrt{3})(\sqrt{5}+\sqrt{7})=\sqrt{10}+\sqrt{14}+\sqrt{15}+\sqrt{21}$$
83. B **85.** A **87.** C **89a.** 2.8% **89b.** $P=\dfrac{A}{(1+r)^t}$
89c. \$2678.02

Lesson 7-5

1.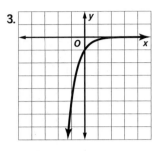
1;
D = {all real numbers};
R = {y | y > 0}; y = 0

3.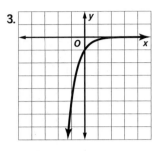
−1;
D = {all real numbers};
R = {y | y < 0}; y = 0

5.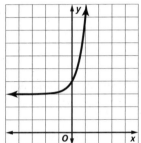
4;
D = {all real numbers};
R = {y | y > 3}; y = 3

7a. D = {t | t ≥ 0}, the number of days is greater than or equal to 0; R = {f(t) | f(t) ≥ 100}, the number of fruit flies is greater than or equal to 100. **7b.** about 198 fruit flies **9.** Yes; the domain values are at regular intervals, and the range values have a common factor of 4.

11.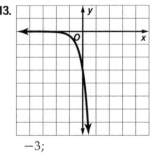
2;
D = {all real numbers};
R = {y | y > 0}; y = 0

13.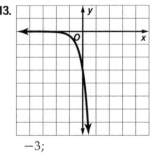
−3;
D = {all real numbers};
R = {y | y < 0}; y = 0

15.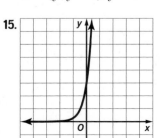
3;
D = {all real numbers};
R = {y | y > 0}; y = 0

17.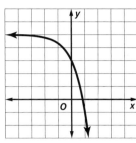

The y-intercept is -3.5; D = all real numbers; $R = \{y \mid y > -4\}$; $y = -4$

19.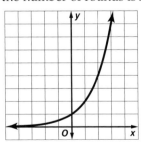

The y-intercept is 3; D = all real numbers; $R = \{y \mid y < 5\}$; $y = 5$

21. No; the domain values are at regular intervals of 4.

$2 \times (-2) = 4$
$-4 \times (-2) = 8$
$8 \times (-2) = -16$
$-16 \times (-2) = 32$

The range values differ by the common factor of -2. The range values do not have a positive common factor.

23. Yes; the domain values are at regular intervals, and the range values have a common factor of 2.

25. This enlargement is about 506% bigger than the original. **27.** exponential **29.** linear **31.** neither **33.** about 75 **35.** $y = 0$ **37.** $y = -3$ **39.** $y = 1$

41. $f(x) = 3(2)^x$

43. Sample answer: The number of teams competing in a basketball tournament can be represented by $y = 2^x$, where the number of teams competing is y and the number of rounds is x.

The y-intercept of the graph is 1. The graph increases quickly for $x > 0$. With an exponential model, each team that joins the tournament will play all of the other teams. If the scenario were modeled with a linear function, each team that joined would play a fixed number of teams.

45. Sample answer: First, look for a pattern by making sure that the domain values are at regular intervals and the range values differ by a common factor.

47. D **49a.** A **49b.** D

1. translated down 1 unit **3.** translated left 1 unit **5.** $g(x) = 2^{x-4}$ **7.** stretched vertically **9.** reflected across y-axis and compressed horizontally **11.** reflected across x-axis and stretched vertically **13.** reflected across y-axis, compressed vertically, and translated up 1 unit

15. For $g(x) = \left(\dfrac{5}{4}\right)^{4x} - 2$, the parent function is $f(x) = \left(\dfrac{5}{4}\right)^x$. $a = 4$, so $f(x)$ is compressed horizontally because $a > 1$ and multiplied before x is evaluated.
$f(x)$ is translated down 2 units because $k = -2$.

17. translated right 2 units **19.** stretched horizontally **21.** stretched vertically **23.** compressed vertically **25.** reflected across x-axis **27.** compressed vertically and translated right 1 unit **29.** reflected across y-axis and compressed horizontally **31.** reflected across y-axis, compressed vertically, and translated left 3 units **33.** compressed horizontally and translated down 1 unit **35.** reflected across x-axis, compressed vertically, and translated left 4 units and down 5 units

37. $f(x + n)$ indicates a horizontal translation of n units. All parent functions $f(x)$ pass through the point $(0, 1)$. $n = -4$ because $g(x)$ passes through $(-4, 1)$.

39. 0.5 **41.** -3 **43.** 3

45. 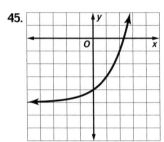 **47.**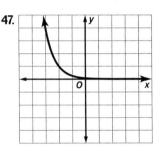

49. 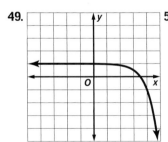 **51.**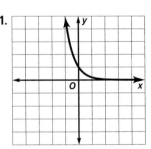

53. $m(x) = f(x) - 2$ **55.** $j(x) = -2f(x + 2)$ **57.** $n(x) = 5f(x)$

59. $j(x), g(x), p(x)$ **61a.** They have the same asymptote and end behavior as $x \to \infty$. For $y = \frac{2^x}{5}$, the function is decreasing as $x \to \infty$. The function $y = -\frac{2^x}{5}$, is increasing as $x \to \infty$. The y-intercepts are 1 and -1, respectively.

61. They have the same asymptote and y- intercept. For $y = \frac{2^x}{5}$, the function is decreasing as $x \to \infty$. The function $y = \frac{2^{-x}}{5}$ is increasing as $x \to \infty$.

63a. The graph is compressed vertically.

63b. The graph becomes more compressed and farther away from the y-axis.

63c. The graph is stretched vertically.

63d. The graph becomes more stretched and closer to the y-axis.

65. For any point (x, y) of the function, the reflected function will have $(-x, y)$.

67. Case 1: As x decreases, y still approaches 0, but as x increases, y goes to negative infinity instead of positive infinity. Case 2: As x decreases, y goes to negative infinity instead of positive infinity, but as x increases, y still approaches 0.

69. Sample answer: $y = -\left(\frac{1}{4}\right)^x - 2$

71. C   **73.** B, C

75. $g(x) = -7\left(\frac{1}{2}\right)^{x+1}$

77a. -2; reflects across the x-axis and vertically stretches the graph

77b. -1; translates the graph left 1 unit

77c. 4; translates the graph up 4 units

Lesson 7-7

1. $y = 0.5(3)^x$ **3.** $y = 4(3)^x$ **5.** initial: 240; rate of change: 65% per week; if this growth rate continues, then the number of likes will continue to increase, exceeding 100,000 in 12 weeks.   **7a.** $y = 2200(0.98)^t$
7b. about 1625   **9.** $y = 0.25(2)^x$ **11.** $y = 10(0.5)^x$
13. $y = -0.5(4)^x$ **15.** $y = 5(2)^x$
17. Initial: 82 grams of food; rate of change: decreasing by 35% per minute; if this decay rate continues, there will be less than a gram of food after 11 minutes.

19 $y = a(1 + r)^t$
   $= 300(1 + 0.05)^5$
   $= 300(1.05)^5$
   ≈ 382.88 or about \$382.88

21. \$3964.93
23. Sample answer: No; she will have about \$199.94 in the account in 4 years.
25. Sample answer: No; the car is worth about \$5774.61.

27 **a.** Write an equation to represent the loss of value: $I = 194.375(1 - 0.0425)^t$
   b. Solve the equation: about \$81,549
29a. $w(t) = 19,000(0.995)^t$ **29b.** $p(t) = 300t$
29c. $C(t) = 300t + 19,000(0.995)^t$; The function represents the number of gallons of water in the pool at any time after the hose is turned on.   **29d.** about 7.3 h
31. about 9.2 yr   **33.** Sample answer: Exponential models can grow without bound, which is usually not the case of the situation that is being modeled. For instance, a population cannot grow without bound due to space and food constraints. Therefore, when using a model, the situation that is being modeled should be carefully considered when used to make decisions.
35. A   **37.** \$240   **39.** A, B, D   **41.** A

Lesson 7-8

1 **a.** The interest rate is 3.1% and the initial investment is \$1, so
   $A(t) = a(1 + r)^t$
   $A(t) = 1(1 + 0.031)^t$
   $A(t) = 1.031^t$
   To write an equivalent function with an exponent of $4t$:
   $A(t) = 1.031^{\left(\frac{1}{4} \cdot 4\right)t}$
   $A(t) = (1.031^{\frac{1}{4}})^{4t}$
   $A(t) \approx (1.0077)^{4t}$
   b. The second equation shows 1.0077, or $1 + 0.0077$, as the base of the exponential expression. So the effective quarterly interest rate is 0.0077, or 0.77%.
   c. Oak Hill Financial; The effective quarterly rate of 0.77% is greater than the 0.7% quarterly rate at First City Bank.

3. World Mutual; The effective monthly interest rate is about 1.19%, which is lower than the monthly rate for Super City Card.

5 Write an equivalent function with an exponent of $12t$:

$$P(t) = 10,200(1.08)^t$$
$$P(t) = 10,200(1.08)^{\left(\frac{1}{12} \cdot 12\right)t}$$
$$P(t) = 10,200(1.08^{\frac{1}{12}})^{12t}$$
$$P(t) \approx 10,200(1.0064)^{12t}$$

The base of the exponential expression is 1.0064 or $1 + 0.0064$, so the effective monthly growth rate is 0.0064 or about 0.64%.

7a. about 25.5%   **7b.** about 19.77 crowns   **9.** No; in the expression 0.987^{12t}, the base of the expression represents a decrease of 1.3%, so the effective monthly decrease is about 1.3%, not 98.7%.   **11.** Sample answer: Any amount can be used for the initial investment, and $1 is convenient. Equivalent expressions used to compare rates of increase or decrease require transformations of the exponential expression, not the coefficient outside the parentheses. The initial amount does not have an effect on the rates you want to compare.   **13.** B, C, E   **15.** B
17. 1.1%

Lesson 7-9

1. Geometric; the common ratio is $\frac{1}{5}$.
3. Arithmetic; the common difference is 3.
5. 160, 320, 640   **7.** $-\frac{1}{16}, -\frac{1}{64}, -\frac{1}{256}$

9. $a_n = -6 \cdot (4)^{n-1}$; -1536 **11.** $a_n = 72 \cdot \left(\frac{2}{3}\right)^{n-1}$; $\frac{4096}{2187}$

13.

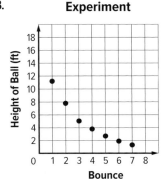

Experiment

15. Arithmetic; the common difference is 10.
17. Geometric; the common ratio is $\frac{1}{2}$.
19. Neither; there is no common ratio or difference.

21 Step 1: Find the common ratio.

$36 \times \frac{1}{3} = 12$

$12 \times \frac{1}{3} = 4$

The common ratio is $\frac{1}{3}$.

Step 2: Multiply each term by the common ratio to find the next three terms.

$4\left(\frac{1}{3}\right) = \frac{4}{3}$

$\left(\frac{4}{3}\right)\left(\frac{1}{3}\right) = \frac{4}{9}$

$\left(\frac{4}{9}\right)\left(\frac{1}{3}\right) = \frac{4}{27}$

So, the next three terms are $\frac{4}{3}, \frac{4}{9}$, and $\frac{4}{27}$.
23. $\frac{25}{4}, \frac{25}{16}, \frac{25}{64}$ **25.** $-2, \frac{1}{4}, -\frac{1}{32}$ **27.** 134,217,728
29. $-1,572,864$ **31.** 19,683   **33a.** the second option
33b. She should choose whichever option would earn her the most money over the summer. For the first option, $30 a week for 9 weeks would yield $270. The second option is a geometric sequence with a common ratio of 2, totaling $511 over 9 weeks. So, although this option starts out slow, it ends up being the most profitable.

35 Divide the 3rd term by the 2nd term to find the common ratio. The common ratio is $\frac{1}{3}$. Substitute 2 for n and $\frac{1}{3}$ for r to find the first term.

$$a_n = a_1 r^{n-1}$$
$$a_2 = a_1\left(\frac{1}{3}\right)^{2-1}$$
$$3 = a_1\left(\frac{1}{3}\right)^1$$
$$9 = a_1$$

The first term is 9. Find the 4th term.

$$a_n = a_1 r^{n-1}$$
$$a_4 = 9\left(\frac{1}{3}\right)^{4-1}$$
$$a_4 = 9\left(\frac{1}{3}\right)^3$$
$$a_4 = \frac{1}{3}$$

The fourth term is $\frac{1}{3}$.

37a.

Richter Number (x)	Increase in Magnitude (y)	Rate of Change (slope)
1	1	—
2	10	9
3	100	90
4	1,000	900
5	10,000	9000

37b.

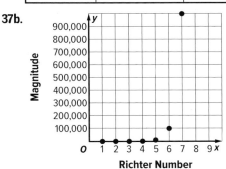

37c. The graph appears to be exponential. The rate of change between any two points does not match any others.   **37d.** $1 \cdot (10)^{x-1} = y$ **39.** Neither; Haro calculated the exponent incorrectly. Matthew did not calculate $(-2)^8$ correctly.   **41.** Sample answer: When graphed, the terms of a geometric sequence lie

on a curve that can be represented by an exponential function. They are different in that the domain of a geometric sequence is the set of natural numbers, while the domain of an exponential function is all real numbers. Thus, geometric sequences are discrete, while exponential functions are continuous. **43.** C
45. B **47.** A, D, E **49a.** 8295 **49b.** 13,054

Lesson 7-10

1. 16, 13, 10, 7, 4 **3.** $a_1 = 1, a_n = a_{n-1} + 5, n \geq 2$
5a. $a_1 = 10, a_n = 0.6a_{n-1}, n \geq 2$
5b. $a_n = 10(0.6)^{n-1}$

7 $a_n = 5n + 8$ is an explicit formula for an arithmetic sequence with $d = 5$ and $a_1 = 5(1) + 8$ or 13. Therefore, $a_1 = 13, a_n = a_{n-1} + 5, n \geq 2$.

9. $a_n = 22(4)^{n-1}$ **11.** 48, -16, 16, 0, 8

13. 12, 15, 24, 51, 132 **15.** $\frac{1}{2}, 2, \frac{7}{2}, 5, \frac{13}{2}$

17. $a_1 = 27, a_n = a_{n-1} + 14, n \geq 2$

19. $a_1 = 100, a_n = 0.8a_{n-1}, n \geq 2$

21. $a_1 = 81, a_n = \frac{1}{3}a_{n-1}, n \geq 2$

23. $a_1 = 3, a_n = 4a_{n-1}, n \geq 2$ **25.** $a_n = 38\left(\frac{1}{2}\right)^{n-1}$

27 **a.** Barbara was the first to receive the message, so the first term is 1. She then forwarded it to 5 of her friends, so the second term is 5. Each of her 5 friends forwarded the message to 5 more friends, so the third term is 5 · 5 or 25. This pattern continues. The fourth term is 25 · 5 or 125, and the fifth term is 125 · 5 or 625. Therefore, the first five terms are 1, 5, 25, 125, and 625.

b. There is a common ratio of 5. The sequence is geometric.
$a_n = r \cdot a_{n-1}$
$a_n = 5a_{n-1}$
The first term a_1 is 1, and $n \geq 2$. So, $a_1 = 1$,
$a_n = 5a_{n-1}, n \geq 2$.

c. We found that $a_5 = 625$.
$a_6 = 5a_{6-1}$
$= 5a_5$
$= 5(625)$ or 3125
$a_7 = 5a_{7-1}$
$= 5a_6$
$= 5(3125)$ or 15,625
$a_8 = 5a_{8-1}$
$= 5a_7$
$= 5(15,625)$ or 78,125

29a. $a_1 = 10, a_n = 1.1a_{n-2}, n \geq 2$ **29b.** 16.1 ft

31. Both; sample answer: The sequence can be written as the recursive formula $a_1 = 2, a_n = (-1)a_{n-1}, n \geq 2$. The sequence can also be written as the explicit formula $a_n = 2(-1)^{n-1}$.

33. False; sample answer: A recursive formula for the sequence 1, 2, 3, ... can be written as $a_1 = 1, a_n = a_{n-1} + 1, n \geq 2$ or as $a_1 = 1, a_2 = 2, a_n = a_{n-2} + 2, n \geq 3$.

35. Sample answer: In an explicit formula, the nth term a_n is given as a function of n. In a recursive formula, the nth term a_n is found by performing operations to one or more of the terms that precede it.
37. C **39a.** B **39b.** 1024

Chapter 7   Study Guide and Review

1. cube root **3.** exponential function **5.** exponential equation **7.** If the domain values are at equal intervals, find the common factors among each consecutive pair of range values. If the common factors are the same, the data are exponential.
9. x^9 **11.** $20a^6b^6$ **13.** $64r^{18}t^6$ **15.** $8x^{15}$

17. $45\pi x^4$ **19.** $\frac{27x^3y^9}{8z^3}$ **21.** $\frac{c^6}{a^3}$ **23.** x^6 **25.** $\frac{6}{yx^3}$

27. 7 **29.** 5 **31.** 64 **33.** 2401 **35.** 5
37. $6|x|y^3\sqrt{y}$ **39.** $3\sqrt{2}$ **41.** $21 - 8\sqrt{5}$ **43.** $\frac{5\sqrt{2}}{|a|}$
45. $-6 - 3\sqrt{5}$ **47.** $-2\sqrt{6} + 11\sqrt{3}$ **49.** $5\sqrt{2} + 3\sqrt{6}$
51. $24\sqrt{10} + 8\sqrt{2} + 6\sqrt{15} + 2\sqrt{3}$
53. y-intercept 1; D = {all real numbers}; R = $\{y \mid y > 0\}$; $y = 0$

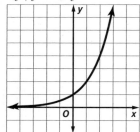

55. y-intercept 3; D = {all real numbers}; R = $\{y \mid y > 2\}$; $y = 2$

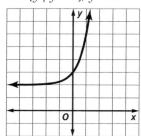

57. about 568 **59.** reflected across the x-axis, compressed vertically, and translated down 2 units
61. $g(x) = -2(4)^x$ **63.** $y = 4 \cdot (3)^x$ **65.** $3053.00
67. For First Bank, the effective yearly rate is about 1.8% which is greater than the yearly rate of 1.5% at Main Street Bank. **69.** about 0.6% **71.** 81, 243, 729
73. $a_n = -1(-1)^{n-1}$
75. $a_n = 256\left(\frac{1}{2}\right)^{n-1}$ **77.** 11, 7, 3, -1, -5
79. $a_1 = 2, a_n = a_{n-1} + 5, n \geq 2$
81. $a_1 = 2, a_n = 2a_{n-1} + 1, n \geq 2$

Chapter 8   Get Ready

1. A factor in front of parentheses multiplies across all terms inside the parentheses. $a(b + c) = ab + ac$

3. Multiply length by width; area $= a(b + 3c)$.   **5.** $-2, 6$

7. Yes; $-18y^7$

Lesson 8-1

1. yes; 3; trinomial   **3.** yes; 2; monomial

5. yes; 5; binomial   **7.** $2x^5 + 3x - 12$; 2

9. $-5z^4 - 2z^2 + 4z$; -5 **11.** $4x^3 + 5$

⑬ $(4 + 2a^2 - 2a) - (3a^2 - 8a + 7) = (2a^2 - 2a + 4) - (3a^2 - 8a + 7) = (2a^2 - 3a^2) + (-2a - (-8a)) + (4 - 7) = -a^2 + 6a - 3$

15. $-8z^3 - 3z^2 - 2z + 13$ **17.** $4y^2 + 3y + 3$

19a. $D(n) = 6n + 14$ **19b.** 116,000 students

19c. 301,000 students   **21.** yes; 0; monomial

23. No; the exponent is a variable.   **25.** yes; 4; binomial   **27.** $7y^3 + 8y$; 7

29. $-y^3 - 3y^2 + 3y + 2$; -1

31. $-r^3 + r + 2$; -1 **33.** $-b^6 - 9b^2 + 10b$; -1

㉟ $(2x + 3x^2) - (7 - 8x^2) = (2x + 3x^2) + (-7 + 8x^2)$
$\qquad = [3x^2 + 8x^2] + 2x + (-7)$
$\qquad = 11x^2 + 2x - 7$

37. $2z^2 + z - 11$ **39.** $-2b^2 + 2a + 9$

41. $7x^2 - 2xy - 7y$ **43.** $3x^2 - rxt - 8r^2x - 6rx^2$

45. quadratic trinomial   **47.** quartic binomial

49. quintic polynomial   **51a.** $s = 0.55t^2 - 0.05t + 3.7$

51b. 3030 students

53a. the area of the rectangle

53b. the perimeter of the rectangle

55. $10a^2 - 8a + 16$

57. $7n^3 - 7n^2 - n - 6$

㊾ **a.** Words: $25 plus $0.35 per mile
Expression: $25 + 0.35m$
The expression is $25 + 0.35m$.

b. $25 + 0.35m = 25 + 0.35(145)$
$\qquad = 25 + 50.75$
$\qquad = 75.75$
The cost is $75.75.

c. $247   **d.** $714

61. Neither; neither of them found the additive inverse correctly. All terms should have been multiplied by -1.   **63.** $6n + 9$ **65.** Sample answer: To add polynomials in a horizontal format, combine like terms. For the vertical format, write the polynomials in standard form, align like terms in columns, and combine like terms. To subtract polynomials in a horizontal format you find the additive inverse of the polynomial you are subtracting, and then combine like terms. For the vertical format, you write the polynomials in standard form, align like terms in columns, and subtract by adding the additive inverse.   **67.** C   **69.** $x^2 + 2x + 9$ **71.** $14x + 7$

Lesson 8-2

1. $-15w^3 + 10w^2 - 20w$

3. $32k^2m^4 + 8k^3m^3 + 20k^2m^2$

⑤ $2ab(7a^4b^2 + a^5b - 2a) = 2ab(7a^4b^2) + 2ab(a^5b) + 2ab(-2a)$
$\qquad = 14a^5b^3 + 2a^6b^2 + (-4a^2b)$
$\qquad = 14a^5b^3 + 2a^6b^2 - 4a^2b$

7. $4t^3 + 15t^2 - 8t + 4$

9. $-5d^4c^2 + 8d^2c^2 - 4d^3c + dc^4$ **11.** 30   **13.** $-\dfrac{20}{9}$

15. 20   **17.** 1   **19.** $f^3 + 2f^2 + 25f$

21. $10j^5 - 30j^4 + 4j^3 + 4j^2$

23. $8t^5u^3 - 40t^4u^5 + 8t^3u$

25. $-8a^3 + 20a^2 + 4a - 12$ **27.** $-9g^3 + 21g^2 + 12$

29. $8n^4p^2 + 12n^2p^2 + 20n^2 - 8np^3 + 12p^2$

㉛ $7(t^2 + 5t - 9) + t = t(7t - 2) + 13$
$7t^2 + 35t - 63 + t = 7t^2 - 2t + 13$
$7t^2 + 36t - 63 = 7t^2 - 2t + 13$
$\qquad 36t - 63 = -2t + 13$
$\qquad 38t = 76$
$\qquad t = 2$

33. $\dfrac{43}{6}$ **35.** $\dfrac{30}{43}$ **37.** $20np^4 + 6n^3p^3 - 8np^2$

39. $-q^3w^3 - 35q^2w^4 + 8q^2w^2 - 27qw$

41a. $100.5 - 0.5h$ **41b.** $94.50

㊸ **a.** $A = \ell w$
$\qquad = (1.5x + 24)x$
$\qquad = 1.5x^2 + 24x$

b. $x(x - 9) = x^2 - 9x$

c. $2(2.5x) = 2(2.5)(36)$
$\qquad = 180$ ft
$2(x + 6) = 2(36 + 6)$
$\qquad = 2(42)$
$\qquad = 84$ ft
Perimeter $= 180 + 84$ or 264 ft

45. Ted; Pearl used the Distributive Property incorrectly.

47. $8x^2y^{-2} + 24x^{-10}y^8 - 16x^{-3}$

49. Sample answer: $3n, 4n + 1$; $12n^2 + 3n$ **51.** B

53a. A   **53b.** C

53c. 20 m²

53d. $30x^3 + 45x^2 + 90x = 240 + 180 + 180 = 600$ m²

53e. 600 m² $\times$ $10/m² = $6000

Lesson 8-3

1. $x^2 + 7x + 10$ **3.** $b^2 - 4b - 21$ **5.** $16h^2 - 26h + 3$
7. $4x^2 + 72x + 320$ **9.** $16y^4 + 28y^3 - 4y^2 - 21y - 6$
11. $10n^4 + 11n^3 - 52n^2 - 12n + 48$ **13.** $2g^2 + 15g - 50$
(15) $(4x + 1)(6x + 3) = 4x(6x) + 4x(3) + 1(6x) + 1(3)$
$$= 24x^2 + 12x + 6x + 3$$
$$= 24x^2 + 18x + 3$$
17. $24d^2 - 62d + 35$ **19.** $49n^2 - 84n + 36$
21. $25r^2 - 49$ **23.** $33z^2 + 7yz - 10y^2$
25. $2y^3 - 17y^2 + 37y - 22$
27. $m^4 + 2m^3 - 34m^2 + 43m - 12$
29. $6b^5 - 3b^4 - 35b^3 - 10b^2 + 43b + 63$
31. $2m^3 + 5m^2 - 4$
33. $4\pi x^2 + 12\pi x + 9\pi - 3x^2 - 5x - 2$
(35) a. $A = \ell w$
$$= (3y + 4)(6y - 5)$$
$$= 3y(6y) + 3y(-5) + 4(6y) + 4(-5)$$
$$= 18y^2 - 15y + 24y - 20$$
$$= 18y^2 + 9y - 20$$
  b. $3y + 4 = 31$
$$3y = 27$$
$$y = 9$$
  So, the width is $6y - 5 = 6(9) - 5$
$$= 54 - 5$$
$$= 49$$
  $A = \ell w$
$$= (31)(49)$$
$$= 1519 \text{ ft}^2$$
37. $a^2 - 4ab + 4b^2$ **39.** $x^2 - 10xy + 25y^2$
41. $125g^3 + 150g^2h + 60gh^2 + 8h^3$
43a. $x > 4$; If $x = 4$, the width of the rectangular sandbox would be zero and if $x < 4$ the width of the rectangular sandbox would be negative.
43b. square   **43c.** 4 ft^2
45. Always; by grouping two adjacent terms, a trinomial can be written as a binomial, the sum of two quantities, and apply the FOIL method. For example, $(2x + 3)(x^2 + 5x + 7) = (2x + 3)[x^2 + (5x + 7)] = 2x(x^2) + 2x(5x + 7) + 3(x^2) + 3(5x + 7)$. Then use the Distributive Property and simplify.
47. Sample answer: $x - 1, x^2 - x - 1$.
$(x - 1)(x^2 - x - 1) = x^3 - 2x^2 + 1$
49. The Distributive Property can be used with a vertical or horizontal format by distributing, multiplying, and combining like terms. The FOIL method is used with a horizontal format. You multiply the first, outer, inner, and last terms of the binomials and then combine like terms. A rectangular method can also be used by writing the terms of the polynomials along the top and left side of a rectangle and then multiplying the terms and combining like terms.
51a. $x^2, (x + 3)^2$ or $x^2 + 6x + 9$ **51b.** 1   **51c.** 1 unit^2, 16 units^2 **53.** D   **55.** $8x^3 + 6x^2 - 3x - 2$
57a. $2x^3 - 11x^2 + 14x - 3$ **57b.** $6x^3 + 16x^2 + 15x - 3$
59. $8x^2 + 32x + 32$

Lesson 8-4

1. $x^2 + 10x + 25$
(3) $(2x + 7y)^2 = (2x)^2 + 2(2x)(7y) + (7y)^2$
$$= 4x^2 + 28xy + 49y^2$$
5. $g^2 - 8gh + 16h^2$ **7a.** $D^2 + 2Dy + y^2$ **7b.** 75%
9. $x^2 - 25$ **11.** $81t^2 - 36$ **13.** $b^2 - 12b + 36$
15. $x^2 + 12x + 36$ **17.** $81 - 36y + 4y^2$
19. $25t^2 - 20t + 4$ **21a.** $(T + t)^2 = T^2 + 2Tt + t^2$
21b. TT: 25%; Tt: 50%; tt: 25%
(23) $(b + 7)(b - 7) = b^2 - (7)^2$
$$= b^2 - 49$$
25. $16 - x^2$ **27.** $9a^4 - 49b^2$
29. $64 - 160a + 100a^2$ **31.** $9t^2 - 144$
33. $9q^2 - 30qr + 25r^2$ **35.** $g^2 + 10gh + 25h^2$
37. $9a^8 - b^2$ **39.** $64a^4 - 81b^6$
41. $\frac{4}{25}y^2 - \frac{16}{5}y + 16$ **43.** $4m^3 + 16m^2 - 9m - 36$
45. $2x^2 + 2x + 5$ **47.** $6x + 3$
49. $c^3 + 3c^2d + 3cd^2 + d^3$
51. $f^3 + f^2g - fg^2 - g^3$ **53.** $n^3 - n^2p - np^2 + p^3$
(55) a. $A = 3.14(r + 9)^2$
$$= 3.14(r^2 + 18r + 81)$$
$$\approx (3.14r^2 + 56.52r + 254.34) \text{ ft}^2$$
  b. $38^2 - (3.14r^2 + 56.52r + 254.34)$
$$= 1444 - 3.14r^2 - 56.52r - 254.34$$
$$\approx (1189.66 - 3.14r^2 - 56.52r) \text{ ft}^2$$
57. Sample answer: $(2c + d)(2c - d)$; The product of these binomials is a difference of two squares and does not have a middle term. The other three do.   **59.** 81
61. Sample answer: To find the square of a sum, apply the FOIL method or apply the pattern. The square of the sum of two quantities is the first quantity squared plus two times the product of the two quantities plus the second quantity squared. The square of the difference of two quantities is the first quantity squared minus two times the product of the two quantities plus the second quantity squared. The product of the sum and difference of two quantities is the square of the first quantity minus the square of the second quantity.
63a. $x^2 + 16x + 16$ **63b.** $25x^2 - 9$
63c. $-21x^2 + 16x + 7$
65. C   **67.** C   **69.** $(x^4 + 4x^2 + 4) \text{ yd}^2$

Lesson 8-5

1. $3(7b - 5a)$ **3.** $gh(10gh + 9h - g)$
(5) $np + 2n + 8p + 16 = (np + 2n) + (8p + 16)$
$$= n(p + 2) + 8(p + 2)$$
$$= (n + 8)(p + 2)$$
7. $(b + 5)(3c - 2)$ **9.** $(3k + 2)(m - 7)$
11. $(5p^3 - 3q)(2q - p)$
13a. $4t(3 - 4t)$ **13b.** 1.04 ft   **13c.** 2 ft
15. $8(2t - 5y)$

17. $2k(k + 2)$ **19.** $2ab(2ab + a - 5b)$

21 $fg - 5g + 4f - 20 = (fg - 5g) + (4f - 20)$
$$= g(f - 5) + 4(f - 5)$$
$$= (g + 4)(f - 5)$$

23. $(h + 5)(j - 2)$ **25.** $(9q - 10)(5p - 3)$
27. $(3d - 5)(t - 7)$ **29.** $(3t - 5)(7h - 1)$
31. $(r - 5)(5b + 2)$ **33.** $gf(5f + g + 15)$
35. $3cd(9d - 6cd + 1)$ **37.** $2(8u - 15)(3t + 2)$
39. $(5p + 2r)(4p + 3)$ **41.** $(3k^2 - 2)(3m - 5)$
43. $(6f + 1)(8g - 3)$
45a. ab **45b.** $(a + 6)(b + 6)$ **45c.** $6(a + b + 6)$
47a. $8t(55 - 2t)$ **47b.** 2800 ft, 2400 ft   **47c.** 3025 ft

49 $72t - 16t^2 = 8t(9) + 8t(-2t)$
$$= 8t(9 - 2t)$$
$$8(3)[9 - 2(3)] = 24(9 - 6)$$
$$= 24(3)$$
$$= 72 \text{ ft}$$
The height of the arrow after 3 seconds is 72 feet.

51a. 3 and -2

51b.

x^2	$+3x$
$-2x$	-6

51c.

	x	$+3$
x	x^2	$+3x$
-2	$-2x$	-6

$(x + 3)(x - 2)$

51d. Sample answer: Place x^2 in the top left-hand corner and place -40 in the lower right-hand corner. Then determine which two factors have a product of -40 and a sum of -3. Then place these factors in the box. Then find the factor of each row and column. The factors will be listed on the very top and far left of the box.

53. Sample answer:
$$4yz^2 + 24z + 5yz + 30 = (4yz^2 + 24z) + (5yz + 30)$$
$$= (4z + 5)(yz + 6)$$
$$4yz^2 + 24z + 5yz + 30 = 4yz^2 + 5yz + 24z + 30$$
$$= (4yz^2 + 5yz) + (24z + 30)$$
$$= (yz + 6)(4z + 5)$$

55. ac^2; $ac^2 (b^2c + a^6b + a^2c^4)$ **57.** $(5x^2 + 2)$ and $(2x + 3)$
59. A   **61.** D
63a. $(3y^2 + 2)(2y + 7)$ **63b.** $(4a + 3y)(3a - 5b)$
63c. $(w - 2)(4w^2 + 3z)$ **63d.** $3mn(5m - 9n)$
63e. $(4y + 5z)(2w + 3x)$ **65.** $x^2 + 2x - 8$

1. $(x + 2)(x + 12)$ **3.** $(n + 7)(n - 3)$ **5.** $2(2x - 3)(x - 6)$
7. $(3x + 2)(x + 5)$ **9.** prime   **11.** $(x + 3)(x + 14)$
13. $(a - 4)(a + 12)$ **15.** $(h + 4)(h + 11)$
17. $(x + 2)(x - 12)$

19   Find two numbers with a sum of 19 and a product of 48. $(2x + 3)(x + 8)$

21. $2(2x + 1)(x + 5)$ **23.** $(2x + 3)(x - 3)$ **25.** prime
27. $3(4x + 3)(x + 5)$ **29.** $(5x + 8)(x + 3)$
31. $(q + 2r)(q + 9r)$ **33.** $(x - y)(x - 5y)$ **35.** $4x + 48$
37. $-(2x + 5)(3x + 4)$ **39.** $-(x - 4)(5x + 2)$
41. prime   **43a.** a^2 and b^2 **43b.** $a^2 - b^2$
43c. width: $a - b$; length: $a + b$ **43d.** $(a - b)(a + b)$
43e. $(a - b)(a + b)$; the figure with area $a^2 - b^2$ and the rectangle with area $(a - b)(a + b)$ have the same area, so $a^2 - b^2 = (a - b)(a + b)$.   **45.** $-15, -9, 9, 15$
47. 7, 12, 15, 16

49   Sometimes; Sample answer: The trinomial $x^2 + 10x + 9 = (x + 1)(x + 9)$ and $10 > 9$. The trinomial $x^2 + 7x + 10 = (x + 2)(x + 5)$ and $7 < 10$.

51. $(4y - 5)^2 + 3(4y - 5) - 70 = [(4y - 5) + 10] \cdot [(4y - 5) - 7] = (4y + 5)(4y - 12) = 4(y + 5)(y - 3)$
53. 4 ft   **55.** B   **57.** A   **59a.** 524; 56, 100
59b. D   **59c.** $(9x - 13)$ ft and $(3x + 1)$ ft
59d. 115 ft by 167 ft; 16 ft by 32 ft   **61.** $x - 1$

1. yes; $(q + 11)(q - 11)$ **3.** no   **5.** yes; $(4m + k^2)(4m - k^2)$
7. yes; $(5x + 6)^2$ **9.** yes; $(y^2 + 1)^2$
11. $(u + 3)(u - 3)(u^2 + 9)$

13 $20r^4 - 45n^4 = 5(4r^4 - 9n^4)$
$$= 5\left((2r^2)^2 - (3n^2)^2\right)$$
$$= 5(2r^2 + 3n^2)(2r^2 - 3n^2)$$

15. $(c + 1)(c - 1)(2c + 3)$
17. $(t + 4)(t - 4)(3t + 2)$
19a. $(4n + 1)^2 - 5^2$
19b. $(4n + 6)$ by $(4n - 4)$
21. $2m(2m - 7)(3m + 5)$ **23.** $3(2x - 7)^2$
25. $3p(2p + 1)(2p - 1)$ **27.** $(a + 7)(a - 7)$
29. $3(m^4 + 81)$
31. $2(a + 4)(a - 4)(6a + 1)$
33. $(24 + x)$ by $(24 - x)$; 96 ft
35. $(x + 2y)(x - 2)(x + 2)$
37. $(r - 6)(r + 6)(2r - 1)$
39. $2cd(c^2 + d^2)(2c - 5)$

41 $x^4 + 6x^3 - 36x^2 - 216x = x^3(x + 6) - 36x(x + 6)$
$= (x^3 - 36x)(x + 6) = x(x^2 - 36)(x + 6) = x(x + 6)$
$(x - 6)(x + 6) = x(x + 6)^2(x - 6)$

43. $(y - 2)(y + 2)(y^2 + 4)(y^2 + 16)$

45a. $h(h - 6)^2$; h, $(h - 6)$, $(h - 6)$ **45b.** Yes; because length cannot be negative, the value of h has to be greater than 6 so that $(h - 6)$ is greater than 0. So the sides of the box are each greater than 6 inches.

47. Lorenzo; sample answer: Checking Elizabeth's answer gives us $16x^2 - 25y^2$. The exponent on x in the final product should be 4.

49. Sample answer: $x^2 - 3x + \frac{9}{4} = 0$; $\left\{\frac{3}{2}\right\}$

51. When the difference of squares pattern is multiplied together using the FOIL method, the outer and inner terms are opposites of each other. When these terms are added together, the sum is zero.

53. $4x^2 + 10x + 4$ because it is the only expression that is not a perfect square trinomial.   **55.** A

57. $|4x + 5|$ **59.** B   **61.** Sample answer: $3x$, $(2x - 7)$, $(2x + 7)$

Chapter 8   Study Guide and Review

1. false; sample answer: $x^2 + 5x + 7$ **3.** true

5. true   **7.** true   **9.** false; difference of squares

11. difference of squares

13. prime polynomial   **15.** $-x^4 + 1$

17. $3x^5 + x^3 - 2x^2 + 6x - 2$ **19.** $-a^2 + 9a - 6$

21. $4x^2 + 4x + 8$ **23.** 1   **25.** $3x^3 + 3x^2 - 21x$

27. $18a^2 + 3a - 10$ **29.** $10x^2 + 29x + 10$

31. $x^2 - 25$ **33.** $25x^2 + 40x + 16$ **35.** $4r^2 + 20rt + 25t^2$

37. $3x^2 - 21$ **39.** $7xy(2x - 3 + 5y)$ **41.** $(a + b)(a - 4c)$

43. $(3a + 5b)(8m - 3n)$ **45.** $(3r^2 + p)(r - 4)$

47. $3f^4g^2(6g^3 - f^2 + 3g)$

49. $(x - 5)(x - 3)$ **51.** $(x - 6)(x + 1)$ **53.** $(x + 10)(x - 5)$

55. $(x + 4)(x + 8)$ **57.** $(x + 10)(x + 1)$

59. $2(2x - 1)(3x + 7)$ **61.** $3(x - 5)(x + 3)$

63. $(4x - 3)(5x + 4)$ **65.** $(3x + 2)(x - 5)$ **67.** $3x + 7$

69. $(8 + 5x)(8 - 5x)$ **71.** $3(x + 1)(x - 1)$

73. $(3x - 5)(3x + 5)$ **75.** prime   **77.** $(2 - 7a)^2$

79. $x^2(x + 4)(x - 4)$ **81.** $-3(x + 2)^2$

Chapter 9   Get Ready

1. A table can organize data so that you can determine points to use for creating a graph.

3. Plot each point given by the (x, y) pairs, then draw a straight line through all the points.

5.

x	y
0	3
1	2
2	1

7. Sample answer: Find two integers that have a sum of b and a product of c.

Lesson 9-1

1.

x	y
−3	0
−2	−6
−1	−8
0	−6
1	0
2	10

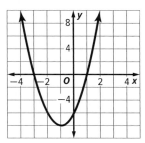

$D = \{-\infty < x < \infty\}; R = \{y \mid y \geq -8\}$

3.

x	y
−1	4
0	−3
1	−8
2	−11
3	−12
4	−11
5	−8
6	−3
7	4

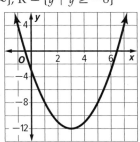

$D = \{-\infty < x < \infty\};$
$R = \{y \mid y \geq -12\}$

5. vertex $(2, 0)$, axis of symmetry $x = 2$, y-intercept 4, zero $(2, 0)$ **7.** vertex $(-2, -1)$, axis of symmetry $x = -2$, y-intercept 3, zeros $(-1, 0)$, $(-3, 0)$ **9.** vertex $(1, 2)$, axis of symmetry $x = 1$, y-intercept -1 **11.** vertex $(2, 1)$, axis of symmetry $x = 2$, y-intercept 5

13 **a.** Because the a value is 3, the graph opens downward and has a maximum.
  b. In this equation $a = -1$, $b = 4$, and $c = -3$.
  $$x = \frac{-b}{2a}$$
  $$= \frac{-4}{2(-1)}$$
  $$= 2$$

  To find the vertex, use the value you found for the x-coordinate of the vertex. To find the y-coordinate, substitute the value for x in the original equation.

$$y = -x^2 + 4x - 3$$
$$= -(2)^2 + 4(2) - 3$$
$$= -4 + 8 - 3$$
$$= 1$$

The maximum is at $(2, 1)$.
  c. The domain of the function is $D = \{-\infty < x < \infty\}$. The range is $R = \{y \mid y \leq 1\}$.

15a. maximum   **15b.** 6   **15c.** $D = \{\text{all real numbers}\}$; $R = \{y \mid y \leq 6\}$

17.

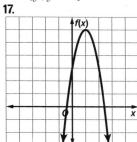

19.

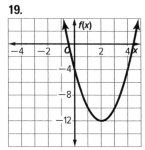

21a.

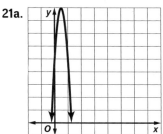

21b. 5 ft   **21c.** 9 ft

23.

x	y
−3	13
−2	7
−1	5
0	7
1	13

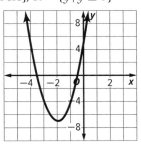

$D = \{\text{all real numbers}\}; R = \{y \mid y \geq 5\}$

25.

x	y
0	5
−1	−4
−2	−7
−3	−4
−4	5

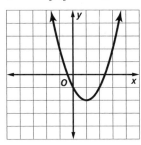

$D = \{-\infty < x < \infty\}; R = \{y \mid y \geq -7\}$

27.

x	y
3	2
2	−1
1	−2
0	−1
−1	2

$D = \{-\infty < x < \infty\}; R = \{y \mid y \geq -2\}$

29. vertex $(2, -4)$, axis of symmetry $x = 2$, y-intercept 0, zeros $(0, 0)$ and $(4, 0)$

31. vertex $(1, 1)$, axis of symmetry $x = 1$, y-intercept 4, no real zeros

33. vertex $(0, 0)$, axis of symmetry $x = 0$, y-intercept 0, zero $(0, 0)$

35 In this equation $a = 2$, $b = 12$, and $c = 10$.
$$x = \frac{-b}{2a} \qquad = \frac{-12}{2(2)}$$
The equation for the axis of symmetry is
$x = -3$.
To find the vertex, use the value you found for the axis of symmetry as the x-coordinate of the vertex. To find the y-coordinate, substitute the value for x in the original equation.
$$y = 2x^2 + 12x + 10$$
$$= 2(-3)^2 + 12(-3) + 10$$
$$= -8$$
The vertex is at $(-3, -8)$.
The y-intercept occurs at $(0, c)$. So, in this case, the y-intercept is 10.

37. vertex $(-3, 4)$, axis of symmetry $x = -3$, y-intercept -5 **39.** vertex $(2, -14)$, axis of symmetry $x = 2$, y-intercept 14 **41.** vertex $(1, -15)$, axis of symmetry $x = 1$, y-intercept -18

43a. maximum **43b.** 9

43c. $D = \{-\infty < x < \infty\}$, $R = \{y \mid y \le 9\}$

45a. minimum **45b.** -48

45c. $D = \{\{-\infty < x < \infty\}$, $R = \{y \mid y \ge -48\}$

47a. maximum **47b.** 33

47c. $D = \{-\infty < x < \infty\}$, $R = \{y \mid y \le 33\}$

49a. maximum **49b.** 4

49c. $D = \{-\infty < x < \infty\}$, $R = \{y \mid y \le 4\}$

51a. maximum **51b.** 3

51c. $D = \{-\infty < x < \infty\}$, $R = \{y \mid y \le 3\}$

53.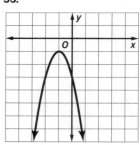

55.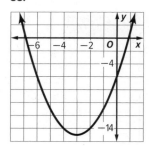

57.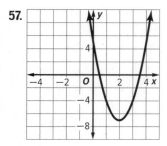

59. 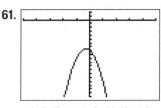 $(-1.25, -0.25)$

$[-5, 5]$ scl:1 by $[-5, 5]$ scl:1

61. 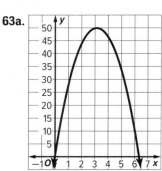 $(-0.3, -7.55)$

$[-5, 5]$ scl: 1 by $[-20, 2]$ scl: 2

63a. Where $h > 0$, the ball is above the ground. The height of the ball decreases as more time passes.

63b. 0 m **63c.** ≈ 50.0 m **63d.** ≈ 6.4 seconds

63e. $D = \{t \mid 0 \le t \le 6.4\}$; $R = \{h \mid 0 \le h \le 50.0\}$

65 **a.** $h = -16t^2 + 90t$
$$= -16(1)^2 + 90(1)$$
$$= 74 \text{ ft}$$
b. $126 = -16t^2 - 90t$
$$0 = -16t^2 - 90t - 126$$
$$0 = (t - 3)(-16t + 42)$$
$$t = 3 \text{ and } t = 2.625$$
c. $h = -16t^2 + 90t$
$$0 = -16t^2 + 90t$$
$$0 = -16t(t - 5)$$
$$t = 0 \text{ and } t = 5.625$$
These represent the time that the ball leaves the ground initially and the time it returns to the ground.

67 **a.**

Equation	Related Function	Zeros	y-Values
$x^2 - x = 12$	$y = x^2 - x - 12$	$-3, 4$	$-3: 8, -6; 4:$ $-6, 8$
$x^2 + 8x = 9$	$y = x^2 + 8x - 9$	$-9, 1$	$-9: 11, -9; 1:$ $-9, 11$
$x^2 = 14x - 24$	$y = x^2 - 14x + 24$	$2, 12$	$2: 11, -9; 12:$ $-9, 11$
$x^2 + 16x = -28$	$y = x^2 + 16x + 28$	$-14, -2$	$-14: 13, -11;$ $-2: -11, 13$

b. Use a graphing calculator to graph.

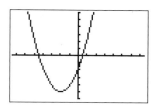

$$y = x^2 - x - 12$$

[−10, 10] scl: 1 by [−10, 10] scl: 1

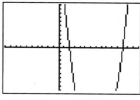

$$y = x^2 + 8x - 9$$

[−15, 15] scl: 2 by [−30, 30] scl: 5

$$y = x^2 - 14x + 24$$

[−10, 15] scl: 1 by [−10, 10] scl: 1

$$y = x^2 + 16x + 28$$

[−15, 10] scl: 1 by [−10, 10] scl: 1

c. Use the table function on the calculator to identify the zeros. The zeros are the values where y is 0. Also list the y-values that are to the left and right of the zeros.

d. The function values have opposite signs just before and just after the zeros.

69. Chase; the lines of symmetry are $x = 2$ and $x = 1.5$.   **71.** $(−1, 9)$; Sample answer: I graphed the points given, and sketched the parabola that goes through them. I counted the spaces over and up from the vertex and did the same on the opposite side of the line $x = 2$.   **73.** Sample answer: The function $y = −x^2 − 4$ has a vertex at $(0, −4)$, but it is a maximum.
75. B   **77.** 3   **79a.** D   **79b.** A   **79c.** C   **79d.** B, D, E, F

Lesson 9-2

1. translated down 11 units   **3.** reflected across the x-axis, translated up 8 units   **5.** reflected across the x-axis, translated left 3 units and stretched vertically   **7.** Because $a = −0.204$, the parent function is reflected across the x-axis and compressed vertically. $h = 6.2$ and $k = 13.8$, so the parent function is translated right 6.2 units and up 13.8 units.
9. $f(x) = 4(x − 1)^2 − 1$, $f(x) = 4x^2 − 8x + 3$

11 The function can be written as $f(x) = ax^2 + k$ where $a = −1$ and $k = −7$. Because $−7 < 0$ and

$−1 < 0$, the graph of $y = −x^2 − 7$ translates the graph of $y = x^2$ down 7 units and reflects it across the x-axis.
13. compressed vertically, translated up 6 units
15. stretched vertically, translated up 3 units
17. translated left 1 unit and up 2.6 units and stretched vertically   **19a.** $c(x)$ is reflected across the x-axis, stretched vertically, and translated right 2.5 units and up 105 units. $p(x)$ is reflected across the x-axis, stretched vertically, and translated right 2.8 units and up 126.5 units.   **19b.** Paulo's   **19c.** Paulo's
21. $f(x) = \frac{4}{9}(x + 1)^2 − 4$, $f(x) = \frac{4}{9}x^2 + \frac{8}{9}x − \frac{32}{9}$
23. $f(x) = \frac{1}{3}(x + 5)^2 − 9$, $f(x) = \frac{1}{3}x^2 + \frac{10}{3}x − \frac{2}{3}$
25 a. The two equations are $h = −16t^2 + 300$ and $h = −16t^2 + 700$.
b. $0 = −16t^2 + 300$, $t \approx 4.3$; $0 = −16t^2 + 700$, $t \approx 6.6$; $6.6 − 4.3 \approx 2.3$ seconds
27a. The graph of $g(x)$ is the graph of $f(x)$ translated 200 yards right and 20 yards up, compressed vertically, and reflected across the x-axis   **27b.** $h(x) = 0.0005(x − 230)^2 + 20$ **29.** Translate the graph of $f(x)$ up 7 units and to the right 2 units.   **31b.** For $f(x) = x^2$, when $b > 1$, the graph is compressed horizontally.   **31c.** For $f(x) = x^2$, when $0 < b < 1$, the graph is stretched horizontally.
31d. For $f(x) = x^2$, when $b < 0$, the squared term offsets the negative. For example, when $b = −3$, $(−3x)^2 = 9x^2$. This is the same transformation as when $b = 3$. Therefore, the sign of b has no effect on the transformation of this function.
33. Graph 1: $y = −\frac{1}{3}x^2$; Graph 2: $y = −2x^2$.   **35.** The graph of $y = 3x^2$ is narrower than the graph of $y = \frac{1}{2}x^2$.
37. Both graphs have the same shape, but the graph of $y = −3x^2$ opens down while the graph of $y = 3x^2$ opens up.   **39.** Both graphs have the same shape, but the graph of $y = x^2 + 2$ is translated up 2 units from the graph of $y = x^2$ and the graph of $y = x^2 − 1$ is translated down 1 unit from the graph of $y = x^2$.
41. Both graphs have the same shape, but the graph of $y = x^2 − 4$ is translated down 4 units from the graph of $y = x^2$ and the graph of $y = (x − 4)^2$ is translated right 4 units from the graph of $y = x^2$.
43. $y = x^2 − 1$ **45.** Sample answer: $f(x) = −\frac{1}{2}x^2$;
47. C   **49.** −4   **51a.** 0.009   **51b.** −7.05   **51c.** −0.071
51d. B, E, F

Lesson 9-3

1. 2, −5

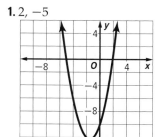

3. −2

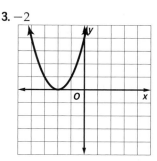

5. −8, 1

7. 5, −5

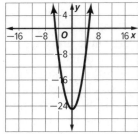

9. about 8.4 seconds

11 Step 1: Graph the related function of $f(x) = x^2 + 2x − 24 = 0$.

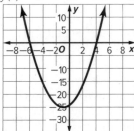

Step 2: The x-intercepts appear to be at −6 and 4, so the solutions are −6 and 4.

13. ∅

15. 1

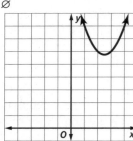
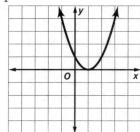

17. ∅

19. −6

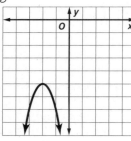
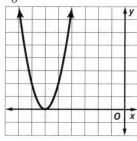

21. 8, −10

23. − 3

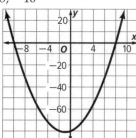
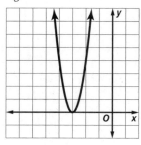

25. 2.2, −4.2

27. 3, −6

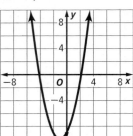
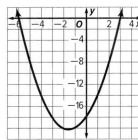

29. about 7.6 seconds　**31.** 1; −2　**33.** 2; −4, −8
35. −3, 4

37 **a.** $h = −16t^2 + 30t + 10$
$0 = −16t^2 + 30t + 10$
Graph the equation and find the x-intercepts.

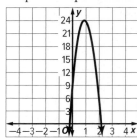

The positive x-intercept appears to be at 2.2, so she is in the air 2.2 seconds.

b. From the graph, she appears to hit a height of 15 feet at 0.2 second and 1.7 seconds.

c. $x = \dfrac{−b}{2a}$

$= \dfrac{−30}{2(−16)}$

$h = (−16)(0.9375)^2 + 30(0.9375) + 10$
$≈ 24$ feet

Her maximum height is about 24 feet, so she gets the bonus points.

39. −2, 1, 4　**41.** Sample answer: A tennis ball being hit in the air; an equation is $h = −16t^2 + 25t + 2$. The ball is in the air for about 1.6 seconds.　**43.** 1.5 and −1.5; Sample answer: Make a table of values for x from −2.0 to 2.0. Use increments of 0.1.　**45a.** B　**45b.** B, E
47a. 2　**47b.** 1　**49c.** C, E　**49.** A, C, E

Lesson 9-4

1 Take the square root of each side of the equation. The square root of x^2 is x. To find the square root of 88, multiply 4 by 22 and take the square root of 4 to get $±2\sqrt{22}$.
$x = ±2\sqrt{22} = ±9.38$　**3.** $x = 3, x = −5$
5. 0.6 second　**7.** −10, 0　**9.** 0, $\dfrac{3}{4}$
11. 1, 9　**13.** 1, $\dfrac{7}{5}$　**15.** $f(x) = −\dfrac{1}{4}x^2 + \dfrac{1}{4}x + 3$　**17.** ±3
19. ±10　**21.** $±6\sqrt{2}$

23. $-5 \pm 2\sqrt{5}$

25. $-2, 6$ **27.** $0, 3$

29. $-2, -\dfrac{1}{2}$

31. $-3, 0$ **33.** $8, 10$

35. $-6, 11$

(37) $16b^2 + 24b + 20 = 15$
$16b^2 + 24b + 5 = 0$
$(4b + 5)(4b + 1) = 0$
$4b + 5 = 0$ or $4b + 1 = 0$
$b = -\dfrac{5}{4}$ $b = -\dfrac{1}{4}$

39. $-\dfrac{3}{4}; -\dfrac{2}{3}$

41. $f(x) = \dfrac{3}{4}x^2 - 3x - 9$

43. 20 feet   **45.** $-9, 1$

47. $f(x) = \dfrac{1}{4}x^2 - 2x + 3$

(49)
$V = \ell wh$
$96 = (h)(h - 2)(h + 8)$
$96 = h^3 + 6h^2 - 16h$
$0 = h^3 + 6h^2 - 16h - 96$
$0 = (h^3 + 6h^2) + (-16h - 96)$
$0 = h^2(h + 6) - 16(h + 6)$
$0 = (h^2 - 16)(h + 6)$
$0 = (h + 4)(h - 4)(h + 6)$
$h + 4 = 0$ or $h - 4 = 0$ or $h + 6 = 0$
$h = -4$ $h = 4$ $h = -6$

Since height cannot be negative, $h = 4$ is the only viable solution. Substitute 4 for h in the length and width. The ballot box is 4 in. high by 12 in. long by 2 in. wide.

51. The quadratic is a perfect square, so I can factor it as $(x + 4)^2 = 0$. Then, I can use the Square Root Property to solve. So, $x = -4$.

53. Because the solutions are $-\dfrac{b}{a}$ and $\dfrac{b}{a}$, $a \neq 0$ and b is any real number.

55. $-\dfrac{1}{2}, 3$

57. $x^2 - 7x = 0$

59a. $h = -16t^2 + 29t + 6$

59b. 2 seconds

61. $-3, 3$ **63.** A, D   **65.** 0

67a. $2x^2 - 16x - 31 = \dfrac{1}{2}(2x)(x + 14)$

67b. 45 inches   **67c.** No; the negative value cannot be used for the solutions since length cannot be negative and $2(-1) = -2$.

(1) Step 1: Find $\dfrac{1}{2}$ of $-18 = -9$.
Step 2: Square the result in step 1: $(-9)^2 = 81$
Step 3: Add the result of step 2 to $x^2 - 18x$:
$x^2 - 18x + 81$
Thus, $c = 81$.

3. $\dfrac{81}{4}$ **5.** $-5.2, 1.2$ **7.** $-2.4, 0.1$ **9.** 8 ft by 18 ft

11. $y = (x + 9)^2 - 45$

13. 169   **15.** $\dfrac{361}{4}$ **17.** $\dfrac{25}{4}$

(19) $x^2 + 6x - 16 = 0$
$x^2 + 6x = 16$
$x^2 + 6x + 9 = 16 + 9$
$(x + 3)^2 = 25$
$x + 3 = \pm 5$
$x = -3 \pm 5$
The solutions are 2 and -8.

21. $-1, 9$ **23.** $-0.2, 11.2$

25. $\varnothing$ **27.** on the 30th and 40th days after purchase

29. 5.3   **31.** $y = (x - 4)^2 - 26$ **33.** $-1, 2$ **35.** $0.2, 0.9$

37a. Earth   **37b.** Earth: 4.9 seconds; Mars: 8.0 seconds

37c. Yes. The acceleration due to gravity is much greater on Earth than it is on Mars. So the time to reach the ground should be much less.

39. $y = 3(x + 4)^2 - 3$ **41a.** $97, -72, -24, 0, 57$

41b. $2, 2, 0, 1, 2$ **41c.** If $b^2 - 4ac$ is negative, the equation has no real solutions. If $b^2 - 4ac$ is zero, the equation has one real solution. If $b^2 - 4ac$ is positive, the equation has 2 real solutions.

41d. 0 because $b^2 - 4ac$ is negative. The equation has no real solutions because taking the square root of a negative number does not produce a real number.

(43) None; sample answer: If you add $\left(\dfrac{b}{2}\right)^2$ to each side of the equation and each side of the inequality, you get $x^2 + bx + \left(\dfrac{b}{2}\right)^2 = c + \left(\dfrac{b}{2}\right)^2$ and $c + \left(\dfrac{b}{2}\right)^2 < 0$. Because the left side of the last equation is a perfect square, it cannot equal the negative number $c + \left(\dfrac{b}{2}\right)^2$. So, there are no real solutions.

45. $y = -(x - 1)^2 + 2$; The function is of the form $y = a(x - h)^2 + k$. Using the coordinates given in the graph, the vertex is $(1, 2)$. So, the function $y = a(x - 1)^2 + 2$. Substitute the coordinates of one of the points to get $-7 = a(-2 - 1)^2 + 2$, or $-7 = 9a + 2$. Solve for a to get $a = -1$. So the function is $y = -(x - 1)^2 + 2$.

47. B   **49.** C   **51.** 4, 6
53. C

Lesson 9-6

1. $-3, 5$ **3.** 6.4, 1.6   **5.** 0.6, 2.5   **7.** $-6, \frac{1}{2}$
9. $\pm\frac{5}{3}$ **11.** -3; no real solutions
13. 0; one real solution   **15.** The discriminant is -14.71, so the equation has no real solutions. Thus, the jaguarundi will not reach a height of 10 feet.

17
$$x^2 + 16 = 0$$
For this equation, $a = 1$, $b = 0$, and $c = 16$.
$$x = \frac{-b \pm \sqrt{b^2 - 4ac}}{2a}$$
$$= \frac{0 \pm \sqrt{(0)^2 - 4(1)(16)}}{2(1)}$$
$$= \frac{\sqrt{-64}}{2}$$
So, there is no real solution. The solution can be written $\varnothing$.

19. $2.2, -0.6$ **21.** $-3, -\frac{6}{5}$ **23.** $0.5, -2$
25. $0.5, -1.2$ **27.** 3   **29.** $-1.2, 5.2$ **31.** $-2, 5$
33. $-6.2, -0.8$ **35.** -0.07; no real solution   **37.** 12.64; two real solutions   **39.** 0; one real solution
41a. 2036   **41b.** Sample answer: No; the parabola has a maximum at about 66, meaning only 66% of the store's customers would ever have the store's loyalty card.   **43.** 0   **45.** 1   **47.** $-1.4, 2.1$

49 **a.** $(20 - 2x)(25 - 7x) = 375$
b. $500 - 50x - 140x + 14x^2 = 375$
$$14x^2 - 190x + 125 = 0$$
$$x = \frac{-b \pm \sqrt{b^2 - 4ac}}{2a}$$
$$= \frac{190 \pm \sqrt{(-190)^2 - 4(14)(125)}}{2(14)}$$
$$= \frac{190 \pm \sqrt{29,100}}{28}$$
$$\approx 0.7 \text{ and } 12.9$$
c. The margins should be 0.7 inches on the sides and $4(0.7)$ or 2.8 inches on the top and $3(0.7)$ or 2.1 inches on the bottom.

51. $k < \frac{9}{40}$ **53.** none   **55.** two   **57.** Sample answer: If the discriminant is positive, the Quadratic Formula will result in two real solutions because you are adding and subtracting the square root of a positive number in the numerator of the expression. If the discriminant is zero, there will be one real solution because you are adding and subtracting the square root of zero. If the discriminant is negative, there will be no real solutions because you are adding and subtracting the square root of a negative number in the numerator of the expression.
59a. D, F   **59b.** B, E   **59c.** Sample answer: She could rewrite the equation so that it equals zero. She could write out the values of a, b, and c.   **61a.** A   **61b.** D
63. B

1. $(3, 0), (0, -3)$ **3.** $(0, 1), (3, -2)$

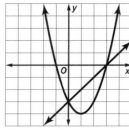   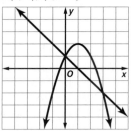

5. $4

7 Substitute one expression for y in the other equation and solve for x.
$$x^2 - 5x + 5 = 5x - 20$$
$$x^2 - 10x + 25 = 0$$
$$(x - 5)^2 = 0$$
$$x - 5 = 0$$
$$x = 5$$
Substitute 5 for x in either equation.
$$y = 5(5) - 20$$
$$y = 5$$
The solution is $(5, 5)$

11. $(5, 0), (2, -3)$ **13.** $(-2, -4), (2, 4)$

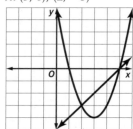   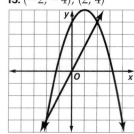

15. $(-1, 6), (0.5, 1.5)$
17 Substitute the first expression for y in the second equation and solve for x.
$$x^2 - 9 = 0$$
$$(x - 3)(x + 3) = 0$$
$$x - 3 = 0 \quad \text{or} \quad x + 3 = 0$$
$$x = 3 \quad \text{or} \quad x = -3$$
Substitute the x-values in either equation.
$$y = (3)^2 \qquad y = (-3)^2$$
$$y = 9 \qquad y = 9$$
The solutions are $(3, 9)$ and $(-3, 9)$.

19. $(-1, 12), (-11, 132)$ **21a.** $(1, 29)$ **21b.** The streets do not intersect because the system has no real solutions.
23a. 2   **23b.** 0   **25.** Sample answer: $y = x^2 + 3$
27. $y = -x^2$ **29.** $(-2.5, -6.25), (1, -1)$ **31.** The solution of the system is $(0, 0)$ which is not the same as no solution.   **33a.** $(2, 1)$ \33b. no solution   **33c.** $(-0.83, 3.34), (4.83, 14.66)$ **35.** A, B   **37.** B, D, F
39a.

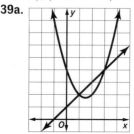

39b. Yes there are two places where they might collide: (3, 4) and (1, 2).   **39c.** Sample answer: $y = 1.75$

Lesson 9-8

1.
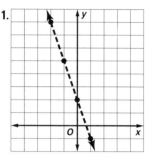
linear

3.
exponential

5. quadratic   **7.** exponential

9. exponential; $y = 3(3)^x$ **11.** linear; $y = \frac{1}{2}x + \frac{5}{2}$
13. linear: $y = 0.5x + 3$

15.
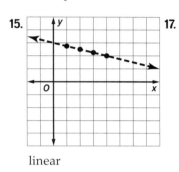
linear

17.
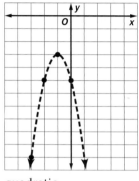
quadratic

19.
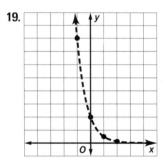
exponential

21 Look for a pattern in the y-values. Start with comparing first differences.

10   2.5   0   2.5   10
  −7.5 −2.5 2.5 7.5

The first differences are not all equal. So, the table of values does not represent a linear function. Find the second differences and compare.

−7.5   −2.5   2.5   7.5
    5   5   5

The second differences are all equal, so the table of values represents a quadratic function.
Write an equation for the function that models the data.

The equation has the form $y = ax^2$. Find the value of a by choosing one of the ordered pairs from the table of values. Let's use (2, 10).

$$y = ax^2$$
$$10 = a(2)^2$$
$$10 = 4a$$
$$\frac{5}{2} = a$$
$$2.5 = a$$

An equation that models the data is $y = 2.5x^2$.
23. exponential; $y = 0.2(5)^x$
25. linear; $y = -5x - 0.25$

27 a. Graph the ordered pairs on a coordinate plane.
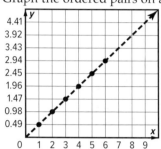
The graph appears to be linear.

b. Look at the first differences of the y-values.

0.49   0.98   1.47   1.96   2.45   2.94
    0.49   0.49   0.49   0.49   0.49

The common difference is 0.49.
The equation is $y = 0.49x$.
c. $y = 0.49x$
    $= 0.49(10)$
    $= \$4.90$

29a.

Time (hour)	0	1	2	3	4
Amount of Bacteria	12	36	108	324	972

29b. exponential   **29c.** $b = 12(3)^t$ **29d.** 78,732
31. Sample answer: $y = 2x^2 - 5$ **33.** $y = 4x + 1$
35. Sample answer: The data can be graphed to determine which function best models the data. You can also find the differences in ratios of the y-values. If the first differences are constant, the data can be modeled by a linear function. If the second differences are constant but the first differences are not, the data can be modeled by a quadratic function. If the ratios are constant, then the data can be modeled by an exponential function.   **37.** B, D, E   **39.** 11   **41.** B

Lesson 9-9

1. $x^2 + x - 8$ **3.** $x^2 - 2x - 9$
5. $x + 1$
7a. $f(t) = 20(1.045)^t$
  b. $g(t) = 42 + 2t$
  c. $v(t) = 20(1.045)^t + 42 + 2t$. It represents the total value of the coins after t years.
  d. $\$76.92$

9 $(f \cdot g)(x) = (3x + 10)(x^2 - 6x - 20)$
$\qquad = 3x^3 - 18x^2 - 6x + 10x^2 - 60x - 20$
$\qquad = 3x^3 - 8x^2 - 66x - 20$
11a. $P(x) = 4 + 0.15x$ **b.** $T(x) = 160 - 10x$
c. $R(x) = -1.5x^2 - 16x + 640$ **d.** \$552
13. $2x^2 - 6x$ **15.** $2^x + x^2 - 7$
17. $-4x^2 - 6x + 18$
19. $-5x^3 + 18x^2 - 11x - 8$
21. $-15x^3 - 15x^2$
23. $25x^3 - 40x^2$
25 **a.** $f(t) = 385 + 360t$
b. $g(t) = 590(1.15)^t$
c. $(f + g)(t) = 590(1.15)^t + 360t + 385$; this represents the total number of subscribers to Aurelio's blogs after t years.
d. $590(1.15)^6 + 360(6) + 385$
$1364.7 + 2160 + 385 \approx 3910$
27. 21 **29.** -4
31a. $f(x) = 10 - 2x$ **31b.** $g(x) = 8 - 2x$
31c. $4x^2 - 36x + 80$; this represents the area of the base.
31d. $D = \{0 < x < 4\}$; to create a square, x must be greater than 0; also, x must be less than half the width of the cardboard.
33. 4 **35.** -2 **37.** -4
39a. $(f - g)(x) = x^2 + 1$; $(g - f)(x) = -x^2 - 1$; the functions are opposites since $(f - g)(x) = -(g - f)(x)$
39b.

39c. The graph of $(g - f)(x)$ is the graph of $(f - g)(x)$ reflected across the x-axis.
39d. Yes; the functions are opposites of each other. This means one graph is a reflection of the other across the x-axis.
41 $(f + g)(x) = f(x) + g(x)$
$\qquad = 3^x + 2(3^x)$
$\qquad = 3(3^x)$
$\qquad = (3^1)(3^x)$
$\qquad = 3^{x+1}$
43. $2(3)^{2x}$
45. sometimes **47.** sometimes **49.** No; $(f \cdot g)(x)$ can be a linear function if one or both of $f(x)$ or $g(x)$ is a constant function.
51a. Mikayla did not distribute the minus sign to every term in $g(x)$. **51b.** $x^2 + 10x + 3$ **53.** B **55.** A, B, E, F
57a. $f(x) = 82,400(1.025)^x$ **57b.** $g(x) = 75,600 - 300x$
57c. $82,400(1.025)^x + 300x - 75,600$; this represents how many more residents live in Oakville than in Elmwood. **57d.** 21,759; there were 21,759 more residents of Oakville than Elmwood in 2016. **59.** C

1. true **3.** false; parabola **5.** false; two **7.** true
9. The solutions of a quadratic equation are the x-intercepts of the graph of the related quadratic function.
11a. minimum **11b.** 0 **11c.** $D = \{-\infty < x < \infty\}$;
$R = \{y \mid y \geq 0\}$ **13a.** minimum **13b.** -4
13c. $D = \{-\infty < x < \infty\}$; $R = \{y \mid y \geq -4\}$
15a. maximum **15b.** 16 **15c.** $D = \{t \mid 0 \leq t \leq 2\}$;
$R = \{h \mid 0 \leq h \leq 16\}$ **17.** translated down 3 units
19. vertical stretch and translated down 18 units
21. vertical compression **23.** reflected across the x-axis, vertically stretched, and translated up 100 units **25.** 3
27. $-4.6, 0.6$ **29.** $-0.8, 3$
31. $-11, 5$ **33.** $-4, 8$ **35.** $-4, 2$ **37.** $1, -7$
39. $10, -2$ **41.** $-0.7, 7.7$ **43.** no solution
45. $-8, 6$ **47.** $-0.7, 0.5$
49. $-5, 1.5$ **51.** $-2.5, 1.5$
53. -4; no solution **55.** $(1, 1), (5, 9)$ **57.** $(0.5, -0.5),$
$(4, 17)$ **59.** $(3, 0)$
61. no solution **63.** exponential; $y = 2^x$
65. linear; $y = 8 - 3x$ **67.** $-6x^2 + x + 4$
69. $4x^2 - x + 2$ **71.** $5x^4 - x^3 - 16x^2 + 3x + 3$

Get Ready

1. Total all the cubes, set that as the denominator; find the total of a specific color, set as the numerator.
3. the Distributive Property   **5.** The range remains the same.   **7.** Substitute a value for x and solve for y or vice versa to generate data points.

Lesson 10-1

1. mean: 4.6; median: 4; mode: none; The mean and median are close enough in value that either of them can represent the data.   **3.** mean
5 a. Canada has more medals than 6 of the countries, and there are 10 countries.

$\frac{6}{10}$ = 60th percentile

b. Yes, the nation with 5 medals below it will be in the 50th percentile. The Netherlands has more medals than 5 countries.

Netherlands = $\frac{5}{10}$ = 50th percentile
7a. mean = \$342.37; median \$329.98; mode = \$259.95
7b. The mean and median are close to the same value. Either measure would be good, but choosing a tablet closest to the median might be best because it is less than the mean.
9. 75th percentile
11 a. Change all the scores to seconds by multiplying the minutes by 60 and then adding the seconds. 20:45 = 20 × 60 + 45 = 1245. Similarly, the other scores in seconds are 1290, 1275, 1352, 1300.
Mean = $\frac{1245 + 1290 + 1275 + 1352 + 1300}{5} = \frac{6462}{5} =$
1292.4. Convert 1292.4 seconds back to minutes by dividing by 60 to get 21.54 minutes. Then convert 0.54 minutes to seconds to multiplying by 60. 0.54 × 60 = 32.4 seconds. So, her mean practice time in minutes and seconds is 21:32.4.
b. Arrange the practice times (in seconds) in order from least to greatest to find the median. 1245, 1275, 1290, 1300, 1352. The middle value is 1290 seconds. Convert it to minutes as seconds as shown in Part a, or 1290 ÷ 60 = 21.5, or 21 minutes 30 seconds. So, the median is 21:30. Since her time is 30 seconds less, the actual race time is exactly 21 minutes.
13a. John: 9.65, Raul: 9.475, Laird, Boris, Han: 9.525; John won.
13b. John: 9.633, Raul: 9.483, Laird, Boris, Han: 9.517; John still won.
13c. Sample answer: For this meet, it didn't matter which method of scoring they used because the overall standings turned out the same.

15. The student is not correct. A counterexample is the set 3 3 3 3 3 3, with mean = 3 and mode = 3.
17. Sample answer: A data set can have zero, one, or more than one mode. This is true because the mode is not dependent on where the data value is positioned in the set or on the distribution of the data values. The mode only depends on how many of a certain data value there are. If no values appear more than once, there will be no mode. If multiple values appear more than once, there will be more than one mode.
19. The percentile rank tells where a score ranks among the other scores. The percent is a comparison of a score to the highest score possible.
21. Sample answer: 40, 40, 50, 50, 93, 95. The mean is 61.3, and the median is 50. If an unusually high test score like 99 is added to the data set, it will not change the median, but the mean will increase by over 5 points.
23. B   **25.** Sample answer: 20, 30, 40, 50, 60   **27.** Sample answer: Mean; The mean is 46.2 people, and the median is 33 people. The mean is closer to the number of participants in the activities.

Lesson 10-2

1.

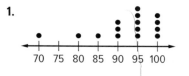

3.

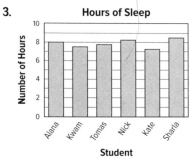

5.

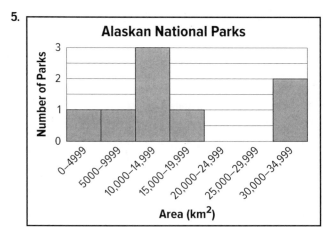

7.

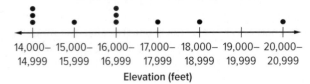

Elevation of Mountains in United States

Elevation (feet)

14,000– 15,000– 16,000– 17,000– 18,000– 19,000– 20,000–
14,999 15,999 16,999 17,999 18,999 19,999 20,999

9.

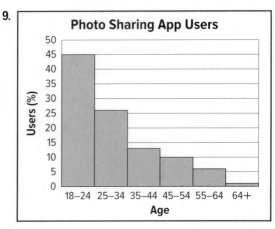

Photo Sharing App Users

Users (%): 18–24, 25–34, 35–44, 45–54, 55–64, 64+
Age

11. continuous;

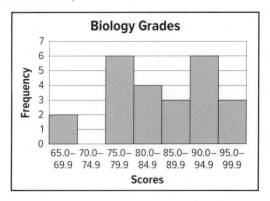

Biology Grades

Frequency

65.0– 70.0– 75.0– 80.0– 85.0– 90.0– 95.0–
69.9 74.9 79.9 84.9 89.9 94.9 99.9
Scores

13. discrete;

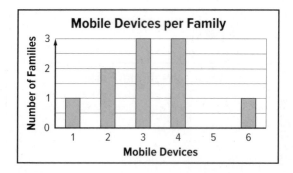

Mobile Devices per Family

Number of Families

1   2   3   4   5   6
Mobile Devices

15.

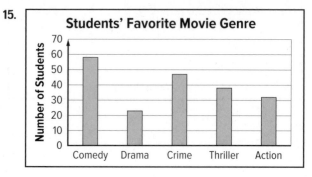

Students' Favorite Movie Genre

Number of Students

Comedy Drama Crime Thriller Action

17a. 56°F   **17b.** 20   **17c.** 11   **19.** B, C, D   **21.** 10; 3; 4; 6; 2; 5; 1 and 7   **23.** The data is categorical, so a bar graph should be used to display the data.   **25a.** 7 in. **25b.** 72–73 in.   **25c.** 35

Lesson 10-3

1. $99.99; $199.99; $299.99; $399.99; $439.99

③ Find the mean.

$$\bar{x} = \frac{10 + 12 + 0 + 6 + 9 + 15 + 12 + 10 + 11 + 20}{10}$$ or 10.5

Find the sum of the square of the differences, $(\bar{x} - x_n)^2$.

$$(10 - 10.5)^2 = 0.25$$
$$(12 - 10.5)^2 = 2.25$$
$$(0 - 10.5)^2 = 110.25$$
$$(6 - 10.5)^2 = 20.25$$
$$(9 - 10.5)^2 = 2.25$$
$$(15 - 10.5)^2 = 20.25$$
$$(12 - 10.5)^2 = 2.25$$
$$(10 - 10.5)^2 = 0.25$$
$$(11 - 10.5)^2 = 0.25$$
$$+ (20 - 10.5)^2 = 90.25$$
$$= 248.50$$

Find the variance.

$$\sigma^2 = \frac{(\bar{x} - x_1)^2 + (\bar{x} - x_2)^2 + ... + (\bar{x} - x_n)^2}{n}$$

$$= \frac{248.50}{10}$$ or 24.85

Find the standard deviation.

$$\sigma = \sqrt{\sigma^2}$$

$$= \sqrt{24.85}$$ or about 4.98

Sample answer: The mean is 10.5 and the standard deviation is about 4.98. The standard deviation is relatively high due to outliers 0 and 20.

5. 62 people; 66 people; 73 people; 82 people; 99 people
7. ≈21.83; 14.5; 82; ≈16.36; 14; mean   **9.** 2.4; Sample answer: With a mean of about $9.29, the standard deviation of about $2.38 suggests that there is a good amount of deviation to the data. This deviation is mostly caused by the outlier of $15.00. If this outlier were removed, the new mean of the data would be about $8.33 with a standard deviation of about $0.49.   **11.** The Wildcats had a mean of about 11.9 fouls per game with a standard deviation of about 2.2 fouls. The Trojans had a mean of about 10.1 fouls per game with a standard deviation of about 2.1 fouls. The data sets had nearly the same variability while the Wildcats had a mean of almost 2 fouls more than the Trojans. The Wildcats coach can conclude that his team consistently has about 2 more fouls than the Trojans.

⑬ **a.** Movie A:

Find the mean.

$$\bar{x} =$$
$$\frac{7 + 8 + 7 + 6 + 8 + 6 + 7 + 8 + 6 + 8 + 8 + 6 + 7 + 7 + 8 + 8}{16}$$
or about 7.2

Find the sum of the square of the differences, $(\bar{x} - x_n)^2$.

$$(7 - 7.2)^2 = 0.04$$
$$(8 - 7.2)^2 = 0.64$$
$$(7 - 7.2)^2 = 0.04$$
$$(6 - 7.2)^2 = 1.44$$
$$(8 - 7.2)^2 = 0.64$$
$$(6 - 7.2)^2 = 1.44$$
$$(7 - 7.2)^2 = 0.04$$
$$(8 - 7.2)^2 = 0.64$$
$$(6 - 7.2)^2 = 1.44$$
$$(8 - 7.2)^2 = 0.64$$
$$(8 - 7.2)^2 = 0.64$$
$$(6 - 7.2)^2 = 1.44$$
$$(7 - 7.2)^2 = 0.04$$
$$(7 - 7.2)^2 = 0.04$$
$$(8 - 7.2)^2 = 0.64$$
$$+ (8 - 7.2)^2 = 0.64$$
$$= 10.44$$

Find the variance.

$$\sigma^2 = \frac{(\bar{x} - x_1)^2 + (\bar{x} - x_2)^2 + \dots + (\bar{x} - x_n)^2}{n}$$

$$= \frac{10.44}{16} \text{ or } 0.6525$$

Find the standard deviation.

$$\sigma = \sqrt{\sigma^2}$$
$$= \sqrt{0.6525} \text{ or about } 0.81$$

Movie B:

Find the mean.

$$\bar{x} =$$
$$\frac{9 + 3 + 8 + 2 + 5 + 10 + 3 + 8 + 10 + 9 + 9 + 10 + 6 + 4 + 9 + 3}{16}$$

or about 6.8

Find the sum of the square of the differences, $(\bar{x} - x_n)^2$.

$$(9 - 6.8)^2 = 4.84$$
$$(3 - 6.8)^2 = 14.44$$
$$(8 - 6.8)^2 = 1.44$$
$$(2 - 6.8)^2 = 23.04$$
$$(5 - 6.8)^2 = 3.24$$
$$(10 - 6.8)^2 = 10.24$$
$$(3 - 6.8)^2 = 14.44$$
$$(8 - 6.8)^2 = 1.44$$
$$(10 - 6.8)^2 = 10.24$$
$$(9 - 6.8)^2 = 4.84$$
$$(9 - 6.8)^2 = 4.84$$
$$(10 - 6.8)^2 = 10.24$$
$$(6 - 6.8)^2 = 0.64$$
$$(4 - 6.8)^2 = 7.84$$
$$(9 - 6.8)^2 = 4.84$$
$$+ (3 - 6.8)^2 = 14.44$$
$$= 131.04$$

Find the variance.

$$\sigma^2 = \frac{(\bar{x} - x_1)^2 + (\bar{x} - x_2)^2 + \dots + (\bar{x} - x_n)^2}{n}$$

$$= \frac{131.04}{16} \text{ or } 8.19$$

Find the standard deviation.

$$\sigma = \sqrt{\sigma^2}$$
$$= \sqrt{8.19} \text{ or about } 2.86$$

Sample answer: Movie A had a mean of about 7.2 with a standard deviation of about 0.81. Movie B had a mean of about 6.8 with a standard deviation of about 2.86. While both movies had a mean rating of close to 7, the ratings for Movie B had a wider range than those for Movie A.

b. Movie A: Sample answer: All the reviews are between 6 and 8. The ratings were consistent amongst the group.
Movie B: Sample answer: The ratings range from 2 to 10 with half of the students rating the movie between 8 and 10.

15. Both; when an outlier is removed, the spread and standard deviation will decrease. When more values that are equal to the mean of a data set are added, the outliers will have less influence.
17. Sample answer: Poll voters to determine if a particular presidential candidate is favored to win the election. Use a stratified random sample to call 100 people throughout the country.
19. Both values are helpful in identifying the measure of spread. The range of a data set is the difference between the highest and lowest values in a set of data. The interquartile range is the difference between Q3, the upper quartile, and Q1, the lower quartile. This value is helpful in identifying outliers.
21. D **23.** A **25.** C **27a.** $433; $215
27b. $181.6, $114.50, $70 **27c.** There is no outlier.
27d. $142.34; a standard deviation of $142.34 is large compared to the mean of $181.60. This suggests that most of the data values are not relatively close to the mean.

Lesson 10-4

1.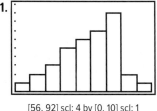

[56, 92] scl: 4 by [0, 10] scl: 1

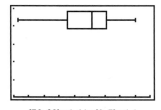

[56, 92] scl: 4 by [0, 5] scl: 1

negatively skewed

3. Sample answer: The distribution is skewed, so use the five-number summary. The range is 92 − 52 or 40. The median is 65, and half of the data are between 59.5 and 74.

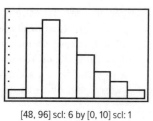

[48, 96] scl: 6 by [0, 10] scl: 1

5.

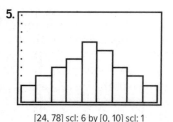

[24, 78] scl: 6 by [0, 10] scl: 1

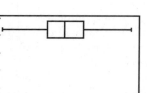

[24, 78] scl: 6 by [0, 5] scl: 1

symmetric

7 Use a graphing calculator to enter the data into L1 and create a histogram. Adjust the window to the dimensions shown. The graph is high on the right. Therefore, the distribution is negatively skewed.

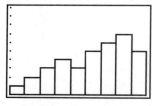

[10, 55] scl: 5 by [0, 10] scl: 1

The distribution is negatively skewed, so use the five-number summary.

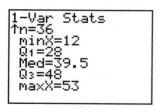

The range is 53 − 12 or 41. The median of the numbers is 39.5, and half the numbers are between 28 and 48.

9. Sample answer: The distribution is symmetric, so use the mean and standard deviation to describe the center and spread. The mean temperature is 52.8° with standard deviation of about 4.22°.

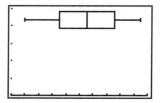

[42, 62] scl: 2 by [0, 5] scl: 1

11 **a.** Enter the list of prices as L1 and create a box-and-whisker plot. Adjust the window to the dimensions shown. The right whisker is longer than the left. The median is closer to the shorter whisker. Thus, the distribution is positively skewed.

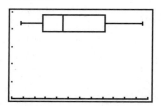

[7, 19] scl: 1 by [0, 5] scl: 1

The distribution is positively skewed, so use the five-number summary.

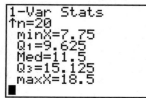

The range of prices is $18.50 − $7.75 or $10.75. The median price is $11.50, and half of the prices are between $9.63 and $15.13.

b. Delete all the entries in L1 that are greater than $15 and create another box-and-whisker plot. The lengths of the whiskers are approximately equal, and the median is in the middle of the data. Thus, the distribution is symmetric.

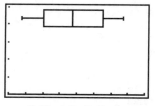

[7, 15] scl: 1 by [0, 5] scl: 1

The distribution is symmetric, so use the mean and standard deviation to describe the center and the spread of the new data.

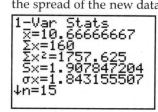

The mean is about \$10.67 with standard deviation of about \$1.84.

13. i **15.** Sample answer: A bimodal distribution is a distribution of data that is characterized by having data divided into two clusters, thus producing two modes, and having two peaks. The distribution can be described by summarizing the center and spread of each cluster of data.

17. Sample answer: In a symmetrical distribution, the majority of the data is located near the center of the distribution. The mean of the distribution is also located near the center of the distribution. Therefore, the mean and standard deviation should be used to describe the data. In a skewed distribution, the majority of the data lies either on the right or left side of the distribution. Because the distribution has a tail or may have outliers, the mean is pulled away from the majority of the data. The median is less affected. Therefore, the five-number summary should be used to describe the data.

19. A **21a.** \$48,000 **21b.** \$60,200 **21c.** \$256,000 is an outlier. **23.** A

Lesson 10-5

1. 3.5, 3.5, 1, 8, 2.4

③ Find the mean, median, mode, range, and standard deviation of the original set.
Mean: 5.75
Median: 5.5
Mode: 3
Range: 10
Standard deviation: 3.14
Multiply the mean, median, mode, range, and standard deviation by 3.
Mean: $3 \times 5.75 \approx 17.3$
Median: $3 \times 5.5 = 16.5$
Mode: $3 \times 3 = 9$
Range: $3 \times 10 = 30$
Standard deviation: $3 \times 3.14 \approx 9.4$

5a. Kyle's Distances

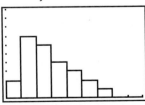

[17, 19.25] scl: 0.25 by [0, 10] scl: 1

Mark's Distances

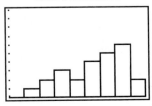

[17, 19.25] scl: 0.25 by [0, 10] scl: 1

Kyle, positively skewed; Mark, negatively skewed

5b. Sample answer: The distributions are skewed, so use the five-number summaries. Kyle's upper quartile is 17.98, while Mark's lower quartile is 18.065. This means that 75% of Mark's distances are greater than 75% of Kyle's distances. Therefore, we can conclude that overall, Mark's distances are higher than Kyle's distances.

7. 60.9, 60, 60, 14, 4.7

9. 22.5, 21.5, no mode, 24, 7.4

11. 36.8, 38, 12, 56, 20.0

13. 26.8, 27.2, 29.6, 10.4, 3.5

15a.

1st Period

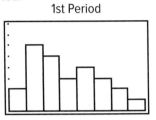

[200, 600] scl: 50 by [0, 8] scl: 1

6th Period

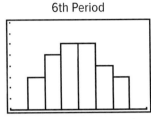

[200, 600] scl: 50 by [0, 8] scl: 1

1st period, positively skewed; 6th period, symmetric

15b. Sample answer: One distribution is symmetric and the other is skewed, so use the five-number summaries. The lower quartile for 1st period is 291 pages, while the minimum for 6th period is 294 pages. This means that the lower 25% of data for 1st period is lower than any data from 6th period. The range for 1st period is $578 - 206$ or 372 pages. The range for 6th period is $506 - 294$ or 212 pages. The median for 1st period is about 351 pages, while the median for 6th period is 392 pages. This means, that while the median for 6th period is greater, 1st period's pages have a greater range and include greater values than 6th period.

⑰ **a.** Enter Leon's times as L1 and Cassie's times as L2. Then use STAT PLOT to create a histogram for each list.

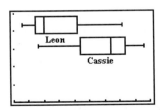

[1.5, 6] scl: 0.5 by [0, 5] scl: 1

For Leon's times, the right whisker is longer than the left, and the median is closer to the left whisker. The distribution is positively skewed. For Cassie's times, the left whisker is longer than the right and the median is closer to the right whisker. The distribution is negatively skewed.

b. One distribution is positively skewed and the other is negatively skewed, so use the five-number summaries to compare the data.

Leon's Times

```
1-Var Stats
↑n=16
 minX=1.8
 Q1=2.2
 Med=2.5
 Q3=3.6
 maxX=5.1
■
```

Cassie's Times

```
1-Var Stats
↑n=16
 minX=2.3
 Q1=3.7
 Med=4.7
 Q3=5.2
 maxX=5.8
```

The lower quartile for Leon's times is 2.2 minutes, while the minimum for Cassie's times is 2.3 minutes. This means that 25% of Leon's times are less than all of Cassie's times. The upper quartile for Leon's times is 3.6 minutes, while the lower quartile for Cassie's times is 3.7 minutes. This means that 75% of Leon's times are less than 75% of Cassie's time. Overall, we can conclude that Leon completed the brainteasers faster than Cassie.

19a. 52.96, 53, 53, 19, 6.08

19b. 47.96, 48, 48, 19, 6.08

21 Find the total cost of all 20 SUVs and luxury cars.
20 × $39,250 = $785,000
Next subtract the price reduction of the 15 SUVs.
$785,000 − (15 × 2000) = $755,000
Finally, divide this new total by the 20 cars.
$755,000 ÷ 20 = $37,750
The new average price for all the vehicles is $37,750.

23. Sample answer: Histograms show the frequency of values occurring within set intervals. This makes the shape of the distribution easy to recognize. However, no specific values of the data set can be identified from looking at a histogram, and the overall spread of the data can be difficult to determine. The box-and-whisker plots show the data divided into four sections. This aids when comparing the spread of one set of data to another. However, the box-and-whisker plots are limited because they cannot display the data any more specifically than showing it divided into four sections.   **25.** Sample answer: When two distributions are symmetric, the first thing to determine is how close the averages are and how spread out each set of data is. The mean and standard deviation are the best values to use for this comparison. When distributions are skewed, we also want to determine the direction and the degree to which it is skewed. The mean and standard deviation cannot provide any information in this regard, but we can get this information by comparing the range, quartiles, and medians found in the five-number summaries. So, if one or both sets of data are skewed, it is best to compare their five-number summaries.   **27a.** A   **27b.** Their means and standard deviations are more appropriate because both distributions are relatively symmetric.   **29a.** C   **29b.** Their five-number summaries are more appropriate because one of the distributions is skewed.

1 The total number of Norman nobles is found in column 2, row 2. 29   **3.** 78

5.

Role	Normans	Saxons	Total
Peasants	34%	30%	64%
Nobles	25%	11%	36%
Total	59%	41%	100%

7. Marginal frequencies are found by adding the totals for each column and row. 34% + 25% = 59%; 30% + 11% = 41%; 34% + 30% = 64%; 25% + 11% = 36%; The marginal frequencies are 64%, 36%, 59%, and 41%.

9.

Gender	Retriever	Small Dog	Total
Female	24	20	44
Male	24	12	36
Total	48	32	80

11. 11%

13. It seems that there is an association; Americans are less likely to wear a full ski suit.

15. 18   **17.** 14   **19.** 11

21.

Meal	Ate Everything	Left Food	Total
Vegetarian	6	18	24
Chicken	18	9	27
Beef	21	9	30
Total	45	36	81

23.

Meal	Ate Everything	Left Food	Total
Vegetarian	7%	22%	30%
Chicken	22%	11%	33%
Beef	26%	11%	37%
Total	56%	44%	100%

25. They are equal.

27. Yes, it appears that there is an association, because the conditional relative frequency of vegetarians eating everything on their plates is much lower than the conditional relative frequency of diners with other meals eating everything on their plates.

29.

Flavor	Waffle Cone	Regular Cone	Total
Strawberry	2	2	4
Vanilla	1	1	2
Chocolate	3	5	8
Total	6	8	14

31. 4%   **33.** 6%

35a.

Safety Gear	Paperwork Ready	Paperwork Not Ready	Total
Have Safety Gear	8	6	14
Do Not Have Safety Gear	4	6	10
Total	12	12	24

35b. 50%

35c. 33% **35d.** 58%

35e.

Safety Gear	Paperwork Ready	Paperwork Not Ready	Total
Have Safety Gear	33%	25%	58%
Do Not Have Safety Gear	17%	25%	42%
Total	50%	50%	100%

35f. 34% **35g.** 50%

37.

Allergy	Eat Gluten	Should Not Eat Gluten	Total
Peanuts	4	3	7
Soy	2	1	3
Milk	1	2	3
Total	7	6	13

Allergy	Eat Gluten	Should Not Eat Gluten	Total
Peanuts	31%	23%	54%
Soy	15%	8%	23%
Milk	8%	15%	23%
Total	54%	46%	100%

39 Relative frequencies give the percentage of each joint data point within the data set. However, the data set may be skewed, as one category in the data set may have only a fraction of the data points that another category has.

41a. A **41b.** C **41c.** B **43a.** B, D, E, G **43b.** 79%
43c. 44% **43d.** tomatoes: 79%; peas: 40%; squash: 40%; Yes, there is an association. A much larger percentage of tomatoes were damaged in the storm **43e.** C

Chapter 10 Study Guide and Review

1. mean **3.** quantitative **5.** A bar graph compares categories of data, while a histogram displays data that have been organized into equal intervals. The mean is the average of the data set, the median is the middle number when written in order, and mode occurs the most often. **9.** 21; 19; 20 **11.** 85.5; 86.6; no mode; Because the data set seems to be spread evenly, both the mean and the median are good measures of center.

13.

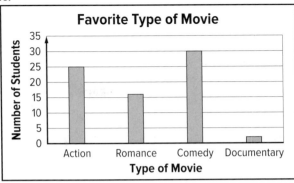

15.

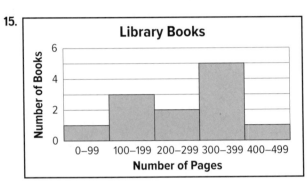

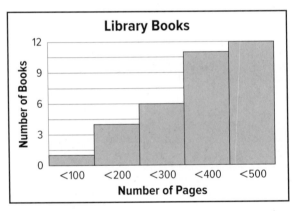

17. 3.4; 1.02; Sample answer: The standard deviation is relatively small compared to the mean. So, Ben mows about the same number of lawns each week. **19.** The day customers had a mean of about 9.6 times per month with a standard deviation of about 3.5. The night customers had a mean of about 11.7 times per month with a standard deviation of about 2.5. The night customers had a higher average and their data values were more consistent.

21. negatively skewed;

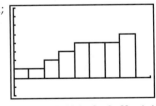

[0, 90] scl: 10 by [−2, 8] scl: 1

23. 22.5, 21.5, no mode, 24, 7.4

25a. Ms. Miller:

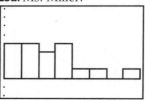

[9, 17] scl: 1 by [−2, 8] scl: 1

Ms. Anderson:

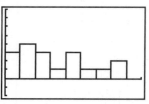

[0, 9] scl: 1 by [−2, 8] scl: 1

25b. Sample answer: Use the five-number summaries. The range for Ms. Miller is 7. The median is 11. Half of the data are between 10 and 12. The range for Ms. Anderson is 7. The median is 2. Half of the data are between 1 and 4. All of the data in Ms. Anderson's distribution are less than all of the data in Ms. Miller's distribution. Therefore, Ms. Miller will more than likely hand out more disciplinary actions than Ms. Anderson.

27a.

Gender	Yes	No	Totals
Boys	8	20	28
Girls	19	15	34
Total	27	35	62

27b.

Gender	Yes	No	Totals
Boys	12.9%	32.3%	45.2%
Girls	30.6%	24.2%	54.8%
Total	43.5%	56.5%	100%

27c.

Gender	Yes	No	Totals
Boys	28.6%	71.4%	100%
Girls	55.9%	44.1%	100%

Gender	Yes	No
Boys	29.6%	57.1%
Girls	70.4%	42.9%
Total	100%	100%

Glossary/Glosario

Multilingual eGlossary

Go to **connectED.mcgraw-hill.com** for a glossary of terms in these additional languages:

Arabic	Haitian Creole	Tagalog
Bengali	Hmong	Urdu
Brazilian Portuguese	Korean	Vietnamese
Chinese	Russian	
English	Spanish	

English / Español

A

absolute error (p. 313) The absolute error of a measurement is equal to one half the unit of measure.

error absoluto El error absoluto de una medida es igual a un medio de la unidad de medida.

absolute value (p. P11) The distance a number is from zero on the number line.

valor aboluto Es la distancia que dista de cero en una recta numerica.

absolute value function (p. 205) A function written as $f(x) = |x|$, in which $f(x) \geq 0$ for all values of x.

función del valor absoluto Una función que se escribe $f(x) = |x|$, donde $f(x) \geq 0$, para todos los valores de x.

accuracy (p. 34) The degree to which a measured value comes close to the actual or desired value.

exactitud El grado de cercanía entre un valor medido y el valor real o deseado.

additive identity (p. 16) For any number a, $a + 0 = 0 + a = a$.

identidad de la adición Para cualquier número a, $a + 0 = 0 + a = a$.

additive inverses (p. P11) Two integers, x and $-x$, are called additive inverses. The sum of any number and its additive inverse is zero.

inverso aditivo Dos enteros x y $-x$ reciben el nobre de inversos aditivos. La suma de cualquier número y su inverso aditivo es cero.

algebraic expression (p. 5) An expression consisting of one or more numbers and variables along with one or more arithmetic operations.

expresión algebraica Una expresión que consiste en uno o más números y variables, junto con una o más operaciones aritméticas.

area (p. P26) The measure of the surface enclosed by a geometric figure.

área La medida de la superficie incluida por una figura geométrica.

arithmetic sequence (p. 190) A numerical pattern that increases or decreases at a constant rate or value. The difference between successive terms of the sequence is constant.

sucesión aritmética Un patrón numérico que aumenta o disminuye a una tasa o valor constante. La diferencia entre términos consecutivos de la sucesión es siempre la misma.

association (p. 247) See correlation.

asociación Véase correlación.

asymptote (p. 430) A line that a graph approaches.

asíntota Una línea a que un gráfico acerca.

augmented matrix (p. 374) A coefficient matrix with an extra column containing the constant terms.

matriz aumentada una matriz del coeficiente con una columna adicional que contiene los términos de la constante.

axis of symmetry (p. 559) The vertical line containing the vertex of a parabola.

bar graph (p. 659) A graphic form using bars to make comparisons of statistics.

base (p. 5) In an expression of the form x^n, the base is x.

best-fit line (p. 259) The line that most closely approximates the data in a scatter plot.

binomial (p. 491) The sum of two monomials.

bivariate data (p. 247) Data with two variables.

boundary (p. 321) A line or curve that separates the coordinate plane into regions.

causation (p. 254) A relationship in which one event causes the other.

center (p. P24) The given point from which all points on the circle are the same distance.

circle (p. P24) The set of all points in a plane that are the same distance from a given point called the center.

circumference (p. P24) The distance around a circle.

closed (p. 30) A set is closed under an operation if for any numbers in the set, the result of the operation is also in the set.

closed half-plane (p. 321) The solution of a linear inequality that includes the boundary line.

coefficient (p. 26) The numerical factor of a term.

coefficient of determination (p. 628) A measure of how well a regression line fits a data set, denoted by r^2.

common difference (p. 190) The difference between successive terms in an arithmetic sequence.

common ratio (p. 462) The ratio of successive terms of a geometric sequence.

complements (p. P33) One of two parts of a probability making a whole.

completing the square (p. 596) To add a constant term to a binomial of the form $x^2 + bx$ so that the resulting trinomial is a perfect square.

---

eje de simetría La recta vertical que pasa por el vértice de una parábola.

gráfico de barra Forma gráfica usando barras para comparar estadísticas.

base En una expresión de la forma x^n, la base es x.

recta de ajuste óptimo La recta que mejor aproxima los datos de una gráfica de dispersión.

binomio La suma de dos monomios.

datos bivariate Datos con dos variables.

frontera Recta o curva que divide el plano de coordenadas en regiones.

causalidad Una relación en la que un suceso causa el otro.

centro Punto dado del cual equidistan todos los puntos de un circulo.

círculo Conjunto de todos los puntos del plano que están a la misma distancia de un punto dado del plano llamado centro.

circunferencia Longitud del contorno de un círculo.

cerrado Un conjunto es cerrado bajo una operación si para cualquier número en el conjunto, el resultado de la operación es también en el conjunto.

mitad-plano cerrado La solución de una desigualdad linear que incluye la línea de límite.

coeficiente Factor numérico de un término.

coeficiente de determinación Una medida de la proximidad o el ajuste de una recta de regresión a un conjunto de datos, expresado como r^2.

diferencia común Diferencia entre términos consecutivos de una sucesión aritmética.

razón común El razón de términos sucesivos de una secuencia geométrica.

complementos Una de dos partes de una probabilidad que forma un todo.

completar el cuadrado Adición de un término constante a un binomio de la forma $x^2 + bx$, para que el trinomio resultante sea un cuadrado perfecto.

compound inequality (p. 309) Two or more inequalities that are connected by the words *and* or *or*.

compound interest (p. 451) A special application of exponential growth.

conditional relative frequency (p. 692) In a two-way frequency table, conditional relative frequencies are found by dividing joint frequencies by row or column totals, which are marginal frequencies.

conjugates (p. 421) Binomials of the form $a\sqrt{b} + c\sqrt{d}$ and $a\sqrt{b} - c\sqrt{d}$.

consecutive integers (p. 96) Integers in counting order.

consistent (p. 339) A system of equations that has at least one ordered pair that satisfies both equations.

constant (pp. 143, 395) A monomial that is a real number.

constant functions (pp. 172, 142) A linear function of the form $y = b$.

constraint (p. 227) A condition that a solution must satisfy.

continuous function (p. 50) A function that can be graphed with a line or a smooth curve.

coordinate (p. P8) The number that corresponds to a point on a number line.

coordinate plane (p. 42) The plane containing the *x*- and *y*-axes.

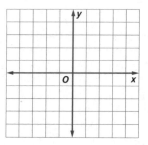

coordinate system (p. 42) The grid formed by the intersection of two number lines: the horizontal axis and the vertical axis.

correlation (p. 247, 254) A relationship between a set of data with two variables.

correlation coefficient (p. 259) A value that shows how close data points are to a line.

cube root (p. 411) If $a^3 = b$, then *a* is the cube root of *b*.

cumulative frequency (p. 659) The sum of frequencies of all preceding events.

desigualdad compuesta Dos o más desigualdades que están unidas por las palabras *y* u *o*.

interés compuesto Aplicación especial de crecimiento exponencial.

frecuencia relativa condicionada En una tabla de frecuencias de doble entrada, las frecuencias relativas condicionadas se calculan dividiendo las frecuencias conjuntas entre los totales de las filas o las columnas, que son las frecuencias marginales.

conjugados Binomios de la forma $a\sqrt{b} + c\sqrt{d}$ and $a\sqrt{b} - c\sqrt{d}$.

enteros consecutivos Enteros en el orden de contar.

consistente Sistema de ecuaciones para el cual existe al menos un par ordenado que satisface ambas ecuaciones.

constante Monomio que es un número real.

función constante Función lineal de la forma $f(x) = b$.

restricción Una condición que una solución debe satisfacer.

función continua Función cuya gráfica puedes ser una recta o una curva suave.

coordenada Número que corresponde a un punto en una recta numérica.

plano de coordenadas Plano que contiene los ejes *x* y *y*.

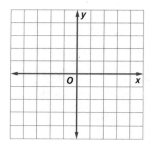

sistema de coordenadas Cuadriculado formado por la intersección de dos rectas numéricas: los ejes *x* y *y*.

correlación Es la relación que hay en un conjunto de datos con dos variables.

coeficiente de correlación Un valor que demostraciones cómo los puntos de referencias cercanos están a una línea.

raíz cúbica Si $a^3 = b$, entonces *a* es la raíz cúbica de *b*.

frecuencia acumulada La suma de las frecuencias de todos los sucesos inferiores.

Glossary/Glosario

debt-to-income ratio (p. 33) The ratio of how much a person owes per month to how much the person earns per month; used to determine if a person qualifies for a loan.

relación deuda-ingresos La razón de la cantidad de dinero que una persona debe por mes a la cantidad de dinero que gana por mes; se usa para determinar si una persona califica para un préstamo.

decreasing (p. 59) The graph of a function goes down on a portion of its domain when viewed from left to right.

decreciente El gráfico de una función va abajo en una porción de su dominio cuando está visto de izquierda a derecha.

defining a variable (p. P5) Choosing a variable to represent one of the unspecified numbers in a problem and using it to write expressions for the other unspecified numbers in the problem.

definir una variable Consiste en escoger una variable para representar uno de los números desconocidos en un problema y luego usarla para escribir expresiones para otros números desconocidos en el problema.

degree of a monomial (p. 491) The sum of the exponents of all its variables.

grado de un monomio Suma de los exponentes de todas sus variables.

degree of a polynomial (p. 491) The greatest degree of any term in the polynomial.

grado de un polinomio El grado mayor de cualquier término del polinomio.

dependent (p. 339) A system of equations that has an infinite number of solutions.

dependiente Sistema de ecuaciones que posee un número infinito de soluciones.

dependent variable (p. 44) The variable in a relation with a value that depends on the value of the independent variable.

variable dependiente La variable de una relación cuyo valor depende del valor de la variable independiente.

diameter (p. P24) The distance across a circle through its center.

diámetro La distancia a través de un círculo a través de su centro.

difference of two squares (p. 539) Two perfect squares separated by a subtraction sign.
$a^2 - b^2 = (a + b)(a - b)$ or
$a^2 - b^2 = (a - b)(a + b)$.

diferencia de cuadrados Dos cuadrados perfectos separados por el signo de sustracción.
$a^2 - b^2 = (a + b)(a - b)$ or
$a^2 - b^2 = (a - b)(a + b)$.

dilation (pp. 182, 572) A transformation that stretches or compresses the graph of a function.

homotecia Transformación que estira ó comprime la gráfica de una función.

dimensional analysis (p. 124) The process of carrying units throughout a computation.

análisis dimensional Proceso de tomar en cuenta las unidades de medida al hacer cálculos.

dimensions (p. 374) The number of rows m and the number of columns n of a matrix written as $m \times n$.

dimension El número de filas, de m, y del número de la columna, n, de una matriz escrita como $m \times n$.

discrete function (p. 50) A function of points that are not connected.

función discreta Función de puntos desconectados.

discriminant (p. 610) In the Quadratic Formula, the expression under the radical sign, $b^2 - 4ac$.

discriminante En la fórmula cuadrática, la expresión debajo del signo radical, $b^2 - 4ac$.

distribution (p. 673) A graph or table that shows the theoretical frequency of each possible data value.

distrubución Un gráfico o una tabla que muestra la frecuencia teórica de cada valor de datos posible.

domain (p. 42) The set of the first numbers of the ordered pairs in a relation.

dominio Conjunto de los primeros números de los pares ordenados de una relación.

dot plot (p. 657) A diagram that shows the frequency of data on a number line; also called a line plot.

diagrama de puntos Un diagrama que muestra la frecuencia de los datos en una recta numérica.

double root (p. 581) The roots of a quadratic function that are the same number.

raíces dobles Las raíces de una función cuadrática que son el mismo número.

E

elements (p. 374) Each entry in a matrix.

elemento Cada entrada de una matriz.

elimination (p. 354) The use of addition or subtraction to eliminate one variable and solve a system of equations.

eliminación El uso de la adición o la sustracción para eliminar una variable y resolver así un sistema de ecuaciones.

end behavior (p. 59) Describes how the values of a function behave at each end of the graph.

comportamiento extremo Describe como los valores de una función se comportan en el cada fin del gráfico.

equally likely (p. P33) The outcomes of an experiment are equally likely if there are n outcomes and the probability of each is $\frac{1}{n}$.

igualmente probablemente Los resultados de un experimento son igualmente probables si hay resultados de n y la probabilidad de cada uno es $\frac{1}{n}$.

equation A mathematical sentence that contains an equal sign, $=$.

ecuación Enunciado matemático que contiene el signo de igualdad, $=$.

equivalent equations (p. 87) Equations that have the same solution.

ecuaciones equivalentes Ecuaciones que poseen la misma solución.

equivalent expressions (p. 16) Expressions that denote the same value for all values of the variable(s).

expresiones equivalentes Expresiones que denotan el mismo valor para todos los valores de la(s) variable(s).

evaluate (p. 10) To find the value of an expression.

evaluar Calcular el valor de una expresión.

exponent (p. 5) In an expression of the form x^n, the exponent is n. It indicates the number of times x is used as a factor.

exponente En una expresión de la forma x^n, el exponente es n. Éste indica cuántas veces se usa x como factor.

exponential decay (p. 430) When an initial amount decreases by the same percent over a given period of time.

desintegración exponencial La cantidad inicial disminuye según el mismo porcentaje a lo largo de un período de tiempo dado.

exponential decay function (p. 430) A function of the form $y = ab^x$ where $a > 0$ and $0 < b < 1$.

función de decaimiento exponencial Una función con la forma $y = ab^x$, donde $a > 0$ y $0 < b < 1$.

exponential equation (p. 413) An equation in which the variables occur as exponents.

ecuación exponencial Ecuación en que las variables aparecen en los exponentes.

exponential function (p. 430) A function that can be described by an equation of the form $y = a^x$, where $a > 0$ and $a \neq 1$.

función exponencial Función que puede describirse mediante una ecuación de la forma $y = a^x$, donde $a > 0$ y $a \neq 1$.

exponential growth (p. 430) When an initial amount increases by the same percent over a given period of time.

crecimiento exponencial La cantidad inicial aumenta según el mismo porcentaje a lo largo de un período de tiempo dado.

exponential growth function (p. 430) A function of the form $y = ab^x$ where $a > 0$ and $b > 1$.

función de crecimiento exponencial Una función con la forma $y = ab^x$, donde $a > 0$ y $b > 1$.

extrema (p. 59) The greatest or least value of a function over an interval.

extremo El valor mayor o el valor menor de una función en un intervalo.

extremes (p. 116) In the ratio $\frac{a}{b} = \frac{c}{d}$, a and d are the extremes.

extremos En la razón $\frac{a}{b} = \frac{c}{d}$, a y d son los extremos.

Glossary/Glosario

factoring (p. 520) To express a polynomial as the product of monomials and polynomials.

factoring by grouping (p. 521) The use of the Distributive Property to factor some polynomials having four or more terms.

factors (p. 5) In an algebraic expression, the quantities being multiplied are called factors.

family of graphs (p. 151) Graphs and equations of graphs that have at least one characteristic in common.

five-number summary (p. 665) The three quartiles and the minimum and maximum values of a data set.

FOIL method (p. 507) To multiply two binomials, find the sum of the products of the First terms, the Outer terms, the Inner terms, and the Last terms.

formula (p. 80) An equation that states a rule for the relationship between certain quantities.

four-step problem-solving plan (p. P5)
Step 1 Understand the problem.
Step 2 Plan the solution.
Step 3 Solve the problem.
Step 4 Check the solution.

frequency table (p. 659) A chart that indicates the number of values in each interval.

function (p. 49) A relation in which each element of the domain is paired with exactly one element of the range.

function notation (p. 52) A way to name a function that is defined by an equation. In function notation, the equation $y = 3x - 8$ is written as $f(x) = 3x - 8$.

Fundamental Counting Principle (p. P34) If an event M can occur in m ways and is followed by an event N that can occur in n ways, then the event M followed by the event N can occur in $m \times n$ ways.

factorización La escritura de un polinomio como producto de monomios y polinomios.

factorización por agrupamiento Uso de la Propiedad distributiva para factorizar polinomios que poseen cuatro o más términos.

factores En una expresión algebraica, los factores son las cantidades que se multiplican.

familia de gráficas Gráficas y ecuaciones de gráficas que tienen al menos una característica común.

resumen de cinco números Los tres cuartiles, el valor mínimo y el valor máximo de un conjunto de datos.

método FOIL Para multiplicar dos binomios, busca la suma de los productos de los primeros (First) términos, los términos exteriores (Outer), los términos interiores (Inner) y los últimos términos (Last).

fórmula Ecuación que establece una relación entre ciertas cantidades.

solución de problemas en cinco pasos
Paso 1 Comprender el problema.
Paso 2 Planifica la solución.
Paso 3 Resuelve el problema.
Paso 4 Examina la solución.

Tabla de frecuencias Tabla que indica el número de valores en cada intervalo.

función Una relación en que a cada elemento del dominio le corresponde un único elemento del rango.

notación funcional Una manera de nombrar una función definida por una ecuación. En notación funcional, la ecuación $y = 3x - 8$ se escribe $f(x) = 3x - 8$.

Principio fundamental de contar Si un evento M puede ocurrir de m maneras y lo sigue un evento N que puede ocurrir de n maneras, entonces el evento M seguido del evento N puede ocurrir de $m \times n$ maneras.

geometric sequence (p. 462) A sequence in which each term after the first is found by multiplying the previous term by a constant r, called the common ratio.

graph (p. P8) To draw, or plot, the points named by certain numbers or ordered pairs on a number line or coordinate plane.

secuencia geométrica Una secuencia en la cual cada término después de que la primera sea encontrada multiplicando el término anterior por un r constante, llamado el razón común.

graficar Marcar los puntos que denotan ciertos números en una recta numérica o ciertos pares ordenados en un plano de coordenadas.

Glossary/Glosario

greatest integer function (p. 197) A step function, written as $f(x) = [\![x]\!]$, where $f(x)$ is the greatest integer less than or equal to x.

La función más grande del número entero Una función del paso, escrita como $f(x) = [\![x]\!]$, donde está el número entero $f(x)$ es el número más grande menos que o igual a x.

H

half-plane (p. 321) The region of the graph of an inequality on one side of a boundary.

semiplano Región de la gráfica de una desigualdad en un lado de la frontera.

histogram (p. 659) A graphical display that uses bars to display numerical data that have been organized into equal intervals.

histograma Una exhibición gráfica que utiliza barras para exhibir los datos numéricos que se han organizado en intervalos iguales.

I

identity (pp. 35, 102) An equation that is true for every value of the variable.

identidad Ecuación que es verdad para cada valor de la variable.

identity matrix (p. 375) A square matrix that, when multiplied by another matrix, equals that same matrix. If A is any $n \times n$ matrix and I is the $n \times n$ identity matrix, then $A \cdot I = A$ and $I \cdot A = A$.

matriz de la identidad Una matriz cuadrada que, cuando es multiplicada por otra matriz, iguala que la misma matriz. Si A es alguna de la matriz $n \times n$ e I es la matriz de la identidad de $n \times n$, entonces $A \cdot I = A$ e $I \cdot A = A$.

inconsistent (p. 339) A system of equations with no ordered pair that satisfy both equations.

inconsistente Un sistema de ecuaciones para el cual no existe par ordenado alguno que satisfaga ambas ecuaciones.

increasing (p. 59) The graph of a function goes up on a portion of its domain when viewed from left to right.

crecciente El gráfico de una función va arriba en una porción de su dominio cuando está visto de izquierda a derecha.

independent (p. 339) A system of equations with exactly one solution.

independiente Un sistema de ecuaciones que posee una única solución.

independent variable (p. 44) The variable in a function with a value that is subject to choice.

variable independiente La variable de una función sujeta a elección.

inequality (p. 289) An open sentence that contains the symbol $<$, $\leq$, $>$, or $\geq$.

desigualdad Enunciado abierto que contiene uno o más de los símbolos $<$, $\leq$, $>$, o $\geq$.

integers (p. P7) The set $\{\ldots, -2, -1, 0, 1, 2, \ldots\}$.

enteros El conjunto $\{\ldots, -2, -1, 0, 1, 2, \ldots\}$.

intercepts (p. 58) Points where the graph intersects an axis.

puntos de corte Los puntos en los que la gráfica se interseca con un eje.

interquartile range (p. 666) The range of the middle half of a set of data. It is the difference between the upper quartile and the lower quartile.

amplitud intercuartílica Amplitude de la mitad central de un conjunto de datos. Es la diferenccia entre el cuartil superior y el inferior.

intersection (p. 309) The graph of a compound inequality containing *and*; the solution is the set of elements common to both inequalities.

intersección Gráfica de una desigualdad compuesta que contiene la palabra y; la solución es el conjunto de soluciones de ambas desigualdades.

interval (p. 179) The points between two points or numbers between two values.

intervalo Los puntos que están entre dos puntos o los números que están entre dos valores.

inverse function (p. 268) A function that undoes the action of another function; can be written as $f^{-1}(x)$ and read as "the inverse of f of x."

función inversa Una función que cancela la acción de otra función; se puede escribir $f^{-1}(x)$ y se lee "el inverso de f de x."

inverse relation (p. 267) The set of ordered pairs obtained by exchanging the x- and y-coordinates of each ordered pair in a relation.

irrational numbers (p. P7) Numbers that cannot be expressed as terminating or repeating decimals.

relación inversa El conjunto de pares ordenados que se obtiene al intercambiar las coordenadas x e y de cada par ordenado de una relación.

números irracionales Números que no pueden escribirse como decimales terminales o periódicos.

J

joint frequency (p. 691) In a two-way frequency table, joint frequencies display the frequency of two conditions happening together.

frecuencia conjunta En una tabla de frecuencias de doble entrada, las frecuencias conjuntas muestran la frecuencia de dos condiciones que ocurren juntas.

L

leading coefficient (p. 492) The coefficient of the term with the highest degree in a polynomial.

coeficiente inicial El coeficiente del término con el grado más alto (el primer coeficiente inicial) en un polinomio.

like terms (p. 25) Terms that contain the same variables, with corresponding variables having the same exponent.

términos semejantes Expresiones que tienen las mismas variables, con las variables correspondientes elevadas a los mismos exponentes.

line of fit (p. 248) A line that describes the trend of the data in a scatter plot.

recta de ajuste Recta que describe la tendencia de los datos en una gráfica de dispersión.

line symmetry (p. 59) A graph possesses line symmetry if one half of the graph is a reflection of the other half across the line.

simetría axial Una gráfica tiene simetría axial si una mitad de la gráfica es una reflexión de la mitad que está del otro lado del eje.

linear equation (p. 143) An equation in the form $Ax + By = C$, with a graph that is a straight line.

ecuación lineal Ecuación de la forma $Ax + By = C$, cuya gráfica es una recta.

linear extrapolation (p. 227) The use of a linear equation to predict values that are outside the range of data.

extrapolación lineal Uso de una ecuación lineal para predecir valores fuera de la amplitud de los datos.

linear function (p. 141) A function with ordered pairs that satisfy a linear equation.

función lineal Función cuyos pares ordenados satisfacen una ecuación lineal.

linear interpolation (p. 249) The use of a linear equation to predict values that are inside the data range.

interpolación lineal Uso de una ecuación lineal para predecir valores dentro de la amplitud de los datos.

linear regression (p. 259) An algorithm to find a precise line of fit for a set of data.

regresión linear Un algoritmo para encontrar una línea exacta del ajuste para un sistema de datos.

linear transformation (p. 680) One or more operations performed on a set of data that can be written as a linear function.

transformación lineal Una o más operaciones que se hacen en un conjunto de datos y que se pueden escribir como una función lineal.

literal equation (p. 123) A formula or equation with several variables.

ecuación literal Un fórmula o ecuación con varias variables.

lower quartile (p. 665) Divides the lower half of the data into two equal parts.

cuartil inferior Éste divide en dos partes iguales la mitad inferior de un conjunto de datos.

mapping (p. 42) Illustrates how each element of the domain is paired with an element in the range.

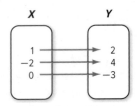

marginal frequency (p. 691) In a two-way frequency table, marginal frequencies are totals for each row or column.

matrix (p. 374) Any rectangular arrangement of numbers in rows and columns.

maximum (p. 559) The highest point on the graph of a curve.

mean (p. 651) The sum of numbers in a set of data divided by the number of items in the data set.

measure of spread (p. 665) A measure of the variation in a data set.

measures of central tendency (p. 651) Numbers or pieces of data that can represent the whole set of data.

measures of position (p. 665) Measures that compare the position of a value relative to other values in a set.

measures of variation (p. 665) Used to describe the distribution of statistical data.

median (p. 651) The middle number in a set of data when the data are arranged in numerical order. If the data set has an even number, the median is the mean of the two middle numbers.

median fit line (p. 262) A type of best-fit line that is calculated using the medians of the coordinates of the data points.

metric (p. 33) A rule for assigning a number to some characteristic or attribute.

minimum (p. 559) The lowest point on the graph of a curve.

mode (p. 651) The number(s) that appear most often in a set of data.

monomial (p. 395) A number, a variable, or a product of a number and one or more variables.

multiplicative identity (p. 17) For any number a, $a \cdot 1 = 1 \cdot a = a$.

aplicaciones Ilustra la correspondencia entre cada elemento del dominio con un elemento del rango.

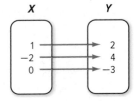

frecuencia marginal En una tabla de frecuencias de doble entrada, las frecuencias marginales son los totales de cada fila o columna.

matriz Disposción rectangular de numeros colocados en filas y columnas.

máximo El punto más alto en la gráfica de una curva.

media La suma de los números de un conjunto de datos dividida entre el numero total de artículos.

medida de dispersión Una medida de la variación en un conjunto de datos.

medidas de tendencia central Números o fragmentos que pueden representar el conjunto de datos total de datos.

medidas de la posición Las medidas que comparar la posición de un valor relativo a otros valores de un conjunto.

medidas de variación Números que se usan para describir la distribución o separación de un conjunto de datos.

mediana El número central de conjunto de datos, una vezque los datos han sido ordenados numéricamente. Si hay un número par de datos, la mediana es el promedio de los datos centrales.

línea apta del punto medio Tipo de mejor-cupo la línea se calcula que usando los puntos medios de los coordenadas de los puntos de referencias.

métrico Una regla para asignar un número a alguna característica o atribuye.

mínimo El punto más bajo en la gráfica de una curva.

moda El número(s) que aparece más frecuencia en un conjunto de datos.

monomio Número, variable o producto de un número por una o más variables.

identidad de la multiplicación Para cualquier número $a \cdot 1 = 1 \cdot a = a$.

multiplicative inverses (pp. P18, 17) Two numbers with a product of 1.

multi-step equation (p. 95) Equations with more than one operation.

inversos multiplicativos Dos números cuyo producto es igual a 1.

mutuamente exclusivos Eventos que no pueden ocurrir simultáneamente.

N

n th root (p. 411) If $a^n = b$ for a positive integer n, then a is an nth root of b.

natural numbers (p. P7) The set $\{1, 2, 3, ...\}$.

negative (p. 59) A function is negative on a portion of its domain where its graph lies below the x-axis.

negative correlation (p. 247) In a scatter plot, as x increases, y decreases.

negative exponent (p. 404) For any real number $a \neq 0$ and any integer n, $a^{-n} = \frac{1}{a^n}$ and $\frac{1}{a^{-n}} = a^n$.

negative number (p. P7) Any value less than zero.

nonlinear function (p. 52) A function with a graph that is not a straight line.

number theory (p. 96) The study of numbers and the relationships between them.

raíz enésima Si $a^n = b$ para cualquier entero positive n, entonces a se llama una raíz enésima de b.

números naturales El conjunto $\{1, 2, 3, ...\}$.

negativo Una función es negativa en una porción de su dominio donde su gráfico está debajo del eje-x.

correlación negativa En una gráfica de dispersión, a medida que x aumenta, y disminuye.

exponiente negativo Para números reales, si $a \neq 0$, y cualquier número entero n, entonces $a^{-n} = \frac{1}{a^n}$ and $\frac{1}{a^{-n}} = a^n$.

número negativo Cualquier valor menor que cero.

función no lineal Una función con un gráfica que no es una línea recta.

teoría del número El estudio de números y de las relaciones entre ellas.

O

odds (p. P35) The ratio of the probability of the success of an event to the probability of its complement.

open half-plane (p. 321) The solution of a linear inequality that does not include the boundary line.

opposites (p. P11) Two numbers with the same absolute value but different signs.

order of magnitude (p. 405) The order of magnitude of a quantity is the number rounded to the nearest power of 10.

order of operations (p. 10)
1. Evaluate expressions inside grouping symbols.
2. Evaluate all powers.
3. Do all multiplications and/or divisions from left to right.
4. Do all additions and/or subtractions from left to right.

ordered pair (p. 42) A set of numbers or coordinates used to locate any point on a coordinate plane, written in the form (x, y).

probabilidades El cociente de la probabilidad del éxito de un acontecimiento a la probabilidad de su complemento.

abra el mitad-plano La solución de una desigualdad linear que no incluya la línea de límite.

opuestos Dos números que tienen el mismo valor absoluto, pero que tienen distintos signos.

orden de magnitud de una cantidad Un número redondeado a la potencia más cercana de 10.

orden de las operaciones
1. Evalúa las expresiones dentro de los símbolos de agrupamiento.
2. Evalúa todas las potencias.
3. Multiplica o divide de izquierda a derecha.
4. Suma o resta de izquierda a derecha.

par ordenado Un par de números que se usa para ubicar cualquier punto de un plano de coordenadas y que se escribe en la forma (x, y).

origin (p. 42) The point where the two axes intersect at their zero points.

origen Punto donde se intersecan los dos ejes en sus puntos cero.

outliers (p. 666) Data that are more than 1.5 times the interquartile range beyond the quartiles.

valores atípicos Datos que distan de los cuartiles más de 1.5 veces la amplitud intercuartílica.

P

parabola (p. 559) The graph of a quadratic function.

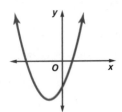

parábola La gráfica de una función cuadrática.

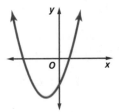

parallel lines (p. 239) Lines in the same plane that do not intersect and either have the same slope or are vertical lines.

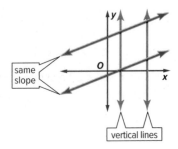

rectas paralelas Rectas en el mismo plano que no se intersecan y que tienen pendientes iguales, o las mismas rectas verticales.

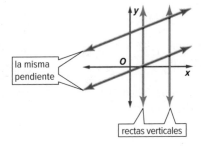

parent function (p. 151) The simplest of functions in a family.

función básica La función más fundamental de un familia de funciones.

percent (p. P20) A ratio that compares a number to 100.

porcentaje Razón que compara un numero con 100.

percent of change (p. 457) When an increase or decrease is expressed as a percent.

porcentaje de cambio Cuando un aumento o disminución se escribe como un tanto por ciento.

percent proportion (p. P20)
$$\frac{\text{part}}{\text{whole}} = \frac{\text{percent}}{100} \text{ or } \frac{a}{b} = \frac{P}{100}$$

proporción porcentual
$$\frac{\text{parte}}{\text{todo}} = \frac{\text{por ciento}}{100} \text{ or } \frac{a}{b} = \frac{P}{100}$$

percentile (p. 653) A measure that tells what percent of a data set are below a reported value in the set.

percentil Una medida que indica qué porcentaje de un conjunto de datos está por debajo de un valor informado del conjunto.

perfect square (p. P7) A number with a square root that is a rational number.

cuadrado perfecto Número cuya raíz cuadrada es un número racional.

perfect square trinomial (p. 540) A trinomial that is the square of a binomial.
$(a + b)^2 = (a + b)(a + b) = a^2 + 2ab + b^2$ or
$(a - b)^2 = (a - b)(a - b) = a^2 - 2ab + b^2$

trinomio cuadrado perfecto Un trinomio que es el cuadrado de un binomio.
$(a + b)^2 = (a + b)(a + b) = a^2 + 2ab + b^2$ or
$(a - b)^2 = (a - b)(a - b) = a^2 - 2ab + b^2$

Glossary/Glosario

perimeter (p. P23) The distance around a geometric figure.

perpendicular lines (p. 240) Lines that intersect to form a right angle.

piecewise-defined function (p. 198) A function that is written using two or more expressions.

piecewise-linear function (p. 197) A function written using two or more linear expressions.

point-slope form (p. 233) An equation of the form $y - y_1 = m(x - x_1)$, where m is the slope and (x_1, y_1) is a given point on a nonvertical line.

polynomial (p. 491) A monomial or sum of monomials.

positive (p. 59) A function is positive on a portion of its domain where its graph lies above the x-axis.

positive correlation (p. 247) In a scatter plot, as x increases, y increases.

positive number (p. P7) Any value that is greater than zero.

power (p. 5) An expression of the form x^n, read x *to the nth power*.

prime polynomial (p. 535) A polynomial that cannot be written as a product of two polynomials with integral coefficients.

principal square root (p. P7) The nonnegative square root of a number.

probability (p. P33) The ratio of the number of favorable equally likely outcomes to the number of possible equally likely outcomes.

product (p. 5) In an algebraic expression, the result of quantities being multiplied is called the product.

proof (p. 21) A proof is a logical argument in which each statement is supported by a true statement.

proportion (p. 115) An equation of the form $\frac{a}{b} = \frac{c}{d}$, where $b, d \neq 0$, stating that two ratios are equivalent.

Pythagorean Theorem If a and b are the measures of the legs of a right triangle and c is the measure of the hypotenuse, then $c^2 = a^2 + b^2$.

perímetro Longitud alrededor una figura geométrica.

recta perpendicular Recta que se intersecta formando un ángulo recto.

función definida por partes Función que se escribe usando dos o más expresiones.

función lineal por partes Función que se escribe usando dos o más expresiones lineal.

forma punto-pendiente Ecuación de la forma $y - y_1 = m(x - x_1)$, donde m es la pendiente y (x_1, y_1) es un punto dado de una recta no vertical.

polinomio Un monomio o la suma de monomios.

positiva Una función es positiva en una porción de su dominio donde su gráfico está encima del eje-x.

correlación positiva En una gráfica de dispersión, a medida que x aumenta, y aumenta.

número positivos Cualquier valor mayor que cero.

potencia Una expresión de la forma x^n, se lee x *a la enésima potencia*.

polinomio primo Polinomio que no puede escribirse como producto de dos polinomios con coeficientes enteros.

raíz cuadrada principal La raíz cuadrada no negativa de un número.

probabilidad La razón del número de maneras en que puede ocurrir el evento al numero de resultados posibles.

producto En una expresión algebraica, se llama producto al resultado de las cantidades que se multiplican.

demostración Una demostración es un argumento lógico en el que cada enunciado se reafirma con un enunciado verdadero.

proporción Ecuación de la forma $\frac{a}{b} = \frac{c}{d}$, donde $b, d \neq 0$, que afirma la equivalencia de dos razones.

Teorema de Pitágoras Si a y b son las longitudes de los catetos de un triángulo rectángulo y si c es la longitud de la hipotenusa, entonces $c^2 = a^2 + b^2$.

quadratic equation (p. 571) An equation of the form $ax^2 + bx + c = 0$, where $a \neq 0$.

quadratic expression (p. 507) An expression in one variable with a degree of 2 written in the form $ax^2 + bx + c$.

Quadratic Formula (p. 606) The solutions of a quadratic equation in the form $ax^2 + bx + c = 0$, where $a \neq 0$, are given by the formula

$$x = \frac{-b \pm \sqrt{b^2 - 4ac}}{2a}.$$

quadratic function (p. 559) An equation of the form $y = ax^2 + bx + c$, where $a \neq 0$.

qualitative data (p. 651) Data that can be categorized, but not measured.

quantitative data (p. 651) Data that can be measured.

quartile (p. 665) The values that divide a set of data into four equal parts.

ecuación cuadrática Ecuación de la forma $ax^2 + bx + c = 0$, donde $a \neq 0$.

expression cuadratica Una expresión en una variable con un grado de 2, escritos en la forma $ax^2 + bx + c$.

Fórmula cuadrática Las soluciones de una ecuación cuadrática de la forma $ax^2 + bx + c = 0$, donde $a \neq 0$, vienen dadas por la fórmula

$$x = \frac{-b \pm \sqrt{b^2 - 4ac}}{2a}.$$

función cuadrática Función de la forma $y = ax^2 + bx + c$, donde $a \neq 0$.

datos cualitativos Los datos que se pueden categorizar, pero no se pueden medir.

datos cuantitativos Los datos que se pueden medir.

cuartile Valores que dividen en conjunto de datos en cuàrto partes iguales.

radical expression (p. 419) An expression that contains a square root.

range (pp. 42, 665) 1. The set of second numbers of the ordered pairs in a relation. 2. The difference between the greatest and least data values.

rate (p. 117) The ratio of two measurements having different units of measure.

rate of change (p. 160) How a quantity is changing with respect to a change in another quantity.

ratio (p. 115) A comparison of two numbers by division.

rational exponent (p. 410) For any positive real number b and any integers m and $n > 1$, $b^{\frac{m}{n}} = \left(\sqrt[n]{b}\right)^m$ or $\sqrt[n]{b^m}$. $\frac{m}{n}$ is a rational exponent.

rational numbers (p. P7) The set of numbers expressed in the form of a fraction $\frac{a}{b}$, where a and b are integers and $b \neq 0$.

rationalizing the denominator (p. 421) A method used to eliminate radicals from the denominator of a fraction.

real numbers (p. P7) The set of rational numbers and the set of irrational numbers together.

expresión radical Expresión que contiene una raíz cuadrada.

rango 1. Conjunto de los segundos números de los pares ordenados de una relación. 2. La diferencia entre los valores de datos más grande o menos.

tasa Razón de dos medidas que tienen distintas unidades de medida.

tasa de cambio Cómo cambia una cantidad con respecto a un cambio en otra cantidad.

razón Comparación de dos números mediante división.

exponent racional Para cualquier número real no nulo b y cualquier entero m y $n > 1$, $b^{\frac{m}{n}} = \left(\sqrt[n]{b}\right)^m$ or $\sqrt[n]{b^m}$. $\frac{m}{n}$ es un exponent racional.

números racionales Conjunto de los números que pueden escribirse en forma de fracción $\frac{a}{b}$, donde a y b son enteros y $b \neq 0$.

racionalizar el denominador Método que se usa para eliminar radicales del denominador de una fracción.

números reales El conjunto de los números racionales junto con el conjunto de los números irracionales.

reciprocal (pp. P18, 17) The multiplicative inverse of a number.

reciproco Inverso multiplicativo de un número.

recursive formula (p. 469) Each term is formulated from one or more previous terms.

fórmula recursiva Cada tórmino proviene de uno o más terminos anteriores.

reflection (p. 184, 574) A transformation where a figure, line, or curve, is flipped across a line.

reflexión Transformación en que cadapunto de una figura se aplica a través de una recta de simetría a su imagen correspondiente.

relation (p. 42) A set of ordered pairs.

relación Conjunto de pares ordenados.

relative error (p. 313) The ratio of the absolute error to the expected measure.

error relativo La razón del error absoluto a la medida esperada.

relative frequency (p. 690) The number of times an event occurred compared to the whole

frecuencia relativa Número de veces que aparece un resultado en un experimento probabilístico.

relative maximum (p. 59) A point on graph is a relative maximum if no other nearby points have a greater y-coordinate.

máximo relativo Un punto de una gráfica es un máximo relativo si no hay puntos cercanos que tengan una coordenada y mayor.

relative minimum (p. 59) A point on graph is a relative minimum if no other nearby points have a lesser y-coordinate.

mínimo relativo Un punto de una gráfica es un mínimo relativo si no hay puntos cercanos que tengan una coordenada y menor.

residual (p. 260) The difference between an observed y-value and its predicted y-value on a regression line.

residual Diferencia entre el valor observado de y y el valor redicho de y en la recta de regression.

root (p. 151) The solutions of a quadratic equation.

raíces Las soluciones de una ecuación cuadrática.

row reduction (p. 375) The process of performing elementary row operations on an augmented matrix to solve a system.

reducción de la fila El proceso de realizar operaciones elementales de la fila en una matriz aumentada para solucionar un sistema.

S

sample space (p. P33) The list of all possible outcomes.

espacio muestral Lista de todos los resultados posibles.

scale (p. 118) The relationship between the measurements on a drawing or model and the measurements of the real object.

escala Relación entre las medidas de un dibujo o modelo y las medidas de la figura verdadera.

scale model (p. 118) A model used to represent an object that is too large or too small to be built at actual size.

modelo a escala Modelo que se usa para representar un figura que es demasiado grande o pequeña como para ser construida de tamaño natural.

scatter plot (p. 247) A scatter plot shows the relationship between a set of data with two variables, graphed as ordered pairs on a coordinate plane.

gráfica de dispersión Es un diagrama que muestra la relación entre un conjunto de datos con dos variables, graficados como pares ordenados en un plano coordenadas.

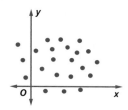

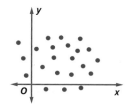

sequence (p. 190) A set of numbers in a specific order.

set-builder notation (p. 152) A concise way of writing a solution set. For example, $\{t \mid t < 17\}$ represents the set of all numbers t such that t is less than 17.

simplest form (p. 25) An expression is in simplest form when it is replaced by an equivalent expression having no like terms or parentheses.

slope (p. 162) The ratio of the change in the y-coordinates (rise) to the corresponding change in the x-coordinates (run) as you move from one point to another along a line.

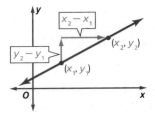

slope-intercept form (p. 171) An equation of the form $y = mx + b$, where m is the slope and b is the y-intercept.

solve an equation (p. 87) The process of finding all values of the variable that make the equation a true statement.

square root (p. P7) One of two equal factors of a number.

standard deviation (p. 667) The square root of the variance.

standard form of a linear equation (pp. 143, 232) The standard form of a linear equation is $Ax + By = C$, where $A \geq 0$, A and B are not both zero, and A, B, and C are integers with a greatest common factor of 1.

standard form of a polynomial (p. 492) A polynomial that is written with the terms in order from greatest degree to least degree.

step function (p. 197) A function with a graph that is a series of horizontal line segments.

substitution (p. 348) Use algebraic methods to find an exact solution of a system of equations.

surface area (p. P31) The sum of the areas of all the surfaces of a three-dimensional figure.

system of equations (p. 339) A set of equations with the same variables.

sucesión Conjunto de números en un orden específico.

notación de construcción de conjuntos Manera concisa de escribir un conjunto solución. Por ejemplo, $\{t \mid t < 17\}$ representa el conjunto de todos los números t que son menores o iguales que 17.

forma reducida Una expresión está reducida cuando se puede sustituir por una expresión equivalente que no tiene ni términos semejantes ni paréntesis.

pendiente Razón del cambio en la coordenada y (elevación) al cambio correspondiente en la coordenada x (desplazamiento) a medida que uno se mueve de un punto a otro en una recta.

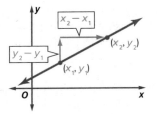

forma pendiente-intersección Ecuación de la forma $y = mx + b$, donde m es la pendiente y b es la intersección y.

resolver una ecuación Proceso en que se hallan todos los valores de la variable que hacen verdadera la ecuación.

raíz cuadrada Uno de dos factores iguales de un número.

desviación típica Calculada como la raíz cuadrada de la varianza.

forma estándar La forma estándar de una ecuación lineal es $Ax + By = C$, donde $A \geq 0$, ni A ni B son ambos cero, y A, B, y C son enteros cuyo máximo común divisor es 1.

forma de estándar de un polinomio Un polinomio que se escribe con los términos en orden del grado más grande a menos grado.

función escalonada Función cuya gráfica es una serie de segmentos de recto.

sustitución Usa métodos algebraicos para hallar una solución exacta a un sistema de ecuaciones.

área de superficie Suma de las áreas de todas las superficies (caras) de una figura tridimensional.

sistema de ecuaciones Conjunto de ecuaciones con las mismas variables.

system of inequalities (p. 376) A set of two or more inequalities with the same variables.

sistema de desigualdades Conjunto de dos o más desigualdades con las mismas variables.

T

term (p. 5) A number, a variable, or a product or quotient of numbers and variables.

término Número, variable o producto, o cociente de números y variables.

terms of a sequence (p. 190) The numbers in a sequence.

términos Los números de una sucesión.

transformation (pp. 181, 571) A movement of a graph on the coordinate plane.

transformación Desplazamiento de una gráfica en un plano de coordenadas.

translation (pp. 181, 571) A transformation where a figure is slid from one position to another without being turned.

translación Transformación en que una figura se desliza sin girar, de una posición a otra.

tree diagram (p. P34) A diagram used to show the total number of possible outcomes.

diagrama de árbol Diagrama que se usa para mostrar el número total de resultados posibles.

trinomials (p. 491) The sum of three monomials.

trinomios Suma de tres monomios.

two-way frequency table (p. 688) A table used to organize the frequency of data in two categories.

tabla de frecuencias de doble entrada Una tabla que se usa para organizar las frecuencias de los datos en dos categorías.

two-way relative frequency table (p. 690) A table used to organize the relative frequency of data in two categories; frequencies may be relative to row totals, column totals, or the total count.

tabla de frecuencias relativas de doble entrada Una tabla que se usa para organizar las frecuencias relativas de los datos en dos categorías; las frecuencias pueden ser relativas a los totales de las filas, a los totales de las columnas o al total general.

U

union (p. 310) The graph of a compound inequality containing or; the solution is a solution of either inequality, not necessarily both.

unión Gráfica de una desigualdad compuesta que contiene la palabra o; la solución es el conjunto de soluciones de por lo menos una de las desigualdades, no necesariamente ambas.

unit analysis (p. 124) The process of including units of measurement when computing.

análisis de la unidad Proceso de incluir unidades de medida al computar.

unit rate (p. 117) A ratio of two quantities, the second of which is one unit.

tasa unitaria Tasa reducida que tiene denominador igual a 1.

upper quartile (p. 665) The median of the upper half of a set of data.

cuartil superior Mediana de la mitad superior de un conjunto de datos.

V

variable (pp. 5, 651) 1. Symbols used to represent unspecified numbers or values. 2. a characteristic of a group of people or objects that can assume different values.

variable 1. Símbolos que se usan para representar números o valores no especificados. 2. una característica de un grupo de personas u objetos que pueden asumir valores diferentes.

variance (p. 667) The mean of the squares of the deviations from the arithmetic mean.

varianza Media de los cuadrados de las desviaciones de la media aritmética.

Glossary/Glosario

variation (p. 665) A measure of how spread out or scattered a data set is.

vertex (pp. 205, 559) The maximum or minimum point of a parabola.

vertex form (p. 576) A quadratic function in the form $f(x) = a(x - h)^2 + k$.

vertical line test (p. 51) If any vertical line passes through no more than one point of the graph of a relation, then the relation is a function.

volume (p. P29) The measure of space occupied by a solid region.

variación Una medida de la dispersión de los datos de un conjunto de datos.

vértice Punto máximo o mínimo de una parábola.

forma de vértice Una función cuadrática de la forma $f(x) = a(x - h)^2 + k$.

prueba de la recta vertical Si cualquier recta vertical pasa por un sólo punto de la gráfica de una relación, entonces la relación es una función.

volumen Medida del espacio que ocupa un solido.

W

whole numbers (p. P7) The set $\{0, 1, 2, 3, ...\}$.

números enteros El conjunto $\{0, 1, 2, 3, ...\}$.

X

x-axis (p. 42) The horizontal number line on a coordinate plane.

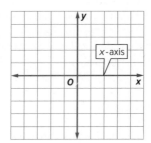

eje x Recta numérica horizontal que forma parte de un plano de coordenadas.

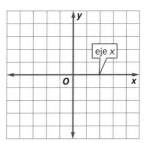

x-coordinate (p. 42) The first number in an ordered pair.

x-intercept (pp. 58, 144) The x-coordinate of a point where a graph crosses the x-axis.

coordenada x El primer número de un par ordenado.

intersección x La coordenada x de un punto donde la gráfica corte al eje de x.

Y

y-axis (p. 42) The vertical number line on a coordinate plane.

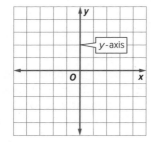

eje y Recta numérica vertical que forma parte de un plano de coordenadas.

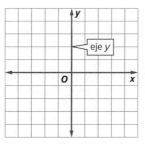

y-coordinate (p. 42) The second number in an ordered pair.

coordenada y El segundo número de un par ordenado.

y-intercept (pp. 58, 144) The *y*-coordinate of a point where a graph crosses the *y*-axis.

intersección y La coordenada *y* de un punto donde la grafica corta al eje de *y*.

Z

zero exponent (p. 403) For any nonzero number a, $a^0 = 1$. Any nonzero number raised to the zero power is equal to 1.

exponente cero Para cualquier número distinto a cero a, $a^0 = 1$. Cualquier número distinto a cero levantado al potente cero es igual a 1.

zeros (p. 151) The *x*-intercepts of the graph of a function; the values of *x* for which $f(x) = 0$.

cero Las intersecciones *x* de la grafica de una función; los puntos *x* para los que $f(x) = 0$.

Index

Index

Index

Index

Q

Index

Index

Symbols

$\neq$	is not equal to	AB	measure of $\overline{AB}$
$\approx$	is approximately equal to	$\angle$	angle
$\sim$	is similar to	$\triangle$	triangle
$>, \geq$	is greater than, is greater than or equal to	$^\circ$	degree
$<, \leq$	is less than, is less than or equal to	π	pi
$-a$	opposite or additive inverse of a	$\sin x$	sine of x
$\lvert a \rvert$	absolute value of a	$\cos x$	cosine of x
$\sqrt{a}$	principal square root of a	$\tan x$	tangent of x
$a : b$	ratio of a to b	$!$	factorial
(x, y)	ordered pair	$P(a)$	probability of a
$f(x)$	f of x, the value of f at x	$P(n, r)$	permutation of n objects taken r at a time
$\overline{AB}$	line segment AB	$C(n, r)$	combination of n objects taken r at a time

Algebraic Properties and Key Concepts

Identity	For any number a, $a + 0 = 0 + a = a$ and $a \cdot 1 = 1 \cdot a = a$.
Substitution (=)	If $a = b$, then a may be replaced by b.
Reflexive (=)	$a = a$
Symmetric (=)	If $a = b$, then $b = a$.
Transitive (=)	If $a = b$ and $b = c$, then $a = c$.
Commutative	For any numbers a and b, $a + b = b + a$ and $a \cdot b = b \cdot a$.
Associative	For any numbers a, b, and c, $(a + b) + c = a + (b + c)$ and $(a \cdot b) \cdot c = a \cdot (b \cdot c)$.
Distributive	For any numbers a, b, and c, $a(b + c) = ab + ac$ and $a(b - c) = ab - ac$.
Additive Inverse	For any number a, there is exactly one number $-a$ such that $a + (-a) = 0$.
Multiplicative Inverse	For any number $\frac{a}{b}$, where $a, b \neq 0$, there is exactly one number $\frac{b}{a}$ such that $\frac{a}{b} \cdot \frac{b}{a} = 1$.
Multiplicative (0)	For any number a, $a \cdot 0 = 0 \cdot a = 0$.
Addition (=)	For any numbers a, b, and c, if $a = b$, then $a + c = b + c$.
Subtraction (=)	For any numbers a, b, and c, if $a = b$, then $a - c = b - c$.
Multiplication and Division (=)	For any numbers a, b, and c, with $c \neq 0$, if $a = b$, then $ac = bc$ and $\frac{a}{c} = \frac{b}{c}$.
Addition (>)*	For any numbers a, b, and c, if $a > b$, then $a + c > b + c$.
Subtraction (>)*	For any numbers a, b, and c, if $a > b$, then $a - c > b - c$.
Multiplication and Division (>)*	For any numbers a, b, and c, 1. if $a > b$ and $c > 0$, then $ac > bc$ and $\frac{a}{c} > \frac{b}{c}$. 2. if $a > b$ and $c < 0$, then $ac < bc$ and $\frac{a}{c} < \frac{b}{c}$.
Zero Product	For any real numbers a and b, if $ab = 0$, then $a = 0$, $b = 0$, or both a and b equal 0.
Square of a Sum	$(a + b)^2 = (a + b)(a + b) = a^2 + 2ab + b^2$
Square of a Difference	$(a - b)^2 = (a - b)(a - b) = a^2 - 2ab + b^2$
Product of a Sum and a Difference	$(a + b)(a - b) = (a - b)(a + b) = a^2 - b^2$

* These properties are also true for $<$, $\geq$, and $\leq$.

Formulas

Slope	$m = \dfrac{y_2 - y_1}{x_2 - x_1}$
Distance on a coordinate plane	$d = \sqrt{(x_2 - x_1)^2 + (y_2 - y_1)^2}$
Midpoint on a coordinate plane	$M = \left(\dfrac{x_1 + x_2}{2}, \dfrac{y_1 + y_2}{2}\right)$
Pythagorean Theorem	$a^2 + b^2 = c^2$
Quadratic Formula	$x = \dfrac{-b \pm \sqrt{b^2 - 4ac}}{2a}$
Perimeter of a rectangle	$P = 2\ell + 2w$ or $P = 2(\ell + w)$
Circumference of a circle	$C = 2\pi r$ or $C = \pi d$

Area

rectangle	$A = \ell w$	**trapezoid**	$A = \frac{1}{2}h(b_1 + b_2)$
parallelogram	$A = bh$	**circle**	$A = \pi r^2$
triangle	$A = \frac{1}{2}bh$		

Surface Area

cube	$S = 6s^2$	**regular pyramid**	$S = \frac{1}{2}P\ell + B$
prism	$S = Ph + 2B$	**cone**	$S = \pi r \ell + \pi r^2$
cylinder	$S = 2\pi rh + 2\pi r^2$		

Volume

cube	$V = s^3$	**regular pyramid**	$V = \frac{1}{3}Bh$
prism	$V = Bh$	**cone**	$V = \frac{1}{3}\pi r^2 h$
cylinder	$V = \pi r^2 h$		

Measures

Metric	Customary
Length	
1 kilometer (km) = 1000 meters (m) 1 meter = 100 centimeters (cm) 1 centimeter = 10 millimeters (mm)	1 mile (mi) = 1760 yards (yd) 1 mile = 5280 feet (ft) 1 yard = 3 feet 1 foot = 12 inches (in.) 1 yard = 36 inches
Volume and Capacity	
1 liter (L) = 1000 milliliters (mL) 1 kiloliter (kL) = 1000 liters	1 gallon (gal) = 4 quarts (qt) 1 gallon = 128 fluid ounces (fl oz) 1 quart = 2 pints (pt) 1 pint = 2 cups (c) 1 cup = 8 fluid ounces
1 kilogram (kg) = 1000 grams (g) 1 gram = 1000 milligrams (mg) 1 metric ton (t) = 1000 kilograms	1 ton (T) = 2000 pounds (lb) 1 pound = 16 ounces (oz)